AF760075

Kiwifruit

Kiwifruit

Botany, Production and Uses

Edited by

Annette C. Richardson

The New Zealand Institute for Plant and Food Research Ltd, Kerikeri, New Zealand

Jeremy N. Burdon

The New Zealand Institute for Plant and Food Research Ltd, Auckland, New Zealand

A. Ross Ferguson

The New Zealand Institute for Plant and Food Research Ltd, Auckland, New Zealand

with the assistance of Ruth Williams

The New Zealand Institute for Plant and Food Research Ltd, Auckland, New Zealand

CABI

CABI is a trading name of CAB International

CABI
Nosworthy Way
Wallingford
Oxfordshire OX10 8DE
UK

Tel: +44 (0)1491 832111
E-mail: info@cabi.org
Website: www.cabi.org

CABI
200 Portland Street
Boston
MA 02114
USA

Tel: +1 (617)682-9015
E-mail: cabi-nao@cabi.org

© CAB International 2024. All rights reserved. No part of this publication may be reproduced in any form or by any means, electronically, mechanically, by photocopying, recording or otherwise, without the prior permission of the copyright owners.

The views expressed in this publication are those of the author(s) and do not necessarily represent those of, and should not be attributed to, CAB International (CABI). Any images, figures and tables not otherwise attributed are the author(s)' own. References to internet websites (URLs) were accurate at the time of writing.
CAB International and, where different, the copyright owner shall not be liable for technical or other errors or omissions contained herein. The information is supplied without obligation and on the understanding that any person who acts upon it, or otherwise changes their position in reliance thereon, does so entirely at their own risk. Information supplied is neither intended nor implied to be a substitute for professional advice. The reader/user accepts all risks and responsibility for losses, damages, costs and other consequences resulting directly or indirectly from using this information.

CABI's Terms and Conditions, including its full disclaimer, may be found at https://www.cabi.org/terms-and-conditions/.

A catalogue record for this book is available from the British Library, London, UK.

Library of Congress Cataloging-in-Publication Data

Names: Richardson, Annette C., editor. | Burdon, Jeremy N., editor. | Ferguson, A. R., editor.
Title: Kiwifruit : botany, production and uses / edited by Annette C. Richardson, Jeremy N. Burdon, A. Ross Ferguson.
Other titles: Botany, production and uses.
Description: Boston, MA : CAB International, [2023] | Series: Botany, production and uses | Includes bibliographical references and index. | Summary: "A comprehensive and authoritative reference text on kiwifruit; their history, plant material, culture, vine physiology, pest and disease control, and fruit utilisation. Particular attention is given to recent threats and opportunities, e.g. climate change, the disease Pseudomonas syringae pv. actninidiae, new genetics, new growing areas and technology advances"-- Provided by publisher.
Identifiers: LCCN 2023032718 (print) | LCCN 2023032719 (ebook) | ISBN 9781800620919 (hardback) | ISBN 9781800620926 (ebk) | ISBN 9781800620933 (epub)
Subjects: LCSH: Kiwifruit.
Classification: LCC QK495.A177 K595 2023 (print) | LCC QK495.A177 (ebook) | DDC 634/.4--dc23/eng/20231004
LC record available at https://lccn.loc.gov/2023032718
LC ebook record available at https://lccn.loc.gov/2023032719

ISBN-13: 9781800620919 (hardback)
9781800620926 (ePDF)
9781800620933 (ePub)

DOI: 10.1079/ 9781800620933.0000

Commissioning Editor: Rebecca Stubbs
Editorial Assistant: Emma McCann
Production Editor: Rosie Hayden

Typeset by Straive, Pondicherry, India
Printed and bound by CPI Group (UK) Ltd, Croydon, CR0 4YY

Contents

List of Contributors vii

Foreword ix

Acknowledgements xi

List of *Actinidia* taxa mentioned xiii

List of kiwifruit cultivars xv

1 **History of Kiwifruit: Evolution of a Global Crop** 1
A. Ross Ferguson, Hongwen Huang and Guglielmo Costa

2 **Global Industry and Markets** 16
Jayne E. Chamberlain

3 **Biology of the Genus *Actinidia*** 29
A. Ross Ferguson

4 **Fruiting Cultivars** 52
A. Ross Ferguson, Jinbao Fang and Guglielmo Costa

5 **Male Cultivars** 72
Alan Seal, Raffaele Testolin and Jinbao Fang

6 **Kiwifruit Rootstocks** 85
Grant Thorp and Andrew M. Barnett

7 **Kiwiberry** 94
Piotr Latocha and Filip Debersaques

8 **Propagation of Kiwifruit Plants** 119
Ed R. Morgan

9 **Genetic Improvement, Kiwifruit Genome and Dioecy** 127
Raffaele Testolin and Mark A. McNeilage

10 Soil, Climate and Site Selection 150
Annette C. Richardson, Roberta M. Gentile and Indrakumar Vetharaniam

11 Vine Nutrition and Water Requirements 164
Cristos Xiloyannis, Bartolomeo Dichio and Alba N. Mininni

12 Vine Management Systems 184
Kevin J. Patterson and Grant Thorp

13 Kiwifruit Pollination 207
Melissa A. Broussard and David E. Pattemore

14 Axillary Bud, Shoot and Flower Development 228
Annette C. Richardson and Erika Varkonyi-Gasic

15 Vine Canopy and Root Physiology 255
Nick Gould and Michael J. Clearwater

16 Fruit Development: Growth, Maturation and Ripening 273
Simona Nardozza and Jeremy N. Burdon

17 Vine and Fruit Diseases 295
Joy L. Tyson, Irene Donati and Kerry R. Everett

18 The Bacterial Canker of Kiwifruit Caused by *Pseudomonas syringae* pv. *actinidiae* 317
Marco Scortichini, Francesco Spinelli and Matt D. Templeton

19 Pests and Their Management 332
David P. Logan, Qiguo Zhuang and Aldo Pollini

20 Postharvest: Fresh Fruit Harvest, Storage and Supply 353
Jeremy N. Burdon and Ruiling Wang

21 Consumer Research into Fresh Kiwifruit: Fruit Quality, Consumer Methods and Future Trends 375
Denise C. Hunter and Anne White

22 Kiwifruit: Future Potential, Risks and New Directions 385
Annette C. Richardson and Jeremy N. Burdon

Index 391

Contributors

Andrew M. Barnett, The New Zealand Institute for Plant and Food Research Ltd, 412 No. 1 Road, RD 2, Te Puke 3182, New Zealand. Email: andrew.barnett@plantandfood.co.nz

Melissa A. Broussard, The New Zealand Institute for Plant and Food Research Ltd, Private Bag 3230, Waikato Mail Centre, Hamilton 3240, New Zealand. Email: melissa.broussard@plantandfood.co.nz

Jeremy N. Burdon, The New Zealand Institute for Plant and Food Research Ltd, Private Bag 92169, Auckland Mail Centre, Auckland 1142, New Zealand. Email: jeremy.burdon@plantandfood.co.nz

Jayne E. Chamberlain, Kiwifruit Breeding Centre, 412 No. 1 Road, RD 2, Te Puke 3182, New Zealand. Email: jayne.chamberlain@next.kiwi

Michael J. Clearwater, School of Science, The University of Waikato, Private Bag 3105, Hamilton 3240, New Zealand. Email: mike.clearwater@waikato.ac.nz

Guglielmo Costa, Alma Mater Studiorum – University of Bologna, Viale G. Fanin 46, Bologna 40127, Italy. Email: guglielmo.costa1@gmail.com

Filip Debersaques, Department of Plants and Crops, Faculty of Bioscience Engineering, Ghent University, Valentin Vaerwyckweg 1, Ghent 9000, Belgium. Email: filip.debersaques@UGent.be

Bartolomeo Dichio, Department of European and Mediterranean Cultures: Architecture, Environment, Cultural Heritage, University of Basilicata, Via Lanera 20, Matera 75100, Italy. Email: bartolomeo.dichio@unibas.it

Irene Donati, Zespri Fresh Produce, Via Del Faggio 1/11, Bologna 40132, Italy. Email: irene.donati@zespri.com

Kerry R. Everett, The New Zealand Institute for Plant and Food Research Ltd, Private Bag 92169, Auckland Mail Centre, Auckland 1142, New Zealand. Email: kerry.everett@plantandfood.co.nz

Jinbao Fang, Zhengzhou Fruit Research Institute, Chinese Academy of Agricultural Sciences, Zhengzhou 450009, Henan, China. Email: fangjinbao@caas.cn

A. Ross Ferguson, The New Zealand Institute for Plant and Food Research Ltd, Private Bag 92169, Auckland Mail Centre, Auckland 1142, New Zealand. Email: ross.ferguson@plantandfood.co.nz

Roberta M. Gentile, The New Zealand Institute for Plant and Food Research Ltd, Private Bag 11600, Palmerston North 4442, New Zealand. Email: roberta.gentile@plantandfood.co.nz

Nick Gould, The New Zealand Institute for Plant and Food Research Ltd, 412 No. 1 Road, RD 2, Te Puke 3182, New Zealand. Email: nick.gould@plantandfood.co.nz

Hongwen Huang, Lushan Botanical Garden, Chinese Academy of Sciences, Jiujiang 332900, Jiangxi, China. Email: huanghw@scbg.ac.cn

Denise C. Hunter, FORWARD Insight and Strategy, Auckland 1245, New Zealand. Email: denise@forward-insight.com

Piotr Latocha, Department of Environmental Protection and Dendrology, Institute of Horticultural Sciences, Warsaw University of Life Sciences – SGGW, Nowoursynowska 166, Warsaw 02-787, Poland. Email: piotr_latocha@sggw.edu.pl

David P. Logan, The New Zealand Institute for Plant and Food Research Ltd, Private Bag 92169, Auckland Mail Centre, Auckland 1142, New Zealand. Current address: School of Agriculture, Food and Wine, The University of Adelaide, Urrbrae 5064, Adelaide, Australia. Email: david.logan@adelaide.edu.au

Mark A. McNeilage, The New Zealand Institute for Plant and Food Research Ltd, Private Bag 92169, Auckland Mail Centre, Auckland 1142, New Zealand. Email: mark.mcneilage@plantandfood.co.nz

Alba N. Mininni, Department of European and Mediterranean Cultures: Architecture, Environment, Cultural Heritage, University of Basilicata, Via Lanera 20, Matera 75100, Italy. Email: alba.mininni@unibas.it

Ed R. Morgan, The New Zealand Institute for Plant and Food Research Ltd, Private Bag 11600, Palmerston North 4442, New Zealand. Email: ed.morgan@plantandfood.co.nz

Simona Nardozza, The New Zealand Institute for Plant and Food Research Ltd, Private Bag 92169, Auckland Mail Centre, Auckland 1142, New Zealand. Email: simona.nardozza@plantandfood.co.nz

David E. Pattemore, The New Zealand Institute for Plant and Food Research Ltd, Private Bag 3230, Waikato Mail Centre, Hamilton 3240, New Zealand. Email: david.pattemore@plantandfood.co.nz

Kevin J. Patterson, The New Zealand Institute for Plant and Food Research Ltd, Private Bag 92169, Auckland Mail Centre, Auckland 1142, New Zealand. Email: kevin.patterson@plantandfood.co.nz

Aldo Pollini, Via Luxemburg 11, Imola 40026, Bologna, Italy. Email: aldopollini.studio@gmail.com

Annette C. Richardson, The New Zealand Institute for Plant and Food Research Ltd, 121 Keri Downs Road, RD1, Kerikeri 0294, New Zealand. Email: annette.richardson@plantandfood.co.nz

Marco Scortichini, Council for Agricultural Research and Economics (CREA) – Research Centre for Olive, Fruit and Citrus Crops, Via di Fioranello 52, Roma I-00134, Italy. Email: marco.scortichini@crea.gov.it

Alan Seal, Kiwifruit Breeding Centre, 412 No. 1 Road, RD 2, Te Puke 3182, New Zealand. Email: alan.seal@next.kiwi

Francesco Spinelli, Department of Agriculture and Food Science, University of Bologna, Viale G. Fanin 44, Bologna 40127, Italy. Email: francesco.spinelli3@unibo.it

Raffaele Testolin, Department of Agricultural, Food, Environmental and Animal Sciences, University of Udine, Via delle Scienze 206, Udine 33100, Italy. Email: raffaele.testolin@uniud.it

Matt D. Templeton, The New Zealand Institute for Plant and Food Research Ltd, Private Bag 92169, Auckland Mail Centre, Auckland 1142, New Zealand. Email: matt.templeton@plantandfood.co.nz

Grant Thorp, The New Zealand Institute for Plant and Food Research Ltd, 412 No. 1 Road, RD 2, Te Puke 3182, New Zealand. Email: grant.thorp@plantandfood.co.nz

Joy L. Tyson, The New Zealand Institute for Plant and Food Research Ltd, Private Bag 92169, Auckland Mail Centre, Auckland 1142, New Zealand. Email: joy.tyson@plantandfood.co.nz

Erika Varkonyi-Gasic, The New Zealand Institute for Plant and Food Research Ltd, Private Bag 92169, Auckland Mail Centre, Auckland 1142, New Zealand. Email: erika.varkonyi-gasic@plantandfood.co.nz

Indrakumar Vetharaniam, The New Zealand Institute for Plant and Food Research Ltd, Private Bag 3230, Waikato Mail Centre, Hamilton 3240, New Zealand. Email: kumar.vetharaniam@plantandfood.co.nz

Ruiling Wang, The New Zealand Institute for Plant and Food Research Ltd, Private Bag 92169, Auckland Mail Centre, Auckland 1142, New Zealand. Email: tina.wang@plantandfood.co.nz

Anne White, The New Zealand Institute for Plant and Food Research Ltd, Private Bag 92169, Auckland Mail Centre, Auckland 1142, New Zealand. Email: anne.white@plantandfood.co.nz

Cristos Xiloyannis, Department of European and Mediterranean Cultures: Architecture, Environment, Cultural Heritage, University of Basilicata, Via Lanera 20, Matera 75100, Italy. Email: cristosxiloyannis15@gmail.com

Qiguo Zhuang, Sichuan Provincial Academy of Natural Resource Sciences, Chengdu 610015, Sichuan, China. Email: qgzhuang@126.com

Foreword

The publication of *Kiwifruit: Botany, Production and Uses* is a welcome addition to the literature on this important fruit crop. It is now 33 years since the publication of *Kiwifruit: Science and Management* and there have been many new developments to this crop over that period. The diversity of new cultivars has increased dramatically, new production methods have emerged, and the area of crop grown worldwide has continued to increase. Most significantly, the industry worldwide has had to counter the bacterial canker disease caused by *Pseudomonas syringae* pv. *actinidiae* which has devastated established plantings in many countries over the past two decades.

Unlike *Kiwifruit: Science and Management*, which was authored only by scientists in New Zealand, this new text brings together the contributions of research scientists and industry personnel from Belgium, China, Italy, Poland and New Zealand, a strong indication of the global research effort that is now focused on kiwifruit and the closely related kiwiberry.

The largest recent change in commercial kiwifruit production has been the rapid adoption of the new yellow-fleshed and red-fleshed cultivars that have emerged from long-term R&D-based breeding programmes. These new cultivars have resulted in markedly enhanced returns for growers through greater consumer appeal and higher yields. Nonetheless, production of green-fleshed kiwifruit, primarily 'Hayward', continues to be important as it remains popular to many consumers. Because of the need to remain commercially profitable with all cultivars, greater attention has had to be given to all aspects of production, including pollination, choice of male cultivars, bud development and flowering, vine management, mineral nutrition management and irrigation control, and fruit maturation. Furthermore, postharvest methods have had to adjust to accommodate the specific needs of these new fruit types. All the new innovations involved have been the focus of considerable R&D and these topics are covered in this new text. The rate of development and the degree of adoption of new cultivars and new growing methods, along with the expansion of underpinning scientific knowledge, has been phenomenal.

Perhaps the biggest change in the science of kiwifruit over the past 30 years has been the development of extensive knowledge at the molecular level of the kiwifruit genome. There is now, for example, a sound understanding of the genetic basis of dioecy, and molecular markers have been developed that allow the specific identification of individual cultivars. Molecular techniques have now become an indispensable component of breeding programmes, allowing enhanced targeting of specific breeding and selection goals. These techniques also affect other areas such as plant disease identification and control.

Notwithstanding the major expansion of kiwifruit plantings globally over the past 30 years, and the impressive orchard gate returns achieved, especially with the new yellow-fleshed cultivars, the industry faces some major challenges moving forward. For example, the increasing urbanization of societies has brought with it a critical shortage of workers who are prepared to work in rural occupations. The development of image-analysis methods coupled with smart computer programming has revolutionized the grading and sorting of many fruit crops, including kiwifruit. Automatic tray filling and other packing techniques are also evolving. However, the development of robotic systems for crop husbandry practices (such as pruning, pollination and fruit thinning) and for harvesting, while mostly still at the prototype stage, will revolutionize the production of this crop in the future and reduce dependency on human labour, as it has already done in the mechanical harvesting of many vegetable, berry and nut crops.

Climate change is also beginning to have an impact on the kiwifruit crop in several different ways. Global temperature increases are already affecting the amount of winter chilling received to break bud dormancy. Higher growing temperatures also affect the progression of harvest maturity and, potentially, fruit quality. The current physiological understanding of these processes is little better than rudimentary. Climate change is bringing an increased frequency of high rainfall and wind events that are unprecedented and influence plant health, crop yield and fruit quality. A consequence of such changes in the environment is that production in some currently established areas may become marginal, requiring expansion into other areas that will be more suitable under the new climatic norm.

There is no doubt that there will continue to be a high rate of innovation in the kiwifruit industry worldwide, providing that investment in R&D is sustained. These innovations, like those of the past 30 years since the publication of *Kiwifruit: Science and Management*, will no doubt warrant a further text.

Ian Warrington
Emeritus Professor I.J. Warrington, CNZM, FRSNZ, FISHS
(co-editor of *Kiwifruit: Science and Management* (1990))

List of *Actinidia* taxa mentioned

We have followed Li, J.-Q., Li, X.-W. and Soejarto, D.D. (2007) Actinidiaceae. In: Wu, Z-Y., Raven, P.H. and Hong, D.-Y. (eds) *Flora of China*, Vol. 12. Science Press, Beijing, and Missouri Botanical Gardens, St Louis, Missouri, pp. 334–360. Plants cited in the literature discussed in the chapters are not necessarily correctly identified.

Actinidia arguta (Sieb. et Zucc.) Planch. ex Miq.
Actinidia arguta var. *hypoleuca* (Nakai) Kitam.
Actinidia arguta var. *purpurea* (Rehder) C.F.Liang ex Q.Q.Chang, now included in *A. arguta* var. *arguta*
Actinidia arguta var. *giraldii* (Diels) Vorosch.
Actinidia chinensis Planch. var. *chinensis*
Actinidia chinensis var. *deliciosa* (A.Chev.) A.Chev.
Actinidia chinensis var. *hispida* C.F.Liang, now synonym of *A. chinensis* var. *deliciosa* (A.Chev.) A.Chev.
Actinidia chrysantha C.F.Liang
Actinidia deliciosa (A.Chev.) C.F.Liang et A.R.Ferguson, now treated as *A. chinensis* var. *deliciosa* (A.Chev.) A.Chev.
Actinidia eriantha Benth.
Actinidia glaucophylla F.Chun now synonym of *Actinidia fortunatii* Finet et Gagnep.
Actinidia grandiflora C.F.Liang
Actinidia guilinensis C.F.Liang synonym of *A. latifolia*
Actinidia hemsleyana Dunn
Actinidia hypoleuca Nakai now treated as *A. arguta* var. *hypoleuca* (Nakai) Kitam.
Actinidia indochinensis Merr.
Actinidia kolomikta (Maxim. et Rupr.) Maxim.
Actindia latifolia (Gardner et Champ.) Merr.
Actinidia longicarpa R.G.Li et M.Y.Liang
Actinidia macrosperma C.F.Liang
Actinidia melanandra Franch.
Actinidia polygama (Sieb. et Zucc.) Maxim.
Actinidia rufa (Sieb. et Zucc.) Planch. ex Miq.
Actinidia sabiifolia Dunn
Actinidia setosa (Li) C.F.Liang et A.R.Ferguson, now sometimes treated as *A. chinensis* var. *setosa* Li
Actinidia valvata Dunn

List of kiwifruit cultivars

Accepted cultivar names are given in bold. Some cultivars, especially Zespri-owned cultivars, are better known by the tradename under which the fruit are sold.

Cultivar	Taxon	Gender
'Abbott'	*A. chinensis* var. *deliciosa*	F
'AC 1536'	*A. chinensis* var. *chinensis*	F
Fruit marketed as Dorì		
'Ac 459 011'	*A. chinensis* var. *chinensis*	F
'Allison'	*A. chinensis* var. *deliciosa*	F
'Ananasnaya'	*A. arguta*	F
'AU Authur'	*A chinensis* var. *deliciosa*	M
'AU Fitzgerald'	*A chinensis* var. *deliciosa*	F
'Autari' (Ac 12.255)	*A chinensis* var. *deliciosa*	M
'Belén' (A 134.16)	*A. chinensis* var. *chinensis*	M
'Bingo'	*A. arguta*	F
Boerica® ('Bo-Erica')	*A. chinensis* var. *deliciosa*	F
'Bohwa' (mis-spelling of '**Pohwa**')	*A. chinensis* var. *deliciosa*	M
'Bounty71' (rootstock)	*A. macrosperma*	
'Boverlig 02' (Earligreen®)	*A. chinensis* var. *deliciosa*	F
'Bruce'	*A. chinensis* var. *chinensis*	M
'Bruno'	*A. chinensis* var. *deliciosa*	F
'Chang Jiang No. 1'	*A. arguta*	F
'Chaohong'	*A. eriantha* × *A. chinensis* var. *chinensis*	M
'Chengxiang'	*A. arguta*	F fruit parthenocarpic
'Cheongsan'	*A. arguta*	F
'Chiak'	*A. arguta*	F
'Chico' **('Hayward')**	*A. chinensis* var. *deliciosa*	F
'Chico-male' (California male)	*A. chinensis* var. *deliciosa*	M
'Chieftain' (M59)	*A. chinensis* var. *deliciosa*	M
'Chuhong'	*A. chinensis* var. *chinensis*	F
'Chunmi-1'	*A. chinensis* var. *deliciosa*	F
'Convi 97.001' (Green Light®)	*A. chinensis* var. *deliciosa*	F
'Cornell'	*A. arguta*	F

Continued

Continued.

Cultivar	Taxon	Gender
'Cuixiang'	*A. chinensis* var. *deliciosa*	F
'Cuiyu'	*A. chinensis* var. *chinensis*	F
'D1 Vitroplant' (D1 Vitroplant®) (rootstock)	*A. chinensis* var. *deliciosa*	M
'Deliwoong'	*A. chinensis* var. *deliciosa*	M
'Don Juan'	*A. arguta*	M
'Donghong'	*A. chinensis* var. *chinensis*	F
'Earp' (MC79)	*A. chinensis* var. *chinensis*	M
'Emihoutao No. 1' ('Jinkui')	*A. chinensis* var. *deliciosa*	F
'Emihoutao No.2' ('Jinnong No. 2')	*A. chinensis* var. *chinensis*	F
'Emihoutao No. 3' ('Jinyang')	*A. chinensis* var. *chinensis*	F
'Fave'	*A. arguta*	M
'Fengyue'	*A. chinensis* var. *chinensis*	F
'Ferguson' (MC110)	*A. chinensis* var. *chinensis*	M
'Fuchu'	*A. rufa*	F
'Ganmi No. 1' ('Zaoxian')	*A. chinensis* var. *chinensis*	F
'Ganmi No. 2' ('Kuimi')	*A. chinensis* var. *chinensis*	F
'Ganmi No. 3' ('Jinfeng')	*A. chinensis* var. *chinensis*	F
'Ganmi No. 6'	*A. eriantha*	F
'Garmrok'	*A. chinensis* var. *deliciosa*	F
'Gemma'	*A. arguta*	M
'Geneva'	*A. arguta*	F
'Golden Dragon'	*A. chinensis* var. *chinensis*	F
'Golden Sunshine'	*A. chinensis* var. *chinensis*	F
'Goldone'	*A. chinensis* var. *chinensis*	F
Green Angel®	*A. chinensis* var. *deliciosa*	F
'Guichang'	*A. chinensis* var. *deliciosa*	F
'H-15'	*A. chinensis* var. *deliciosa*	F
'Haegeum'	*A. chinensis* var. *chinensis*	F
'Haewon'	*A. chinensis* var. *deliciosa*	F
'Halla Gold'	*A. chinensis* var. *chinensis*	F
'Hayward'	*A. chinensis* var. *deliciosa*	F
'Hayward Behi'	*A. chinensis* var. *deliciosa*	F
'Hayward' Clone 8	*A. chinensis* var. *deliciosa*	F
'Hayward' Clone K	*A. chinensis* var. *deliciosa*	F
'Hayward' Clone Maeba®	*A. chinensis* var. *deliciosa*	F
'Hayward' Kramer	*A. chinensis* var. *deliciosa*	F
'HFR18'	*A. chinensis* var. *chinensis*	F
'Hongshi No. 2'	*A. chinensis* var. *chinensis*	F
'Hongyang'	*A. chinensis* var. *chinensis*	F
'Hort16A'	*A. chinensis* var. *chinensis*	F
Fruit marketed as Zespri Gold™ Kiwifruit		
'Hort22D'	*A. chinensis* var. *chinensis*	F
'Hortgem Rua'	*A. arguta*	F
'Hortgem Tahi'	*A. arguta*	F
'Huate' ('White')	*A. eriantha*	F
'Huayou'	*A. chinensis* var. *chinensis*	F
'Issai'	*A. arguta*	F
Different genotypes may be known by this name		
'Jecy Gold'	*A. chinensis* var. *chinensis*	F
'Jenny'	*A. chinensis* var. *deliciosa*	MF
'Jinkui' **('Emihoutao No. 1')**	*A. chinensis* var. *deliciosa*	F
'Jinmin'	*A. chinensis* var. *deliciosa*	F
'Jinmei'	(*A. eriantha* × *A. chinensis* var. *chinensis*) × *A. chinensis* var. *chinensis*	F

Continued

Continued.

Cultivar	Taxon	Gender
'Jinshi No. 1'	*A. chinensis* var. *chinensis*	F
'Jintao' (WIB 6)	*A. chinensis* var. *chinensis*	F
'Jinxia'	*A. chinensis* var. *chinensis*	F
'Jinyan' (M3)	*A. eriantha* × *A. chinensis* var. *chinensis*	F
'Jinyang' **('Emihoutao No. 3')**	*A. chinensis* var. *chinensis*	F
'Jinyang No. 1'	*A. chinensis* var. *chinensis*	F
'Jinyu'	*A. chinensis* var. *chinensis*	F
'Jinyuan'	(*A. chinensis* var. *chinensis* × *A. eriantha)* × *A. chinensis var. chinensis*	F
'Joker'	*A. arguta*	M
'Jumbo'	*A. arguta*	F
'K56'		M
'Kaimai' (rootstock)	*A. hemsleyana* × *A. eriantha*[1]	
'Ken's Red'	*A. arguta* × *A. melanandra*	F
'Kochi'	*A. arguta*	F
'Koryoku'	*A. chinensis* var. *deliciosa*	F
'Kosui'	*A. rufa* × *A. chinensis* var. *chinensis*[1]	F
'KPSC'	*A. kolomikta*	F
'Kuilv'	*A. arguta*	F
'LD133'	*A. arguta*	F
'Longcheng No. 2'	*A. arguta*	F
'Lushanxiang'	*A. chinensis* var. *chinensis*	F
'M36'	*A. chinensis* var. *deliciosa*	M
'M43'	*A. chinensis* var. *deliciosa*	M
'M56'	*A. chinensis* var. *deliciosa*	M
'M91'	*A. chinensis* var. *chinensis*	M
'MacLoughlin' (M352)	*A. chinensis* var. *chinensis*	M
'Male'	*A. arguta*	M
'Matua'	*A. chinensis* var. *deliciosa*	M
'Meader'	*A. arguta*	M
'Meader Male'	*A. arguta*	M
'Meris'	*A. chinensis* var. *deliciosa*	F
'Meteor' (CK3)	*A. chinensis* var. *chinensis*	M
'Miliang' (rootstock)	*A. chinensis* var. *deliciosa*	M
'Miliang No. 1'	*A. chinensis* var. *deliciosa*	F
'Mitsuko'	*A. arguta*	F
'Monty'	*A. chinensis* var. *deliciosa*	F
'Moshan No. 4'	*A. chinensis* var. *deliciosa*	M
'Muller' (MC482)	*A. chinensis* var. *chinensis*	M
'Nijitaro'	*A. chinensis* var. *chinensis*	M
'No. 1 Maoxiong'	*A. eriantha*	M
'Nongda Jinmi'	*A. chinensis* var. *chinensis*	F
'Nongda Mixiang'	*A. chinensis* var. *deliciosa*	F
'Nongda Yuxiang'	*A. chinensis* var. *deliciosa*	F
'Nostino'	*A. arguta*	M
'Pohwa'	*A. chinensis* var. *deliciosa*	M
Pukekohe Dwarf	*A. chinensis* var. *deliciosa*	sterile
'Qihong'	*A. chinensis* var. *chinensis*	F
'Qinmei'	*A. chinensis* var. *deliciosa*	F
Rainbow Red ('Collective name used for red-fleshed *A. chinensis* fruit')	*A. chinensis* var. *chinensis*	F
Red5	*A. chinensis* var. *chinensis*	F

Continued

Continued.

Cultivar	Taxon	Gender
'Red Queen'	*A. chinensis* var. *chinensis* × *A. chinensis* var. *deliciosa*	F
'Redvita'	*A. chinensis* var. *chinensis*	F
'Renact'	*A. chinensis* var. *deliciosa*	MF
'Ruiyu'	*A. chinensis* var. *deliciosa*	F
'Russell'	*A. chinensis* var. *chinensis*	M
'Sanuki Gold'	*A. chinensis* var. *chinensis*	F
'Sanuki kafunriki'	*A. chinensis* var. *chinensis*	M
'Senty'	*A. kolomikta*	F
'SF0612M'	*A. chinensis* var. *chinensis*	M
'Skelton A19' Fruit marketed as EnzaGold™	*A. chinensis* var. *chinensis*	F
'Soreli' (Ac 171.76)	*A. chinensis* var. *chinensis*	F
'Sparkler' (CK2)	*A. chinensis* var. *chinensis*	M
'Summerkiwi 3373'	*A. chinensis* var. *deliciosa*	F
'Summerfaenza'	*A. chinensis* var. *deliciosa*	M
'Taihang-Xiongying'	*A. chinensis* var. *deliciosa*	M
'Tomua'	*A. chinensis* var. *deliciosa*	F
'Tomuri'	*A. chinensis* var. *deliciosa*	M
Top Star®	*A. chinensis* var. *deliciosa*	F
'Tsechelidis' (Megakiwi™)	*A. chinensis* var. *deliciosa*	F
'Weiki'	*A. arguta*	F
'WF 1'	*A. eriantha*	M
'Wilkins Super'	*A. chinensis* var. *deliciosa*	F
'Wuzhi No. 3'	*A. chinensis* var. *chinensis*	F
'Wuzhi-7'	*A. chinensis* var. *chinensis*	F
'Xiangji'	*A. chinensis* var. *deliciosa*	F fruit parthenocarpic
'Xuguan'	*A. chinensis* var. *deliciosa*	F
'Xuxiang'	*A. chinensis* var. *deliciosa*	F
'Y368'	*A. chinensis* var. *chinensis*	F
'Y374'	*A. chinensis* var. *chinensis*	F
'Yangshi Jinhong No. 50'	*A. chinensis* var. *chinensis*	F
'Yuxiang'	*A. chinensis* var. *deliciosa*	F
'Z487'	*A. chinensis* var. *chinensis*	F
'Z1 Vitroplant' (Vip Zedone®)	*A. chinensis* var. *deliciosa* × *A. arguta* 'Gemma'	unstated
'Zes006'	*A. chinensis* var. *chinensis*	F
'Zes007'	*A. chinensis* var. *deliciosa*	F
'Zes008' Fruit marketed as Zespri RubyRed™ Kiwifruit	*A. chinensis* var. *chinensis*	F
'Zesh004' Fruit marketed as Zespri Sweet Green™ Kiwifruit	*A. chinensis* var. *deliciosa* × *A. chinensis* var. *chinensis*	F
'Zesy002' (Gold3) Fruit marketed as Zespri SunGold™ Kiwifruit	*A. chinensis* var. *chinensis*	F
'Zesy003' (Gold9) Fruit marketed as Zespri Charm™ Kiwifruit	*A. chinensis* var. *chinensis*	F
'Zhongmi No. 2'	*A. chinensis* var. *deliciosa*	F
'Zuva' (A 125.93)	*A. chinensis* var. *chinensis*	M
'Zhuxi No. 2'	*A. chinensis* var. *deliciosa*	F
'74-49'	*A. arguta*	F
'74-46'	*A. arguta*	M

[1]Probable taxon.

1 History of Kiwifruit: Evolution of a Global Crop

A. Ross Ferguson[1]*, Hongwen Huang[2] and Guglielmo Costa[3]
[1]The New Zealand Institute for Plant and Food Research Ltd, Auckland, New Zealand; [2]Lushan Botanical Garden, Chinese Academy of Sciences, Jiujiang, Jiangxi, China; [3]Alma Mater Studiorum – University of Bologna, Bologna, Italy

1.1 Introduction

In the *New Zealand Oxford Dictionary* of 2005, kiwifruit are described as having 'fruit with a thin, hairy skin, green flesh and black seeds'. This narrow definition is because the name kiwifruit had originally been devised in New Zealand in 1959 for fruit of the cultivar 'Hayward' of *Actinidia chinensis* var. *deliciosa* (Ferguson, 2020a). 'Hayward' kiwifruit have a hairy skin and brilliant, green fruit flesh. The new name was quickly accepted in export markets and for many years was essentially restricted to 'Hayward' fruit as the world kiwifruit industry was then totally dominated by this cultivar. From 2000 onwards, the name became extended to *A. chinensis* var. *chinensis* when fruit of that variety, with less hairy skins and fruit flesh of different colours, became available in international trade. The name is now sometimes even used for the whole of the genus *Actinidia*, both the fruit and the vines. The name kiwiberry is used for fruit of *A. arguta* and some related species (see Chapter 7).

Actinidia chinensis comes from China. There, the name mihoutao (Fig. 1.1) has generally been used for *Actinidia* fruit, particularly fruit of *A. chinensis*. Mihoutao translates as macaque monkey peach since the ripe fruit are eaten by the monkeys in the mountains (Fig. 1.2). In Chinese botanical literature, mihoutao is usually taken as referring to the whole genus *Actinidia* (mihoutao shu) and *A. chinensis* as zhonghua mihoutao (Chinese mihoutao). Other names have been used in China for *A. chinensis*, but mainly only in restricted areas. The species is variable, particularly in the hairiness of the fruit, as indicated by Chinese names such as yangtao (sun or goat peach) given to variants with smooth, less hairy fruit and mao yangtao for those with hairier fruit (Dunn, 1911; Wilson, 1913).

Early plant collectors in China and botanists were also aware of this natural variation in *A. chinensis* wild populations (Ferguson, 1990a). Fairchild (1913) published a photograph taken by the plant collector E.H. Wilson, labelling it 'A plate of yang taw fruits. One of at least two distinct strains of *Actinidia chinensis* Planch, which have been observed in China'. The fruit shown are clearly hairy, very similar to the 'Hayward' fruit produced by many commercial orchards today.

The consequent, rather convoluted, nomenclatural history of the two main variants ('distinct strains') of *A. chinensis* in mainland China is described in Chapter 3. Currently, they are considered as varieties of the one species, the less

*Corresponding author: ross.ferguson@plantandfood.co.nz

© CAB International 2024. *Kiwifruit: Botany, Production and Uses* (eds A.C. Richardson, J.N. Burdon and A.R. Ferguson)
DOI: 10.1079/9781800620933.0001

猕猴桃

Fig. 1.1. The simplified Chinese characters for mihoutao (monkey peach). (Photo: The New Zealand Institute for Plant and Food Research Ltd)

Fig. 1.2. Macaque monkey eating a kiwifruit; hence the Chinese name mihoutao or monkey peach. (Photo: Aneesh Thomas, EyeEM)

Fig. 1.3. The hairs on the skin of *Actinidia chinensis* var. *chinensis* fruit (*left*) are typically much finer and softer than those on the skin of *A. chinensis* var. *deliciosa* fruit (*right*). (Photo: The New Zealand Institute for Plant and Food Research Ltd)

hairy as *A. chinensis* var. *chinensis* and the hairier as *A. chinensis* var. *deliciosa* (Fig. 1.3). The other obvious difference for consumers is in the colour of the fruit flesh: the fruit flesh of *A. chinensis* var. *chinensis* is usually golden yellow, although occasionally green or sometimes red to varying extents, whereas that of *A. chinensis* var. *deliciosa* is almost always green, but rarely with some red. Other morphological differences between the two varieties are described in Chapter 3, as are their overlapping geographical distributions.

The two varieties of *A. chinensis* also differ in the place and timing of their initial domestication. The pathways of their domestication are classic examples of domestication by a single introduction of wild germplasm in the case of *A. chinensis* var. *deliciosa* and subsequently by multiple selections from wild germplasm of both *A. chinensis* var. *chinensis* and *A. chinensis* var. *deliciosa*.

The type specimen and the first illustration in the European scientific literature were of *A. chinensis* var. *chinensis* (Fig. 1.4) but it was *A. chinensis* var. *deliciosa* that was domesticated first. Its

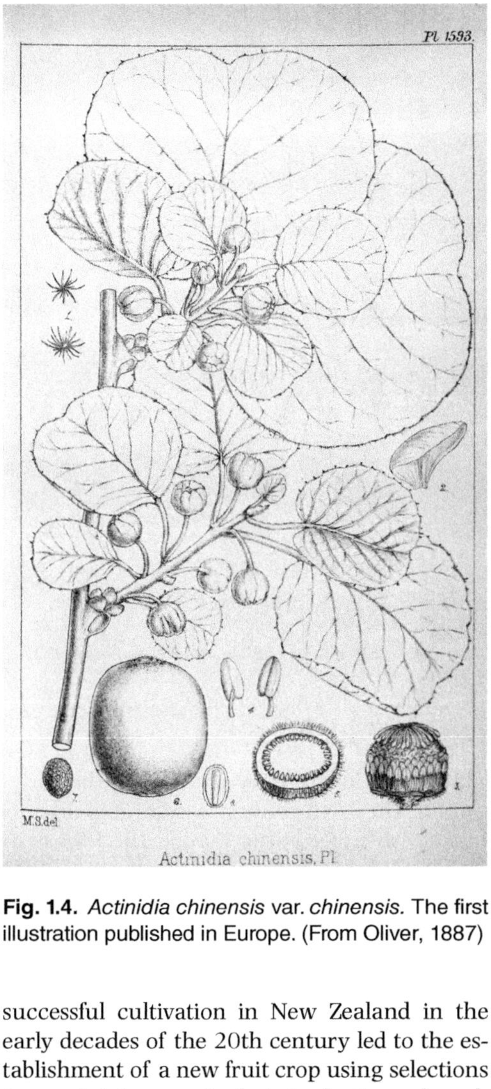

Fig. 1.4. *Actinidia chinensis* var. *chinensis.* The first illustration published in Europe. (From Oliver, 1887)

successful cultivation in New Zealand in the early decades of the 20th century led to the establishment of a new fruit crop using selections descended from a single introduction of seed. This seems to have been largely a matter of luck, although the expertise and efforts of early nurserymen, orchardists and agricultural advisors should not be dismissed. Probably of greatest consequence was the selection of the cultivar 'Hayward', which eventually became the only important fruiting kiwifruit cultivar grown in New Zealand. It was adopted by other countries as they started growing kiwifruit. For many years, the kiwifruit was the cultivar 'Hayward'. 'Most consumers are not even aware that other varieties exist' (Belrose, Inc., 1998).

Half a century after plants of *A. chinensis* var. *deliciosa* first fruited in New Zealand, the success of the expanding New Zealand kiwifruit industry encouraged detailed evaluation of the *Actinidia* germplasm resources in China. Multiple selections from the wild of superior genotypes of both *A. chinensis* var. *chinensis* and *A. chinensis* var. *deliciosa* provided most of the cultivars now grown commercially in China. This was the result of a long-term vision and a well-formulated strategy (Huang, 2022) based on the knowledge that kiwifruit cultivation could be profitable. China brought *A. chinensis* var. *chinensis* into cultivation and thus pioneered the production of yellow-fleshed and red-fleshed kiwifruit.

1.2 The Wild Mihoutao in China

Despite extensive forest clearance, wild *Actinidia* vines are still abundant throughout much of China (Fig. 1.5), even if the distributions of many species are now discontinuous (Huang, 2016). Sporadic attempts were made to cultivate *Actinidia* species, starting about 1200 years ago (Ferguson, 1990b; Cui *et al.*, 2002), but generally, fruit were simply collected from wild plants. Thus, all descriptions in the classic pharmacopoeias (Bencao) and encyclopaedias of the appearance and use of mihoutao as medicine or as fruit for consumption (Fig. 1.6) describe them as being wild plants in the mountains. Thus, the *Bencao Yanyi* [Development of materia medica] compiled by the pharmacist Kou Zongshi between 1111 and 1116 AD (Song Dynasty) described the mihoutao thus: 'The branches are slender and supple, the plant ... supports itself by growing over trees. It is found along pathways in the hills, deep in the mountains. The fruit are much eaten by monkeys' (Yan, 1981). European and American travellers in China in the 19th and early 20th centuries confirmed that mihoutao were not then cultivated in China.

1.3 Domestication and Cultivation of Green-fleshed Kiwifruit, *Actinidia chinensis* var. *deliciosa*

1.3.1 The first plants outside China

The first serious attempts to cultivate *A. chinensis* were not in China but in other countries. At the

Fig. 1.5. Wild kiwifruit or mihoutao. *Actinidia chinensis* var. *deliciosa* growing over trees in Sichuan, China. (Photo: The New Zealand Institute for Plant and Food Research Ltd)

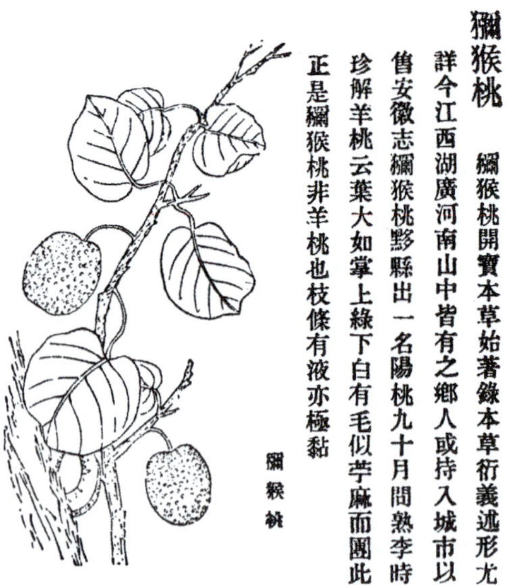

Fig. 1.6. An early Chinese illustration of the mihoutao (*Actinidia chinensis*). The text quotes the *Bencao Gangmu* of Li Shizhen (1518–1593): mihoutao are common in the mountains of Jiangxi, Huguang (largely modern Hubei and Hunan) and Henan and peasants collect the fruit to bring to town to sell in the markets. (From Wu, 1848)

end of the 19th century and during the early years of the 20th century seed and plants of *A. chinensis* were sent to Europe, the USA and New Zealand (Ferguson and Bollard, 1990; Ferguson, 2004). Some introductions, all apparently unsuccessful, were of *A. chinensis* var. *chinensis*; most introductions of *A. chinensis* var. *deliciosa* were only marginally more successful.

In 1904, a London nursery firm, James Veitch & Sons Ltd, advertised plants for sale of what we now know as *A. chinensis* var. *deliciosa*. The seed from which the plants had been raised had been collected by E.H. Wilson (Fig. 1.7). The advertisement said that the vine produced 'edible fruits having the size of walnuts, and the flavour of ripe gooseberries' (James Veitch & Sons, 1904). This seems to be the first time that the flavour of kiwifruit was likened to that of the European gooseberry and is possibly the origin of the former name in English, 'Chinese gooseberry'. The first plants sold by James Veitch & Sons all proved to be male (Fig. 1.8). Eventually, after further introductions of female plants, fruit was occasionally produced in Britain and the USA, but in those countries, *A. chinensis* var. *deliciosa* came, for many years, to be valued mainly as an ornamental plant.

Fig. 1.7. The plant collector E.H. Wilson at the time of his trips to China. (From Anon., 1905)

Fig. 1.8. Staminate flowers of *Actinidia chinensis* var. *deliciosa* from a plant obtained in 1905 from James Veitch & Sons and raised from seed gathered in Hubei, China, by E.H. Wilson. (From Sprague, 1914)

1.3.2 The first plants in New Zealand

One successful introduction of *A. chinensis* var. *deliciosa* seed to New Zealand in 1904 ultimately led to the establishment of a new fruit crop (Ferguson and Bollard, 1990; Ferguson, 2004). In 1903, Isabel Fraser (Fig. 1.9) had visited her sister Katie Fraser, a missionary teacher at Ichang (now Yichang), Hubei Province, China. Ichang was the overwintering base for E.H. Wilson on his plant-collecting trips to China. In 1900, he had introduced the fruit of *A. chinensis* to the foreign residents of Ichang '... with whom it found immediate favour' (Wilson, 1913). When Isabel Fraser returned to New Zealand, it is very likely that her seed of *A. chinensis* var. *deliciosa* came directly or indirectly from Wilson. The fruit from which the seed had been extracted probably originated not from around Ichang itself but from further inland in Sichuan.

The seeds were given to Alexander Allison, a farmer in Whanganui, New Zealand. He successfully raised both male and female plants, and by 1910 his vines were producing fruit (Ferguson and Bollard, 1990). These are the first known

Fig. 1.9. Isabel Fraser. After visiting her sister in Ichang, China, in 1903, she returned to New Zealand in 1904 with seed of *Actinidia chinensis* var. *deliciosa*. (From Anon., 1910)

fruit of *A. chinensis* var. *deliciosa* to be produced outside China. Nurserymen became interested and by 1917 plants were being sold to the public under the name 'Chinese gooseberry', a name that was used for the next 40–50 years. A commercial orchard of 14 vines was established in Whanganui in the late 1920s followed by a larger orchard there of about 60 vines. Fruit were being sold in other parts of New Zealand by the mid-1930s. This was the small beginning of a new commercial fruit crop (Fig. 1.10).

1.3.3 Cultivar selection

For many years, New Zealand orchardists were simply encouraged to propagate female Chinese gooseberry strains with large oval or elongated fruit. One selection with large oval fruit was named 'Hayward' after its selector Hayward Wright (Fig. 1.11), an Auckland nurseryman; another was named 'Bruno' after Bruno Just, another nurseryman (Mouat, 1958). These and other selections can be traced back to one male and two female vines (Ferguson and Bollard, 1990) descended from the original *A. chinensis* var. *deliciosa* seed that had arrived in New Zealand in 1904, a very narrow genetic base for a developing industry.

1.3.4 The development of management practices

Successful introduction is only the first step in the development of a new commercial crop. Growers must learn how to manage the plant, select best forms of the plant and succeed in producing profitable crops that can be harvested at the optimal time, stored, if necessary, then marketed (Ferguson and Bollard, 1990). The management techniques being developed (Fig. 1.12) were initially specific to the conditions in New Zealand.

Fig. 1.10. Whanganui, New Zealand: fruit from the first commercial kiwifruit orchard in the world. (From Whelan, 1935)

1.3.5 The consequences of fruit exports

Chinese gooseberries began to be planted in different parts of New Zealand to provide fruit for local consumption. Storage experiments indicated that export of the fruit by sea to overseas markets on the other side of the world was a realistic possibility. Initially, export was

Fig. 1.11. Hayward Wright with canes of his large-fruited selection of *Actinidia chinensis* var. *deliciosa*, later named 'Hayward' in his honour. (From Rice, 1930)

Fig. 1.12. Innovation in vine management. A 21st century orchard in the Bay of Plenty, New Zealand. The aim is to separate replacement canes from the fruiting canopy below. (Photo: The New Zealand Institute for Plant and Food Research Ltd)

a means of using fruit surplus to local requirements. 'Hayward' fruit reached the markets in better condition than did fruit of the other cultivars then grown (Ferguson and Bollard, 1990). As exports became more important, the clear preference of the overseas markets encouraged new plantings of 'Hayward' or reworking of existing orchards to 'Hayward'. By 1975, only 'Hayward' fruit were being recommended for export. The following year, exports exceeded local consumption for the first time. In 1965, the total area of Chinese gooseberry orchards in New Zealand was about 100 ha; this increased to about 1000 ha in 1975 and to 15,000 by 1985. It was the adoption of 'Hayward' that allowed the successful expansion of an industry producing fruit primarily for export.

1.3.6 A change of name – kiwifruit

The name Chinese gooseberry was used in New Zealand until the first exports of fruit to the USA. A new name was then considered necessary (Ferguson, 2020a). 'Kiwifruit' was chosen in 1959 because the bird, the kiwi (*Apteryx*), is the national emblem of New Zealand, and the people of New Zealand are commonly known as kiwis (Fig. 1.13). By 1970, all exports of kiwifruit from New Zealand went under that name. The international success of exports meant that the name became accepted, and as other countries started growing kiwifruit, they too adopted the new name.

1.3.7 Cultivation of kiwifruit spreads to other countries

Sustained promotion of kiwifruit in California, Europe and Japan encouraged orchardists in those places to consider growing kiwifruit. While kiwifruit may have been novel, a demand for the fruit had been created and New Zealand orchardists had shown that kiwifruit could be cropped profitably. Interest was also stimulated by publicity in horticultural journals and

Fig. 1.13. A new name: kiwifruit. Cover of the advertising booklet *Kiwifruit: A Strange New Zealand Phenomenon*. ([New Zealand] Kiwifruit Export Promotion Committee, 1979)

nursery catalogues. Kiwifruit went from being a crop peculiar to New Zealand to one of increasing worldwide distribution. The cultivars selected in New Zealand were accepted and the management practices adopted. As 'Hayward' was becoming the only green-fleshed kiwifruit cultivar grown in New Zealand, it soon predominated elsewhere.

This spread of kiwifruit around the world can be illustrated by the evolution of the Italian kiwifruit industry (Ferguson and Bollard, 1990; Testolin and Ferguson, 2009; Testolin, 2015; Costa *et al.*, 2018). Experimental plantings were established in 1966/67 but the first commercial plantings came about five years later. At first, many female cultivars such as 'Abbott', 'Bruno' and 'Monty' were introduced as well as 'Hayward', but 'Hayward' quickly dominated. Many initial plantings failed or produced low fruit crops because growers were inexperienced and lacked knowledge or because marginal land was used. Once more suitable areas were chosen and growers became more aware of the plant's requirements, the industry developed rapidly, almost doubling in area every two years (Costa *et al.*, 1992). Initially, the small quantities of marketable fruit returned very high prices to orchardists, and this encouraged more planting. Another advantage was that, at that stage, 'Hayward' was substantially free of disease.

The management techniques, support structures and methods of training and pruning were largely those developed in New Zealand (Testolin and Ferguson, 2009; Testolin, 2015) (Fig. 1.14). However, in New Zealand, the bulk of kiwifruit orchards (> 80%) is in one part of the country, the Bay of Plenty. Kiwifruit cultivation in Italy extends throughout the length of the country with more diverse environmental conditions. Therefore, some management practices had to be modified (Costa *et al.*, 2018). The most striking difference between the two countries is the structure of the industries, particularly in marketing. The New Zealand industry is unusually unified with most exports under the control of one marketing organization, Zespri Group Limited. The Italian kiwifruit industry is much less tightly organized with multiple exporters.

In 1978, there were about 600–800 ha of kiwifruit planted in Italy, increasing to ~10,000 ha in 1988 and 27,000 ha in 2008, before decreasing to 24,500 ha in 2020 (Costa and Testolin, 2022). At one stage, kiwifruit production in Italy exceeded that of New Zealand, but from 2015 to 2020, there was a decline in annual production from 575,096 tonnes to only 305,356 tonnes, despite the total area of plantings remaining essentially unchanged (Costa and Testolin, 2022). 'Hayward' remains the most widely grown kiwifruit cultivar in Italy. Towards the end of the 20th century other *A. chinensis* var. *deliciosa* cultivars owned by nurseries or consortia began to be planted and, more recently, cultivars *of A. chinensis* var. *chinensis*.

There are extensive kiwifruit orchards in Chile and in northern hemisphere countries such as France, Greece, Iran, Portugal and the USA (see Chapter 2). In all these countries, 'Hayward' still predominates. 'Hayward' is even important in China, the original home of kiwifruit.

Fig. 1.14. A protected (covered) kiwifruit orchard in Emilia-Romagna, Italy, just after budbreak in spring. (Photo: U. Palara, Italy)

1.4 Domestication and Cultivation of *Actinidia chinensis* var. *chinensis* and *A. chinensis* var. *deliciosa* in China

1.4.1 First kiwifruit orchards

In 1957, the Beijing Botanic Garden, Chinese Academy of Sciences, grew plants of *A. chinensis* var. *deliciosa* raised from seed collected in the Qinling Mountains, Shaanxi, and followed this in 1961 with plants of *A. chinensis* var. *chinensis* from seed collected in the Funiu Mountains, Henan (Huang *et al.*, 2003). These seem to be the first modern attempts to cultivate kiwifruit in China. This was also the start of the domestication of *A. chinensis* var. *chinensis*, even if the first plantings were valuable mainly for scientific research.

The first small commercial kiwifruit orchards in China were established around 1980 in Chibi County, Hubei, and then in Dujiangyan, Sichuan. The cultivars used (*A. chinensis* var. *deliciosa*) and orchard management techniques adopted were based on those used in New Zealand. Possibly the first ever commercial orchard in China, indeed the world, of *A. chinensis* var. *chinensis* was established around 1980 by transplantation of wild vines to an orchard in Xixia, Henan.

1.4.2 Evaluation of wild *Actinidia* germplasm in China

The success of kiwifruit in New Zealand stimulated Chinese horticulturists to initiate a national germplasm survey of wild *Actinidia* resources. The immediate aim was to develop cultivars by selection of superior genotypes to replace the cultivar ‘Hayward’ (Cui *et al.*, 2002; Huang and Ferguson, 2007; Huang, 2012, 2016; Huang and Zhong, 2018). The ultimate aim was a Chinese kiwifruit industry that was competitive with international kiwifruit industries. The plan also emphasized the development of orchard, storage and marketing techniques suited to Chinese conditions, of processed products and possible medical uses.

Starting in 1974, there were extensive surveys of wild *Actinidia* germplasm resources. There are many millions of plants of both *A. chinensis* var. *chinensis* and *A. chinensis* var. *deliciosa* throughout China. In 1974, the Guangxi Forest Institute started surveying the *Actinidia* resources in the mountainous areas of Guangxi Province. In 1975, the Zhengzhou Fruit Institute, Chinese Academy of Agricultural Sciences, collaborated with the Xixia Forest Institute in an extensive survey of *Actinidia* resources in Xixia County, Henan. This survey was extended to other parts of Henan in co-operation with other research institutes during 1976–78. Likewise, in 1978, the Central China Agricultural College at Yichang, Hubei, started surveying the *Actinidia* resources in the mountainous areas of Hubei Province. Eventually, more than 1400 large-fruited plants of *A. chinensis* var. *chinensis* and *A. chinensis* var. *deliciosa* were selected directly from the wild for further study and cultivar development. It is a striking example of fruit crop improvement being based on multiple selections from wild populations of germplasm from widely separated parts of the natural distribution (Huang *et al.*, 2003). The selections varied greatly in many commercial traits (Huang, 2016), including fruit size, vitamin C content, soluble solids content and suitability for storage. Some selections were cold-hardy, drought-tolerant or could adapt readily to the different environments in many parts of China.

1.4.3 Selections of *Actinidia chinensis* var. *deliciosa* in China

A dozen cultivars of *A. chinensis* var. *deliciosa* were initially selected from wild germplasm, including 'Qinmei', 'Guichang' and 'Miliang No. 1'. These, as well as 'Hayward', were important in the initial stages of commercial kiwifruit production in China (Huang and Ferguson, 2001, 2007; Huang, 2016; Huang and Zhong, 2018) and many are still important (see Chapter 4). Some subsequent selections were one generation removed from the wild. In 2020, cultivars of *A. chinensis* var. *deliciosa* comprised about half of Chinese kiwifruit plantings (Zhong *et al.*, 2022; Chapter 4). All these cultivars have green-fleshed fruit.

1.4.4 Selections of *Actinidia chinensis* var. *chinensis* in China

Surveys of wild *Actinidia* resources in China quickly revealed the potential of *A. chinensis* var. *chinensis* genotypes. As early as 1985–90, several cultivars selected from wild populations were released. Nearly all these had yellow-fleshed fruit. Since then, many more cultivars of *A. chinensis* var. *chinensis* have been authorized or registered in China than have cultivars of *A. chinensis* var. *deliciosa*.

In 2002, only about 20% of the total Chinese area planted in kiwifruit was of cultivars of *A. chinensis* var. *chinensis* (Huang *et al.*, 2003). By 2020, this had risen to almost 50%, one-third of which was in yellow-fleshed cultivars and two-thirds in red-fleshed cultivars (Zhong *et al.*, 2022). The domestication of *A. chinensis* var. *chinensis* resulted in China being the first country to have extensive plantings of green-, yellow- and red-fleshed kiwifruit. 'Jinyan' (described in Chapter 4) remains exceptional in that it is the first kiwifruit cultivar resulting from interspecific hybridization to be planted on a commercial scale.

Red-fleshed kiwifruit are uniquely important in China. 'Hongyang', the first red-fleshed cultivar, was selected in Sichuan from seedlings raised from seed collected from wild *A. chinensis* var. *chinensis* populations in Henan (Wu and Li, 1993; Wang *et al.*, 2003). By 2020, it was the most widely planted kiwifruit cultivar in China (66,700 ha), nearly a quarter of the total area of kiwifruit. Its fruit has proved very popular, receiving about three times the wholesale price of 'Hayward' fruit. Its popularity is probably largely due to its perceived sweetness.

1.4.5. Expansion of the Chinese kiwifruit industry

In 1980, there was only about 1 ha of cultivated kiwifruit in China. That year, France had about

600 ha, California and Italy each had an estimated 1000 ha and New Zealand had 3500 ha (Warrington, 1990). Subsequently, the areas planted in kiwifruit in Italy and New Zealand increased rapidly but were soon dwarfed by those in China (Fig. 1.15): 4000 ha in 1990, 45,000 ha in 1998, 57,400 ha in 2002, 70,000 ha in 2011, 286,500 ha in 2020 (Huang and Ferguson, 2001; Huang *et al.*, 2003; Zhong *et al.*, 2022).

A striking feature of the evolution of the Chinese kiwifruit industry since 1980 is the frequent changes in the cultivars grown. There has been a much greater use of cultivars of *A. chinensis* var. *chinensis*, particularly of the red-fleshed cultivars 'Hongyang' and 'Donghong', probably in response to the market prices realized. Perceived fruit quality and market returns have probably also influenced plantings of *A. chinensis* var. *deliciosa* cultivars.

Another notable difference is that kiwifruit production in China is almost entirely for local consumption, with exports being insignificant. In New Zealand and Chile, and, to a lesser extent, other countries, kiwifruit are produced for export.

1.5 *Actinidia chinensis* var. *chinensis* in Countries Other than China

1.5.1 Yellow-fleshed kiwifruit

Until 2000, consumers throughout most of the world had little choice in the kiwifruit they could buy. The only real choice was the country of origin, as different countries in both hemispheres were, by then, producing and marketing 'Hayward' kiwifruit. To consumers, 'Hayward' *was* the kiwifruit. Only in China were fruit of different kiwifruit cultivars readily available, but these were not offered outside China and were not generally known.

Fig. 1.15. Successful kiwifruit cultivation near Chengdu, Sichuan, China. (Photo: C.H. Cheng, Kiwifruit Breeding Centre, New Zealand)

Zespri GOLD™ Kiwifruit were the first yellow-fleshed kiwifruit on the international market, available from 2000. They were fruit of a new *A. chinensis* var. *chinensis* cultivar, 'Hort16A', bred by HortResearch in New Zealand (Muggleston *et al.*, 1998). Limited quantities of seed and budwood of *A. chinensis* var. *chinensis* that had been introduced from China to New Zealand were the starting material for a large breeding programme. 'Hort16A' kiwifruit were novel, differing from 'Hayward' fruit in that they were ovoid with a protruding distal end (beak), the fruit skin was covered by soft fine hairs, rather like the fuzz on a peach, the internal flesh was a beautiful clear yellow (if the fruit were harvested at the right maturity), and the flavour was very different, being sweeter, more aromatic and less tart. 'Hort16A' was also novel in that it was the first commercially successful kiwifruit cultivar to come from a planned breeding programme. It was a protected cultivar, which meant that it could be legally grown only by growers licensed by Zespri, the New Zealand owner. Initially, 'Hort16A' was grown only in New Zealand, but plantings were established under licence in northern hemisphere countries such as France, Italy, South Korea and Japan, to enable year-round marketing of the fruit. This is increasingly the trend with other protected kiwifruit cultivars.

Selection of a new cultivar is only the first step (Martin and Luxton, 2005). Suitable vine management techniques had to be developed, including the selection of suitable pollenizer cultivars since 'Hort16A' flowers some weeks earlier than 'Hayward'. Harvest maturity indices needed to take into account the fruit flesh changing from green to yellow only late in the season. Procedures were also required for the handling and storage of the fruit since they were very different from those of 'Hayward'.

'Hort16A' proved very successful, and the fruit fetched very high prices; it was very profitable for growers. By 2010 it accounted for a third of the New Zealand kiwifruit industry's total sales. Unfortunately, it was also very susceptible to bacterial canker of kiwifruit, caused by *Pseudomonas syringae* pv. *actinidiae* (Psa). Within a couple of years of this disease arriving in New Zealand, 'Hort16A' could no longer be grown commercially. Fortunately, the New Zealand kiwifruit breeding programme had produced a seedling selected on fruit weight, shape and taste (Ferguson, 2020b). This cultivar, 'Zesy002', turned out to be much less susceptible to Psa. It was chosen as a replacement for 'Hort16A', and its fruit marketed as Zespri SunGold™ Kiwifruit. New Zealand is currently the only kiwifruit-producing country in which yellow-fleshed kiwifruit are more important than green-fleshed kiwifruit. In 2022, almost two-thirds, by volume, of the kiwifruit exported from New Zealand were yellow-fleshed and as they fetched much higher prices, they were much more profitable for growers than 'Hayward'. Kiwifruit is now the most valuable of New Zealand's horticultural exports.

Yellow-fleshed *A. chinensis* var. *chinensis* cultivars are also important in Italy (Costa and Testolin, 2022); in 2020, yellow-fleshed cultivars made up about 15% of kiwifruit plantings but 25% of total kiwifruit production. The most widely cultivated is 'Zesy002', grown under licence for Zespri. The second most widely planted yellow-fleshed cultivar is a Chinese cultivar, 'Jintao' (fruit marketed by Jingold S.p.A.). There are also plantings of one other Chinese cultivar, 'Jinyan', and two that are the result of Italian kiwifruit breeding programmes, 'Soreli' and 'AC 1536' (fruit marketed by Consorzio Dorì Europe S.r.l). Limited areas of these yellow-fleshed kiwifruit cultivars have been established in other countries such as Chile and South Africa.

'Jintao' is now, after 'Zesy002', the most widely planted yellow-fleshed kiwifruit cultivar outside China. Many other countries, such as Japan and South Korea, which established kiwifruit industries based on 'Hayward', are also exploring the potential of yellow-fleshed kiwifruit.

1.5.2 Red-fleshed kiwifruit

There are limited plantings in Italy of several red-fleshed cultivars of Chinese origin including 'Hongyang' and 'Donghong'. In South Korea, the red-fleshed cultivar 'Redvita' has been registered (Kwack *et al.*, 2017). In New Zealand, 'Zes008' (Zespri RubyRed™ Kiwifruit), the result of a breeding programme, is being commercialized with the beginning of exports. It is likely that cultivation of red-fleshed kiwifruit will become more general because of keen consumer demand.

1.6 Cultivation of Other *Actinidia* Species

1.6.1 Kiwiberries and related species

The first *Actinidia* species to be cultivated outside their natural distributions were *A. arguta* (kiwiberries), *A. polygama* and *A. kolomikta* (Huang and Ferguson, 2007). Initially, they were valued mainly as ornamentals. Hobbyists in Europe and North America grew vines for fruit but the first commercial kiwiberry orchards seem to be those established in Oregon, USA (Finn, 1999). Since then, orchards have been established in many countries, mainly in China (see Chapter 7).

1.6.2 *Actinidia eriantha*

Superior *A. eriantha* genotypes have been selected and commercial orchards established in Zhejiang, China and South Korea. The cultivar 'Jinyan' is a hybrid between *A. eriantha* and *A. chinensis* var. *chinensis* (see Chapter 4).

1.7 Evolution of Kiwifruit as a Crop

The development of the green-fleshed kiwifruit (*A. chinensis* var. *deliciosa*) as a crop is a classic example of plant domestication from one sampling of wild germplasm. The industry that developed in New Zealand was based on what appears to have been a single introduction of seed from China at the beginning of the 20th century.

The development of the Chinese kiwifruit industry followed a very different pattern. Systematic evaluation of wild germplasm led to the selection and cultivation of superior genotypes of both *A. chinensis* var. *chinensis* and *A. chinensis* var. *deliciosa* with green, yellow or red fruit flesh. This is a rare example of crop development being based on multiple selections from wild germplasm.

Cultivation of yellow-fleshed kiwifruit, *A. chinensis* var. *chinensis*, started with the first commercial orchard in China around 1980. International trade in yellow-fleshed kiwifruit started in 2000 with the commercialization of 'Hort16A', bred in New Zealand. Subsequent loss of 'Hort16A' as a commercial cultivar from disease demonstrates the very real dangers of an industry being based on only a very small number of cultivars. 'Hort16A' was soon replaced by 'Zesy002' and now exports from New Zealand of yellow-fleshed kiwifruit are more important than those of green-fleshed kiwifruit. Yellow-fleshed kiwifruit are likewise becoming more important in other countries.

The first red-fleshed kiwifruit cultivar, 'Hongyang', was registered in 1997 and by 2020 was the most widely planted of all kiwifruit cultivars in China. 'Hongyang' and most of the red-fleshed kiwifruit cultivars grown in China are very closely related but together account for more than 40% of all Chinese kiwifruit plantings. Again, there is an obvious danger in such extensive plantings having so limited a genetic base. Red-fleshed kiwifruit are currently uniquely important in China, but production is starting in other countries.

'Jinyan', the first commercially important interspecific *Actinidia* cultivar (*A. eriantha* × *A. chinensis* var. *chinensis*) was authorized as a new kiwifruit cultivar by the Chinese National Forestry Cultivar Approval Committee in 2010.

We can reasonably expect, soon, sweeter, green-fleshed cultivars; red-fleshed cultivars with larger fruit that are more consistently red; hermaphrodite, self-setting cultivars; more cultivars coming from interspecific crosses. Perhaps green-fleshed kiwifruit will steadily become less important. Certainly, the dominance of 'Hayward' is being rapidly eroded. Thus, the evolution of the kiwifruit as a crop plant continues; indeed, it appears to be accelerating.

References

Anon. (1905) Plant collecting in China. *The Gardeners' Chronicle*, 3rd Series 37, 113–114.
Anon. (1910) [Photograph of Isabel Fraser]. *The Adastrian* ((Annual Magazine of the Girls' College, Wanganui) 13.

Belrose, Inc. (1998) *World Kiwifruit Review 1998*. Belrose, Inc., Pullman, Washington.
Costa, G. and Testolin, R. (2022) Recent changes in the kiwifruit (*Actinidia* spp.) industry since its introduction: Italy as a case study. *Acta Horticulturae* 1332, 1–10. DOI: 10.17660/ActaHortic.2022.1332.1
Costa, G., Testolin, R. and Biasi, R. (1992) Situazione attuale e prospettive della coltura dell'actinidia in Italia. *Rivista di Frutticoltura e di Ortofloricoltura* 54(9), 7–16.
Costa, G., Ferguson, R., Huang, H. and Testolin, R. (2018) Main changes in the kiwifruit industry since its introduction: present situation and future. *Acta Horticulturae* 1218, 1–16. DOI: 10.17660/ActaHortic.2018.1218.1
Cui, Z., Huang, H. and Xiao, X. (2002) *Actinidia in China*. Agricultural Science and Technology Press, Beijing.
Dunn, S.T. (1911) A revision of the genus *Actinidia*, Lindl. *Journal of the Linnean Society of London, Botany* 38, 394–410 + plate. DOI: 10.1111/j.1095-8339.1911.tb02326.x
Fairchild, D. (1913) Some Asiatic Actinidias. In: Circular 110 (Miscellaneous Paper), Bureau of Plant Industry, United States Department of Agriculture. Government Printing Office, Washington, DC, pp. 7–12 + 2 plates.
Ferguson, A.R. (1990a) Botanical nomenclature: *Actinidia chinensis*, *Actinidia deliciosa*, and *Actinidia setosa*. In: Warrington, I.J. and Weston, G.C. (eds) *Kiwifruit: Science and Management*. Ray Richards Publisher in association with the New Zealand Society for Horticultural Science, Auckland, New Zealand, pp. 36–57.
Ferguson, A.R. (1990b) The kiwifruit in China. In: Warrington, I.J. and Weston, G.C. (eds) *Kiwifruit: Science and Management*. Ray Richards Publisher in association with the New Zealand Society for Horticultural Science, Auckland, New Zealand, pp. 155–164.
Ferguson, A.R. (2004) 1904 – the year that kiwifruit (*Actinidia deliciosa*) came to New Zealand. *New Zealand Journal of Crop and Horticultural Science* 32, 3–27. DOI: 10.1080/01140671.2004.9514276
Ferguson, A.R. (2020a) Changing names: tamarillo and kiwifruit. *New Zealand Garden Journal* 23(1), 4–8.
Ferguson, A.R. (2020b) Plant Raisers' Award: Russell George Lowe. *New Zealand Garden Journal* 23(1), 19.
Ferguson, A.R. and Bollard, E.G. (1990) Domestication of the kiwifruit. In: Warrington, I.J. and Weston, G.C. (eds) *Kiwifruit: Science and Management*. Ray Richards Publisher in association with the New Zealand Society for Horticultural Science, Auckland, New Zealand, pp. 165–246.
Finn, C. (1999) Temperate berry crops. In: Janick, J. (ed.) *Perspectives on New Crops and New Uses*. ASHS Press, Alexandra, Virginia, pp. 324–344.
Huang, H. (2012) [*Germplasm Resources and Cultivars of Actinidia in China*.] China Forest Publishing House, Beijing.
Huang, H. (2016) *Kiwifruit: The Genus* Actinidia. Elsevier, London.
Huang, H. (2022) Discovery and domestication of new fruit trees in the 21st century. *Plants* 11, 2107. DOI: 10.3390/plants 11162107
Huang, H. and Ferguson, A.R. (2001) Review: kiwifruit in China. *New Zealand Journal of Crop and Horticultural Science* 29, 1–14. DOI: 10.1080/01140671.2001.9514154
Huang, H. and Ferguson, A.R. (2007) Genetic resources of kiwifruit: domestication and breeding. *Horticultural Reviews* 33, 1–121. DOI: 10.1002/9780470168011.ch1
Huang, H. and Zhong, C. (2018) [*Forty Years of Actinidia Research and Kiwifruit Industry in China*.] China University of Science and Technology Press, Hefei, China.
Huang, H., Wang, S., Jiang, Z., Zhang, Z. and Gong, J. (2003) Exploration of *Actinidia* genetic resources and development of kiwifruit industry in China. *Acta Horticulturae* 610, 29–43. DOI: 10.17660/ActaHortic.2003.610.2
James Veitch & Sons Ltd (1904) *Novelties Offered by James Veitch & Sons Ltd*, Royal Exotic Nursery, Chelsea, SW. James Veitch & Sons Ltd, Chelsea, London, UK. Available at: Special Collections & Archives Research Center, http://scarc.library.oregonstate.edu/omeka/items/show/231 (accessed 31 August 2022).
Kwack, Y.-B., Kim, H.-L., Lee, J.-H., Chung, K.H. and Chae, W.B. (2017) 'Redvita': a yellow-fleshed kiwifruit with red color around the core. *Horticultural Science and Technology* 35, 387–391. DOI: 10.12972/kjhst.20170041
Martin, R.A. and Luxton, P. (2005) The successful commercialisation of Zespri™ Gold Kiwifruit. *Acta Horticulturae* 694, 35–40. DOI: 10.17660/ActaHortic.2005.694.2
Mouat, H.M. (1958) New Zealand varieties of yang-tao or Chinese gooseberry. *New Zealand Journal of Agriculture* 97, 161–165.
Muggleston, S., McNeilage, M., Lowe, R. and Marsh, H. (1998) Breeding new kiwifruit cultivars: the creation of Hort16A and Tomua. *The Orchardist of New Zealand* 71(8), 38–40.

Oliver, D. (1887) *Actinidia chinensis*, Planchon. *Hooker's Icones Plantarum* 16, Ser. 3, 6, plate 1593.
Rice, W.H. (1930) Chinese gooseberry. The New Zealand Smallholder 12, 276, 277.
Sprague, T.A. (1914) *Actinidia chinensis. Curtis's Botanical Magazine* 140, Tab. 8536. Image available at http://plantillustrations.org/illustratration.php?id_illustration=7242 (accessed 6 June 2023).
Testolin, R. (2015) Kiwifruit (*Actinidia* spp.) in Italy: the history of the industry, international scientific cooperation and recent advances in genetics and breeding. *Acta Horticulturae* 1096, 47–61. DOI: 10.17660/ActaHortic.2015.1096.2
Testolin, R. and Ferguson, A.R. (2009) Kiwifruit (*Actinidia* spp.) production and marketing in Italy. *New Zealand Journal of Crop and Horticultural Science* 37, 1–32. DOI: 10.1080/01140670909510246
Wang, M., Li, M. and Meng, A. (2003) Selection of a new red-fleshed kiwifruit cultivar 'Hongyang'. *Acta Horticulturae* 610, 115–117. DOI: 10.17660/ActaHortic.2003.610.13
Warrington, I.J. (1990) Areas and trends of kiwifruit production in New Zealand and around the world. In: Warrington, I.J. and Weston, G.C. (eds) *Kiwifruit: Science and Management*. Ray Richards Publisher in association with the New Zealand Society for Horticultural Science, Auckland, New Zealand, pp. 512–525.
Whelan, J.W. (1935) Chinese gooseberry (*Actinidia chinensis*). *The Guide* (Horticulture Division, Department of Agriculture, New Zealand) 4, 326–329.
Wilson, E.H. (1913) *A Naturalist in Western China with Vasculum, Camera and Gun*. Methuen & Co, Ltd, London.
Wu, B.-L. and Li, X.-D. (1993) 'Hongyang' – a red-fleshed, top quality and good storing kiwifruit cultivar. *China Fruits* 1993(4), 15, 27.
Wu, Q.-J. (1848 [1965]) *Zhiwu Mingshi Tukao [Illustrated account of flora]*. Taipei, Taiwan, Shangwu.
Yan, J.Y. (1981) Histoire d'*Actinidia chinensis* Planch. et conditions actuelles de sa production à l'étranger. *Journal d'Agriculture Tropicale et de Botanique Appliquée* 28, 281–290. DOI: 10.3406/jatba.1981.3847
Zhong, C., Huang, W., Wang, Z., Li, L., Li, D. *et al.* (2022) The breeding progress and development status of the kiwifruit industry in China. *Acta Horticulturae* 1332, 445–454. DOI: 10.17660/ActaHortic.2022.1332.59

2 Global Industry and Markets

Jayne E. Chamberlain
Kiwifruit Breeding Centre, Te Puke, New Zealand

2.1 Introduction

The global kiwifruit sector has experienced significant growth in the first 20 years of this century, both in increased supply and increased value. The total fresh fruit category experienced strong growth, with volumes increasing by ~54% from 2000 to 2020. However, global kiwifruit supply more than doubled over the same period. What is most impressive for kiwifruit is that the value of fruit traded increased over fourfold from 2000 to 2020.

This chapter looks in more detail at the production trends for kiwifruit against total fresh fruit supply, considers the leading producing countries, the importance of export trading for the kiwifruit sector and how the percentage mix of kiwifruit cultivars in production is changing. With the impressive increase in value generated from the global kiwifruit sector over the past 20 years, the chapter also outlines what is driving increased demand, the potential risks to kiwifruit demand and supply, and the key market drivers and opportunities for the sector. The global kiwifruit industry is focused on large fruited *Actinidia chinensis* var. *chinensis* and *A. chinensis* var. *deliciosa* cultivars. For details on kiwiberry (*A. arguta*), a small proportion of the overall kiwifruit sector/market, see Chapter 7.

2.2 Global Kiwifruit Supply

2.2.1 What is total global kiwifruit production compared with total fresh fruit supply?

Global production of both total fresh fruit and kiwifruit increased during the period 2000–2020 (Fig. 2.1). Total fresh fruit increased from 577 million metric tonnes (Mt) in 2000 to 887 Mt in 2020, an increase of ~54% (FAO, 2022). Over the same time period, kiwifruit production increased from 1.87 Mt in 2000 to 4.4 Mt in 2020, an increase of ~136%. As a proportion of the total fresh fruit supply, kiwifruit increased from 0.32% in 2000 to 0.50% in 2020. Other fresh fruits that had greater production growth than kiwifruit over this same period were blueberries (303%) and avocados (199%) (FAO, 2022).

2.2.2 Which are the leading countries producing kiwifruit?

China is the biggest producer of kiwifruit with ~50% of the total kiwifruit produced in 2020. In 2020, production in China was more than three times greater than that of the second-ranked

Address for correspondence: jayne.chamberlain@next.kiwi

© CAB International 2024. *Kiwifruit: Botany, Production and Uses* (eds A.C. Richardson, J.N. Burdon and A.R. Ferguson)

DOI: 10.1079/9781800620933.0002

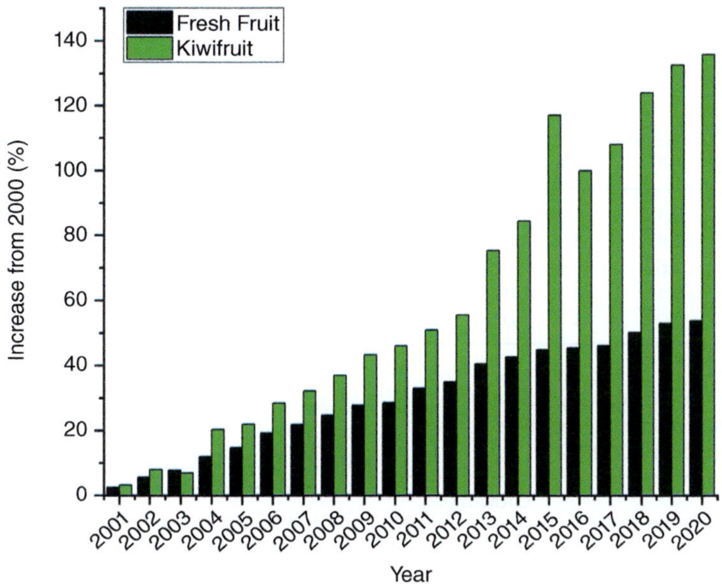

Fig. 2.1. Percentage increase in global production of fresh fruit and kiwifruit since 2000. Baselines in 2000: fresh fruit 577 Mt; kiwifruit 1.87 Mt. (From FAO, 2022)

producing country, New Zealand, which had a 14% share. Italy was the third-ranked country with a ~12% share (Fig. 2.2).

According to FAO data, this represents a change in rankings as, historically (2017 and prior), Italy ranked second in total kiwifruit production, ahead of New Zealand (Table 2.1). Europe has experienced poor weather conditions in recent years which has affected growth in Italian production (*FreshPlaza*, 2021; Table 2.2). Italy has also had challenges, with vines being removed due to kiwifruit vine decline syndrome (also called *moria* or *Moria del Kiwi* in Italy) and the ongoing management of the bacterial disease caused by *Pseudomonas syringae* pv. *actinidiae* (Psa) (Costa and Testolin, 2022). This is whilst New Zealand kiwifruit production has continued to increase (Table 2.2) owing to Zespri Group Limited (marketer and exporter of New Zealand kiwifruit) releasing commercial licences for orchardists to grow their proprietary *A. chinensis* var. *chinensis* 'Zesy002' cultivar (fruit sold as Zespri SunGold™ Kiwifruit). China's production is gradually expanding (Table 2.2), mainly due to the increasing production of new cultivars and with the execution of a comprehensive, targeted, poverty-alleviation national task force that has made many remote mountainous areas suitable for kiwifruit planting (Zhong *et al.*, 2022). One of the countries with the largest growth in production is Greece, where kiwifruit production increased by 88% between 2016 and 2020 (Table 2.2), albeit from a relatively low production base. This was largely due to growth in plantings of *A. chinensis* var. *deliciosa* 'Hayward' kiwifruit and improved management practices.

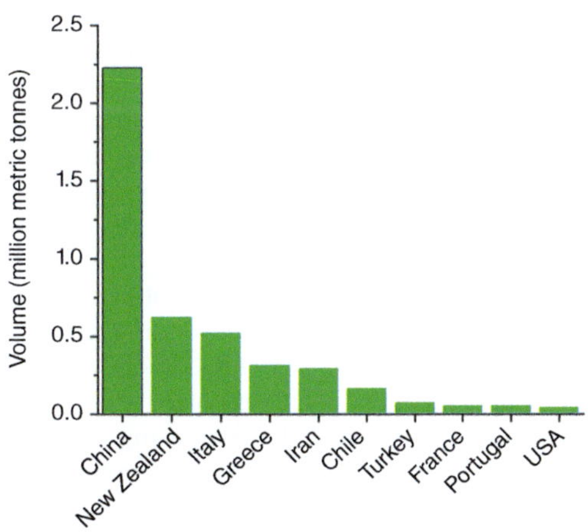

Fig. 2.2. Production volumes (Mt) of the top ten kiwifruit-producing countries in 2020. (From FAO, 2022)

Table 2.1. Production of kiwifruit (in Mt) and ranking of the top five producing countries between 2016 and 2020. (From FAO, 2022)

	2016		2017		2018		2019		2020	
Country	Volume (Mt)	Rank	Volume (Mt)	Rank	Volume (Mt)	Rank	Volume (Mt)	Rank	Volume (Mt)	Rank
China	2.12	1	2.13	1	2.11	1	2.20	1	2.23	1
New Zealand	0.47	3	0.44	3	0.60	2	0.60	2	0.62	2
Italy	0.52	2	0.54	2	0.56	3	0.52	3	0.52	3
Greece	0.16	5	0.20	5	0.27	5	0.29	5	0.31	4
Iran	0.28	4	0.29	4	0.29	4	0.29	4	0.29	5

Table 2.2. Production growth (by fruit volume) for the top five kiwifruit producing countries between 2016 and 2020. (From FAO, 2022)

Country	Growth (%)
China	5
New Zealand	34
Italy	0
Greece	88
Iran	3

Yellow-fleshed kiwifruit cultivars, e.g. 'Zesy002', are also being trialled in Greece to determine if standards can be met to commercialize these cultivars. It is worth mentioning Iran, in fifth position (Fig. 2.2; Table 2.2), where production increased by around 3% between 2016 and 2020 (FAO, 2022).

Data on production of kiwifruit can differ from that presented above, depending on the source. For instance, based on CSO-Italy data, Costa and Testolin (2022) reported that Italy produced 0.42 Mt in 2016 decreasing to 0.31 Mt in 2020, an approximate 30% reduction. Similarly, for China, a survey in 2021 by the Kiwifruit Section of the Chinese Society for Horticultural Science (Zhong *et al.*, 2022) found China's production increased to 3.33 Mt in 2020 from 2.4 Mt in 2016, an approximate 40% increase; significantly higher than the 5% increase calculated from FAO data.

2.2.3 Has the mix of commercial kiwifruit cultivars changed?

Historically, 'Hayward' was the dominant kiwifruit cultivar. However, over the last 5–10 years, the cultivar mix being grown has changed in each of the top three kiwifruit producing countries. In China, there seems to have been a shift to more red-fleshed plantings and in Italy and New Zealand, yellow-fleshed kiwifruit has had the biggest proportional increases. In all these countries, the green-fleshed cultivar 'Hayward' had the biggest proportional decreases. In the following section, these production trends are investigated for China, New Zealand and Italy.

China

In 2020, green-fleshed kiwifruit cultivars still had the largest planted area in China at 142,867 ha (~50% of total kiwifruit plantings), with red-fleshed cultivars at 93,267 ha (~33%) and yellow-fleshed cultivars at 45,267 ha (~16%) (Zhong *et al.*, 2022).

It is interesting to note the changes in percentage of plantings attributable to different cultivars between 2011 and 2020 (Table 2.3). There has been a decline in 'Hayward', with 'Hongyang' and 'Donghong', both red-fleshed cultivars, and 'Xuxiang' and 'Cuixiang', both green-fleshed cultivars, having the biggest increases. In 2020, 'Hongyang' accounted for the largest proportion at 23% of estimated kiwifruit plantings in China.

New Zealand

Zespri's yellow-fleshed cultivar, 'Zesy002', has increased in plantings and overall production, relative to 'Hayward', since 2017 (Table 2.4). 'Zesy002' is more productive than 'Hayward', so although the area planted in 'Zesy002' was smaller, its overall fruit production exceeded that of 'Hayward' in 2020. New Zealand's area of

Table 2.3. Change in the planted area of the main kiwifruit cultivars in China between 2011 and 2020. (From Zhong *et al.*, 2022)

		Planted area			
		2011		2020	
Cultivar	Flesh colour	ha	%	ha	%
'Hongyang'	Red	10,710	15.3	66,667	23.3
'Xuxiang'	Green	6,930	9.9	53,333	18.6
'Cuixiang'	Green	0	0	20,000	7.0
'Donghong'	Red	0	0	20,000	7.0
'Hayward'	Green	21,420	30.6	16,667	5.8
'Guichang'	Green	<1,400	<2	16,000	5.6
'Jinyan'	Yellow	3,150	4.5	14,000	4.9
'Miliang No. 1'	Yellow/Green	2,420	3.6	10,000	3.5
'Jintao'	Yellow	<350	<0.5	7,333	2.6
'Qinmei'	Green	5,040	7.2	5,330	1.9
'Ruiyu'	Green	0	0	4,011	1.4
'Jinkui'	Green	5,040	7.2	2,664	0.9
'Cuiyu'	Yellow	<1,400	<0.2	2,664	0.9

Note: The total kiwifruit planting area in China was 70,000 ha in 2011 and 286,500 ha in 2020.

Table 2.4. Proportion of producing land area and fruit production of kiwifruit by flesh colour in New Zealand from 2017 to 2020. (From Zespri Group Limited, 2021a)

	Producing area (%)				Production volume (%)			
Flesh colour	2017	2018	2019	2020	2017	2018	2019	2020
Green	63	61	58	54	57	56	49	45
Yellow	37	39	42	46	43	44	51	55

land producing green-fleshed cultivars ('Hayward' and 'Zesh004', marketed as Zespri Sweet Green™ Kiwifruit) decreased between 2017 and 2020 by about 10%, primarily due to the replacement of 'Hayward' with 'Zesy002'. 'Zesh004' accounts for only a small proportion of the land used to produce kiwifruit in New Zealand, at 48 ha in 2020, from a total across all cultivars of 13,334 ha (Zespri Group Limited, 2021a). Zespri released licences for a proprietary, red-fleshed kiwifruit (*A. chinensis* var. *chinensis* 'Zes008', marketed as Zespri RubyRed™ Kiwifruit) in 2020, with small volumes available from 2022.

Italy

Consistent with China and New Zealand, the proportion of green kiwifruit production has decreased in Italy (Table 2.5). Production of yellow-fleshed kiwifruit has increased steadily from 2015, and by 2020 constituted 24% of all Italian kiwifruit production. 'Zesy002' dominates the plantings of yellow-fleshed kiwifruit, accounting for an estimated 77% of the planted area of yellow-fleshed vines in 2020 (3085 ha) with lesser amounts of 'Jintao', 'Soreli', 'AC 1536' (Dorì) and 'Jinyan' (Costa and Testolin, 2022). 'Hayward' is still the most commonly grown green-fleshed kiwifruit. Others include 'Summerkiwi 3373', 'Boverlig 02' (Earligreen®) and 'Convi 97.001' (Green Light®). More recently, around 140 ha of red-fleshed cultivars have been planted (Costa and Testolin, 2022).

2.2.4 Which countries trade the most kiwifruit?

New Zealand is the largest exporter of kiwifruit. In 2020, calculated from FAO data, 39% (0.57 Mt) of the total global exported volume came from New Zealand (~93% of total kiwifruit produced in New Zealand). New Zealand's focus is on exporting all kiwifruit produced that meets their Class 1 export quality standards. In 2020, Italy accounted for 19% of global kiwifruit exports, Greece 11% and Chile 10%. Overall, these top four countries accounted for nearly 79% of all exported kiwifruit. China exported a very small volume, accounting for only 1% of global exports in 2020, indicating that most of the large volume of kiwifruit they produce is consumed domestically (Goljic, 2021). When compared with total global fruit exports, kiwifruit accounted for 1.8% of all fresh fruit exported in 2020 (FAO, 2022).

In terms of imports, according to FAO data, Belgium was the biggest importer of kiwifruit in 2020, accounting for 11% of the global total, with Spain at 9%. China and Japan were equal third, accounting for 8% each of the total global imported volumes (FAO, 2022). However, Belgium should be considered a re-exporter as large volumes of New Zealand kiwifruit are sent to the Zeebrugge port in Belgium and then redistributed across Europe, the UK and Ireland. Potentially, a proportion of the import volumes attributable to Spain could also have been re-exported across mainland Europe.

One-third of the kiwifruit volume produced worldwide is exported compared with 9% of the total for all fresh fruit production. This illustrates the importance of trade for the kiwifruit sector, particularly for southern hemisphere suppliers in New Zealand and Chile (Table 2.6).

2.2.5 What are the global risks for kiwifruit production?

For the continued supply of kiwifruit, four key risks have been identified (Zespri Group Limited, 2021b, c) and are discussed below.

Biosecurity. There is a broad range of pests and diseases that may affect crop quality, volume and even market access in the future. The bacterial disease Psa effectively wiped out the kiwifruit cultivar 'Hort16A' in New Zealand over the period 2010–2013. The brown marmorated stink bug *Halyomorpha halys* also has the potential to cause significant fruit loss (up to 30%) from feeding damage and subsequent storage rot (Kiwifruit Vine Health, 2019).

Climate change. It is likely that a higher frequency of warm winters, more climate variability and extreme weather events will occur in the coming years, affecting how and where kiwifruit are grown. Temperature increases will impact productivity of green-fleshed cultivars, which generally require more chilling for sufficient bud break and flowering.

Table 2.5. Proportion of production by flesh colour for kiwifruit grown in Italy between 2015 and 2020. (From Costa and Testolin, 2022)

	Proportion of production (%)					
Flesh colour	2015	2016	2017	2018	2019	2020
Green	97	94	90	85	81	76
Yellow	3	6	10	15	19	24
Red (starburst)	0	0	0	0	0	0.1

Table 2.6. Proportion of total fruit production (Mt) exported for all fresh fruit and kiwifruit in 2020. (From FAO, 2022)

Worldwide fresh fruit volumes			Worldwide kiwifruit volumes		
Total volume (Mt)	Export volume (Mt)	Export (%)	Total volume (Mt)	Export volume (Mt)	Export (%)
887.03	83.02	9	4.41	1.47	33

Non-fresh fruit export figures were removed from the FAO dataset.

Labour and on-orchard costs. In many countries, minimum wages are increasing and as kiwifruit production increases so does demand for seasonal labour. To continue the growth in global kiwifruit supply, a focus is needed on how to attract labour to the industry, in the short term, and finding longer-term technological solutions to help relieve labour shortages.

Logistics. With exports being an important part of the kiwifruit industry, difficulties in logistics can cause fruit loss, increased costs and a sub-optimal mix of markets to which kiwifruit is exported.

2.3 Global Kiwifruit Demand

Significant growth has been observed in the production and global supply of kiwifruit. This section focuses on kiwifruit value trends and projections, including key risks to demand.

2.3.1 What are the values of the global fruit and kiwifruit markets?

Revenue in the fruit sector was estimated at US$630 billion in 2022. The market is expected to grow annually at a compound annual growth rate (CAGR) of 5–6% in 2022–2026. Globally, most revenue is generated in China (US$89 billion in 2022; *Statista*, 2022). However, most of China's fruit is produced and consumed locally (Goljic, 2021).

For kiwifruit, the only data source available estimated that global market revenue amounted to US$7.6 billion in 2018, with China accounting for an estimated US$3.9 billion of this (Global Trade, 2019). The global kiwifruit market is projected to grow at a CAGR of 5.8% during 2020–2025 (Mordor Intelligence, 2022).

This sub-section focuses on the export value of kiwifruit and how this compares with all fruit. The export value of kiwifruit was estimated at US$3.3 billion in 2020, compared with US$84 billion for all fresh fruit. Although kiwifruit makes up a small percentage of total fruit volume exported (~1.8%), in terms of traded fruit value kiwifruit accounts for about 3.9% of fresh fruit export value and was the sixth most valuable traded fresh fruit category in 2020, behind citrus, bananas, grapes, apples and cherries (FAO, 2022).

Similar increases in export volume have been seen across all fresh fruit and kiwifruit (Table 2.7). However, while all fresh fruit has significantly increased in value, kiwifruit has had exceptional increases in value with a percentage increase of 434% between 2000 and 2020 (Table 2.7).

For comparison, other fresh fruit categories that had larger percentage increases in export value over the same period (and had similar or greater export volumes than kiwifruit in 2020) include: lemons and limes (447%), mangoes, mangosteens and guava (729%) and cherries (1039%) (FAO, 2022). Each of these fruit categories added value beyond just increasing supply. Each of the top four kiwifruit-exporting countries also increased value beyond volume over the period 2000–2020 (Table 2.8).

In 2020, New Zealand contributed 52% of the global value generated from kiwifruit exports. This was considerably above Italy, which, in second place, contributed 16% of the value from all kiwifruit exports that year.

Table 2.7. Worldwide growth in volume (Mt) and value (US$ billion) of all fruit and kiwifruit exports from 2000 to 2020. (From FAO, 2022)

	2000		2020		Increase	
Category	Volume (Mt)	Value (US$B)	Volume (Mt)	Value (US$B)	Volume (%)	Value (%)
All fresh fruit exports	43	19.8	83	84.1	94	325
Kiwifruit exports	0.77	0.61	1.47	3.3	91	434

Non-fresh fruit export figures were removed from the FAO dataset.

2.3.2 How is the kiwifruit sector increasing global trade value?

Zespri case study

As New Zealand is the biggest exporter of kiwifruit, this sub-section reviews the key factors that have influenced how this country has increased its traded value in kiwifruit.

This case study is based on figures and information from Zespri Group Limited, the primary exporter of New Zealand kiwifruit. Zespri is 100% New Zealand grower-owned and currently markets only whole fresh kiwifruit. Zespri saw increases in both volume and value in the 2020/21 season (Table 2.9). However, Zespri also procures kiwifruit from outside New Zealand, and this contributes to their total revenue and overall returns to New Zealand growers. Revenue increased over the period 2014/15 to 2020/21 (Fig. 2.3).

The main countries that Zespri exports to saw substantial growth over the three years from 2017 to 2019, particularly through increased value (Table 2.10), suggesting demand and prices continue to rise. It is interesting to note that while the volume that Zespri exported to Taiwan decreased, returns still increased between 2017 and 2019.

Four factors driving increases in demand for and value of New Zealand kiwifruit exports are described below.

A globally consistent brand identity and delivering to consumer preferences. The Zespri brand was developed in 1997 when New Zealand kiwifruit growers decided to work together and unite under one brand. The brand was refreshed in 2020 along with the launch of Zespri's sustainability commitments (Zespri Group Limited, 2021b).

Premium quality kiwifruit. The Zespri system is key in ensuring the brand is supported by great-tasting and high-quality kiwifruit. This system contains the knowledge, policies and processes that embody the Zespri brand across the supply chain, ensuring consistency to support the consumer promise for which the brand stands. This includes a robust, scientifically based quality assurance system and proprietary cultivars. There are minimum taste standards, by cultivar, for harvesting the fruit and growers are paid an incentive to ensure the best-tasting fruit is produced and then marketed (NZKGI, 2018).

Product mix is important. Zespri currently markets two kiwifruit cultivars under two different growing regimes: (i) 'Hayward' as Zespri Green™ Kiwifruit and Zespri Organic Green™ Kiwifruit; and (ii) 'Zesy002' as Zespri SunGold™ Kiwifruit and Zespri Organic SunGold Kiwifruit. There are also very small volumes of another green-fleshed cultivar, 'Zesh004', which is sold as Zespri Sweet

Table 2.8. Growth in volume (Mt) and value (US$ billion) of kiwifruit exports from the top four exporter countries from 2000 to 2020. (From FAO, 2022)

	2000		2020		Increase	
Country	Volume (Mt)	Value (US$B)	Volume (Mt)	Value (US$B)	Volume (%)	Value (%)
New Zealand	0.25	0.27	0.57	1.73	129	546
Italy	0.27	0.17	0.27	0.53	0	204
Greece	0.03	0.01	0.17	0.20	432	1207
Chile	0.11	0.06	0.15	0.20	31	225

Table 2.9. Increase in volume in trays (where 1 tray = ~3.6 kg) and value of Zespri kiwifruit between the 2019/20 and 2020/21 seasons. (From Zespri Group Limited, 2021a.)

Factor	Change	Value/volume
Global sales	+14%	US$2.31 billion
Global volume (trays sold)	+10%	181.5 million trays

Figures converted to US$ at exchange rate of NZ$ = 0.6455 US$.

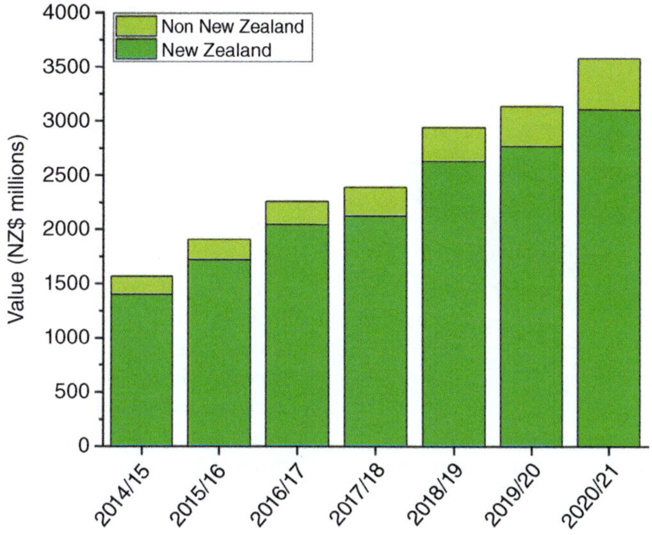

Fig. 2.3. Global sales (NZ$ millions) of New Zealand-grown and non-New Zealand-grown kiwifruit by Zespri in the 2014/15 to 2020/21 seasons. (Data from Zespri Group Limited, 2021a)

Table 2.10. Volume and value of Zespri kiwifruit export markets from 2017 to 2019. (From New Zealand Horticulture Export Authority, 2022)

	Volume (tonnes)				Value (US$ billion)			
Export market	2017	2018	2019	Change	2017	2018	2019	Change
Japan	96,122	130,550	125,521	31%	0.256	0.381	0.433	69%
European Union	204,370	214,749	225,330	10%	0.330	0.364	0.411	25%
Mainland China	101,457	107,558	112,905	11%	0.295	0.329	0.365	24%
Taiwan	35,360	35,155	33,289	–6%	0.089	0.092	0.095	7%
South Korea	28,196	33,829	38,664	37%	0.059	0.097	0.082	38%
USA	22,119	30,126	29,100	32%	0.042	0.059	0.077	83%

Years changed from source to more accurately reflect year grown; source based on financial year ending June. Figures converted to US$ at exchange rate of NZ$ = 0.6455 US$.

Green™ Kiwifruit. Volumes and value have been increasing for Zespri SunGold™ Kiwifruit as more commercial licences are released to growers and consumer demand increases. Controlling supply (which Zespri can do as owner of the plant variety rights for 'Zesy002') to align with demand is essential in ensuring sustainable premium returns. Zespri SunGold™ Kiwifruit is predicted to continue to be a key driver of growth (Fig. 2.4) for Zespri, both from New Zealand- and non-New Zealand-grown fruit. The recently released cultivar 'Zes008', marketed as Zespri RubyRed™ Kiwifruit, will start having small commercial volumes sold from 2022.

Further opportunities to increase demand. Zespri currently sells kiwifruit in over 50 countries. However, there are opportunities to open new markets while continuing to grow and strengthen relationships with existing markets. There are also opportunities to increase awareness and demand by focusing on strong, targeted marketing campaigns that support the Zespri brand, and continuing to

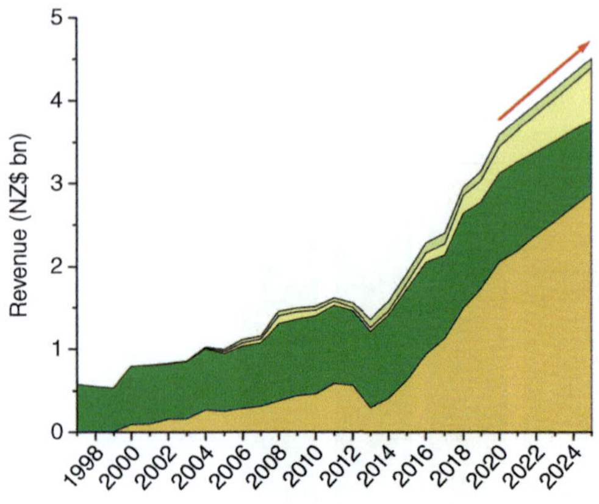

Fig. 2.4. Zespri's current and predicted (arrow) revenue based on fruit flesh colour for kiwifruit grown in New Zealand (Yellow ■, Green ■) and non-New Zealand grown kiwifruit (Yellow ■, Green ■). (From an unpublished Zespri Overview Presentation 2021). There was a dip in revenue in 2013, which was due to *Pseudomonas syringae* pv. *actinidiae* (Psa) as the New Zealand kiwifruit industry moved from growing 'Hort16A' to establishing 'Zesy002', a cultivar more tolerant of the disease caused by Psa.

innovate by developing new/unique kiwifruit cultivars. The Kiwifruit Breeding Centre, a 50/50 joint venture between The New Zealand Institute for Plant and Food Research Ltd and Zespri, was established in 2021 to drive greater innovation in kiwifruit breeding and to create healthier, better-tasting and more sustainability-focused cultivars.

As the biggest producer and consumer of kiwifruit, China has a strong influence on future trends in the global kiwifruit sector. It is difficult to find detailed datasets on China's kiwifruit sales and consumption year on year. However, based on historical trends, the growth in their own supply of kiwifruit and the increases in volume and value from exporting countries sending kiwifruit to China, such as New Zealand (Table 2.10), suggest demand remains strong.

Predicting future demand for kiwifruit in China is extremely difficult. Anecdotally, there have been comments in the media that kiwifruit is on a strong growth trajectory in China owing to the trend towards buying healthier foods (Neo, 2021).

2.3.3 What are the global risks for kiwifruit demand and value?

There are four key risks for the global kiwifruit sector in continuing to increase consumer demand and value:

Economic recession. Varying levels of economic recession can impact consumers' disposable income and their willingness to pay a premium on certain products.

Market access. Restrictions into premium markets could reduce the overall profitability of the sector. This could include increased tariffs as well as restrictions resulting from pests or diseases.

Consumer preference. Remaining relevant to consumers and delivering to their expectations is essential for continued growth. For example, consumers could move away from whole fresh kiwifruit as their need for convenience increases, if no alternative solutions are provided.

Category reputation. Perceived or real damage to the category's reputation could reduce demand for global kiwifruit. For example, a non-compliance incident that affected the kiwifruit industry's reputation or misleading claims made regarding attributes of the fruit.

2.4 Global Kiwifruit Brands and Products

This section introduces the key kiwifruit brands in the global marketplace. This is not intended as an exhaustive list and should not be confused with kiwifruit fruiting cultivars (see Chapter 4). There has been an increase in brands to market fresh kiwifruit over recent years.

2.4.1 What are the leading kiwifruit brands?

Zespri (New Zealand). The most established kiwifruit brand is Zespri (www.zespri.com), developed in 1997. For more details refer to the Zespri Case Study (Section 2.3.2).

Jingold (Italy). The Jingold brand (www.jingold.it/en/) was established in 2012. Jingold was founded from the decision to obtain the exclusive rights for the production and commercialization of the 'Jintao' yellow-fleshed kiwifruit in 2001. Jingold currently markets yellow, green, red and organic kiwifruit.

Mighties® (California, USA). Sun Pacific (www.sunpacific.com) launched the Mighties® brand in 2014, although their first investment in kiwifruit was in 1997. Sun Pacific now grows 50% of the North American kiwifruit supply (predominately 'Hayward') and plans to treble this by 2025. Sun Pacific reported there was double-digit growth in their kiwifruit sales the first year after the Mighties® brand was launched.

Sweeki® (Italy). Sweeki® was created in 2015 by the Origine Group to market Italian kiwifruit globally (www.sweeki.co). Sweeki® also markets kiwifruit sourced from Chile to support its strategy of supplying kiwifruit 12 months a year. The company markets Sweeki® Green, Sweeki® Gold and Sweeki® Organic Kiwifruit.

Kiwi Passion™ (Italy). The Kiwi Passion™ consortium is a collaboration among Italian and other European companies to protect and develop the best kiwifruit cultivars (www.kiwipassion.it/en). The consortium currently has Red Passion™ and Gold Passion™ as part of their portfolio.

Oscar® (France). For 35 years the Oscar® brand has been recognized in more than 35 countries marketing the 'Hayward' cultivar (www.oscar.kiwi/en). In 2015, Oscar created a new identity with a new logo and the slogan 'Be Kiwi, Be Happy'.

2.4.2 Processed kiwifruit products

The global fruit and vegetable processing market was predicted to be worth US$314 billion in 2022, a 3.3% growth in value from 2021. The annualized market size growth from 2017 to 2022 was estimated at 2.4% (IBISWorld, 2021).

As outlined earlier in this chapter, fresh kiwifruit's growth and increase in value has been substantial over the last 20 years. By comparison, the production of processed, green-fleshed kiwifruit products remains low, at around 5% of the crop. Green kiwifruit are not usually processed as the chlorophyll responsible for the green colour is destroyed and the characteristic flavour is lost during processing. Yellow-fleshed kiwifruit have become an alternative for food processors. Processing of 'Jintao' has shown good results with the yellow colour surviving well in processed products such as juices and jams. However, high prices for fresh, yellow-fleshed kiwifruit lead to limited processing volumes and options (Guroo *et al.*, 2017).

2.5 What Are the Key Future Trends and Opportunities for the Kiwifruit Sector?

This section focuses on seven key trends that will create opportunities for the global kiwifruit sector to continue to realize substantial growth through expanding production and increasing value.

2.5.1 Health

In research conducted by Zespri to understand why consumers buy kiwifruit, the top three decision-making factors were taste and quality (20%), health (20%) and convenience (16%) (Fig. 2.5).

With taste and quality being important for any food group, health provides an opportunity to differentiate kiwifruit. Kiwifruit is one of the most nutritionally dense fruit among the most popular fruit categories, although nutritional values can differ depending on the kiwifruit cultivar. The high amount of vitamin C in kiwifruit is the primary driver of its high nutrient adequacy score; the nutrient adequacy score is defined as the mean of percent daily values for 16 nutrients, based on 100 g of food (Darmon *et al.*, 2005). Other key nutrients include fibre, folate, potassium and vitamin E. Kiwifruit also contains a wide variety of antioxidants, phytonutrients and enzymes with digestive benefits, while being low in calories (Zespri International, 2016).

It has been reported that there has been an increase in demand for kiwifruit in China, the largest consumer of kiwifruit, where it is well known for being high in vitamin C and dietary

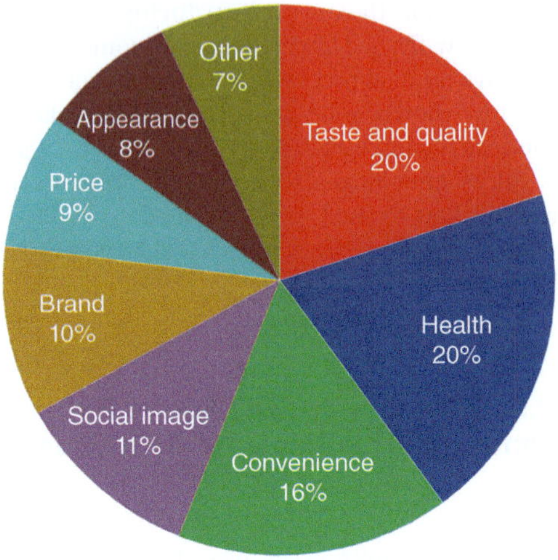

Fig. 2.5. Consumer decision making factors for buying kiwifruit – findings from Zespri research. (From NZKGI, 2019)

fibre as well as having other health benefits (Neo, 2021). Zespri has also reported that the global COVID-19 pandemic has helped to drive further demand for kiwifruit as consumers are looking for healthy and nutritious food choices (Burke, 2021).

2.5.2 Ageing population

The number of people aged over 60 is expected to double by 2050 (WHO, 2021), surpassing those aged 10–24. This is an opportunity for kiwifruit because in most markets, as people age, they spend more and consume more fruit (Zespri Group Limited, 2021b).

2.5.3 Eating occasions

Snacking is an increasing trend (Oller, 2021; Zespri Group Limited, 2021b) and fruit is the most consumed product at snacking occasions (Mintel, 2019). There is an opportunity for kiwifruit as a healthy snack, meeting consumers' increased health focus and desire for healthy eating owing to health issues such as rising rates of diabetes and obesity (Zespri Group Limited, 2021b).

2.5.4 Sustainability

Consumers are increasingly focused on sustainability whilst regulatory changes in this area are also ongoing (Zespri Group Limited, 2021b,c). As a natural product, there is an opportunity for the global kiwifruit industry to capitalize on this trend by ensuring growing techniques are as sustainable as possible and use of plastic packaging is reduced.

2.5.5 New retail channels

As worldwide internet access and adoption are rapidly increasing, the number of digital buyers keeps climbing each year (Pasquali, 2022). In 2021, retail e-commerce sales amounted to approximately US$5.2 trillion worldwide. This figure is forecast to grow by 56%, reaching about US$8.1 trillion by 2026 (Chevalier, 2022). Being able to distribute quality kiwifruit through this rapidly growing retail channel is important for continued growth in revenue.

2.5.6 Convenience

With convenience being the third-ranked reason consumers buy kiwifruit (Fig. 2.5), the industry needs to continue to innovate with new cultivars or delivery formats that support convenience. Important aspects of convenience are ensuring kiwifruit is available to consumers all year round, that it is ready to eat and readily available across a range of retail channels and formats.

2.5.7 New product innovation

Realizing the potential growth of the kiwifruit sector requires continuing innovation to develop new/unique kiwifruit cultivars that are healthier, better-tasting, more sustainability-focused, and meet specific consumer-demand niches.

References

Burke, P. (2021) Covid boosts kiwifruit demand. Available at: www. ruralnewsgroup.co.nz/rural-news/rural-general-news/covid-boosts-kiwifruit-demand (accessed 12 January 2022).

Chevalier, S. (2022) Global retail e-commerce sales 2014–2026 *Statista* 21 September 2022. Available at: https://www.statista.com/statistics/379046/worldwide-retail-e-commerce-sales/ (accessed 21 November 2022).

Costa, G. and Testolin, R. (2022) Recent changes in the kiwifruit (*Actinidia* spp.) industry since its introduction: Italy as a case study. *Acta Horticulturae* 1332, 1–10. DOI: 10.17660/ActaHortic.2022.1332.1

Darmon, N., Darmon, M., Maillot, M. and Drewnowski, A. (2005) A nutrient density standard for vegetables and fruits: nutrients per calorie and nutrients per unit cost. *Journal of the American Dietetic Association* 105, 1881–1887. DOI: 10.1016/j.jada.2005.09.005

FAO (2022) FAOStat. Available at: https://www.fao.org/faostat/en/#data (accessed 29 January 2022).

FreshPlaza (2021) Global overview kiwi. *FreshPlaza* 10 September 2021. Available at: www.freshplaza.com/article/9354005/global-overview-kiwi/ (accessed 12 January 2022).

Global Trade (2019) Global kiwifruit market 2019 – New Zealand and Italy are the leading exporters of kiwi. Available at: www.globaltrademag.com/global-kiwi-fruit-market-2019-new-zealand-and-italy-are-the-leading-exporters-of-kiwi-fruits (accessed 12 January 2022).

Goljic, D. (2021) 34 fruit consumption statistics for a refreshed 2022. *Supplements 101*, 25 December 2021. Available at www.supplements101.net/fruit-consumption-statistics/ (accessed 12 January 2022).

Guroo, I., Wani, S.A., Wani, S.M., Ahmad, M., Mir, S.A. *et al.* (2017) A review of production and processing of kiwifruit. *Journal of Food Processing & Technology* 8, 699. DOI: 10.4172/2157-7110.1000699

IBISWorld (2021) Global fruit & vegetable processing: market size 2005–2027. Available at: https://www.ibisworld.com/global/market-size/global-fruit-vegetable-processing/ (accessed January 2022).

Kiwifruit Vine Health (2019) Fact sheet: brown marmorated stink bug. *Kiwifruit Vine Health*, October. Available at: www.kvh.org.nz/vdb/document/104414 (accessed 14 January 2022).

Mintel (2019) Better for You Snacking - US - 2019 : Consumer market research report. Mintel.com

Mordor Intelligence (2022) Kiwifruit market – growth, trends, Covid-19 impact and forecasts (2022–2027). Available at: www.mordorintelligence.com/industry-reports/kiwi-fruit-market (accessed 14 January 2022).

Neo, P. (2021) Global kiwifruit king: demand in China rising rapidly due to premiumisation and health trends. *Nutra Ingredients-Asia*, 11 October. Available at: www.nutraingredients-asia.com/Article/2021/09/27/Global-kiwifruit-king-Demand-in-China-rising-rapidly-due-to-premiumisation-and-health-trends (accessed 14 January 2022).

New Zealand Horticulture Export Authority (2022) Kiwifruit and kiwiberry. Available at: www.hea.co.nz/2012-05-11-03-05-28/kiwifruit-trade (accessed 29 January 2022).

NZKGI (2018) Business. In: *Kiwifruit Book 2018*. New Zealand Kiwifruit Growers Inc., Mount Maunganui, New Zealand, pp. 73–98. Available at: www.nzkgi.org.nz/wp-content/uploads/2018/12/2018-Kiwifruit-Book.pdf (accessed 14 September 2022).

Oller, S. (2021) How the pandemic is reshaping the snacking occasion. *FOODDIVE* 12 April 2021. Available at: www.fooddive.com/news/how-the-pandemic-is-reshaping-the-snacking-occasion/597757/ (accessed 14 January 2022).

Pasquali, M. (2022) E-commerce worldwide – statistics & facts. *Statista*, 17 November. Available at: www.statista.com/topics/871/online-shopping/#dossier-chapter1 (accessed 21 November 2022).

Statista (2022) Fresh fruits worldwide. Available at: www.statista.com/outlook/cmo/food/fruits-nuts/fresh-fruits/worldwide (accessed 29 January 2022).

WHO (2021) Aging and health. World Health Organisation, 4 October. Available at: www.who.int/news-room/fact-sheets/detail/ageing-and-health (accessed 12 January 2022).

Zespri Group Limited (2021a) *Annual Report 2020/21*. Available at: www.zespri.com/en-NZ/annualreports (accessed 14 January 2022).

Zespri Group Limited (2021b) *Zespri 5 Year Outlook December 2020*. Available at: https://www.zespri.com/content/dam/zespri/nz/outlooks/2021-Outlook.pdf (accessed 22 November 2022).

Zespri Group Limited (2021c) *Zespri 5 Year Outlook December 2021*. Available at: https://www.zespri.com/content/dam/zespri/nz/outlooks/Zespri-5YearOutlook-2022.pdf (accessed 22 November 2022).

Zespri International (2016) *Zespri® Scientific Pack: Nutritional & Health Attributes of Kiwifruit*. Available at: www.zespri.eu/_hcp/_img/Downloads/Zespri%20Kiwifruit%20Scientific%20Pack_ENG.pdf (accessed 14 January 2022).

Zhong, C., Huang, W., Wang, Z., Li, L., Li, D. *et al.* (2022) The breeding progress and development status of the kiwifruit industry in China. *Acta Horticulturae* 1332, 445–454. DOI: 10.17660/ActaHortic.2022.1332.59

3 Biology of the Genus *Actinidia*

A. Ross Ferguson
The New Zealand Institute for Plant and Food Research Ltd, Auckland, New Zealand

3.1 Introduction

The kiwifruit of commerce are mostly large-fruited selections of *Actinidia chinensis* var. *chinensis* or *A. chinensis* var. *deliciosa* (see Chapters 1 and 4). Kiwiberries (see Chapter 7) are selections of *A. arguta* or are hybrids between *A. arguta* and *A. melanandra* and, possibly, *A. kolomikta*. However, there are many other species in the genus *Actinidia* (Actinidiaceae) showing great diversity in many characters. Even individual *Actinidia* species can be highly heterozygous as indicated by their phenotypic diversity and by molecular studies.

3.2 The Family Actinidiaceae

The family Actinidiaceae is usually placed in the Ericales, sometimes the Theales. The family has three genera: *Saurauia* Willd., *Clematoclethra* (Franch.) Maxim. and *Actinidia* Lindl. (Dressler and Bayer, 2004):

- *Saurauia* has some 350 species of trees and shrubs in tropical and subtropical Asia and America. The stamens are connected at their base with the petals.
- *Clematoclethra* consists of perhaps a single very variable species in western and central China. It is a woody climber. The stamens, usually ten but sometimes more, are free, not connected to the petals. The ovary has up to five locules with a single style.
- *Actinidia* currently has 54 species of woody climbers, mostly in China. The flowers have many free stamens, and the ovaries have many locules with the stylar branches distinct. The generic name *Actinidia* is based on the characteristic radiating styles of the pistillate flower (Fig. 3.1).

3.3 The Genus *Actinidia*

The early taxonomic history of the genus is described in Dunn (1911). Subsequently, the genus was revised by Li (1952), Liang (1984) and, most recently, Li *et al.* (2007), who recognized 54 species and 36 infraspecific varieties.

Earlier authors divided *Actinidia* into infrageneric sections and series based on morphological features such as the type of hair, the shape of the ovary, the presence of lenticels on the fruit and whether the pith was lamellate or solid (Li, 1952; Liang, 1984; Huang and Ferguson,

---

Address for correspondence: ross.ferguson@plantandfood.co.nz

© CAB International 2024. *Kiwifruit: Botany, Production and Uses* (eds A.C. Richardson, J.N. Burdon and A.R. Ferguson)
DOI: 10.1079/9781800620933.0003

Fig. 3.1. Dissected gynoecium of a flower of *Actinidia chinensis* var. *deliciosa* 'Bruno'. (Photo: The New Zealand Institute for Plant and Food Research Ltd)

2007) but this was not attempted by Li *et al.* (2007). Li and Li (2010) provisionally accepted the divisions of Li (1952) because it 'reflects the morphological differentiation among *Actinidia* species' but revised the sectional nomenclature. Although the section *Leiocarpae* Dunn appears to form a coherent grouping, the other sections variously proposed seem artificial, particularly as molecular evidence indicates that they are polyphyletic (Huang and Ferguson, 2007; Li and Li, 2010; Tang *et al.*, 2019). Further work is required.

3.3.1 *Actinidia* species

It is difficult to circumscribe *Actinidia* species. The plants, like many other climbers, are notoriously variable in morphology and, according to Li (1952), 'It appears that *Actinidia* is in general more variable in the individual plant than most other ligneous vines'. Another problem is that taxonomists may vary in their approach. The difficulties in separating many *Actinidia* taxa and the frequent nomenclatural changes mean that identification of plants used in earlier published studies is often doubtful. Individual plants in a germplasm collection may not be typical of the taxon.

Detailed studies of wild populations show that the morphological characters used to define an *Actinidia* species can vary across its geographic distribution. At the extremes of the geographical distribution, differences can be clear-cut and even sufficient to divide a species into infraspecific taxa, but across the distribution range, changes in diagnostic characters may be only gradual and not always correlated with each other. Li (1952) describes such forms of *A. arguta* as passing 'into each other imperceptibly by inter-grading forms. Clearly this [*A. arguta*] is another of the very widespread, more or less polymorphic, species characteristic of the genus *Actinidia*.'

Leaves on male and female plants can differ (Li, 1952), at times even resulting in male and female plants being placed in different species (Cuong *et al.*, 2007). Leaves can vary with time of emergence or on different types of shoot of the one plant. Dunn (1911) describes how:

> The long arching shoots which appear during the spring and summer months give rise the following year to secondary branches bearing leaves which are often strikingly different in shape and character from those of the primary stems – a point which will explain some anomalies in the series of specimens in herbaria.

This is why leaves can vary even in a single herbarium specimen (Huang, 2016a). Individual herbarium specimens may not accurately reflect the variability within taxa and some taxa have been described on single or fragmentary specimens.

Not only are individual *Actinidia* species often variable, but there can also be no clear boundaries between species. Furthermore, breeding barriers between species are not always sufficient to produce complete reproductive isolation; many *Actinidia* species coexist in the wild and hybrid swarms are often observed. Controlled interspecific crosses in *Actinidia* are often successful in producing thousands of seedlings (Ferguson and Wu, 2016).

Some molecular studies indicate that more than half of all *Actinidia* species may have arisen from hybridization at least once during their evolutionary history (Chat *et al.*, 2004; Li and Liu, 2016). However, Tang *et al.* (2019) concluded that as molecular studies of relationships of species within *Actinidia* may have been limited by the use of a few genotypes for each taxon and insufficient molecular markers, more work is required.

The variability within species and the frequent lack of clear boundaries between species have been attributed to habitat diversification owing to the broken topography of southern China encouraging allopatric speciation (Liu *et al.*, 2022), the frequent sympatric distribution of taxa, and polyploidization and recurrent inter-taxon and interploidy hybridization (Huang and Ferguson, 2007; Liu *et al.*, 2015; Li and Liu, 2016).

Another difficulty is that taxonomists vary in their treatment of the variability in *Actinidia*: some such as Liang (1984) are 'splitters', wanting to name even minor variants on a few specimens; others such as Li *et al.* (2007) are 'lumpers', preferring broader species concepts, accepting that some species are morphologically quite variable.

The net result of all these difficulties and the recent, much more intensive, studies of wild populations in China is that there have been frequent changes in the nomenclature of *Actinidia* taxa. This is illustrated by the nomenclatural changes in the *A. arguta* complex and the *A. chinensis* complex, as these are commercially the most important and the subject of most research.

Changes in nomenclature in the A. arguta *complex*

A. arguta is widely distributed geographically and is variable, particularly in leaf and fruit characteristics. The confusing variation in *A. arguta* and related species has been treated in different ways.

Liang (1984) accepted five varieties within *A. arguta*, but the characteristics he used to distinguish infraspecific varieties were not always correlated or clearly discontinuous. Hence, Li *et al.* (2007) and Li *et al.* (2009) considered the species in China should be split into only two varieties, *A. arguta* var. *arguta* and *A. arguta* var. *giraldii*.

An apparently distinct variant with purple fruit had been variously treated as a separate species, *A. purpurea* (Rehder, 1915), or as a variety, *A. arguta* var. *purpurea* (Liang, 1984). Li *et al.* (2007) and Li *et al.* (2009) considered that the distinguishing features used were not sufficiently reliable or well correlated to justify separating a distinct taxon: they included such plants with purple fruit in *A. arguta* var. *arguta*. Kiwiberries with different coloured skin or flesh are commercially important so this change should be noted.

A separate change affects what was thought to be a closely related species in Japan, *A. hypoleuca* (Nakai, 1933), now reduced to *A. arguta* var. *hypoleuca*.

Another closely related species is *A. melanandra*. Liang (1984) divided *A. melanandra* into five varieties, whereas Li *et al.* (2007) recognized only two. Furthermore, *A. melanandra* is separated from *A. arguta* by the lower surface of the leaves being glaucous, but the difference is not always consistent. Li *et al.* (2009) and Liu *et al.* (2017) have suggested that *A. melanandra* could be merged with *A. arguta*. Fortunately, if the two species are eventually combined, the epithet *arguta* has priority.

Changes in nomenclature in the A. chinensis *complex*

The *A. chinensis* complex contains the plants generally known as kiwifruit. Morphological variants within *A. chinensis* were quickly recognized by early plant collectors (see Chapter 1); the problem has been how to treat this variation taxonomically. The type specimen was probably collected near the coast, where the fruit have only fine soft hairs, easily shed, whereas the first plants to be grown outside China and which formed the initial basis for the kiwifruit industry were collected far inland: these plants had fruit covered by long stiff hairs, usually persistent. The distinction was formalized by Chevalier, who named the 'hairy' variant as *A. chinensis* var. *deliciosa* (Chevalier, 1940, 1941), although his papers largely went unnoticed. Later, Liang (1975), unaware of Chevalier's earlier papers, called the 'hairy' variant *A. chinensis* var. *hispida*. Another variant from Taiwan with even stiffer fruit hairs had earlier been separated by Li (1952) as *A. chinensis* var. *setosa*.

Gui (1981) proposed that, because of differences in morphology, geographic distribution and ploidy, the varieties *A. chinensis* var. *chinensis* and *A. chinensis* var. *hispida* be raised to specific status, and this was effected in 1984 by Liang and Ferguson (1984, 1986), who also acknowledged that the epithet *deliciosa* had priority over *hispida*. Liang and Ferguson therefore proposed the new combination *A. deliciosa*. They also raised the variant from Taiwan to specific status as *A. setosa*.

Subsequent detailed studies of the wild populations in China have shown that the distinction between the two mainland variants is not that clear-cut. *A. chinensis* occurs more to the east, whereas *A. deliciosa* occurs more to the west and at higher altitudes (Liang, 1975; Ferguson, 1990a). Where they occur together, *A. chinensis* is at lower altitudes, *A. deliciosa* above at higher altitudes, but with partially sympatric distributions (Huang and Liu, 2014). At the geographic extremes, the two variants are clearly distinct but there is a progression in morphology across the geographic range. Where they overlap, there is extensive gene flow (Liu *et al.*, 2008; Li *et al.*, 2010a; Liu *et al.*, 2015) despite differences in ploidy.

Li *et al.* (2007) therefore decided that it was better to treat the two variants as varieties of the one species, rather than as separate species. In this volume, we have followed the classification used in Volume 12 of the *Flora of China* and the 'hairy' variant is accordingly accepted as *A. chinensis* var. *deliciosa*. Li *et al.* (2007) also reduced *A. setosa* to a variety of *A. chinensis*, but others, e.g. Hsieh *et al.* (2011) and Tang *et al.* (2019), still consider it a separate species.

It is necessary to be aware of such changes: the literature on 'Hayward', the most studied kiwifruit cultivar, can refer to it as belonging to *Actinidia chinensis* Planch., *A. chinensis* var. *hispida* C.F.Liang, *A. deliciosa* (A.Chev.) C.F.Liang et A.R.Ferguson, or as now, *A. chinensis* var. *deliciosa* (A.Chev.) A.Chev.

The *A. chinensis* complex might be enlarged by incorporating what appear to be closely related species. Nevertheless, further changes to the taxonomic status, and thus the nomenclature, of the commercially important kiwifruit taxa seem unlikely.

3.3.2 Distinguishing *Actinidia chinensis* var. *chinensis* and *A. chinensis* var. *deliciosa*

The main morphological differences (Li *et al.*, 2010b; Huang, 2016a) can be summarized.

- *Trichomes (hairs).* In variety *chinensis*, young shoots and petioles are either hairless or covered in sparse, soft, white hairs, soon largely lost, and young fruits have dense fine hairs, readily lost, so that the mature fruit are essentially hairless. In variety *deliciosa*, the hairs of young shoots are often initially bright red and strigose fading to gingery brown and are persistent. The fruit are densely hispid, the yellow-brown hairs being retained at maturity. Descriptions of the fruit skin usually make it obvious as to which variety is being discussed: smooth skin or with very fine hairs means *A. chinensis* var. *chinensis*; hairy skin means *A. chinensis* var. *deliciosa*.
- *Overwintering buds.* In variety *chinensis*, these are obvious and exposed, covered only by bud scales, whereas those of variety *deliciosa* are buried deeply in the bark of the leaf axil.
- *Leaves.* Those of variety *chinensis* are smaller and softer than those of variety *deliciosa* and the lower leaf surface is covered by white stellate hairs, whereas in variety *deliciosa* the hairs are pale yellow-brown.
- *Flowers.* Pistillate flowers of variety *chinensis* are smaller than those of variety *deliciosa*, 4.0 cm vs. ~5.5 cm across, and have fewer styles. Staminate flowers of variety *chinensis* are likewise smaller than those of variety *deliciosa*, ~2.5 cm vs. ~4.0 cm across, and have fewer stamens, ~45 vs. ~150 or more.
- *Fruit shape.* Fruit of variety *chinensis* tend to be shorter, more spherical and less cylindrical than those of variety *deliciosa*. However, fruit of wild vines can vary greatly in shape (Hu *et al.*, 2021). Differences in fruit size and shape are less obvious in commercial cultivars, which have been selected, in part, on fruit size.
- *Ripe fruit flesh colour.* The fruit flesh of variety *chinensis* is usually, although not always, yellow; that of variety *deliciosa* is usually green.

In addition, diploid plants will have been classified as *A. chinensis* var. *chinensis*, hexaploid plants as *A. chinensis* var. *deliciosa* and tetraploid plants usually as *A. chinensis* var. *chinensis*.

3.3.3 Geographic distribution of *Actinidia* species

The centre of the genus is in southern China between the Yangzi river (Chang Jiang) and the Pearl river (Zhu Jiang), between approximately 25° and 30° N (Liang, 1983; Huang and Ferguson, 2007; Huang, 2016a) although species are found as far as 50° N or just south of the equator. *Actinidia* species are generally found in moist, sheltered conditions, often shady areas under the tree canopy, and seem to grow best in mountainous regions, the broken topography providing different habitats and microclimates. They can also be found along the banks of streams, on forest edges, and in sparse forest or scrub. Most taxa are endemic to the Chinese mainland, but a few extend to Taiwan and neighbouring countries and only two are not found in China (Huang, 2016a).

Some species are very widely distributed throughout much of China, and in Siberia and Japan. In the far north, they may be found at sea level, while at the southern extremes they may be found at 3000 m above sea level or higher (Huang and Ferguson, 2007). Other species have very narrow distributions, both geographically and in altitude, and are sometimes endemic to a single province or even a single county. To some extent, current discontinuous distributions are probably due to deforestation and habitat loss (Cui *et al.*, 2002).

Many species overlap geographically with at least one other species, sometimes even growing together up a tree. Within the same biogeographical region, their vertical distributions may overlap to different extents (Li and Liu, 2016). Such sympatric distribution increases the probability of interspecific introgression resulting in natural hybrid zones.

3.3.4 Chromosome numbers and ploidy of *Actinidia* species

Actinidia chromosomes are very small and numerous (He *et al.*, 2005) (Fig. 3.2). The basic chromosome number is $n = 29$ (Huang and Ferguson, 2007; Ferguson and Huang, 2016), a high number that indicates probable whole genome duplication (McNeilage and Considine, 1989; He *et al.*, 2005; Shi *et al.*, 2010; Wang *et al.*, 2018). Diploid *Actinidia* species ($2n = 58$) can therefore be considered as cryptic polyploids or rediploidized palaeopolyploids.

Within the genus there is extensive variation in ploidy. *Actinidia* taxa at the same ploidy level seem to vary little in genome size, and polyploidization does not seem to have resulted in a significant loss of nuclear DNA. This allows the use of flow cytometry to estimate ploidy in *Actinidia* genotypes; it is now realistic to measure ploidy of every individual in large breeding populations (Ferguson *et al.*, 2009) or in the wild (Wang, Z. *et al.*, 2021).

Chromosome numbers or determinations of ploidy by flow cytometry have been published for about 30 *Actinidia* species. Many species are consistently diploid; a few consistently tetraploid ($2n = 4x = 116$). At least ten species contain ploidy races (Ferguson and Huang, 2016) (Fig. 3.3). That so many species contain ploidy races, despite rather limited testing, suggests that more

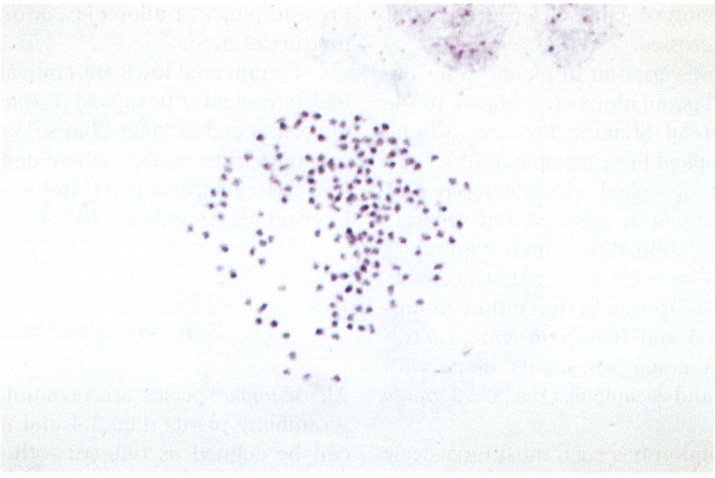

Fig. 3.2. Chromosome squash from hexaploid *Actinidia chinensis* var. *deliciosa*, $2n = 6x = 174$. Squash by Yan Guijun. (Photo: The New Zealand Institute for Plant and Food Research Ltd)

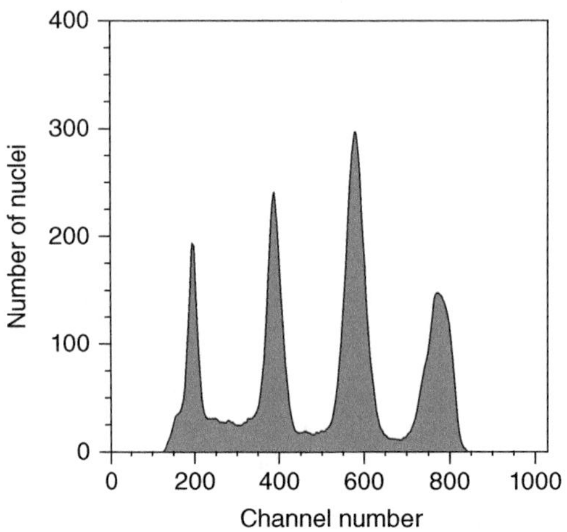

Fig. 3.3. Ploidy races in *Actinidia* are easily distinguished by flow cytometry. Peaks from left to right are *Actinidia arguta* var. *hypoleuca*, 2*x*; *A. arguta* var. *arguta* 4*x*, *A. arguta* var. *arguta* 'Issai' 6*x* and *A. arguta* var. *arguta* 8*x*. Plant samples were co-chopped and DNA content measured by a Beckman Coulter FC500 MPL flow cytometer. Ploidies were confirmed by chromosome counts. (Unpublished data of A.R. Ferguson, A.M. Duffy and J.-L. Zhang, The New Zealand Institute for Plant and Food Research Ltd)

ploidy races might be revealed by further studies. With the exception, to some extent, of the *A. chinensis* complex, ploidy races cannot usually be unequivocally distinguished morphologically. It is not safe to assume the ploidy of any individual *Actinidia* genotype, either from the wild or from interploid crosses.

The greatest variation in ploidy so far observed is in wild populations of *A. arguta*. In the Daba mountains of Shaanxi Province, China, there were tetraploid (4*x*), hexaploid (6*x*), octaploid (8*x*) and decaploid (10*x*) cytotypes of *A. arguta* var. *arguta* in adjacent but separate populations, the populations often containing sympatric ploidy races (Li *et al.*, 2013). Likewise, Zhang *et al.* (2017) found in the Qinling mountains of Shaanxi and Henan Provinces tetraploids (4*x*) of *A. arguta* var. *arguta*, along with octaploids (8*x*) and decaploids (10*x*) of *A. arguta* var. *giraldii*.

Ploidy variation has been most intensively studied in the *A. chinensis* complex. The two varieties can be readily distinguished at the geographic extremes of the complex, but morphologically intermediate forms can be recognized where the two varieties overlap. The morphological transition is accompanied by a transition in ploidy (Zeng *et al.*, 2009; Li *et al.*, 2010a; Liu *et al.*, 2011; Liu *et al.*, 2015). Across the geographic distribution of the complex, plants morphologically identifiable as belonging to *A. chinensis* var. *chinensis* were diploid (2*x*) or tetraploid (4*x*) with a few triploids (3*x*); those identifiable as *A. chinensis* var. *deliciosa* were tetraploid (4*x*) or hexaploid (6*x*) with a small number of octaploids (8*x*) (Wang Z. *et al.*, 2021). All diploids were identifiable as belonging to *A. chinensis* var. *chinensis*, all hexaploids to *A. chinensis* var. *deliciosa*. Tetraploids could be identified as belonging to either variety, but some were morphologically intermediate.

This may be an oversimplification: Xia *et al.* (2023) reported a wild plant of unstated origin, identified as belonging to *A. chinensis* var. *deliciosa*, as diploid. *A. chinensis* var. *setosa* (syn. *A. setosa*) is also diploid (Yan *et al.*, 1997).

Production of numerically unreduced 2*n* gametes could contribute to the numerous ploidy races (Yan *et al.*, 1997; Seal *et al.*, 2012; Fraser and McNeilage, 2016). Unreduced gametes can be useful in breeding programmes to cross ploidy barriers. Somatic doubling (Datson and Ferguson, 2011) may also occur in wild plants. Polyploid races within a taxon may not be monophyletic (Wang *et al.*, 2022). Thus, the origin of the polyploids in the *A. chinensis* complex is still debatable as to whether they are autoploids or alloploids, or quite possibly a mixture of both.

Commercial kiwifruit cultivars can be diploid, tetraploid or hexaploid (Li *et al.*, 2010b) (see Chapters 4 and 5). Many Chinese kiwifruit cultivars originate from regions where diploid/tetraploid and tetraploid/hexaploid forms of *A. chinensis* intermingle (Li and Liu, 2016).

3.4 Growth Habit

All *Actinidia* species are perennial climbing or scrambling plants (Fig. 3.4 and Fig. 3.5). They can be defined as twiners – they have stems which, in the wild, twine or twist to climb. They do not have tendrils, spines, thorns, adhesive pads or adventitious roots to assist climbing, but

Fig. 3.4. Wild *Actinidia chinensis* var. *chinensis* vine growing up trees in Jiangxi Province, China. (Photo: C.H. Cheng, Kiwifruit Breeding Centre, New Zealand)

Fig. 3.5. Wild *Actinidia chinensis* var. *chinensis* vine scrambling over rocks beside a stream, Jiangxi Province, China. (Photo: C.H. Cheng, Kiwifruit Breeding Centre, New Zealand)

instead rely on the interlacing of shorter shoots and the twisting (usually dextrally) of the more vigorous shoots. The species vary in vigour, particularly in the height to which they can grow, many of them up to 5–7 m, often enveloping whole trees, or, once above the canopy, spreading from tree to tree. However, in clearings where there are no trees, or in the far north, or at the extremes of their altitudinal range, even rampant growers can be reduced to forming vast thickets, rather like brambles.

The commercially important *Actinidia* species are all vigorous and a critical step in transforming the wild, uncontrolled climber to a tamed, cultivated crop plant was learning how to develop an orderly vine structure and to establish a balance between vegetative growth and crop production. The next requirement was to achieve the most profitable compromise between fruit number, fruit size and fruit dry matter content (see Chapter 12). Vine management and pruning are major expenses in commercial kiwifruit orchards and need to consider the growth habit of the vines. The methods used are largely based on experience with the cultivars grown commercially in New Zealand: 'Hayward' (*A. chinensis* var. *deliciosa*), and 'Hort16A' and 'Zesy002' (both *A. chinensis* var. *chinensis*). Methods may have to be modified for other cultivars and different growing conditions.

3.5 Vine Structure

Most of the information published is of cultivated kiwifruit vines which can consist of several parts or genotypes. If they have been propagated vegetatively, then the roots and the above-ground parts are all one genotype, usually a selected male or female cultivar. Frequently, however, they consist of a selected scion cultivar grafted onto either a seedling rootstock or a selected, propagated rootstock. They may, if an orchard has been reworked to a new scion cultivar, even have three parts: the roots, an interstock and a scion. Many of the studies on orchard-grown kiwifruit vines, especially those from New Zealand, are of 'Hayward' scions grafted onto 'Bruno' (*A. chinensis* var. *deliciosa*) seedling rootstocks.

3.5.1 Roots

The root systems of cultivated kiwifruit can be massive: 16-year-old plants can have roots spreading through 80 m^3 soil per vine, although this will depend on soil conditions (McAneney and Judd, 1983), and the below-ground dry weight of the vine can almost equal the above-ground dry weight (Hughes *et al.*, 1992). The root system consists of long, thickened roots, which comprise most of the dry weight, and fibrous roots (Lemon and Considine, 1993). The fine roots, which make up about 8% of the standing biomass of the vine, can turn over at extraordinarily high rates (Reid *et al.*, 1993). The rooting pattern of kiwifruit seems to be uneven, less uniform throughout the volume of soil than most other horticultural trees and vines (Hughes *et al.*, 1992) (see Chapters 11 and 15).

Older roots of *A. chinensis* var. *deliciosa* are unlike those of many other plants in that the cortex and the endodermis are retained during secondary root thickening (Lemon and Considine, 1993). This is also the case in the roots of *A. chinensis* var. *chinensis* and five other *Actinidia* species (Wang *et al.*, 1994). The root anatomy of the *Actinidia* taxa studied does, however, vary in the cross-sectional area of the xylem vessels and the cross-sectional area of fibres and idioblasts in the phloem. These differences may affect the behaviour of different taxa as rootstocks.

3.5.2 Stems

The stems are relatively narrow in cross-section, compared with those of self-supporting woody plants. The xylem vessels of lianes such as kiwifruit are often exceptionally wide to allow large amounts of water to be conducted through the stems (Ferguson, 1984). The vessels in the roots and stem of *A. chinensis* var. *deliciosa* can, accordingly, be very large (McAneney and Judd, 1983; Condon, 1991a), especially in the outer growth rings (see Chapter 15).

3.5.3 Buds

Growth occurs in spring from axillary buds formed the previous season (Brundell, 1975a; Walton *et al.*, 1997). These are generally in the axils of leaves distal on the shoot to the last flowering axil. A bud primordium starts developing nearly a year before the differentiated shoot emerges. By midsummer, the bud has largely completed development and is dormant by winter. In taxa that grow in areas with colder winters, e.g. *A. arguta*, *A. polygama* and *A. chinensis* var. *deliciosa*, the dormant bud is usually buried deep in the swollen leaf base. In some taxa from warmer areas, e.g. *A. chinensis* var. *chinensis*, the buds are much more exposed and covered only by bud scales. There is limited information on the overwintering buds of different *Actinidia* taxa (Condon, 1991b; Snowball, 1997a; Huang, 2016a).

The time of spring budbreak varies with taxon (Li *et al.*, 1987), between different genotypes within the one taxon (Lee *et al.*, 2020), and from year to year. Not all buds break, especially those on long, vigorous shoots. Percentage budbreak is an important characteristic to be considered in choosing commercial cultivars (see Chapter 14).

3.5.4 Leaves

The leaves of *Actinidia* species are simple, usually with a long petiole, and are generally dentate to varying degrees. They vary in size, shape, texture and hairiness across different species, within one species and often in a single plant (Dunn, 1911; Huang, 2016a). Most *Actinidia* species are deciduous but some from warmer areas are variously described as semi-deciduous or even evergreen (Huang, 2016a). However, the species in commercial cultivation are all deciduous.

Differences between the leaves of *A. chinensis* var. *chinensis*, *A. chinensis* var. *deliciosa* and *A. chinensis* var. *setosa* are summarized in Ferguson (1990a). Photographs of the stellate hairs on the lower surface of *A. chinensis* var. *deliciosa* leaves are in Ferguson (1990b). The very limited information on *Actinidia* leaf anatomy is summarized in Ferguson (1984, 1990b) and Oláh *et al.* (1997).

3.5.5 Flowering shoots

Actinidia seedlings do not usually flower until they are at least 3–4 years old, female seedlings at an older age than male seedlings. Flowering shoots develop in the same way as purely vegetative shoots. Floral evocation occurs by late summer in *A. chinensis* var. *deliciosa* and floral differentiation only in the following spring (Brundell, 1975b; Walton *et al.*, 1997) (see Chapter 14).

Flowers are produced towards the base of lateral shoots that develop in the current season, from buds formed in leaf axils of shoots grown the previous season. Shoots that develop early in the season may produce flowers, while those breaking later are usually vegetative (Grant and Ryugo, 1982). Some vines occasionally produce flowers later in the season. The male cultivar ‘Chaohong’, a hybrid between *A. eriantha* and *A. chinensis* var. *chinensis*, regularly flowers several times during the growing season (Wang *et al.*, 2009; Lei *et al.*, 2010).

Flowers are usually in small cymes or pseudoumbels. The number of flowers in an inflorescence depends on the species and gender; male plants usually have more flowers per inflorescence (Snowball, 1997b; Huang, 2016a). Single flowers in *A. chinensis* var. *deliciosa* female cultivars result from abortion of lateral flower buds (Engin *et al.*, 2011). Fruiting cultivars with single flowers are often preferred as this reduces fruit-to-fruit variation and the risk of proximity damage, mainly wind rub.

Flowering of the different *Actinidia* species, when grown at the same site in New Zealand, occurs over some three months. Diploid *A. chinensis* var. *chinensis*, *A. indochinensis* and *A. kolomikta* are amongst the first to flower (mid- to late October, southern hemisphere), *A. latifolia* not until the following mid-January. A similar sequence, although more compact, has been observed in Guilin, Guangxi, China (Li *et al.*, 1987). In New Zealand, diploid *A. chinensis* var. *chinensis* generally flowers 1–2 weeks before tetraploid *A. chinensis* var. *chinensis*, which, in turn, flowers before hexaploid *A. chinensis* var. *deliciosa*. A similar sequence has been observed in kiwifruit cultivars in Wuhan, Hubei, China (Zhong *et al.*, 2015) and South Korea (Lee *et al.*, 2020). This may simply reflect the different geographical origins in China (and climatic conditions) of the ploidy races (Huang and Ferguson, 2007). Zhong *et al.* (2015) also found that the duration of flowering of male cultivars was consistently longer than that of female cultivars, but this could be just the result of cultivar selection.

Even within the one taxon and ploidy level, e.g. a diploid *A. chinensis* var. *chinensis* breeding population, there can be considerable variation in the time taken to reach 50% flowering (~6 weeks) and in the duration of flowering (Cheng *et al.*, 2006). Flowering time was highly heritable and should be amenable to selection. Selections of *A. chinensis* var. *deliciosa* can likewise differ consistently in time of flowering (Brundell, 1975c); 'Hayward' is amongst the last to flower. The duration of flowering can be greatly affected by environment and can vary from season to season (McPherson *et al.*, 1994).

In commercial orchards of 'Hayward', nearly every flower on a female vine that is pollinated forms a fruit (Palmer-Jones and Clinch, 1974), and unless disease intervenes, there is usually limited fruit drop before harvest. Crop load in commercial orchards is thus largely determined by the number of flowers. In most other fruit crops only a small proportion of flowers set fruit that develop fully.

3.5.6 Flowers and dioecism

Details of flower structure are given in Dickison (1972), McGregor (1976), Schmid (1978), Ferguson (1984), Hopping (1990), Howpage *et al.* (1998) and Fraser and McNeilage (2016).

Actinidia flowers are radially symmetrical. In most species, the flowers are cup-shaped, but in some the petals are strongly reflexed and revolute. The largest flowers in the genus include those of *A. chinensis* and *A. eriantha*, up to ~5 cm across; the smallest, in *A. sabiifolia*, are only ~1 cm across. Pistillate flowers are generally larger than staminate flowers (Huang, 2016a).

Amongst 60 selections of the *A. chinensis* complex, chosen partly on fruit size, increases in ploidy were significantly correlated with increased flower size: diploid cultivars of *A. chinensis* var. *chinensis* had smaller flowers than tetraploid cultivars, which, in turn, had smaller flowers than hexaploid *A. chinensis* var. *deliciosa* cultivars (Zhong *et al.*, 2015).

The flowers usually have five sepals and five petals, but there can be more or fewer. In many species, the petals are different shades of white, ageing in the case of *A. chinensis* var. *deliciosa* to a dirty cream, but in some species they can be yellow, rose-pink to red, greenish-grey or partially coloured. The anthers are light yellow to golden yellow, dark brown or purple-to-black, depending on the species and, in some taxa, geographic location. No nectar is produced (Schmid, 1978).

All *Actinidia* species are thought to be dioecious (Fig. 3.6 and Fig. 3.7), i.e. there are female plants with pistillate flowers and male plants with staminate flowers. *Actinidia* species may appear androdioecious, with staminate and apparently hermaphrodite flowers on separate plants, but functionally they are cryptically dioe-

Fig. 3.6. Dioecism in *Actinidia*. Flowers of *A. chinensis* var. *deliciosa*. *Left:* pistillate flower; *right:* staminate flower. (Photos: The New Zealand Institute for Plant and Food Research Ltd)

Fig. 3.7. Dioecism in *Actinidia*. Multi-branched inflorescences of *A. latifolia*. *Left:* pistillate flowers; *right:* staminate flowers. (Photos: The New Zealand Institute for Plant and Food Research Ltd)

cious (Schmid, 1978; McNeilage, 1991; Kawagoe and Suzuki, 2004). Female plants produce flowers that are morphologically perfect with well-developed syncarpous ovaries containing hundreds of ovules, up to 1500 in *A. chinensis* var. *deliciosa* cultivars, and fully formed stamens, but the pollen produced is sterile. After pollination, the ovules develop into seeds and the ovary into a fruit, i.e. the flowers are functionally pistillate. The possible evolution of dioecy in *Actinidia* is discussed by Fraser and McNeilage (2016).

The lower parts of the carpels of the gynoecium in pistillate flowers fuse to form a superior ovary. The number of carpels depends on the species: many have 20–30, but *A. sabiifolia* has only ~10, whereas *A. eriantha* can have almost 40 carpels. Amongst selections of the *A. chinensis* complex, chosen partly on fruit size, diploid selections had fewer carpels than tetraploid or hexaploid selections (Zhong *et al.*, 2015). The resultant ovary can be conoid, globose, ovoid, cylindrical or bottle-shaped, pubescent or glabrous, depending on the species. The fusion of the carpel bases results in a central core of parenchymatous tissue, the columella (Schmid, 1978), at the top of which, just below the separation of the styles, is a compitum (termed a pollen tube distributor cup by Howpage *et al.*, 1998). This appears to distribute pollen tubes evenly to the different carpels of the ovary, even when pollination is limited to only a few stigmata. The upper parts of the carpels form the styles: these are free, mostly reflexed, radiating and persistent, elongating after flowering. The upper parts of the styles become broadened into V-shaped stigmata.

The filaments of pistillate flowers are short, resulting in the stamens being tucked neatly under the radiating styles. This means that any pollinators collecting the pollen must first clamber over the styles, thereby transferring pollen. Failure of pollen maturation in pistillate flowers occurs very late in flower development (Brundell, 1975b; Schmid, 1978; Messina, 1993; Fraser and McNeilage, 2016). Pollen development seems normal up to the microspore stage. Then the nucleus does not undergo mitosis but instead the cytoplasm condenses, and the cells die as a result of programmed cell death. The pollen, although empty, is still collected by bees.

Staminate flowers have numerous stamens producing viable pollen but have only rudimentary ovaries lacking ovules, and the vestigial styles are short and without stigmatic papillae. Staminate flowers do not set seed, nor do their ovaries develop into fruit. Ovary development stops much earlier than the corresponding failure of pollen development in pistillate flowers (Caporali *et al.*, 2019). Most species have numerous stamens, generally 30–50, but flowers of *A. kolomikta* can have only 16 stamens whereas staminate flowers of *A. eriantha* can have 200 (Huang, 2016a). Among 60 selections of the *A. chinensis* complex, diploid cultivars had markedly fewer stamens than hexaploid cultivars (Zhong *et al.*, 2015). The flowers of two female hexaploid cultivars, 'Xuguan' and 'Xuxiang', have more than 300 stamens.

Both staminate and pistillate flowers produce scent (volatile organic compounds) which attracts honey bees (Twidle *et al.*, 2015). The volatiles produced vary between *Actinidia* taxa and between genotypes within a taxon (Twidle *et al.*, 2017, 2018; Stasiak and Latocha, 2020).

Dioecism is not absolute: fruiting (andromonoecious) or inconstant males occasionally occur (see Chapter 9). The *A. chinensis* var. *deliciosa* cultivar 'Tsechelidis' appears to be an inconstant female (gynomonoecious) with flowers of varying pollen fertility (Sakellariou *et al.*, 2016). Gender inconstancy has also been occasionally recorded in other *Actinidia* species (Hirsch *et al.*, 1990; Tang and Jiang, 1995; Cui *et al.*, 2002; Matsumoto *et al.*, 2013).

Gender inconstancy indicates that commercial hermaphrodite kiwifruit cultivars are a possibility. Ideally, these should be self-setting. Several hermaphrodite cultivars of *A. chinensis* var. *deliciosa* have been released but their fruit are small (see Chapter 4). Flowers of the large-fruited cultivar 'Tsechelidis' are sometimes both hermaphrodite and self-setting (Sakellariou *et al.*, 2016). All the flowers of a hermaphrodite of *A. chinensis* var. *deliciosa*, produced by crossing a female with a fruiting male, have fertile pollen and set large fruit after selfing (McNeilage and Steinhagen, 1998), although not of adequate commercial quality. This vine was used by Akagi *et al.* (2019) to establish that two Y-chromosome-encoded genes determine sex in kiwifruit (see Chapters 9 and 14).

Pollination is not always required for fruit set. 'Xiangji', a selection from the wild of *A. chinensis* var. *deliciosa*, if not pollinated, can produce parthenocarpic fruit lacking seed. Pollination results in fruit containing seed (Pei *et al.*, 2011). 'Chengxiang', a selection from the wild in China of *A. arguta*, can also produce parthenocarpic fruit lacking seed (Jia *et al.*, 2022). There are also contradictory reports that Japanese selections of *A. arguta* are self-setting, but in at least one case, fruit containing no seed were produced parthenocarpically and retained until harvest (Mizugami *et al.*, 2007).

Dioecism has many consequences. *Actinidia* plants are obligatorily outcrossing. This, together with probable chromosome duplication, the high chromosome number allowing extensive recombination (Fraser and McNeilage, 2016), and the frequent polyploidy, particularly in the *A. chinensis* complex, helps maintain heterozygosity. As well as the genetic variation in numerous phenotypic characters (Testolin *et al.*, 1995; Zhu *et al.*, 2002), there is also remarkably high molecular heterozygosity (Huang and Ferguson, 2007).

Dioecism creates problems in breeding programmes (see Chapter 9). It also has practical implications for growers (see Chapter 5). About 10% of each commercial kiwifruit orchard must be dedicated to pollenizer (male) vines. If orchards consist of only female vines relying on mechanical pollination, then pollen must either be purchased or pollen orchards established. Normally, only pollen of hexaploid *A. chinensis* var. *deliciosa* is available commercially.

3.5.7 Pollination

Actinidia flowers have characteristics typical of both insect-pollinated and wind-pollinated flowers (Craig *et al.*, 1988). Insects are attracted to the flowers by the conspicuously coloured petals and the scent and pollen produced by both staminate and pistillate flowers. However, the lack of nectar or pollen kit (a sticky covering on the surface of the pollen grains) is more typical of wind-pollinated flowers. Some species, e.g. *A. kolomikta* and *A. polygama*, have leaves that are brightly variegated in spring and these may attract pollinating insects to the plants (Koller, 1990). The situation in the wild with scattered *Actinidia* plants is very different to commercial kiwifruit orchards where growers cannot rely solely on wind to effect pollination (see Chapter 13).

In the *A. chinensis* var. *deliciosa* cultivars 'Bruno', 'Hayward' and 'Monty', fruit weight at harvest is related to the number of seeds (Hopping, 1976a; Hopping and Hacking, 1983). A 100 g 'Hayward' kiwifruit contains, on average, about 1100 seeds. The link between seed number and fruit size indicates the importance of effective pollination. Fruit of plants held in germplasm repositories are likely to be larger than fruit of the same genotype in the wild because of better pollination (Ferguson, 1991).

3.5.8 Fruit

Only a few *Actinidia* taxa have fruit of immediate commercial potential; most have small fruit between 5 g and 20 g (Huang, 2016a) and some are less than 1 g. The largest fruit are those of some genotypes of *A. chinensis* var. *chinensis* and *A. chinensis* var. *deliciosa*.

Actinidia fruit are defined botanically as berries, i.e. fleshy fruits without a pit or stone, produced by a single flower containing one ovary. In *Actinidia*, the ovary consists of many fused carpels (Fig. 3.8). The ovary wall develops into the pericarp, and the fertilized ovules into seeds embedded in the inner flesh. The fleshy fruit indicates that *Actinidia* have evolved to favour seed dispersal by birds and other frugivores such as monkeys (Fraser and McNeilage, 2016). Most detailed descriptions of the structure of *Actinidia* fruit are of the fruit of *A. chinensis* var. *deliciosa* (Habart, 1974; Hopping, 1976b, 1990; Schmid, 1978; Ferguson, 1984; Nardozza *et al.*, 2011; Guo *et al.*, 2013) (see Chapter 16).

The fruit can be divided into four parts:

- *Skin (epicarp)*, with or without hairs. The skin can be an epidermis consisting of living cells with a thick cuticle (e.g. *A. arguta*) or a thick periderm consisting of one or more layers of dead, collapsed cells, as in *A. chinensis* (Hallett and Sutherland, 2005, 2007). Hairs, if present on the skin surface, vary from the fine, soft hairs of most *A. chinensis* var. *chinensis* genotypes, often readily lost at maturity, to the persistent, stiff, coarse hairs of most *A. chinensis* var. *deliciosa* genotypes, or the long, woolly hairs of *A. eriantha* (White, 1986). Just below the periderm there may be several layers of thick-walled cells forming a hypodermis.
- *Outer pericarp*, consisting of loosely organized parenchymatous cells, individual or small groups of large cells in a matrix of densely packed smaller cells (Hallett *et al.*, 1992). In most species, there are mucilaginous cells containing calcium oxalate crystals (raphides). There may also be stone cells. Chloroplasts, when present, are similar to those in the leaves (Possingham *et al.*, 1980).
- *Inner pericarp*. Between the locules are septa consisting of elongated cells with thicker walls, and within the locules there are two radial rows of seed supported in a matrix of thin-walled cells. The structure of the seed of *A. chinensis* var. *deliciosa* is summarized in Ferguson (1984). The locules can contain particularly numerous oxalate raphides (e.g. *A. arguta*, *A. chinensis* var. *chinensis* and *A. eriantha*), but in *A. macrosperma*, the raphides are concentrated in the seed and very few are present in the locules (Rassam *et al.*, 2007). The chloroplasts of the inner pericarp are very different to those in the outer pericarp (Possingham *et al.*, 1980), with a proliferation of inter-granal membranes and with the grana and inter-granal membranes extending the length of the chloroplasts.
- *Core (columella)*, consisting of uniform parenchyma cells and secondary vascular tissue (Guo *et al.*, 2013). There can be a woody spike at the stalk end.

Fig. 3.8. A backlit equatorial slice of an *Actinidia chinensis* var. *deliciosa* 'Hayward' fruit, derived from a syncarpous ovary containing, in this case, 40 carpels. (Photo: The New Zealand Institute for Plant and Food Research Ltd)

In fruit of 24 genotypes of *A. chinensis* var. *deliciosa*, the outer pericarp occupied ~50–55% of the volume of the fruit, the inner pericarp ~40% and the core ~5–10% (Nardozza *et al.*, 2011). Fruit of *A. chinensis* var. *chinensis* 'Zesy002' have a visibly small core (see Chapter 16).

3.5.9 Fruit diversity

There is great variation in fruit characteristics amongst *Actinidia* species (Fig. 3.9 and Fig. 3.10) (Huang *et al.*, 1983; Li *et al.*, 1985; Ferguson, 1991; Huang *et al.*, 2004; Nishiyama, 2007; Henare, 2015; Huang, 2016a, b). Ripe *Actinidia* fruit can vary noticeably in texture (Schröder

Fig. 3.9. Diversity in *Actinidia* fruit colour. (Photo: The New Zealand Institute for Plant and Food Research)

Fig. 3.10. Diversity in *Actinidia* fruit colour. Segregation in fruit colour in F_1 progeny from an *Actinidia macrosperma* × *A. melanandra* cross. (Photo: The New Zealand Institute for Plant and Food Research Ltd)

et al., 2018). Usually, however, only a few individual genotypes have been examined, not necessarily typical of a particular taxon. Considerable variation within taxa has been revealed by studies of the fruit of numerous genotypes of *A. arguta*, *A. chinensis* var. *chinensis*, *A. chinensis* var. *deliciosa* and *A. eriantha*, but the plants studied (Li *et al.*, 2010b; Henare, 2015; Lee *et al.*, 2020; Wang, S. *et al.*, 2021) were often cultivars selected for fruit size in addition to other characteristics.

Nevertheless, there is also evidence of considerable infrataxon fruit diversity in wild germplasm. The common conclusion of such studies is that the wide variation observed provides ample potential for selection of new, better cultivars (see Chapter 9). Song *et al.* (2022) assessed fruit of 67 genotypes of wild *A. arguta* from northern China. Fruit weight ranged from 3.5 g to 24.2 g (although this might also have been affected by pollination and growing conditions), soluble solids content (SSC) at 'commercial maturity' from 11.9 to 20.6 °Brix, titratable acid from 0.15% to 1.2%, and vitamin C content from 3.51 to 230.8 mg 100 g^{-1} fresh weight. There were also differences in mineral content, and electronic tongue analysis revealed significant differences in fruit sourness, sweetness, bitterness and umami.

He *et al.* (2020) found significant variation among 145 genotypes of wild *A. eriantha* from Jiangxi Province, China, in fruit sugar, dry matter, soluble solids, carotenoid, chlorophyll, ascorbic acid contents and titratable acidity. For example, soluble solids ranged from 6.0% to 17.0% and dry matter 6.6% to 23.1%.

Similar diversity has been demonstrated in wild populations of *A. chinensis* var. *chinensis* (Hu *et al.*, 2021). Sui *et al.* (2013) found significant variation in fruit shape, fruit hairs and the intensity and spread of red pigmentation and background flesh colour in the red-fleshed fruit of 56 accessions of *A. chinensis* var. *chinensis* and *A. chinensis* var. *deliciosa*.

F_1 populations of breeding programmes can also show considerable variation. For example, in *A. chinensis* var. *chinensis*, Cheng *et al.* (2004) found significant variation in all characters measured such as vitamin C (ascorbate) content, titratable acidity, SSC, fruit weight and dry matter content. There was more than a tenfold range in vitamin C content, and nearly a nine-fold range in fruit weight.

Ever since the discovery that the fruit of kiwifruit (*A. chinensis* var. *deliciosa*) are exceptionally rich in vitamin C (Randoin and Boisselot, 1941; Keys, 1942), they have been promoted for their possible health benefits. There is enormous variation in fruit vitamin C content between species, from about 4 mg 100 g^{-1} fresh weight in *A. henryi* to more than 2000 mg 100 g^{-1} fresh weight (> 2% fresh weight) in some genotypes of *A. latifolia*. There can also be considerable variation within taxa or between cultivars (Ferguson, 1984, 1991; Ferguson and MacRae, 1992; Huang *et al.*, 2004; Huang and Ferguson, 2007; Nishiyama, 2007; Huang, 2016a, b).

The colour of the outer and inner pericarps is often used to classify commercial kiwifruit cultivars as green-fleshed, yellow-fleshed or red-fleshed (Fig. 3.11) (see Chapters 4 and 16). The

Fig. 3.11. Diversity in *Actinidia* fruit flesh colour. Equatorial cross-sections. (Photo: The New Zealand Institute for Plant and Food Research Ltd)

colour of the fruit flesh of most *Actinidia* species is hidden until the fruit is cut open but is obvious in kiwiberries and related species that have thin, transparent skins.

The pericarp of all *Actinidia* fruit is green to very dark green early in development and in most taxa remains green even when the fruit are fully ripe. This is due to the retention of chloroplasts containing chlorophyll (Possingham *et al.*, 1980; McGhie and Ainge, 2002; Nishiyama *et al.*, 2005; Montefiori *et al.*, 2009) (see Chapter 16). Changes in colour late in development are the result of chlorophyll degradation unmasking other pigments already present or the synthesis of new pigments. Thus, the ripe fruit flesh of most genotypes of *A. chinensis* var. *chinensis* is yellow-green to yellow, depending on the extent to which chlorophyll is lost (Montefiori *et al.*, 2016), whereas fruit of *A. chinensis* var. *deliciosa* and a few genotypes of *A. chinensis* var. *chinensis* retain their green flesh. Differences in the final fruit flesh colour of the various cultivars are commercially important (see Chapter 4).

The skin and fruit flesh of *A. melanandra* and some *A. arguta* genotypes change from green to light red or purplish-red when ripe owing to the synthesis of anthocyanins (Peng *et al.*, 2019). The flesh of the mature fruit of some genotypes of *A. chinensis* var. *chinensis* and *A. chinensis* var. *deliciosa* can contain anthocyanins, but the distribution and intensity of the red pigmentation varies (Sui *et al.*, 2013) (see Chapters 9 and 16).

The flavour of most *Actinidia* fruit can be described as unexceptional, bland, acid-sweet or somewhat unpleasant. Unripe fruit of *A. macrosperma*, *A. polygama* and *A. valvata* are mouth-numbing with an intense chili flavour; the fruit of species such as *A. globosa* and *A. grandiflora* are described as tasting very acidic (Huang *et al.*, 2003); the fruit of individual genotypes of *A. arguta*, *A. chinensis* (both var. *chinensis* and var. *deliciosa*) and *A. kolomikta* can have an excellent flavour. Flavour is the sum of taste and aroma (volatile organic constituents). Taste largely depends on the relative amounts of acids and sugars in the fruit (Marsh *et al.*, 2003) and fruit of different *Actinidia* species contain differing ratios of sugars and acids (Huang *et al.*, 1983; Li *et al.*, 1985; Huang *et al.*, 2004; Huang, 2016b). Most research on kiwifruit aromas has analysed in detail the volatiles from only a small number of cultivars such as 'Hayward' and 'Hort16A' (Garcia *et al.*, 2012). Volatiles allowed four yellow-fleshed cultivars of *A. chinensis* var.

chinensis to be distinguished from two green-fleshed cultivars of *A. chinensis* var. *deliciosa* (Cozzolino *et al.*, 2020), whereas this was not possible with 20 cultivars of *A. chinensis* var. *chinensis* and var. *deliciosa* with different flesh colours (green, yellow and red) (Lan *et al.*, 2021). This emphasizes the dangers of extrapolating from the behaviour of a small number of genotypes.

3.5.10 Time of fruit harvest

Fruit of 14 *Actinidia* species grown at Guilin, Guangxi Province, China, were considered ready for harvest over about ten weeks, from mid-September to late November (northern hemisphere) (Huang *et al.*, 1983). At Te Puke, New Zealand (southern hemisphere), fruit of different species are ready for harvest from February to June. The date depends to some extent on when the fruit are considered 'ready for harvest' (see Chapters 16 and 20). Fruit of *A. arguta* var. *arguta* cultivars start to ripen on the vine about 100 days after flowering whereas fruit of 'Hayward' reach an SSC of 6.2 °Brix about 150 days after flowering and, if there are no frosts, slowly soften on the vines, being retained until the following spring.

3.6 Climatic Requirements

Cultivated kiwifruit vines are relatively demanding in their climatic requirements (Lötter, 1984) and are restricted to temperate regions. They need shelter from strong winds and have a high requirement for water. Vines require a long frost-free period from budbreak to fruit harvest and winters that are not too cold, but that provide sufficient winter chilling for adequate budbreak and flowering. Excessive sun can damage the fruit but internal vine shading from too dense a canopy can reduce flower evocation (see Chapter 14).

Actinidia species, because the genus is so widespread geographically, are likely to differ in their climatic requirements. Their response to winter cold is a useful example. Those that come from the far north can survive severe winter cold, e.g. *A. kolomikta* can withstand temperatures below −35°C (Evreinoff, 1949). *A. arguta* and *A. polygama* are likewise much hardier than *A. chinensis* var. *deliciosa* (Chat, 1995). It is reasonable to expect cultivars of *A. chinensis* var. *chinensis* and *A. chinensis* var. *deliciosa* to differ in their responses to climatic factors because of the different ecological habitats from which the two varieties derive in China.

In the *Actinidia* species studied, some winter chilling is required to break dormancy and promote the formation of flower buds. The chilling requirements (time at low temperatures) for floral emergence are significantly greater for hexaploid *A. chinensis* var. *deliciosa* than diploid *A. chinensis* var. *chinensis*, with those for tetraploid *A. chinensis* var. *chinensis* being intermediate (Wang *et al.*, 2017; Zhao *et al.*, 2017). This is consistent with the geographic distribution of the ploidy races of the *A. chinensis* complex. There is, however, considerable variation in chilling requirements amongst the diploids (see Chapters 10 and 14).

3.7 The Importance of Diversity

Genetic diversity provides the raw material for breeding and selection programmes. Although many of the present kiwifruit cultivars have excellent qualities, the aim of kiwifruit breeding programmes is to develop cultivars with even better qualities, introduce new types of fruit to meet consumer demand for novelty, introduce convincing resistance to serious diseases such as bacterial canker of kiwifruit caused by *Pseudomonas syringae* pv. *actinidiae* (Psa), and respond to likely changes in climatic conditions. It is not just the fruit attributes that are important. A good example would be the use of *A. rufa*, a species that occurs in Japan, Korea and Taiwan. *A. rufa* has very low chilling requirements and fruits even when minimum winter temperatures remain above 15°C. It is also tolerant of hot, dry summers. It has therefore been crossed with diploid *A. chinensis* var. *chinensis* with the aim of producing kiwifruit better adapted to warm conditions (Kataoka *et al.*, 2014). Some genotypes of *A. rufa* also have strong tolerance to the highly pathogenic Psa biovar 3. *A. rufa* is now being used in breeding programmes to introduce tolerance to Psa (Kisaki *et al.*, 2019).

Actinidia species are diverse. They are highly heterozygous, as indicated by their phenotypic diversity and as revealed by molecular studies. There is great diversity within infraspecific taxa such as *A. chinensis* var. *chinensis* and *A. chinensis* var. *deliciosa*. Even existing selections of the *A. chinensis* complex have a high level of mean genetic diversity (Li *et al.*, 2014). The potential diversity within many other *Actinidia* taxa is often not appreciated because so few representatives have been studied in detail. Worse, most detailed research has so far considered only a very small number of selected genotypes. The great diversity in the genus *Actinidia* makes it dangerous to generalize. Each selected cultivar, each individual genotype, is possibly unique in its attributes, its behaviour and its requirements.

References

Akagi, T., Pilkington, S.M., Varkonyi-Gasic, E., Henry, I.M., Sugano, S.S. *et al.* (2019) Two Y-chromosome-encoded genes determine sex in kiwifruit. *Nature Plants* 5, 801–809. DOI: 10.1038/s41477-019-0489-6

Brundell, D.J. (1975a) Flower development of the Chinese gooseberry (*Actinidia chinensis* Planch.). I. Development of the flowering shoot. *New Zealand Journal of Botany* 13, 473–483. DOI: 10.1080/0028825X.1975.10430338

Brundell, D.J. (1975b) Flower development of the Chinese gooseberry (*Actinidia chinensis* Planch.). II. Development of the flower bud. *New Zealand Journal of Botany* 13, 485–496. DOI: 10.1080/0028825X.1975.10430339

Brundell, D.J. (1975c) Quantitative aspects of flowering in the Chinese gooseberry (*Actinidia chinensis* Planch.). *New Zealand Journal of Agricultural Research* 18, 371–374. DOI: 10.1080/00288233.1975.10421061

Caporali, E., Testolin, R., Pierce, S. and Spada, A. (2019) Sex change in kiwifruit (*Actinidia chinensis* Planch.): a developmental framework for the bisexual to unisexual floral transition. *Plant Reproduction* 32, 323–330. DOI: 10.1007/s00497-019-00373-w

Chat, J. (1995) Cold hardiness within the genus *Actinidia*. *HortScience* 30, 329–332. DOI: org/10.21273/HORTSCI.30.2.329

Chat, J., Jáuregui, B., Petit, R.J. and Nadot, S. (2004) Reticulate evolution in kiwifruit (*Actinidia*, Actinidiaceae) identified by comparing their maternal and paternal phylogenies. *American Journal of Botany* 91, 736–747. DOI: 10.3732/ajb.91.5.736

Cheng, C.H., Seal, A.G., Boldingh, H.L., Marsh, K.B., MacRae, E.A. *et al.* (2004) Inheritance of taste characters and fruit size and number in a diploid *Actinidia chinensis* (kiwifruit) population. *Euphytica* 138, 185–195. DOI: 10.1023/B:EUPH.0000046802.28347.41

Cheng, C.H., Seal, A.G., Murphy, S.J. and Lowe, R.G. (2006) Variability and inheritance of flowering time and duration in *Actinidia chinensis*. *Euphytica* 147, 395–402. DOI: 10.1007/s10681-005-90936-9

Chevalier, A. (1940) Sur des lianes fruitières intéressantes: les *Actinidia*. *Revue de Botanique Appliquée et d'Agriculture Tropicale* 20, 10–15.

Chevalier, A. (1941) Un *Actinidia* à fruits comestibles intéressant pour la France. (*A. chinensis* Planch. var. *deliciosa* Chev.) *Revue de Botanique Appliquée et d'Agriculture Tropicale* 21, 240–244.

Condon, J.M. (1991a) Aspects of kiwifruit stem structure in relation to transport. *Acta Horticulturae* 297, 419–426. DOI: 10.17660/ActaHortic.1992.297.55

Condon, J.M. (1991b) Aspects of comparative vegetative morphology as an aid to *Actinidia* taxonomy. PhD thesis, University of Canterbury, New Zealand.

Cozzolino, R., De Giulio, B., Petriccione, M., Martignetti, A., Malorni, L. *et al.* (2020) Comparative analysis of volatile metabolites, quality and sensory attributes of *Actinidia chinensis* fruit. *Food Chemistry* 316, 126340. DOI: 10.1016/j.foodchem.2020.126340

Craig, J.L., Stewart, A.M., Pomeroy, N., Heath, A.C.G. and Goodwin, R.M. (1988) A review of kiwifruit pollination: Where to next? *New Zealand Journal of Experimental Agriculture* 16, 385–399. DOI: 10.1080/03015521.1988.10425667

Cui, Z., Huang, H. and Xiao, X. (2002) *Actinidia in China*. Agricultural Science and Technology Press, Beijing.

Cuong, N.M., Soejarto, D.D. and Li, J. (2007) A taxonomic revision of Actinidiaceae of Vietnam. *Blumea* 52, 209–243. DOI: 10.3767/000651907X608981

Datson, P.M. and Ferguson, A.R. (2011) *Actinidia*. In: Kole, C. (ed.) *Wild Crop Relatives: Genomic and Breeding Resources, Tropical and Subtropical Fruits*. Springer-Verlag, Berlin, Heidelberg, Germany, pp. 1–20. DOI: 10.1007/978-3-642-20447-0_1

Dickison, W.C. (1972) Observations on the floral morphology of some species of *Saurauia*, *Actinidia*, and *Clematoclethra*. *Journal of the Elisha Mitchell Scientific Society* 88, 43–54.

Dressler, S. and Bayer, C. (2004) Actinidiaceae. In: Kubitzki K. (ed.) *Flowering Plants. Dicotyledons. The Families and Genera of Vascular Plants*, Vol. 6. Springer, Berlin, Heidelberg, Germany, pp. 14–19. DOI: 10.1007/978-3-662-07257-8_3

Dunn, S.T. (1911) A revision of the genus *Actinidia*, Lindl. *Journal of the Linnean Society of London, Botany* 38, 394–410 + plate. DOI: 10.1111/j.1095-8339.1911.tb02326.x

Engin, H., Gökbayrak, Z. and Dardeniz, A. (2011) Flower aberrations in kiwifruit (*Actinidia deliciosa*). *European Journal of Horticultural Science* 76(3), 91–94.

Evreinoff, V.A. (1949) Notes sur les variétiés d'*Actinidia*. *Revue Horticole* 106 (N.S. 24), 69–70.

Ferguson, A.R. (1984) Kiwifruit: a botanical review. *Horticultural Reviews* 6, 1–64. DOI: 10.1002/9781118060797.ch1

Ferguson, A.R. (1990a) Botanical nomenclature: *Actinidia chinensis*, *Actinidia deliciosa*, and *Actinidia setosa*. In: Warrington, I.J. and Weston, G.C. (eds) *Kiwifruit: Science and Management*. Ray Richards Publisher in association with the New Zealand Society for Horticultural Science, Auckland, New Zealand, pp. 36–57.

Ferguson, A.R. (1990b) Stem, branches, leaves and roots of the kiwifruit vine. In: Warrington, I.J. and Weston, G.C. (eds) *Kiwifruit: Science and Management*. Ray Richards Publisher in association with the New Zealand Society for Horticultural Science, Auckland, New Zealand, pp. 58–70.

Ferguson, A.R. (1991) Kiwifruit (*Actinidia*). *Acta Horticulturae* 290, 603–653. DOI: 10.17660/ActaHortic.1991.290.14

Ferguson, A.R. and Huang, H. (2016) Cytology, ploidy and ploidy manipulation. In: Testolin, R., Huang, H.-W. and Ferguson, A.R. (eds) *The Kiwifruit Genome. Compendium of Plant Genomes*. Springer Nature, Cham, Switzerland, pp. 55–63. DOI: 10.1007/978-3-319-32274-2_5

Ferguson, A.R. and MacRae, E.A. (1992) Vitamin C in *Actinidia*. *Acta Horticulturae* 297, 481–487. DOI: 10.17660/ActaHortic.1992.297.63

Ferguson, A.R. and Wu, J.-H. (2016) Interploid and interspecific hybridization for kiwifruit improvement. In: Mason, A.S. (ed.) *Polyploidy and Hybridization for Crop Improvement*. CRC Press, Boca Raton, Florida, pp. 339–363. DOI: 10.1201/9781315369259

Ferguson, A.R., Zhang, J.-L., Duffy, A.M., Beatson, R.A., Cheng, C.-H. *et al.* (2009) Ploidia e uso della citometria a flusso nel miglioramento genetico dell'actinidia. [Ploidy and the use of flow cytometry in kiwifruit breeding] *Italus Hortus* 16(5), 78–83.

Fraser, L.G. and McNeilage, M.A. (2016) Reproductive biology. In: Testolin, R., Huang, H.-W. and Ferguson, A.R. (eds) *The Kiwifruit Genome. Compendium of Plant Genomes*. Springer Nature, Cham, Switzerland, pp. 65–84. DOI: 10.1007/978-3-319-32274-2_6

Garcia, C.V., Quek, S.-Y., Stevenson, R.J. and Winz, R.A. (2012) Kiwifruit flavour: a review. *Trends in Food Science & Technology* 24, 82–91. DOI: 10.1016/j.tifs.2011.08.012

Grant, J.A. and Ryugo, K. (1982) Influence of developing shoots on flowering potential of dormant buds of *Actinidia chinensis*. *HortScience* 17, 977–978. DOI: 10.21273/HORTSCI.17.6.977

Gui, Y.-L. (1981) [A comparative morphological observation of *Actinidia chinensis* Planch. var. *chinensis* and *A. chinensis* Planch. var. *hispida* C.F. Liang.] *Acta Phytotaxonomica Sinica* 19, 304–307 + plate.

Guo, X.-M., Xiao, X., Wang, G.-X. and Gao, R.-F. (2013) Vascular anatomy of kiwi fruit and its implications for the origin of the carpels. *Frontiers in Plant Science* 4, 391. DOI: 10.3389/fpls.2013.00391

Habart, J.L. (1974) La baie de l'*Actinidia chinensis* Planch. var. *chinensis*. *Fruits* 29, 191–207.

Hallett, I.C. and Sutherland, P.W. (2005) Structure and development of kiwifruit skins. *International Journal of Plant Sciences* 166, 693–704. DOI: 10.1086/431232

Hallett, I.C. and Sutherland, P.W. (2007) Kiwifruit skins: the fruit's natural packaging. *Acta Horticulturae* 753, 89–95. DOI: 10.17660/ActaHortic.2007.753.8

Hallett, I.C., MacRae, E.A. and Wegryzyn, T.F. (1992) Changes in kiwifruit cell wall ultrastructure and cell packing during postharvest ripening. *International Journal of Plant Sciences* 153, 49–60. DOI: 10.1086/297006

He, Y., Li, Z., Liao, G., Chen, L., Zhong, M. *et al.* (2020) Variation in fruit quality within wild *Actinidia eriantha* germplasm. *New Zealand Journal of Crop and Horticultural Science* 48, 153–163. DOI: 10.1080/01140671.2020.1761405

He, Z.-C., Li, J.-Q., Cai, Q. and Wang, Q. (2005) The cytology of *Actinidia*, *Saurauia* and *Clematoclethra* (Actinidiaceae). *Botanical Journal of the Linnean Society* 147, 369–374. DOI: 10.1111/j.1095-8339.2005.00372.x

Henare, S.J. (2015) The nutritional composition of kiwifruit (*Actinidia* spp.) In: Simmonds, M.S.J. and Preddy, V.R. (eds) *Nutritional Composition of Fruit Cultivars*. Academic Press, San Diego, California, pp. 337–370. DOI: 10.1016/B978-0-12-408117-8.00015-5

Hirsch, A.-M., Fortune, D. and Blanchet, P. (1990) Study of dioecism in kiwifruit, *Actinidia deliciosa*, Chevalier. *Acta Horticulturae* 282, 367–376. DOI: 10.17660/ActaHortic.1990.282.46

Hopping, M.E. (1976a) Effect of exogenous auxins, gibberellins, and cytokinins on fruit development in Chinese gooseberry (*Actinidia chinensis* Planch.). *New Zealand Journal of Botany* 14, 69–75. DOI: 10.1080/0028825X.1976.10428652

Hopping, M.E. (1976b) Structure and development of fruit and seeds in Chinese gooseberry (*Actinidia chinensis* Planch.). *New Zealand Journal of Botany* 14, 63–68. DOI: 10.1080/0028825X.1976.10428651

Hopping, M.E. (1990) Floral biology, pollination, and fruit set. In: Warrington, I.J. and Weston, G.C. (eds) *Kiwifruit: Science and Management*. Ray Richards Publisher in association with the New Zealand Society for Horticultural Science, Auckland, New Zealand, pp. 71–96.

Hopping, M.E. and Hacking, N.J.A. (1983) A comparison of pollen application methods for the artificial pollination of kiwifruit. *Acta Horticulturae* 139, 41–50. DOI:10.17660/ActaHortic.1983.139.5

Howpage, D., Vithanage, V. and Spooner-Hart, R. (1998) Pollen tube distribution in the kiwifruit (*Actinidia deliciosa* A. Chev. C. F. Liang) pistil in relation to its reproductive process. *Annals of Botany* 81, 697–703. DOI: 10.1006/anbo.1998.0615

Hsieh, T.-Y., Ku, S.-M., Chien, C.-T. and Liou, Y.-T. (2011) Classifier modelling and numerical taxonomy of *Actinidia* (Actinidiaceae) in Taiwan. *Botanical Studies* 52, 337–357.

Hu, G.-M., Xia, W.-J., Zheng, L., Rao, H.-K., Lei, M. *et al.* (2021) Investigation and fruit genetic diversity analysis of wild *Actinidia* germplasm resources in Tongshan County, Hubei Province. *Plant Science Journal* 39, 620–631. DOI: 10.11913/PSJ.2095-0837.2021.60620

Huang, H. (2016a) *Kiwifruit: The Genus ACTINIDIA*. Elsevier, London.

Huang, H. (2016b) Genetic resources. In: Testolin, R., Huang, H.-W. and Ferguson, A.R. (eds) *The Kiwifruit Genome. Compendium of Plant Genomes*. Springer Nature, Cham, Switzerland, pp. 15–36. DOI: 10.1007/978-3-319-32274-2_2

Huang, H. and Ferguson, A.R. (2007) Genetic resources of kiwifruit: domestication and breeding. *Horticultural Reviews* 33, 1–121. DOI: 10.1002/9780470168011.ch1

Huang, H. and Liu, Y. (2014) Natural hybridization, introgression breeding, and cultivar improvement in the genus *Actinidia*. *Tree Genetics & Genomes* 10, 1113–1122. DOI: 10.1007/s11295-014-0771-8

Huang, H., Wang, S., Jiang, Z., Zhang, Z. and Gong, J. (2003) Exploration of *Actinidia* genetic resources and development of kiwifruit industry in China. *Acta Horticulturae* 610, 29–43. DOI: 10.17660/ActaHortic.2003.610.2

Huang, H., Wang, Y., Zhang, Z., Jiang, Z. and Wang, S. (2004) *Actinidia* germplasm resources and kiwifruit industry in China. *HortScience* 39, 1165–1172. DOI: 10.21273/HORTSCI.39.6.1165

Huang, Z.-F., Liang, M.-Y., Huang, C.-G. and Li, R.-G. (1983) A preliminary study on the character and nutritive composition of *Actinidia* fruits. *Guihaia* 3, 53–56, 66.

Hughes, K.A., de Willigen, P., Gandar, P.W. and Clothier, B.E. (1992) Kiwifruit root systems: structure and function. *Acta Horticulturae* 297, 383–389. DOI: 10.17660/ActaHortic.1992.297.51

Jia, Y., Wu, Y.-P., Wang, F.-W., Zhang, L., Yu, G. *et al.* (2022) Full-length transcriptome sequencing analysis and characterization of gene isoforms involved in flavonoid biosynthesis in the seedless kiwifruit cultivar 'Chengxiang' (*Actinidia arguta*). *Diversity* 14, 424. DOI: 10.3390/d14060424

Kataoka, I., Matsumoto, H., Kawano, A., Beppu, K., Ohtani, M. *et al.* (2014) Selection of low-chill kiwifruit adapting to warm climate by utilizing *Actinidia rufa* native to southwestern part of Japan. *Acta Horticulturae* 1059, 85–88. DOI: 10.17660/ActaHort.2014.1059.9

Kawagoe, T. and Suzuki, N. (2004) Cryptic dioecy in *Actinidia polygama*: a test of the pollinator attraction hypothesis. *Canadian Journal of Botany* 82, 214–218. DOI: 10.1139/b03-150

Keys, O.H. (1942) Vitamin C in apples and other materials. *New Zealand Journal of Science and Technology* Section B, 24, 146–148.

Kisaki, G., Shimagami, T., Matsudaira, K., Tsugi, Y., Moriguchi, K. *et al.* (2019) A kiwifruit cultivar crossbred with *Actinidia chinensis* and *Actinidia rufa* has practical tolerance to *Pseudomonas syringae* biovar 3. *Journal of Plant Pathology* 101, 1211–1214. DOI: 10.1007/s42161-019-00349-9

Koller, G.L. (1990) Kolomikta Kiwi. *Arnoldia* 50(1), 36–40.

Lan, T., Gao, C., Yuan, Q., Wang, J., Zhang, H. *et al.* (2021) Analysis of the aroma chemical composition of commonly planted kiwifruit cultivars in China. *Foods* 10(7), 1645. DOI: 10.3390/foods10071645

Lee, M., Kim, H.L., Rhee, H.-C., Kwack, Y.-B. and Kumarihami, H.M.P.C. (2020) Evaluation of the genetic resources of kiwifruit with multivariate analysis. *Horticultural Science and Technology* 38, 569–581. DOI: 10.7235/HORT.20200053

Lei, Z., Li, Z., Wang, Y., Jiang, Z., Wang, S. *et al.* (2010) Vitamin C, flower color and ploidy variation of hybrids from a ploidy-unbalanced *Actinidia* interspecific cross and SSR characterization. *Euphytica* 175, 133–143. DOI: 10.1007/s10681-010-0194-z

Lemon, C.W. and Considine, J.A. (1993) Anatomy and histochemistry of the root system of the kiwifruit vine, *Actinidia deliciosa* var. *deliciosa*. *Annals of Botany* 71, 117–129. DOI: 10.1006/anbo.1993.1015

Li, D. and Liu, Y. (2016) Gene introgression from wild relatives. In: Testolin, R., Huang, H.-W. and Ferguson, A.R. (eds) *The Kiwifruit Genome. Compendium of Plant Genomes*. Springer Nature, Cham, Switzerland, pp. 237–248. DOI: 10.1007/978-3-319-32274-2_18

Li, D., Liu, Y., Zhong, C. and Huang, H. (2010a) Morphological and cytotype variation of wild kiwifruit (*Actinidia chinensis* complex) along an altitudinal and longitudinal gradient in central-west China. *Botanical Journal of the Linnean Society* 164, 72–83. DOI: 10.1111/j.1095-8339.2010.01073-x

Li, D., Zhong, C., Liu, Y. and Huang, H. (2010b) Correlation between ploidy level and fruit characters of the main kiwifruit cultivars in China: implication for selection and improvement. *New Zealand Journal of Crop and Horticultural Science* 38, 137–145. DOI: 10.1080/01140671.2010.482966

Li, D., Liu, Y., Li, X., Rao, J., Yao, X. *et al.* (2014) Genetic diversity in kiwifruit polyploid complexes: insights into cultivar evaluation, conservation, and utilization. *Tree Genetics & Genomes* 10, 1451–1463. DOI: 10.1007/s11295-014-0773-6

Li, H.-L. (1952) A taxonomic review of the genus *Actinidia*. *Journal of the Arnold Arboretum* (Harvard University) 33, 1–61.

Li, J.-Q., Li, X.-W. and Soejarto, D.D. (2007) Actinidiaceae. In Wu, Z-Y., Raven, P.H. and Hong, D.-Y. (eds) *Flora of China*, Vol. 12. Science Press, Beijing, and Missouri Botanical Gardens, St Louis, Missouri, pp. 334–360.

Li, R.-G., Huang, C.-G., Liang, M.-Y. and Huang, Z.-F. (1985) Investigation of germplasm resources of genus *Actinidia* in Guangxi. *Guihaia* 5, 253–267.

Li, R.-G., Liang, M.-Y., Huang, C.-G., Hu, S.-H. and Huang, Z.-F. (1987) Observation and analysis of Actinidias planted in Guilin Germplasm Orchard. *Guihaia* 7, 325–331.

Li, X., Li, J. and Soejarto, D.D. (2009) Advances in the study of the systematics of *Actinidia* Lindley. *Frontiers of Biology in China* 4, 55–61. DOI: 10.1007/s11515-008-0110-2

Li, X.-W. and Li, J.-Q. (2010) A review of the intrageneric taxonomy and nomenclature of *Actinidia* (Actinidiaceae). *Annales Botanici Fennici* 47, 106–108. DOI: 10.5735/085.047.0292

Li, Z.-Z., Man, Y.-P., Lan, X.-Y. and Wang, Y.-C. (2013) Ploidy and phenotype variation of a natural *Actinidia arguta* population in the east of Daba Mountain located in a region of Shaanxi. *Scientia Horticulturae* 161, 259–265. DOI: 10.1016/j.scienta.2013.07.008

Liang, C.-F. (1975) Classification of *Actinidia chinensis* Planch. *Acta Phytotaxonomica Sinica* 13, 32–35 + plate.

Liang, C.-F. (1983) On the distribution of Actinidias. *Guihaia* 3, 229–248.

Liang, C.-F. (1984) Actinidiaceae. In: Feng, K.-M. (ed.) *Flora Reipublicae Popularis Sinicae*. Science Press, Beijing 49(2), 196–268, 309–324.

Liang, C.-F. and Ferguson, A.R. (1984) Emendation of the Latin name of *Actinidia chinensis* Planch. var. *hispida* C.F.Liang. *Guihaia* 4, 181–182.

Liang, C.-F. and Ferguson, A.R. (1986) The botanical nomenclature of the kiwifruit and related taxa. *New Zealand Journal of Botany* 24, 183, 184. DOI: 10.1080/0028825X.1986.10409728

Liu, Y., Li, D., Yan, L. and Huang, H. (2015) The microgeographical patterns of morphological and molecular variation of a mixed ploidy population in the species complex *Actinidia chinensis*. *PLOS ONE* 10(2), e0117596. DOI: 10.1371/journal.pone.0117596

Liu, Y., Li, D., Zhang, Q., Song, C., Zhong, C. *et al.* (2017) Rapid radiations of both kiwifruit hybrid lineages and their parents shed light on a two-layer mode of species diversification. *New Phytologist* 215, 877–890. DOI: 10.1111/nph.14607

Liu, Y., Yu, W., Wu, B. and Li, J. (2022) Patterns of genomic divergence in sympatric and allopatric speciation of three Mihoutao (*Actinidia*) species. *Horticulture Research* 9, uhac054. DOI: 10.1093/hr/uhac054/6541630

Liu, Y.F., Li, D.W. and Huang, H. (2011) Cytotype variation in the natural populations of *Actinidia chinensis* complex in China. *Acta Horticulturae* 913, 157–162. DOI: 10.17660/ActaHortic.2011.913.18

Liu, Y.-L., Li, Z.-Z., Jiang, Z.-W., Liu, Y.-F. and Huang, H.-W. (2008) Genetic structure and hybridization introgression in natural populations of two closely related *Actinidia* species, *A. chinensis* and *A. deliciosa*. *Chinese Journal of Plant Ecology* 32, 704–708. DOI: 10.3773/j.issn.1005-264x.2008.03.021

Lötter, J. de V. (1984) An evaluation of the climatic suitability of various areas in Southern Africa for commercial production of Hayward kiwifruit (*Actinidia chinensis*). *Deciduous Fruit Grower* 34, 122–130.

Marsh, K., Rossiter, K., Lau, K., Walker, S., Gunson, A. *et al.* (2003) The use of fruit pulp to explore flavour in kiwifruit. *Acta Horticulturae* 610, 229–237. DOI: 10.17660/ActaHortic.2003.610.30

Matsumoto, H., Beppu, K. and Kataoka, I. (2013) Identification of hermaphroditism and self-fruitfulness in the wild *Actinidia* found in warm region of Japan. *Horticultural Research* (Japan) 12, 361–366. DOI: 10.2503/hrj.12.361

McAneney, K.J. and Judd, M.J. (1983) Observations on kiwifruit (*Actinidia chinensis* Planch.) root exploration, root pressure, hydraulic conductivity, and water uptake. *New Zealand Journal of Agricultural Research* 26, 507–510. DOI: 10.1080/00288233.1983.10427029

McGhie, T.K. and Ainge, G.D. (2002) Color in fruit of the genus *Actinidia*: carotenoid and chlorophyll compositions. *Journal of Agricultural and Food Chemistry* 50, 117–121. DOI: 10.1021/jf010677l

McGregor, S.E. (1976) *Insect Pollination of Cultivated Crop Plants. Agriculture Handbook* 496. Agriculture Research Service, U.S. Department of Agriculture, Washington DC.

McNeilage, M.A. (1991) Gender variation in *Actinidia chinensis*, the kiwifruit. *Sexual Plant Reproduction* 4, 267–273. DOI: 10.1007/BF00200546

McNeilage, M.A. and Considine, J.A. (1989) Chromosome studies in some *Actinidia* taxa and implications for breeding. *New Zealand Journal of Botany* 27, 71–81. DOI: 10.1080/0028825X.1989.10410145

McNeilage, M.A. and Steinhagen, S. (1998) Flower and fruit characteristics in a kiwifruit hermaphrodite. *Euphytica* 101, 69–72. DOI: 10.1023/A:1018398924819

McPherson, H.G., Hall, A.J. and Stanley, C.J. (1994) Seasonal and regional variation in budbreak and flowering of kiwifruit vines in New Zealand. *New Zealand Journal of Crop and Horticultural Science* 22, 263–276. DOI: 10.1080/01140671.1994.9513835

Messina, R. (1993) Microsporogenesis in male-fertile cv. Matua and male-sterile cv. Hayward of *Actinidia deliciosa* var. *deliciosa* (kiwifruit). *Advances in Horticultural Science* 7, 77–81.

Mizugami, T., Kim, J.G., Beppu, K., Fukuda, T. and Kataoka, I. (2007) Observation of parthenocarpy in *Actinidia arguta* selection 'Issai'. *Acta Horticulturae* 753, 199–203. DOI: 10.17660/ActaHortic.2007.753.22

Montefiori, M., McGhie, T.K., Hallett, I.C. and Costa, G. (2009) Changes in pigments and plastid ultrastructure during ripening of green-fleshed and yellow-fleshed kiwifruit. *Scientia Horticulturae* 119, 377–387. DOI: 10.1016/j.scienta.2008.08.022

Montefiori, M., Pilkington, S.M., Davies, K.M. and Allan, A.C. (2016) Genetics of pigment biosynthesis and degradation. In: Testolin, R., Huang, H.-W. and Ferguson, A.R. (eds) *The Kiwifruit Genome. Compendium of Plant Genomes*. Springer Nature, Cham, Switzerland, pp. 149–161. DOI: 10.1007/978-3-319-32274-2_12

Nakai, T. (1933) Notulae ad plantas Japoniae & Koreae. XLIII. *Botanical Magazine* Tokyo, 47, 235–267. DOI: 10.15281/jplantres1887.40.161

Nardozza, S., Hallett, I.C., McCartney, R., Richardson, A.C., MacRae, E.A. *et al.* (2011) Is fruit anatomy involved in variation in fruit starch concentration between *Actinidia deliciosa* genotypes? *Functional Plant Biology* 38, 63–75. DOI: 10.1071/FP10158

Nishiyama, I. (2007) Fruits of the *Actinidia* genus. *Advances in Food and Nutrition Research* 52, 293–324. DOI: 10.1016/S1043-4526(06)52006-6

Nishiyama, I., Fukuda, T. and Oota, T. (2005) Genotypic differences in chlorophyll, lutein, and β-carotene contents in the fruits of *Actinidia* species. *Journal of Agricultural and Food Chemistry* 53, 6403–6407. DOI: 10.1021/jf050785y

Oláh, R., Masarovičová, E., Šamaj, J., Ovečka, M., Borák, M. *et al.* (1997) Anatomical and morphological parameters of leaves and leaf petioles of *Actinidia deliciosa*. *Biologia Plantarum* 39, 271–280. DOI: 10.1023/A:1000361408183

Palmer-Jones, T. and Clinch, P.G. (1974) Observations on the pollination of Chinese gooseberries variety 'Hayward'. *New Zealand Journal of Experimental Agriculture* 2, 455–458. DOI: 10.1080/03015521.1974.10427714

Pei, C.-J., Liu, S.-B., Xiang, Y.-P. and Zhang, J.-X. (2011) Brief report on the selection and cultivation of the seedless kiwifruit 'Xiangji' (*Actinidia deliciosa*). *Acta Horticulturae* 913, 131–134. DOI: 10.17660/ActaHortic.2011.913.15

Peng, Y., Lin-Wang, K., Cooney, J.M., Wang, T., Espley, R.V. *et al.* (2019) Differential regulation of the anthocyanin profile in purple kiwifruit (*Actinidia* species). *Horticulture Research* 6, 3. DOI: 10.1038/s41438-018-0076-4

Possingham, J.V., Coote, M. and Hawker, J.S. (1980) The plastids and pigments of fresh and dried Chinese gooseberries (*Actinidia chinensis*). *Annals of Botany* 45, 529–533. DOI: 10.1093/oxfordjournals.aob.a085856

Randoin, L. and Boisselot, J. (1941) Détermination de la valeur antiscorbutique du fruit comestible d'*Actinidia chinensis* Planchon, source exceptionnellement riche en vitamin C. *Comptes Rendus de la Société de Biologie, Paris* 135, 209–212.

Rassam, M., Bulley, S.M. and Laing, W.A. (2007) Oxalate and ascorbate in *Actinidia* fruit and leaves. *Acta Horticulturae* 753, 479–485. DOI: 10.17660/ActaHortic.2007.753.62

Rehder, A. (1915) Dilleniaceae. *Actinidia* Lindl. In: Sargent, C.S. (ed.) *Plantae Wilsonianae: An Enumeration of the Woody Plants Collected in Western China for the Arnold Arboretum of Harvard University During the Years 1907, 1908, and 1910 by E.H. Wilson* (Arnold Arboretum Publication No. 4). The University Press, University, Cambridge, Massachusetts, Part 2, Vol. 2, pp. 378–386.

Reid, J.B., Sorensoen, J. and Petrie, R.A. (1993) Root demography in kiwifruit (*Actinidia deliciosa*). *Plant, Cell & Environment* 16, 949–957. DOI: 10.1111/j.1365-3040.1993.tb00518.x

Sakellariou, M.A., Mavromatis, A.G., Adimargono, S., Nakabayashi, K. and Nakas, C. (2016) Agronomic, cytogenetic and molecular studies on hermaphroditism and self-compatibility in the Greek kiwifruit (*Actinidia deliciosa*) cultivar 'Tsechelidis'. *Journal of Horticultural Science and Technology* 91, 2–13. DOI: 10.1080/14620316.2015.1108542

Schmid, R. (1978) Reproductive anatomy of *Actinidia chinensis* (Actinidiaceae). *Botanischer Jahrbücher für Systematik, Planzengeshichte und Pflanzengeographie* 100, 149–195.

Schröder, R., Virgin-Harris, T., Beresford, M.K. and Hallett, I.C. (2018) Kiwifruit texture – a pilot study. *Acta Horticulturae* 1218, 171–175. DOI: 10.17660/ActaHortic.2018.1218.23

Seal, A.G., Ferguson, A.R., de Silva, H.N. and Zhang, J.-L. (2012) The effect of 2*n* gametes on sex ratios in *Actinidia*. *Sexual Plant Reproduction* 25, 197–203. DOI: 10.1007/s00497-012-0191-6

Shi, T., Huang, H. and Barker, M.S. (2010) Ancient genome duplications during the evolution of kiwifruit (*Actinidia*) and related Ericales. *Annals of Botany* 106, 497–504. DOI: 10.1093/aob/mcq129

Snowball, A.M. (1997a) Axillary shoot bud development in selected *Actinidia* species. *New Zealand Journal of Crop and Horticultural Science* 25, 233–242. DOI: 10.1080/01140671.1997.9514012

Snowball, A.M. (1997b) Seasonal cycle of shoot development in selected *Actinidia* species. *New Zealand Journal of Crop and Horticultural Science* 25, 221–231. DOI: 10.1080/01140671.1997.9514011

Song, M., Xu, H., Xin, G., Liu, C., Sun, X. *et al.* (2022) Comprehensive evaluation of *Actinidia arguta* fruit based on the nutrition and taste: 67 germplasm native to northeast China. *Food Science and Human Wellness* 11, 393–404. DOI: 10.1016/j.fshw.2021.11.020

Stasiak, A. and Latocha, P. (2020) Comparative analysis of volatile compounds in flowers of different *Actinidia* species. *Plants* 9, 1675. DOI: 10.3390/plants9121675

Sui, L., Liu, Y., Zhong, C. and Huang, H. (2013) Geographical distribution and morphological diversity of red-fleshed kiwifruit germplasm (*Actinidia chinensis* Planchon) in China. *Genetic Resources and Crop Evolution* 60, 1873–1883. DOI: 10.1007/s10722-013-9961-8

Tang, P., Xu, Q., Shen, R. and Yao, X. (2019) Phylogenetic relationship in *Actinidia* (Actinidiaceae) based on four noncoding chloroplast DNA sequences. *Plant Systematics and Evolution* 305, 787–796. DOI: 10.1007/s00606-019-01607-0

Tang, S. and Jiang, S. (1995) Detection and observation on the bud mutation of *Actinidia chinensis* Planch. *Acta Horticulturae* 403, 71–73. DOI: 10.17660/ActaHortic.1995.403.13

Testolin, R., Cipriani, G. and Costa, G. (1995) Estimate of variance components and heritability of characters in kiwifruit (*Actinidia deliciosa*). *Acta Horticulturae* 403, 182–188. DOI: 10.17660/ActaHortic.1995.403.35

Twidle, A.M., Mas, F., Harper, A.R., Horner, R.M., Welsh, T.J. *et al.* (2015) Kiwifruit flower odor perception and recognition by honey bees, *Apis mellifera*. *Journal of Agricultural and Food Chemistry* 63, 5597–5602. DOI: 10.1021/acs.jafc.5b01165

Twidle, A.M., Suckling, D.M., Seal, A.G., Fedrizzi, B., Pilkington, L.I. *et al.* (2017) Identification of *in situ* flower volatiles from kiwifruit (*Actinidia chinensis* var. *deliciosa*) cultivars and their male pollenisers in a New Zealand orchard. *Phytochemistry* 141, 61–69. DOI: 10.1016/j.phytochem.2017.05.011

Twidle, A.M., Barker, D., Seal, A.G., Fedrizzi, B. and Suckling, D.M. (2018) Identification of floral volatiles and pollinator responses in kiwifruit cultivars, *Actinidia chinensis* var. *chinensis*. *Journal of Chemical Ecology* 44, 406–415. DOI: 10.1007/s10886-018-0936-2

Walton, E.F., Fowke, P.J., Weis, K. and McLeay, P.L. (1997) Shoot axillary bud morphogenesis in kiwifruit (*Actinidia deliciosa*). *Annals of Botany* 80, 13–21. DOI: 10.1006/anbo.1996.0381

Wang, J.-P., Yu, J.-G., Li, J., Sun, P.-C., Wang, L. *et al.* (2018) Two likely auto-tetraploidization events shaped kiwifruit genome and contributed to establishment of the Actinidiaceae family. *iScience* 7, 230–240. DOI: 10.1016/j.isci.2018.08.003

Wang, S., Huang, C., Tao, J., Zhong, M., Qu, X. *et al.* (2017) Evaluation of chilling requirement of kiwifruit (*Actinidia* spp.) in south China. *New Zealand Journal of Crop and Horticultural Science* 45, 289–298. DOI: 10.1080/01140671.2017.1334670

Wang, S., Qiu, Y. and Zhu, F. (2021) Kiwifruit (*Actinidia* spp.): a review of chemical diversity and biological activities. *Food Chemistry* 350, 128469. DOI: 10.1016/foodchem.2020.128469

Wang, S.-M., Jiang, Z.-W., Zhong, C.-H., Zhang, Z.-H., Gong, J.-J. *et al.* (2009) A new ornamental *Actinidia* Lindely cultivar 'Chaohong'. *Acta Horticulturae Sinica* 5, 773.

Wang, Z., Zhong, C., Li, D., Yan, C., Yao, X. *et al.* (2021) Cytotype distribution and chloroplast phylogeography of the *Actinidia chinensis* complex. *BMC Plant Biology* 21, 325. DOI: 10.1186/s12870-021-03099-y

Wang, Z., Hu, G., Li, Z., Zhong, C. and Yao, X. (2022) Characterizing tetraploid populations of *Actinidia chinensis* for kiwifruit genetic improvement. *Plants* 11, 1154. DOI: 10.3390/plants11091154

Wang, Z.-Y., Gould, K.S. and Patterson, K.J. (1994) Comparative root anatomy of five *Actinidia* species in relation to rootstock effects on kiwifruit flowering. *Annals of Botany* 73, 403–413. DOI: 10.1006/anbo.1994.1050

White, J. (1986) Morphology of the fruit hairs of *Actinidia deliciosa* var. *deliciosa*, *Actinidia eriantha* and *Actinidia rufa*. *New Zealand Journal of Botany* 24, 415–423. DOI: 10.1080/0028825X.1986.10409819

Xia, H., Deng, H., Li, M., Xie, Y., Lin, L. *et al.* (2023) Chromosome-scale genome assembly of a natural diploid kiwifruit (*Actinidia chinensis* var. *deliciosa*). *Scientific Data* 10, 92. DOI: 10.1038/s41597-023-02006-4

Yan, G., Ferguson, A.R., McNeilage, M.A. and Murray, B.G. (1997) Numerically unreduced (2*n*) gametes and sexual polyploidization in *Actinidia*. *Euphytica* 96, 267–272. DOI: 10.1023/A:1003037920247

Zeng, H., Li, D.-W. and Huang, H.-W. (2009) Distribution pattern of ploidy variation of *Actinidia chinensis* and *A. deliciosa*. *Plant Science Journal (Journal of Wuhan Botanical Research)* 27, 312–317.

Zhang, Y., Zhong, C., Liu, Y., Zhang, Q., Sun, X. *et al.* (2017) Agronomic trait variations and ploidy differentiation of kiwiberries in northwest China: implication for breeding. *Frontiers in Plant Science* 8, 711. DOI: 10.3389/fpls.2017.00711

Zhao, T., Li, D., Li, L., Han, F., Liu, X. *et al.* (2017) The differentiation of chilling requirements of kiwifruit cultivars related to ploidy variation. *HortScience* 52, 1676–1679. DOI: 10.21273/HORTSCI12410-17

Zhong, C.H., Li, D.W., Han, F., Wang, S.M. and Huang, H.W. (2015) Impacts of polyploidy on blooming and flower morphological traits of kiwifruit cultivars. *Acta Horticulturae* 1096, 229–240. DOI: 10.17660/ActaHortic.2015.1096.25

Zhu, D., Lawes, G.S. and Gordon, I.L. (2002) Estimates of genetic variability and heritability for vegetative and reproductive characters of kiwifruit (*Actinidia deliciosa*). *Euphytica* 124, 93–98. DOI: 10.1023/A:1015662523096

4 Fruiting Cultivars

A. Ross Ferguson[1]*, Jinbao Fang[2] and Guglielmo Costa[3]
[1]*The New Zealand Institute for Plant and Food Research Ltd, Auckland, New Zealand;* [2]*Zhengzhou Fruit Research Institute, Chinese Academy of Agricultural Sciences, Zhengzhou, Henan, China;* [3]*Alma Mater Studiorum – University of Bologna, Bologna, Italy*

4.1 Introduction

This chapter considers the fruiting cultivars of *Actinidia chinensis* var. *chinensis*, *A. chinensis* var. *deliciosa*, *A. eriantha* and their hybrids. These cultivars produce the fruit that are recognizably 'kiwifruit'. Kiwiberries are discussed in Chapter 7.

Several apparently hermaphrodite cultivars of *A. chinensis* var. *deliciosa* have been reported, e.g. 'Renact' and 'Jenny'. These seem to be useful only for home gardeners, and all current commercial cultivars of *A. chinensis* and *A. eriantha* are dioecious. Thus, both male (pollenizer) and female (fruiting) cultivars are important to growers. Male cultivars are discussed in Chapter 5. Only the female cultivars that produce fruit attract the attention of marketers, retailers and consumers.

Consumers are seldom interested in the botanical classification of the fruiting vine, whether it be a cultivar of *A. chinensis* var. *chinensis*, *A. chinensis* var. *deliciosa* or *A. eriantha*. Instead, for the consumer, it is the fruit attributes (Huang, 2022) that are important.

4.2 What Makes the Ideal Fruiting Female Kiwifruit Cultivar?

The fruit of a successful cultivar must meet the requirements of the grower, the marketer, the retailer and the consumer. Kiwifruit industries rely on repeat purchases by the consumer: consumer satisfaction is therefore critical. Only the grower is concerned with the vine that produces the fruit, but vine characteristics are nevertheless important. If a cultivar cannot produce a commercially profitable crop, the quality of the fruit is essentially irrelevant. Handling the fruit must also be profitable for the marketer and retailer.

'[T]he attributes we need for a successful ... [kiwifruit] cultivar [are] great taste, good yield, size and storage, Psa [*Pseudomonas syringae* pv. *actinidiae*] tolerance, strong health attributes, consumer liking ...' (Zespri Group Ltd, 2015). These attributes would be considered desirable by all those growing kiwifruit commercially, but the priorities may vary between different producing countries.

Growing a new cultivar is always risky, even after careful assessment under semi-commercial conditions. Reworking an orchard to a new cultivar entails a considerable loss of grower income for several years. A new cultivar must have significantly improved agronomic traits, or its fruit must attain a premium price in the marketplace if replacement of the existing cultivar is to be justified. Susceptibility to diseases and pests is also important, as demonstrated by the impact of bacterial canker of kiwifruit caused by Psa (see Chapter 18).

*Corresponding author: ross.ferguson@plantandfood.co.nz

© CAB International 2024. *Kiwifruit: Botany, Production and Uses* (eds A.C. Richardson, J.N. Burdon and A.R. Ferguson)

DOI: 10.1079/9781800620933.0004

4.2.1 Psa tolerance

Until recently, disease resistance of a kiwifruit cultivar was a relatively minor consideration. Now, resistance to, or tolerance of, Psa is one of the most important priorities in choosing which kiwifruit cultivar to grow.

Some diploid cultivars of *A. chinensis* var. *chinensis* seem particularly susceptible. The New Zealand-bred cultivar 'Hort16A' was eliminated as a commercial cultivar in Italy and New Zealand because of the disease. The Chinese cultivar 'Hongyang' is also very susceptible, and this may eventually limit its cultivation in China. Many other kiwifruit cultivars can be affected, but generally to a lesser extent. Management practices can be modified to reduce the effects of infection, but the best long-term solution is to select cultivars that can be grown successfully and produce profitable crops even in the presence of Psa.

4.2.2 Fruit size, shape, skin and handling

Larger kiwifruit tend to fetch higher prices in the market. Cultivars with an average fruit size appreciably below about 100 g are at a disadvantage as the 'tail' of small fruit cannot be sold as Class I or Class II fruit under current United Nations Economic Commission for Europe standards, prepared with *A. chinensis* var. *deliciosa* 'Hayward' fruit in mind (UNECE, 2017). Fruit size is strongly affected by crop load, amongst other factors, and growers need to determine the most profitable compromise between fruit number, fruit size and fruit quality (as determined by dry matter content) (see Chapter 12).

Cultivars that produce large numbers of 'flat' fruit that are markedly oval in equatorial cross-section are also at a disadvantage as the minimum ratio of the lesser to greater equatorial diameters is defined for Class I fruit. Fasciated fruit, formed by the fusion of terminal and lateral flowers, can be even larger, but have little commercial value.

Fruit shape and skin robustness can affect vine management and handling procedures. The sharp 'beak' or stylar protrusion of 'Hort16A' fruit meant that they had to be handled more carefully to avoid fruit-to-fruit damage (Patterson *et al.*, 2003). Conversely, a marked indentation at the stylar end of the fruit can provide a refuge for pests. The tough peridermal skins covered in long hairs, as in *A. chinensis* var. *deliciosa*, provide the fruit with more protection than those of *A. chinensis* var. *chinensis* with their much softer, finer hairs. Fruit of *A. chinensis* var. *chinensis* need to be protected during both growth on the vines and handling procedures during and after harvest. The techniques suitable for 'Hayward' are not necessarily suitable for all kiwifruit cultivars. Fruit shape can also affect packaging efficiency in pocket packs.

The skin of ripe fruit of *A. chinensis* is usually light brown but is sometimes reddish-brown to yellow-green or even dark green (Huang *et al.*, 1983). Skin colour can be darkened by exposure to the sun. Skin colour of ripe fruit is not greatly affected by the colour of the underlying fruit flesh.

4.2.3 Fruit flesh colour

The fruit flesh of ripe kiwifruit can be green, yellow or red (Huang, 2016a) (Fig. 4.1). Flesh colour is important as kiwifruit are often used as food garnishes or decoration.

The brilliant emerald green of green-fleshed kiwifruit is part of the consumer appeal, as long as the green does not fade too much during fruit development or storage. Yellow flesh also has its appeal if it is truly yellow. Yellow-fleshed kiwifruit are that colour because of an accumulation of β-carotene (Ampomah-Dwamena *et al.*, 2009) and the loss of chlorophyll (Montefiori *et al.*, 2009) (see Chapters 9 and 16). Partial loss of

Fig. 4.1. The three main categories of commercial kiwifruit: green-fleshed (*Actinidia chinensis* var. *deliciosa* 'Hayward'), yellow-fleshed (*A. chinensis* var. *chinensis* 'Zesy002') and red-fleshed (*A. chinensis* var. *chinensis* 'Zes008'). Other red-fleshed cultivars usually are only partially red, with a starburst of red around the core. (Photo: The New Zealand Institute for Plant and Food Research Ltd)

chlorophyll results in the flesh being lime-green or greenish-yellow. Red fruit flesh occurs in *A. melanandra* and some *A. arguta* genotypes but is less common, occurring only occasionally in most other *Actinidia* species. It is found in only a few cultivars of *A. chinensis* and can be somewhat inconsistent, being affected by environmental conditions and vine management (Wang *et al.*, 2012; Nardozza *et al.*, 2015). The patterns of red pigment distribution throughout the fruit can vary, sometimes restricted to the locules of the inner pericarp (termed 'starburst'), e.g. 'Hongyang', and sometimes present throughout both the inner and outer pericarps (termed 'block red'), e.g. 'Zes008', both cultivars of *A. chinensis* var. *chinensis*.

Consumers may learn to associate and then expect certain fruit flavours with a particular flesh colour (Lavin and Lawless, 1998): more acidic, 'fresh' flavours with green flesh; sweeter, less acidic, sometimes more 'subtropical' flavours with yellow flesh; more intense sweetness, even berry-like flavours with red flesh. Of course, flesh colour is not visible until the fruit is cut open. Consumers are therefore likely to choose fruit based on external cues, such as fruit shape or the hairiness of the fruit skin, and promotional advertizing.

4.2.4 Fruit flavour and texture

Consumers can vary markedly in their kiwifruit flavour preferences (Harker *et al.*, 2007). Preference maps resulting from consumer studies reveal that overall preferences seem to be based on a liking for acidic, green, grassy flavours, typical of the fruit of 'Hayward' or the sweet, 'subtropical' flavours belonging to 'Hort16A'. Some consumers liked both, but many showed a distinct preference for one or the other. It is often stated that Asian consumers tend to prefer sweeter, less acidic fruit than Europeans, but this does not seem to have been established, except anecdotally, for kiwifruit. However, this is plausible as Chinese consumers are prepared to pay considerably higher prices for 'Hongyang' kiwifruit, which have a noticeably sweeter flavour as well as red fruit flesh (Ferguson, 2015; Zhang *et al.*, 2020).

Comparisons of the fruit of different cultivars is often difficult because individual fruit may differ, and consumer preferences are strongly affected by the ripe fruit soluble solids content (SSC) and the firmness of the fruit when eaten (Jaeger *et al.*, 2003; see Chapter 21). Fruit of different cultivars mature and are ready for harvest at different times.

Kiwifruit are normally harvested firm and allowed to ripen and soften before consumption. The different tissues in some genotypes can soften at different rates (Jackson and Harker, 1997; White *et al.*, 2005; Jeffrey *et al.*, 2013). This can leave a hard core when the rest of the fruit is ready for eating, a definite disadvantage. A hard woody spike at the stalk end of the fruit is also a disadvantage. Ripe *Actinidia* fruit can vary noticeably in texture (Schröder *et al.*, 2018; Jaeger *et al.*, 2019), e.g. smooth, mushy, mealy, dry, juicy.

4.2.5 Convenience

Kiwifruit are often considered as inconvenient fruit because it is difficult to judge their stage of ripeness, and cutlery (knives and spoons) is required (Harker *et al.*, 2007). Fruit of *A. chinensis* var. *chinensis*, which lack the stiff, coarse hairs of *A. chinensis* var. *deliciosa* fruit, could be considered more convenient, as could fruit of *A. eriantha* that can be peeled (Wu *et al.*, 2009), or kiwiberries as they have smooth, hairless and edible skins (see Chapter 7).

4.2.6 Harvesting

Kiwifruit are harvested while still firm and then stored until marketed. The time of harvest is determined by the maturity of the fruit, maturity being the physiological state reached when the fruit can be ripened to meet consumer satisfaction. Optimum time of harvest depends on when the fruit are to be marketed (Burdon, 2019). The time of maturity may vary with the cultivar, and the criteria for determining the maturity at which harvest can be carried out may also vary. The criteria used with 'Hayward' are not necessarily suitable for other cultivars (see Chapter 20).

Fruit of 'Hayward' mature relatively late in the season. Later maturity and harvesting can be an advantage or a disadvantage for the grower. In parts of China with higher summer temperatures, later harvesting can be an advantage as the fruit contain less heat at harvest, but further north or at higher altitudes in China, early-maturing cultivars could be preferred. Likewise, in Italy, late maturity can be a disadvantage because of the risks of autumnal frosts or heavy rain in some districts. This has encouraged the selection of Italian cultivars that can be harvested earlier in the season, although a stronger incentive was probably the higher prices obtained for early-marketed fruit. In China, it is an advantage if fruit can be harvested and are ready for sale before important holidays that occur in autumn such as National Day (1 October).

4.2.7 Storage

The remarkable storage potential of 'Hayward' kiwifruit is one of the main reasons why fruit of this cultivar were the only kiwifruit exported from New Zealand from 1975–2000 (Ferguson and Bollard, 1990). Storage life remains a top priority for new kiwifruit cultivars in New Zealand, as the industry there exists primarily for export, and there is a desire for extended marketing. In countries less reliant on exporting, long storage life is less critical although still a great advantage.

Assessment of storage life is more than just measuring firmness or textural characteristics (see Chapter 20). The loss of flavour, production of off-flavours, development of storage-related disorders and shelf life after being removed from storage must also be considered. Conditions that suit 'Hayward' fruit are not necessarily suitable for fruit of other cultivars: storage temperatures may vary, e.g. 0°C, or just below, for 'Hayward' and a degree or two above 0°C for fruit of some *A. chinensis* var. *chinensis* cultivars. Storage temperatures are a compromise between slowing ripening to prolong the storage life and increasing the risk of chilling damage. Cultivar differences in susceptibility to chilling damage can be important.

4.2.8 Health attributes

Kiwifruit are rich in bioactive compounds which have a number of nutritional and health benefits (Singletary, 2012; Stonehouse *et al.*, 2013; Henare, 2015; Richardson *et al.*, 2018; Cundra *et al.*, 2020). High ascorbic acid (vitamin C) content appears to be the most obvious nutritional benefit of eating fresh kiwifruit. Marketing advertisements frequently stress the vitamin C content, which for most commercial kiwifruit cultivars is at least 80–120 mg 100 g^{-1} fresh weight. 'Hayward' fruit contain ~85 mg 100 g^{-1} fresh weight; some cultivars, especially those of *A. chinensis* var. *chinensis*, considerably more. A single 100 g fruit of most cultivars therefore contains enough vitamin C to meet the daily requirement of most adults. Hence, most kiwifruit cultivars would meet the requirements for health claims in the European Union (EU) for vitamin C. Even cultivars with less vitamin C than 'Hayward' would still be a very good sources of vitamin C, better than most other fruit available commercially.

Initially, most studies of the health attributes of kiwifruit compared in detail only green-fleshed kiwifruit (almost certainly 'Hayward') and one yellow-fleshed cultivar (usually 'Hort16A'). An exception was Nishiyama (2007), who analysed several species and a range of genotypes. More recently, there have been studies, particularly from China, of the differences among cultivars in constituents that may affect human health (e.g. Zhang *et al.*, 2020).

Kiwifruit are one of the more common elicitors of food allergies, and the frequency of the allergy seems to be increasing (Wang *et al.*, 2019). Allergens are present in both green-fleshed kiwifruit (*A. chinensis* var. *deliciosa*, usually referring to 'Hayward') and yellow-fleshed cultivars (*A. chinensis* var. *chinensis*, usually 'Zesy002' or 'Hort16A'). Relative allergenicity does not seem to have so far been considered in cultivar selection.

4.2.9 Vine attributes

Attributes that affect the ability to produce a profitable crop yield can also determine the choice of cultivar. There may be resistance or susceptibility to diseases other than Psa or to pests. There may be climatic or edaphic responses. Cultivars differ in their requirement for winter chilling: profitable cultivation of 'Hayward' in parts of New Zealand requires the use of dormancy-breaking chemicals (see Chapter 14). Climatic change may render continued cultivation of

'Hayward' uneconomic in many of the areas where it is currently grown, particularly in the absence of dormancy-breaking chemicals (Vetharaniam *et al.*, 2022). The timing of budbreak and flowering can be important if there is a risk of spring frosts.

In kiwifruit orchards, every flower usually sets a fruit. The number of flowers therefore determines the potential crop. Not all buds break in spring and percentage budbreak and percentage floral budbreak (the proportion of resultant shoots that do carry flowers) are important factors in assessing selections as potential cultivars. Single flowers are preferred. Inflorescences consisting of a king (terminal) flower and lateral flowers are a disadvantage because the fruit from lateral flowers are usually smaller. Furthermore, lateral flowers usually open after king flowers, leading to differences in fruit age and maturity. An extended flowering period is a disadvantage. In an infructescence, two or more fruit in close proximity may suffer skin damage caused by rubbing together. Lateral fruit may have to be removed by thinning. Flowers should also be self-grooming, i.e. the petals should fall soon after fruit set as retention of dead petals can encourage diseases such as *Sclerotinia*.

The fruit must stay on the vine until harvest. Some selections are commercially non-viable because, under certain environmental conditions, they can suffer extensive premature fruit drop.

Cultivars may differ in their response to training and pruning. 'Hayward' was initially thought to be more difficult to manage than some other cultivars grown at that time, but this handicap was outweighed by the superior qualities of its fruit. Cultivars may also differ in their compatibility with rootstocks.

4.2.10 Ploidy

Kiwifruit cultivars that belong to *A. chinensis* var. *chinensis* are diploid or tetraploid (Li *et al.*, 2010, 2014) and cultivars that belong to *A. chinensis* var. *deliciosa* are hexaploid. There is a positive correlation between cultivar ploidy and fruit weight, but it should be remembered that cultivars of any ploidy have been selected for fruit size. Fruit shape, SSC, vitamin C and total soluble acids are not correlated with ploidy (Li *et al.*, 2010). Knowledge of the ploidy of a fruiting cultivar is important in choosing compatible pollenizer cultivars (see Chapter 5). There is an apparently consistent pattern with *A. chinensis* var. *chinensis* that tetraploid cultivars are more tolerant to Psa than diploid cultivars (Wang *et al.*, 2022).

4.2.11 Novelty

In the international food market, there is a demand by retailers for new and innovative products (Jaeger *et al.*, 2003). The New Zealand kiwifruit industry, facing increased competition towards the end of the 20th century, supported breeding programmes to develop new kiwifruit cultivars. Now there are programmes in many countries seeking new and novel cultivars by breeding or by selection from wild germplasm.

Novelty, both the different flesh colour and the sweeter flavour, was probably an important factor in the ready uptake of 'Hort16A' fruit by consumers. Red-fleshed kiwifruit are also likely to appeal to consumers because of their novelty. Consumer studies indicate that rather than improving on existing kiwifruit cultivars there may be advantages in choosing selections with fruit that have novel flavours (Harker *et al.*, 2007).

4.3 Cultivar Promotion

The success of a new cultivar depends on its promotion and marketing as well as the attributes of the vine and its fruit (O'Rourke, 2011). The cost of this promotion, brand protection and marketing of the fruit can be many times the cost of the original research to develop a new cultivar (Martin and Luxton, 2005). Successful marketing could explain the ready acceptance of 'Zesy002' fruit when 'Hort16A' was lost to disease.

Branding of kiwifruit started when 'Hayward' fruit were the only kiwifruit in international trade (Ferguson, 2011). The cultivar was not protected: anybody could grow it and use the name 'kiwifruit'. At that time, 'The kiwifruit industry [was] unique among global fruit industries in being so totally dominated by one variety, the Hayward. Most consumers [were] not even aware that other varieties exist' (Belrose Inc., 1998). In 1970, New Zealand was responsible for 95%

of the total world production of kiwifruit but by 1997 this had fallen to only 25% and was continuing to decrease. Competition from producers in other countries encouraged the development of the Zespri brand with the aim of distinguishing Zespri kiwifruit from those of competitors and to achieve higher prices (Beverland, 2001).

There is an increasing tendency for newer cultivars to be protected by plant variety rights (plant breeding rights) or plant patents and made available only to consortia which limit the areas planted (and thus the production), provide technical advice and make exclusive marketing arrangements (Costa *et al.*, 2018).

The dominant consortium is Zespri Group Limited which manages the bulk of exports of New Zealand-produced fruit as well as some production in Europe and Asia. In New Zealand, kiwifruit are produced primarily for export, meaning that Zespri Group Limited determines the fruit of which cultivars it will accept for export and thus the range of kiwifruit cultivars grown in New Zealand.

In Italy and elsewhere, there is an increasing number of consortia that own or have the propagation rights of some cultivars and also manage promotion, consultancy and marketing activities (Testolin *et al.*, 2023). Such consortia include Zespri Fresh Produce Italy S.r.l. ('Zesy002'), Jingold SpA (Boerica®, 'Jintao', 'Jinyan' and 'Donghong' (Oriental Red®)), Consorzio Dorì Europe Srl (Dorì ('AC 1536') and 'Meris') and New Plant (Italia) ('Ac 459 011'). Consortia are sometimes established at the instigation of the owners of nurseries, with the intention of promoting the sale of plants, or individual nursery owners may have propagation rights, e.g. Zanzi Vivai (Green Angel® and 'HFR18').

Consortia frequently want to ensure year-round supply of their products to the market, and this requires fruit production in both the northern and southern hemisphere. Jingold SpA has licensed production of 'Jintao' in both Italy and Chile. Zespri Group Limited has arranged for its protected cultivars to be grown under licence in the northern hemisphere (Italy, France, Japan and South Korea) as well as New Zealand (Zespri Group Ltd, 2021).

The area planted in a new cultivar will depend on the strength of such consortia, the number of their members and on their promotion, not necessarily on the fruit qualities of the cultivar. In China, there was initially a preference for local provincial selections (Huang and Ferguson, 2001), but now a few cultivars, such as 'Hongyang', have become widespread throughout different provinces (Zhong *et al.*, 2022).

4.4 Fruiting Cultivars Grown

Selectors of kiwifruit cultivars and marketers are increasingly emphasizing the flesh colour of the ripe fruit, particularly as the colour is often associated with differences in flavour. Cultivar names or marketing tradenames often include:

- green, e.g. Earligreen®, Green Angel®, Green Light®, Zespri Sweet Green™ Kiwifruit;
- gold, e.g. EnzaGold™, Jingold™ Kiwifruit, Zespri SunGold™ Kiwifruit, ('yellow' might be a more accurate descriptor than 'gold' but 'gold' implies value);
- red, e.g. Rainbow Red, Red Passion®, 'Red Queen', 'Redvita', Zespri RubyRed™ Kiwifruit;
- jin (Mandarin for gold), e.g. 'Jinfeng', 'Jintao', 'Jinyang', Jingold™ Kiwifruit; and
- hong (Mandarin for red), e.g. 'Chuhong', 'Donghong', 'Hongyang'.

The most important of the kiwifruit cultivars currently grown, previously grown and some promising new cultivars are therefore listed below by flesh colour and by country of origin, not by botanical variety or ploidy. Most kiwifruit cultivars developed in China are grown only in that country. A small number of Chinese cultivars and of those developed outside China are grown in both the northern and southern hemisphere to facilitate year-round marketing.

The list is not exhaustive as there are at least 150 registered or authorized kiwifruit cultivars from China alone (Huang, 2013; Zhong *et al.*, 2022) and possibly another 100 cultivars or named selections from outside China. The areas planted in each cultivar are taken from Costa and Testolin (2022), Zespri Group Ltd (2021) and Zhong *et al.* (2022) and usually include the relevant pollenizer cultivars. The average size of the fruit and other fruit quality parameters will depend on the age of the vine, growing conditions and crop management. More information on individual cultivars is available in Ferguson (1997, 2020, 2022), Ferguson *et al.* (2012), Huang (2013, 2016b), Li *et al.* (2010) and

Wang *et al.* (2021) or in specific citations. Information on ploidy is provided where known.

4.4.1 Green-fleshed kiwifruit cultivars

All cultivars, unless stated, belong to *A. chinensis* var. *deliciosa* and are hexaploid. The few less widely grown exceptions are the tetraploid *A. chinensis* var. *chinensis* cultivar 'Ac 459 011', promoted in Italy, the tetraploid hybrid 'Zesh004' from New Zealand, and some tetraploid, green-fleshed selections of *A. chinensis* var. *chinensis* from China, e.g. 'Wuzhi No. 3'.

Global

'Hayward' (Figs 4.2, 4.3) is still the most widely grown green-fleshed cultivar in most countries where kiwifruit are grown commercially, except China. When kiwifruit were initially grown in New Zealand, there were a number of different pistillate selections, all descended from the introduction of kiwifruit seed in 1904. Eventually, as kiwifruit orchards were planted to produce fruit primarily for export rather than for local sale, 'Hayward' became the only viable commercial cultivar (see Chapter 1). It was sent to the USA in 1935 and for many years was known there as 'Chico'.

The fruit are large and have greenish-brown skin with stiff brown hairs. The fruit flesh is green, and the core is oval, pale yellow and often fluted. The flavour is fresh and somewhat acidic. A very important attribute is the remarkably long storage life: the fruit can be held in storage for many months and still ripen to a state acceptable to consumers.

Although the success of kiwifruit can be ascribed to the choice of 'Hayward', it does have some limitations, listed by Ferguson *et al.* (1990):

- The vine is slower to come into full production; it does not respond well to some pruning systems and has a high requirement for winter chilling. Crop yields can be low.
- Under some growing conditions, it can be affected by Psa.
- It flowers relatively late in the season, later than many commonly used pollenizers.
- The fruit are ovoid, not round. There is a tendency to produce flat and fasciated fruit. The 'Hayward mark', a protrusion on the surface of the fruit, is a common fault.
- The fruit core can be large and sometimes has a hard woody spike at the stalk end.
- The fruit mature late in the season.
- The flavour is too sharp or too acidic for some consumers.
- The vitamin C content, about 85 mg 100 g^{-1} fresh weight, although high compared with that of most other commercially available fruit, is relatively low for a kiwifruit cultivar.

Fig. 4.2. *Actinidia chinensis* var. *deliciosa* 'Hayward' fruit on the vine. (Photo: The New Zealand Institute for Plant and Food Research Ltd)

Fig. 4.3. Green flesh of *Actinidia chinensis* var. *deliciosa* 'Hayward' kiwifruit. (Photo: The New Zealand Institute for Plant and Food Research Ltd)

Despite these potential disadvantages, 'Hayward' has proved to be a very successful cultivar (Thorp *et al.*, 1990) and is still the mainstay of kiwifruit production in most parts of the world.

Clones of 'Hayward' claimed to have fewer fruit defects include Clone 8, Clone K and Clone Maeba® from Italy, and Clone Kramer from New Zealand. Bud mutations of 'Hayward' have also been identified such as Boerica® with more cylindrical fruit, 'Wilkins Super' with more regular fruit and 'Meris' with fewer flat fruit and triples. 'Boverlig 02' (Earligreen®) and 'Convi 97.001' (Green Light®) are considered ready for harvest earlier than 'Hayward', avoiding autumnal frosts. Consumers have been generally unable to detect differences between the fruit of bud mutations of 'Hayward' and 'standard' 'Hayward'. Hoped-for price premiums for individual clones did not eventuate and reworking of existing 'Hayward' orchards could not be justified so plantings have remained small.

China

In 2020, green-fleshed cultivars accounted for 142,870 ha, nearly half the total Chinese plantings of kiwifruit. The situation in China is very different to that in other kiwifruit-producing countries: there is a range of cultivars, and the mix and leading cultivars have changed considerably since kiwifruit cultivation began. Initially, commercial orchards were established using cultivars from New Zealand, particularly 'Hayward', before a Chinese cultivar, 'Qinmei', predominated. Then 'Hayward became more widely planted and in 2020 it was still the third most widely planted green-fleshed kiwifruit cultivar (16,670 ha) in China. Large plantings of new green-fleshed cultivars, especially 'Xuxiang', 'Cuixiang', and 'Guichang', have been made (Huang and Ferguson, 2001; Zhong *et al.*, 2022). 'Xuxiang', now the most widely planted green-fleshed cultivar in China, was apparently selected from open-pollinated seed of 'Hayward'.

Other commonly grown and some promising new green-fleshed kiwifruit cultivars grown in China are listed below, with area planted for the more established cultivars. Many of these cultivars were selections from the wild but some resulted from kiwifruit improvement programmes or deliberate planned crosses. They vary in vine characteristics such as vigour, resistance to environmental stress and time of flowering and of fruit harvest, although timing will depend on the region in China where they are grown. The fruit vary in shape, size, flavour (as indicated by the SSC and the ratio of sugars to acids when eating ripe) and vitamin C content (Xie *et al.*, 2019; Zhang *et al.*, 2020).

'Cuixiang' (20,000 ha in plantings, 2020) was selected from the wild in the Qinling Mountains by the Xi'an Kiwifruit Research Institute and the Agricultural Experimental Station of Zhouzhi County, Shaanxi Province, and was released in 2008. Plantings have expanded very rapidly and by 2020, this was the second most widely planted green-fleshed cultivar in China. The vine is vigorous and productive but often requires fruit thinning. In the Xi'an area, Shaanxi Province, the fruit mature early in the season and are harvested in late August to mid-September. The fruit are obovoid, oval in cross-section, with a pointed stylar end, and have an average weight of 90 g. The skin is thick, yellowish-brown with sparse, brown, stiff, readily shed hairs. The flesh is dark green, with a pleasant aromatic flavour.

'Cuiyu' (2700 ha, 2020) is exceptional in that it is the only widely planted tetraploid, green-fleshed cultivar of *A. chinensis* var. *chinensis*: all other green-fleshed cultivars belong to *A. chinensis* var. *deliciosa* or *A. eriantha* and the fruit flesh of most *A. chinensis* var. *chinensis* genotypes is varying shades of yellow when ripe. 'Cuiyu' was selected from the wild by the Hunan Horticultural Research Institute and registered in 2001. The vine is moderately vigorous and tolerant of adverse conditions. It produces mostly single flowers. In Changsha, Hunan Province, the fruit are harvested in early October. The fruit are conical, with an average weight of 90 g. The greenish-brown skin is smooth and hairless, with downy hairs that are readily shed.

'Emihoutao No. 1' ('Jinkui') (2660 ha, 2020) was selected by the Fruit and Tea Research Institute, Hubei Academy of Agricultural Sciences, from open-pollinated seedlings of 'Zhuxi No. 2' (a selection from the wild). It was registered in 1993. The vine is vigorous, very productive and tolerant of environmental stress. The fruit can be harvested in late October to early November and are similar to those of 'Hayward', being cylindrical, oval in cross-section and wider towards the slightly concave stylar end. The fruit are large, weighing, on average, more than 100 g. They have a yellowish-brown skin with readily shed brownish hairs and the fruit flesh is bright green. The fruit store well.

'Guichang' (16,000 ha, 2020) is the fourth most widely planted green-fleshed cultivar in China. It was selected from the wild in Ziyun County, Guizhou, by the Guizhou Fruit Research Institute. The vine is moderately vigorous and very productive under different growing conditions, being adaptable and tolerant of stress. The fruit, which are harvested in late September to early October, are elongated cylinders, like fruit of the cultivar 'Bruno', with an average weight of 85 g. The stylar end is rounded and slightly convex. The brown skin is covered with stiff, greyish-brown hairs and the flesh is light green.

'Huate' ('White') (1053 ha, 2020) is a particularly large-fruited diploid *A. eriantha* cultivar, with an average fruit weight of 94 g; most *A. eriantha* genotypes have much smaller fruit. It was selected from the wild in Wenzhou, Zhejiang Province, by the Institute of Horticulture, Zhejiang Academy of Agricultural Sciences, and plant variety rights were granted in 2008. In southern Zhejiang, the fruit can be harvested in late October. The fruit are long cylinders, and the skin, which is readily peeled, is covered with the long white hairs typical of the species. The flesh is dark green. The vitamin C content is very high at 687 mg 100 g^{-1} fresh weight, about eight times that of 'Hayward' fruit.

'Miliang No.1' (10,000 ha, 2020) was selected from the wild near Miliang, western Hunan Province, by Jishou University and registered in 1995. The vine is very vigorous, resistant to environmental stress and very productive. It is not seriously affected by pests or diseases. The flowers are mainly singles. The fruit are harvested in late October but have a relatively short storage life. The fruit are elongated cylinders, with a pointed stylar end, and an average fruit weight of 95 g. The brown skin is covered with long, stiff hairs and the flesh is yellowish-green.

'Nongda Mixiang' was selected from seedlings of 'Xuguan' (an open-pollinated seedling of 'Hayward') by the Northwest Agriculture and Forestry University, Yangling, and registered in March 2015. In the Guanzhong area of Shaanxi Province, the fruit are harvested in mid- to late October. The fruit are elongated cylinders. The fruit size is uniform with an average weight of 98 g. The skin is brown with short, soft hairs and the flesh is yellowish-green.

'Nongda Yuxiang' was also selected from seedlings of 'Xuguan' by the Northwest Agriculture and Forestry University, Yangling, and registered in December 2016. The vine is vigorous and resistant to environmental stress. In the Guanzhong area of Shaanxi Province, fruit are harvested in early October. The fruit are elongated cylinders with an average weight of 110 g. The fruit skin is brown with stiff hairs. The flesh is light green to yellow-green, the core is small, and the flesh is smooth and sweet. The vitamin C content is 252 mg 100 g^{-1} fresh weight, which is high for a green-fleshed kiwifruit.

'Qinmei' (5300 ha, 2020) was selected from the wild by the Shaanxi Provincial Fruit Research Institute and the Zhouzhi County Kiwifruit Experiment Station and registered in 1986. The vine is vigorous and highly resistant to environment stresses, regularly producing heavy crops. In Shaanxi Province, the fruit are harvested in October. The fruit (Fig. 4.4) are ellipsoid to ovoid with a distinct beak at the stylar end, weighing, on average, 100 g. The greenish-brown skin is covered with dense, stiff, yellow-brown hairs, which are readily shed. The flesh is light green. At one stage, 'Qinmei' was the most widely grown green-fleshed kiwifruit cultivar in China, but the inferior flavour and quality of the fruit flesh means that it now receives only low prices in the market.

'Ruiyu' (4000 ha, 2020) was selected from a cross of 'Qinmei' × 'K56' by the Shaanxi Rural Science and Technology Development Center and the Shaanxi Bairui Kiwifruit Research Institute and was registered in January 2015. In Shaanxi, north of the Qinling Mountains, the fruit are harvested in late September. They are elongated cylinders and somewhat flattened

Fig. 4.4. Fruit of *Actinidia chinensis* var. *deliciosa* 'Qinmei'. (Photo: The New Zealand Institute for Plant and Food Research Ltd)

laterally, oval or almost round in cross-section. The average fruit weight is 90 g. The skin is brown and is covered with yellow-brown hairs. The ripe fruit SSC is high at 17.3–19.6%, the total acid content is low at 0.91% and the vitamin C content is 140 mg 100 g^{-1} fresh weight. 'Ruiyu' is exceptional amongst green-fleshed cultivars for its high SSC and its high sugar:acid ratio (Zhang *et al.*, 2020). It is a good example of the Chinese trend towards selection of kiwifruit with markedly sweet flavours. The fruit fetch particularly high prices on the Chinese market.

'Xuxiang' (53,350 ha, 2020), a seedling, from Beijing Bontanical Garden, possibly 'Hayward' open-pollinated, selected at Xuhou Fruit Orchard, Jiangsu, China, is currently the most widely planted of all green-fleshed kiwifruit in China, accounting for almost one-fifth of all kiwifruit plantings. It was registered in 1990. The vine is moderately vigorous and productive. In Xuzhou, the fruit are harvested in late October. The fruit are short cylinders with a slight stylar protrusion. The average fruit weight is 90 g. The yellowish-green skin is covered with stiff, yellow-brown hairs and the flesh is green. The flavour is sweeter than that of 'Hayward' and the fruit fetch particularly high prices.

'Zhongmi No. 2' was selected from a cross of 'Miliang No. 1' × 'Taihang-Xiongying' by the Zhengzhou Fruit Research Institute, Chinese Academy of Agricultural Sciences, and was registered in 2014. The cultivar is vigorous. In Zhengzhou, Henan Province, fruit are harvested in late September to early October. The fruit are ovoid, with a slightly concave stylar end; the average weight is 108 g. The fruit have a green-brown skin and stiff, grey-brown hairs, while the flesh is bright green.

Italy

'Summerkiwi 3373' (800 ha, globally, in 2020, mostly in Italy) came from a cross with 'Hayward'. Its fruit mature earlier in the season.

'Ac 459 011', recently released, is unusual in that it belongs to *A. chinensis* var. *chinensis*, but the fruit flesh remains green when eating-ripe. It is a product of a joint breeding and selection programme of the Universities of Bologna and Udine and New Plant (Italy).

Japan

'Koryoku' is an open-pollinated seedling of 'Hayward'. The fruit flesh is dark green with a high SSC when ripe. The fruit are harvested two or three weeks earlier than 'Hayward'.

New Zealand

Three green-fleshed cultivars have been released, with limited commercial production.

'Tomua' was selected from a cross between 'Hayward' and an early-flowering male. It was promoted as an early-maturing complement to 'Hayward'.

'Zesh004' came from a cross of *A. chinensis* var. *deliciosa* × *A. chinensis* var. *chinensis*. The fruit has a sweet, green flesh when harvested early but a reduced green colour with increased maturity. Tetraploid.

'Zes007' fruit has several desirable characteristics, including a sweet 'tropical' flavour.

4.4.2 Yellow-fleshed kiwifruit cultivars

Almost all yellow-fleshed kiwifruit cultivars belong to *A. chinensis* var. *chinensis* and are diploid or tetraploid. There are exceptions such as the interspecific hybrids 'Jinyan' and 'Jinyuan' from China.

Yellow-fleshed kiwifruit are uniquely important in New Zealand, where more than half the kiwifruit exports are of the yellow-fleshed cultivar 'Zesy002'. China has the most abundant plantings (45,267 ha) of many yellow-fleshed cultivars, but they account for only 15.8% of the total Chinese kiwifruit plantings. In Italy, yellow-fleshed cultivars also make up about 15% of the total area in kiwifruit. In 2021, there were also minor plantings, mainly in Chile, Japan, South Korea and the USA. Some of the cultivars planted in these countries have been introduced from China and Italy; others are the result of planned breeding programmes.

China

Initially, Chinese yellow-fleshed kiwifruit cultivars were selections from the wild but, more recently, some have resulted from breeding programmes. The fruit tend to be sweeter than those of green-fleshed cultivars and the vitamin C content is particularly high, up to 250 mg 100 g^{-1} fresh weight. The main production regions for yellow-fleshed cultivars are Sichuan, Henan, Hubei, Jiangxi and Anhui Provinces. Some of the more widely grown cultivars and some promising selections are listed.

'Fengyue' was selected from the wild by the Hunan Horticultural Research Institute and was registered in 2000. The vine is moderately vigorous. In the Changsha area, Hunan Province, fruit harvest is in mid- to late September. The fruit are ovoid or nearly round, with a very small stylar protrusion. The average fruit weight is 90 g. The greenish-brown skin has greyish-white hairs, and the fruit flesh is yellow-green or golden yellow, smooth and juicy. Diploid.

'Ganmi No. 2' ('Kuimi') was selected from the wild in Fengxin County, Jiangxi Province, by the Institute of Horticultural Research, Jiangxi Academy of Agricultural Sciences. It was registered in 1985 and renamed 'Ganmi No. 2' in 1992. The vine is moderately vigorous but highly productive, producing mainly single flowers. In the Nanchang area of Jiangxi, the fruit are harvested in mid-September. The fruit are a flattened spheroid, a very characteristic apple shape, with an average weight of 100 g. The skin is yellowish-brown to dark brown and covered with dense, short, fine hairs, while the flesh is yellow to greenish-yellow. The storage life and shelf life are limited. 'Ganmi No. 2' was one of the first Chinese kiwifruit cultivars to be grown outside China and is known by names such as Apple Sensation, Kamitsu or Turandot. Now it is not widely grown in China. Tetraploid.

'Ganmi No. 3' ('Jinfeng') was selected from the wild in Fengxin County, Jiangxi Province, by the Institute of Horticultural Research, Jiangxi Academy of Agricultural Sciences. It was registered in 1985 and renamed 'Ganmi No.3' in 1992. The cultivar is very vigorous and productive. In the Nanchang region, the fruit are harvested in mid- to late September. The fruit are uniform, ellipsoid, rather bluntly pointed at the stylar end, and have an average weight of about 90 g. The skin is yellowish-brown to dark brown, with dense, readily shed, short hairs, and the flesh is bright yellow. Tetraploid.

'Huayou' was selected from seedlings of fruit collected from the wild, tested by the Rural Science and Technology Development Center and the Zhouzhi Kiwifruit Experiment Station, and registered in 2007. The vine is vigorous and, in Shaanxi Province, fruit are harvested in late September. The fruit are ellipsoid with a pronounced stylar protrusion; average fruit weight 95 g. The skin has sparse, short, readily shed, downy hairs. The flesh is light yellow-green to yellow with a sweet flavour.

'Jinmei' was selected from a cross between 'Jinyan' (see below) and a male whose female relatives have red flesh. 'Jinyuan' (see below) came from the same cross by the Wuhan Botanical Garden, Chinese Academy of Sciences. 'Jinmei' was registered in 2014. The cultivar is vigorous. In the Wuhan area of Hubei Province, fruit are harvested from late September to early October. The fruit are long and flattened ovoid with a rounded, concave stylar end. The skin is greenish-brown with short hairs, and the flesh is yellow. Tetraploid.

'Jinshi No. 1' was selected from the wild in Jiangxi Province by the Sichuan Provincial Academy of Natural Resource Sciences and registered in 2014. In Sichuan Province, at an altitude of 700 m, fruit are harvested in early November. The fruit are long trapezoidal, elongated oval in cross-section, with a blunt stylar end and rounded shoulders. The average fruit weight is 86 g. The skin is yellow-brown with moderately dense, short, yellow-brown hairs which are easily removed, and the flesh is yellow.

'Jintao' (7300 ha, 2020) was selected from the wild in Wuning County, Jiangxi Province, by the Wuhan Botanical Garden, Chinese Academy of Sciences. It was registered in 2005. The vine is moderately to strongly vigorous and highly productive. In the Wuhan area of Hubei Province, fruit are harvested in mid- to late September. The long cylindrical fruit (Fig. 4.5), which are borne mainly as singles, are round in

Fig. 4.5. *Actinidia chinensis* var. *chinensis* 'Jintao', a Chinese selection first commercialized in Italy. (Photo: R. Testolin, University of Udine, Italy)

Fig. 4.6. *Actinidia eriantha* × *A. chinensis* var. *chinensis* 'Jinyan': the first commercial *Actinidia* interspecific hybrid to be grown commercially. (Photo: Z. Jiang, Wuhan Botanical Garden, Chinese Academy of Sciences, Hubei, China)

cross-section and the stylar end is almost flat. The skin is yellowish-brown with almost no hair at maturity, and the flesh is yellow-green to yellow. The vitamin C content is high at about twice that of 'Hayward'. The fruit store well, better than the fruit of many other *A. chinensis* var. *chinensis* cultivars. 'Jintao' is also planted in Italy and other countries. Tetraploid.

'Jinyan' (14,000 ha, 2020) was selected from a cross of *A. eriantha* × *A. chinensis* var. *chinensis* by the Wuhan Botanical Garden, Chinese Academy of Sciences. It was approved by the Hubei Provincial Forest Cultivar Appraisal Committee in 2006 and as an authorized new kiwifruit cultivar by the National Forestry Cultivar Approval Committee in 2010. It is the first commercial cultivar developed by interspecific crossing, although the fruit more closely resemble other genotypes of *A. chinensis* var. *chinensis* rather than *A. eriantha*. The vine is vigorous, precocious and productive. In Wuhan, Hubei Province, the fruit are harvested late in the season, in late October to early November. The fruit (Fig. 4.6) are elongated cylinders with an average weight of 100 g. The stylar end is slightly concave. The skin is yellowish-brown with short hairs, and the flesh is brilliant yellow and juicy. Surprisingly, the vitamin C content is only 105 mg 100 g^{-1} fresh weight, rather low for an *A. chinensis* var. *chinensis* genotype and considerably lower than the average for *A. eriantha* genotypes. Tetraploid.

'Jinyuan' is also a selection from a cross by the Wuhan Botanical Garden, of 'Jinyan' and an *A. chinensis* male whose female siblings had red flesh. It was registered in 2012. The cultivar is vigorous. In Wuhan, Hubei Province, fruit harvest is in late September to early October. The fruit are short cylinders, with a flattened or slightly concave stylar end, and have an average weight of 84 g. The skin is yellow-brown, covered with dense, short hairs, and the flesh is golden or orange yellow. 'Jinyuan' fruit have a long storage life. Tetraploid.

'Lushanxiang' was selected from the wild in Wuning County, Jiangxi Province, by the Lushan Botanical Garden, Chinese Academy of Sciences and was registered in 1985. The cultivar is vigorous and productive. At Lushan, the fruit are harvested in mid-October. The fruit are elongated cylinders, slightly concave at the stylar end, with an average fruit weight of 90 g. The skin is light yellow to yellow-brown, covered with sparse, fine, easily shed yellow hairs, and the flesh is light yellow. 'Lushanxiang' was also one of the first Chinese cultivars to be grown outside China and was given names such as Elizabeth, First Emperor and Yellow Joy. It is now relatively unimportant in China. Tetraploid.

'Nongda Jinmi' was selected by the Northwest A & F University from a cross of 'Jinnong No. 2' ('Emihoutao No. 2') × male of 'Jinyang No. 1' and was registered in 2016. The cultivar is vigorous and productive. In the Guanzhong region of Shaanxi Province, fruit are harvested in early September. The fruit are nearly cylindrical with a greenish-brown skin with sparse fine hairs. The average fruit weight is 82 g. The flesh is greenish-yellow to yellow and is sweet, smooth and juicy. The vitamin C content is high at 204 mg 100 g^{-1} fresh weight.

Italy

In 2020, about 15% of all kiwifruit plantings in Italy were of yellow-fleshed cultivars (Costa and Testolin, 2022), producing about a quarter of the total kiwifruit crop. Most plantings were of 'Zesy002' from New Zealand grown under licence, replacing 'Hort16A', which also has yellow-fleshed fruit (see below). Two of the other yellow-fleshed cultivars grown in Italy came from China – 'Jintao' and 'Jinyan'. All cultivars have fruit with bright yellow flesh, are sweet and are ready for harvest up to a month before those of 'Hayward'.

'AC 1536' (Dorì) (330 ha, 2020) is a product of a breeding programme at the Universities of Bologna and Udine. The vine is moderately vigorous and resistant to winter cold. At Bologna, the fruit are harvested in the first half of September, a month before other yellow-fleshed cultivars grown in Italy. It has large oblong fruit with an almost flat, slightly concave stylar end and an average weight of 105 g. The fruit skin is yellow-brown at harvest with fine downy hairs that are readily shed. The fruit flesh is a brilliant yellow and the fruit are sweet. Tetraploid.

'Jintao' (1281 ha, 2020) is described in Section 4.4.2, China.

'Jinyan' (281 ha, 2020) is described in Section 4.4.2, China.

'Soreli' (332 ha, 2020) is a product of a breeding programme at the University of Udine. It is highly productive. The fruit are long, oblong-ovoid and large, weighing on average more than 100 g. The fruit skin is bright brown and essentially hairless at harvest, and the fruit flesh is brilliant yellow. The fruit can be harvested about a month before 'Hayward'. The fruit are sweet with a tropical flavour, but the storage life is short. 'Soreli' is under the control of Variety Innovation B.V. (formerly Biogold). Tetraploid.

'Zesy002' (3085 ha, 2020) is grown under licence from Zespri Group Ltd and is described in Section 4.4.2, New Zealand.

Japan

'Sanuki Gold' is a tetraploid seedling of the well-known Chinese cultivar 'Ganmi No. 2'. It has very large fruit with bright yellow flesh that are high in vitamin C and sugar (Fukuda *et al.*, 2007).

South Korea

Several yellow-fleshed cultivars have been released, e.g. 'Haegeum' (Cho *et al.*, 2011), but do not appear to be widely grown.

New Zealand

The New Zealand Institute for Plant and Food Research Ltd (Plant & Food Research) and its predecessors, with support from Zespri Group Limited, have had a large programme of kiwifruit breeding for more than 30 years, recently separated into a joint venture, the Kiwifruit Breeding Centre.

'Hort16A' was the first major success of the breeding programme (Muggleston *et al.*, 1998) (Fig. 4.7). It was the first widely grown yellow-fleshed kiwifruit cultivar outside China. It was also the first yellow-fleshed kiwifruit to be introduced into international trade in 2000. Within a few years it became the most important and most profitable New Zealand kiwifruit cultivar. Plantings were established in the northern hemisphere (mainly Italy) under licence to provide year-round supplies of fruit. The fruit have an average weight of 95–100 g, and the vines produce high yields. The fruit are ovoid, tapering towards the stylar end with a prominent beak. The skin is green-brown to brown, depending on sun exposure, and is covered by soft, silky, easily shed hairs. The flesh is lemon yellow to brighter yellow and has a strongly aromatic, tropical,

Fig. 4.7. *Actinidia chinensis* var. *chinensis* 'Hort16A', the first yellow-fleshed kiwifruit in international trade. It was also the first commercial kiwifruit cultivar produced by a breeding programme. (Photo: The New Zealand Institute for Plant and Food Research Ltd)

sweet flavour. Unfortunately, 'Hort16A' proved to be particularly susceptible to Psa, and almost all 'Hort16A' orchards were eventually removed in both Italy and New Zealand. It is no longer a viable commercial cultivar. Diploid.

'Zesy002' (6200 producing ha in New Zealand, 4000 ha planted in northern hemisphere, 2020) was also developed by Plant & Food Research. Initially, it was selected for other attributes, but proved to be reasonably tolerant of Psa. It was therefore chosen as the replacement for 'Hort16A' and the transition to 'Zesy002' was achieved remarkably quickly, despite the flavour being different. This confirms that there is no single acceptable flavour for kiwifruit. The fruit are naturally large, with an average weight > 100 g, and the vines can bear good crop loads. The fruit are ovoid with a flat, slightly concave stylar end. The skin is yellow-green to dark brown, depending on sun exposure, and is covered by short, soft, easily shed hairs. The fruit are harvested in April in New Zealand. Tetraploid.

'Zesy003' was released at the same time as 'Zesy002' to find a replacement for 'Hort16A' in response to Psa but was dropped in favour of 'Zesy002'.

Other yellow-fleshed kiwifruit cultivars have been developed in New Zealand by D. Skelton, but are not widely grown there, including 'Skelton A19', 'Y368' (SweetGold kiwifruit), 'Y374' (Kiwikiss™ Gold kiwifruit), and 'Z487' (FirstGold kiwifruit). 'Skelton A19' and 'Y368' are grown on a small scale in South Africa and South America.

4.4.3 Red-fleshed kiwifruit cultivars

Red-fleshed kiwifruit are becoming very popular because of their colour and perceived sweetness (Zhang *et al.*, 2020). Fruit categorized as having red flesh, may be bi-coloured, with the red pigmentation limited to the locules of the inner pericarp, e.g. 'Hongyang', or have red pigmentation in both the inner and outer pericarp, e.g. 'Zes008'.

Red pigmentation sometimes occurs in the ripe fruit of wild genotypes of both *A. chinensis* var. *chinensis* and *A. chinensis* var. *deliciosa*. All the red-fleshed cultivars currently grown belong to *A. chinensis* var. *chinensis*, and are probably diploid, except for 'Chuhong' which is tetraploid (D.-W. Li, Wuhan, 2023, personal communication).

China

Red-fleshed kiwifruit are uniquely important in China. Consumer demand and the consequent high prices obtained have strongly encouraged planting, even in unsuitable areas, so that in 2020, red-fleshed cultivars accounted for nearly one-third (93,270 ha) of all Chinese kiwifruit plantings. 'Hongyang', the most important cultivar, is the most widely planted kiwifruit cultivar in China and many of the other red-fleshed cultivars grown are related to it, e.g. 'Donghong'. Many red-fleshed cultivars are particularly susceptible to Psa, the fruit are relatively small, the red pigmentation is often inconsistent and influenced by growing conditions and the fruit storage life is relatively short. However, the returns to growers compensate initially for these disadvantages.

'Chuhong' was selected from the wild in Xupu County, Hunan Province, by the Hunan Horticultural Research Institute and was registered in 2005. The vine is vigorous. The flowers are mostly singles. In the Changsha area of Hunan Province, the fruit can be harvested in early September. The fruit are elongated oval or oblate, with an average fruit weight of 70–80 g. The skin is dark green and hairless. The outer pericarp is yellowish-green with red pigments in the locules of the inner pericarp. Tetraploid.

'Donghong' (20,000 ha, 2020) was selected from F_1 seedlings of open-pollinated 'Hongyang' by the Wuhan Botanical Garden, Chinese Academy of Sciences, and was registered in 2011. The vine is moderately vigorous. In Wuhan, Hubei Province, fruit can be harvested in late August to early September. The fruit are elongated cylinders with a rounded stylar end. The average weight is about 70–75 g, somewhat greater than that of 'Hongyang'. The fruit skin is greenish-brown and hairless. The outer pericarp is golden yellow and the inner pericarp is red.

'Hongyang' (also known as Red Sun No. 1 or RS1) (66,700 ha, 2020) was selected from a seedling population, raised from seed collected in the wild in Henan Province, by the Sichuan Provincial Natural Resources Institute and

Cangxi County Agricultural Bureau. It was registered in 1997. At Cangxi, Sichuan Province, fruit are harvested in early September. The fruit are cylindrical and short obovoid with a concave stylar end. The fruit (Figs 4.8, 4.9 and 4.10) are relatively small, with an average fruit weight of 65 g. The skin is green or greenish-brown and has short hairs which can be easily removed. The outer fruit flesh is yellow-green, the inner pericarp is red, and the core is white. The flavour is particularly sweet, with a high sugar and low acid content. 'Hongyang' is intolerant of high summer temperatures and drought, and the development of red colouration can be reduced by environmental conditions, sometimes almost completely disappearing (Zhong *et al.*, 2007).

Fig. 4.8. *Actinidia chinensis* var. *chinensis* 'Hongyang' fruit on the vine; the most widely planted kiwifruit cultivar in China. (Photo: C.H. Cheng, Kiwifruit Breeding Centre, New Zealand)

'Hongshi No. 2' was selected from a cross of 'Hongyang' × SF0612M by the Sichuan Provincial Natural Resources Institute and registered in 2014. The vine is vigorous. In the Deyang region of Sichuan Province, the fruit can be harvested in mid-September. The fruit are broadly ellipsoid, with rounded shoulders and a deeply concave stylar end. The average fruit weight is 78 g. The skin is greenish-brown with sparse, short, yellow hairs. The outer pericarp is yellow-green, and the inner pericarp is red.

'Qihong' is a bud mutation of 'Hongyang' found in 2011 by the Northwest Agriculture and Forestry University, Yanling. It was registered in March 2014. In the Guanzhong region of Shaanxi Province, the fruit can be harvested in late September. The fruit are nearly cylindrical,

Fig. 4.9. The delicate skin of *Actinidia chinensis* var. *chinensis* 'Hongyang' fruit is protected by bagging in a Chinese kiwifruit orchard. (Photo: S. Max, Zespri International Ltd, New Zealand)

Fig. 4.10. Equatorial slices of the fruit of *Actinidia chinensis* var. *chinensis* 'Hongyang' with a starburst of red in the locules of the inner pericarp contrasting with the greenish-yellow outer pericarp. (Photo: The New Zealand Institute for Plant and Food Research Ltd)

Fig. 4.11. Fruit of *Actinidia chinensis* var. *chinensis* 'Zes008'. (Photo: The New Zealand Institute for Plant and Food Research Ltd)

with a distinctive protuberance at the stylar end. The average fruit weight is about 100 g, considerably larger than that of 'Hongyang'. The skin is brown and hairless. The outer pericarp is yellow-green with red colour around the small, oval core.

'Yangshi Jinhong No. 50' was selected by Yang Shangmou in 1999 from a cross of 'Hongyang' × *A. chinensis* var. *chinensis* male. The vine is moderately vigorous. In Xixia County, Henan Province, the fruit can be harvested in early October. The fruit are cylindrical with a slightly concave or flattened stylar end. The average fruit weight is 90 g. The skin is smooth and brownish-yellow. The outer pericarp is yellow and the inner pericarp red.

Italy

Small plantings (about 140 ha in 2020) of red-fleshed kiwifruit have been established in Italy. Several cultivars are being tested, e.g. 'Donghong' (Oriental Red®), 'HFR18' and 'Red Queen' (RK2018). All appear to be related to the Chinese red-fleshed cultivar, 'Hongyang', and are being promoted by consortia.

South Korea

'Redvita', a yellow-fleshed kiwifruit with red in the locules of the inner pericarp, was registered in 2015 and released to growers the following year. With an average weight of 97 g, the fruit are larger than fruit of 'Hongyang' (Kwack *et al.*, 2017). Tetraploid.

New Zealand

'Zes008' is the result of a long-term breeding programme by Plant & Food Research. Commercial production started in 2020 with the planting of 150 ha. The fruit mature early in the season and in New Zealand are harvested in March. The fruit are elliptical and oblate in cross-section with a slightly concave stylar end (Fig. 4.11). The average fruit weight, 74 g, is small, about the same as that of 'Hongyang'. The skin is greenish-brown with short, soft, reddish-brown hairs mainly at the stylar end. The softness of the skin means that the fruit require careful handling. The fruit are very sweet, and the texture is smooth. The red pigmentation is intensely dark in the locules of the inner pericarp but is spread throughout the whole pericarp, although the intensity in the outer pericarp can vary from fruit to fruit. This intensity and distribution of the pigmentation distinguishes 'Zes008' from many other red-fleshed kiwifruit cultivars. Diploid.

4.5 Current and Future Trends

The first 20 years of the 21st century have seen many changes in the kiwifruit cultivars grown:

1. Yellow-fleshed kiwifruit arrived in the international marketplace. Kiwifruit were no longer simply fruit of the 'Hayward' cultivar – hairy outside and green-fleshed inside. The change was most dramatic in New Zealand, where yellow-fleshed kiwifruit have become more important (and more profitable for growers) than the original green-fleshed kiwifruit, 'Hayward'.
2. A red-fleshed cultivar, 'Hongyang', has become the most widely planted kiwifruit cultivar in China. Red-fleshed kiwifruit have also started to be available internationally. Some commercially important cultivars, such as 'Zesy002', are the results of planned breeding programmes, rather than selections from the wild or seedling selections.
3. The first widely grown interspecific hybrid cultivar became important in China: 'Jinyan', a cross of *A. eriantha* × *A. chinensis* var. *chinensis*.
4. Branding of kiwifruit became widespread.
5. Many newer cultivars are protected by plant variety rights, and their propagation and the marketing of their fruit are restricted to consortia.
6. The Psa epidemic showed how much kiwifruit cultivars can vary in their susceptibility to a particular disease. It also demonstrated the dangers of industries relying on only one or two cultivars. As a result of the breeding programme, New Zealand had 'Zesy002' available to replace 'Hort16A'. However, many red-fleshed kiwifruit are descended from 'Hongyang', which is very susceptible to Psa, again indicating the dangers of a narrow genetic base.

In China, the rapid changes in cultivar composition as kiwifruit plantings have expanded are likely to continue. Changes will be driven by the considerable differences in the prices obtained in the Chinese market for kiwifruit of different cultivars. Internationally, plantings of yellow-fleshed and red-fleshed cultivars will probably increase. Although 'Hayward' has been the leading kiwifruit cultivar for nearly 50 years, its dominance is being increasingly threatened by yellow- and red-fleshed cultivars and new green-fleshed cultivars with fruit that are sweeter and less acidic. The list above of the more important commercial kiwifruit cultivars is very likely to change.

Commercial hermaphrodite kiwifruit cultivars are becoming a realistic possibility. The *A. chinensis* var. *deliciosa* selection, 'Jenny', has only small fruit in the absence of a male pollenizer, but sets more fruit if a pollenizer is present. Another selection of *A. chinensis* var. *deliciosa*, 'Renact', a seedling from open-pollination of 'Hayward', produced, in the absence of a pollenizer, fruit approximately 70% the size of 'Hayward' fruit.

However, choosing a cultivar is only the first step. It can be dangerous to predict the quality of fruit of mature cropping vines from the fruit of the first one or two harvests from a young developing vine. Semi-commercial trials are necessary to confirm the initial promise. It is then necessary to develop management techniques, the handling and postharvest procedures that will allow profitable production of fruit that will appeal to marketers, retailers and consumers. Acceptance of the fruit of new cultivars by consumers will depend on the reliability of supplies of consistently good quality fruit and promotion in the markets.

References

Ampomah-Dwamena, C., McGhie, T., Wibisono, R., Montefiori, M., Hellens, R.P. *et al.* (2009) The kiwifruit lycopene beta-cyclase plays a significant role in carotenoid accumulation in fruit. *Journal of Experimental Botany* 60, 3765–3779. DOI: 10.1093/jxb/erp218

Belrose, Inc. (1998) *World Kiwifruit Review*, 1998 edn. Belrose Inc., Pullman, Washington.

Beverland, M. (2001) Creating value through brands: the ZESPRI™ kiwi fruit case. *British Food Journal* 103, 383–399. DOI: 10.1108/00070700110400389

Burdon, J. (2019) Kiwifruit biology: the commercial implications of fruit maturation. *Horticultural Reviews* 46, 385–421. DOI: 10.1002/9781119521082.ch8

Cho, Y.-S., Cho, H.-S., Park, M.-Y., Ma, K.-C., Lim, D.-G. *et al.* (2011) A selection of Korean gold kiwifruit 'Haegeum' and its performances. *Acta Horticulturae* 913, 249–253. DOI: 10.17660/ActaHortic.2011.913.32

Costa, G. and Testolin, R. (2022) Recent changes in the kiwifruit (*Actinidia* spp.) industry since its introduction: Italy as a case study. *Acta Horticulturae* 1332, 1–16. DOI: 10.17660/ActaHortic.2022.1332.1

Costa, G., Ferguson, R., Huang, H. and Testolin, R. (2018) Main changes in the kiwifruit industry since its introduction: present situation and future. *Acta Horticulturae* 1218, 1–16. DOI: 10.17660/ActaHortic.2018.1218.1
Cundra, L.B., D'Souza, S.M., Parekh, J. and Johnson, D.A. (2020) The impact of kiwifruit on human health. *Gut and Gastroenterology* 3(1), 1–10.
Ferguson, A.R. (1997) Kiwifruit (Chinese gooseberry). In: Neff, M.W. and Brooks, R.M. (eds) *The Brooks and Olmo Register of Fruit & Nut Varieties*, 3rd edn. ASHS Press, Alexandria, Virginia, pp. 319–323.
Ferguson, A.R. (2011) Kiwifruit: evolution of a crop. *Acta Horticulturae* 913, 31–42. DOI: 10.17660/ActaHortic.2011.913.1
Ferguson, A.R. (2015) Kiwifruit in the world – 2014. *Acta Horticulturae* 1096, 33–46. DOI: 10.17660/ActaHortic.2015.1096.1
Ferguson, A.R. (2020) Kiwifruit. In: Gasic, K., Preece, J.E. and Karp, D. (eds) Register of new fruit and nut cultivars. List 50. *HortScience* 55, 1186–1190. DOI: 10.21273/HORTSCI50register-20
Ferguson, A.R. (2022) Kiwifruit. In: Preece, J.E. and Karp, D. (eds) Register of new fruit and nut cultivars. List 51. *HortScience* 57, 1215–1216. DOI: 10.21273/HORTSCI.57.9.1174
Ferguson, A.R. and Bollard, E.G. (1990) Domestication of the kiwifruit. In: Warrington, I.J. and Weston, G.C. (eds) *Kiwifruit: Science and Management*. Ray Richards Publisher in association with the New Zealand Society for Horticultural Science, Auckland, New Zealand, pp. 165–246.
Ferguson, A.R., Seal, A.G. and Davison, R.M. (1990) Cultivar improvement, genetics and breeding of kiwifruit. *Acta Horticulturae* 282, 335–348. DOI: 10.17660/ActaHortic.1990.282.43
Ferguson, A.R., Huang, H.-W. and Testolin, R. (2012) Kiwifruit. In: Finn, C.E. and Clark, J.R. (eds) Register of new fruit and nut cultivars. List 46. *HortScience* 47, 547–550. DOI: 10.21273/HORTSCI.47.5.536
Fukuda, T., Suezawa, K. and Katagiri, T. (2007) New kiwifruit cultivar 'Sanuki Gold'. *Acta Horticulturae* 753, 243–246. DOI: 10.17660/ActaHortic.2007.753.28
Harker, F.R., Jaeger, S.R., Lau, K. and Rossiter, K. (2007) Consumer perceptions and preferences for kiwifruit: a review. *Acta Horticulturae* 753, 81–88. DOI: 10.17660/ActaHortic.2007.753.7
Henare, S.J. (2015) The nutritional composition of kiwifruit (*Actinidia* spp.). In: Simmonds, M.S.J. and Preddy, V.R. (eds) *Nutritional Composition of Fruit Cultivars*. Academic Press, San Diego, California, pp. 337–370. DOI: 10.1016/B978-0-12-408117-8.00015-5
Huang, H. (2013) [*Actinidia Germplasm Resources in China*] China Forestry Publishing House, Beijing. [In Chinese]
Huang, H. (2016a) Genetic resources. In: Testolin, R., Huang, H.-W. and Ferguson, A.R. (eds) *The Kiwifruit Genome. Compendium of Plant Genomes*. Springer Nature, Cham, Switzerland, pp. 15–36. DOI: 10.1007/978-3-319-32274-2_2
Huang, H. (2016b) *Kiwifruit: The Genus ACTINIDIA*. Elsevier, London.
Huang, H. (2022) Discovery and domestication of new fruit trees in the 21st century. *Plants* 11, 2107. DOI: 10.3390/plants11162107
Huang, H. and Ferguson, A.R. (2001) Kiwifruit in China. *New Zealand Journal of Crop and Horticultural Science* 29, 1–14. DOI: 10.1080/01140671.2001.9514154
Huang, Z.-F., Liang, M.-Y., Huang, C.-G. and Li, R.-G. (1983) A preliminary study on the character and nutritive composition of *Actinidia* fruits. *Guihaia* 3, 53–60, 66.
Jackson, P.J. and Harker, F.R. (1997) Changes in firmness of the outer pericarp, inner pericarp, and core of *Actinidia* species during ripening. *New Zealand Journal of Crop and Horticultural Science* 25, 185–189. DOI: 10.1080/01140671.1997.9514005
Jaeger, S.R., Rossiter, K.L., Wismer, W.V. and Harker, F.R. (2003) Consumer-driven product development in the kiwifruit industry. *Food Quality and Preference* 14, 187–198. DOI: 10.1016/s0950-3293(02)00053-8
Jaeger, S.R., Hunter, D.C., Vidal, L., Chheang, S.L., Ares, G. *et al.* (2019) Sensory product characterization by consumers using check-all-that-apply questions: investigations linked to term development using kiwifruit as a case study. *Journal of Sensory Studies* 34, e12490. DOI: 10.1111/joss.12490
Jeffery, P.B., East, A.R., Jabbar, A., Heyes, J.A. and Bollen, A.F. (2013) Characterisation, measurement and factors influencing columella softening in kiwifruit. *Acta Horticulturae* 1012, 1487–1491. DOI: 10.17660/ActaHortic.2013.1012.201
Kwack, Y.-B., Kim, H.-L., Lee, J.-H., Chung, K.-H. and Chae, W.B. (2017) 'Redvita': a yellow-fleshed kiwifruit with red color around the core. *Horticultural Science and Technology* 35, 387–391. DOI: 10.12972/kjhst.20170041
Lavin, J.G. and Lawless, H.T. (1998) Effects of color and odor on judgements of sweetness among children and adults. *Food Quality and Preference* 9, 283–289. DOI: 10.1016/S0950-3293(98)00009-3

Li, D., Zhong, C., Liu, Y. and Huang, H. (2010) Correlation between ploidy level and fruit characters of the main kiwifruit cultivars in China: implications for selection and improvement. *New Zealand Journal of Crop and Horticultural Science* 38, 137–145. DOI: 10.1080/01140671.2010.482966

Li, D., Liu, Y., Li, X., Rao, J., Yao, X. *et al.* (2014) Genetic diversity in kiwifruit polyploidy complexes: insights into cultivar evaluation, conservation, and utilization. *Tree Genetics & Genomes* 10, 1451–1463. DOI: 10.1007/s11295-014-0773-6

Martin, R.A. and Luxton, P. (2005) The successful commercialisation of Zespri™ Gold Kiwifruit. *Acta Horticulturae* 694, 35–40. DOI: 10.17660/ActaHortic.2005.694.2

Montefiori, M., McGhie, T.K., Hallett, I.C. and Costa, G. (2009) Changes in pigments and plastid ultrastructure during ripening of green-fleshed and yellow-fleshed kiwifruit. *Scientia Horticulturae* 119, 377–387. DOI: 10.1016/j.scienta.2008.08.022

Muggleston, S., McNeilage, M., Lowe, R. and Marsh, H. (1998) Breeding new kiwifruit cultivars: the creation of Hort16A and Tomua. *The Orchardist of New Zealand* 71(8), 38–40.

Nardozza, S., Kashuba, P., McCaughan, L., Philippe, M., Wohlers, M. *et al.* (2015) Leaves are important to obtain consistent red flesh pigmentation in *Actinidia chinensis* fruit. *Scientia Horticulturae* 197, 496–503. DOI: 10.1016/j.scienta.2015.10.008

Nishiyama, I. (2007) Fruits of the *Actinidia* genus. *Advances in Food and Nutrition Research* 52, 293–324. DOI: 10.1016/S1043-4526(06)52006-6

O'Rourke, D. (2011) Dealing with changing world markets for kiwifruit. *Acta Horticulturae* 913, 45–49. DOI: 10.17660/ActaHortic.2011.913.2

Patterson, K., Burdon, J. and Lallu, N. (2003) 'Hort16A' kiwifruit: progress and issues with commercialisation. *Acta Horticulturae* 610, 267–273. DOI: 10.17660/ActaHortic.2003.610.35

Richardson, D.P., Ansell, J. and Drummond, L.N. (2018) The nutritional and health attributes of kiwifruit: a review. *European Journal of Nutrition* 57, 2659–2676. DOI: 10.1007/s00394-018-1627-z

Schröder, R., Virgin-Harris, T., Beresford, M.K. and Hallett, I.C. (2018) Kiwifruit texture – a pilot study. *Acta Horticulturae* 1218, 171–175. DOI: 10.17660/ActaHortic.2018.1218.23

Singletary, K. (2012) Kiwifruit: overview of potential health benefits. *Nutrition Today* 47(3), 133–147. DOI: 10.1097/NT.0b013e31825744bc

Stonehouse, W., Gammon, C.S., Beck, K.L., Conlon, C.A., von Hurst, P.R. *et al.* (2013) Kiwifruit: our daily prescription for health. *Canadian Journal of Physiology and Pharmacology* 91, 442–447. DOI: 10.1139/cjpp-2012-0303

Testolin, R., Cipriani, G. and Costa, G. (2023) Kiwi verde, giallo o rosso sapersi distinguere fa la differenza sul mercato. *Rivista di Frutticoltura e di Ortofloricoltura*, 87(4), Insert, x–xv. Available at: https://rivistafrutticoltura.edagricole.it/colture/kiwi/kiwi-verde-giallo-rosso-sapersi-distinguere-sul-mercato/ (accessed 9 July 2023).

Thorp, T.G., Zhang, J. and Lay-Yee, M. (1990) Horticultural characteristics of seven pistillate and three staminate New Zealand cultivars of kiwifruit. *New Zealand Journal of Crop and Horticultural Science* 18, 233–240. DOI: 10.1080/01140671.1990.10428100

UNECE (United Nations Economic Commission for Europe) (2017) *UNECE Standard FFV-46 Concerning the Marketing and Commercial Quality Control of Kiwifruit*, 2017 edn. United Nations, New York and Geneva, Switzerland.

Vetharaniam, I., van den Dijssel, C., Stanley, J., Richardson, A. and Müller, K. (2022) Projecting suitability of kiwifruit by chill requirements for different locations in New Zealand under climate change. *Acta Horticulturae* 1332, 455–462. DOI: 10.17660/ActaHortic.2022.1332.60

Wang, J., Vanga, S.K., McCusker, C. and Raghavan, V. (2019) A comprehensive review on kiwifruit allergy: pathogenesis, diagnosis, management, and potential modification of allergens through processing. *Comprehensive Reviews in Food Science and Food Safety* 18, 500–513. DOI: 10.1111/1541-4337.12426

Wang, S., Qiu, Y. and Zhu, F. (2021) Kiwifruit (*Actinidia* spp.): a review of chemical diversity and biological activities. *Food Chemistry* 350, 128469. DOI: 10.1016/j.foodchem.2020.128469

Wang, Y.-C., Zhang, L., Man, Y.-P., Li, Z.-Z. and Qin, R. (2012) Phenotypic characterization and simple sequence repeat identification of red-fleshed kiwifruit germplasm accessions. *HortScience* 47, 992–999. DOI: 10.21273/HORTSCI.47.8.992

Wang, Z., Hu, G., Li, Z., Zhong, C. and Yao, X. (2022) Characterizing tetraploid populations of *Actinidia chinensis* for kiwifruit genetic improvement. *Plants* 11, 1154. DOI: 10.3390/plants11091154

White, A., de Silva, H.N., Requejo-Tapia, F. and Harker, F.R. (2005) Evaluation of softening characteristics of fruit from 14 species of *Actinidia*. *Postharvest Biology and Technology* 35, 143–151. DOI: 10.1016/j.postharvbio.2004.08.004

Wu, Y.J., Xie, M., Zhang, Q.C., Jiang, G.H., Zhang, H.Q. *et al.* (2009) Characteristics of 'White': a new easy-peel cultivar of *Actinidia eriantha*. *New Zealand Journal of Crop and Horticultural Science* 37, 369–373. DOI: 10.1080/01140671.2009.9687592

Xie, Q., Zhang, H., Yan, F., Yan, C., Wei, S. *et al.* (2019) Morphology and molecular identification of twelve commercial varieties of kiwifruit. *Molecules* 24, 888. DOI: 10.3390/molecules24050888

Zespri Group Ltd (2015) *Annual Review 2014/2015*. Zespri Group Limited, Mt Maunganui, New Zealand.

Zespri Group Ltd (2021) *Annual Report 2020/21*. Zespri Group Limited, Mt Maunganui, New Zealand. Available at: https://www.zespri.com/content/dam/zespri/nz/annual-reports/Annual_Report_2020_21.pdf (accessed 9 June 2023).

Zhang, H., Zhao, Q., Lan, T., Geng, T., Gao, C. *et al.* (2020) Comparative analysis of physiochemical characteristics, nutritional and functional components and antioxidant capacity of fifteen kiwifruit (*Actinidia*) cultivars – comparative analysis of fifteen kiwifruit (*Actinidia*) cultivars. *Foods* 9, 1267. DOI: 10.3390/foods9091267

Zhong, C.-H., Wang, Z.-Y., Bu, F.-W. and Peng, D.-F. (2007) Studies on the mechanism of red coloration in 'Chuhong' kiwifruit cultivar. In: Huang, H.-W. (ed.) *Advances in Actinidia Research (IV)*. Science Press, Beijing, China, pp. 213–218.

Zhong, C.H., Huang, W., Wang, Z., Li, L., Li, D. *et al.* (2022) The breeding progress and development status of the kiwifruit industry in China. *Acta Horticulturae* 1332, 445–454. DOI: 10.17660/ActaHortic.2022.1332.59

5 Male Cultivars

Alan Seal[1]*, Raffaele Testolin[2] and Jinbao Fang[3]
[1]*Kiwifruit Breeding Centre, Te Puke, New Zealand;* [2]*University of Udine, Udine, Italy;* [3]*Zhengzhou Fruit Research Institute, Chinese Academy of Agricultural Sciences, Zhengzhou, Henan, China*

5.1 Significance of Males

All known *Actinidia* species are functionally dioecious – they have separate female and male vines. While the flowers on female vines have male parts (stamens) and female parts (ovaries, styles and stigmata), their pollen is sterile (Fig. 5.1). Flowers on male vines usually have only rudimentary non-functional female parts (Fig. 5.1), but their stamens produce copious amounts of viable pollen (Fig. 5.2). To achieve fruit set, viable pollen from male vines needs to be transferred to the stigmata of flowers on female vines (see Chapter 13). Pollination by wind and wild pollinators alone is usually insufficient to produce a commercial crop consistently (Testolin *et al.*, 1991), so pollen transfer is usually facilitated by introducing fans or bees into the orchard at flowering time (Goodwin *et al.*, 2013) or by hand pollination. In commercial kiwifruit orchards (*A. chinensis* var. *chinensis* or *A. chinensis* var. *deliciosa*), male vines are usually planted among the female vines, either singly at regular intervals, or as entire rows alternating with female rows. Specific male cultivars, selected by breeders and growers to suit each particular female cultivar, are usually clonally propagated for pollination. Male cultivars need to flower in synchrony with each chosen female cultivar, produce numerous flowers that attract pollinators and release large amounts of fertile and compatible pollen.

While pollination is essential to achieve fruit set, it also can have important effects on vine productivity and fruit quality. Fruit weight is a key component of kiwifruit yield and marketability, and therefore can greatly affect growers' profitability. The effectiveness of pollination determines seed set, and within each female kiwifruit cultivar there is a moderate to strong relationship between seed set and fruit weight at harvest (Goodwin *et al.*, 2013). Fruit containing few seeds tend to be very small, and if the seeds are distributed unevenly within the fruit, they may also be mis-shapen. The number of seeds required to produce fruit of commercially acceptable size varies among female cultivars and is dependent on management factors such as crop load (Hopping and Hacking, 1983). To produce a fruit of more than 100 g, at least 1000 seeds may be required in the hexaploid cultivar 'Hayward' (*A. chinensis* var. *deliciosa*) (Hopping and Hacking, 1983; Testolin *et al.*, 1991), whereas only about half that number is needed in the tetraploid cultivar 'Zesy002' (*A. chinensis* var. *chinensis*) (Seal *et al.*, 2017). Pollen from different male cultivars can vary in its ability to set seed and, therefore, to size the fruit of a given female cultivar (McNeilage *et al.*, 1992; Chen *et al.*,

*Corresponding author: alan.seal@next.kiwi

© CAB International 2024. *Kiwifruit: Botany, Production and Uses* (eds A.C. Richardson, J.N. Burdon and A.R. Ferguson)

DOI: 10.1079/9781800620933.0005

Fig. 5.1. *Left:* open pistillate flowers of diploid (*top*) and tetraploid (*centre*) *Actinidia chinensis* var. *chinensis*, and hexaploid *A. chinensis* var. *deliciosa* (*bottom*), cut in half to show stamens, large ovary and styles. *Right:* staminate flowers of diploid (*top*) and tetraploid (*centre*) *Actinidia chinensis* var. *chinensis*, and hexaploid *A. chinensis* var. *deliciosa* (*bottom*), cut in half to show long stamens and rudimentary ovary. Bar = 1 cm. (Photo: Kiwifruit Breeding Centre, New Zealand)

1996; Qi *et al.*, 2007; Seal *et al.*, 2017). As well as fruit weight, other vine productivity and fruit quality traits can be affected by the choice of pollen donor. Variation in male performance is particularly common when males differ in ploidy from the recipient females (Seal *et al.*, 2013a; Chen *et al.*, 2019; Oh *et al.*, 2021).

In summary, male vines are a vital component of current commercial kiwifruit growing systems. High-performing male cultivars can enhance the productivity and quality, and hence the profitability, of female cultivars.

5.2 Key Attributes

5.2.1 Synchrony with females

Male cultivars need to overlap in their flowering period with each chosen female. The relative timing and length of the flowering period of individual females and males can vary in response to the environment (Testolin *et al.*, 1995). Therefore, growers are often advised to grow at least two male cultivars with slightly different flowering periods to ensure coverage of the entire female flowering period. Synchrony can also be affected by the use of budbreak enhancers, different rootstocks and vine management.

5.2.2 Ploidy

Within *A. chinensis*, there is a general relationship between ploidy and flowering time: diploid cultivars tend to flower before tetraploids, and tetraploids before hexaploids. Thus, diploid female cultivars such as 'Hongyang' and 'Zes008' require early flowering, usually diploid males, whereas hexaploid female cultivars such as

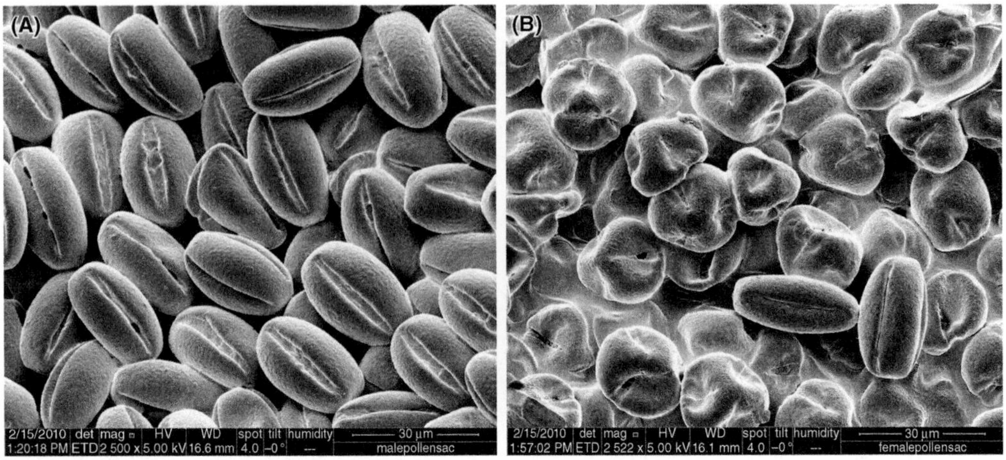

Fig. 5.2. Scanning electron micrographs of pollen grains from flower buds of a male vine (**A**) and a female vine (**B**) of *Actinidia chinensis* var. *deliciosa*. Note the mostly shrivelled, non-functional grains from the female flower bud. (Photos: The New Zealand Institute for Plant and Food Research Ltd)

'Hayward' and 'Jinkui' require late flowering hexaploid males to ensure coincident flowering. Tetraploid female cultivars such as 'Zesy002' and 'Jintao' are usually matched with tetraploid males. However, there is a degree of overlap in flowering between diploids and tetraploids, and between tetraploids and hexaploids. With the release of a wider range of female and male cultivars, it is possible that interploid pollination could occur either inadvertently or deliberately. Moreover, only pollen from hexaploid males is currently available for use in artificial pollination systems, as the pollen yield from lower ploidy males is considered uneconomic. Interploid pollination (Fig. 5.3) can affect many seed and fruit traits (Chen *et al.*, 2019; Oh *et al.*, 2021). Seed number, individual and total seed weight, fruit set, fruit weight and, in red-fleshed kiwifruit, even the intensity of red pigmentation can be affected (Seal *et al.*, 2013a, 2016, 2018). Such effects, when present, are usually unfavourable, but can be beneficial (Seal *et al.*, 2013b).

Kiwiberries (*A. arguta*) can be pollinated by males of *A. arguta* or *A. chinensis* (Latocha *et al.*, 2021). Pollination of three female tetraploid kiwiberry cultivars by six tetraploid male *A. arguta* cultivars and the hexaploid *A. chinensis* var. *deliciosa* male, 'Tomuri', generated some differences in mean seed number but very few significant differences in fruit set or weight (Stasiak *et al.*, 2021).

5.2.3 Flower number

Although essential, male cultivars occupy space in the canopy that could otherwise be occupied by fruitful female vines. Therefore, an ideal male vine would produce an abundance of flowers in a very constrained area. A high budbreak, high proportion of floral buds, short internode length and numerous flowers per shoot would be desirable. Kiwifruit inflorescences are compound structures (cymes) consisting of several branches (dichasia), each supporting a terminal 'king' flower and sometimes a variable number of additional lateral flowers. Whereas lateral flowers on a female dichasium are generally deemed undesirable and removed, multiple flowers on each dichasium on a male inflorescence can be advantageous as they may extend the flowering period.

5.2.4 Pollen yield and fertility

The amount of fertile pollen per flower is clearly an important attribute of male cultivars. There is a general association between

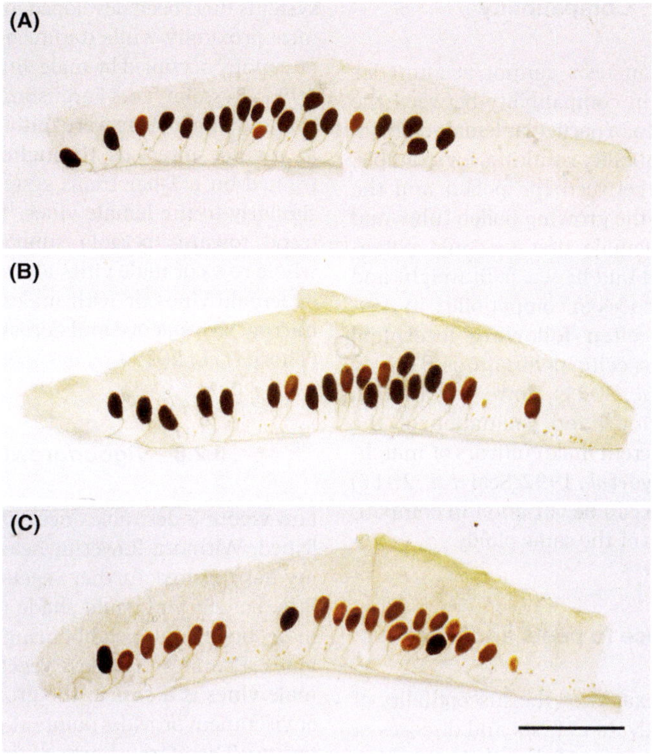

Fig. 5.3. Individual locules from mature fruit of *Actinidia chinensis* var. *chinensis* 'Hort16A' following pollination with pollen from (**A**) diploid *Actinidia chinensis* var. *chinensis* 'Bruce'; (**B**) a tetraploid *A. chinensis* var. *chinensis* seedling; and (**C**) a commercial mix of pollen from hexaploid *A. chinensis* var. *deliciosa*. Stylar end to the left. The frequency of well-formed (black) seeds decreased with increasing ploidy of the pollen parent. However, in 'Hort16A' this had no negative effects on fruit weight or dry matter content. Bar = 1 cm. Reprinted from: Pollen parent effects on fruit attributes of diploid *Actinidia chinensis* 'Hort16A' kiwifruit. (A.G. Seal, J.K. Dunn, Y.L. Jia, *New Zealand Journal of Crop and Horticultural Science*, 2013. Copyright ©The Royal Society of New Zealand, reprinted by permission of Taylor & Francis Ltd, http://www.tandfonline.com on behalf of The Royal Society of New Zealand)

pollen yield of *A. chinensis* males and ploidy: hexaploid males tend to produce more pollen per flower than tetraploids, and tetraploids more than diploids, but there can be significant variation in the pollen yield per flower even among males of the same ploidy (Hopping, 1986; Gonzalez *et al.*, 1994). Pollen viability can also vary widely among male genotypes. New methods that incorporate automated image analysis to distinguish hollow, non-viable pollen grains from those that are potentially fertile have been successfully tested on other plant species. Such techniques, if applicable to kiwifruit, might improve the accuracy and efficiency of pollen viability testing (Ascari *et al.*, 2020), but there is no guarantee the apparently fully developed grains will germinate (Wizenberg *et al.*, 2021). *In vitro* germination tests can provide estimates of pollen fertility (Alves *et al.*, 2018; Chen *et al.*, 2019). Oh *et al.* (2020) assessed pollen production by *A. chinensis* var. *deliciosa* 'Chieftain' and 'Bohwa' (a misspelling of 'Pohwa') in two consecutive years. There appeared to be only minor differences between years in the amount, viability and germinability *in vitro* of the pollen from each cultivar. There have been no comprehensive studies of the variation in pollen yield and fertility among flowers within male vines, during the flowering season, or in different environments.

5.2.5 Compatibility

In vitro germination tests cannot account for possible variation in compatibility between the pollen and the pistil of a particular female cultivar. Variation in compatibility resulting, for example, from interactions between the pollen and the stigma or between the growing pollen tubes and the surrounding female tissues, could potentially affect seed set and hence, fruit weight and other traits. Differences in compatibility in *Actinidia* arise most often following interploid (Fig. 5.3) or interspecific pollinations (Harvey *et al.*, 1992; Oh *et al.*, 2021). However, variation in seed set following hand pollination of the same female by different male cultivars of matching ploidy (McNeilage *et al.*, 1992; Seal *et al.*, 2017) suggests there also can be variation in compatibility among males of the same ploidy.

5.2.6 Resistance to pests and diseases

Few studies have examined the susceptibility of male kiwifruit cultivars to pests and diseases or environmental stresses. Males are just as vulnerable as females to infection by *Pseudomonas syringae* pv. *actinidiae* (Psa), with male vines of *A. chinensis* var. *deliciosa* particularly susceptible to floral bud rot which can greatly reduce their performance as pollenizers. Susceptible males can also be an unwanted source of inoculum in the orchard. Similarly, male cultivars whose flowers retain their senescent tissues long after flowering might harbour *Sclerotinia sclerotiorum*, which could cause fruit rot or skin damage in neighbouring female vines. Ideally, male cultivars should shed their senescent floral tissues soon after flowering.

5.2.7 Suitability for growing systems

Deployment of male kiwifruit vines within the orchard is critical for successful pollination. The spatial distribution and size of male vines need careful consideration because the fruit set and fruit size of female vines depends on their proximity to the nearest males (Testolin, 1991). Many different arrangements and management systems have been developed in attempts to optimize proximity while minimizing the total area of canopy occupied by male vines (Testolin *et al.*, 1991; Testolin and Ferguson, 2009). In New Zealand, male vines were initially planted singly at regular intervals throughout the orchard, trained on a T-bar trellis system and managed similarly to the female vines. There has been a trend towards pergola support systems with whole rows of male vines alternating with rows of female vines or with males trained to form narrow strips above and across the female rows (Thorp *et al.*, 2011).

5.2.8 Vigour/growth habit

Low vigour is desirable once a male vine is established. Within a flowering season, once flowering has finished, further vegetative growth is of little benefit and could shade the next season's male buds and neighbouring fruiting female vines. Pruning of excess vegetative growth on male vines is a cost to the grower and in a Psa environment provides points of entry for the bacterium. Therefore, short self-terminating shoots and a growth habit well suited to a low-vigour management system would be advantageous.

5.2.9 Attractiveness to pollinators

Unless artificial pollination is being used, male flowers need to attract pollinators such as honey bees or bumble bees. It is important that the pollinators do not show a strong preference for either male or female flowers (known as 'floral constancy'), as they need to move frequently between the two to ensure effective pollen transfer. As honey bees tend to show a distinct preference for female flowers (Goodwin and Steven, 1993; Goodwin *et al.*, 2013), floral constancy could be an important attribute, albeit one that is difficult to quantify accurately. Kiwifruit flowers do not produce nectar. Recent studies of commercial male and female cultivars in New Zealand have identified some of the main floral volatile organic compounds to which honey bees respond (Twidle *et al.*, 2017, 2018). However, the presence and concentrations of these compounds have yet to be linked to bee pollination behaviour.

5.2.10 Suitability for artificial pollination

In locations or seasons in which pollination by insects or wind might be inadequate, stored pollen can be mechanically applied to female flowers. There are some differences in the criteria for selecting male cultivars for artificial versus insect-mediated pollination. Clearly, synchrony of flowering is not so important. Early flowering, in advance of the females, could be advantageous by allowing fresh pollen to be used in the same season, avoiding the need for storage of frozen pollen. A series of males with different flowering times might help spread the labour and reduce the impact of adverse weather events while harvesting flower buds for pollen extraction. While attractiveness to insects might not be essential, supplementary pollen may act as an attractant to enhance insect-mediated pollination, and insects attracted to pollen applied in the leaf canopy could assist in its redistribution to female flowers within the orchard. Pollen yield and fertility are particularly important criteria for selecting males for artificial pollination. Therefore, males with numerous large flowers and high pollen yields are highly desirable. Pollen stored until the following spring must retain high fertility when frozen at about −18°C (Bomben *et al.*, 1999). There may be other selection criteria associated with the processing and application of pollen, such as ease of release from anthers and suitability for use dry or in liquid suspension in various mechanical application devices (Tacconi *et al.*, 2016). Ideally, pollen should be sourced from male vines free of Psa, as contaminated pollen can transmit the disease to healthy vines (Donati *et al.*, 2018).

5.3 Cultivars

Most of the male kiwifruit cultivars currently widely grown are listed in Table 5.1. The first cultivated female kiwifruit vines in New Zealand (*A. chinensis* var. *deliciosa*) were most likely pollinated by male siblings that happened to flower at the appropriate time. As the kiwifruit industry developed, specific male clones were selected and clonally propagated by growers. In 1968, 'Matua' and 'Tomuri' were among the first of these male cultivars to be named (Ferguson and Bollard, 1990; Thorp *et al.*, 1990), followed by the 'M Series' males including 'M56' and 'M59', which were released as 'Chieftain' in 1989. Some of these males, especially 'Chieftain', are still cultivated in New Zealand and other countries for pollination of 'Hayward'.

The diploid, yellow-fleshed female *A. chinensis* var. *chinensis* 'Hort16A', registered in 1995, flowered much earlier than any of the available hexaploid *A. chinensis* var. *deliciosa* male cultivars. Two floriferous diploid *A. chinensis* var. *chinensis* males, 'Meteor' (commonly known as CK3) and 'Sparkler' (CK2), were released to cover the flowering period of 'Hort16A'. A third diploid male, 'Bruce', was released in 2007. 'Hort16A' and all these diploid males later succumbed to Psa. 'Russell', a diploid male with more tolerance to Psa infection, is currently being used for pollination of the recently released red-fleshed female 'Zes008' (*A. chinensis* var. *chinensis*).

After 2010, 'Hort16A' was replaced in New Zealand by the tetraploid female *A. chinensis* var. *chinensis* 'Zesy002'. A tetraploid male cultivar, 'M91', was released for pollination. Recently, four new tetraploid males, 'Earp' (MC79), 'Ferguson' (MC110), 'MacLoughlin' (MC352) and 'Muller' (MC482), were released to allow growers to select those best suited to their specific location and management system. Some growers also use 'Earp', 'MacLoughlin' or 'Muller' to pollinate late-opening flowers of 'Zes008'. Stored pollen from *A. chinensis* var. *deliciosa* males is used for supplementary artificial pollination of 'Zesy002'. All the above-named cultivars were selected by the New Zealand Institute for Plant and Food Research Ltd or its predecessor organizations.

While many countries initially relied on imported New Zealand male cultivars for kiwifruit pollination, some have since selected their own male cultivars to suit local female cultivars and conditions. Among the first was 'Chico-Male' (also known as California Male), selected in the USA for pollination of 'Hayward' (Ferguson and Bollard, 1990). The tetraploid male *A. chinensis* var. *chinensis* 'Moshan No. 4' was selected by the Wuhan Institute of Botany in 1984 from plants growing in the wild in Wuning County, Jiangxi

Table 5.1. Widely grown male cultivars of *Actinidia chinensis*.

Cultivar name (common name)	Ploidy	Year released	Source	Taxon[1]	Comments	Flowering time[2]
'Russell'	2*x*	2019	Seedling population, New Zealand	a	Psa-tolerant; selected to match 'Zes008'	Early
'Nijitaro'	2*x*	2019	Seedling population, Japan	a	Selected to match the flowering time of 'Hongyang'	Early
'Belén' (A 134.16)	4*x*	2001	Seedling population, Italy	a	Selected to match the flowering time of 'Jintao'; also used for 'Soreli', 'AC 1536'	Mid
'Moshan No. 4' (S-SV-AC-016-2006)	4*x*	2006	Selected from the wild, China	a	Long flowering period; used for pollination of many tetraploid cultivars	Mid–late
'M91'	4*x*	2010	Seed accession, New Zealand	a	Selected to match flowering time of 'Zesy002'	Mid–late
'Earp' (MC79)	4*x*	2019	Seedling population, New Zealand	a	Selected to match early flowers of 'Zesy002'; sometimes used to cover late flowers of 'Zes008'	Early–mid
'Ferguson' (MC110)	4*x*	2019	Seedling population, New Zealand	a	Selected to match early flowers of 'Zesy002'; sometimes used to cover late flowers of 'Zes008'	Early–mid
'MacLoughlin' (MC352)	4*x*	2019	Seedling population, New Zealand	a	Selected to match early flowers of 'Zesy002'; sometimes used to cover late flowers of 'Zes008'	Early–mid
'Muller' (MC482)	4*x*	2019	Seedling population, New Zealand	a	Selected to match early flowers of 'Zesy002'; sometimes used to cover late flowers of 'Zes008'	Early–mid
'Sanuki kafunriki'	4*x*	2017	Seedling population, Japan	a	Selected to match the flowering time of 'Sanuki Gold'	Mid
'Zuva' (A 125.93)	4*x*	2019	Seedling population, Italy	a	Selected to match the flowering time of 'Soreli'	Mid
'Matua'	6*x*	1968	Seedling on grower's property, New Zealand	b	Long flowering period; used for pollination of 'Hayward'	Mid
'Tomuri'	6*x*	1968	Seedling on grower's property, New Zealand	b	Still used for pollination of 'Hayward' in some locations	Late
'Chieftain'	6*x*	1989	Seedling on grower's property, New Zealand	b	Still widely used internationally for pollination of 'Hayward', 'Qinmei', 'Jinkui', 'Haegeum'; also used for artificial pollination of many cultivars including 'Zesy002'	Late
'Pohwa'	6*x*	1996	Seedling population, South Korea	b	High pollen yield, used for artificial pollination of hexaploid and tetraploid females including 'Garmrok', 'Goldone', 'Haegeum', 'Hayward'	Late

Table 5.1. Continued.

Cultivar name (common name)	Ploidy	Year released	Source	Taxon[1]	Comments	Flowering time[2]
'Autari' (Ac 12.255)	6*x*	1998	Seedling population, Italy	b	Selected to match the flowering time of 'Hayward' in Italy	Mid–late
'Deliwoong'	6*x*	2014	Seedling population, South Korea	b	Selected to match the mid-flowering time of 'Garmrok', but also used for artificial pollination of many hexaploid and tetraploid females including 'Garmrok', 'Goldone', 'Haegeum', 'Hayward'	Mid
'M36'	6*x*	2015	Seedling population, New Zealand	b	High pollen yield; suitable for pollen farming	Mid
'M43'	6*x*	2015	Seedling population, New Zealand	b	High pollen yield; suitable for pollen farming	Mid

[1]Taxon: a *Actinidia chinensis* var. *chinensis*; b *A. chinensis* var. *deliciosa*.
[2]Flowering time is in relation to other genotypes of the same ploidy.
Psa, *Pseudomonas syringae* pv. *actinidiae*

Province, China (Huang *et al.*, 2006; Ferguson *et al.*, 2012). It is renowned for its extended flowering period and is used for pollination of many tetraploid and some early flowering hexaploid female cultivars, including the tetraploid *A. eriantha* × *A. chinensis* hybrid 'Jinyan'.

In Italy, the hexaploid *A. chinensis* var. *deliciosa* male cultivar 'Autari' was selected by the Università degli Studi di Udine from seedlings from a cross between 'Hayward' and M3h, a fruiting (inconstant) male form of 'Matua' (Messina *et al.*, 1990; Testolin *et al.*, 1995). It is used mainly for pollination of 'Hayward'. An earlier-flowering hexaploid male cultivar, 'Summerfaenza', is used for pollination of 'Summerkiwi 3373'. It was selected by V. Ossani from Ravenna, Italy, from seedlings of a cross between 'Hayward' and ML2, another fruiting male (Ferguson *et al.*, 2012).

The tetraploid male cultivar *A. chinensis* var. *chinensis* 'Belén', released in 2001, is used for pollination of the tetraploid female cultivars 'Jintao', 'Soreli' and 'AC 1536'. It originated as a seedling from open pollination of the Chinese female cultivar 'Ganmi No. 1' (Ferguson *et al.*, 2012). Another tetraploid male cultivar, 'Zuva', was released in 2019 to match the flowering time of 'Soreli' (R. Testolin, Italy, 2021, personal communication). Both these male cultivars were selected by the Università degli Studi di Udine, Italy, from seedling populations.

In Japan, an early-flowering diploid male cultivar, 'Nijitaro', has been selected for pollination of the red-fleshed diploid female cultivar 'Hongyang' ('Rainbow Red') (Murakami *et al.*, 2019). A tetraploid male cultivar 'Sanuki kafunriki' has been released for pollination of the tetraploid female cultivar 'Sanuki Gold'. Most kiwifruit orchards in Japan are hand pollinated. About 20% of the pollen is sourced from New Zealand, USA and Chile (K. Suezawa, Japan, 2021, personal communication).

In South Korea, artificial pollination is common. The hexaploid male cultivars 'Chieftain' and 'Pohwa', which were released in 1996, are the main cultivars used for pollination of several tetraploid and hexaploid female cultivars. The high pollen yield of 'Pohwa' (16 mg per bud) makes it well suited for artificial pollination. Similarly, 'Deliwoong', released in 2014 to match the mid-flowering female cultivar 'Garmrok', is also used for artificial pollination (Y.-B. Kwack, South Korea, 2021, personal communication).

The release of the HortGem series of four female cultivars of *A. arguta* in New Zealand was accompanied by the release of three unnamed *A. arguta* male selections (Williams *et al.*, 2003). Several named *A. arguta* male cultivars have been selected and used elsewhere (see Chapter 7).

In China, male cultivars for pollination of female *A. eriantha* cultivars have been selected from wild *A. eriantha* seedlings. The male cultivar 'No. 1 Maoxiong' is recommended for pollination of 'Huate' ('White') (Wu *et al.*, 2009), and male 'WF 1' for pollination of 'Ganmi No. 6' (Liao *et al.*, 2022).

5.4 Selection and Breeding

Most male cultivars have been selected directly from the wild, from plants derived from seed of fruit collected from the wild, from seedlings grown on growers' properties, or from male seedlings segregating in breeding populations designed to produce new female cultivars. These approaches have been successful but rely on the chance discovery of male seedlings with the required characteristics.

Ideally, a male selection programme should take account of all the key traits discussed above. The timing and duration of flowering are easily measured, and flower density can be estimated by counting buds on floral shoots on a measured length of one-year-old wood. The amount of pollen (by weight) produced by a sample of buds can be readily measured, and pollen quality can be evaluated based on *in vitro* staining or germination. These variables can be combined to estimate the amount of germinable pollen per unit length of one-year-old wood, providing a useful index for discriminating among male candidates (Gonzalez *et al.*, 1994). Following these assessments, the most promising selections can then be propagated for more detailed field evaluation, potentially on growers' properties and in multiple locations. Flowering time and duration can be confirmed, and fruit set, seed set and fruit quality of the intended recipient female cultivars evaluated following hand pollinations.

There have been very few sustained attempts to breed specifically for enhanced male characteristics over multiple generations. Estimates of the heritability of the timing and duration of flowering have varied among populations (Testolin *et al.*, 1995; Marsh *et al.*, 1999; Zhu *et al.*, 2002; Cheng *et al.*, 2006; Testolin *et al.*, 2011; Pinto *et al.*, 2018). Traits related to the number of flowers on male vines (such as flowers per shoot, number of shoots, percentage of floral shoots) have consistently shown low heritability (Testolin *et al.*, 1995; Zhu *et al.*, 2002; Pinto *et al.*, 2018) and may be slow to improve by selecting parents based on their individual performance. There appear to have been no studies of the heritability of pollen yield and fertility, although both traits are known to vary among male genotypes (Gonzalez *et al.*, 1994; Chen *et al.*, 2019). Resistance to Psa is independent of vine gender and appears to be under polygenic control with moderate to high heritability (Cheng, 2014). Little is known about the inheritance of resistance to other pests, diseases or environmental stresses in kiwifruit. The publication of the genome sequence of the diploid *A. chinensis* var. *chinensis* male 'Russell' – the first of any male kiwifruit cultivar – should lead to a much more detailed understanding of the genetic control of male characteristics (Tahir *et al.*, 2022).

5.5 Future Prospects

Climate change, in particular increasing winter temperature, is predicted to adversely affect the viability of 'Hayward' kiwifruit production in some regions unless effective mitigation measures are implemented (Tait *et al.*, 2018). Higher winter temperatures reduce the number of flowers produced by vines and can affect the timing and duration of flowering in male and female cultivars. In some locations in New Zealand and other countries, but not in Europe, the budbreak-enhancing chemical, hydrogen cyanamide, is often used to offset potential reductions in flower numbers resulting from insufficient winter chilling, especially of the 'Hayward' cultivar. However, this practice is likely soon to be phased out, and alternative treatments may be less effective. Higher temperatures in spring could affect pollen viability (Seyrek *et al.*, 2016). In *A. chinensis* var. *deliciosa*, pollen germination *in vitro* declines rapidly at temperatures above 28°C (Jansson and Warrington, 1988). Climate change could also affect male performance through increased vine vigour, the emergence of new pests or diseases, or an increasing incidence of drought or flooding. New male cultivars with a lower winter-chilling requirement and enhanced tolerance to environmental stresses are desirable.

Radical changes in production systems in future, such as greenhouse cultivation or planar cordon systems used for apple (Tustin *et al.*, 2021), might require male cultivars with different vine architecture and flowering habits. In several woody crops, clonal rootstocks have been selected that enhance the performance of grafted scions. This represents a largely unexploited opportunity in kiwifruit. It may be possible to modify the timing, duration and density of flowering and the growth habit of male cultivars by grafting them onto specific clonal rootstocks.

New technologies such as gene editing are being used to identify and modify the genes that control flowering in kiwifruit. It is becoming possible to modify key floral characteristics such as the timing and duration of flowering, and even gender. Recent research has shown that sex determination in kiwifruit appears to be controlled by a small number of genes, and that females and males can be converted to self-fertile hermaphrodites (Akagi *et al.*, 2019; De Mori *et al.*, 2021; Varkonyi-Gasic *et al.*, 2021). As hermaphrodites may offer advantages to breeders and growers, they are likely to become commonplace in the future. Several other fruit crops, including grapes, strawberries and papaya, have transitioned from dioecy to hermaphroditism during domestication. Should kiwifruit follow a similar course, male cultivars may no longer be needed for commercial production.

5.6 Conclusions

Male cultivars can vary in terms of their coincidence in flowering period with the target female cultivar and the quantity, quality and compatibility of their pollen. This variation can affect the productivity and fruit quality of the female cultivar. In particular, pollen from males differing in ploidy from the female cultivar can affect fruit quality. While the opportunistic selection of superior males has been successful, the variation in key characteristics specific to male performance is yet to be exploited by targeted multi-generational breeding. The management of male vines within orchards can be expected to continue to evolve as new production systems are explored. The application of new technologies based on a detailed understanding of the genetic control of flowering has the potential to radically modify male phenology. The development of hermaphrodite cultivars through conventional breeding or the use of gene-editing technologies could eventually remove the need for male cultivars.

References

Akagi, T., Pilkington, S.M., Varkonyi-Gasic, E., Henry, I.M., Sugano, S.S. *et al.* (2019) Two Y-chromosome-encoded genes determine sex in kiwifruit. *Nature Plants* 5(8), 801–809. DOI: 10.1038/s41477-019-0489-6

Alves, J.P.M., Rodrigues, J.R.O. and Oliveira, C.M.M.S. (2018) Fertility of female ('Hayward', 'Earligreen', 'Tsechelidis' and 'Soreli') and male ('Autari', 'M56', 'Chieftain', 'P1'and 'Belén') kiwifruit in the north of Portugal. *Acta Horticulturae* 1218, 391–396. DOI: 10.17660/ActaHortic.2018.1218.54

Ascari, L., Novara, C., Dusio, V., Oddi, L. and Siniscalco, C. (2020) Quantitative methods in microscopy to assess pollen viability in different plant taxa. *Plant Reproduction* 33(3), 205–219. DOI: 10.1007/s00497-020-00398-6

Bomben, C., Malossini, C., Cipriani, G. and Testolin, R. (1999) Long term storage of kiwifruit pollen. *Acta Horticulturae* 498, 105–108. DOI: 10.17660/ActaHortic.1999.498.12

Chen, L., Wang, S.-Y., Zhong, M., Huang, C.-H., Liao, G.-L. *et al.* (2019) Effects of pollens from the 10 selected *Actinidia* male genotypes on 4 commercial planting kiwifruit female cultivars in Southern China. *New Zealand Journal of Crop and Horticultural Science* 47(3), 155–169. DOI: 10.1080/01140671.2019.1616569

Chen, Q.H., Zhang, Z.H., Qin, Z.Q. and Jiang, Y.C. (1996) Studies on male species selection and pollen induction of Jinkui kiwifruit. *China Fruits* 2, 23–24.

Cheng, C.-H. (2014) Inheritance of resistance to *Pseudomonas syringae* pv. *actinidiae* and genetic correlations with fruit characters in a diploid *Actinidia chinensis* (kiwifruit) population. *Euphytica* 198, 305–315. DOI: 10.1007/s10681-014-1107-3

Cheng, C.H., Seal, A.G., Murphy, S.J. and Lowe, R.G. (2006) Variability and inheritance of flowering time and duration in *Actinidia chinensis* (kiwifruit). *Euphytica* 147(3), 395–402. DOI: 10.1007/s10681-005-9036-9

De Mori, G., Zaina, G., Franco-Orozco, B., Testolin, R., De Paoli, E. *et al.* (2021) Targeted mutagenesis of the female-suppressor *SyGI* gene in tetraploid kiwifruit by CRISPR/CAS9. *Plants* 10, 62. DOI: 10.3390/plants10010062

Donati, I., Cellini, A., Buriani, G., Mauri, S., Kay, C. *et al.* (2018) Pathways of flower infection and pollen-mediated dispersion of *Pseudomonas syringae* pv. *actinidiae*, the causal agent of kiwifruit bacterial canker. *Horticulture Research* 5, 56. DOI: 10.1038/s41438-018-0058-6

Ferguson, A.R. and Bollard, E.G. (1990) Domestication of the kiwifruit. In: Warrington, I.J. and Weston, G.C. (eds) *Kiwifruit: Science and Management.* Ray Richards Publisher in association with the New Zealand Society for Horticultural Science, Auckland, New Zealand, pp. 165–246.

Ferguson, A.R., Huang, H.W. and Testolin, R. (2012) Kiwifruit. In: Finn, C.E. and Clark, J.R. (eds) Register of new fruit and nut cultivars. List 46. *HortScience* 47, 536–562. DOI: 10.21273/HORTSCI.47.5.536

Gonzalez, M., Coque, M. and Herrero, M. (1994) Pollinator selection in kiwifruit (*Actinidia deliciosa*). *Journal of Horticultural Science* 69(4), 697–702. DOI: 10.1080/14620316.1994.11516502

Goodwin, R.M. and Steven, D. (1993) Behaviour of honey bees visiting kiwifruit flowers. *New Zealand Journal of Crop and Horticultural Science* 21(1), 17–24. DOI: 10.1080/01140671.1993.9513741

Goodwin, R.M., McBrydie, H.M. and Taylor, M.A. (2013) Wind and honey bee pollination of kiwifruit (*Actinidia chinensis* 'HORT16A'). *New Zealand Journal of Botany* 51(3), 229–240. DOI: 10.1080/0028825X.2013.806934

Harvey, C.F., Fraser, L.G. and Kent, J. (1992) *Actinidia* seed development in interspecific crosses. *Acta Horticulturae* 297, 71–78. DOI: 10.17660/ActaHortic.1992.297.7

Hopping, M.E. (1986) Kiwifruit. In: Monselise, S.P. (ed.) *Handbook of Fruit Set and Development.* CRC Press, Boca Raton, Florida, pp. 217–232. DOI: 10.1201/9781351073042

Hopping, M.E. and Hacking, N.J.A. (1983) A comparison of pollen application methods for the artificial pollination of kiwifruit. *Acta Horticulturae* 139, 41–50. DOI: 10.17660/ActaHortic.1983.139.5

Huang, H., Zhang, Z., Jiang, Z., Cheng, Z., Chen, X. *et al.* (2006) A new male cultivar of *Actinidia chinensis* 'Moshan 4'. *Acta Horticulturae Sinica* 33(5), 1163.

Jansson, D.M. and Warrington, I.J. (1988) The influence of temperature during floral development and germination *in vitro* on the germinability of kiwifruit pollen. *New Zealand Journal of Experimental Agriculture* 16, 225–230. DOI: 10.1080/03015521.1988.10425644

Latocha, P., Debersaques, F. and Hale, I. (2021) *Actinidia arguta* (Kiwiberry): botany, production, genetics, nutritional value, and postharvest handling. *Horticultural Reviews* 48, 37–152. DOI: 10.1002/9781119750802.ch2

Liao, G., Jiang, Z., He, Y., Zhong, M., Huang, C. *et al.* (2022) The comprehensive evaluation analysis of the fruit quality in *Actinidia eriantha* pollinated with different pollen donors based on the membership function method. *Erwerbs-Obstbau* 64(1), 91–96. DOI: 10.1007/s10341-021-00620-3

Marsh, H., McNeilage, M. and Gordon, I. (1999) Breeding value of parents in kiwifruit (*Actinidia deliciosa*). *Acta Horticulturae* 498, 85–92. DOI: 10.17660/ActaHortic.1999.498.9

McNeilage, M.A., Seal, A.G., Steinhagen, S. and McGowan, J. (1992) Evaluation of kiwifruit pollinizers. *Acta Horticulturae* 297, 277–282. DOI: 10.17660/ActaHortic.1992.297.36

Messina, R., Vischi, M., Marchetti, S., Testolin, R. and Milani, N. (1990) Observations on subdioeciousness and fertilization in a kiwifruit breeding program. *Acta Horticulturae* 282, 377–385. DOI: 10.17660/ActaHortic.1990.282.47

Murakami, S., Taneishi, M., Suzuki, K., Sasaki, T. and Hashimoto, N. (2019) Selection of new pollinizer variety 'Nijitaro' for 'Rainbow Red' kiwifruit and development of techniques for its use. *Horticultural Research (Japan)* 18(2), 117–125. DOI: 10.2503/hrj.18.117

Oh, E.U., Jeong, S.Y., Kang, H.H. and Song, K.J. (2020) Characterization of pollen development in staminate kiwifruit (*Actinidia* sp.) cultivars. *Horticultural Science and Technology* 38, 1–8. DOI: 10.7235/HORT.20200001

Oh, E.U., Jeong, S.Y., Kim, J.Y. and Song, K.J. (2021) Response of fruit set and quality and seed formation to ploidy levels of pollen donor in yellow-fleshed kiwifruits. *Horticulture, Environment, and Biotechnology* 62(1), 9–15. DOI: 10.1007/s13580-020-00293-z

Pinto, C., Testolin, R. and Cipriani, G. (2018) Analysis of components of genetic variance and breeding value of kiwifruit (*Actinidia chinensis* Planch.) using best linear unbiased prediction (BLUP) and pedigree information. *Acta Horticulturae* 1218, 73–79. DOI: 10.17660/ActaHortic.2018.1218.9

Qi, X.J., Han, L.X., Li, M., Xu, S.K., Zhu, Y.S. *et al.* (2007) Studies on pollen xenia of kiwifruit. *Journal of Fruit Science* 24, 774–777.

Seal, A.G., Dunn, J.K., De Silva, H.N., McGhie, T.K. and Lunken, R.C.M. (2013a) Choice of pollen parent affects red flesh colour in seedlings of diploid *Actinidia chinensis* (kiwifruit). *New Zealand Journal of Crop and Horticultural Science* 41(4), 207–218. DOI: 10.1080/01140671.2013.803129

Seal, A.G., Dunn, J.K. and Jia, Y.L. (2013b) Pollen parent effects on fruit attributes of diploid *Actinidia chinensis* 'Hort16A' kiwifruit. *New Zealand Journal of Crop and Horticultural Science* 41(4), 219–229. DOI: 10.1080/01140671.2013.803130

Seal, A.G., McGhie, T., Boldingh, H., Rees, J., Blackmore, A. *et al.* (2016) The effect of pollen donor on fruit weight, seed weight and red colour development in *Actinidia chinensis* 'Hort22D'. *New Zealand Journal of Crop and Horticultural Science* 44(1), 1–12. DOI: 10.1080/01140671.2015.1100127

Seal, A.G., Clark, C.J., Sharrock, K.R., de Silva, H.N., Jaksons, P. *et al.* (2017) Choice of pollen donor affects weight but not composition of *Actinidia chinensis* var. *chinensis* 'Zesy002' (Gold3) kiwifruit. *New Zealand Journal of Crop and Horticultural Science* 46(2), 133–143. DOI: 10.1080/01140671.2017.1365732

Seal, A., Blackmore, A., Jaksons, P., McAtee, P. and Cheng, C.H. (2018) Pollen donor effects on the expression of red flesh colour in kiwifruit. *Acta Horticulturae* 1218, 373–379. DOI: 10.17660/ActaHortic.2018.1218.52

Seyrek, U.A., Tao, J., Huang, C., Lang, B., Zhong, M. *et al.* (2016) Effects of heat stress on germination and tube growth of stored pollen in *Actinidia eriantha*. *New Zealand Journal of Crop and Horticultural Science* 44(2), 152–163. DOI: 10.1080/01140671.2016.1159580

Stasiak, A., Latocha, P. and Bieniasz, M. (2021) Effect of genetically diverse pollen on pollination, pollen tube overgrow, fruit set and morphology of kiwiberry (*Actinidia arguta*). *Agronomy* 11(9), 1814. DOI: 10.3390/agronomy11091814

Tacconi, G., Michelotti, V., Cacioppo, O. and Vittone, G. (2016) Kiwifruit pollination: the interaction between pollen quality, pollination systems and flowering stage. *Journal of Berry Research* 6(4), 417–426. DOI:10.3233/JBR-160138

Tahir, J., Crowhurst, R., Deroles, S., Hilario, E., Deng, C. *et al.* (2022) First chromosome-scale assembly and deep floral-bud transcriptome of a male kiwifruit. *Frontiers in Genetics* 13, 852161. DOI: 10.3389/fgene.2022.852161

Tait, A., Paul, V., Sood, A. and Mowat, A. (2018) Potential impact of climate change on Hayward kiwifruit production viability in New Zealand. *New Zealand Journal of Crop and Horticultural Science* 46(3), 175–197. DOI: 10.1080/01140671.2017.1368672

Testolin, R. (1991) Male density and arrangement in kiwifruit orchards. *Scientia Horticulturae* 48, 41–50. DOI: 10.1016/0304-4238(91)90151-N

Testolin, R. and Ferguson, A.R. (2009) Kiwifruit (*Actinidia* spp.) production and marketing in Italy. *New Zealand Journal of Crop and Horticultural Science* 37(1), 1–32. DOI: 10.1080/01140670909510246

Testolin, R., Vizzotio, G. and Costa, G. (1991) Kiwifruit pollination by wind and insects in Italy. *New Zealand Journal of Crop and Horticultural Science* 19(4), 381–384. DOI: 10.1080/01140671.1991.10422880

Testolin, R., Cipriani, G. and Costa, G. (1995) Estimate of variance components and heritability of characters in kiwifruit (*Actinidia deliciosa*). *Acta Horticulturae* 403, 182–188. DOI: 10.17660/ActaHortic.1995.403.35

Testolin, R., Cipriani, G., Fiori, G., Piccinini, L. and Costa, G. (2011) Evaluation of parental genotypes as a tool for planning the mating design in kiwifruit breeding programs. *Acta Horticulturae* 913, 213–219. DOI: 10.17660/ActaHortic.2011.913.27

Thorp, T.G., Zhang, J. and Lay-Yee, M. (1990) Horticultural characteristics of seven pistillate and three staminate New Zealand cultivars of kiwifruit. *New Zealand Journal of Crop and Horticultural Science* 18(4), 233–240. DOI: 10.1080/01140671.1990.10428100

Thorp, T.G., Barnett, A.M. and Blattmann, P. (2011) What are the productivity limits for 'Hayward' kiwifruit? *Acta Horticulturae* 913, 419–424. DOI: 10.17660/ActaHortic.2011.913.56

Tustin, D.S., Breen, K.C. and van Hooijdonk, B.M. (2021) Light utilisation, leaf canopy properties and fruiting responses of narrow-row, planar cordon apple orchard planting systems: a study of the productivity of apple. *Scientia Horticulturae* 294, 110778. DOI: 10.1016/j.scienta.2021.110778

Twidle, A.M., Suckling, D.M., Seal, A.G., Fedrizzi, B., Pilkington, L.I. *et al.* (2017) Identification of in situ flower volatiles from kiwifruit (*Actinidia chinensis* var. *deliciosa*) cultivars and their male pollenisers in a New Zealand orchard. *Phytochemistry* 141, 61–69. DOI: 10.1016/j.phytochem.2017.05.011

Twidle, A.M., Barker, D., Seal, A.G., Fedrizzi, B. and Suckling, D.M. (2018) Identification of floral volatiles and pollinator responses in kiwifruit cultivars, *Actinidia chinensis* var. *chinensis*. *Journal of Chemical Ecology* 44(4), 406–415. DOI: 10.1007/s10886-018-0936-2

Varkonyi-Gasic, E., Wang, T., Cooney, J., Jeon, S., Voogd, C. *et al.* (2021) Shy Girl, a kiwifruit suppressor of feminization, restricts gynoecium development via regulation of cytokinin metabolism and signalling. *New Phytologist* 230(4), 1461–1475. DOI: 10.1111/nph.17234

Williams, M.H., Boyd, L.M., McNeilage, M.A., MacRae, E.A., Ferguson, A.R. *et al.* (2003) Development and commercialization of 'Baby Kiwi' (*Actinidia arguta* Planch.). *Acta Horticulturae* 610, 81–86. DOI: 10.17660/ActaHortic.2003.610.8

Wizenberg, S.B., Dang, M. and Campbell, L.G. (2021) No correlation between pollen fertility and viability: differential measures of male gametophytic fitness in *Cannabis sativa* L. *bioRxiv* DOI: 10.1101/2021.11.04.467333

Wu, Y.J., Xie, M., Zhang, Q.C., Jiang, G.H., Zhang, H.Q. *et al.* (2009) Characteristics of 'White': a new easy-peel cultivar of *Actinidia eriantha*. *New Zealand Journal of Crop and Horticultural Science* 37(4), 369–373. DOI: 10.1080/01140671.2009.9687592

Zhu, D., Lawes, G.S. and Gordon, I.L. (2002) Estimates of genetic variability and heritability for vegetative and reproductive characters of kiwifruit (*Actinidia deliciosa*). *Euphytica* 124(1), 93–98. DOI: 10.1023/A:1015662523096

6 Kiwifruit Rootstocks

Grant Thorp* and Andrew M. Barnett
The New Zealand Institute for Plant and Food Research Ltd, Te Puke, New Zealand

6.1 Introduction

The recent domestication of kiwifruit (*Actinidia* spp.) has involved little development of rootstocks through breeding and selection compared with other temperate tree and vine crops. With kiwifruit seed being plentiful and easy to germinate, initial plantings were as grafted seedlings. In later years, plants were propagated as own-rooted cuttings or *in vitro* using tissue culture. As kiwifruit production has intensified across a number of regions and countries, often on very difficult soils, pest and disease pressure has intensified. This has increased the interest in breeding and selecting rootstocks that thrive under these conditions.

6.2 Seedling Rootstocks

Seedling rootstocks were the preferred option during the early stages of domestication in New Zealand. A set of *A. chinensis* var. *deliciosa* scion cultivars had been identified (Thorp *et al.*, 1990) and 'Bruno' gained prominence as a reliable seed source for propagating rootstocks. 'Bruno' seeds were easy to germinate and were reputed to produce a uniform line of plants for grafting with *A. chinensis* var. *deliciosa* 'Hayward' (Lawes, 1990).

6.3 Clonal Rootstocks

Vegetative cuttings and tissue culture have become popular methods to produce clonal 'Hayward' plants on their own roots (Biasi *et al.*, 1990; Lawes, 1990; Loreti, *et al.*, 1992; Monastra and Testoni, 1992; Zuccherelli, 1994). Rootstocks derived from tissue culture can be micro-grafted with the desired scion cultivar (Bao *et al.*, 2020). More recent studies have further developed *in vitro* propagation methods for *A. arguta* (Hameg *et al.*, 2020) and *A. chinensis* var. *deliciosa* (Akbaş *et al.*, 2007). Micro-propagated 'Hayward' plants have remained the preferred rootstock in Italy, even now when new scion cultivars of *A. chinensis* var. *chinensis* are available. Growers argue that if the new scion cultivars fail to perform, they still have the 'Hayward' rootstock to bring into production.

Comparisons of 'Hayward' propagated on clonal 'Hayward' and on 'Bruno' seedling rootstocks have often been inconclusive owing to the lack of genuine replication of plants derived from seedlings (Loreti *et al.*, 1992; Monastra and Testoni, 1992; Cruz-Castillo *et al.*, 1997; Díaz *et al.*,1997). Early studies reported that micropropagated plants were more vigorous and produced fewer flowers when young compared with cutting or grafted seedling vines; however,

*Corresponding author: grant.thorp@plantandfood.co.nz

© CAB International 2024. *Kiwifruit: Botany, Production and Uses* (eds A.C. Richardson, J.N. Burdon and A.R. Ferguson)
DOI: 10.1079/9781800620933.0006

the cumulative yields were similar over seven years (Monastra and Testoni, 1992). While vines propagated from cuttings can be slower to establish than grafted vines (Loreti *et al.*, 1992), cutting-grown vines are reported to have higher cumulative yields than grafted plants (Díaz *et al.*,1997).

6.3.1 'D1 Vitroplant' (D1 Vitroplant®)

Apart from seedling rootstocks and clonally propagated 'Hayward' on its own roots, most kiwifruit rootstocks are a different *Actinidia* species from the scion cultivar. D1 Vitroplant® from Italy is an exception (Zuccherelli, 1994). It was selected from a population of *A. chinensis* var. *deliciosa* seedlings and grows well in high pH, calcareous soils, increasing iron (Fe) uptake by the roots and preventing Fe-deficiency-induced leaf chlorosis when grafted with 'Hayward' (Rombolà *et al.*, 2002). *A. chinensis* var. *chinensis* plants appear to be especially susceptible to Fe-induced leaf chlorosis (Thorp *et al.*, 2021) and should respond extremely well on this rootstock when grown in alkaline soils.

6.3.2 'Miliang'

'Miliang' (*A. chinensis* var. *deliciosa*), a seedling of 'Miliang No. 1', has been widely used as a rootstock in Chinese commercial orchards because it has high tolerance to soilborne diseases (Li, D. *et al.*, 2021).

6.3.3 'Z1 Vitroplant' (Vip Zedone®)

'Z1 Vitroplant' rootstock is a hybrid of *A. chinensis* var. *deliciosa* (P1) and *A. arguta* 'Gemma' that is claimed to show good tolerance to Fe deficiency and possibly to kiwifruit vine decline syndrome (KVDS; also called *moria* or *Moria del Kiwi* in Italy) (see Section 6.6.1) (Zuccherelli, 1994; Iliescu *et al.*, 2021).

6.3.4 'Kaimai'

The first interspecific rootstock (where the rootstock is a different species to the scion) to be commercialized in New Zealand was 'Kaimai' (possibly *A. hemsleyana* × *A. eriantha*). A male vine, noted for its ability to reduce floral abortion in buds and thereby double flower numbers on vines grafted with 'Hayward' scions, created significant interest in the 1990s (Wang *et al.*, 1994). Less floral abortion, however, meant more lateral flowers and fruit that required additional thinning. Fruit from lateral flowers are generally smaller and, more often, mis-shapen than fruit produced from king (terminal) flowers and so are the first to be thinned during crop management. Having excess lateral flowers significantly increased thinning costs for growers and 'Kaimai' rootstock fell out of favour.

6.3.5 'Bounty71'

Apart from clonal 'Hayward', the most important clonal rootstock at this time is 'Bounty71', a selection of *A. macrosperma* (Figs 6.1 and 6.2) (Donati *et al.*, 2020; Mian *et al.*, 2022). Identified as a rootstock that promoted higher flower numbers and larger fruit sizes when compared with 'Bruno' seedling rootstock (Zhong *et al.*, 2018),

Fig. 6.1. Graft union between *Actinidia chinensis* var. *chinensis* ('Zesy002') on top and *A. macrosperma* ('Bounty71') rootstock below. (Photo: The New Zealand Institute for Plant and Food Research Ltd)

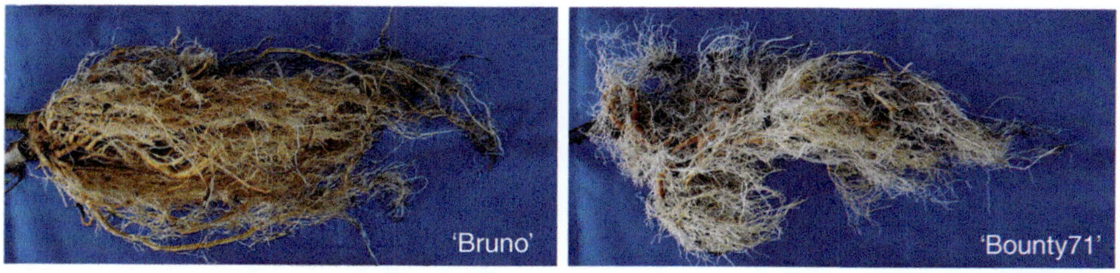

Fig. 6.2. Comparison of *Actinidia chinensis* var. *deliciosa* ('Bruno' seedling) and *A. macrosperma* ('Bounty71' clonal) root systems of young, ungrafted potted plants. (Photos: The New Zealand Institute for Plant and Food Research Ltd)

recent reports from Italian researchers indicate that 'Bounty71' rootstock is likely to be tolerant of drought (Zhong *et al.*, 2018), intermittent waterlogging (Donati *et al.*, 2020) and will survive in KVDS-inducing soils (Mian *et al.*, 2022).

6.3.6 Potential of other *Actinidia* species as rootstocks

The rich genetic diversity of *Actinidia* species provides an abundance of materials for new rootstock selection.

A study seeking frost-tolerant rootstocks compared growth responses of 'Hayward' grafted onto clonal plants of *A. arguta*, *A. kolomikta*, *A. polygama* and *A. chinensis* var. *deliciosa* × *A. arguta* hybrids (Chartier and Blanchet, 1997). The frost tolerance of interspecific rootstocks, even those of *A. arguta*, the hardy kiwifruit, was greatly reduced once grafted with 'Hayward' and active scion shoot growth had commenced.

Wang *et al.* (1994) compared flowering and fruiting responses of a range of interspecific combinations, including 'Hayward', 'Kaimai', *A. eriantha*, *A. rufa* and *A. chinensis* var. *chinensis*, grafted with 'Hayward'. 'Kaimai' increased flowering by 110%, with smaller increases of 73% and 30% for vines on *A. eriantha* and *A. rufa* rootstocks, respectively. Nitta and Ogasawara (1999) found 'Hayward' vines were more productive with *A. rufa* as a rootstock than with *A. polygama* and suggested this was because *A. rufa* was better adapted to drought and transient waterlogging. A similar result was found for *A. rufa* grafted with a scion of *A. chinensis* var. *chinensis* compared with *A. chinensis* var. *chinensis* or *A. arguta* rootstocks (Suezawa *et al.*, 2018).

A comparison of eight interspecific rootstocks with *A. chinensis* var. *chinensis* 'Hort16A' as scion examined rootstock effects on vegetative vigour, flowering and fruit quality responses (Clearwater *et al.*, 2004, 2006, 2007; Thorp *et al.*, 2007). Rootstocks included clonal plants of 'Bounty71', 'Hayward' Clone Kramer, *A. eriantha*, *A. hemsleyana* × *A. eriantha*, *A. chrysantha*, *A. kolomikta* 'KPSC' and 'Senty', and *A. polygama*. There were obvious differences in scion vigour ranging from extreme dwarfism on the two *A. kolomikta* and *A. polygama* rootstocks to slight to no vigour reduction, compared with the control 'Hayward' rootstock (Fig. 6.3) (Clearwater *et al.*, 2006). Low-vigour rootstocks produced vine canopies with short shoots with small leaves and few long and medium shoots. Control of vine vigour appeared to correlate with the timing of sap flow in spring, with root pressure increasing earlier in spring on high-vigour rootstocks compared with the low-vigour rootstocks in this trial. High vigour was also correlated with high concentrations of phosphorus and potassium in the fruit but results for calcium and magnesium were inconsistent. Fruit on vines with 'Bounty71', a moderately-high-vigour rootstock, and *A. polygama*, a low-vigour rootstock, had low calcium and magnesium concentrations which correlated with higher levels of postharvest disorders such as physiological pitting, compared with the other rootstocks studied (Thorp *et al.*, 2007). Few data were provided on the quality of the graft union in these studies. In any rootstock studies, it can be difficult to separate genuine effects of rootstock from the indirect effects of graft incompatibility.

6.4 Graft Compatibility

Interspecific grafting between different species has been successful in a number of fruit crops, including kiwifruit (Wang *et al.*, 1994). A basic requirement is that the scion and rootstock

Fig. 6.3. Effect of interspecific scion/rootstock combinations on kiwifruit vine vigour as described by Clearwater *et al.* (2006). *Actinidia chinensis* var. *chinensis* 'Hort16A' grafted onto *A. chinensis* var. *deliciosa* 'Hayward' rootstocks were high vigour and produced mainly long and medium shoots (*left*); 'Hort16A' grafted onto *A. kolomikta* rootstocks were extremely low vigour and produced mainly short and medium shoots (*right*). (Photos: The New Zealand Institute for Plant and Food Research Ltd)

species are graft-compatible. Early studies using reciprocal grafting between 'Hayward' and clonal plants of *A. arguta*, *A. kolomikta*, *A. polygama* and *A. chinensis* var. *deliciosa* × *A. arguta* hybrids, with open-pollinated seedlings of 'Hayward' as controls, were undertaken in France (Chartier and Blanchet, 1997). Graft unions between 'Hayward' and *A. arguta* were short-lived and for unions between 'Hayward' and *A. kolomikta* only a few plants survived after eight years. Combinations with the *A. chinensis* var. *deliciosa* × *A. arguta* hybrids and *A. polygama* grafted onto 'Hayward' were successful.

More comprehensive studies examined graft compatibility across a range of *Actinidia* species (Li, D. *et al.*, 2021). All grafts were successful within the different ploidy levels (diploid, tetraploid and hexaploid). Incompatibility was observed only among species of the *Leiocarpae*, with clusters of necrotic layers and abnormal vascular connections forming at the graft union. Grafting survival rate was generally positively correlated with genetic affinity between the scion and rootstock species, but there were exceptions with genotypes within the same species showing variable grafting success. The *A. macrosperma* genotypes tested in this study showed relatively poor grafting success.

6.5 Screening for Difficult Soils and Drought

An important criterion when screening for new rootstocks is the productivity of the grafted vine when grown in difficult soils and drought. Most studies with kiwifruit have focused on identifying mechanisms conferring tolerance to transient waterlogging and high pH, calcareous soils, although some researchers have also equated drought tolerance with tolerance to transient waterlogging (Nagata *et al.*, 1997; Nitta and Ogasawara, 1999).

6.5.1 Waterlogged soils and drought

The ability of plants to recover from anoxia caused by transient waterlogging is usually tested by placing young plants in a glasshouse with their roots subjected to anoxia, either by being grown in solution without aeration or in soil with the roots completely immersed in water for specified periods (Miller *et al.*, 1997; Li *et al.*, 2020; Beppu *et al.*, 2022). Treatments are typically assessed by observing leaf and root damage and measuring stomatal conductance and leaf gas exchange during the treatment and recovery phases.

Miller *et al.* (1997) compared seedlings of *A. chinensis* var. *deliciosa*, *A. chinensis* var. *chinensis*, *A. eriantha*, *A. rufa* and *A. hemsleyana* grown in either solution culture or a peat/sand mix and subjected to periods of anoxia. *A. hemsleyana* showed symptoms of decline within 3 h of treatment, while *A. chinensis* var. *deliciosa*, *A. chinensis* var. *chinensis* and *A. eriantha* did not show symptoms until three days after the start of anoxia, and *A. rufa* developed hardly any symptoms at all. The recovery phase, however, did not show a similar pattern with *A. hemsleyana* plants quickly producing new shoot growth. *A. chinensis* var.

chinensis, *A. eriantha* and, to a lesser extent, *A. chinensis* var. *deliciosa* did not resume growth, while *A. rufa*, despite showing no visible symptoms, never returned to normal rates of stomatal conductance.

Nagata *et al.* (1997) also reported that *A. rufa* rootstocks grafted with *A. chinensis* var. *deliciosa* ('Hayward' and 'Koryoku') conferred strong tolerance to both transient waterlogging and drought stress compared with *A. chinensis* var. *deliciosa* rootstocks. Nitta and Ogasawara (1999) reported more damage from drought stress with 'Hayward' grafted onto *A. polygama* than onto *A. rufa* or clonal 'Hayward' rootstocks and recommended *A. rufa* as a potential rootstock for kiwifruit in Japan. It has also been suggested that *A. rufa* rootstocks are more tolerant to *Phytopythium helicoides* and possibly *Phytopythium vexans* root rots, which are prevalent in Japan, than *A. chinensis* var. *deliciosa* or *A. chinensis* var. *chinensis* rootstocks (Yano *et al.*, 2011). Suezawa *et al.* (2018) also recommended *A. rufa* rootstocks for kiwifruit in Japan. Working with kiwifruit growing in a rice paddy field with poor drainage, they found that *A. chinensis* var. *chinensis* 'Sanuki Gold' performed best in terms of vine growth and yield when grafted onto *A. rufa* 'Fuchu' compared with 'Bruno', 'Sanuki Gold', *A. arguta* 'Kochi' and *A. rufa* × *A. chinensis* var. *chinensis* 'Kosui' rootstocks.

Comparison between 'Hayward' and *A. macrosperma* genotype Fuchu revealed 'Hayward' to be highly susceptible to anoxia, with reduced photosynthesis and stomatal conductance and high rates of leaf damage and defoliation when subjected to 13 days of waterlogging, whereas *A. macrosperma* plants did not exhibit these symptoms (Beppu *et al.*, 2022). These authors suggested that roots on *A. macrosperma* plants might be able to eliminate or detoxify substances formed during periods of anoxia.

Similar projects comparing *A. valvata* with 'Hayward' plants reported that the *A. valvata* genotype KR5 maintained higher root activity, stronger photosynthetic capacity and sustained less damage from anoxia during waterlogging stress than 'Hayward' plants (Li, Z. *et al.*, 2021; Bai *et al.*, 2022). The KR5 plants were able to alleviate active oxygen damage to roots during waterlogging stress, possibly through differences in sucrose and lysine metabolism by the roots (Li, Z. *et al.*, 2021). Six key genes involved in the waterlogging response were also identified (Bai *et al.*, 2022). It could therefore be possible to use morphological, metabolic and molecular responses in a screening programme for tolerance to anoxia.

Zhong *et al.* (2018) used *in vitro* methods to compare the drought tolerance of genotypes of five *Actinidia* species (*A. macrosperma*, *A. longicarpa*, *A. chinensis* var. *deliciosa* ('Hayward'), *A. hemsleyana* and *A. valvata*) under tissue culture conditions. They concluded that *A. macrosperma* and *A. valvata* were the most resilient genotypes and warranted further exploration to identify sources of drought-tolerant rootstocks in breeding programmes. Morphological measures such as plant fresh weight, height and leaf number and a drought damage index based on leaf condition and shoot-tip abortion were suggested as rapid and effective screening techniques once plantlets had been decanted from *in vitro* culture.

6.5.2 Alkaline, calcareous soils

Fe deficiency, and associated leaf and fruit chlorosis, is a major constraint to kiwifruit productivity in producing regions with high pH, calcareous soils, including regions in China (Wang *et al.*, 2019), Greece (Loupassaki *et al.*, 1997) and Italy (Rombolà and Tagliavini, 2006). The newer, yellow-fleshed cultivars of *A. chinensis* var. *chinensis* appear to be especially sensitive (Wang *et al.*, 2019; Thorp *et al.*, 2021).

Current methods to reduce the impact of high pH soils on plant growth and productivity mainly rely on Fe chelate treatments (Tagliavini and Rombolà, 2001; Rombolà and Tagliavini, 2006). These are expensive, and applications made in one year do not prevent chlorosis in the next (Tagliavini *et al.*, 2000). More sustainable solutions include intercropping vines with cover crops that secrete phytosiderophores into the soil, which assists Fe uptake by the vine (Briat *et al.*, 2015). The development of rootstocks tolerant of high pH soils, however, might offer the best and most sustainable solution (Viti *et al.*, 1990; Rombolà *et al.*, 2002; Rombolà and Tagliavini, 2006).

There have been no large-scale screening trials reported for rootstocks tolerant to Fe deficiency in *Actinidia*. Studies on mechanisms and preventative measures for chlorosis in kiwifruit have either studied clonal 'Hayward' vines alone

(Vizzotto *et al.*, 1997) or involved comparisons with 'D1 Vitroplant' (Viti *et al.*, 1990; Rombolà *et al.*, 2002) or advanced selections of *A. chinensis* var. *deliciosa* (Vizzotto *et al.*, 1999).

6.6 Resistance to Pests and Diseases

While the ultimate measure of success for a kiwifruit rootstock is that it improves the production efficiency and product quality of the scion cultivar, a critical factor in this success is the ability of the rootstock to withstand soilborne pests and diseases. Important among these are KVDS and *Meloidogyne* spp. (root-knot nematodes). Rootstocks should also have low susceptibility to infection by *Pseudomonas syringae* pv. *actinidiae* (Psa).

6.6.1 Kiwifruit vine decline syndrome (KVDS)

KVDS is a relatively new disease complex that has become a major concern for kiwifruit industries in Europe, especially in the Verona Province of northern Italy (Sorrenti *et al.*, 2016; Donati *et al.*, 2020; Savian *et al.*, 2020). Waterlogging has been implicated as a causal factor along with a complex of soilborne pathogens (see Chapter 17). Current recommendations are for growers to optimize their irrigation systems (Sorrenti *et al.*, 2016) but ultimately the solution may come from the development of rootstocks that can survive in KVDS-inducing soils. Genotypes of *A. arguta* and *A. macrosperma*, including 'Bounty71', survived in KVDS-inducing soils in Italy, whereas 'Hayward' and another genotype of *A. arguta* did not (Mian *et al.*, 2022).

6.6.2 Root-knot nematodes

Root-knot nematodes (*Meloidogyne* spp.) are plant-parasitic and relatively common in the main producing regions for kiwifruit, with *M. incognita* the most common species (Grandison 1990; Haygood *et al.*, 1990; Watson *et al.*,1992; Knight, 2001; Tao *et al.*, 2017). The impact of mild infestation on kiwifruit vine productivity is not always clear but heavy infestation can cause significant damage, as found in the sandy coastal soils of South Carolina, USA (Haygood *et al.*, 1990). While preventative measures can be employed to inhibit infestations of orchards, development of resistant rootstocks would provide a long-term and sustainable option for growers. Research on genetic resistance of *Actinidia* spp. to root-knot nematode has found several examples of resistance among genotypes of *A. chinensis* var. *chinensis* and *A. chinensis* var. *deliciosa* that would be of interest to breeding programmes (Nicotra *et al.*, 2003).

6.6.3 *Pseudomonas syringae* pv. *actinidiae* (Psa)

Psa is the most devastating disease of kiwifruit in the major producing regions (Vanneste, 2017) (see Chapter 18). Though not a soilborne disease, kiwifruit rootstocks can become infected with Psa through the systemic movement of the disease from a point of infection above ground down through the trunk, past the graft union, and into the rootstock (Tyson *et al.*, 2014). While Psa infection might not damage the root system, it is important to have a tolerant or resistant rootstock in case the scion cultivar dies and regrafting is required.

A preliminary study on the possible transfer of protection against Psa from Gui-1, a Psa-tolerant rootstock of *A. guilinensis* (now included in *A. latifolia*), to 'Hongyang' (*A. chinensis* var. *chinensis*), a susceptible scion cultivar, has been reported (Wang *et al.*, 2021). Evidence included the rate of lesion development in wood after inoculation, from the rootstock and scion on grafted and non-grafted plants. Results were not always consistent, and the authors discussed the possibility of there being an indirect effect of the rootstock improving scion vigour and the stronger root system supplying more water, nutrients and minerals, rather than a direct effect of the rootstock transferring resistance to the scion.

6.7 Future Research

Kiwifruit is a vigorously growing vine which requires regular and detailed pruning, so a rootstock to reduce vegetative vigour and increase carbon partitioning to fruit, without compromising fruit yield and quality, is desired. Interspecific rootstocks, for example 'Bounty71', that partly satisfy these requirements have been released to growers. Rootstocks that inhibit shoot growth and thus reduce pruning costs for male vines are also desirable as long as they do not negatively affect flowering and pollen production/viability. However, the more important challenge for breeders of kiwifruit rootstocks is to identify/develop rootstocks that perform in difficult soils. Rootstocks that are tolerant of anoxia and/or drought would most likely provide a sustainable solution for KVDS, as would rootstocks tolerant of the disease complex associated with KVDS. Rootstocks that are tolerant of alkaline and calcareous soils would reduce the impact of the associated leaf chlorosis and provide substantial and sustainable improvements in vine productivity in regions with these soil types. Importantly, each of these breeding objectives requires close collaboration between breeders, physiologists and pathologists to ensure objectives are focused on primary mechanisms rather than secondary effects.

References

Akbaş, F.A., Işikalan, Ç., Namli, S. and Başaran, D. (2007) Micropropagation of kiwifruit (*Actinidia deliciosa*). *International Journal of Agriculture & Biology* 9(3), 489–493.

Bai, D., Li, Z., Gu, S., Li, Q., Sun, L. *et al.* (2022) Effects of kiwifruit rootstocks with opposite tolerance on physiological responses of grafting combinations under waterlogging stress. *Plants* 11, 2098. DOI: 10.3390/plants11162098

Bao, W.-W., Zhang, X.-C., Zhang, A.-L., Zhao, L., Wang, Q.-C. *et al.* (2020) Validation of micrografting to analyze compatibility, shoot growth, and root formation in micrografts of kiwifruit (*Actinidia* spp.). *Plant Cell, Tissue and Organ Culture* 140(1), 209–214. DOI: 10.1007/s11240-019-01723-2

Beppu, K., Ogihara, Y., Ohtani, M. and Kataoka, I. (2022) Comparison of waterlogging tolerance between *Actinidia macrosperma* and *Actinidia deliciosa*. *Acta Horticulturae* 1332, 219–226. DOI: 10.17660/ActaHortic.2022.1332.29

Biasi, R., Marino, G. and Costa, G. (1990) Propagation of Hayward (*Actinidia deliciosa*) from soft and semi-hardwood cuttings. *Acta Horticulturae* 282, 243–250. DOI: 10.17660/ActaHortic.1990.282.31

Briat, J.-F., Dubos, C. and Gaymard, F. (2015) Iron nutrition, biomass production, and plant product quality. *Trends in Plant Science* 20(1), 33–40. DOI: 10.1016/j.tplants.2014.07.005

Chartier, J. and Blanchet, P. (1997) Reciprocal grafting compatibility of kiwifruit and frost hardy *Actinidia* species. *Acta Horticulturae* 444, 149–154. DOI: 10.17660/ActaHortic.1997.444.21

Clearwater, M.J., Lowe, R.G., Hofstee, B.J., Barclay, C., Mandemaker, A.J. *et al.* (2004) Hydraulic conductance and rootstock effects in grafted vines of kiwifruit. *Journal of Experimental Botany* 55(401), 1371–1382. DOI: 10.1093/jxb/erh137

Clearwater, M.J., Seleznyova, A.N., Thorp, T.G., Blattmann, P., Barnett, A.M. *et al.* (2006) Vigor-controlling rootstocks affect early shoot growth and leaf area development of kiwifruit. *Tree Physiology* 26(4), 505–515. DOI: 10.1093/treephys/26.4.505

Clearwater, M.J., Blattmann, P., Luo, Z. and Lowe, R.G. (2007) Control of scion vigour by kiwifruit rootstocks is correlated with spring root pressure phenology. *Journal of Experimental Botany* 58(7), 1741–1751. DOI: 10.1093/jxb/erm029

Cruz-Castillo, J.G., Lawes, G.S., Woolley, D.J. and Ganesh, S. (1997) Evaluation of rootstock and 'Hayward' scion effects on field performance of kiwifruit vines using a multivariate analysis technique. *New Zealand Journal of Crop and Horticultural Science* 25(3), 273–282. DOI: 10.1080/01140671.1997.9514016

Díaz, M.B., Ciordia, M., Garcia, J.C. and Garcia, J. (1997) Performance of kiwifruit plant material propagated by different methods. *Acta Horticulturae* 444, 155–162. DOI: 10.17660/ActaHortic.1997.444.22

Donati, I., Cellini, A., Sangiorgio, D., Caldera, E., Sorrenti, G. *et al.* (2020) Pathogens associated to kiwifruit vine decline in Italy. *Agriculture* 10(4), 119. DOI: 10.3390/agriculture10040119

Grandison, G.S. (1990) Nematodes and kiwifruit. In: Warrington, I.J. and Weston, G. (eds) *Kiwifruit: Science and Management*. Ray Richards Publisher in association with the New Zealand Society for Horticultural Science, Auckland, New Zealand, pp. 413–419.

Hameg, R., Arteta, T.A., Landin, M., Gallego, P.P. and Barreal, M.E. (2020) Modeling and optimizing culture medium mineral composition for *in vitro* propagation of *Actinidia arguta*. *Frontiers in Plant Science* 11, 554905. DOI: 10.3389/fpls.2020.554905

Haygood, R.A., Saunders, J.A. and Miller, R.W. (1990) Widespread occurrence of *Meloidogyne incognita* on kiwifruit in the coastal areas of South Carolina. *Plant Disease* 74(1), 81. DOI: 10.1094/PD-74-0081F

Iliescu, L.M., Alexandru, C.M. and Stănică, F. (2021) 'Z1 Vitroplant' – valuable rootstock for kiwifruit cultivars – grafting results. *Scientific Papers. Series B, Horticulture* 65(1), 99–110.

Knight, K.W.L. (2001) Plant parasitic nematodes associated with six subtropical crops in New Zealand. *New Zealand Journal of Crop and Horticultural Science* 29(4), 267–275. DOI: 10.1080/01140671.2001.9514188

Lawes, G.S. (1990) Propagation of kiwifruit. In: Warrington, I.J. and Weston, G. (eds) *Kiwifruit: Science and Management*. Ray Richards Publisher in association with the New Zealand Society for Horticultural Science. Auckland, New Zealand, pp. 297–321.

Li, D., Han, F., Liu, X., Lv, H., Li, L. *et al.* (2021) Localized graft incompatibility in kiwifruit: analysis of homografts and heterografts with different rootstock & scion combinations. *Scientia Horticulturae* 283, 110080. DOI: 10.1016/j.scienta.2021.110080

Li, Z., Zhong, Y., Bai, D., Lin, M., Qi, X. *et al.* (2020) Comparative analysis of physiological traits of three *Actinidia valvata* Dunn genotypes during waterlogging and post-waterlogging recovery. *Horticulture, Environment, and Biotechnology* 61(5), 825–836. DOI: 10.1007/s13580-020-00276-0

Li, Z., Bai, D., Zhong, Y., Abid, M., Qi, X. *et al.* (2021) Physiological responses of two contrasting kiwifruit (*Actinidia* spp.) rootstocks against waterlogging stress. *Plants* 10, 2586. DOI: 10.3390/plants10122586

Loreti, F., Piccotino, D., Baroni, G. and Xiloyannis, C. (1992) Effect of propagation technique on vegetative growth and fruiting in kiwifruit. *Acta Horticulturae* 297, 183–191. DOI: 10.17660/ActaHortic.1992.297.23

Loupassaki, M.H., Lionakis, S.M. and Androulakis, I.I. (1997) Iron deficiency in kiwi and its correction by different methods. *Acta Horticulturae* 444, 267–271. DOI: 10.17660/ActaHortic.1997.444.41

Mian, G., Cipriani, G., Saro, S., Martini, M. and Ermacora, P. (2022) Evaluation of germplasm resources for resistance to kiwifruit vine decline syndrome (KVDS). *Acta Horticulturae* 1332, 125–130. DOI: 10.17660/ActaHortic.2022.1332.17

Miller, S.A., Smith, G.S. and Cookson, J.H. (1997) The effects of transient waterlogging on different species of *Actinidia*. *Acta Horticulturae* 444, 323–327. DOI: 10.17660/ActaHortic.1997.444.50

Monastra, F. and Testoni, A. (1992) Horticultural performance and quality characteristics of fruit from kiwifruit plants (cv. Hayward) obtained by *in-vitro* propagation, by cuttings and by grafting. *Acta Horticulturae* 297, 197–203. DOI:10.17660/ActaHortic.1992.297.25

Nagata, K., Morinaga, K. and Ikeda, F. (1997) Photosynthetic ability of kiwifruit (*Actinidia deliciosa*) and characteristics of Shima-sarunashi (*Actinidia rufa*) rootstock. *Bulletin of the Shikoku National Agricultural Experiment Station (Japan)* 61, 159–166.

Nicotra, A., Simeone, A.M. and De Vito, M. (2003) Research on kiwifruit source of genetic resistance to root-knot and lesion nematodes. *Acta Horticulturae* 610, 449–453. DOI: 10.17660/ActaHortic.2003.610.60

Nitta, H. and Ogasawara, S. (1999) Characteristics of 'Hayward' kiwifruit vines grown on their own roots or grafted onto *Actinidia polygama* or *Actinidia rufa*. *Acta Horticulturae* 498, 319–324. DOI: 10.17660/ActaHortic.1999.498.38

Rombolà, A.D. and Tagliavini, M. (2006) Iron nutrition of fruit crops. In: Barton, L.L. and Abadía, J. (eds) *Iron Nutrition in Plants and Rhizospheric Microorganisms*. Springer, Dordrecht, The Netherlands, pp. 61–83. DOI: 10.1007/1-4020-4743-6_3

Rombolà, A.D., Brüggemann, W., López-Millán, A.F., Tagliavini, M., Abadía, J. *et al.* (2002) Biochemical responses to iron deficiency in kiwifruit (*Actinidia deliciosa*). *Tree Physiology* 22(12), 869–875. DOI: 10.1093/treephys/22.12.869

Savian, F., Ginaldi, F., Musetti, R., Sandrin, N., Tarquini, G. *et al.* (2020) Studies on the aetiology of kiwifruit decline: interaction between soil-borne pathogens and waterlogging. *Plant and Soil* 456, 113–128. DOI: 10.1007/s11104-020-04671-5

Sorrenti, G., Toselli, M., Reggidori, G., Spinelli, F., Tosi, L. *et al.* (2016) Implicazioni della gestione idrica nella 'moria del kiwi' del veronese. *Rivista di Frutticoltura e di Ortofloricoltura* 78(3), 45–51.

Suezawa, K., Fukuda, T., Mizutani, R., Yamashita, T., Otani, M. *et al.* (2018) Field performance of tetraploid *Actinidia chinensis* 'Sanuki Gold' on *Actinidia rufa* rootstocks. *Acta Horticulturae* 1218, 413–418. DOI: 10.17660/ActaHortic.2018.1218.57

Tagliavini, M. and Rombolà, A.D. (2001) Iron deficiency and chlorosis in orchard and vineyard ecosystems. *European Journal of Agronomy* 15(2), 71–92. DOI: 10.1016/S1161-0301(01)00125-3

Tagliavini, M., Abadía, J., Rombolà, A.D., Abadía, A., Tsipouridid, C. *et al.* (2000) Agronomic means for the control of iron deficiency chlorosis in deciduous fruit trees. *Journal of Plant Nutrition* 23(11), 2077–2022. DOI: 10.1080/01904160009382161

Tao, Y., Xu, C., Yuan, C., Wang, H., Lin, B. *et al.* (2017) *Meloidogyne aberrans* sp. nov. (Nematoda: Meloidogynidae), a new root-knot nematode parasitizing kiwifruit in China. *PLOS ONE* 12(8), e0182627. DOI: 10.1371/journal.pone.0182627

Thorp, T.G., Zhang, J. and Lay Yee, M. (1990) Horticultural characteristics of seven pistillate and three staminate New Zealand cultivars of kiwifruit. *New Zealand Journal of Crop and Horticultural Science* 18, 233–240. DOI: 10.1080/01140671.1990.10428100

Thorp, T.G., Boyd, L.M., Barnett, A.M., Lowe, R.G., Hofstee, B.J. *et al.* (2007) Effect of inter-specific rootstocks on inorganic nutrient concentrations and fruit quality of 'Hort16A' kiwifruit (*Actinidia chinensis* Planch. var. *chinensis*). *The Journal of Horticultural Science and Biotechnology* 82(6), 829–838. DOI: 10.1080/14620316.2007.11512314

Thorp, T.G., Barnett, A.M., Civolani, C., Spinelli, R., Lallu, N. *et al.* (2021) Iron chelate applications improve productivity and fruit quality at harvest and during cold storage of *Actinidia chinensis* var. *chinensis* ('Hort16A') kiwifruit growing in high-pH calcareous soils. *European Journal of Horticultural Science* 86(6), 609–619. DOI: 10.17660/eJHS.2021/86.6.4

Tyson, J.L., Curtis, C.L., Manning, M.A., Rees-George, J., Snelgar, W.P. *et al.* (2014) Systemic movement of *Pseudomonas syringae* pv. *actinidiae* in kiwifruit vines in New Zealand. *New Zealand Plant Protection* 67, 41–47. DOI: 10.30843/nzpp.2014.67.5720

Vanneste, J.L. (2017) The scientific, economic, and social impacts of the New Zealand outbreak of bacterial canker of kiwifruit (*Pseudomonas syringae* pv. *actinidiae*). *Annual Review of Phytopathology* 55, 377–399. DOI: 10.1146/annurev-phyto-080516-035530

Viti, R., Xiloyannis, C., Trinci, M. and Ragone, A.F. (1990) Effect of calcareous soil on vegetative growth of own rooted and grafted kiwi fruit trees. *Acta Horticulturae* 282, 209–216. DOI: 10.17660/ActaHortic.1990.282.27

Vizzotto, G., Matosevic, I., Pinton, R., Varanini, Z. and Costa, G. (1997) Iron deficiency responses in roots of kiwi. *Journal of Plant Nutrition* 20(2-3), 327–334. DOI: 10.1080/01904169709365253

Vizzotto, G., Pinton, R., Bomben, C., Cesco, S., Varanini, Z. *et al.* (1999) Iron reduction in iron-stressed plants of *Actinidia deliciosa* genotypes: involvement of PM Fe(III)-Chelate reductase and H^+-ATPase activity. *Journal of Plant Nutrition* 22(3), 479–488. DOI: 10.1080/01904169909365645

Wang, F.-M., Li, J.-W., Ye, K.-Y., Gong, H.-J., Liu, P.-P. *et al.* (2021) Preliminary report on the improved resistance towards *Pseudomonas syringae* pv. *actinidiae* of cultivated kiwifruit (*Actinidia chinensis*) when grafted onto wild *Actinidia guilinensis* rootstock in vitro. *Journal of Plant Pathology* 103(1), 51–54. DOI: 10.1007/s42161-020-00719-8

Wang, N., Yao, C., Li, M., Li, C., Liu, Z. *et al.* (2019) Anatomical and physiological responses of two kiwifruit cultivars to bicarbonate. *Scientia Horticulturae* 243, 528–536. DOI: 10.1016/j.scienta.2018.09.011

Wang, Z.-Y., Gould, K.S. and Patterson, K.J. (1994) Comparative root anatomy of five *Actinidia* species in relation to rootstock effects on kiwifruit flowering. *Annals of Botany* 73(4), 403–413. DOI: 10.1006/anbo.1994.1050

Watson, R.N., Wilson, E.A. and Marsden, R.S. (1992) Distribution of plant-parasitic nematodes in the rhizosphere of kiwifruit. *Acta Horticulturae* 297, 537–543. DOI: 10.17660/ActaHortic.1992.297.70

Yano, T., Shimizu, S., Miyoshi, T., Miyata, N., Immon, K. *et al.* (2011) Tolerant *Actinidia* species to *Pythium helicoides* and *P. vexans* causing root rot. *Acta Horticulturae* 913, 517–523. DOI: 10.17660/ActaHortic.2011.913.69

Zhong, Y.-P., Li, Z., Bai, D.-F., Qi, X.-J. and Chen, J.-Y. (2018) In vitro variation of drought tolerance in five *Actinidia* species. *Journal of the American Society for Horticultural Science* 143(3), 226–234. DOI: 10.21273/JASHS04399-18

Zuccherelli, G. (1994) *L'Actinidia e i Nuovi Kiwi*. Fruitticoltura Moderna. Edagricole – Edizioni Agricole, Bologna, Italy.

7 Kiwiberry

Piotr Latocha[1]* and Filip Debersaques[2]
[1]*Warsaw University of Life Sciences – SGGW, Warsaw, Poland;*
[2]*Ghent University, Ghent, Belgium*

7.1 Introduction

For many centuries, wild kiwiberries have been harvested for both food and medicinal uses (Huang, 2016). Attractive features include their convenient size, thin edible peel, exotic flavour and rich nutritional composition. They meet the consumer's desire for novel foods (Harker *et al.*, 2007; Latocha and Jankowski, 2011; Latocha *et al.*, 2011; Vanhonacker and Debersaques, 2018). However, due to differences between kiwiberry and kiwifruit, including their morphology, cultivation, marketability, perishability and edibility, distinct cultivation, postharvest and marketing strategies have been developed for this fruit. The name kiwiberry is now widely used for *Actinidia arguta* and hybrids in European Union countries, the USA and New Zealand, reflecting the similarities of these species with other berries.

7.2 Kiwiberry Botany

Kiwiberry is largely *A. arguta* or hybrids with other *Actinidia* species (see Table 7.1). According to the latest revision of the *Actinidia* genus, *A. arguta* is divided into two varieties, *A. arguta* var. *arguta* and *A. arguta* var. *giraldii* (Huang, 2016). Like most *Actinidia* species, *A. arguta* originated from Asia and is one of the most widespread, being found in many parts of China, far-eastern Russia, the Korean peninsula and Japan. Kiwiberry vines naturally have a wide elevation range, from 500 to nearly 2700 m.a.s.l. (Kolbasina *et al.*, 2008; Huang, 2016). It has been suggested that climate change will have a positive effect on the distribution of kiwiberry in China, where the suitability of the northeast region will gradually increase and the distribution of *A. arguta* will move closer to the centre of its natural range (Ma *et al.*, 2021).

Studies on wild *A. arguta* populations indicate a high variability in ploidy, ranging from diploid to decaploid (Li *et al.*, 2013; Zhang *et al.*, 2017), with tetraploid genotypes most common ($2n=4x=232$; Kataoka *et al.*, 2010). The presence of wild kiwiberry genotypes across a breadth of habitats and ploidy levels suggests large genetic diversity. Kiwiberry has many biological similarities with other *Actinidia* species. It is a vigorous climber with twining shoots, has single leaves with long petioles, and flowers and fruit develop from one-year-old buds. Morphological variability within kiwiberry includes variation in the colour of one-year-old shoots (Fig. 7.1), shape, serration and size of leaves (Fig. 7.2), size, shape and colour of the fruit skin

*Corresponding author: piotr_latocha@sggw.edu.pl

© CAB International 2024. *Kiwifruit: Botany, Production and Uses* (eds A.C. Richardson, J.N. Burdon and A.R. Ferguson)
DOI: 10.1079/9781800620933.0007

Fig. 7.1. *Actinidia arguta* one-year-old shoots. (Photo: P. Latocha)

Fig. 7.2. *Actinidia arguta* leaf diversity. (Photo: P. Latocha)

and pulp (Fig. 7.3) and variation in flower number. There are also differences in tolerance to low temperatures in winter.

Kiwiberry cultivars are dioecious, with male and female plants. The vines have a dense, expansive root system located near the soil surface (Hughes *et al.*, 1992). The bark of one-year-old shoots is brown to light grey and smooth with a lamellar core (Fig. 7.1). Buds exhibit axillary structures of both basal buds and simple primordia (Walton and Wu, 1999), although earlier studies suggested otherwise (Snowball, 1997a). Kiwiberry plants produce both terminated and non-terminated shoots (Fig. 7.4) (Ferguson, 1984; Jackson, 2000), with 63–79% of all shoots being terminated on female plants and 94% on male plants (Stasiak *et al.*, 2020). Kiwiberry plants produce both floral and vegetative shoots, with floral shoots accounting for 50–64% of all shoots on female plants (Tiyayon and Strik, 2003a) and 44–100% on male plants (Stasiak *et al.*, 2020). According to Snowball (1997b), flowers can originate from nodes 1–26 on non-terminating shoots while Tiyayon and Strik (2003b) found that nodes 6–26 are usually floral, although there is year-to-year variation.

Kiwiberry leaves may vary considerably in shape, from oval to oblong or even elongated, and leaf size depends on the origin and ploidy of the plant (Fig. 7.2). Plants of higher ploidy levels tend to have larger leaves than diploids (Kataoka *et al.*, 2010). Kiwiberry leaves are much smaller and almost hairless compared with kiwifruit leaves. The veins of the lower surface and petiole are covered with simple hairs (Huang, 2016).

Flower development in kiwiberry occurs over a two-year cycle (Walton and Wu, 1999; Kolbasina *et al.*, 2008) with floral evocation in spring/summer of the first season and floral differentiation in spring of the second season (see Chapter 14). The length of the flowering period of male and female plants differs, owing to the size of the inflorescences. Female plants usually flower for 6–8 days (Skrypchenko and Moroz, 2002) and males for up to 13 days (Kim *et al.*, 2016). The number of flowers is influenced by cultural and environmental factors such as shading; shaded plants are less productive (Tiyayon and Strik, 2004).

Kiwiberry flowers are small, up to 2.5 cm in diameter. Male (staminate) flowers are usually smaller than female (pistillate) flowers. Flowers have four to seven cream-white uniform petals (Fig. 7.5). Male flowers do not have a pistil with characteristically arranged stigmas, but only stamens with dark brown or purple anthers. A single male flower can produce up to ~4 million pollen grains (Stasiak *et al.*, 2020); however, there is large variation between genotypes. While structurally perfect, female flowers are functionally female because the anthers contain non-viable (sterile) pollen grains, which is known as cryptic dioecy (Kawagoe and Suzuki, 2004).

Kiwiberry flowers emerge from the leaf axils, either individually or in cymes, and the proportion of individual flowers or cymes varies among genotypes (Fig. 7.5). Male plants may have up to 15 flowers of a similar size per cyme while female plants usually have one to three, with the king (terminal) flower sometimes slightly larger than the lateral flowers. A single mature female plant may produce more than 11,000 flowers (Tiyayon and Strik, 2003b).

Kiwiberries are much smaller (6–20 g) than kiwifruit (60–120 g) and have delicate, edible skins. They can be oblong-ovoid, cylindrical-ellipsoid, nearly spherical, acorn-like, or flattened in shape with skin colour ranging from green, yellow-green, green with a red or brownish blush, to deep burgundy-red and even dark purple (Fig. 7.3). Depending on genotype, the ripe fruit pulp colour can be yellow to dark green, reddish-rose, dark red or even green with a red core.

Kiwiberry fruit develop very quickly, reaching nearly 80% of their final size 30–40 days after anthesis (Hassal *et al.*, 1998). During this period, the delicate skin of the fruit is very sensitive to damage from strong wind and temperature and humidity fluctuations, which may result in microcracks and russeting of the skin. Fruit reaches physiological maturity when the seeds change colour from white to brown. Most genotypes develop the typical skin colour (blush) before reaching physiological maturity but those with fully red fruit usually colour later. Fruit coloration depends on light, but the mechanism remains unclear (Huang *et al.*, 2021).

The ripening of individual kiwiberry fruit on a vine is not synchronized during the harvest period. Kiwiberry ripening physiology is similar to that of kiwifruit (see Chapter 16). During fruit maturation, soluble solids content (SSC) and dry matter increase, reaching 15–25% and 17–25.5%, respectively (Bieniek, 2012; Wojdyło

Fig. 7.3. *Actinidia arguta* fruit diversity. (Photo: P. Latocha)

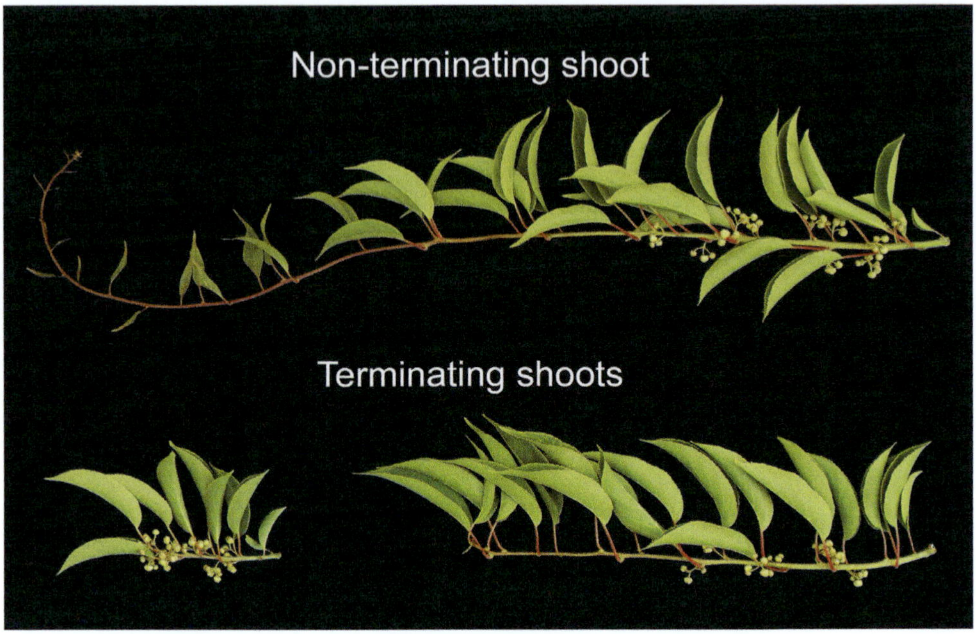

Fig. 7.4. *Actinidia arguta* shoots in summer. (Photo: P. Latocha)

Fig. 7.5. Male (*left*) and female (*right*) *Actinidia arguta* flowers. (Photo: P. Latocha)

et al., 2017). The softening of kiwiberry fruit is faster than that of the larger kiwifruit. Within seven days of initiation of softening at 20°C, fruit firmness can drop below 0.5 N (White *et al.*, 2005; Sutherland *et al.*, 2017), and therefore kiwiberry shelf life is relatively short.

7.3 Cultivars

In many places around the world, kiwiberry cultivation has developed based on local selections (Latocha *et al.*, 2021) (Fig. 7.6). Cultivars are either *A. arguta* or its hybrids (Table 7.1). Most cultivars are freely available to growers, but some have plant variety protection. Due to high natural variability, fruit of a kiwiberry cultivar can differ markedly in shape, skin and pulp colour as well as taste (Debersaques *et al.*, 2019).

In addition to the female cultivars there are also many male cultivars. The main male cultivars in the USA are 'Meader Male', 'Fave', 'Cornell' and Robert Smith's 74 series (Jackson, 2000). In Europe, 'Male', 'Joker', 'Nostino', '74-46', 'Meader' and 'Don Juan' are dominant (Skrypchenko and Latocha, 2017; Debersaques *et al.*, 2019).

Fig. 7.6. Commercial kiwiberry (*Actinidia arguta*) cultivars: 'Mitsuko', 'Ananasnaya', 'Weiki', 'Longcheng No. 2', 'Kuilv', 'Bingo', 'Issai', 'Hortgem Rua', 'Jumbo', 'Ken's Red' (*A. arguta* × *A. melanandra*), 'Hortgem Tahi', 'Geneva'. (Photos: P. Latocha, with permission from I. Kataoka ('Mitsuko'), J. Machado ('Hortgem Tahi', 'Hortgem Rua'), H. Qin ('Longcheng No. 2', 'Kuilv').

7.4 Commercial Fruit Production

For successful kiwiberry production, orchard location, vine architecture, trellis structure, plant maintenance, fruit harvest and postharvest handling are of crucial importance. The growth period of kiwiberry vines is short, with the annual growth cycle completed in approximately 150 frost-free days (Strik and Hummer, 2006). The fruit generally reaches physiological maturity prior to autumn frost. However, the dates of the phenological phases and their relative durations are strongly dependent on climatic conditions, particularly temperature (Kolbasina *et al.*, 2008).

7.4.1 Orchard site selection

An important factor in establishing a new orchard is temperature. Dormant kiwiberry

Table 7.1. The most common commercially grown female kiwiberry cultivars and the characteristics of the ripe fruit.

Cultivar name (species)	Shape	Weight (g)	Skin colour	Pulp colour	References
Cultivars popular in Europe and North and South America					
'Ananasnaya' (= 'Anna', 'Ananasnaja') (AA)	Oval	6–15	Green, strong red blush	Green	Strik and Hummer (2006); Debersaques *et al.* (2019); Strik and Davis (2021)
'Bingo' (AA)	Ovate-oblong, slightly flattened	8–14	Green, reddish-pink blush	Yellowish-green	Latocha (2012)
'Geneva' (AA)	Round, slightly flattened	6–12	Green, light reddish-brown blush	Green	Debersaques *et al.* (2019); Strik and Davis (2021)
'Issai'[1] (AA x AP, hexaploid)	Oblong, laterally flattened fruit	3–9	Green	Green	Mizugami *et al.* (2007); Debersaques *et al.* (2019)
'Jumbo' (= 'Jumbo Verde') (AA)	Oblong, laterally flattened	15–19	Green	Green to yellow-green	Debersaques *et al.* (2019)
'Ken's Red' (AA x AM)	Cylindrical, tapers to a point at the apex	8–12	Green to reddish-brown	Green, brownish-red core	Debersaques *et al.* (2019)
'Weiki' (= 'Bayern-kiwi') (AA)	Oval	5–12	Green, strong red or reddish-brown blush	Green	Debersaques *et al.* (2019)
Cultivars popular in Asia, Australia and Oceania					
'Chang Jiang No. 1' (AA)	Long, cylindrical, slightly oval	16	Green	Green	Liu and Liu (2015)
'Kuilv' ('Jian Feng Kui Lü', 'Kuilu', '8025') (AA)	Oval	~15	Deep green	Deep green	Debersaques *et al.* (2019)
'LD133' (AA)	Round, laterally flattened	~20	Deep green	Deep green	Debersaques *et al.* (2019)
'Longcheng No. 2' (AA)	Elongated	~18	Green, slightly blushing	Green	Debersaques *et al.* (2019)
'Cheongsan' (AA)	Oblong	7–11	Green	Green	Han *et al.* (2019)
'Chiak' (AA)	Oblong	12–16	Green	Green	Jo *et al.* (2007)
'Mitsuko' (AA, hexaploid)	Obovate	12–16	Green	Green	Kim *et al.* (2012)
'74-49' (AA)	Roundish, slightly flattened	6–12	Green, reddish-brown blush	Green	Latocha (2007)
'Hortgem Tahi' (AA)	Spheroid	10–12	Green	Medium green	Williams *et al.* (2003); Cossio *et al.* (2015)
'Hortgem Rua' (AA)	Oblong, stylar end strongly protruding to a point	14–18	Green, brown-green	Green, red core	Williams *et al.* (2003); Cossio *et al.* (2015)

AA, *Actinidia arguta*; AM, *Actinidia melanandra*; AP, *Actinidia polygama*
[1]There are a number of different selections of different origins named 'Issai' (Mizugami *et al.*, 2007).

vines require only about 200 h below 7°C to break winter dormancy (Phivnil *et al.*, 2004). While dormant kiwiberry plants are resistant to temperatures below −30°C (Latocha, 2007; Kataoka *et al.*, 2010), most cultivars start to break bud very early and are sensitive to spring frosts (Van Labeke *et al.*, 2015) (Fig. 7.7) when temperatures of −1°C for as little as 30 minutes can significantly injure young shoots (Strik and Davis, 2021). About 50% of buds break in spring, independent of growing conditions or cultivar (Snowball, 1997b; Tiyayon and Strik, 2003b). Mature kiwiberry plants that sustain frost damage in spring can readily produce vegetative shoots from the remaining dormant buds, whereas young kiwiberry plants are more susceptible to frost injury and can be completely killed by spring frosts.

For good yields and fruit quality, vines need fertile soil with high organic matter content, a well-structured A-horizon and good drainage in the B- and C-horizons (Rothgerber, 1992). The optimum soil pH is between 5.6 and 6.5 (Whitworth, 2017; Strik and Davis, 2021). As kiwiberry plants are susceptible to verticillium wilt, sites with a history of *Verticillium* should be avoided (Strik and Davis, 2021).

7.4.2 Trellis structure and planting design

Kiwiberry is extremely vigorous and requires a permanent trellis to support vines that are very susceptible to wind damage (Calvez *et al.*, 2015). The pergola (Fig. 7.8) and T-bar trellis systems (Fig. 7.9) used for kiwifruit are most commonly used for kiwiberry, but a hedgerow system (Fig. 7.10), similar to that used for grapes, is used in some European countries (Debersaques *et al.*, 2019).

Vines on a pergola and T-bar are better protected from wind damage and sunburn than vines on a hedgerow system (Latocha, 2010), are less exposed to ground frosts, produce higher yields, and are easier to protect from disease. Whitworth (2017) concluded that plants on pergola trellises have larger, less sunburned fruit and are less wind-damaged than those on T-bar support systems. However, as the pergola system hinders air movement, increased humidity may lead to fungal infections at high temperatures (e.g. flyspeck (*Schizothyrium pomi* (Mont.) v. Arx)).

Depending on the trellis structure, plant density can vary from 400 to 1250 vines ha^{-1} (Debersaques *et al.*, 2019). Planting distances vary from 3–4.5 m within rows and 3.4–4.5 m

Fig. 7.7. Spring frost protection for regions with colder climates. (Photo: F. Debersaques)

Fig. 7.8. Pergola system for kiwiberry production in Portugal. (Photo: P. Latocha)

Fig. 7.9. T-bar system for kiwiberry production in Portugal. (Photo: P. Latocha)

Fig. 7.10. Hedgerow system for kiwiberry production in Germany. (Photo: P. Latocha)

between rows (Hastings and Hale, 2019; Strik and Davis, 2021).

Male plants need to be evenly distributed among female plants in the orchard for pollination. The ratio of male:female plants is commonly 1:8 (Konrad and Willging, 2011; Calvez *et al.*, 2015; Debersaques *et al.*, 2015; Strik and Davis, 2021) or 1:6–8 (Hastings and Hale, 2019).

7.4.3 Growing conditions and maintenance

Kiwiberry vines are vigorous, particularly males, and require regular pruning (Reich, 2004; Hastings and Hale, 2019). To maintain high production, kiwiberry vines are pruned in both summer and winter (Snowball, 1997b; Reich, 2004). While female plants are mainly pruned during winter dormancy, male plants are pruned just after flowering with minimal winter pruning to achieve maximum flowering (Reich, 2004).

For a T-bar system, mature plants should have one or two permanent, straight trunks (more trunks in colder countries), two permanent cordons (leaders) and 8–12 temporary (annual) fruiting laterals (canes), which emerge from the cordons. These canes are trained perpendicular to the trellis wires (Reich, 2004). Canes that fruited in the previous year are removed during winter pruning and selected replacement canes are trained across the trellis to fruit in the current season (Debersaques *et al.*, 2019). Medium-vigour canes are recommended as the optimal wood type so fruit will ripen more evenly (Medic *et al.*, 2021). Summer pruning to maintain an open canopy during the fruit ripening period will improve fruit quality (colour, taste, antioxidant content; Figiel-Kroczyńska *et al.*, 2021).

In a pergola system, with, typically, a continuous overhead canopy, trunks from the vines are trained to a wire along which cordons are grown in both directions. Fruiting canes are trained across the outer wires of the pergola to the centre of the row, with plants nearly meeting in the middle of the rows (Strik and Davis, 2021).

7.4.4 Pollination

Male plants that produce abundant, good quality pollen and flower at the same time as female cultivars are needed for high yields of quality fruit (Latocha *et al.*, 2021). Some male genotypes more effectively pollinate specific female varieties (e.g. greater pollen tube overgrowth, fertilization, fruit set) (Stasiak *et al.*, 2021). Effective pollination is essential for both yield and fruit quality, as seed number affects fruit shape, size and chemical composition (the xenia phenomenon; Tiyayon and Strik, 2003b; Stasiak *et al.*, 2019). A minimum ambient temperature of 14°C is needed for successful pollen tube germination. At the optimum temperature of 25°C, pollen germination may exceed 70% after 24 h (Kim *et al.*, 2016).

Actinidia plants can be both insect- and wind-pollinated. Although *Actinidia* flowers produce no nectar, insects are important pollen vectors. Commercially, wind pollination is generally supplemental to insect pollination (Fraser and McNeilage, 2016). Wind pollination can be effective for kiwiberry when pollinator activity is low (Jackson, 2000). However, other research has found that wind pollination results in a small number of commercially unacceptable fruit (Gonzalez *et al.*, 1998).

A. arguta flowers may be more attractive to insects as they contain more aromatic compounds that attract pollenizers (about 80) than *A. chinensis* var. *deliciosa* (22), *A. kolomikta* (16–17) and *A. polygama* (11) (Matich *et al.*, 2003; Stasiak and Latocha, 2020). Kiwiberry pollen is also a good source of potassium (K) and fat for bees (Ghosh and Jung, 2017). Male *A. arguta* flowers produce viable pollen during the first 2–3 days after anthesis, while female flowers remain receptive for 4–9 days after anthesis (Debersaques *et al.*, 2019; Strik and Davis, 2021).

Under natural conditions, pollen vectors are mainly honey bees (*Apis mellifera* L.) and bumble bees (*Bombus* spp.), although other insects including hoverflies (Syrphidae family), ladybird beetles (Coleoptera order) and solitary bees (Hymenoptera order) may distribute kiwiberry pollen (Howpage *et al.*, 2001; Debersaques *et al.*, 2019). According to Goodwin and Steven (1993), honey bees are not optimal pollinators for kiwiberry because they can show a preference for pistillate flowers. Debersaques *et al.* (2019) concluded that honey bees are more active during midday hours and do not fly in cloudy or rainy weather, while bumble bees are active throughout the day and fly in cooler temperatures and cloudy conditions. However, an excess of bumble bees may result in flower destruction before anthesis or damage to the ovary, resulting in cosmetic fruit skin defects.

Artificial pollination of kiwiberry with *A. chinensis* var. *deliciosa* pollen was less effective than pollination by insects or manual application of *A. arguta* pollen under optimal conditions (Jackson, 2000; Stasiak *et al.*, 2021). However, *A. arguta* pollen is not commercially available and there is a risk of *Pseudomonas syringae* pv. *actinidiae* (Psa) transfer using *A. chinensis* var. *deliciosa* pollen (Tontou *et al.*, 2014). Artificial pollination can improve the balance between fruit quantity and quality (Debersaques *et al.*, 2015) and is the only option for commercial production of some hybrid genotypes in greenhouses (e.g. 'Issai'; Mertens, 2012).

Under optimal conditions, kiwiberry fruit set is high, usually resulting in low fruit weight. Thinning flower buds before flowering increased lateral fruit volume and king fruit volume by 18% and 27%, respectively (Pescie and Strik, 2004). However, such treatment is labour-intensive on an orchard scale.

7.4.5 Irrigation and nutrition

Water supply

Kiwiberry vines have high water requirements during the growing season (Debersaques *et al.*, 2015) and irrigation may be needed to ensure good plant establishment and fruit production (Hastings and Hale, 2019). Water deficiency can cause poor growth, lack of flower production and increased susceptibility to cold damage (Hastings and Hale, 2019), while excess water can cause crown or root rot, especially in young plants (Strik and Davis, 2021).

Plant and soil management (e.g. planting design, summer pruning, mulching) and rainfall determine the frequency and amount of irrigation to apply (Hastings and Hale, 2019). Average requirements for mature vines during the growing season vary from 15 L m^{-2} per week (Debersaques *et al.*, 2015) to 28 L m^{-2} per week (Konrad and Willging, 2011). It is important to begin irrigating when soil-water content falls below 60% of field capacity (Huang, 2016).

Drip irrigation is efficient (Huang, 2016) for young vines before the full lateral root system of mature vines has developed (Debersaques *et al.*, 2015; Hastings and Hale, 2019). Under-vine micro-sprinklers are used to reach the spreading root system of mature vines (Fig. 7.11A). Drip irrigation and micro-sprinklers provide the possibility for simultaneous fertigation (Debersaques *et al.*, 2019).

Where there is a risk of spring frost, an overhead sprinkler system can protect young growth (Fig. 7.11b) (Debersaques *et al.*, 2019; Strik and Davis, 2021). Impact sprinklers are commonly used, but mini- or micro-sprinkler systems may be used (e.g. in windless weather; Hastings and Hale, 2019). Systems may be configured to enable rapid conversion from above-vine for frost protection to under-vine for irrigation (Debersaques *et al.*, 2019). Irrigation water for kiwifruit should have low boron (B) (< 0.25 mg L^{-1}), Cl (< 70 mg L^{-1}) and sodium (< 50 mg L^{-1}) (Strik and Davis, 2021). Excess calcium carbonate can cause fruit spotting when applied through overhead irrigation (Debersaques *et al.*, 2019).

Nutrient management

To ensure a healthy, high-yielding kiwiberry orchard, a sustainable fertilization programme is needed to meet plant requirements (Debersaques *et al.*, 2015). The programme should be based on soil nutrient analysis relative to kiwiberry requirements, with fertilizer making up for any deficiencies. Unfortunately, limited information is available regarding fertilization of kiwiberry vines (Stefaniak *et al.*, 2017a; Vance and Strik, 2018). Therefore, current recommendations are largely based on those for kiwifruit production (Strik and Davis, 2021). However, some nutrient recommendations for kiwiberry have been published (Table 7.2).

Kiwiberry roots, especially in young plants, are very sensitive to fertilizer burn or high salts (electrical conductivity) (Strik and Davis, 2021). Foliar nutrient application of a standard NPK fertilizer with micronutrients can increase B, iron (Fe) and zinc (Zn) concentrations in the leaves and improve yield and storage life (Decorte, 2013; Decorte and Debersaques, 2018).

Late and excessive nitrogen (N) applications can cause vigorous vegetative growth that is unable to harden off before winter (Hastings and Hale, 2019; Strik and Davis, 2021). Split applications with the last N application prior to (Debersaques *et al.*, 2012, 2019) or just after anthesis (Whitworth, 2017) are recommended. Also recommended are split applications of equal amounts of N throughout the season, with rates varying with vine age (Hastings and Hale, 2019).

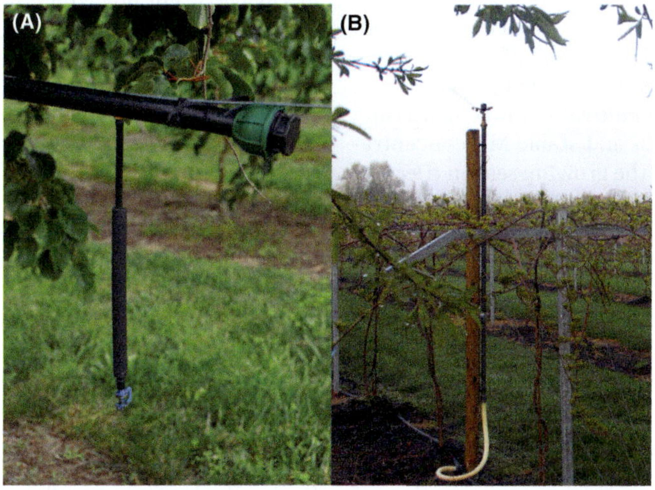

Fig. 7.11. (**A**) Under-vine micro-sprinkler. (**B**) Overhead sprinkler for frost protection. (Photo: P. Latocha and F. Debersaques)

Table 7.2. Nutrient recommendations for young and mature kiwiberry vines (kg ha^{-1}).

Nitrogen		Phosphorus		Potassium		Manganese		
Young	Mature	Young	Mature	Young	Mature	Young	Mature	References
60	80	20	50	30	110			Debersaques *et al.* (2012)
50	50	10	30	70	70	5	10	Konrad and Willging (2011)
		60	65	90	150			Hastings and Hale (2019)

There is little information on kiwiberry micronutrient requirements and toxicity. Some reports suggest sensitivity of the crop to high B (Sotiropoulos *et al.*, 2002) and high Cl (Strik and Davis, 2021). Others suggest low concentrations of Cl are beneficial for the vegetative growth of *A. arguta* (Debersaques *et al.*, 2011). Decorte *et al.* (2015) found varietal differences in Cl requirements.

Based on available research, kiwiberry is very sensitive to manganese (Mn) deficiency (Decorte *et al.*, 2018). The same authors observed varietal differences in foliar magnesium (Mg) and phosphorus (P) concentrations, suggesting that there are opportunities at the cultivar level for fine tuning of fertilization management.

Optimal leaf nutrient ranges for kiwiberry pre-flowering, from flowering until 50 days after full flowering and from 50 days after full flowering until harvest have been published (Decorte *et al.*, 2018). Compared with kiwifruit, there is a higher B, higher Cl and lower Zn optimal level in kiwiberry leaves. Differences in seasonal time courses of a range of nutrients in leaves of kiwiberry cultivars showed increasing copper (Cu) and Zn and stable Mn concentrations at the end of the growing season (Decorte *et al.*, 2015). Leaf N and K concentrations decreased and calcium (Ca) and Mg concentrations increased throughout the season (Stefaniak *et al.*, 2019). For best nutrient management practices, Vance and Strik (2018) recommended maintaining detailed annual data on leaf tissue analysis, as well as yield results and plant observations (shoot growth, leaf deficiency and toxicity symptoms).

Latocha *et al.* (2015a) reported lower K and Cu concentrations and higher concentrations of P, Ca, Zn and Fe for ripe kiwiberry fruit than 'Hayward' kiwifruit.

7.4.6 Pests and diseases

In countries where kiwifruit is grown, specific pests and diseases are found in both kiwifruit and kiwiberry. The major pests and fungal diseases of kiwiberry are given in Tables 7.3 and 7.4, respectively. In countries where kiwiberry is a recent crop, vines are relatively free of pests and diseases (Debersaques *et al.*, 2012) and are suitable for organic cultivation.

Considerable attention is being given to bacterial diseases such as Psa. To date, Psa has had no real economic impact on kiwiberry (Vanneste *et al.*, 2014). However, Psa strains have been isolated from kiwiberry plants, so a potential risk exists (Liu *et al.*, 2016). Polyploid selections of *A. arguta* showed reduced susceptibility to Psa compared with diploid selections (Cotruţ *et al.*, 2013).

There is limited literature on viral diseases in kiwiberry, but one study reported cucumber necrosis virus was found in *A. arguta* plants (Blouin *et al.*, 2013).

7.4.7 Harvest and postharvest handling

Harvesting at the correct time is essential for optimal fruit quality and storage (Fisk *et al.*, 2006a). The lack of uniformity in the ripening of fruit within a vine (Williams *et al.*, 2003; Strik and Davis, 2021) poses a challenge to producers and marketers alike (Hastings and Hale, 2019).

For most kiwiberry cultivars, there are no clear morphological parameters such as colour that reliably indicate fruit maturity (Strik and Davis, 2021). This complicates identifying the optimum harvest time, which, in turn, affects the storage, shelf life and taste of the fruit (Oh

Table 7.3. Common insect pests of kiwiberry.

Pest	Organism	Symptoms	Possible economic impact	References
Armoured scale	*Hemiberlesia lataniae*; *Hemiberlesia rapax*	Reduced plant vigour	Reduced yield and fruit quality (Quarantine pest in New Zealand)	Mauchline and Hill (2005)
Cicadas	*Amphipsalta zelandica*	Linked to sooty mould	Reduced vine growth and health	McKenna *et al.* (2009)
Japanese beetle	*Popillia japonica*	Feeding damage on foliage, immature and ripe berries	Can affect fruit skin quality	Knodel *et al.* (2017); Hastings and Hale (2019)
Leafroller	*Ctenopseustis* sp.; *Cnephasia jactatana*	Mis-shapen young fruit	Fruit quality issue (Quarantine pest in New Zealand)	Steven (1992); McKenna *et al.* (2009)
Mealybug	*Pseudococcus longispinus*	Honeydew	Fruit quality issue (Quarantine pest in New Zealand)	McKenna *et al.* (2009); Martin (2019)
Passionvine hopper	*Scolypopa australis*	Linked to sooty mould	Fruit quality issue	Logan *et al.* (2002); McKenna *et al.* (2009); MAF (2010)
Spotted wing drosophila	*Drosophila suzukii*	Larvae feed on the pulp	Not a problem when the fruit is harvested firm	Debersaques *et al.* (2019)
Thrips	*Heliothrips haemorrhoidalis*	Fruit skin damaged by superficial silver or bronze scarring	Fruit quality issue	McKenna *et al.* (2009)

Table 7.4. Common fungal diseases of kiwiberry.

Fungal disease	Organism	Symptoms	Possible economic impact	References
Anthracnose	*Colletotrichum gloeosporioides*	Watery lesions, developing into brown or black spots and leaf drop	A decline in fruit quality and yield	Deng *et al.* (2017)
Armillaria	*Armillaria mellea*	Interwoven filamentous mycelium through darkened cortical tissue	Can be fatal to vines	Huang (2016)
Black rot	*Alternaria alternata*	Black spots around the calyx of fruit	Not common	Kim *et al.* (2017)
Brown spot	*Alternaria alternata*	Fruit brown spot	Reduced yield and marketability	Ruan *et al.* (2022)
Flyspeck	*Schizothyrium pomi*	Patches of discrete small (< 0.5 mm) black spots on the fruit skin	Fruit quality issue	McKenna *et al.* (2012); Debersaques *et al.* (2019)
Sooty mould/ sooty splotch	*Peltaster fructicola*; *Geastrumia polystigmatis*	Black powdery coating on the surface of fruit	Fruit quality issue	McKenna *et al.* (2009); McKenna *et al.* (2012)
Grey mould	*Botrytis cinerea*	Surface depressions, flesh decay, deformation and drying	Fruit quality issue	Romanazzi (2009)
Leaf spot and blight	*Pestalotiopsis bicilia*	Irregular brown spots and blight symptoms on the margins of infected leaves	No data available	Kim *et al.* (2017)
Phytophthora	*Phytophthora* spp.	Root rot, collar rot or crown rot	Can be fatal to vines	Latorre *et al.* (1991); Akıllı *et al.* (2011)
Powdery mildew	*Uncinula actinidiae*	Circular whitish powdery mycelia and conidia on leaves	No data available	Kim *et al.* (2017)
Sclerotinia rot	*Sclerotinia nivalis*	Begins as black-brown watery soft rot expanding to girdling of basal stems	Can be fatal to vines	Wang *et al.* (2016)
Verticillium wilt	*Verticillium albo-atrum*	Reddish-brown discolouration of the xylem, followed by sudden wilting and dieback of plants	Can be fatal to vines	Auger *et al.* (2009)

et al., 2017). Fruit harvested too early does not store well or achieve a sufficiently high SSC when ripe. Fruit harvested too late is too soft and is susceptible to postharvest handling damage (Williams *et al.*, 2003). The change in seed colour from white to brown/black is a marker of the physiological maturity of the fruit (Latocha *et al.*, 2021) and is a minimum prerequisite for harvesting (Hastings and Hale, 2019). After this coloured-seed stage, the fruit SSC is used as a harvest-time determinant with a range of values between 6.5 and 14 °Brix suggested (Fisk *et al.*, 2006a; Debersaques *et al.*, 2015, 2019; Han *et al.*, 2019; Strik and Davis, 2021). Other harvest indicator methods, such as iodine staining of starch, proved impractical for growers (Fisk *et al.*, 2006b). At 6.5–8 °Brix SSC the skin and pulp colour of red kiwiberry cultivars may not have fully developed and delaying harvest is suggested until fully coloured (Hassal *et al.*, 1998). However, this delay can allow fruit softening, resulting in reduced long-term storage and shelf life.

A more recent development has been the extension of the understanding of the relationship between at-harvest dry matter and ripe fruit SSC, previously described for kiwifruit, to give an at-harvest estimate of the consumer acceptability of the fruit (Hunter *et al.*, 2020) (see Chapter 21).

Other biochemical changes in the fruit have been investigated as harvest indicators (Kim *et al.*, 2014; Oh *et al.*, 2017), although these parameters proved less applicable in practice than in the laboratory. More recently, Mendes da Silva *et al.* (2020) suggested more study be undertaken on composite colorimetric parameters (colour changes). Combinations of these parameters appear to be more sensitive to changes in fruit ripening and less cultivar-specific than chemical composition and can potentially be used as new harvesting and maturity quality indicators. However, this research is in its early stages and the development of the harvesting index requires further research on a larger number of samples.

Giuggioli *et al.* (2021) determined that the perception of taste and the kiwiberry sensory attributes are key factors for consumer satisfaction. They proposed the first version of a standardized sensory terminology for kiwiberry to distinguish between high- and low-quality fruit in terms of sensory attributes in relation to a maturity index.

Kiwiberry harvest is labour-intensive because either each individual fruit is harvested from the vine by hand (Strik and Davis, 2021) or individual fruit are cut from the shoots of fruiting branches that had been removed from the vine and taken to a cooler place (Debersaques *et al.*, 2019).

Kiwiberry fruit have a short storage life with rapid softening and desiccation being the most challenging commercial limitations (Hassal *et al.*, 1998; Fisk *et al.*, 2006a; Latocha *et al.* 2014; Strik and Davis, 2021). Kiwiberry storability (up to 4–8 weeks) is much shorter than that of kiwifruit (up to 6 months). The fast postharvest decline in kiwiberry may also reduce nutritional values (Kim, A.-N. *et al.*, 2018).

As with most fruit, temperature plays a key role in kiwiberry perishability. After picking, the fruit must be cooled as quickly as possible to remove field heat and maximize storage life (Strik and Davis, 2021). Shock cooling (cooled to 3–4°C within 2 h of harvest) of kiwiberry improves the storage quality and extends shelf life (Figiel-Kroczyńska and Ochmian, 2021). For optimal storage life, good quality pre-cooled berries should be stored at 0–2°C with a relatively humidity of 90% or higher (Latocha *et al.*, 2021). Cold storage of still-firm fruit harvested early but mature (i.e. brown/black seeds, 6.5–7 °Brix) slows down the ripening process, resulting in fruit with the desired taste and aroma (Latocha *et al.*, 2014). During cold storage in air, fruit firmness will decrease rapidly within the first seven days (Krupa *et al.*, 2011). For the most effective long-term storage, refrigeration supplemented by a controlled atmosphere of low oxygen (O_2) and high carbon dioxide (CO_2) is necessary. Kiwiberries can be successfully stored at 1.5% O_2 and 1.5% CO_2 for five or even eight weeks depending on the cultivar (Krupa *et al.*, 2011; Latocha *et al.*, 2014; Szpadzik *et al.*, 2021). A higher CO_2 concentration (5–10%) can prolong storage up to 12 weeks (Krupa and Tomala, 2021), although CO_2 > 15% can reduce fruit quality.

Treatments applied prior to cold storage, including organic acids (oxalic, salicylic, acetylsalicylic) and calcium chloride, resulted in minor improvements in fruit storability but some did affect fruit chemical composition (e.g. total phenolics and vitamin C content) (Stefaniak *et al.*, 2017b). Treatment with 1-methylcyclopropene (1-MCP) significantly improved kiwiberry storability (Fisk *et al.*, 2008; Wang, R.C. *et al.*, 2015; Wang, Y. *et al.*, 2015; Stefaniak *et al.*, 2017b).

Water loss resulting in fruit shrivel is a major postharvest risk. Lidded plastic polyethylene terephthalate (PET) clamshells extended fruit storage by as much as 60 days compared with those without lids (Giuggioli *et al.*, 2019a), while kiwiberry could be stored for 7–10 weeks in high-vent clamshell containers or 14 weeks in low-vent packaging combined with an edible coating (Fisk *et al.*, 2008). Further positive effects of edible coatings of packaging on fruit storability have been reported (Strik and Hummer, 2006; Kim, J.G. *et al.*, 2018; Jamróz *et al.*,

2019) while not affecting consumer acceptance (Fisk *et al.*, 2008).

While in warehouse storage, covering pallets of fruit with stretch-wrap film helped maintain fruit quality, particularly water content and textural properties (Giuggioli *et al.*, 2019b). Mathematical models to predict the shelf life of kiwiberries stored under different conditions, considering relevant pre-harvest factors, have been investigated (Giuggioli *et al.*, 2017). However, varietal differences make this difficult and, to date, there is no universal model for determining kiwiberry shelf life.

Developing new postharvest fruit treatment (e.g. 1-MCP) or packaging technologies (e.g. controlled and modified atmosphere packaging or other active packaging systems including the application of chemical sensors and biosensors) extends the shelf life of fruit and increasingly enables transport over longer distances (Kirtil and Oztop, 2016; Kuswandi and Moradi, 2019).

Little research exists on kiwiberry storage diseases, although limited reports suggest considerable spoilage is caused by *Botryosphaeria dothidea*, *Diaporthe actinidiae* and *Botrytis cinerea* (Koh et al., 2003) and the occurrence of postharvest fruit rot on kiwiberry at the picking scar (Kim *et al.*, 2017).

7.5 Fruit Quality and Bioactive Composition

Fruit quality is largely determined by the physical parameters and chemical composition of the fruit. The relevant physical parameters include fruit shape, size, skin quality, colour and firmness, while key chemical components include sugar content, acidity and concentration of bioactive ingredients, all of which depend on genetic characteristics (specific for cultivar) and cultivation practices. Russeting is a common problem when growing in changing weather conditions (e.g. temperature fluctuations during early fruit development). Like kiwifruit, kiwiberries are a valuable source of nutrients that are important for our health and well-being (Latocha, 2017; Pinto *et al.*, 2020; Latocha *et al.*, 2021).

Kiwiberry is a good source of beneficial vitamins, minerals and compounds essential to good health (reviewed by Latocha *et al.*, 2021). These include vitamins A, C and B_3, *myo*-inositol, lutein, beta-carotene, chlorophyll and phenolics. Kiwiberry is also a good source of K, Ca, P, Mg, Fe, Zn, Cu and Mn, with higher concentrations in red fruit than green fruit (Skrypchenko and Moroz, 2002). An advantage of kiwiberry over kiwifruit is its delicate edible skin, which can contain up to three times more vitamin C and up to 15 times more phenolic compounds than the flesh (Kim *et al.*, 2009; Latocha *et al.*, 2015b). Kiwiberry also provides soluble and insoluble dietary fibre and a range of 19 amino acids. Therefore, kiwiberry is often considered as a 'superfood' or 'health food'.

In addition to healthful compounds, kiwiberry, like kiwifruit, contains compounds considered harmful – mostly oxalate in both soluble and insoluble forms. The oxalate concentration in kiwiberry is cultivar-dependent (Zhang *et al.*, 2021) and can be lower, similar or higher than in kiwifruit (Watanabe and Takahashi, 1998; Rassam *et al.*, 2007; Nguyen and Savage, 2013).

7.6 Uses

A. arguta has only recently been cultivated commercially in many countries around the world (Latocha *et al.*, 2018). Due to its costly, labour-intensive cultivation, current commercial production is focused on supplying fresh fruit as this form best demonstrates the nutritional and taste qualities of kiwiberry. In addition to direct consumption as fresh fruit, kiwiberries can be added to various dishes or salads. Unfortunately, the availability of fresh kiwiberries is limited by their short season and short storage life.

As with other fruits, some fruit is deformed or damaged, sometimes comprising up to 40% of total production (Hastings and Hale, 2019) and could be used for processed products. These include preserves (juices, jams, compotes), fillings for candies or chocolates, additions to fruit yoghurt and dried forms (Wang and Li, 2018) (Fig. 7.12). As examples, fortified drinks, dried fruits, fruit vinegars and fruit powders are already produced in China (Liu and Liu, 2015; Wang and Li, 2018).

Juices and dried fruit are the most common processed kiwiberry products. Kiwiberry juice has an intense flavour and aroma, with

Fig. 7.12. Kiwiberry processed products. (Photo: P. Latocha)

palatability characteristics that may limit its acceptability (Kostyra *et al.*, 2021). Thus, kiwiberry juice products are either diluted (smoothie or nectar) or mixed with juices from other widely available fruits. The juice blends preferred by consumers were kiwiberry with > 40% apple and kiwiberry with > 30% pear (Kostyra *et al.*, 2021).

Kiwiberries are often a component of Chinese and Korean traditional medicine treatments for digestive problems, dyspepsia, rheumatism, haemorrhoids, diuresis, diabetes mellitus and osteoporosis (Park *et al.*, 2011; Ravipati *et al.*, 2012). Despite this pedigree in traditional medicine, there are few published studies on the effects of kiwiberry on human health. To date, kiwiberry has been reported to exhibit anti-inflammatory (Choi *et al.*, 2008; An *et al.*, 2016), anti-tumour (Zuo *et al.*, 2012; Nishimura *et al.*, 2016), anti-fatigue (Liu and Liu, 2016), anti-amnesic (Ha *et al.*, 2015), anti-oxidant (An *et al.*, 2016; Leontowicz *et al.*, 2016a; Latocha, 2017) and anti-microbial (Wang *et al.*, 2022) properties, as well as showing some positive effects against hypercholesterolaemia (Leontowicz *et al.*, 2016b). Recent studies indicate that kiwiberry may slow the progression of neurodegeneration in Alzheimer's and Parkinson's diseases (Ha *et al.*, 2020; Kitamura *et al.*, 2021).

7.7 Future Perspective

As production and sales volumes of kiwiberry have increased substantially over the last decade, its transition from a marginal, novelty fruit to a commercial crop appears very likely. Consequently, the increasing quantity of kiwiberry fruit on the market requires intensive promotion among consumers in order to maintain a balance in supply and demand. Although kiwiberry is often consumed locally, new postharvest technologies now allow long-distance transport.

Further work will be required in terms of variety improvement. Breeding targets should include reducing the large variation in picking time, delaying budbreak, increasing spring frost resistance as well as drought and heat tolerance.

Continued efforts will need to be made to further develop the market and product utilization, in addition to the work underway to better understand all stages of kiwiberry production, harvesting and processing.

References

Akıllı, S., Serçe, Ç.U., Katırcıoğlu, Y.Z., Karakaya, A. and Maden, S. (2011) Involvement of *Phytophthora citrophthora* in kiwifruit decline in Turkey. *Journal of Phytopathology* 159, 579–581. DOI: 10.1111/j.1439-0434.2011.01798.x

An, X., Lee, S.G., Kang, H., Heo, H.J., Cho, Y.-S. *et al.* (2016) Antioxidant and anti-inflammatory effects of various cultivars of kiwi berry (*Actinidia arguta*) on lipopolysaccharide-stimulated RAW 264.7 cells. *Journal of Microbiology and Biotechnology* 26, 1367–1374. DOI: 10.4014/jmb.1603.03009

Auger, J., Pérez, I., Fullerton, R.A. and Esterio, M. (2009) First report of verticillium wilt of gold kiwifruit, *Actinidia chinensis* cv. Hort 16A, caused by *Verticillium albo-atrum* in Chile. *Plant Disease* 93, 553. DOI: 10.1094/PDIS-93-5-0553B

Bieniek, A. (2012) Yield, morphology and biological value of fruits of *Actinidia arguta* and *Actinidia purpurea* and some of their hybrid cultivars grown in north-eastern Poland. *Acta Scientiarum Polonorum Hortorum Cultus* 11, 117–130. Available at: https://czasopisma.up.lublin.pl/index.php/asphc/article/view/3088 (accessed 11 July 2023).

Blouin, A.G., Pearson, M.N., Chavan, R.R., Woo, E.N.Y., Lebas, B.S.M. *et al.* (2013) Viruses of kiwifruit (*Actinidia* species). *Journal of Plant Pathology* 95, 221–235. DOI: stable/23721513

Calvez, E., Neuberger, S., Wagner, K., Schoedl-Hummel, K. and Debersaques, F. (2015) Management strategies for minikiwi (*Actinidia arguta*): practical experiences of European growers. *Acta Horticulturae* 1096, 451–454. DOI: 10.17660/ActaHortic.2015.1096.54

Choi, J.J., Park, B., Kim, D.H., Pyo, M.-Y., Choi, S. *et al.* (2008) Blockade of atopic dermatitis-like skin lesions by DA-9102, a natural medicine isolated from *Actinidia arguta*, in the Mg-deficiency induced dermatitis model of hairless rats. *Experimental Biology and Medicine* 233, 1026–1034. DOI: 10.3181/0801-RM-19

Cossio, F., Debersaques, F. and Latocha, P. (2015) Kiwiberry (*Actinidia arguta*): new perspectives for a great future. *Acta Horticulturae* 1096, 423–434. DOI: 10.17660/ActaHortic.2015.1096.51

Cotruţ, R., Renzi, M., Taratufolo, M.C., Mazzaglia, A., Balestra, G. *et al.* (2013) *Actinidia arguta* ploidy level variation in relation to *Pseudomonas syringae* pv. *actinidiae* susceptibility. *Lucrări Ştiinţifice, Universitatea de Ştiinţe Agricole si Medicină Veterinară 'Ion Ionescu de la Brad' Iaşi, Iaşi, Romania. Seria Agronomie* 56, 29–38.

Debersaques, F. and Mekers, O. (2011) Influence of chlorine on shoot growth of kiwiberry (*Actinidia arguta* (Sieb. et Zucc.) Planch. ex Miq.). *Acta Horticulturae* 913, 345–351. DOI: 10.17660/ActaHortic. 2011.913.45

Debersaques, F., Mekers, O. and De Kezel, C. (2012) *Van Kiwi tot Kiwibes: Telershandleiding*. Hogeschool Gent, Gent, Belgium. [In Dutch]

Debersaques, F., Mekers, O., Decorte, J., Van Labeke, M.C., Schoedl-Hummel, K. *et al.* (2015) Challenges faced by commercial kiwiberry (*Actinidia arguta* Planch.) production. *Acta Horticulturae* 1096, 435–442. DOI: 10.17660/ActaHortic.2015.1096.52

Debersaques, F., Latocha, P. and Decorte, J. (2019) *Kiwiberry* Actinidia arguta *Grower's Manual*. Polish Dendrology Society, Warsaw. ISBN 978-83-938299-7-2

Decorte, J. (2013) Evolutie van de vruchtgroei, bladen vruchtsamenstelling van kiwibes (*Actinidia arguta* Planch). MS thesis, Ghent University College, Gent, Belgium. [In Dutch]

Decorte, J. and Debersaques, F. (2018) Positive effects of foliar nutrition on yield and storage life of kiwiberry. *Acta Horticulturae* 1218, 579–585. DOI: 10.17660/ActaHortic.2018.1218.79

Decorte, J., Debersaques, F. and Mekers, O. (2015) Seasonal patterns of leaf mineral content of kiwiberry cultivars. *Acta Horticulturae* 1096, 471–477. DOI: 10.17660/ActaHortic.2015.1096.58

Decorte, J., Vereecke, D. and Debersaques, F. (2018) Mineral status of kiwiberry. *Acta Horticulturae* 1218, 435–441. DOI: 10.17660/ActaHortic.2018.1218.60

Deng, J.C., Guan, Y.M., Wu, L.J. and Zhang, Y.Y. (2017) First report of anthracnose caused by *Colletotrichum gloeosporioides* on *Actinidia arguta* in China. *Plant Disease* 101, 1032. DOI: 10.1094/PDIS-01-16-0103-PDN

Ferguson, A.R. (1984) Kiwifruit: a botanical review. *Horticultural Reviews* 6, 1–64. DOI: 10.1002/9781118060797.ch1

Figiel-Kroczyńska, M. and Ochmian, I. (2021) Effect on phytochemical content and microbial contamination of Actinidia fruit after shock cooling and storage. *Acta Universitatis Cibiniensis. Series E: Food Technology* 25, 155–166. DOI: 10.2478/aucft-2021-0015

Figiel-Kroczyńska, M., Ochmian, I., Lachowicz, S., Krupa-Małkiewicz, M., Wróbel, J. *et al.* (2021) Actinidia (mini kiwi) fruit quality in relation to summer cutting. *Agronomy* 11(5), 964. DOI: 10.3390/agronomy11050964

Fisk, C.L., McDaniel, M.R., Strik, B.C. and Zhao, Y. (2006a) Physicochemical, sensory, and nutritive qualities of hardy kiwifruit (*Actinidia arguta* 'Ananasnaya') as affected by harvest maturity and storage. *Journal of Food Science* 71, S204–S210. DOI: 10.1111/j.1365-2621.2006.tb15642.x

Fisk, C.L., Strik, B.C. and Zhao, Y. (2006b) Iodine staining does not indicate harvest maturity of 'Ananasnaya' hardy kiwifruit berries. *HortTechnology* 16(4), 655–658. DOI: 10.21273/HORTTECH.16.4.0655

Fisk, C.L, Silver, A.M., Strik, B.C. and Zhao, Y. (2008) Postharvest quality of hardy kiwifruit (*Actinidia arguta* 'Ananasnaya') associated with packaging and storage conditions. *Postharvest Biology and Technology* 47, 338–345. DOI: 10.1016/j.postharvbio.2007.07.015

Fraser, L.G. and McNeilage, M.A. (2016) Reproductive biology. In: Testolin, R., Huang, H.-W. and Ferguson, A.R. (eds) *The Kiwifruit Genome. Compendium of Plant Genomes.* Springer Nature, Cham, Switzerland, pp. 65–84. DOI: 10.1007/978-3-319-32274-2_6

Ghosh, S. and Jung, C. (2017) Nutritional value of bee-collected pollens of hardy kiwi, *Actinidia arguta* (Actinidiaceae) and oak, *Quercus* sp. (Fagaceae). *Journal of Asia-Pacific Entomology* 20, 245–251. DOI: 10.1016/j.aspen.2017.01.009

Giuggioli, N.R., Girgenti, V., Baudino, C., Giacalone, G. and Chiabrando, V. (2017) Prediction models for the softening of new commercial *Actinidia* cultivars. *Journal of Food Processing and Preservation* 41(6), e13245. DOI: 10.1111/jfpp.13245

Giuggioli, N.R., Baudino, C., Briano, R. and Peano, C. (2019a) Quality of packed baby kiwi cultivar 'Hortgem Tahi®' and 'Hortgem Rua®'. *Progress in Nutrition* 21(2), 440–448. DOI: 10.23751/pn.v21i2.7042

Giuggioli, N.R., Briano, R., Baudino, C. and Peano, C. (2019b) Post-harvest warehouse management of *Actinidia arguta* fruits. *Polish Journal of Food and Nutrition Sciences* 69, 63–70. DOI: 10.31883/pjfns-2019-0006

Giuggioli, N.R., Peano, C., Brondino, L. and Mendes da Silva, T. (2021) Qualitative preliminary approach for the development of a sensory vocabulary for *Actinidia arguta* fruits. *Applied Sciences* 11(20), 9361. DOI: 10.3390/app11209361

Gonzalez, M.V., Coque, M. and Herrero, M. (1998) Influence of pollination systems on fruit set and fruit quality in kiwifruit (*Actinidia deliciosa*). *Annals of Applied Biology* 132, 349–355. DOI: 10.1111/j.1744-7348.1998.tb05210.x

Goodwin, R.M. and Steven, D. (1993) Behaviour of honey bees visiting kiwifruit flowers. *New Zealand Journal of Crop and Horticultural Science* 21, 17–24. DOI: 10.1080/01140671.1993.9513741

Ha, J.S., Jin, D.E., Park, S.K., Park, C.H., Seung, T.W. *et al.* (2015) Antiamnesic effect of *Actinidia arguta* extract intake in a mouse model of TMT-induced learning and memory dysfunction. *Evidence-based Complementary and Alternative Medicine* 2015, 876484. DOI: 10.1155/2015/876484

Ha, J.S., Kang, J.Y., Kang, J.E., Park, S.K., Kim, J.M. *et al.* (2020) Pentacyclic triterpenoid-rich fraction of the hardy kiwi (*Actinidia arguta*) improves brain dysfunction in high fat diet-induced obese mice. *Scientific Reports* 10, 5788. DOI: 10.1038/s41598-020-62810-5

Han, N., Park, H., Kim, C.-W., Kim, M.-S. and Lee, U. (2019) Physiochemical quality of hardy kiwifruit (*Actinidia arguta* L. cv. Cheongsan) during ripening is influenced by harvest maturity. *Forest Science and Technology* 15, 187–191. DOI: 10.1080/21580103.2019.1658646

Harker, F.R., Jaeger, S.R., Lau, K. and Rossiter, K. (2007) Consumer perceptions and preferences for kiwifruit: a review. *Acta Horticulturae* 753, 81–88. DOI: 10.17660/ActaHortic.2007.753.7

Hassal, A.K., Pringle, G.J. and MacRae, E.A. (1998) Development, maturation, and postharvest responses of *Actinidia arguta* (Sieb. et Zucc.) Planch. ex Miq. fruit. *New Zealand Journal of Crop and Horticultural Science* 26, 95–108. DOI: 10.1080/01140671.1998.9514046

Hastings, W. and Hale, I. (2019) Growing kiwiberries in New England: a guide for regional producers. Available at: http://www.noreastkiwiberries.com (accessed 30 November 2021).

Howpage, D., Spooner-Hart, R.N. and Vithanage, V. (2001) Influence of honey bee (*Apis mellifera*) on kiwifruit pollination and fruit quality under Australian conditions. *New Zealand Journal of Crop and Horticultural Science* 29, 51–59. DOI: 10.1080/01140671.2001.9514160

Huang, H. (2016) *Kiwifruit: The Genus ACTINIDIA*. Elsevier, London.

Huang, H., Abid, M., Lin, M., Wang, R., Gu, H. *et al.* (2021) Comparative transcriptome analysis of different *Actinidia arguta* fruit parts reveals difference of light response during fruit coloration. *Biology* (Basel) 10(7), 648. DOI: 10.3390/biology10070648

Hughes, K.A., de Willigen, P., Gandar, P.W. and Clothier, B.E. (1992) Kiwifruit root system: structure & function. *Acta Horticulturae* 297, 383–389. DOI: 10.17660/ActaHortic.1992.297.51

Hunter, D.C., Pidakala, P.P.B., Baylis, E.M., Wohlers, M.W., Barnett, A.M. *et al.* (2020) Physico-chemical attributes influence consumer preferences for kiwiberries (*Actinidia arguta* 'Hortgem Tahi'). *New Zealand Journal of Crop and Horticultural Science* 48, 143–152. DOI: 10.1080/01140671.2020.1759111

Jackson, D. (2000) Sare Grant Project no. FNE00-307: Hardy kiwi pollenation and production. Final report. Available at: https://cdn.sare.org/wp-content/uploads/20171204134814/904717pdf-of-final-report-307.pdf (accessed 30 November 2021).

Jamróz, E., Kopel, P., Juszczak, L., Kawecka, A., Bytešníková, Z. *et al.* (2019) Development of furcellaran-gelatin films with Se-AgNPs as an active packaging system for extension of mini kiwi shelf life. *Food Packaging and Shelf Life* 21, 100339. DOI: 10.1016/j.fpsl.2019.100339

Jo, Y.S., Ma, K.C., Cho, H.S., Park, J.O., Kim, S.C. *et al.* (2007) 'Chiak', a new selection of *Actinidia arguta*. *Acta Horticulturae* 753, 259–262. DOI: 10.17660/ActaHortic.2007.753.31

Kataoka, I., Mizugami, T., Kim, J.G., Beppu, K., Fukuda, T.S. *et al.* (2010) Ploidy variation of hardy kiwifruit (*Actinidia arguta*) resources and geographic distribution in Japan. *Scientia Horticulturae* 124, 409–414. DOI: 10.1016/j.scienta.2010.01.016

Kawagoe, T. and Suzuki, N. (2004) Cryptic dioecy in *Actinidia polygama*: a test of the pollinator attraction hypothesis. *Canadian Journal of Botany* 82(2), 214–218. DOI: 10.1139/b03-150

Kim, A.-N., Kim, H.-J., Chun, J., Heo, H.J., Kerr, W.L. *et al.* (2018) Degradation kinetics of phenolic content and antioxidant activity of hardy kiwifruit (*Actinidia arguta*) puree at different storage temperatures. *LWT – Food Science and Technology* 89, 535–541. DOI: 10.1016/j.lwt.2017.11.036

Kim, C.-W., Oh, S.-I., Kim, M.-J. and Park, Y. (2014) Optimal harvest time by the seasonal fruit quality and ripening characteristics of hardy kiwifruit in Korea. *Journal of Korean Society of Forest Science* 103(3), 353–358. DOI: 10.14578/JKFS.2014.103.3.353

Kim, G.H., Kim, D.R., Park, S.-Y., Lee, Y.S., Jung, J.S. *et al.* (2017) Incidence rates of major diseases of kiwiberry in 2015 and 2016. *The Plant Pathology Journal* 33(4), 434–439. DOI: 10.5423/PPJ.NT.01.2017.0010

Kim, J.G., Beppu, K. and Kataoka, I. (2009) Varietal differences in phenolic content and astringency in skin and flesh of hardy kiwifruit resources in Japan. *Scientia Horticulturae* 120, 551–554. DOI: 10.1016/j.scienta.2008.11.032

Kim, J.G., Beppu, K. and Kataoka, I. (2012) Physical and compositional characteristics of 'mitsuko' and local hardy kiwifruits in Japan. *Horticulture, Environment, and Biotechnology* 53(1), 1–8. DOI: 10.1007/s13580-012-0066-7

Kim, J.G., Park, Y., Shin, M.H., Muneer, S., Lerud, R. *et al.* (2018) Application of NIR-Spectroscopy to predict the harvesting maturity, fruit ripening and storage ability of Ca-chitosan treated baby kiwifruit. *Journal of Stored Products and Postharvest Research* 9(4), 44–53. DOI: 10.5897/JSPPR2018.0257

Kim, J.-H., Park, Y. and Kim, S.-H. (2016) Flowering time, amount of pollen and characteristics of pollen germination of *A. arguta* male flower. *Journal of Apiculture* 31(4), 373–378. DOI: 10.17519/apiculture.2016.11.31.4.373l

Kirtil, E. and Oztop, M.H. (2016) Controlled and modified atmosphere packaging. In: *Reference Module in Food Science*. Elsevier. DOI: 10.1016/B978-0-08-100596-5.03376-X

Kitamura, Y., Sakanashi, M., Ozawa, A., Saeki, Y., Nakamura, A. *et al.* (2021) Protective effect of *Actinidia arguta* in MPTP-induced Parkinson's disease model mice. *Biochemical and Biophysical Research Communications* 555, 154–159. DOI: 10.1016/j.bbrc.2021.03.116

Knodel, J.J., Elhard, C. and Beauzay, P.B. (2017) Integrated pest management of Japanese beetle in North Dakota. E1631 (revised) North Dakota State University Extension Service. Available at: https://www.ag.ndsu.edu/publications/crops/integrated-pest-management-of-japanese-beetle-in-north-dakota (accessed 30 November 2021).

Koh, Y.-J., Lee, J.G., Hur, J.-S. and Jung, J.-S. (2003) Incidences and causal agents of postharvest fruit rots in kiwifruits in Korea. *Research in Plant Disease* 9, 196–200. DOI: 10.5423/RPD.2003.9.4.196 [In Korean]

Kolbasina, E.I., Solovyova, L.V., Tulnova, N.N., Kozak, N.V., Skripchenko, N.V. *et al.* (2008) *Cultured Flora of Russia: Volume Actinidia. Schisandra*. GNU MOS VSTISP Russian Agricultural Academy, Moscow. [In Russian]

Konrad, P. and Willging, C. (2011) Kulturblatt Mini-kiwi. Bildungs- und Beratungszentrum, Arenenberg. Gemüse- und Beerenbau. Available at: https://www.srf.ch/sendungen/content/download/7798945/file/KB_Mini-Kiwi.pdf (accessed 30 November 2021).

Kostyra, E., Król, K., Knysak, D., Piotrowska, A., Żakowska-Biemans, S. *et al.* (2021) Characteristics of volatile compounds and sensory properties of mixed organic juices based on kiwiberry fruits. *Applied Sciences* 11, 529. DOI: 10.3390/app11020529

Krupa, T., Latocha, P. and Liwińska, A. (2011) Changes of physicochemical quality, phenolics and vitamin C content in hardy kiwifruit (*Actinidia arguta* and its hybrid) during storage. *Scientia Horticulturae* 130(2), 410–417. DOI: 10.1016/j.scienta.2011.06.044

Krupa, T. and Tomala, K. (2021) Effect of oxygen and carbon dioxide concentration on the quality of minikiwi fruits after storage. *Agronomy* 11, 2251. DOI: 10.3390/agronomy11112251

Kuswandi, B. and Moradi, M. (2019) Sensor trends in beverages packaging. In: Grumezescu, A.M. and Holban, A.M. (eds) *Trends in Beverage Packaging 16: The Science of Beverages*. Woodhead Publishing, Elsevier, Duxford, UK, pp. 279–302. DOI: 10.1016/B978-0-12-816683-3.00010-4

Latocha, P. (2007) The comparison of some biological features of *Actinidia arguta* cultivars fruit. *Annals of Warsaw University of Life Sciences – SGGW, Horticulture and Landscape Architecture* 28, 105–109.

Latocha, P. (2010) Morfologia oraz wartość użytkowa owoców aktinidii ostrolistnej i mieszańcowej [Morphology and the use value of fruit of hardy kiwifruit and its hybrid] *Actinidia arguta* (Siebold et Zucc.) Planch ex. Miq., *A. arguta* × *A. purpurea* Rehd. Postdoctoral dissertation, Wieś Jutra, Warsaw. [In Polish]

Latocha, P. (2012) Some morphological and biological features of 'Bingo' – new hardy kiwifruit cultivar from Warsaw University of Life Sciences (WULS) in Poland. *Rocznik Polskiego Towarzystwa Dendrologicznego* 60, 61–67.

Latocha, P. (2017) The nutritional and health benefits of kiwiberry (*Actinidia arguta*) – a review. *Plant Foods for Human Nutrition* 72(4), 325–334. DOI: 10.1007/s11130-017-0637-y

Latocha, P. and Jankowski, P. (2011) Genotypic difference in postharvest characteristics of hardy kiwifruit (*Actinidia arguta* and its hybrids) as a new commercial crop. Part II: Consumer acceptability and its main drivers. *Food Research International* 44, 1946–1955. DOI: 10.1016/j.foodres.2011.01.032

Latocha, P., Jankowski, P. and Radzanowska, J. (2011) Genotypic difference in postharvest characteristics of hardy kiwifruit (*Actinidia arguta* and its hybrids) as a new commercial crop. Part I: Sensory profiling and physicochemical differences. *Food Research International* 44, 1936–1945. DOI: 10.1016/j.foodres.2011.01.033

Latocha, P., Krupa, T., Jankowski, P. and Radzanowska, J. (2014) Changes in postharvest physicochemical and sensory characteristics of hardy kiwifruit (*Actinidia arguta* and its hybrid) after cold storage under normal versus controlled atmosphere. *Postharvest Biology and Technology* 88, 21–33. DOI: 10.1016/j.postharvbio.2013.09.005

Latocha, P., Debersaques, F. and Decorte, J. (2015) Varietal differences in mineral composition of kiwiberry (*Actinidia arguta* Siebold et Zucc.) Planch. ex. Miq. *Acta Horticulturae* 1096, 479–486. DOI: 10.17660/ActaHortic.2015.1096.59

Latocha, P., Łata, B. and Stasiak, A. (2015) Phenolics, ascorbate and the antioxidant potential of kiwiberry vs. common kiwifruit: the effect of cultivar and tissue type. *Journal of Functional Foods* 19, 155–163. DOI: 10.1016/j.jff.2015.09.024

Latocha, P., Vereecke, D. and Debersaques, F. (2018) Kiwiberry commercial production – What stage are we at? *Acta Horticulturae* 1218, 559–564. DOI: 10.17660/ActaHortic.2018.1218.76

Latocha, P., Debersaques, F. and Hale, I. (2021) *Actinidia arguta* (kiwiberry): botany, production, genetics, nutritional value, and postharvest handling. *Horticultural Reviews* 48, 37–152. DOI: 10.1002/9781119750802.ch2

Latorre, B.A., Alvarez, C. and Ribeiro, O.K. (1991) Phytophthora root rot of kiwifruit in Chile. *Plant Disease* 75(9), 949–952. DOI: 10.1094/PD-75-0949

Leontowicz, H., Leontowicz, M., Latocha, P., Jesion, I., Park, Y.-S. *et al.* (2016a) Bioactivity and nutritional properties of hardy kiwi fruit *Actinidia arguta* in comparison with *Actinidia deliciosa* 'Hayward' and *Actinidia eriantha* 'Bidan'. *Food Chemistry* 196, 281–291. DOI: 10.1016/j.foodchem.2015.08.127

Leontowicz, M., Leontowicz, H., Jesion, I., Bielecki, W., Najman, K. *et al.* (2016b) *Actinidia arguta* supplementation protects aorta and liver in rats with induced hypercholesterolemia. *Nutrition Research* 36(11), 1231–1242. DOI: 10.1016/j.nutres.2016.09.010

Li, Z.-Z., Man, Y.-P., Lan, X.-Y. and Wang, Y.-C. (2013) Ploidy and phenotype variation of natural *Actinidia arguta* population in the east of Daba Mountain located in a region of Shaanxi. *Scientia Horticulturae* 161, 259–265. DOI: 10.1016/j.scienta.2013.07.008

Liu, P., Xue, S., He, R., Hu, J., Wang, X. *et al.* (2016) *Pseudomonas syringae* pv. *actinidiae* isolated from non-kiwifruit plant species in China. *European Journal of Plant Pathology* 145(4), 743–754. DOI: 10.1007/s10658-016-0863-4

Liu, Y.H. and Liu, C.J. (2015) Current situation and potential of wild *Actinidia arguta* resources in northeastern China. *Acta Horticulturae* 1096, 443–449. DOI: 10.17660/ActaHortic.2015.1096.53

Liu, Y.H. and Liu, C.J. (2016) Antifatigue and increasing exercise performance of *Actinidia arguta* crude alkaloids in mice. *Journal of Food and Drug Analysis* 24, 738–745. DOI: 10.1016/j.jfda.2016.03.001

Logan, D.P., Allison, P.A. and Stannard, K. (2002) Selection of wild hosts for feeding by passion vine hopper, *Scolypopa australis* (Walker) (Hemiptera: Ricaniidae) in the Bay of Plenty. *New Zealand Plant Protection* 55, 368–373. DOI: 10.30843/nzpp.2002.55.3934

Ma, Y., Lu, X., Li, K., Wang, C., Guna, A. *et al.* (2021) Prediction of potential geographical distribution patterns of *Actinidia arguta* under different climate scenarios. *Sustainability* 13(6), 3526. DOI: 10.3390/su13063526

MAF (Ministry of Agriculture and Forestry, New Zealand) (2010) Horticulture and arable monitoring 2010. Bay of Plenty kiwifruit. Ministry of Agriculture and Forestry, Wellington. ISBN 978-0-478-36388-3. Available at: www.mpi.govt.nz/dmsdocument/5470-farm-monitoring-report-2010-horticulture-monitoring-bay-of-plenty-kiwifruit (accessed 30 November 2021).

Martin, N.A. (2019) Long-tailed mealybug – *Pseudococcus longispinus*. Interesting insects and other invertebrates. *New Zealand Arthropod Factsheet Series* No. 160. Available at: http://nzacfactsheets.landcareresearch.co.nz/Index.html (accessed 16 September 2022).

Matich, A.J., Young, H., Allen, J.M., Wang, M.Y., Fielder, S. *et al.* (2003) *Actinidia arguta*: volatile compounds in fruit and flowers. *Phytochemistry* 63(3), 285–301. DOI: 10.1016/s0031-9422(03)00142-0

Mauchline, N.A. and Hill, M.G. (2005) Settlement of armoured scale insects on fruit of commercial *Actinidia* spp. *New Zealand Plant Protection* 58, 294–298. DOI: 10.30843/nzpp.2005.58.4266

McKenna, C.E., Dobson, S.J. and Phare, J.M. (2009) The insect pest complex of *Actinidia arguta* kiwifruit. *New Zealand Plant Protection* 62, 262–267. DOI: 10.30843/nzpp.2009.62.4793

McKenna, C.E., Fullerton, R.A., Dobson, S.J. and Jia, Y. (2012) Control of flyspeck (*Schizothyrium pomi*) on *Actinidia arguta* kiwifruit. *New Zealand Plant Protection* 65, 213–217. DOI: 10.30843/nzpp.2012.65.5368

Medic, A., Hudina, M. and Veberic, R. (2021) The effect of cane vigour on the kiwifruit (*Actinidia chinensis*) and kiwiberry (*Actinidia arguta*) quality. *Scientific Reports* 11, 12749. DOI: 10.1038/s41598-021-92161-8

Mendes da Silva, T., Briano, R., Peano, C. and Giuggioli, N.R. (2020) The use of a new explanatory methodology to assess maturity and ripening indices for kiwiberry (*Actinidia arguta*): preliminary results. *Postharvest Biology and Technology* 163, 111122. DOI: 10.1016/j.postharvbio.2020.111122

Mertens, S. (2012) Kunstmatige bestuiving in de teelt van kiwibes (*A. arguta*). MSc thesis, Hogeschool, Gent, Belgium. [In Dutch]

Mizugami, T., Kim, J.G., Beppu, K., Fukuda, T. and Kataoka, I. (2007) Observation of parthenocarpy in *Actinidia arguta* selection 'Issai'. *Acta Horticulturae* 753, 199–203. DOI: 10.17660/ActaHortic.2007.753.22

Nguyễn, H.V. and Savage, G.P. (2013) The effects of temperature and pH on the extraction of oxalate and pectin from the green kiwifruit (*Actinidia deliciosa* L.), golden kiwifruit (*Actinidia chinensis* L), kiwiberry (*Actinidia arguta*) and persimmon (*Diospyros kaki*). *International Journal of Food Science and Technology* 48, 794–800. DOI: 10.1111/ijfs.12029

Nishimura, M., Okimasu, Y., Miyake, N., Tada, M., Hida, R. *et al.* (2016) Inhibitory effect of *Actinidia arguta* on mutagenesis, inflammation and two-stage mouse skin tumorigenesis. *Genes and Environment* 38, 25. DOI: 10.1186/s41021-016-0053-9

Oh, S.B., Muneer, S., Kwack, Y.-B., Shin, M.H. and Kim, J.G. (2017) Characteristic of fruit development for optimal harvest date and postharvest storability in 'Skinny Green' baby kiwifruit. *Scientia Horticulturae* 222, 57–61. DOI: 10.1016/j.scienta.2017.05.003

Park, H.M., Son, M.-W., Kim, D., Kim, S.-H., Kim, S.-H. *et al.* (2011) Fatty acid components of hardy kiwifruit (*Actinidia arguta*) as IL-4 production inhibitor. *Biomolecules & Therapeutics* 19(1), 126–133. DOI: 10.4062/biomolther.2011.19.1.126

Pescie, M.A. and Strik, B.C. (2004) Thinning before bloom affects fruit size and yield of hardy kiwifruit. *HortScience* 39(6), 1243–1245. DOI: 10.21273/HORTSCI.39.6.1243

Phivnil, K., Beppu, K., Mochioka, R., Fuluda, T. and Kataoka, I. (2004) Low-chill trait for endodormancy completion in *Actinidia arguta* Planch. (Sarunashi) and *A. rufa* (Shima-sarunashi), indigenous *Actinidia* species in Japan and their interspecific hybrids. *Journal of Japanese Society for Horticultural Science* 73(3), 244–246. DOI: 10.2503/jjshs.73.244

Pinto, D., Delerue-Matos, C. and Rodrigues, F. (2020) Bioactivity, phytochemical profile and pro-healthy properties of *Actinidia arguta*: a review. *Food Research International* 136, 109449. DOI: 10.1016/j.foodres.2020.109449

Rassam, M., Bulley, S.M. and Laing, W.A. (2007) Oxalate and ascorbate in *Actinidia* fruit and leaves. *Acta Horticulturae* 753, 479–484. DOI: 10.17660/ActaHortic.2007.753.62
Ravipati, A.S., Zhang, L., Koyyalamudi, S.R., Jeong, S.C., Reddy, N. *et al.* (2012) Antioxidant and anti-inflammatory activities of selected Chinese medicinal plants and their relation with antioxidant content. *BMC Complementary and Alternative Medicine* 12, 173. DOI: 10.1186/1472-6882-12-173
Reich, L. (2004) Hardy kiwifruit: emerald gems. *New York Berry News* 3(12). Available at: https://cpb-us-e1.wpmucdn.com/blogs.cornell.edu/dist/0/7265/files/2016/12/hardykiwi-1i4i0gk.pdf (accessed 30 November 2021).
Romanazzi, G. (2009) Gray mold infection of *Actinidia arguta* in Italy. *Plant Disease* 93, 1221. DOI: 10.1094/PDIS-93-11-1221A
Rothgerber, P. (1992) *Actinidia arguta* thesis. *Kiwifruit Enthusiasts Journal* 6, 109–122.
Ruan, R., Xi, D., Zhang, C., Huang, K., Luo, H. *et al.* (2022) *Alternaria alternata* causing brown spot on kiwiberry (*Actinidia arguta*) in China. *Crop Protection* 152, 105857. DOI: 10.1016/j.cropro.2021.105857
Skrypchenko, N.W. and Latocha, P. (2017) The genesis and current state of *Actinidia* collection in M.M. Grishko National Botanical Garden in Ukraine. *Polish Journal of Natural Sciences* 32(3), 513–525.
Skrypchenko, N.W. and Moroz, P.A. (2002) *Aktinidia (sorty, wyraszcziwania, rozmnożenia)*. M.M. Grishko National Botanic Garden, National Academy of Sciences of Ukraine, Kiev. [In Ukrainian]
Snowball, A.M. (1997a) Axillary shoot bud development in selected *Actinidia* species. *New Zealand Journal of Crop and Horticultural Science* 25, 233–242. DOI: 10.1080/01140671.1997.9514012
Snowball, A.M. (1997b) Seasonal cycle of shoot development in selected *Actinidia* species. *New Zealand Journal of Crop and Horticultural Science* 25, 221–231. DOI: 10.1080/01140671.1997.9514011
Sotiropoulos, T.E., Therios, I.N., Dimassi, K.N., Bosabalidis, A. and Kofidis, G. (2002) Nutritional status, growth, CO_2 assimilation, and leaf anatomical responses in two kiwifruit species under boron toxicity. *Journal of Plant Nutrition* 25, 1249–1261. DOI: 10.1081/PLN-120004386
Stasiak, A. and Latocha, P. (2020) Comparative analysis of volatile compounds in flowers of different *Actinidia* species. *Plants* 9, 1675. DOI: 10.3390/plants9121675
Stasiak, A., Latocha, P., Drzewiecki, J., Hallmann, E., Najman, K. *et al.* (2019) The choice of female or male parent affects some biochemical characteristics of fruit or seed of kiwiberry (*Actinidia arguta*). *Euphytica* 215, 52. DOI: 10.1007/s10681-019-2375-8
Stasiak, A., Łata, B., Bieniasz, M. and Latocha, P. (2020) Morphological variation of male *A. arguta* plants affects their flowering potential and pollen efficiency. *Horticultural Sciences* (Prague) 47(2), 100–109. DOI: 10.17221/37/2019-HORTSCI
Stasiak, A., Latocha, P. and Bieniasz, M. (2021) Effect of genetically diverse pollen on pollination, pollen tube overgrow, fruit set and morphology of kiwiberry (*Actinidia arguta*). *Agronomy* 11, 1814. DOI: 10.3390/agronomy11091814
Stefaniak, J., Stasiak, A., Latocha, P. and Łata, B. (2017a) Effect of nitrogen fertilization on *Actinidia arguta* plants vigour and soil characteristics. *Journal of International Scientific Publications – Agriculture & Food* 5, 314–323.
Stefaniak, J., Sawicka, M., Krupa, T., Latocha, P. and Łata, B. (2017b) Effect of kiwiberry pre-storage treatments on the fruit quality during cold storage. *Zemdirbyste-Agriculturae* 104(3), 235–242. DOI: 10.13080/z-a.2017.104.030
Stefaniak, J., Stasiak, A., Latocha, P. and Łata, B. (2019) Seasonal changes in macronutrients in the leaves and fruit of kiwiberry: nitrogen level and cultivar effects. *Communications in Soil Science and Plant Analysis* 50(22), 2913–2926. DOI: 10.1080/00103624.2019.1689255
Steven, D. (1992) Pest Lepidoptera of kiwifruit In New Zealand. *Acta Horticulturae* 297, 531–536. DOI: 10.17660/ActaHortic.1992.297.69
Strik, B.C. and Davis, A.J. (2021) *Growing Kiwifruit: A Guide to Kiwiberries and Fuzzy Kiwifruit for Pacific Northwest Producers*. PNW 507. A Pacific Northwest Extension Publication. Available at: https://catalog.extension.oregonstate.edu/sites/catalog/files/project/pdf/pnw507.pdf (accessed 30 November 2021).
Strik, B.C. and Hummer, K.E. (2006) 'Ananasnaya' hardy kiwifruit. *Journal of the American Pomological Society* 60(3), 106–112.
Sutherland, P.W., Fullerton, C.G., Schröder, R. and Hallett, I.C. (2017) Cell wall changes in *Actinidia arguta* during softening. *Scientia Horticulturae* 226, 173–183. DOI: 10.1016/j.scienta.2017.08.027
Szpadzik, E., Zaraś-Januszkiewicz, E. and Krupa, T. (2021) Storage quality characteristic of two minikiwi fruit (*Actinidia arguta* (Siebold & Zucc.) Planch. ex Miq.) cultivars: 'Ananasnaya' and 'Bingo' – a new one selected in Poland. *Agronomy* 11, 134. DOI: 10.3390/agronomy11010134
Tiyayon, C. and Strik, B. (2003a) Effect of fruiting cane origin on fruitfulness of hardy kiwifruit, *Actinidia arguta*. *New Zealand Journal of Crop and Horticultural Science* 31, 179–186. DOI: 10.1080/01140671.2003.9514250

Tiyayon, C. and Strik, B. (2003b) Flowering and fruiting morphology of hardy kiwifruit, *Actinidia arguta*. *Acta Horticulturae* 610, 171–176. DOI: 10.17660/ActaHortic.2003.610.23

Tiyayon, C. and Strik, B. (2004) Influence of time of overhead shading on yield, fruit quality, and subsequent flowering of hardy kiwifruit, *Actinidia arguta*. *New Zealand Journal of Crop and Horticultural Science* 32, 235–241. DOI: 10.1080/01140671.2004.9514301

Tontou, R., Giovanardi, D. and Stefani, E. (2014) Pollen as a possible pathway for the dissemination of *Pseudomonas syringae* pv. *actinidiae* and bacterial canker of kiwifruit. *Phytopathologia Mediterranea* 53(2), 333–339. DOI: 10.14601/Phytopathol_Mediterr-13757

Van Labeke, M.-C., Soleme, J. and Debersaques, F. (2015) Cold acclimation patterns of *Actinidia arguta* cultivars in a temperate climate zone. *Acta Horticulturae* 1096, 467–470. DOI: 10.17660/ActaHortic.2015.1096.57

Vance, A.J. and Strik, B.C. (2018) Seasonal changes in leaf nutrient concentration of male and female hardy kiwifruit grown in Oregon. *European Journal of Horticultural Sciences* 83, 247–258. DOI: 10.17660/eJHS.2018/83.4.5

Vanhonacker, F. and Debersaques, F. (2018) How to position kiwiberry in the Flemish market based upon consumer research. *Acta Horticulturae* 1218, 67–71. DOI: 10.17660/ActaHortic.2018.1218.8

Vanneste, J.L., Cornish, D.A., Yu, J. and Stokes, C.A. (2014) First report of *Pseudomonas syringae* pv. *actinidiae*: the causal agent of bacterial canker of kiwifruit on *Actinidia arguta* vines in New Zealand. *Plant Disease* 98(3), 418. DOI: 10.1094/PDIS-06-13-0667-PDN

Walton, E.F. and Wu, R.-M. (1999) Buds on *Actinidia arguta* shoots do contain axillary meristems. *New Zealand Journal of Crop and Horticultural Science* 27, 181–185. DOI: 10.1080/01140671.1999.9514094

Wang, C.W., Ai, J., Qin, H.Y., Yang, Y.M., Fan, S.T. *et al.* (2016) First report of *Sclerotinia nivalis* causing sclerotinia rot on hardy kiwifruit (*Actinidia arguta*) in China. *Plant Disease* 100(9), 1952. DOI: 10.1094/PDIS-08-15-0899-PDN

Wang, H., Quan, H., Sun, T., Wang, Z. and Yang, Y. (2022) Chemical composition, antimicrobial, and antioxidant cytotoxic activities of essential oil from *Actinidia arguta*. *Archives of Microbiology* 204, 239. DOI: 10.1007/s00203-022-02775-3

Wang, R.C., Zeng, Z.L., Pang, L.P., Yang, X.X. and Shi, H. (2015) Effects of postharvest treatments on storability of *Actinidia arguta* kiwifruit. *Acta Horticulturae* 1096, 461–466. DOI: 10.17660/ActaHortic.2015.1096.56

Wang, Y., Xu, F., Feng, X. and MacArthur, R.L. (2015) Modulation of *Actinidia arguta* fruit ripening by three ethylene biosynthesis inhibitors. *Food Chemistry* 173, 405–413. DOI: 10.1016/j.foodchem.2014.10.044

Wang, Y.C. and Li, Z.Z. (2018) The door is opening for commercialization of Ruanzao kiwifruit (*Actinidia arguta*) in China. *Acta Horticulturae* 1218, 587–592. DOI: 10.17660/ActaHortic.2018.1218.80

Watanabe, K. and Takahashi, B. (1998) Determination of soluble and insoluble oxalate contents in kiwifruit (*Actinidia deliciosa*) and related species. *Journal of the Japanese Society for Horticultural Science* 67(3), 299-305. DOI: 10.2503/jjshs.67.299

White, A., de Silva, H.N., Requejo-Tapia, C. and Harker, F.R. (2005) Evaluation of softening characteristics of fruit from 14 species of *Actinidia*. *Postharvest Biology and Technology* 35, 143–151. DOI: 10.1016/j.postharvbio.2004.08.004

Whitworth, J. (2017) Kiwifruit production in Oklahoma. Oklahoma Cooperative Extension Service Fact Sheets. HLA-6294. Available at: http://factsheets.okstate.edu/documents/hla-6249-kiwifruit-production-in-oklahoma/ (accessed 30 November 2021).

Williams, M.H., Boyd, L.M., McNeilage, M.A., MacRae, E.A., Ferguson, A.R. *et al.* (2003) Development and commercialization of 'Baby Kiwi' (*Actinidia arguta* Planch.). *Acta Horticulturae* 610, 81–86. DOI: 10.17660/ActaHortic.2003.610.8

Wojdyło, A., Nowicka, P., Oszmiański, J. and Golis, T. (2017) Phytochemical compounds and biological effects of *Actinidia* fruits. *Journal of Functional Foods* 30, 194–202. DOI: 10.1016/j.jff.2017.01.018

Zhang, J., Tian, J., Gao, N., Gong, E.S., Xin, G. *et al.* (2021) Assessment of the phytochemical profile and antioxidant activities of eight kiwi berry (*Actinidia arguta* (Siebold & Zuccarini) Miquel) varieties in China. *Food Science & Nutrition* 9, 5616–5625. DOI: 10.1002/fsn3.2525

Zhang, Y., Zhong, C., Liu, Y., Zhang, Q., Sun, X. *et al.* (2017) Agronomic trait variations and ploidy differentiation of kiwiberries in northwest China: implication for breeding. *Frontiers in Plant Science* 8, 711. DOI: 10.3389/fpls.2017.00711

Zuo, L.-L., Wang, Z.-Y., Fan, Z.-L., Tian, S.-Q. and Liu, J.-R. (2012) Evaluation of antioxidant and antiproliferative properties of three *Actinidia* (*Actinidia kolomikta*, *Actinidia arguta*, *Actinidia chinensis*) extracts *in vitro*. *International Journal of Molecular Sciences* 13, 5506–5518. DOI: 10.3390/ijms13055506

8 Propagation of Kiwifruit Plants

Ed R. Morgan
The New Zealand Institute for Plant and Food Research Ltd, Palmerston North, New Zealand

8.1 Introduction

The supply of high-quality kiwifruit plants for growers and breeding programmes requires reliable propagation technologies. As the industry has developed, so too have propagation practices. Orchards may comprise fruiting cultivars that have been grafted onto seedlings or propagated from cuttings, or tissue-cultured plants that have been grown on their own roots or grafted onto clonal rootstocks. Work to develop clonal rootstocks is expected to result in fruiting cultivars being grafted to rootstocks specifically bred for this purpose.

Since the review of Lawes (1990), much has been learnt of the requirements for sexual and asexual propagation of quality plants of *Actinidia* species. Some of these learnings are reviewed here.

8.2 Seed Propagation

Seeds are used to produce plants used as rootstocks and for breeding. Seed storage can also provide a cost-effective method for conservation of genetic diversity or to provide resilience in breeding programmes against biotic or abiotic threats. The seed used to propagate rootstock plants is typically from a parent for which seed germination is high and results in vigorous seedlings. Better understanding of germination requirements is needed for breeding and conservation programmes. Poor germination (~18% germination) is cited as a factor in the endangered status of *A. chrysantha* in the wild (Mo *et al.*, 2016). In general, seed germination can be influenced by factors such as the seed-pollen parent combination, seed maturity, seed dormancy, seed handling (including of the fruit from which seed is harvested), exposure to extreme temperatures, prolonged suboptimal storage and germination conditions.

An 80–120 g 'Hayward' kiwifruit (*A. chinensis* var. *deliciosa)* typically contains ~900–1400 seeds (Hopping, 1976), with seed viability (seeds with viable embryos) affected by factors including pollination and compatibility of the parents. Hybridization between even the closely related *A. chinensis* var. *deliciosa* and *A. chinensis* var. *chinensis* can result in poor seed set or non-viable or empty seed (Harvey *et al.*, 1992), an observation possibly explained by ploidy differences between parents.

Low germination in *Actinidia* is typically attributed to seed dormancy, i.e. the absence of germination in a specified time period under conditions otherwise favourable for growth (Baskin

*Corresponding author: ed.morgan@plantandfood.co.nz

© CAB International 2024. *Kiwifruit: Botany, Production and Uses* (eds A.C. Richardson, J.N. Burdon and A.R. Ferguson)
DOI: 10.1079/9781800620933.0008

and Baskin, 2004). Protocols to overcome seed dormancy in kiwifruit typically include low-temperature stratification with or without subsequent alternating temperatures (Smith and Toy, 1967; Lawes and Anderson, 1980; Wu and Datson, 2015; Windauer *et al.*, 2016; Sekhukhune *et al.*, 2018a; Maghdouri *et al.*, 2021) or use of plant-growth regulators, e.g. gibberellic acid (GA_3), as a substitute or a supplement for stratification temperature treatments (Lawes and Anderson, 1980; Wu and Datson, 2015; Sekhukhune *et al.*, 2018a, b). The finding that low-temperature stratification of seed followed by alternating temperatures promoted germination of *A. chinensis* var. *deliciosa* 'Abbott' and 'Bruno' seed (Lawes and Anderson, 1980) provided a sound basis for large-scale commercial propagation. Seed dormancy can also be regulated by the seed coat (Windauer *et al.*, 2016).

A set of experiments on *A. chinensis* var. *deliciosa* seed germination using seed from a 'Bruno' × 'Tomuri' family was reported by Windauer *et al.* (2016). Maximum germination (~80%) was observed when at least three weeks of stratification in darkness at 2°C or 5°C was followed by alternating temperatures (20/30°C), also in darkness; stratification at 12°C gave low germination (16%) even when followed by alternating temperatures. Seed stratified at 2°C and 5°C followed by constant 25°C also exhibited poor germination. Seed could be stratified within intact fruit held for four weeks at 2°C (this was as effective as treating isolated seed), though seed had to be extracted from fruit tissues before exposure to alternating temperatures for germination. That the presence of fruit tissues on seed inhibited germination had been noted by Wu and Datson (2015). Light quality can impact germination, with a 20-minute pulse of far-red light inhibiting germination, but this effect could be overcome by 20 minutes of red light 1 h after the far-red treatment (Windauer *et al.*, 2016). Removing the seed coat from seed extracted from freshly harvested fruit and incubating in alternating temperatures in darkness gave 40% germination, as did sulphuric acid scarification of seed; intact seed did not germinate (Windauer *et al.*, 2016).

Breaking seed dormancy alone will not meet the requirements of large-scale propagation programmes as the challenges of overcoming non-uniform germination and poor seedling vigour remain. Efficiencies may be gained by applying a 'stratification thermal time' model to dormancy release, allowing users to predict the proportion of a seed population that will germinate following stratification (Lastuvka *et al.*, 2021), thus helping propagators achieve more uniform germination. Additionally, attention must be paid to the influence of growing conditions on seedling growth and survival (e.g. Lawes, 1990; Maghdouri *et al.*, 2021).

8.3 Vegetative Propagation

Kiwifruit cultivars for commercial production are clonally propagated by cuttings or micropropagation (discussed later) and there has been significant investment by the nursery industry to successfully refine protocols for large-scale propagation.

Kiwifruit has been successfully propagated using hardwood, semi-hardwood and softwood cuttings of different species and genotypes. For all three types of cuttings, treatment with indole-3-butyric acid (IBA) has been shown to increase rooting rates. For example, 'Bruno' cuttings collected in December (southern hemisphere, Brazil) gave 58% rooting with 4000 mg L^{-1} IBA, which was the highest rate tested (Guasso *et al.*, 2021). When propagating four genotypes of *A. chinensis* var. *deliciosa* ('Abbott', 'Hayward', 'Hayward Behi' and a male genotype), the highest rate of IBA tested, 1000 mg L^{-1}, gave the highest percentage of rooting and the male genotype gave a higher percentage of rooting compared with the three female cultivars (Arfaei *et al.*, 2021).

There are abundant reports of softwood propagation with plant-growth regulators, cutting type and size, time of cutting, genotype, and growing medium among the factors investigated. 'Hayward' cuttings taken at 15-day intervals through the growing season (mid-May to September, northern hemisphere) were treated with 0–6000 mg L^{-1} IBA (Biasi *et al.*, 1990). The best rooting percentage, root number and length were in cuttings taken in July and August (northern hemisphere) and treated with 2000, 4000 or 6000 mg L^{-1} IBA. Similar results were reported using 'Hayward' and 'Matua' semi-hardwood cuttings, where the best results were obtained with cuttings taken in July and

August (northern hemisphere) and treated with 4000 and 6000 mg L^{-1} IBA (Zenginbal and Özcan, 2014). Similar rooting percentages were reported for both cultivars.

Hardwood cuttings can also be used for propagation. Rooting was evaluated on cuttings with two or three buds taken in January and February (northern hemisphere, Turkey) and treated with IBA at different rates. The best results for 'Hayward' were from cuttings taken in January with three buds, and the best results for 'Matua' were from the larger cuttings. The best rooting was seen on those cuttings treated with higher rates of IBA (4000 and 6000 mg L^{-1}) (Zenginbal and Özcan, 2013).

Cutting propagation has been reported for species other than *A. chinensis* var. *deliciosa*: *A. arguta* (Beyl *et al.*, 1995; Chesoniene and Paal, 1998; Ivánek, 2006; Peticilă *et al.*, 2016) and *A. rufa* (Nagata *et al.*, 1997) are among other species reported to have been propagated by cuttings.

8.4 Grafting

It is possible to achieve a high success rate (> 95%) when grafting kiwifruit (Lawes, 1990). Increasing interest in using attributes of other species, e.g. waterlogging tolerance found in a species such as *A. macrosperma* (Kataoka *et al.*, 2021), will require better understanding of the factors regulating grafting success. There has been, however, little investigation of the development of the graft union in *Actinidia* and there appears to have been little investigation of interspecies grafting (heterografting).

Grafting method (cleft grafting by hand and machine, whip and splice) and date of grafting of 'Hayward' scions to seedling rootstocks were investigated (Zenginbal *et al.*, 2006). All methods and graft times were successful, though the best results were observed for whip and splice grafting undertaken on 1 February (in Turkey). Another study (Sedaghathoor and Noie, 2016) investigated factors affecting grafting success of 'Hayward' onto 'Matua' and 'Bruno'. Graft union success was 100% for cleft grafting using 'Bruno' and tongue grafting using 'Matua'.

The anatomy of graft formation in *A. chinensis* var. *deliciosa* following chip-budding, tongue and whip grafting has been investigated (Karadeniz, 1998). The callus bridge between stock and scion was formed by 14 days, cambial tissue was differentiated at 21 days and the union between stock and scion was continuous by 90 days after grafting. *In vitro* micro-grafting was used to graft the *A. chinensis* var. *deliciosa* cultivars 'Yuxiang', 'Jinmin' and 'Hayward' and the *A. chinensis* var. *chinensis* cultivar 'Qihong' as scions onto the *A. valvata* selection MX1 (Bao *et al.*, 2020). Callus formation had initiated from rootstock parenchyma cells three days after micro-grafting; new cambial cells had initiated from the callus at about seven days; and vascular bundles connecting scion and rootstock were apparent at 14 days and were fully developed in the *in vitro* plants at 21 days. Almost all micro-grafts of the four scion cultivars developed into *in vitro* plantlets with good root systems (Bao *et al.*, 2020).

Graft incompatibility in kiwifruit seems to have received little attention. As breeders and growers explore the diversity within the genus, graft compatibility is likely to need attention. In general, homografts combining scion and rootstock of the same species are compatible but work with other genera suggests that increasing taxonomic distance between graft partners reduces the rate of success (see Chapter 6).

8.5 *In Vitro* Propagation

In vitro (tissue culture) propagation has been used to propagate plants for commercial production but has additional applications in generating and maintaining high-health plants that can be used for nursery mother stock. Lawes (1990) anticipated that *in vitro* protocols would have applications to support breeding or propagation of kiwifruit and examples are listed by Kumar and Sharma (2002). There have been many publications describing *in vitro* culture of *Actinidia* genotypes so this discussion will mostly focus on work describing trends across multiple genotypes.

In vitro plant performance depends on factors such as genotype, the physiological maturity and general health of the explant used to initiate cultures, medium composition and *in vitro* growth conditions. Most genotypes will grow at some level *in vitro* (Kumar and Sharma, 2002) and in many cases the same protocol can be used for a diverse range of genotypes. When developing a cryopreservation protocol for multiple *Actinidia* genotypes across four species (*A. chinensis* var.

chinensis (three genotypes), *A. chinensis* var. *deliciosa* 'Hayward', *A. macrosperma* (three genotypes), *A. arguta* and *A. polygama*), the same protocols were used to produce experimental plants and recover plants following freezing for each genotype (Pathirana *et al.*, 2021). However, different genotypes may have different requirements for optimal or even satisfactory growth.

In vitro cultures are typically initiated using explants that comprise axillary buds or shoot tips sourced from fresh spring growth of field- or greenhouse-grown plants. Initiation of cultures from field-grown material brings risk of contamination (Debenham *et al.*, 2016; Hameg *et al.*, 2017). Sprouting dormant winter-collected budwood indoors will provide explants for culture initiation (Prado *et al.*, 2005; Wu, 2018; Pathirana *et al.*, 2021). Even after surface sterilization, explants may carry endogenous microorganisms that do not readily grow on standard propagation media, or only grow slowly (Orlikowska *et al.*, 2017) – endogenous contaminating microorganisms are discussed further in Section 8.6. Newly initiated explants can be grown on agar-solidified medium comprising MS salts (Murashige and Skoog, 1962) and B5 vitamins (Gamborg *et al.*, 1968), supplemented with IBA, GA_3 and 6-benzylaminopurine (BAP), before transfer to multiplication medium (Pathirana *et al.*, 2021).

A range of salt formulations has been used to provide the macro- and micronutrients needed for *in vitro* plant growth since the initial publication by Harada (1975). There has been little optimization of the nutrient medium for kiwifruit and MS medium is still widely used, though B5 and the N6 medium of Chu *et al.* (1975) have been used in different applications (Kumar and Sharma, 2002). When trying to improve protocols for establishing *in vitro* cultures, Debenham *et al.* (2016) screened two genotypes each of *A. chinensis* and *A. polygama* on LP medium (Quoirin and Lepoivre, 1977), woody plant medium (Lloyd and McCown, 1980), and B5 and MS macronutrient salt formulations. They concluded that although there were differences in plant performance, with smaller and paler leaves on LP and B5 media, this did not limit culture initiation success. As their goal was not to optimize growth of individual genotypes but to initiate cultures of large numbers of genotypes, they persisted with MS medium (Debenham *et al.*, 2016). MS medium provides a good starting point for *in vitro* work though it is not considered ideal, with Moncaleán *et al.* (2003) describing it as 'supra-optimal' for 'Hayward' except with respect to phosphate. Several authors have used MS salts at half-strength (e.g. Kim *et al.*, 2007; Pathirana *et al.*, 2021).

In vitro propagation of *A. arguta* 'Issai' was evaluated in response to six different nutrient formulations previously used for kiwifruit (Hameg *et al.*, 2018). Plants were evaluated in terms of shoot number, shoot length, leaf number and leaf area. The medium of Standardi (1981) gave the best or equal-best results for the parameters measured; B5 medium gave the poorest results. Even so, Standardi (1981) medium produced some undesirable callus growth. A modelling approach was used to predict an optimal nutrient salt combination for *A. arguta* (Hameg *et al.*, 2020). Despite the observation of Kumar and Sharma (2002) that kiwifruit plants are not very specific with respect to nutrient medium, the findings of Hameg *et al.* (2018, 2020) show opportunities may exist to improve *in vitro* plant performance.

Following propagation, tissue-culture shoots are typically transferred to an auxin-supplemented medium for root initiation, though this is not always necessary. Most (70%) of the *in vitro* shoots of *A. macrosperma*, *A. polygama*, *A. melanandra* and *A. arguta* genotypes transferred to an MS medium supplemented with IBA, BAP and GA_3 produced roots (Debenham *et al.*, 2016). By comparison, only 2% of the *A. chinensis* var. *chinensis* and *A. chinensis* var. *deliciosa* genotypes tested produced roots (Debenham *et al.*, 2016). When transferred to MS medium supplemented with 3 mg L^{-1} IBA, 65% (of 299) of *A. chinensis* var. *chinensis* genotypes and 60% (of 169) of *A. chinensis* var. *deliciosa* genotypes had roots after four weeks. Root initiation on this medium was also observed on small numbers of *A. eriantha*, *A. glaucophylla*, *A. kolomikta*, *A. latifolia*, *A. melanandra* and *A. polygama* genotypes tested (Debenham *et al.*, 2016). Another protocol for root initiation was that of Prado *et al.* (2005) who reported that immersing the bases of *in vitro* shoots of *A. chinensis* var. *deliciosa* in IBA for 15 s resulted in 100% root initiation after 30 days.

The success of root initiation may be affected by the plant growth regulators used for shoot propagation. The cytokinin 6-(3-hydroxybenzyl) adenine (meta-Topolin, mT) was compared with BAP and zeatin for micropropagation of *A. chinensis* var. *chinensis* 'Zes006' (Saeiahagh *et al.*, 2019). Plants propagated on mT-supplemented medium had significantly higher shoot numbers, shoot weights, leaf numbers and leaf areas. Additionally, shoots produced on mT-supplemented medium readily produced roots, facilitating transfer to the greenhouse. By comparison, none of the shoots from BAP- or zeatin-supplemented media produced roots.

A concern for plant propagators is the genetic integrity of clonally propagated plants. The somaclonal variation (Larkin and Scowcroft, 1981) sometimes seen in plants produced using *in vitro* culture can be detected by DNA-based techniques (e.g. Palombi and Damiano, 2002; Prado *et al.*, 2005; Saeiahagh *et al.*, 2019). Although they detected a very low rate of somaclonal variation using simple sequence repeat primers, Saeiahagh *et al.* (2019) considered that better methods are needed for more precise detection of any changes. It should be noted here that plants propagated by tissue culture may display physical differences to cutting-propagated plants. For example, Piccotino *et al.* (1997) noted that leaves of micropropagated plants displayed juvenile characteristics in their first year of growth.

8.6 Disease-free Plants

Clonally propagated crops, including those produced through tissue culture, are a risk for disease transmission because their microbiome reflects their growing and propagation history. Movement of vegetative material risks spreading pathogens such as *Pseudomonas syringae* pv. *actinidiae* (Psa). Field-grown Psa-infected 'Hayward' plants can remain asymptomatic for four years after inoculation (Tyson *et al.*, 2018a), highlighting the need for pathogen testing of mother stock plants.

Although *in vitro* plants are often considered 'sterile', they can asymptomatically host endophytic bacteria that will not grow on the culture medium and thus are not easily detected (Orlikowska *et al.*, 2017). Psa did not readily grow on the surface of an MS-based medium used for kiwifruit but rapidly colonized all parts of plants following inoculation, including in cultivars considered resistant/tolerant to Psa (Tyson *et al.*, 2018b). When *in vitro* micropropagated shoots of 'Hayward' were inoculated with Psa, the bacteria persisted through multiple subcultures (Minardi *et al.*, 2019). Minardi *et al.* (2019) noted that although plants became symptomatic, control *in vitro* plantlets that were not inoculated sometimes showed unusual features on the foliage that could be mistaken for symptoms.

Data from Tyson *et al.* (2018b) provided a simple protocol for verifying that *in vitro* plants have a very low likelihood of carrying Psa. Adding peptone (3 g L^{-1}) to MS-based medium promoted outgrowth of Psa onto the culture medium from plants inoculated *in vitro*, providing a visual screen for the presence of the bacteria. This coupled with PCR testing of samples from plants grown on peptone-supplemented medium provided confidence of a very low likelihood of Psa in the *in vitro* plants. In the event that pathogen-free plants cannot be identified, technologies to rid plants of pathogens are becoming available.

Cryotherapy has been used to eliminate viruses and other pathogens from *in vitro* plant material. A droplet vitrification method used for cryopreservation (Mathew *et al.*, 2018) was used to eliminate pseudomonad bacteria from kiwifruit plantlets (Pathirana *et al.*, 2019). Plants of *A. chinensis* var. *chinensis* 'Hort16A' infected with *Pseudomonas fluorescens* and *A. chinensis* var. *deliciosa* 'Hayward' infected with *P. viridiflava* were used. Following immersion in liquid nitrogen for 1 h, plantlets regenerated from the shoot tips were screened for the bacteria. All the regenerated 'Hort16A' plants were free of *P. fluorescens* and 80% of the regenerated 'Hayward' plants were free of *P. viridiflava* (Pathirana *et al.*, 2019).

8.7 Into the Future

There are many elements to successful propagation and delivery of kiwifruit plants to growers who, in addition to utilizing more of the diversity available within the genus, are also growing in a greater range of environments. The use of more species in breeding for fruiting or rootstock cultivars will require new knowledge of how to deliver these new plants to growers.

Seed germination in kiwifruit has only been explored in a few species and increased knowledge of germination requirements is needed. There appear to be multiple seed dormancy mechanisms in *Actinidia* species, but this requires further investigation to ascertain the seed germination requirements of a greater range of species.

Propagation by cuttings and grafting are well understood by the nursery industry, with multiple *Actinidia* species and genotypes having been propagated by cuttings. The use of multiple species in breeding for both fruiting and rootstock cultivars suggests there will be a need for a better understanding of the physiology of grafting in the *Actinidia* genus.

In vitro propagation has roles in the generation, maintenance, propagation and movement of pathogen-free plant material. Protocols have been developed for many species. A better knowledge of the *in vitro* nutritional requirements for *Actinidia* propagation can be expected to enhance other *in vitro* work with kiwifruit. Little work has been undertaken into the risks potentially posed by somaclonal variation.

Clonal propagation of plants brings the risk of transfer of a range of pathogens. It is essential that propagation begins with pathogen-free plants that are monitored for health status. *In vitro* technologies offer protocols for pathogen removal and to maintain plants in isolation to minimize opportunities for reinfection.

References

Arfaei, L., Alizadeh, M. and Zarei, H. (2021) Kiwifruit (*Actinidia deliciosa*) propagation: influence of stock plant sexuality on root development. *International Journal of Minor Fruits, Medicinal and Aromatic Plants* 7(2), 68–72.

Bao, W.-W., Zhang, X.-C., Zhang, A-L., Lei, Z., Wang, Q.-C. *et al.* (2020) Validation of micrografting to analyze compatibility, shoot growth, and root formation in micrografts of kiwifruit (*Actinidia* spp.). *Plant Cell, Tissue and Organ Culture* 140, 209–214. DOI: 10.1007/s11240-019-01723-2

Baskin, J.M. and Baskin, C.C. (2004) A classification system for seed dormancy. *Seed Science Research* 14, 1–16. DOI: 10.1079/ssr2003150

Beyl, C.A., Ghale, G. and Zhang, L.J. (1995) Characteristics of hardwood cuttings influence rooting of *Actinidia arguta* (Siebold & Zucc.) Planch. *HortScience* 30, 973–976. DOI: 10.21273/hortsci.30.5.973

Biasi, R., Marino, G. and Costa, G. (1990) Propagation of Hayward (*Actinidia deliciosa*) from soft and semi-hardwood cuttings. *Acta Horticulturae* 282, 243–250. DOI: 10.17660/ActaHortic.1990.282.31

Chesoniene, L. and Paal, T, (1998) Winter hardiness of *Actinidia* Lindl. cultivars propagated by softwood cuttings. *Metsanduslikud Uurimused* 30, 35–39.

Chu, C.-C., Wang, C-.C., Sun, C.-S., Chen, H., Yin, K.-C. *et al.* (1975) Establishment of an efficient medium for anther culture of rice through comparative experiments on nitrogen sources. *Scientia Sinica* 18, 659–668. DOI: 10.1360/ya1975-18-5-659

Debenham, M.C., Seelye, J.F. and Mullan, A.C. (2016) An in vitro repository for clonal kiwifruit. *Acta Horticulturae* 1113, 93–97. DOI: 10.17660/ActaHortic.2016.1113.13

Gamborg, O.L., Miller, R.A. and Ojima, K. (1968) Nutrient requirements of suspension cultures of soybean root cells. *Experimental Cell Research* 50, 151–158. DOI: 10.1016/0014-4827(68)90403-5

Guasso, L.Z., Marodin, F.A., Altmann, T., da Silveira, S.V. and Dutra de Souza, P.V. (2021) Propagation of 'Bruno' kiwifruit: influence of cutting collection time and indolebutyric acid. *Semina-Ciencias Agrárias* 42, 2203–2215. DOI: 10.5433/1679-0359.2021v42n4p2203

Hameg, R., Gallego, P.P. and Barreal, M.E. (2017) *In vitro* establishment and multiplication of hardy kiwi (*Actinidia arguta* 'Issai'). *Acta Horticulturae* 1187, 51–57. DOI 10.17660/ActaHortic.2017.1187.6

Hameg, R., Arteta, T., Gallego, P.P. and Barreal, M.E. (2018) Selecting an efficient proliferation medium for *Actinidia arguta* 'Issai' explants. *Acta Horticulturae* 1218, 565–572. DOI: 10.17660/ActaHortic.2018.1218.77

Hameg, R., Arteta, T.A., Landin, M., Gallego, P.P. and Barreal, M.E. (2020) Modeling and optimizing culture medium mineral composition for *in vitro* propagation of *Actinidia arguta*. *Frontiers in Plant Science* 11, 554905. DOI: 10.3389/fpls.2020.554905

Harada, H. (1975) *In vitro* organ culture of *Actinidia chinensis* PL as a technique for vegetative multiplication. *Journal of Horticultural Science* 50, 81–83. DOI: 10.1080/00221589.1975.11514606

Harvey, C.F., Fraser, L.G. and Kent, J. (1992) *Actinidia* seed development in interspecific crosses. *Acta Horticulturae* 297, 71–78. DOI: 10.17660/ActaHortic.1992.297.7

Hopping, M.E. (1976) Effect of exogenous auxins, gibberellins, and cytokinins on fruit development in Chinese gooseberry (*Actinidia chinensis* Planch.). *New Zealand Journal of Botany* 14, 69–75. DOI: 10.1080/0028825X.1976.10428652

Ivánek, I. (2006) The effects of growth stimulators on *Actinidia arguta* shoot cuttings. *Agricultura Tropica et Subtropica* 39, 8–10.

Karadeniz, T. (1998) Anatomical and histological investigations on the chip-budding, tongue and whip grafting methods in kiwifruit (*Actinidia deliciosa* A. Chev.). *Bahce* 26, 93–103.

Kataoka, I., Yamada, T., Fukuda, T., Ohtani, M., Suezawa, K. *et al.* (2021) Waterlogging tolerance of *Actinidia macrosperma* and its application to kiwifruit by grafting *Actinidia macrosperma*. *Horticultural Research (Japan)* 20(3), 265–271. DOI: 10.2503/hrj.20.265

Kim, M., Kim, S.-C., Moon, D.-Y. and Song, K.J. (2007) Rapid shoot propagation from micro-cross sections of kiwifruit (*Actinidia deliciosa* cv. 'Hayward'). *Journal of Plant Biology* 50, 681–686. DOI: 10.1007/bf03030613

Kumar, S. and Sharma, D.R. (2002) *In vitro* propagation of kiwifruit. *Journal of Horticultural Science & Biotechnology* 77, 503–508. DOI: 10.1080/14620316.2002.11511530

Larkin, P.J. and Scowcroft, W.R. (1981) Somaclonal variation – a novel source of variability from cell cultures for plant improvement. *Theoretical and Applied Genetics* 60, 197–214. DOI: 10.1007/bf02342540

Lastuvka, M., Benech-Arnold, R. and Windauer, L. (2021) A stratification thermal time-based model as a tool for designing efficient methodologies to overcome seed dormancy constraints to kiwifruit seedling production. *Scientia Horticulturae* 277, 109796. DOI: 10.1016/j.scienta.2020.109796

Lawes, G.S. (1990) Propagation of kiwifruit. In: Warrington, I.J. and Weston, G.C. (eds) *Kiwifruit: Science and Management*. Ray Richards Publisher in association with the New Zealand Society for Horticultural Science, Auckland, New Zealand, pp. 297–321.

Lawes, G.S. and Anderson, D.R. (1980) Influence of temperature and gibberellic acid on kiwifruit (*Actinidia chinensis*) seed germination. *New Zealand Journal of Experimental Agriculture* 8, 277–280. DOI: 10.1080/03015521.1980.10426272

Lloyd, G. and McCown, B.H. (1980) Commercially-feasible micropropagation of mountain laurel, *Kalmia latifolia*, by use of shoot-tip culture. *Combined Proceedings – International Plant Propagators' Society* 30, 421–427.

Maghdouri, M., Ghasemnezhad, M., Rabiei, B., Golmohammadi, M. and Atak, A. (2021) Optimizing seed germination and seedling growth in different kiwifruit genotypes. *Horticulturae* 7(9), 314. DOI: 10.3390/horticulturae7090314

Mathew, L., McLachlan, A., Jibran, R., Burritt, D.J. and Pathirana, R. (2018) Cold, antioxidant and osmotic pre-treatments maintain the structural integrity of meristematic cells and improve plant regeneration in cryopreserved kiwifruit shoot tips. *Protoplasma* 255, 1065–1077. DOI: 10.1007/s00709-018-1215-3

Minardi, P., Ardizzi, S. and Lucchese, C. (2019) A reliable method for *Pseudomonas syringae* pv. *actinidiae* detection in kiwifruit explants during in vitro propagation. *Acta Horticulturae* 1243, 103–108. DOI: 10.17660/ActaHortic.2019.1243.17

Mo, Q.-H., Li, J.-W., Gong, H.-J., Ye, K.-Y., Jiang, Q.-S. *et al.* (2016) Reproductive biology of endangered plant *Actinidia chrysantha*. *Guangxi Zhiwu/Guihaia* 36, 640–645.

Moncaleán, P., Cañal, M.J., Fernández, H., Fernández, B. and Rodríguez, A. (2003) Nutritional and gibberellic acid requirements in kiwifruit vitroponic cultures. *In Vitro Cellular & Developmental Biology-Plant* 39, 49–55. DOI: 10.1079/ivp2002371

Murashige, T. and Skoog, F. (1962) A revised medium for rapid growth and bio assays with tobacco tissue cultures. *Physiologia Plantarum* 15, 473–497. DOI: 10.1111/j.1399–3054.1962.tb08052.x

Nagata, K., Morinaga, K. and Ikeda, F. (1997) Photosynthetic ability of kiwifruit (*Actinidia deliciosa*) and characteristics of Shima-sarunashi (*Actinidia rufa*) rootstock. *Bulletin of the Shikoku National Agricultural Experiment Station* 61, 159–166.

Orlikowska, T., Nowak, K. and Reed, B. (2017) Bacteria in the plant tissue culture environment. *Plant Cell, Tissue and Organ Culture* 128, 487–508. DOI: 10.1007/s11240-016-1144-9

Palombi, M.A. and Damiano, C. (2002) Comparison between RAPD and SSR molecular markers in detecting genetic variation in kiwifruit (*Actinidia deliciosa* A. Chev.). *Plant Cell Reports* 20, 1061–1066. DOI: 10.1007/s00299-001-0430-z

Pathirana, R., Mathew, L., Jibran, R., Hunter, D.A. and Morgan, E.R. (2019) Cryopreservation and cryotherapy research on horticultural crops in New Zealand. *Acta Horticulturae* 1234, 29–35. DOI: 10.17660/ActaHortic.2019.1234.4

Pathirana, R., Mathew, L. and McLachlan, A. (2021) A simplified method for high recovery of kiwifruit (*Actinidia* spp.) shoot tips after droplet vitrification cryopreservation suitable for long-term conservation. *Plant Cell, Tissue and Organ Culture* 144, 97–102. DOI: 10.1007/s11240-020-01860-z

Peticilă, A.G., Madjar, R., Scăeteanu, G.V. and Călin, C. (2016) Rooting evaluation for green cuttings at *Actinidia* sp. under rhizogene hormonal treatments. *Romanian Biotechnological Letters* 21, 11160–11165.

Piccotino, D., Massai, R., Dichio, B. and Nuzzo, V. (1997) Morphologicial and anatomical modifications induced by *in vitro* propagation of kiwifruit plants. *Acta Horticulturae* 444(1), 127–132. DOI: 10.17660/ActaHortic.1997.444.17

Prado, M.J., Herrera, M.T., Vázquez, R.A., Romo, S. and González, M.V. (2005) Micropropagation of two selected male kiwifruit and analysis of genetic variation with AFLP markers. *HortScience* 40, 740–746. DOI: 10.21273/hortsci.40.3.740

Quoirin, M. and Lepoivre, P. (1977) Improved media for in vitro culture of *Prunus* sp. *Acta Horticulturae* 78, 437–442. DOI: 10.17660/ActaHortic.1977.78.54

Saeiahagh, H., Mousavi, M., Wiedow, C., Bassett, H.B. and Pathirana, R. (2019) Effect of cytokinins and sucrose concentration on the efficiency of micropropagation of 'Zes006' *Actinidia chinensis* var. *chinensis*, a red-fleshed kiwifruit cultivar. *Plant Cell, Tissue and Organ Culture* 138, 1–10. DOI: 10.1007/s11240-019-01597-4

Sedaghathoor, S. and Noie, M. (2016) Study on different grafting methods of kiwifruit 'Hayward' on the 'Matua' and 'Bruno' rootstocks. *Fruits* 71, 275–280. DOI: 10.1051/fruits/2016024

Sekhukhune, M.K., Nikolova, R.V. and Maila, M.Y. (2018a) Effect of cold stratification and gibberellic acid on in vitro germination of *Actinidia arguta* and *Actinidia chinensis*. *Acta Horticulturae* 1204, 65–76. DOI: 10.17660/ActaHortic.2018.1204.9

Sekhukhune, M.K., Maila, M.Y., Nikolova, R.V. and Mphosi, M.S. (2018b) Preliminary studies on in vivo germination of *Actinidia arguta* and *Actinidia chinensis*. *Acta Horticulturae* 1204, 123–131. DOI: 10.17660/ActaHortic.2018.1204.16

Smith, R.L. and Toy, S.J. (1967) Effects of stratification and alternating temperatures on seed germination of the Chinese gooseberry, *Actinidia chinensis* Planch. *Proceedings of the American Society for Horticultural Science* 90, 409–412.

Standardi, A. (1981) Micropropagazione dell'*Actinidia chinensis* Pl. mediante coltura in vitro di apici meristematici. *Rivista di Frutticoltura e Ortofloricoltura* 43(1), 23–27.

Tyson, J.L., Manning, M.A., Mellow, K.D. and Vergara, M.J. (2018a) *Pseudomonas syringae* pv. *actinidiae* survival in point-inoculated kiwifruit vines. *New Zealand Plant Protection* 71, 45–50. DOI: 10.30843/nzpp.2018.71.136

Tyson, J.L., Vergara, M.J., Butler, R.C., Seelye, J.F. and Morgan, E.R. (2018b) Survival, growth and detection of *Pseudomonas syringae* pv. *actinidiae* in *Actinidia in vitro* cultures. *New Zealand Journal of Crop and Horticultural Science* 46, 319–333. DOI: 10.1080/01140671.2017.1414064

Windauer, L.B., Insausti, P., Biganzoli, F., Benech-Arnold, R. and Izaguirre, M.M. (2016) Dormancy and germination responses of kiwifruit (*Actinidia deliciosa*) seeds to environmental cues. *Seed Science Research* 26, 342–350. DOI: 10.1017/s0960258516000192

Wu, J.H. (2018) Establishment of in vitro culture systems for breeding new types of kiwifruit and for *Actinidia* genomics studies. *Acta Horticulturae* 1224, 195–201. DOI: 10.17660/ActaHortic.2018.1224.26

Wu, J.H. and Datson, P.M. (2015) Development of a method for in vitro germination of *Actinidia* seedlings to enable Psa bioassays and ploidy manipulation. *Acta Horticulturae* 1096, 383–388. DOI: 10.17660/ActaHortic.2015.1096.45

Zenginbal, H. and Özcan, M. (2013) The effects of different treatments on propagation by hardwood cuttings in Hayward and Matua kiwifruit cultivars. *Anadolu Tarim Bilimleri Dergisi* 28(3), 115–125.

Zenginbal, H. and Özcan, M. (2014) The effects of cutting time, bud number and IBA concentration on the cutting on rooting of kiwifruit. *Anadolu Tarim Bilimleri Dergisi* 29, 1–11. DOI: 10.7161/anajas.2014.29.1.1

Zenginbal, H., Özcan, M. and Demir, T. (2006) An investigation on the propagation of kiwifruit (*Actinidia deliciosa*, A. Chev.) by grafting under Turkey ecological conditions. *International Journal of Agricultural Research* 1, 597–602. DOI: 10.3923/ijar.2006.597.602

9 Genetic Improvement, Kiwifruit Genome and Dioecy

Raffaele Testolin[1]* and Mark A. McNeilage[2]
[1]*University of Udine, Udine, Italy;* [2]*The New Zealand Institute for Plant and Food Research Ltd, Auckland, New Zealand*

9.1 The Goals of Kiwifruit Breeding: A Quick Overview

Kiwifruit (*Actinidia chinensis*) is one of the few species domesticated in the 20th century. Originating in China, it was introduced into New Zealand in 1904 and after a few decades of limited cultivation, it became a fruit crop symbol of that country. The export of fruit to Europe in the 1960s aroused the interest of French and Italian fruit growers, who developed the industry in Europe, followed by growers in many other countries. Now kiwifruit represents ~0.2% of the total world fruit industry (FAOSTAT, 2020).

For nearly 40 years, world kiwifruit production was based on a single green-fleshed pistillate cultivar, 'Hayward', of a single variety, *A. chinensis* var. *deliciosa*. 'Hayward' dominated in New Zealand from the early 1970s and was the basis of the kiwifruit industry there and in most other countries for 35–40 years. 'Hort16A' (*A. chinensis* var. *chinensis*), the first yellow-fleshed cultivar developed outside China, started to become commercially important in the early 2000s (see Chapter 1).

Since the 1970s, Chinese scientists have intensively explored the natural genetic resources of *Actinidia*, with hundreds of accessions now stored in several repositories in China (Huang and Ferguson, 2007; Huang, 2016a, b). Only a very limited number of these accessions are available outside China because of restrictions imposed by the Chinese government following the Convention on Biological Diversity of 1992 (also known as the Biodiversity Treaty of Rio (https://www.cbd.int/rio/ - accessed 12 July 2023)) and the Nagoya Protocol of 2010 (https://www.cbd.int/abs - accessed 12 July 2023), which came into effect in 2014. These were designed to promote the conservation of biodiversity and ensure the sustainable use and equitable sharing of genetic resources (https://treaties.un.org/ - accessed 12 July 2023).

The narrow genetic diversity of kiwifruit in worldwide cultivation jeopardizes the survival of the crop in most countries through vulnerability to pests and diseases. Little can be done to combat this in the absence of adequate genetic resources (Hanley, 2018). The spectacular genetic diversity of the genus *Actinidia* (Cui *et al.*, 2002; Huang and Ferguson, 2007; Huang 2016a, b) holds the potential to help breeders meet market demand for fruit that are novel in terms of appearance and quality, and to improve fruit quality and storage, mitigate the impact of fruit allergens, achieve hermaphrodite selections,

*Corresponding author: raffaele.testolin@uniud.it

© CAB International 2024. *Kiwifruit: Botany, Production and Uses* (eds A.C. Richardson, J.N. Burdon and A.R. Ferguson)
DOI: 10.1079/9781800620933.0009

and adapt the crop to different climates and soils (see Chapters 3 and 4).

9.2 Reproductive Biology of Kiwifruit and Constraints for the Breeder

9.2.1 Dioecy

All *Actinidia* species are functionally dioecious, with male and female flowers carried separately on different plants (Ferguson, 1984, 1991; Huang and Ferguson, 2007). Male individuals bear staminate flowers with numerous stamens producing viable pollen and have only a rudimentary ovary that lacks ovules. Female individuals bear pistillate flowers with a well-developed ovary containing hundreds of ovules, and have stamens that produce sterile pollen (McNeilage, 1991a) (see Chapter 3).

Dioecy means that crosses are possible only between genotypes of the different sexes, leading to three severe constraints for breeders:

- direct crosses between selected female parents are not possible;
- when crossing a female and male, the genetic contribution of the male parent to the fruit characteristics of the progeny is not known and only a prior progeny test carried out on male genotypes can provide this information (Testolin *et al.*, 2011; Pinto *et al.*, 2018a);
- in a typical F_1 population, dioecism results in a final female-to-male ratio of 1:1, with male offspring representing a waste of land and resources to breeders who are usually only interested in selecting fruit-bearing individuals (Hanley, 2018).

9.2.2 Ploidy and breeding

Diploids (2*x*), tetraploids (4*x*), hexaploids (6*x*), octaploids (8*x*) and decaploids (10*x*) have been found in wild *Actinidia* populations and many taxa contain ploidy races (Ferguson *et al.*, 1997) (see Chapter 3). Ploidy races within a taxon are not necessarily distinguishable morphologically and the ploidy of any genotype in a kiwifruit improvement programme should be checked.

Differences in ploidy can complicate attempts to cross species or even ploidy races within a species. Pollination may fail or embryos may abort, probably because of endosperm failure. However, *Actinidia* fruit, particularly if the flowers are hand-pollinated, can contain hundreds of seeds, and even if most seeds from an interploidy cross fail, a few viable seeds can result.

It is important to check the ploidy of the resultant seedlings because they may not be at the expected ploidy level. For example, when Seal *et al.* (2012) crossed a hexaploid *A. chinensis* var. *deliciosa* female with 24 diploid *A. chinensis* var. *chinensis* males, only 17 of the 24 crosses produced viable seed. Three-quarters of the resultant seedlings were tetraploid as expected, but there were also triploids, pentaploids and hexaploids as well as aneuploids and one mixoploid. The pentaploids could have resulted from fertilization by numerically unreduced gametes from the male diploids. Other examples of interploid crosses producing seedlings at a range of ploidies are given in Ferguson and Wu (2016). The plants produced may not be useful: in interploid crosses within the *A. chinensis* complex, triploids are usually sterile, whereas pentaploids can set fruit that contain very few seeds.

Ploidy in *Actinidia* can be manipulated in several ways (Ferguson and Wu, 2016) but use of numerically unreduced gametes has proved particularly useful in interploidy crosses where it is possible to select the desired cytotypes from hundreds of seedlings using flow cytometry (Ferguson *et al.*, 2009). Numerically unreduced gametes are produced by many different *Actinidia* genotypes (Seal *et al.*, 2012) and can be used to transfer desirable characters from lower to higher ploidy levels. The New Zealand breeding programme that produced the *A. chinensis* var. *chinensis* cultivar 'Zesy002' involved the selection of tetraploids produced from numerically unreduced gametes of diploid females crossed with tetraploid males (Hanley, 2018).

9.2.3 Long juvenility

In most controlled crosses within the *A. chinensis* complex, 90% of seedlings progress through juvenility and flower within 4–5 years (Testolin *et al.*, 1995a; Ferguson *et al.*, 1996). Other *Actinidia* species can have a shorter juvenile phase (Ferguson *et al.*, 1996). Males tend to flower earlier, so that the 1:1 female:male ratio expected in ordinary crosses is seldom reached before the population is culled.

Forcing continuous growth does not lead to earlier flowering; however, interrupting growth by cold treatment without pruning reduced the length of juvenility in several cross combinations (Davies, 1992). Recently, CRISPR/Cas9-mediated genome editing of the *CENTRORADIALIS* (CEN)-like genes transformed a tissue-cultured kiwifruit explant into a compact plant with rapid terminal flower and fruit development (Varkonyi-Gasic *et al.*, 2018). Such induced precocity could be a valuable means to accelerate breeding programmes (Herath *et al.*, 2023).

9.2.4 The large area required to accommodate cross progeny

In commercial kiwifruit orchards, each vine occupies 5–30 m^2. Breeders tend to reduce the space dedicated to seedlings (Fig. 9.1), considering that after 2–3 years of observation the original seedlings will be cleared, except for those selected for advanced evaluation (Ferguson *et al.*, 1996). Nevertheless, the space allocated to individual vines is large compared, for instance, to the progeny of annual plants. In fact, it is difficult to accommodate more than 600 kiwifruit seedlings per 1000 m^2. This means that kiwifruit breeders are usually forced to work with small families (100–1000 individuals/cross combinations) and this in turn limits the possibility of finding rare recombinants. Moreover, there is an unwritten rule that 'one cross be cleared for each new cross', just to remind breeders that orchard land is not unlimited. Nevertheless, traditional breeding of kiwifruit usually requires large numbers of seedlings to be planted; e.g. over the last decade the New Zealand kiwifruit breeding programme has involved planting out many thousands of seedlings each year.

9.3 Traditional Breeding

9.3.1 Domestication and pre-breeding

Kiwifruit domestication was the result of the fortunate introduction in 1904 of kiwifruit seed (*A. chinensis* var. *deliciosa*) into New Zealand by the school headmistress Isabel Fraser after a visit

Fig. 9.1. Although breeders grow seedlings at close spacings (1 m or less within row), the orchard space required for cross families is still very large.

to her sister in Yichang, China (Ferguson and Bollard, 1990) (see Chapter 1). From these seeds, in one or perhaps two generations, 'Hayward' was selected together with other cultivars now maintained in the repositories for their historical significance (Ferguson and Bollard, 1990; Ferguson *et al.*, 1996). The rest of the story of the early domestication is one of mutant selection, mainly 'Hayward' mutants, such as the hairless Top Star®, the early-maturing 'Convi 97.001' (Green Light®) and 'Boverlig 02' (Earligreen®), the cylindrical 'Wilkins Super' and Boerica®, and the large 'Tsechelidis' (Ferguson, 1991; Ferguson and Seal, 2008; Sotiropoulos *et al.*, 2009). Few of these are grown extensively.

Starting in 1974, the Chinese government promoted large-scale selection of elite genotypes from wild material (Huang and Ferguson, 2007; Huang, 2016a; Zhong *et al.*, 2022) (see Chapters 1 and 4). This was not breeding as such but had the merit of creating a large genetic reservoir prior to the expansion of agriculture and land development that has reduced the original abundance of wild *Actinidia* species.

9.3.2 Breeding of the *A. chinensis* complex

The longest and most successful kiwifruit breeding programme began around 1979 with controlled cross-pollinations of *A. chinensis* var. *deliciosa* by New Zealand's Department of Scientific and Industrial Research (DSIR) and subsequently the Horticulture and Food Research Institute of New Zealand Ltd, and most recently The New Zealand Institute for Plant and Food Research Ltd (Plant & Food Research; Hanley, 2018). In 2021, Plant & Food Research and Zespri Group Limited jointly formed a separate company, The Kiwifruit Breeding Centre (KBC), to run this breeding programme. In 1977, the New Zealand programme accelerated following the introduction of yellow-fleshed accessions of *A. chinensis* var. *chinensis* from China. As a result, the yellow-fleshed *A. chinensis* var. *chinensis* 'Hort16A' was commercialized by Zespri Group Limited as the first product from a major breeding programme and the first globally available yellow-fleshed cultivar. Since then, the New Zealand breeding programme has produced several other commercial cultivars including: 'Zesy002', a Psa (*Pseudomonas syringae* pv. *actinidiae*)-tolerant, yellow-fleshed cultivar as a replacement for 'Hort16A'; 'Zes008', a red-fleshed cultivar; and the associated pollenizers.

Other breeding programmes started in different countries have produced green-fleshed cultivars such as 'Koryoku' in Japan, 'Haewon' in South Korea and 'Summerkiwi 3373' in Italy, and yellow-fleshed cultivars such as 'Sanuki Gold' in Japan, 'Haegeum', 'Jecy Gold' and 'Halla Gold' in South Korea, and 'Soreli' and 'AC 1536' in Italy (Ferguson *et al.*, 2012; Kim *et al.*, 2012). The small number of new cultivars selected during the development of kiwifruit breeding programmes up to the end of the 20th century is probably the result of the narrow genetic base available to breeders outside China. However, it may also be the direct result of marketing strategies, with marketers often preferring a limited range of standardized products.

Since 1974, the richness of *Actinidia* genetic resources in China has allowed the development of 156 new kiwifruit selections, mainly selected from the wild. Of these, 79 were of *A. chinensis* var. *chinensis*, 33 of *A. chinensis* var. *deliciosa*, 31 of *A. arguta*, 5 of *A. eriantha* and 8 were interspecific hybrids (Zhong *et al.*, 2022). 'Hongyang' and 'Donghong' (red-fleshed), and 'Xuxiang', 'Cuixiang' and 'Guichang' (green-fleshed) are some of the more successful cultivars developed in China, all of which, except the red-fleshed cultivars, are grown exclusively there (Zhong *et al.*, 2022), where they are appreciated for their sweetness (see Chapter 4).

9.3.3 Other *Actinidia* species

A. arguta, *A. eriantha* and *A. kolomikta* are the three other *Actinidia* species used to produce cultivars for commercial production. Cultivars from these species are the result of limited breeding programmes in the former USSR, China, the USA, New Zealand, France, Italy and Japan (Huang, 2016a; Zhong *et al.*, 2022) (see Chapter 7).

9.3.4 Interspecific *Actinidia* hybrids and somatic hybridization

The recognized pioneer of interspecific hybridization in China is Wang Shengmei, who crossed

A. eriantha × *A. chinensis* in 1984. The eventual result was the interspecific hybrid 'Jinyan', released by the Institute of Botany of Wuhan, Chinese Academy of Sciences in 2006 (Zhong *et al.*, 2012).

Interspecific crosses were also attempted at the Institute of Botany (China), Plant & Food Research (New Zealand), the University of Paris VI (France), the University of Udine (Italy) and, to a minor extent, in other countries. These studies demonstrated that distant *Actinidia* species could be crossed, even species of the two most divergent taxonomic sections, the *Leiocarpae* and the remainder of the genus (Ke *et al.*, 1992; Hirsch *et al.*, 2001; reviewed in Ferguson and Wu, 2016). Embryo rescue of immature seeds can also be valuable in rescuing particular cross combinations (Hirsch *et al.*, 2001).

Somatic hybridization, the fusion of protoplasts from different genotypes, remained, until recently, a little exploited approach to pool genes from different *Actinidia* species, ploidies, and even individuals of the same sex. Methods of somatic embryogenesis followed by protoplast isolation from multiple *Actinidia* tissue types have, however, been developed and successful fusion has been reported (reviewed in Hanley, 2018; Rey *et al.*, 2020).

9.3.5 Selection procedures

The controlled cross

Before flowers open, both male and female parents are isolated with paper bags to avoid contamination. Pollen is extracted from male flowers that were collected at the balloon stage (just before flowers open) and stored at 4°C until use. Flowers of female plants are hand pollinated by brushing pollen collected from male flowers onto the female styles. Emasculation of female flowers is not required because pollen of female flowers is sterile. After petal fall and fruit set, the paper bags are removed and replaced with net bags until fruit harvest. In Italy, fruit are collected over-ripe, seeds are extracted, dried at room temperature for two days and stored in paper bags at 4°C until midwinter, when seeds are moistened overnight and sown in the greenhouse (see Chapter 8).

Seedling growth

Seeds are sown in trays filled with commercial potting mix (usually ~60% blond peat, ~40% black peat), covered with a thin layer of fine vermiculite and moistened. Seeds need to overcome dormancy to germinate. Controlled growth chambers are expensive, and in Italy (northern hemisphere) alternating day-night temperatures can instead be achieved by storing the stacked trays, covered with polyethylene, for 45 days against a shaded north-facing wall – this will allow the day-night alternating temperatures to overcome seed dormancy. Trays are then transferred to an unheated greenhouse (up to 28°C by day and down to 10°C at night) for germination. Young seedlings with two leaves are transplanted to larger-hole trays and subsequently into 1 L pots and grown in a heated greenhouse. The following season, seedlings are transplanted to the open field at within-row spacing of 1 m or less.

Fruiting selections

Female seedlings are first observed as single plants for two or more fruiting seasons. Superior genotypes are then evaluated as replicated plants and at different locations for three or more cropping seasons. In both cases, the breeder observes variables such as phenological traits (e.g. bud break, flowering time, ripening time) and reproductive traits (e.g. number of flowers per cyme, cropping, fruit size, shape, quality, flavour and storage). Susceptibility to pests and diseases must also be considered. Of course, particular attention is dedicated to the key traits that represent the objectives of the specific breeding programme.

In commercial breeding programmes, a key step is an overall assessment that considers all selection criteria simultaneously. This can help reduce the number of seedlings on which to carry out the full set of field observations. Once field observations have been collected, it is possible to apply a multivariate function that combines the main observed traits, each multiplied by a weighted coefficient that determines the relative importance attributed by the breeder to that variable. Such an approach produces a ranking list of the best selections. The equation calculated for each individual has the form:

$$S1 = k1{*}var1 + k2{*}var2 + \ldots kn{*}varn$$

where SI is the selection index and k1, k2 ... kn are the relative weights assigned to the variables var1, var2 ... varn, respectively. The variables are first normalized (i.e. transformed to the range from 0 to 1). Variables with a negative biological meaning (e.g. susceptibility to a disease) are, in turn, transformed as 1− varx.

These two steps form the basis of the evaluation track for new selections. However, the best selections are further evaluated in 'demonstration orchards', where the selections, grown on small commercial blocks of 10–100 plants, are shown to growers to get their feedback and test commercial suitability. Finally, selections can enter the market or – better – undergo a large consumer test in key world markets because consumer expectations are sometimes unpredictable.

Selection cannot be considered complete until storage trials are performed to identify fruit that can be stored for up to six months (to ensure a full year of supply to markets with product from both hemispheres) and that remain firm when shipped over long distances.

Selections with commercial potential are often legally protected through plant-variety rights (see Chapter 4).

Pollenizers

The evaluation of pollenizers differs from that of female cultivars as different aspects are tested. As well as observations on the time and duration of flowering, it is necessary to select male genotypes for which flowering overlaps with the female cultivars for which they are selected as pollenizers. The analysis of pollen germination or viability and testing of 'fruit sizing' and 'seed setting', through controlled pollination of the female cultivars, is also important (McNeilage *et al.*, 1992). Fruit sizing is the capacity of the male to contribute to good-sized fruit, while seed setting is the number of seeds produced per fruit; individual fruit weight in kiwifruit is highly correlated with seed number (McPherson *et al.*, 2001). These features are considered important for a good pollenizer (see Chapter 5).

Rootstocks

Selection of rootstocks is also different, though there have been very few rootstock breeding programmes (see Chapter 6). A rootstock is often needed to correct a specific weakness found in other cultivars (poor adaptability to heavy soils, poor resistance to chlorosis, sensitivity to a given soilborne pathogen etc.). In such cases, a preliminary screening of a germplasm collection can result in the selection of genotypes carrying the trait of interest.

The candidate rootstocks are vegetatively propagated and grafted cultivars are trialled under different soil and climatic conditions. An essential requirement is graft compatibility, which is not guaranteed if a rootstock and scion belong to genetically unrelated species. Graft incompatibility may take many years to be obvious. Hence rootstock selection can be prolonged, taking 20 years or more.

9.4 Population Genetics Underlying Breeding Programmes

9.4.1 Mating designs for the study of population genetics

Dioecy means that only female × male crosses are possible. This reduces the types of mating design suitable for kiwifruit breeding programmes. In practice, the North Carolina Mating II (NCM2) design is the only one commonly used, in which a number of female genotypes (n) are crossed with a number of male genotypes (m), where m and n can be different. Such a mating design allows simultaneous estimates of the heritability of traits and the breeding value of both female and male parents, by analysis of general and specific combining abilities.

9.4.2 Heritability of traits

The heritability of traits was the first genetic analysis inferred through controlled crosses. When it was carried out, it was considered key information for the choice of selection procedure. Traits with low narrow sense heritability ($h^2 < 0.2$) are difficult to improve and, in any case, would require preliminary selection amongst the cross families before proceeding to selection within families, while traits with high heritability ($h^2 > 0.5$) are easily improved even with selection limited to progeny within a cross family. In kiwifruit, heritability appears reasonably high for ripening time and fruit weight, but lower for other characters such as flowering time and crop yield (Table 9.1). It is worth noting that the variation in heritability reported by different authors for the same trait depends on the mating design, the genotypes selected and their range of variation in the trait studied (Pinto *et al.*, 2018a). This, in turn, means that the estimate of heritability must be considered for the particular genotypes tested and caution must be taken in extrapolating the information to other sets of parents.

9.4.3 Combining ability and progeny testing

Breeding for quantitative traits requires the preliminary evaluation of general and specific combining ability in the parental genotypes (Testolin *et al.*, 2011). General combining ability is the average genetic performance of a given genotype when crossed with a number of other genotypes; it represents the capacity of a given parent to transfer to the progeny the improved characters that it carries. Specific combining ability is the genetic performance of the same genotype in each individual cross combination. It is possible to evaluate the combining ability of parents and carry out practical selection at the same time (Testolin *et al.*, 2011; Pinto *et al.*, 2018a).

9.5 Molecular Approaches and Genome-based Breeding

'... amazingly, there once was a time when it was not possible to sequence DNA. To a molecular biologist working today it is almost unimaginable that research could progress without DNA sequencing' (Brown, 1994). Indeed, genetics and breeding progressed considerably in the past without any information on DNA, but it is clear that information on the sequence of genes and their organization in the genome greatly facilitates the work of breeders.

9.5.1 Molecular markers, genetic maps and marker-assisted breeding

In kiwifruit, as in many other species, sequences of specific genes became available only recently. Initially, it was easier to search for molecular markers, i.e. sequences that have unknown function but are easy to identify, associated to phenotypic traits. These markers allowed the trait segregation in the progeny to be followed by simply analysing for the marker associated with that trait.

Several association maps were produced for diploid *A. chinensis* var. *chinensis*. Initially, they were sparsely saturated (Testolin *et al.*, 2001; Fraser *et al.*, 2009), but with the advent of the genotyping by sequencing technologies, the maps produced were so deeply saturated with thousands of single nucleotide polymorphism (SNP) markers that they became the main basis for the assembly of kiwifruit genomes (Huang *et al.*, 2013; Scaglione *et al.*, 2015a; Zhang *et al.*, 2015).

Genetic maps constructed from diploid parents can also provide useful information for polyploids, although specific maps for the many tetraploid accessions of *A. chinensis* var. *chinensis* and for hexaploid *A. chinensis* var. *deliciosa* would be highly desirable. For a number of reasons

Table 9.1. Estimates of narrow-sense heritability for kiwifruit traits from published data obtained from different sets of cross parents.

	Narrow sense heritability (h^2) cited in:				
Trait	Beatson (1991)	Testolin *et al.* (1995b)	Zhu *et al.* (2002)	Marsh *et al.* (2003)	Pinto *et al.* (2018a)
Time of bud burst	–	0.62 ± 0.65	–	–	0.38 ± 0.08
Flowering time	–	0.27 ± 0.39	0.11 ± 0.10	–	0.14 ± 0.04
Ripening time	0.61 ± 0.22	0.22 ± 0.20	–	–	
Fruit weight	0.81 ± 0.31	0.27 ± 0.23	0.52 ± 0.08	0.78 ± 0.59	0.32 ± 0.07
Crop yield	0.01 ± 0.14	–	0.46 ± 0.11	0.23 ± 0.20	0.05 ± 0.02

linked to the homeologous chromosomes that compose the polyploid genomes and the biallelic status of SNP markers, genetic maps for polyploid kiwifruit might provide only approximate results. They have been attempted in hexaploid *A. chinensis* var. *deliciosa*. One such linkage map consisted of 3686 and 3940 markers organized in 183 and 176 linkage groups for the female and male parents, respectively (Popowski *et al.*, 2021). Linkage maps have been used to determine the position of several quantitative trait loci that control traits of interest for breeding, such as time of bud break, time of flowering, number of flowers per cyme, fruit weight, dry matter accumulation and storage firmness (Scaglione *et al.*, 2015b; Pinto *et al.*, 2018a, b).

9.5.2 Genome sequencing and gene annotation

DNA sequencing, genome assembly and gene annotation are the ultimate tools for a breeder, as they make available the entire list of genes and their position in the genome.

Diploid kiwifruit plants have large genomes (1C = 0.7 pg ≈ 670 Mb) (Hirsch *et al.*, 2001), owing to the three putative whole-genome duplications during evolution (Huang *et al.*, 2013; Wu *et al.*, 2019). The first duplication, common to all eudicots, occurred some 150 million years ago (mya), while the second and third occurred ~50–57 mya and ~18–20 mya, respectively (Wang *et al.*, 2018). Genome sequencing of the large genome therefore required considerable investment in funds, DNA-sequencing technologies and bioinformatics expertise.

The first kiwifruit genome published was of the diploid *A. chinensis* var. *chinensis* cultivar 'Hongyang', which has red fruit flesh (Huang *et al.*, 2013). However, the genome contained large gaps, many misplacements and some 163.7 unplaced Mb, as revealed two years later by a well-saturated genetic map (Scaglione *et al.*, 2015a) (Table 9.2). These problems were overcome in the publication of a second assembly of the diploid *A. chinensis* var. *chinensis* selection Red5, where a higher sequencing coverage, an assembly that left unplaced only 5.9 Mb, a better annotation and a manual re-annotation provided a more precise genome assembly and gene prediction (Pilkington *et al.*, 2018). A further sequencing of 'Hongyang' (Wu *et al.*, 2019) and of genotypes of *A. eriantha* (Tang *et al.*, 2019; Yao *et al.*, 2022) contributed to a better understanding of the intra- and interspecific differences in gene composition.

The 29 chromosomes of the haploid kiwifruit series are cumulatively composed of 0.5–0.6 × 10^9 bases. These encode approximately 33,000 genes which represent 5–6% of the estimated genome size, a ratio typical of most land plants sequenced to date (Pilkington *et al.*, 2018).

Genes are expected to show allelic variations that are the basis of kiwifruit genetic diversity. It is this diversity that breeders seek to exploit in their search for genotypes that combine the best alleles for desired traits (Schaffer and Pilkington, 2016). This search is just beginning, and the many genetic variants associated with previously unknown/unexploited traits represent the dedication of kiwifruit geneticists and explorers of *Actinidia* genetic diversity. A few examples of the markers used in the genetic selections, such as for gender (Grainger, 2011), are given in Section 9.7.

Table 9.2. Comparison of the main features of the three published *Actinidia chinensis* var. *chinensis* genomes sequenced and assembled.

Assembly features	Huang *et al.* (2013)	Pilkington *et al.* (2018)	Wu *et al.* (2019)
Assembled genome size (Mb)	452.4	548.0	640.8
Unassigned sequences (Mb)[a]	163.7	5.9	13.1
Total genome size (Mb)	616.1	553.9	653.9
Number of scaffolds	7698	3933	2366
Scaffold N50 (kb)	646.8	575.6	1430.0
Predicted genes (n)	39,040	33,257	40,464

[a]Sequences that could not be assigned to the 29 pseudomolecules representing the chromosomes.

9.5.3 Genomic selection

Genomic selection (GS) is a development of marker-assisted selection (MAS). While MAS is applied to a single character and works with single pairs of marker-trait associations, GS screens the linkage disequilibria of many SNP markers (tens of thousands) regularly scattered along the whole genome, with any phenotypic trait scored in the same individual. The sum of all these linkage disequilibrium scores produces a genome-wide estimate of the breeding value of individuals (Pinto *et al.*, 2018a).

In a typical GS implementation (Fig. 9.2), a relatively small training population, representative of the entire population on which selection will be carried out, is genotyped, analysed with the set of markers and phenotyped for any trait of interest. The prediction of the effect of markers on each trait is calculated from a mathematical model (Meuwissen *et al.*, 2001). The model, once validated, can be used to estimate the breeding value of any genotype of the breeding population, on which only the marker profile need be scored (Hanley, 2018).

GS has been attempted with kiwifruit (Cheng *et al.*, 2019). The training population consisted of 78 female individuals of a red-fleshed diploid *A. chinensis* var. *chinensis* population and was validated on a further ten female individuals. A strong correlation, 0.91, was recorded in predicting fruit firmness, suggesting that males could be selected from this population that would improve fruit storage in progeny.

GS in kiwifruit is not easy and needs to be set up properly, considering the very limited germplasm resources available, the strong population structures and familial relatedness and the ploidy races common in *Actinidia*, where allele dosage confounds functional polymorphism (Cheng *et al.*, 2016). However, GS certainly represents the future of kiwifruit breeding for quantitative traits.

9.6 Transformation and New Breeding Technologies

9.6.1 Genetically modified kiwifruit

The first report of kiwifruit transformation was that of Matsuta *et al.* (1990). This described the basic components of the protocol – the cassette, the plasmid and the use of *Agrobacterium* as vector – and was intended as a tool for studies of functional genomics. The aversion of people in the western world to genetically modified organisms and the questionable results obtained suggested that the technique should be temporarily abandoned. Although Wang *et al.* (2006) subsequently revived the technique, new breeding technologies promised a 'softer' approach to the modification of the DNA sequence which did not leave foreign DNA in the cells.

9.6.2 Genome editing and new genomic technologies

The first reports of gene editing in *Actinidia* date from 2018. Varkonyi-Gasic *et al.* (2018) accelerated

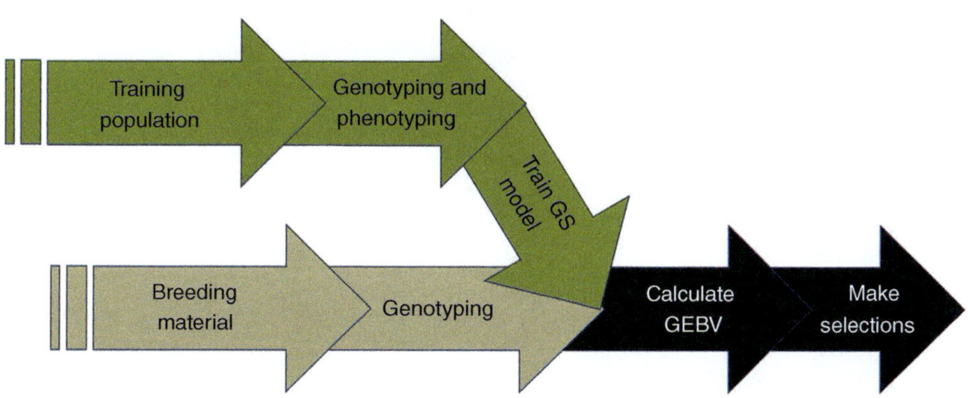

Fig. 9.2. Genomic selection (GS). (GEBV, genome-wide estimate of breeding value) (Adapted from Heffner *et al.*, 2009)

kiwifruit flowering by using CRISPR/Cas9 editing technology to manipulate the *CEN* genes. The implications on prolonged seedling juvenility are discussed in Section 9.2. The same CRISPR/Cas9 editing technology was used to identify the genes controlling dioecy (Akagi *et al.*, 2018, 2019; De Mori *et al.*, 2021) (see Chapter 14). The consequent use in the production of hermaphrodite selections is discussed in Section 9.7.1.

Subsequently, Voogd *et al.* (2021), by using CRISPR/Cas-guided editing, were able to demonstrate that a MADS-box gene with similarity to the *FLOWERING LOCUS C* (*FLC*) of *Arabidopsis* is induced by cold and correlates with epigenetic changes that control bud break in kiwifruit. The knockout of *AcBFT2*, orthologous to *Arabidopsis FLOWERING LOCUS T* (*FT*) family, by CRISPR/Cas9-targeted mutagenesis reduced plant dormancy resulting in an ever-growing phenotype with no adverse effect on flowering (Herath *et al.*, 2022).

These are the first applications of new genomic technologies, but there are laboratories where genes and transcription factors are being sought (Brian *et al.*, 2021) and where these recombinant technologies are already routinely applied; results will soon become known to the scientific community.

9.7 Breeding for Specific Objectives

9.7.1 Hermaphrodite cultivars

Dioecism is not absolute, and inconstant males, also referred to as fruiting males, have been identified among male pollenizers in commercial kiwifruit orchards and germplasm repositories (Fig. 9.3) (McNeilage, 1991a, b). These individuals have bisexual flowers characterized by small ovaries with fewer carpels than those of typical pistillate flowers, fewer ovules per carpel and shorter, thinner styles with small stigmata. They bear small fruit (20–40 g) with only a few dozen seeds (McNeilage, 1991a; Testolin *et al.*, 1995a). However, sexual expression is labile and inconstant and there is a mosaic of staminate, bisexual and mixed inflorescences (McNeilage, 1991b).

Hermaphroditism in *Actinidia* seems erratic but McNeilage and Steinhagen (1998) identified a very rare hermaphrodite of *A. chinensis* var. *deliciosa* within the progeny derived from a cross

Fig. 9.3. An example of an inconstant or 'fruiting' male. These appear irregularly among the progeny in controlled crosses. Hermaphrodite flowers are often interspersed with staminate flowers.

involving an inconstant male parent. In this case, hermaphroditism was stable and transmissible. The difficulty is combining hermaphroditism with all the other vine and fruit attributes required. The production of commercially selectable hermaphrodite individuals through conventional crosses has not yet been reported (see Chapter 4).

In *Actinidia* there is strict genetic control of sex expression by an active Y system, with diploids having homogametic females (XX) and heterogametic males (XY). The same system operates at different ploidy levels, e.g. hexaploid females are XXXXXX and hexaploid males XXXXXY (Testolin *et al.*, 1995a). This is shown by disomic inheritance and the 1:1 sex ratios in the progeny of most controlled crosses at different ploidy levels. The elucidation of the genetic control of gender in kiwifruit (Akagi *et al.*, 2018, 2019) has opened the way to obtaining hermaphrodites through transformation. Two genes control dioecy in *Actinidia*: *SuF* (also known as *Shy Girl* or *SyGI*) regulates the response to cytokinins and suppresses ovary development; *M* (also known as *Friendly Boy* or *FrBy*) is a fasciclin-like gene strongly expressed in the anther cells of the tapetum that ensures the completion of microsporogenesis, independently of *SyGI* (Fig. 9.4). Both genes are located on the Y chromosome (chromosome 25 of the assembly of Huang *et al.*, 2013) within the interval of 500–800 kb.

Hermaphrodite plants could result either by a deletion/silencing of *SyGI* on the Y chromosome that prevents the suppression of the ovary development or by recombination bringing *FrBy* from the Y to the X chromosome. Hermaphrodite genotypes are more readily achieved by silencing *SyGI* in male genotypes, as has already been accomplished by CRISPR/Cas9-based genome-editing technology (Akagi *et al.*, 2019; De Mori *et al.*, 2021).

9.7.2 Fruit flesh colour

Chlorophylls, carotenoids and anthocyanins are responsible for fruit flesh colour in kiwifruit.

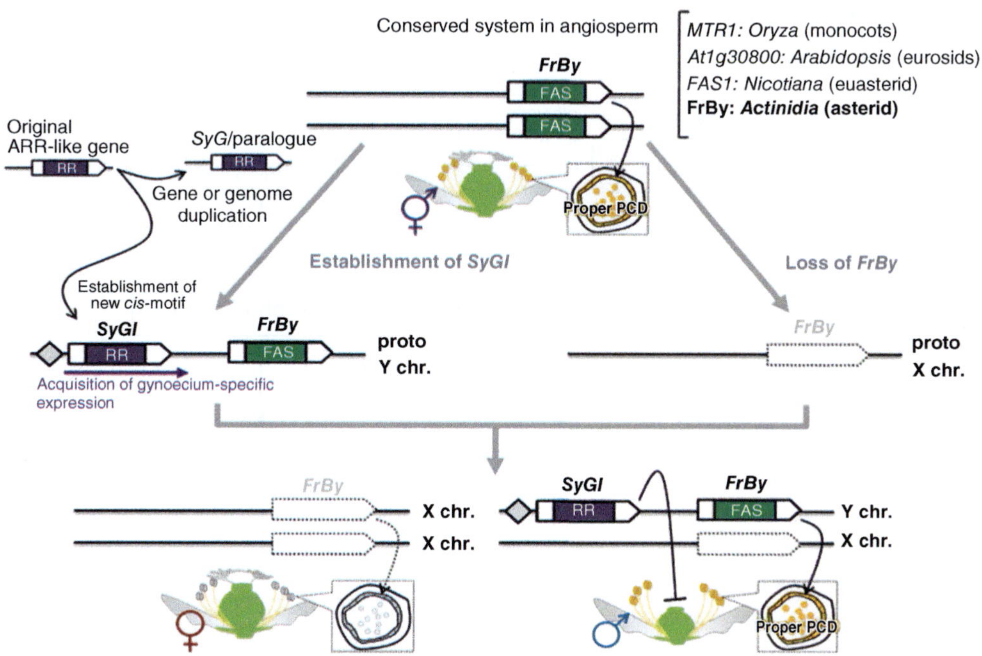

Fig. 9.4. (*Top*) Evolution of sex determining region in kiwifruit; (*bottom*) the Y and X chromosomes in female (XX) and male (XY) individuals of today. (chr., chromosome; FAS, FASCIATA, a class of genes that control the evolution of apical meristems in plants; *FrBy*, *Friendly Boy* gene; PCD, programmed cell death; RR, response regulator gene, a member of the gene family that plays a pivotal role in regulating the developmental responses of plants; *SyGI*, *Shy Girl* gene) (From Akagi *et al.*, 2019)

Breeders are determined to exploit the spectacular colour diversity in *Actinidia* germplasm (see Chapters 3 and 16) by developing novel cultivars whose flesh colour might strongly attract consumers. The manipulation of fruit flesh colour either through MAS of enzyme-coding genes that drive the pathway in favour of one class of pigment or through the more promising new genomic techniques is currently being researched.

Green

Kiwifruit were initially known for their emerald green fruit flesh, a colour symbolic of health and almost uniquely retained when the fruit are ripe. The fruit of all *Actinidia* species accumulate chlorophylls in their flesh, with concentrations varying from one species to another: *A. eriantha* probably accumulates the highest concentrations of chlorophylls, while genotypes of the *A. chinensis* complex cover a range of concentrations.

Species that show different flesh colours (yellow, orange, red etc.) when ripe generally undergo a degreening process, characterized by degradation of chlorophylls (Pilkington *et al.*, 2012; Montefiori *et al.*, 2016). The pathways of degradation of chlorophylls a and b are shown in Fig. 9.5. The end products are finally stored in the vacuole as non-fluorescent chlorophyll catabolites.

Yellow and orange

Carotenoids and xanthophylls are responsible for yellow to orange and occasionally reddish fruit flesh. Chlorophyll degradation unmasks the carotenoids whose synthesis is initiated and sometimes increases during the last phases of fruit development (Fig. 9.6) (Ampomah-Dwamena *et al.*, 2009; Pilkington *et al.*, 2012). The biosynthetic pathway is shown in Fig. 9.7. Carotenoids are the dominant pigments in most yellow-fleshed

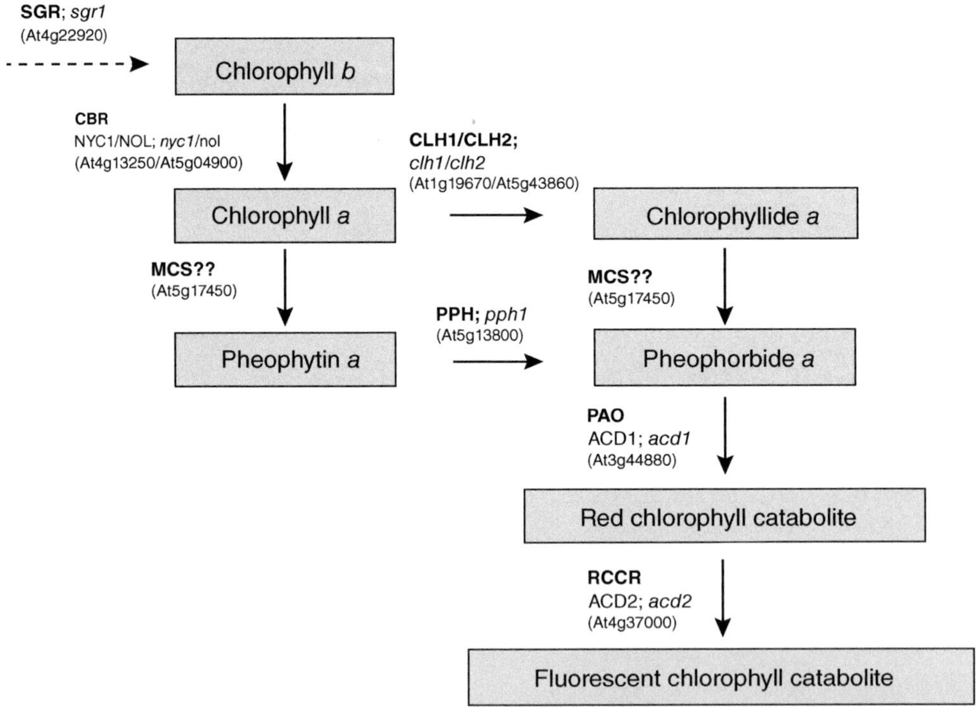

Fig. 9.5. Chlorophyll degradation pathway. (ACD1/2, accelerated cell death 1/2; CBR, chlorophyll *b* reductase; CLH1/2, chlorophyllase 1/2; MCS, metal chelating substance; NOL, NYC1-like; NYC1, non-yellow colouring 1; PAO, pheophorbide *a* oxygenase; PPH, pheophorbide pheophytin hydrolase; RCCR, red chlorophyll catabolite reductase; SGR, stay-green) The orthologous *Arabidopsis* gene is indicated in brackets. (From Montefiori *et al.*, 2016)

Fig. 9.6. Kiwifruit flesh colours and changes during fruit development. (**A**) In this specific example, the red colour of the inner pericarp appears early, 65 days after pollination (DAP), but there are genotypes in which the red colour appears very late in the season and, often, after a period of cooler night temperatures. (**B**) Degradation of chlorophylls, which allows the appearance of the yellow colour of flesh; depends also on genotype either in time and intensity. (**C**) Green-fleshed kiwifruit genotypes do not undergo degreening to the same extent. (Adapted from Liu *et al*., 2017)

A. chinensis var. *chinensis* cultivars, while xanthophylls and their derivatives contribute to the orange colour of *A. polygama* fruit (Montefiori *et al.*, 2009a).

Red

Anthocyanins are generally responsible for red colours in kiwifruit (Montefiori *et al.*, 2009b). They are occasionally synthesized in flower ovaries at the beginning of the season but disappear to reappear again in the later phases of fruit development.

Red-fleshed kiwifruit genotypes accumulate anthocyanins (Fig. 9.8); the presence of the glycosyltransferase F3GT1 enzyme is required for the synthesis of cyanidin 3-O-xylo-galactoside, which is the dominant red pigment in kiwifruit (Montefiori *et al.*, 2011). Degreening, accompanied by synthesis of anthocyanins, results in yellow/red fruit flesh, while a lack of degreening results in green/red fruit flesh (Figs 9.6 and 9.9).

The synthesis of red pigments in kiwifruit is dependent on carbohydrate supply (Nardozza *et al.*, 2015a, 2020) and may be modulated by temperature (Man *et al.*, 2015). Regulatory genes include members of the MYB gene family (Montefiori *et al.*, 2016), which promote the synthesis of the glycosyltransferase enzyme (Li *et al.*, 2017). MYBs and other transcription factors, together with miRNAs that negatively regulate anthocyanin biosynthesis by repressing MYB expression (Huang *et al.*, 2021), are the key genes that might be conveniently manipulated by breeders to enhance the synthesis of red pigments in kiwifruit.

9.7.3 Carbohydrate accumulation

Carbohydrate accumulation is necessary for the high sugar content in kiwifruit. Consumers in different parts of the world tend to differ in their preferences: Asian populations tend to prefer sweeter fruit whereas western populations tend to prefer a balance between sweetness and acidity (Junge *et al.*, 2020). Nevertheless, breeders aim for a high fruit sugar content since that is correlated with consumer liking (see Chapter 21).

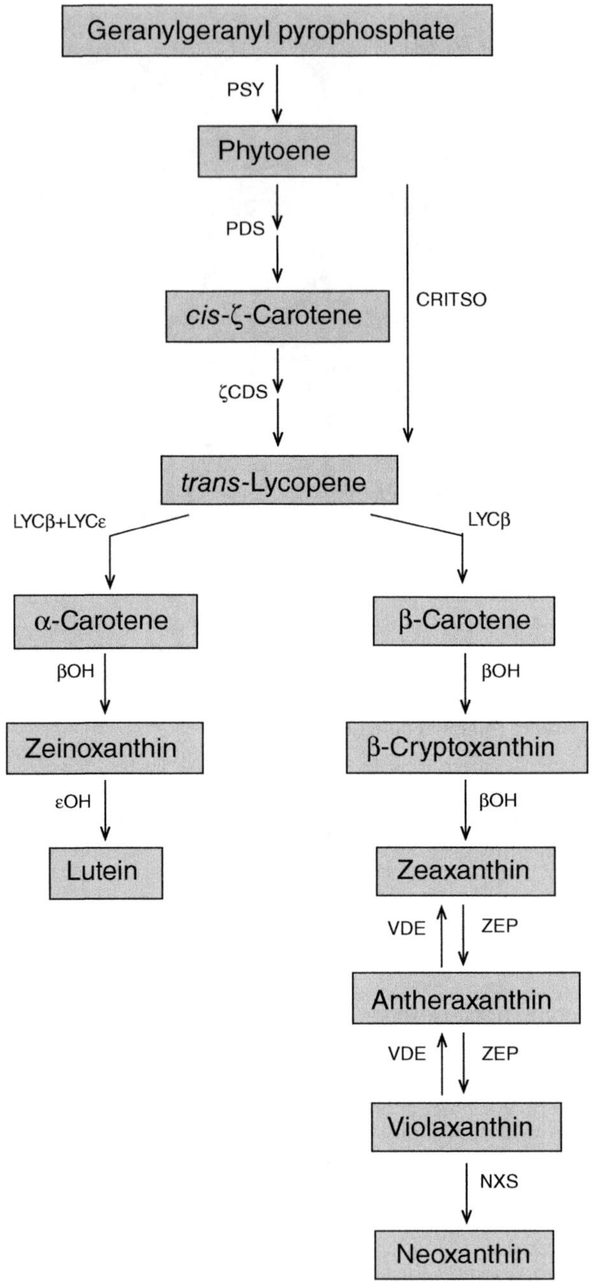

Fig. 9.7. Carotenoid biosynthetic pathway. (βOH, β-ring carotene hydroxylase; CRITSO, carotenoid isomerase; εOH, ε-ring carotene hydroxylase; LCYβ, lycopene β-cyclase; LCYε, lycopene ε-cyclase; NXS, neoxanthin synthase; PDS, phytoene desaturase; PSY, phytoene synthase; VDE, violaxanthin de-epoxidase; ζCDS, ζ-carotene desaturase; ZEP, zeaxanthin epoxidase) (From Montefiori *et al.*, 2016)

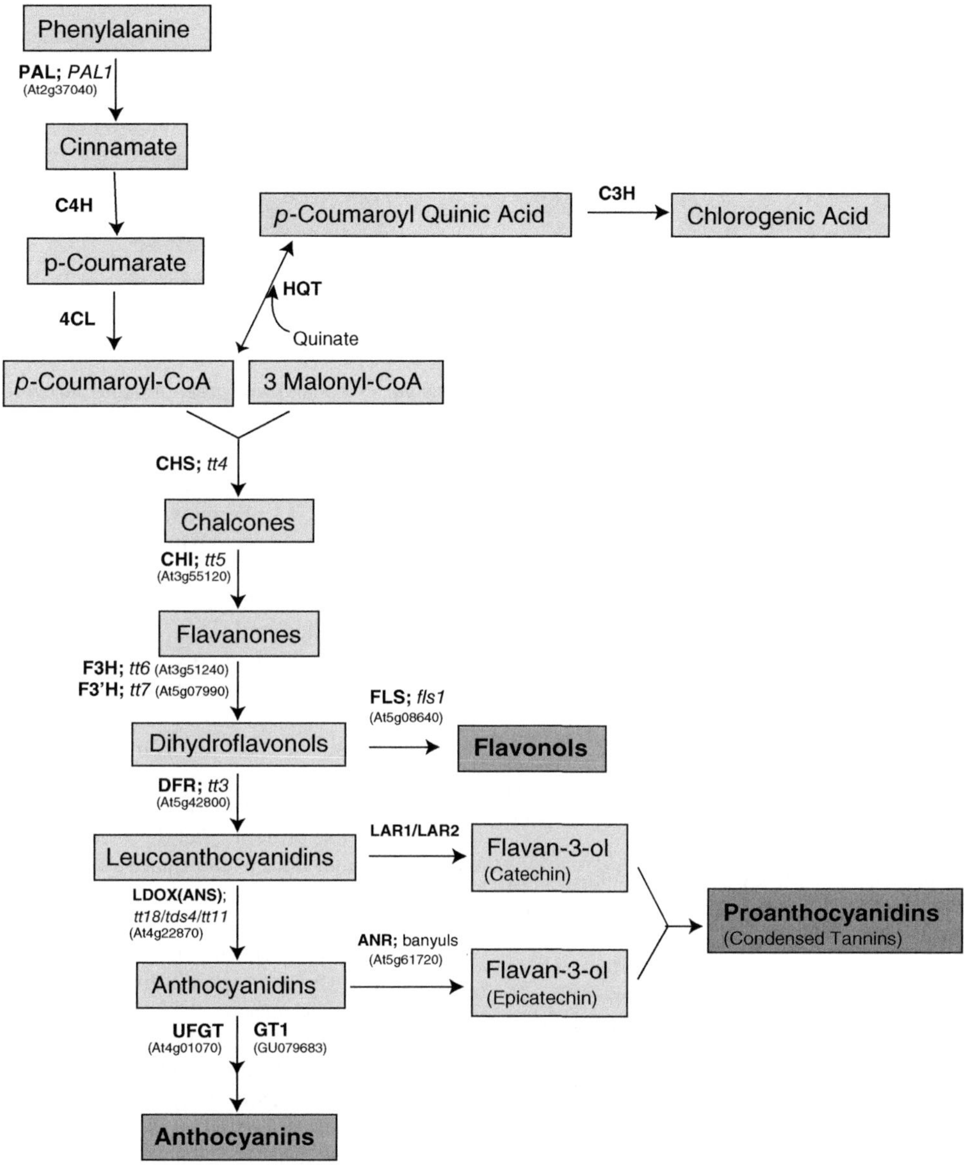

Fig. 9.8. Flavonoid and anthocyanin biosynthetic pathway. (ANR, anthocyanidin reductase; 4CL, 4-coumaryl CoA:ligase; C3H, p-coumarate 3'-hydroxylase; C4H, coumarate 4-hydroxylase; CHI, chalcone isomerase; CHS, chalcone synthase; DFR, dihydroflavonol 4-reductase; F3H flavanone 3-hydroxylase; F3'H, flavonoid 3'-hydroxylase; FLS, flavonol synthase; GT1, anthocyanidin 3-O-glucosyltransferase 1; HQT, hydroxycinnamoyl-CoA quinate transferase; LAR, leucoanthocyanidin 4-reductase; LDOX(ANS), leucoanthocyanidin dioxygenase (anthocyanidin synthase); PAL, phenylalanine ammonia lyase; UFGT, UDP-glucoside: flavonoid glucosyltransferase) The orthologous *Arabidopsis* genes are indicated within the brackets. (From Montefiori *et al.*, 2016)

Fig. 9.9. Two typical bicoloured kiwifruit (*Actinidia chinensis* var. *chinensis*) with green-red and yellow-red flesh. (Photos: (*left*) R. Testolin; (*right*) The New Zealand Institute for Plant and Food Research Ltd)

Starch accumulation and degradation to sugar (see Chapter 16) are under a complex biosynthetic pathway (MacRae *et al.*, 1992) and are therefore considered by breeders to be under polygenic control. Selection follows the rules of the quantitative trait loci analysis.

9.7.4 Skin peelability

Among the many *Actinidia* species, *A. eriantha* has been identified as having an 'easy-to-peel' phenotype that increases the level of convenience associated with kiwifruit by making fruit easier to eat (Huang and Ferguson, 2007; Huang, 2016a).

New Zealand scientists have attempted to understand what controls the easy-to-peel phenotype by investigating chemical and biochemical differences in cell walls, as well as differential expressions of putative genes that would control the easy-to-peel versus poor-peeling genotypes (Atkinson *et al.*, 2009). The lack of isogenic genotypes carrying the contrasting phenotypes has made this difficult and may impede progress in transferring the easy-to-peel trait to other *Actinidia* species.

9.7.5 Overcoming the impact of kiwifruit allergens

Kiwifruit is a highly allergenic food (Wang *et al.*, 2019), and Japanese legislation recommends labelling the presence of kiwifruit in any food. Annotation of the kiwifruit genome allowed the identification of 13 allergens, but the list may still be incomplete (reviewed in Bulley and Laing, 2016). Kiwifruit can trigger a range of allergic symptoms from mild local symptoms of mouth and throat inflammation to the more severe anaphylaxis. Kiwifruit allergy can increase with sensitization to other related allergens such as birch pollen, fruit such as avocados and bananas and legumes such as peanuts (Wang *et al.*, 2019).

Four kiwifruit allergenic proteins, namely Act d 1 (actinidin, a cysteine protease), Act d 2 (thaumatin-like protein), Act d 5 (kiwellin) and Act d 11 (major latex protein), are allergens typical of ripening fruit and are the most critical for kiwifruit consumers (Le *et al.*, 2013). Act d refers to allergens from *A. chinensis* var. *deliciosa*, and Act c to the corresponding allergens from *A. chinensis* var. *chinensis*.

Actinidins are the best known and were the first to be discovered (Fine, 1981); there are many isoforms. Together with many kiwellin isoforms, they may constitute up to 50% of the soluble proteins of the fruit flesh (Bulley and Laing, 2016).

Breeding kiwifruit cultivars with low allergenicity requires several steps. The first is cooperation of kiwifruit geneticists with clinical allergists to better identify the relevant allergens, narrowing down the potential candidates to work with. The second step requires the evaluation of allelic variants and isoforms of the principal allergens and their promoters. It is known from the literature that 'Hongyang', the popular red-fleshed *A. chinensis* var. *chinensis* cultivar, has a low actinidin content because *Act c 1* is not expressed in mature fruit, and this makes 'Hongyang' a potential hypoallergenic cultivar. The promoter of its *Act c 1* gene would represent the first potential target of genetic manipulation (Huang *et al.*, 2013).

This work is unlikely to be easy or to resolve the problem. The large number of individual allergens probably precludes breeding non-allergenic kiwifruit cultivars (Bulley and Laing, 2016).

9.7.6 Fruit storage

Fruit softening is controlled by hormonal and developmental cues, causing an upregulation of cell wall-associated enzymes that break down the complex sugar matrices in the cell wall (Redgwell *et al.*, 1991; Fullerton *et al.*, 2020). The major components of fruit cell walls are pectins, hemicelluloses and cellulose. Pectins are considered to play the key role in maintaining the structural integrity of the cell wall. Most changes during softening involve pectin, xyloglucan and compounds of pectin side-chains, influencing their interactions with cellulose and thereby weakening the cell wall (Redgwell *et al.*, 1992).

The pathway of softening is rather complex (Atkinson and Schröder, 2016). However, a comparison at the microscopic and molecular level of 'fast'- and 'slow'-softening kiwifruit genotypes revealed little difference in the time course of enzymatic activities; cell walls of the fast-ripening genotype seemed simply more loosened, possibly giving cell wall degrading enzymes more space to move and act on their substrates, resulting in a faster softening rate (Fullerton *et al.*, 2020). A recent transcriptomic analysis reported that the transcription factor AdZAT5, which binds and regulates the effector genes pectate lyase and β-galactosidase, may be the key factor for pectin degradation (Zhang *et al.*, 2022). The manipulation of this transcription factor, through stable transformation, would be a way to slow down fruit softening; alternatively, breeders need to use phenotypic selection of progeny derived from controlled crosses.

9.7.7 Resistance to pests and diseases

The first severe disease kiwifruit growers have had to cope with is the bacterial canker caused by *Pseudomonas syringae* pv. *actinidiae* (Psa) biovar 3, which probably originated in China (see Chapter 18). This may have spread through the import of pollen and/or plant material.

Surveys on potential sources of resistance revealed that several species, such as *A. arguta*, *A. polygama*, *A. macrosperma* and *A. guilinensis*, have genotype-wide resistance to the pathogen (Li *et al.*, 2004; Nardozza *et al.*, 2015b; Wang *et al.*, 2020). Genetic analyses to find resistance-related loci have found most of the candidates belong to receptor-like serine/threonine kinase and related genes that are known to play a significant role in plant immunity, especially in early pathogen recognition and activation of plant defence machinery (Fraser *et al.*, 2015; Tahir *et al.*, 2020).

Geneticists prefer this kind of resistance, i.e. based on host-pathogen recognition and programmed cell death, because it is easy to deal with, recognize phenotypically, map, and identify the DNA sequence of the gene involved. However, this kind of resistance is easily overcome by the pathogen, and pathologists prefer less-specific resistance mechanisms that are capable of operating over a large spectrum of strains, albeit not always as efficiently.

From a breeder's point of view, it is more urgent to find sources of resistance to Psa and other diseases in germplasm collections, to ascertain their heritability and to introgress them into the elite cultivars rather than investing in molecular investigation of the underpinning genes.

9.7.8 Rootstocks

Several rootstocks have been developed to meet the needs of orchardists (see Chapter 6). 'Kaimai', a hybrid of uncertain origin, is probably the first example (Wang *et al.*, 1994) but others followed, such as 'D1 Vitroplant', which is tolerant to calcareous soils (Testolin and Ferguson, 2009), 'Bounty71' (Mian *et al.*, 2022), and others. None of these has had extraordinary success.

Breeding for rootstocks is difficult as cultivars, particularly those obtained from interspecific hybridization, may reveal grafting incompatibility many years after orchard establishment. As a consequence, the breeder is committed to test their selection for a long time (20 years) prior to releasing a rootstock. However, the successes of several popular rootstocks, e.g. for apple and peach, have been rewarding.

9.8 Future Perspectives

The new breeding technologies, including cisgenesis, which involves transformation with genes isolated from sexually compatible species, and genome editing, provide great opportunities for reshaping existing elite kiwifruit cultivars to meet the expectations of marketers and consumers (Wang and Gleave, 2012). It is claimed that better fruit quality, more appealing flesh colours, longer storage and shelf life, reduced allergen content and modification of other traits could easily be achieved by editing sequences of individual genes and transcription factors (Chen *et al.*, 2019; Brian *et al.*, 2021).

GS, which makes use of large sets of molecular markers and training populations to produce models for the molecular-based estimation of breeding value of individuals, is now being attempted with kiwifruit following its use with crops such as apples (Testolin *et al.*, 2011; Cheng *et al.*, 2016).

Interspecific hybridization is another field that may prove productive in the near future. Many interspecific crosses have been attempted to produce novel fruits and rootstocks (Li and Liu, 2016). However – and this is a strong message that this chapter wishes to convey – the exploration and exploitation of genetic resources must remain the major commitments of kiwifruit geneticists and breeders.

Chinese scientists, often supported by western colleagues and institutions, made commendable progress by the end of the 20th century in securing into their germplasm repositories an appreciable representation of kiwifruit genetic diversity (Huang, 2016a, b), but further work is required. Mining genetic resources in the natural reservoirs still needs to be encouraged because further diversity is required to cope with climate change and, especially, diseases, which may threaten the world kiwifruit industry.

Much more emphasis is now being placed on sustainable methods of production, especially those with minimal use of insecticides and fungicides (Steven *et al.*, 1997), but the search for resistance to pests and diseases has not yet achieved the attention it deserves. An environmentally sustainable kiwifruit industry should be the mandatory requirement for future kiwifruit breeding.

References

Akagi, T., Henry, I.M., Ohtani, H., Morimoto, T., Beppu, K. *et al.* (2018) A Y-encoded suppressor of feminization arose via lineage-specific duplication of a cytokinin response regulator in kiwifruit. *Plant Cell* 30, 780–795. DOI: 10.1105/tpc.17.00787

Akagi, T., Pilkington, S.M., Varkonyi-Gasic, E., Henry, I.M., Sugano, S.S. *et al.* (2019) Two Y-chromosome-encoded genes determine sex in kiwifruit. *Nature Plants* 5, 801–809. DOI: 10.1038/s41477-019-0489-6

Ampomah-Dwamena, C., McGhie, T., Wibisono, R., Montefiori, M., Hellens, R.P. *et al.* (2009) The kiwifruit lycopene beta-cyclase plays a significant role in carotenoid accumulation in fruit. *Journal of Experimental Botany* 60, 3765–3779. DOI: 10.1093/jxb/erp218

Atkinson, R.G. and Schröder, R. (2016) Genetics of fruit softening. In: Testolin, R., Huang, H.-W. and Ferguson, A.R. (eds) *The Kiwifruit Genome*. Springer International Publishing, Cham, Switzerland, pp. 205–218. DOI: 10.1007/978-3-319-32274-2_16

Atkinson, R.G., Sharma, N.N., Hallett, I.C., Johnston, S.L. and Schröder, R. (2009) *Actinidia eriantha*: a parental species for breeding kiwifruit with novel peelability and health attributes. *New Zealand Journal of Forestry Science* 39, 207–216.

Beatson, R.A. (1991) Inheritance of fruit characters in *Actinidia deliciosa*. *Acta Horticulturae* 297, 79–86. DOI: 10.17660/ActaHortic.1992.297.8

Brian, L., Warren, B., McAtee, P., Rodrigues, J., Nieuwenhuizen, N. *et al.* (2021) A gene expression atlas for kiwifruit (*Actinidia chinensis*) and network analysis of transcription factors. *BMC Plant Biology* 21, 121. DOI: 10.1186/s12870-021-02894-x

Brown, T.A. (1994) *DNA Sequencing: The Basics*. Oxford University Press, Oxford, UK.

Bulley, S.M. and Laing, W. (2016) The kiwifruit allergome. In: Testolin, R., Huang, H.-W. and Ferguson, A.R. (eds) *The Kiwifruit Genome*. Springer International Publishing, Cham, Switzerland, pp. 219–235. DOI: 10.1007/978-3-319-32274-2_17

Chen, K., Wang, Y., Zhang, R., Zhang, H. and Gao, C. (2019) CRISPR/Cas genome editing and precision plant breeding in agriculture. *Annual Review of Plant Biology* 70, 667–697. DOI: 10.1146/annurev-arplant-050718-100049

Cheng, C.-H., Datson, P.M. and Hanley, Z. (2016) Genome-based breeding. In: Testolin, R., Huang, H.-W. and Ferguson, A.R. (eds) *The Kiwifruit Genome*. Springer International Publishing, Cham, Switzerland, pp. 249–257. DOI: 10.1007/978-3-319-32274-2_19

Cheng, C.-H., Datson, P.M., Hilario, E., Deng, C.H., Manako, K.I. *et al.* (2019) Genomic predictions in diploid *Actinidia chinensis* (kiwifruit). *European Journal of Horticultural Science* 84(4), 213–217. DOI: 10.17660/eJHS.2019/84.4.3

Cui, Z., Huang, H. and Xiao, X. (2002) *Actinidia in China*. Agricultural Science and Technology Press, Beijing.

Davies, L.J. (1992) Early induction of flowering in kiwifruit seedlings. *Acta Horticulturae* 297, 231–235. DOI: 10.17660/ActaHortic.1992.297.30

De Mori, G., Zaina, G., Franco-Orozco, B., Testolin, R., De Paoli, E. *et al.* (2021) Targeted mutagenesis of the female-suppressor *SyGI* gene in tetraploid kiwifruit by CRISPR/CAS9. *Plants* 10, 62. DOI: 10.3390/plants10010062

FAOSTAT (2020) Food and Agriculture Organization of the United Nations (FAO). FAOSTAT Database. Available at: http://faostat.fao.org/ (accessed 11 March 2022).

Ferguson, A.R. (1984) Kiwifruit: a botanical review. *Horticultural Reviews* 6, 1–64. DOI: 10.1002/9781118060797.ch1

Ferguson, A.R. (1991) Kiwifruit (*Actinidia*). *Acta Horticulturae* 290, 603–653. DOI: 10.17660/ActaHortic.1991.290.14

Ferguson, A.R. and Bollard, E.G. (1990) Domestication of the kiwifruit. In: Warrington, I.J. and Weston, G.C. (eds) *Kiwifruit: Science and Management*. Ray Richards Publisher in association with the New Zealand Society for Horticultural Science, Auckland, New Zealand, pp. 165–246.

Ferguson, A.R. and Seal, A.G. (2008) Kiwifruit. In: Hancock, J.F. (ed.) *Temperate Fruit Crop Breeding: Germplasm to Genomics*. Springer, Dordrecht, The Netherlands, pp. 235–264.

Ferguson, A.R. and Wu, J.-H. (2016) Interploid and interspecific hybridization for kiwifruit improvement. In: Mason, A.S. (ed.) *Polyploidy and Hybridization for Crop Improvement.* CRC Press, Boca Raton, Florida, pp. 339–363. DOI: 10.1201/9781315369259-18

Ferguson, A.R., Seal, A.G., McNeilage, M.A., Fraser, L.G., Harvey, C.F. *et al.* (1996) Kiwifruit. In: Janick, J. and Moore, J.N. (eds) *Fruit Breeding. Vol. 2: Vine and Small Crops*. John Wiley & Sons, Inc., New York, pp. 371–417.

Ferguson, A.R., O'Brien, I.E.W. and Yan, G.J. (1997) Ploidy in *Actinidia. Acta Horticulturae* 444, 67–71. DOI: 10.17760/ActaHortic.1997.444.7

Ferguson, A.R., Zhang, J.-L., Duffy, A.M., Beatson, R.A., Cheng, C.-H. *et al.* (2009) Ploidia e uso della citometria a flusso nel miglioramento genetico dell'actinidia. [Ploidy and the use of flow cytometry in kiwifruit breeding.] *Italus Hortus* 16(5), 78–83.

Ferguson, A.R., Huang, H.W. and Testolin, R. (2012) Kiwifruit. *HortScience* 47, 547–550. DOI: 10.21273/HORTSCI.47.5.536

Fine, A.J. (1981) Hypersensitivity reaction to kiwi fruit (Chinese gooseberry, *Actinidia chinensis*). *Journal of Allergy and Clinical Immunology* 68(3), 235–237. DOI: 10.1016/0091-6749(81)90189-5

Fraser, L.G., Tsang, G.K., Datson, P.M., De Silva, H.N., Harvey, C.F. *et al.* (2009) A gene-rich linkage map in the dioecious species *Actinidia chinensis* (kiwifruit) reveals putative X/Y sex-determining chromosomes. *BMC Genomics* 10, 102. DOI: 10.1186/1471-2164-10-102

Fraser, L.G., Datson, P.M., Tsang, G.T., Manako, K.I., Rikkerink, E.K. *et al.* (2015) Characterisation, evolutionary trends and mapping of putative resistance and defence genes in *Actinidia* (kiwifruit). *Tree Genetics & Genomes* 11, 21. DOI: 10.1007/s11295-015-0846-1

Fullerton, C.G., Prakash, R., Ninan, A.S., Atkinson, R.G., Schaffer, R.J. *et al.* (2020) Fruit from two kiwifruit genotypes with contrasting softening rates show differences in the xyloglucan and pectin domains of the cell wall. *Frontiers in Plant Science* 11, 964. DOI: 10.3389/fpls.2020.00964

Grainger, A.R. (2011) Plant & Food Research: New Zealand kiwifruit breeding programme. *Acta Horticulturae* 913, 59–61. DOI: 10.17660/ActaHortic.2011.913.4

Hanley, Z. (2018) Kiwifruit (*Actinidia* spp.) breeding. In Al-Khayri, J., Jain, S.M. and Johnson, D.V. (eds) *Advances in Plant Breeding Strategies: Fruits*. Springer International Publishing AG, Cham, Switzerland, pp. 377–401. DOI: 10.1007/978-3-319-91944-7_10

Heffner, E.L., Sorrells, M.E. and Jannink, J.-L. (2009) Genetic selection for crop improvement. *Crop Science* 49, 1–12. DOI: 10.2135/cropsci2008.08.0512
Herath, D., Voogd, C., Mayo-Smith, M., Yang, B., Allan, A.C. *et al.* (2022) CRISPR-Cas9-mediated mutagenesis of kiwifruit *BFT* genes results in an evergrowing but not early flowering phenotype. *Plant Biotechnology Journal* 20, 2064–2076. DOI: 10.1111/pbi.13888
Herath, D., Wang, T., Voogd, C., Peng, Y., Douglas, M. *et al.* (2023) Strategies for fast breeding and improvement of *Actinidia* species. *Horticulture Research* 10, uhad016. DOI: 10.1093/hr/uhad016
Hirsch, A.M., Testolin, R., Brown, S., Chat, J., Fortune, D. *et al.* (2001) Embryo rescue from interspecific crosses in the genus *Actinidia* (kiwifruit). *Plant Cell Reports* 20, 508–516. DOI: 10.1007/s002990100340
Huang, H.-W. (2016a) *Kiwifruit: The Genus ACTINIDIA*. Elsevier, London.
Huang, H.-W. (2016b) Genetic resources. In: Testolin, R., Huang, H.-W. and Ferguson, A.R. (eds) *The Kiwifruit Genome*. Springer International Publishing, Cham, Switzerland, pp. 15–36. DOI: 10.1007/978-3-319-32274-2_2
Huang, H.-W. and Ferguson, A.R. (2007) Genetic resources of kiwifruit: domestication and breeding. *Horticultural Reviews* 33, 1–121. DOI: 10.1002/9780470168011.ch1
Huang, H.W., Abid, M., Lin, M., Wang, R., Gu, H. *et al.* (2021) Comparative transcriptome analysis of different *Actinidia arguta* fruit parts reveals difference of light response during fruit coloration. *Biology* 10(7), 648. DOI:10.3390/biology10070648
Huang, S., Ding, J., Deng, D., Tang, W., Sun, H. *et al.* (2013) Draft genome of the kiwifruit *Actinidia chinensis*. *Nature Communications* 4, 2640. DOI: 10.1038/ncomms3640
Junge, J.Y., Bertelsen, A.S., Mielby, L.A., Zeng, Y., Sun, Y.-X. *et al.* (2020) Taste interactions between sweetness of sucrose and sourness of citric and tartaric acid among Chinese and Danish consumers. *Foods* 9, 1425. DOI: 10.3390/foods9101425
Ke, S.-Q., Huang, R.-H., Wang, S.-M., Xiong, Z.-T. and Wu, Z.-W. (1992) Studies on interspecific hybrids of *Actinidia*. *Acta Horticulturae* 297, 133–139. DOI: 10.17660/ActaHortic.1992.297.15
Kim, S.-C., Song, E.-U. and Kim, C.-H. (2012) A new kiwifruit variety, 'Halla Gold' with high soluble solids content and early harvesting. *Korean Journal of Horticultural Science and Technology* 30, 334–337.
Le, T.-M., Bublin, M., Breiteneder, H., Fernández-Rivas, M., Asero, R. *et al.* (2013) Kiwifruit allergy across Europe: clinical manifestation and IgE recognition patterns to kiwifruit allergens. *Journal of Allergy and Clinical Immunology* 131(1), 164–171. DOI: 10.1016/j.jaci.2012.09.009
Li, B., Xia, Y., Wang, Y., Qin, G. and Tian, S. (2017) Characterization of genes encoding key enzymes involved in anthocyanin metabolism of kiwifruit during storage period. *Frontiers in Plant Sciences* 8, 341. DOI: 10.3389/fpls.2017.00341
Li, D. and Liu, Y. (2016) Gene introgression from wild relatives. In: Testolin, R., Huang, H.-W. and Ferguson, A.R. (eds) *The Kiwifruit Genome*. Springer International Publishing, Cham, Switzerland, pp. 237–248. DOI: 10.1007/978-3-319-32274-2_18
Li, M., Tan, J.-I., Li, Y., Cheng, E.-Y., Han, X. *et al.* (2004) Resistance of different kiwifruit cultivars to kiwifruit bacterial canker caused by *Pseudomonas syringae* pv. *actinidiae* and the cluster analysis. *Plant Protection* 30(5), 51–54. [In Chinese]
Liu, Y., Zhou, B., Qi, Y., Chen, X., Liu, C. *et al.* (2017) Expression differences of pigment structural genes and transcription factors explain flesh coloration in three contrasting kiwifruit cultivars. *Frontiers in Plant Science* 8, 1507. DOI: 10.3389/fpls.2017.01507
MacRae, E., Quick, W.P., Benker, C. and Stitt, M. (1992) Carbohydrate metabolism during postharvest ripening in kiwifruit. *Planta* 188, 314–323. DOI: 10.1007/BF00192797
Man, Y.-P., Wang, Y.-C., Li, Z.-Z., Jiang, Z.-W., Yang, H.-L. *et al.* (2015) High-temperature inhibition of biosynthesis and transportation of anthocyanins results in the poor red coloration in red-fleshed *Actinidia chinensis*. *Physiologia Plantarum* 153, 565–583. DOI: 10.1111/ppl.12263
Marsh, H.D., Paterson, T., Seal, A.G. and McNeilage, M.A. (2003) Heritability estimates in kiwifruit. *Acta Horticulturae* 622, 221–229. DOI: 10.17660/ActaHortic.2003.622.19
Matsuta, N., Iketani, H. and Hayashi, T. (1990) Effect of acetosyringone on kiwifruit transformation. *Japanese Journal of Breeding* 40, 184–185.
McNeilage, M.A. (1991a) Gender variation in *Actinidia deliciosa*, the kiwifruit. *Sexual Plant Reproduction* 4, 267–273. DOI: 10.1007/BF00200546
McNeilage, M.A. (1991b) Sex expression in fruiting male vines of kiwifruit. *Sexual Plant Reproduction* 4, 274–278. DOI: 10.1007/BF00200547
McNeilage, M.A. and Steinhagen, S. (1998) Flower and fruit characters in a kiwifruit hermaphrodite. *Euphytica* 101, 69–72. DOI: 10.1023/A:1018398924819

McNeilage, M.A., Seal, A.G., Steinhagen, S. and McGowan, J. (1992) Evaluation of kiwifruit pollinizers. *Acta Horticulturae* 297, 277–282. DOI: 10.17660/ActaHortic.1992.297.36

McPherson, H.G., Richardson, A.C., Snelgar, W.P., Patterson, K.J. and Currie, M.B. (2001) Flower quality and fruit size in kiwifruit (*Actinidia deliciosa*). *New Zealand Journal of Crop and Horticultural Science* 29, 93–101. DOI: 10.1080/01140671.2001.9514167

Meuwissen, T.H.E., Hayes, B.J. and Goddard, M.E. (2001) Prediction of total genetic value using genome-wide dense marker maps. *Genetics* 157, 1819–1829. DOI: 10.1093/genetics/157.4.1819

Mian, G., Cipriani, G., Saro, S., Martini, M. and Emacora, P. (2022) Potential of different *Actinidia* genotypes as resistant rootstocks for preventing kiwifruit vine decline syndrome. *Horticulturae* 8, 627. DOI: 10.3390/horticulturae8070627

Montefiori, M., McGhie, T.K., Hallett, I.C. and Costa, G. (2009a) Changes in pigments and plastid ultrastructure during ripening of green-fleshed and yellow-fleshed kiwifruit. *Scientia Horticulturae* 119, 377–387. DOI: 10.1016/j.scienta.2008.08.022

Montefiori, M., Comeskey, D.J., Wohlers, M. and McGhie, T.K. (2009b) Characterization and quantification of anthocyanins in red kiwifruit (*Actinidia* spp.). *Journal of Agriculture and Food Chemistry* 57, 6856–6861. DOI: 10.1021/jf900800z

Montefiori, M., Espley, R.V., Stevenson, D., Cooney, J., Datson, P.M. *et al.* (2011) Identification and characterisation of F3GT1 and F3GGT1, two glycosyltransferases responsible for anthocyanin biosynthesis in red-fleshed kiwifruit (*Actinidia chinensis*). *Plant Journal* 65, 106–118. DOI: 10.1111/j.1365-313X.2010.04409.x

Montefiori, M., Pilkington, S.M., Davies, K.M. and Allan, A.C. (2016) Genetics of pigment biosynthesis and degradation. In: Testolin, R., Huang, H.W. and Ferguson, A.R. (eds) *The Kiwifruit Genome*. Springer International Publishing AG, Cham, Switzerland, pp. 149–161. DOI: 10.1007/978-3-319-32274-2_12

Nardozza, S., Kashuba, P., McCaughan, L., Philippe, M., Wohlers, M. *et al.* (2015a) Leaves are important to obtain consistent red flesh pigmentation in *Actinidia chinensis* fruit. *Scientia Horticulturae* 197, 496–503. DOI: 10.1016/j.scienta.2015.10.008

Nardozza, S., Martinez-Sanchez, M., Curtis, C., Datson, P.M. and Montefiori, M. (2015b) Screening *Actinidia* germplasm for different levels of tolerance, or resistance, to Psa (*Pseudomonas syringae* pv. *actinidiae*). *Acta Horticulturae* 1096, 351–355. DOI: 10.17660/ActaHortic.2015.1096.40

Nardozza, S., Boldingh, H.L., Kashuba, P., Feil, R., Jones, D. *et al.* (2020) Carbon starvation reduces carbohydrate and anthocyanin accumulation in red-fleshed fruit via trehalose 6-phosphate and MYB27. *Plant, Cell & Environment* 43, 819–835. DOI: 10.1111/PCE. 13699

Pilkington, S.M., Montefiori, M., Jameson, P.E. and Allan, A.C. (2012) The control of chlorophyll levels in maturing kiwifruit. *Planta* 236, 1615–1628. DOI: 10.1007/s00425-012-1723-x

Pilkington, S.M., Crowhurst, R., Hilario, E., Nardozza, S., Fraser, L. *et al.* (2018) A manually annotated *Actinidia chinensis* var. *chinensis* (kiwifruit) genome highlights the challenges associated with draft genomes and gene prediction in plants. *BMC Genomics* 19, 257. DOI: 10.1186/s12864-018-4656-3

Pinto, C., Testolin, R. and Cipriani, G. (2018a) Analysis of components of genetic variance and breeding value of kiwifruit (*Actinidia chinensis* Pl.) using best linear unbiased prediction (BLUP) and pedigree information. *Acta Horticulturae* 1218, 73–79. DOI 10.17660/ActaHortic.2018.1218.9

Pinto, C., Testolin, R., Cipriani, G., Mesa, K., Meneses, C. *et al.* (2018b) Exploratory QTL analysis for fruit-related traits in kiwifruit (*Actinidia chinensis* Pl.). *Acta Horticulturae* 1218, 101–107. DOI: 10.17660/ActaHortic.2018.1218.12

Popowski, E., Thomson, S.J., Knäbel, M., Tahir, J., Crowhurst, R.N. *et al.* (2021) Construction of a high-density genetic map for hexaploid kiwifruit (*Actinidia chinensis* var. *deliciosa*) using genotyping by sequencing. *G3 Genes|Genomes|Genetics* 11, jkab142. DOI: 10.1093/g3journal/jkab142

Redgwell, R.J., Melton, L.D. and Brasch, D.J. (1991) Cell-wall polysaccharides of kiwifruit (*Actinidia deliciosa*): effect of ripening on the structural features of cell-wall materials. *Carbohydrate Research* 209, 191–202. DOI: 10.1016/0008-6215(91)80156-H

Redgwell, R.J., Melton, L.D. and Brasch, D.J. (1992) Cell wall dissolution in ripening kiwifruit (*Actinidia deliciosa*): solubilization of the pectic polymers. *Plant Physiology* 98, 71–81. DOI: 10.1104/pp.98.1.71

Rey, M., Ferradás, Y., Martínez, Ó. and González, M.V. (2020) Actinidiaceae 1.1 *Actinidia* spp. Kiwifruit. In: Litz, R.E., Pliego-Alfaro, F. and Ignacio Hormaza, J. (eds) *Biotechnology of Fruit and Nut Crops*, 2nd edn. CAB International, Wallingford, UK, pp. 1–18. DOI: 10.1079/9781780648279.0001

Scaglione, D., Fornasiero, A., Pinto, C., Cattonaro, F., Spadotto, A. *et al.* (2015a) A RAD-based linkage map of kiwifruit (*Actinidia chinensis* Pl.) as a tool to improve the genome assembly and to scan the genomic region of the gender determinant for the marker-assisted breeding. *Tree Genetics & Genomes* 11, 115. DOI: 10.1007/s11295-015-0941-3

Scaglione, D., Fornasiero, A., Cattonaro, F., Mesa, K., Pinto, C. *et al.* (2015b) A RAD-based linkage map of kiwifruit (*Actinidia chinensis* Pl.) and map position of genetic determinants of traits of interest to breeders. *Acta Horticulturae* 1096, 191–198. DOI: 10.17660/ ActaHortic.2015.1096.20

Schaffer, R.J. and Pilkington, S.M. (2016) Gene family prediction and annotation. In: Testolin, R., Huang, H.W. and Ferguson, A.R. (eds) *The Kiwifruit Genome*. Springer International Publishing AG, Cham, Switzerland, pp. 125–134. DOI: 10.1007/978-3-319-32274-2_10

Seal, A.G., Ferguson, A.R., de Silva, H.N. and Zhang, J.-L. (2012) The effect of 2*n* gametes on sex ratios in *Actinidia*. *Sexual Plant Reproduction* 25, 197–203. DOI: 10.1007/s00497-012-0191-6

Sotiropoulos, T., Koukourikou-Petridou, M., Petridis, A., Stylianidis, D., Almaliotis, D. *et al.* (2009) 'Tsechelidis' kiwifruit. *HortScience* 44, 466–468. DOI: 10.21273/HORTSCI. 44.2.466

Steven, D., Barnett, S.W., Stevens, P.S. and McKenna, C.E. (1997) Changing pest control on New Zealand kiwifruit. *Acta Horticulturae* 444, 765–772. DOI: 10.17660/actahortic.1997.444.117

Tahir, J., Brendolise, C., Hoyte, S., Lucas, M., Thomson, S. *et al.* (2020) QTL mapping for resistance to cankers induced by *Pseudomonas syringae* pv. *actinidiae* (Psa) in a tetraploid *Actinidia chinensis* kiwifruit population. *Pathogens* 9, 967. DOI: 10.3390/pathogens9110967

Tang, W., Sun, X., Yue, J., Tang, X., Jiao, C. *et al.* (2019) Chromosome-scale genome assembly of kiwifruit *Actinidia eriantha* with single-molecule sequencing and chromatin interaction mapping. *GigaScience* 8, giz027. DOI: 10.1093/gigascience/giz027

Testolin, R. and Ferguson, A.R. (2009) Kiwifruit (*Actinidia* spp.) production and marketing in Italy. *New Zealand Journal of Crop and Horticultural Science* 37, 1–32. DOI: 10.1080/01140670909510246

Testolin, R., Cipriani, G. and Costa, G. (1995a) Sex segregation ratio and gender expression in the genus *Actinidia*. *Sexual Plant Reproduction* 8, 129–132. DOI: 10.1007/BF00242255

Testolin, R., Cipriani, G. and Costa, G. (1995b) Estimate of variance components and heritability of characters in kiwifruit (*Actinidia deliciosa*). *Acta Horticulturae* 403, 182–188. DOI: 10.17660/ActaHortic.1995. 403.35

Testolin, R., Huang, W.G., Lain, O., Messina, R., Vecchione, A. *et al.* (2001) A kiwifruit (*Actinidia* spp.) linkage map based on microsatellites and integrated with AFLP markers. *Theoretical and Applied Genetics* 103, 30–36. DOI: 10.1007/s00122-001-0555-z

Testolin, R., Cipriani, G., Fiori, G., Piccinini, L. and Costa, G. (2011) Evaluation of parental genotypes as a tool for planning the mating design in kiwifruit breeding programs. *Acta Horticulturae* 913, 213–219. DOI: 10.17660/ActaHortic.2011.913.27

Varkonyi-Gasic, E., Wang, T., Voogd, C., Jeon, S., Drummond, R.S.M. *et al.* (2018) Mutagenesis of kiwifruit *CENTRORADIALIS*-like genes transforms a climbing woody perennial with long juvenility and axillary flowering into a compact plant with rapid terminal flowering. *Plant Biotechnology Journal* 17, 869–880. DOI: 10.1111/pbi.13021

Voogd, C., Brian, L.A., Wu, R., Wang, T., Allan, A.C. *et al.* (2021) A MADS-box gene with similarity to *FLC* is induced by cold and correlated with epigenetic changes to control budbreak in kiwifruit. *New Phytologist* 233, 2111–2126. DOI: 10.1111/nph.17916

Wang, F.-M., Mo, Q.-H., Ye, K.-Y., Gong, H.-J., Qi, B.-B. *et al.* (2020) Evaluation of the wild *Actinidia* germplasm for resistance to *Pseudomonas syringae* pv. *actinidiae*. *Plant Pathology* 69, 979–989. DOI: 10.1111/ppa.13184

Wang, J., Vanga, S.K., McCusker, C. and Raghavan, V. (2019) A comprehensive review on kiwifruit allergy: pathogenesis, diagnosis, management, and potential modification of allergens through processing. *Comprehensive Reviews in Food Science and Food Safety* 18, 500–513. DOI: 10.1111/1541-4337.12426

Wang, J.-P., Yu, J.-G., Li, J., Sun, P.-C., Wang, L. *et al.* (2018) Two likely auto-tetraploidization events shaped kiwifruit genome and contributed to establishment of the Actinidiaceae family. *iScience* 7, 230–240. DOI: 10.1016/j.isci.2018.08.003

Wang, T. and Gleave, A.P. (2012) Applications of biotechnology in kiwifruit (*Actinidia*). In: Agbo, E.C. (ed.) *Innovations in Biotechnology*. IntechOpen Ltd, London. DOI: 10.5772/28673

Wang, T., Ran, Y., Atkinson, R.G., Gleave, A.P. and Cohen, D. (2006) Transformation of *Actinidia eriantha*: a potential species for functional genomics studies in *Actinidia*. *Plant Cell Reports* 25, 425–431. DOI: 10.1007/s00299-005-0080-7

Wang, Z-Y., Patterson, K.J., Gould, K.S. and Lowe, R.G. (1994) Rootstock effects on budburst and flowering in kiwifruit. *Scientia Horticulturae* 57, 187–199. DOI: 10.1016/0304-4238(94)90140-6

Wu, H., Ma, T., Kang, M., Ai, F., Zhang, J. *et al.* (2019) A high-quality *Actinidia chinensis* (kiwifruit) genome. *Horticulture Research* 6, 117. DOI: 10.1038/s41438-019-0202-y

Yao, X., Wang, S., Wang, Z., Li, D., Jiang, Q. *et al.* (2022) The genome sequencing and comparative analysis of a wild kiwifruit *Actinidia eriantha*. *Molecular Horticulture* 2, 123. DOI: 10.1186/s43897-022-00034-z

Zhang, Q.Y., Liu, C., Liu, Y., VanBuren, R., Yao, X. *et al.* (2015) High-density interspecific genetic maps of kiwifruit and the identification of sex-specific markers. *DNA Research* 22(5), 367–375. DOI: 10.1093/dnares/dsv019

Zhang, Q.-Y., Ge, J., Liu, X.-C., Wang, W.-Q., Liu, X.-F. *et al.* (2022) Consensus co-expression network analysis identifies AdZAT5 regulating pectin degradation in ripening kiwifruit. *Journal of Advanced Research* 40, 59–68. DOI: 10.1016/j.jare.2021.11.019

Zhong, C., Wang, S., Jiang, Z. and Huang, H.W. (2012) 'Jinyan', an interspecific hybrid kiwifruit with brilliant yellow flesh and good storage quality. *HortScience* 47, 1187–1190. DOI: 10.21273/HORTSCI.47.8.1187

Zhong, C., Huang, W., Wang, Z., Li, L., Li, D. *et al.* (2022) The breeding progress and development status of the kiwifruit industry in China. *Acta Horticulturae* 1332, 445–454. DOI: 10.17660/ActaHortic.2022.1332.59

Zhu, D., Lawes, G.S. and Gordon, I.L. (2002) Estimates of genetic variability and heritability for vegetative and reproductive characters of kiwifruit (*Actinidia deliciosa*). *Euphytica* 124, 93–98. DOI: 10.1023/A:1015662523096

10 Soil, Climate and Site Selection

Annette C. Richardson[1], Roberta M. Gentile[2]* and Indrakumar Vetharaniam[3]
[1]*The New Zealand Institute for Plant and Food Research Ltd, Kerikeri, New Zealand;* [2]*The New Zealand Institute for Plant and Food Research Ltd, Palmerston North, New Zealand;* [3]*The New Zealand Institute for Plant and Food Research Ltd, Hamilton, New Zealand*

10.1 Introduction

The growth and productivity of kiwifruit vines are determined by the local environment, particularly the soil, climate and site characteristics, and each of these varies across the world. Historically, kiwifruit orchards have generally been developed in areas where there are few limitations. However, as competition for growing sites increases, soils degrade and erode and the climate changes, optimizing each of these factors is becoming increasingly difficult.

Complex interactions between vine genetics, vine management and the environment contribute to the commercial success of kiwifruit. This chapter reviews the current understanding of the importance of soils (physical, chemical and biological properties), climate (temperature, rainfall, wind and hail) and site selection on the production of kiwifruit around the world.

10.2 Soil

10.2.1 Soil depth and potential rooting depth

Kiwifruit vines of *Actinidia chinensis* var. *deliciosa* 'Hayward' can produce extensive root systems, with each vine exploring up to 80 m^3 in deep soils while producing very high root-length densities (up to 7 cm cm^{-3}) near the base of the vine (Hughes *et al.*, 1992) (see Chapter 15). While a deeper soil allows strong root systems to develop and provide resilience against drought, kiwifruit vines can perform well in a range of soil depths. For example, *A. chinensis* var. *chinensis* 'Haegeum' vines have been grown commercially in < 30 cm soil in South Korea (Cho *et al.*, 2013) and in New Zealand, 'Hayward' has been grown in soil depths between 50 cm (Northland) and 4 m (Bay of Plenty) (McAneney and Judd, 1983; Greaves, 1985). Greater soil depth provides vines with more soil to exploit for water and nutrients, but root growth competes with canopy and fruit production and may need to be controlled via girdling or root pruning (Black *et al.*, 2012; Currie *et al.*, 2018). Shallow soils control root vigour but have limited water storage during drought, and smaller root systems have fewer reserves to counter stress or damage (e.g. hypoxia and erosion). A shallow root depth can be improved by mounding, addition of organic amendments or intensive soil-water and nutrition management (see Chapter 11).

*Corresponding author: roberta.gentile@plantandfood.co.nz

© CAB International 2024. *Kiwifruit: Botany, Production and Uses* (eds A.C. Richardson, J.N. Burdon and A.R. Ferguson)

DOI: 10.1079/9781800620933.0010

10.2.2 Soil physical properties

Water-holding capacity and waterlogging

'Hayward' and *A. chinensis* var. *deliciosa* 'Abbott' vines are extremely sensitive to waterlogging and require a well or perfectly drained soil (Savé and Serrano, 1986; Smith *et al.*, 1990; Reid and Petrie, 1991; Reid *et al.*, 1991, 1992) (Fig. 10.1). Symptoms of waterlogging may only become evident after high rainfall (Hughes and Wilde, 1989). Some vines are more sensitive than others, as vines with many roots can survive loss of a large proportion of the root system (Reid and Petrie, 1991). In an experiment involving 'Hayward' vines grown with different water table depths (0–150 cm), vine mortality was higher and yield lower in plants grown with a water table depth of 0–50 cm, compared with plants grown with deeper water tables (Magliulo *et al.*, 1991). 'Hayward' vines in well-drained soil had root-length densities > 2 cm cm^{-3} in twice as much soil volume as vines in a poorly drained soil (Hughes and Wilde, 1989). Poor soil drainage and structure, as well as inappropriate soil management, caused early vine decline (Reid *et al.*, 1991). Waterlogged vines were found in soils with brownish-grey and olive-grey mottling in the B horizon, typical of reducing conditions, and higher bulk densities (Hughes and Wilde, 1989).

The relatively high consumption of oxygen (9.1×10^{-10} mol $m^{-1}s^{-1}$) and low porosity of kiwifruit roots means that even short periods of low oxygen concentrations in the root zone seriously affect the physiology and growth of kiwifruit vines (Smith *et al.*, 1989). At this rate of oxygen consumption, a mature kiwifruit vine could deplete the oxygen supply in a well aerated soil in about 5 h. Root anoxia had an impact on the growth of 'Hayward' vines in controlled conditions within one day (Smith *et al.*, 1990) and vines did not recover once oxygen was resupplied to the roots (see Chapter 15).

There is some debate about the relationship between waterlogging and the impacts of root disease. *Phytophthora* spp. and other pathogenic fungi have been associated with waterlogged 'Hayward' vines but were not thought to be the primary cause of vine decline (Reid *et al.*, 1991). In Italy, kiwifruit vine decline syndrome (KVDS) is often observed in poorly drained soils that are susceptible to waterlogging and associated with *Phytophthora* spp. infections, although KVDS can occasionally occur in well-drained soils (Savian *et al.*, 2020) (see Chapter 17).

Soils with a heavy subsoil or hard pan require soil improvement including installation of drainage. After flooding, Reid *et al.* (1991) recommended remedial tillage in many 'Hayward' orchards to reduce compaction and improve aeration of soils. In South Korea, sub-soiling in early

Fig. 10.1. Kiwifruit vines are extremely sensitive to waterlogging. Vines in the poorly drained centre of this orchard block have died, while those in the better-drained edges of the block remain alive. (Photo: The New Zealand Institute for Plant and Food Research Ltd)

spring using pressured air at a depth of ~60 cm, 1.8 m from 'Haegeum' vines, reduced soil resistance by 30–50% and increased soil porosity by 16% (Cho *et al.*, 2013). Addition of organic carbon to soils may also improve soil microporosity (Deurer *et al.*, 2009; Müller *et al.*, 2019) and drainage. A comparison of several *Actinidia* species found that *A. rufa*, *A. macrosperma* and *A. valvata* were the most resistant to waterlogging and may be suitable as rootstocks (see Chapter 6).

10.2.3 Soil chemical properties

Soil pH

Kiwifruit are grown on soils with a wide pH range: neutral to alkaline (pH 6.9–8.5) in Italy, Greece and China, and neutral to acidic (pH 4.8–6.8) in France, New Zealand, Spain and Portugal (Barber *et al.*, 1986; Costa *et al.*, 1992; Tagliavini and Marangoni, 2002; Lema *et al.*, 2010; Otero *et al.*, 2010; Wang *et al.*, 2019). A soil pH of 5.6–6.0 has been recommended for kiwifruit (Strik and Davis, 2021).

Nutrient toxicities and deficiencies

Kiwifruit are prone to several nutrient deficiencies and toxicities arising from soil- and irrigation-water properties, as well as nutrient applications. Deficiencies of potassium, magnesium and nitrogen (N) have been observed in 'Hayward' vines in New Zealand (Smith *et al.*, 1987a, b). Iron deficiencies occur in vines grown on alkaline, calcareous soils in Italy and China as well as in vines in Greece and South Korea (Costa *et al.*, 1992; Zhang *et al.*, 2003; Koh *et al.*, 2007). Zinc deficiency has been observed in vines in Greece and China and chlorine deficiency in China (Costa *et al.*, 1992; Zhang *et al.*, 2003). Calcium (Ca) deficiency was found in vines on slightly acidic soils in Spain (Otero *et al.*, 2010) and Ca toxicity on alkaline calcareous soils in China (Wang *et al.*, 2019). Manganese deficiency was found in vines grown on soils with a pH above 6.8 in New Zealand and South Korea and manganese toxicity in those grown on acidic, poorly drained soils in New Zealand (Asher *et al.*, 1984). Sodium and boron (B) toxicities in vines have been associated with high nutrient concentrations in irrigation water (Barber *et al.*, 1986; Sotiropoulos *et al.*, 2006) or high B in soils (Tsadilas *et al.*, 1997). Soil pH and nutrient imbalances can be modified through lime and fertilizer applications.

Soil salinity

Kiwifruit vines are particularly sensitive to soil salinity and several studies have been carried out on small kiwifruit plants and in controlled environments, e.g. cuttings (Chartzoulakis *et al.*, 1995), seedlings (Klages *et al.*, 1999; Sotiropoulos *et al.*, 2005) and tissue-cultured material (Tian *et al.*, 2011; Zhong *et al.*, 2019). In these studies, salinity reduced shoot growth, leaf area development, photosynthesis and root activity (Chartzoulakis *et al.*, 1995; Zhong *et al.*, 2019) (see Chapter 15). Kiwifruit root systems can play a vital role in salinity tolerance and an *A. valvata* rootstock was found to be more resistant than those from other *Actinidia* species (Zhong *et al.*, 2019; Abid *et al.*, 2020).

10.2.4 Soil biological properties

Soil organic carbon and nitrogen

Soil organic carbon (SOC) and N are important for plant, soil and ecosystem health. SOC concentrations vary temporally and spatially with the rates of carbon inputs and SOC mineralization, and vary between years, between growing regions and within orchards (Rahman *et al.*, 2014). SOC is influenced by management practices. Topsoil (0–15 cm) SOC and N contents were greater under organic compared with conventional kiwifruit management systems, and greater under *A. chinensis* var. *chinensis* 'Hort16A' vines than 'Hayward' vines (Carey *et al.*, 2009). In New Zealand, kiwifruit production increased SOC and N stocks at 1.5–2.0 m depth without changing topsoil stocks, compared with soil under pasture (Gentile *et al.*, 2021).

Soil management systems

Orchard management systems can influence the soil food web, fauna populations and physical and chemical properties. Long-term studies have shown that soils in organic orchards are more suitable for kiwifruit vines than those in conventional orchards, with lower bulk densities, greater porosity and aggregation, better water-holding capacity, higher pH and cation exchange capacity, more N and SOC, and higher Ca and magnesium, but lower phosphate levels (Benge *et al.*, 2007; Carey *et al.*, 2009). In Spain, an organic orchard had vines with better leaf B concentrations and higher fruit Ca, potassium, phosphorus and B concentrations than conventionally grown vines (Lago *et al.*, 2015).

Organic orchards had soils with more diverse food webs, greater bacterial diversity, larger numbers of omnivorous nematodes and earthworms, particularly large, long-lived anecic worms that make permanent vertical burrows in the soil and incorporate SOC, and higher SOC and N retention (Richards *et al.*, 2007; Carey *et al.*, 2009; Castro *et al.*, 2018, 2019; Lago *et al.*, 2019). Soils in conventional orchards had less complex food webs involving small, short-lived, surface-living, epigeic earthworms, as well as microorganisms capable of rapid mineralization of SOC and more mesofauna (mites, collembola and macro-arthropods) (Richards *et al.*, 2007; Carey *et al.*, 2009; Castro *et al.*, 2015, 2018, 2019; Lago *et al.*, 2019). More diverse soil communities may have greater resilience to climate change (Gibb and Cunningham, 2013; Castro *et al.*, 2015, 2018).

The biological properties of orchard soils can be altered through addition of amendments. Microbial diversity can be increased by adding bacteria (Fan *et al.*, 2016). SOC can be increased through adding pasture or sawdust, increasing soil microbial mass, and boosting populations of nematodes and collembola (Wardle *et al.*, 2001). Addition of organic manures increases rhizosphere microbial diversity and network complexity (Liu *et al.*, 2020a). Trace elements (e.g. zinc, nickel, cobalt, copper) also improve the microbial content and stability of soils (Liu *et al.*, 2020b).

10.3 Climate

10.3.1 Wind

Kiwifruit vines are very susceptible to damage from high winds and extreme storms, particularly in the maritime climate of New Zealand (McAneney *et al.*, 1984; McAneney and Judd, 1987). Wind can damage growing shoots and reduce the flowering/fruiting potential of the current crop as well as buds for the following season (Hathaway, 1990). Leaf damage reduces photosynthate production and protection of fruit from sunburn. High winds may increase plant water loss and reduce insect pollinator activity (Strik and Davis, 2021). Downgrading of fruit because of wind-induced skin blemishes is a common problem for growers (McAneney *et al.*, 1984).

Wind damage can be mitigated by using shelterbelts. Protection may extend up to 25 times the height of the shelter, with the best protection provided by shelters with an optical density of 40–80% (Gloyne, 1955; Bird *et al.*, 1993; Řeháček, 2017). Orchard shelters can reduce wind run by 6% for every metre of shelter height and increase soil and maximum air temperatures, but not minimum air temperatures, by 1°C in New Zealand (McAneney *et al.*, 1990). In a continental climate, shelters increased the maximum temperature but decreased the minimum temperature, thereby increasing the possibility of frost damage, particularly at the bottom of slopes (Guyot and Seguin, 1975). Orchard shelters are useful to prevent drift of crop protection sprays.

Artificial shelterbelts are expensive but provide immediate shelter for crops without the negative effects of shade, competition for water and nutrients, management requirements and harbouring of pests and diseases associated with live shelter belts (Hathaway, 1990). Artificial shelters have also been used between rows to help establish young vines, and over the top of blocks to reduce wind turbulence (Fig. 10.2). A wide range of fast-growing tree species, including deciduous *Populus* and *Salix* spp. and evergreen *Pinus radiata*, *Eucalyptus* spp. *Cryptomeria japonica* and *Casuarina* spp., have been used as living shelter for orchards (Fig. 10.3). Traditionally, ten-metre-high shelterbelts were spaced 35–50 m apart and orientated at right angles to the prevailing wind (Hathaway, 1990). Where possible, this results in long narrow blocks (30–50 m wide and 80–150 m long) orientated north to south to avoid shading, but this is dependent on orchard topography and boundaries.

10.3.2 Temperature

Kiwifruit production requires adequate warmth for vine growth, fruit development and maturation as well as cool temperatures in winter (winter chilling) to encourage vine productivity.

Fig. 10.2. Artificial shelter can be used between rows to help the establishment of young vines. (Photo: The New Zealand Institute for Plant and Food Research Ltd)

Requirements for warmth are often expressed as growing degree days (GDD) and Salinger and Kenny (1995) suggested that 'Hayward' required accumulation of 1100 GDD between budbreak and harvest using a base of 10°C. However, there is no defined upper limit for excessive temperatures. 'Hayward' also needs 225–240 frost-free days to develop and mature a crop (Strik and Davis, 2021). The effects and importance of temperature on kiwifruit development vary during the seasons (Tait *et al.*, 2018) and these effects are described below.

Winter

Kiwifruit, like other deciduous plants, require a period of cool temperatures or 'winter chilling' to break winter dormancy, ensuring high budbreak and flowering and hence high fruit yields (McPherson *et al.*, 1994). Insufficient winter chill can result in a reduced amount, or delayed and/or extended periods of, budbreak and flowering (McPherson *et al.*, 1994; Snelgar *et al.*, 2008). Winter chilling requirements also vary substantially between *Actinidia* species and cultivars (Wall *et al.*, 2008; Wang *et al.*, 2017; Zhao *et al.*, 2017) (see Chapter 14).

Climate change predictions suggest kiwifruit growers could be vulnerable to lack of adequate chill in some current kiwifruit growing regions (Cradock-Henry, 2017), and Tait *et al.* (2018) found that production of 'Hayward' may become non-viable in the main growing regions of New Zealand by the end of this century. However, more recent modelling showed that climate change impacts may not be as severe, and identified locations providing high levels of chill that are, or will become, suitable for kiwifruit (Vetharaniam *et al.*, 2022a, b). Application of budbreak enhancers can counteract some effects of warmer winters, but are ineffective if winters are too warm (Allan, 2004). Alternatively, selection and adoption of new low-chill cultivars could mitigate against decreased winter chill.

Kiwifruit vines become acclimated to low temperatures in winter, but severe winter frosts can damage vines, with long-term impacts on vine health and productivity. On young plants, Pyke *et al.* (1986) found that temperatures below −7°C killed some dormant kiwifruit vines in late autumn, but frosts of −9°C and −11°C killed 100% of one-year-old plants in the same month. However, in early winter, no plants were killed at −7°C, and 17%, 33% and 67% of one-year-old plants were killed by frosts of −9°C, −11°C and −13°C, respectively. For mature plants, Weet (1979) found that fully dormant kiwifruit plants in California survived −17.5°C and in Italy, Testolin and Messina (1987) found that 'Hayward' vines survived temperatures of −18°C, with some cane damage. Male plants may be more susceptible to frost damage than female plants (Strik and Davis, 2021).

Spring

Temperatures during very early spring, 40 days before budbreak, can influence the start of budbreak (Warrington and Stanley, 1986). Once budbreak has occurred, young buds and shoots are more sensitive to cold temperatures. Hewett and Young (1981) found that buds, flowers and shoots were not damaged at −1°C but could be severely damaged at or below −1.5°C after as little as 30 minutes. Cultivars that break bud in early spring are more likely

Fig. 10.3. Living shelter, e.g. *Salix* spp., is less costly but slower to establish and requires more management inputs than artificial shelter. (Photo: The New Zealand Institute for Plant and Food Research Ltd)

to suffer frost damage. Kiwifruit vines may be protected from frost via windmills, helicopters, frost fans and overhead irrigation (Perry, 1998) (Fig. 10.4). The risk of bacterial canker of kiwifruit caused by *Pseudomonas syringae* pv. *actinidiae* (Psa) infection is also increased by winter and spring frosts (Ferrante and Scortichini, 2014) (see Chapter 18).

The duration of the period from budbreak to flowering is highly correlated with temperature, and this explains why vines at warm sites are slow to reach budbreak but catch up with

Fig. 10.4. Young buds on cultivars that develop early in spring are extremely sensitive to frost but can be protected using overhead irrigation. (Photo: The New Zealand Institute for Plant and Food Research Ltd)

vines at cooler sites by flowering (McPherson *et al.*, 1992, 2001). Shoot growth, leaf and canopy expansion in spring are key determinants of carbon supply for vine and fruit development throughout the season. For 'Hayward', shoot and leaf development had a broad temperature optimum of 20–30°C (Morgan *et al.*, 1985). Shoot growth, leaf-area development and carbon capture of 'Hayward' vines were substantially increased with a 28/22°C day compared with a 17/12°C day (Greer and Jeffares, 1998). Increasing temperatures from 22/13°C to 30/25°C during summer and early autumn resulted in a ten-fold increase in shoot growth on 'Hayward' vines (Richardson *et al.*, 2004). Maximum root growth occurred at 20°C, with a marked reduction in root growth below 20°C and little or no root growth occurring at 10°C (Smith *et al.*, 1989); root temperatures above 25°C reduce both root and shoot growth.

Summer

During early summer, temperature influences the effectiveness of pollination, fruit set and rapid fruit growth. High temperatures reduce pollen viability and the period over which the stigmas are receptive to pollen. Temperature also influences the activity and effectiveness of insect pollinators (see Chapter 13). Warmer temperatures during early fruit development can speed up fruit growth, but effects do not persist until harvest (Richardson *et al.*, 2004; Snelgar *et al.*, 2005). High temperatures (up to 30°C) later in summer, during fruit starch accumulation, can also increase fruit and shoot growth but at the expense of dry matter accumulation in fruit (Richardson *et al.*, 2004) and red fruit flesh colour (Man *et al.*, 2015). Extreme temperatures, up to and exceeding 40°C, can negatively affect vine performance, causing premature cessation of fruit growth (see Chapter 15).

Autumn

During autumn, temperature influences fruit maturation, with lower temperatures increasing the conversion of starch to sugars in kiwifruit, advancing the maturation of fruit (Burdon *et al.*, 2016). Cooler preharvest temperatures (below 7°C) may reduce the susceptibility of fruit to low temperature breakdown (Burdon *et al.*, 2007).

Preharvest frost can lead to significant leaf drop and fruit damage (Pyke *et al.*, 1986) and increase the risk of Psa infection to vines (Ferrante and Scortichini, 2014).

10.3.3 Hail

Kiwifruit vines are very susceptible to hail damage of the leaves, shoots and fruit. Hail is a particular problem in Italy and Greece (Basile *et al.*, 2012; Roussos *et al.*, 2022) where hail netting is used to prevent damage to vines. Additional advantages of hail netting include moderating the effects of high temperatures on vines (Roussos *et al.*, 2022) and improving vine development, fruit size and fruit quality (Basile *et al.*, 2012, 2014).

10.4 Orchard Site

10.4.1 Water supply

Kiwifruit vine water requirements depend on conditions that affect daily vine water loss (including radiation, wind and rainfall) as well as soil root zone storage and the efficiency of the irrigation system (see Chapter 11). The upper range of daily vine water use is 4.8–6.1 mm day^{-1} (120–152 L d^{-1} for a 5 × 5 m vine footprint); however, water use varies throughout vine development (Judd *et al.*, 1986). Kiwifruit vines are highly sensitive to water deficit, with an immediate reduction in fruit growth and vine development when vines become stressed (see Chapter 15). Prolonged water deficit may cause sunburn, softening and drop of fruit, leaf drop and loss of production in the following season (Prendergast *et al.*, 1987). Excessive water application damages vines and contributes to nutrient leaching into waterways. Water for overhead irrigation may be required for frost protection.

To provide adequate irrigation/frost protection for kiwifruit vines, sufficient water storage or a reliable external supply is required to irrigate vines during periods of drought or frost, and this can be calculated from site weather data. Irrigation water should be of high quality and free of particles that can block irrigation emitters. It must meet chemical guidelines, e.g. those for horticulture in California (chloride < 70 ppm, bicarbonate < 200 ppm, B < 0.25 ppm, sodium < 50 ppm, electrical conductivity (EC), EC × 10 < 0.75) (Strik and Cahn, 1998). High Ca levels in water can cause spotting on fruit if overhead irrigation is used (Strik and Davis, 2021). Water quality may be improved by using filtration or additives.

10.4.2 Orchard position, aspect and slope

Orchard position, aspect and slope all influence the orchard microclimate, vine phenology and vine management. Orchards located near the sea, e.g. in New Zealand, tend to have more moderate climates (cooler in summer and warmer in winter) than those in continental locations like China and Europe (Barden and Neilsen, 2003). In the southern hemisphere, south-facing slopes are cooler, and north-facing slopes warmer, with the reverse in the northern hemisphere, while the east–west aspect can affect exposure to morning or afternoon sun. Temperatures decrease with both increasing latitude and altitude, and this may influence the length of seasons and increase the effect of aspect (Strik and Davis, 2021). The prevailing wind direction also affects site temperature and rainfall. Row orientation influences the amount of shading vines receive, with a north–south orientation preferred to minimize shading effects.

Kiwifruit orchards are generally planted on flat to moderately sloping sites. Slopes of greater than 30° are not safe for orchard machinery (Rowland *et al.*, 2016) and have a greater risk of erosion and slips after intense rainfall. Kiwifruit vines are grown on supporting trellis structures, which further limits their ability to be grown on sloping sites. Variation in temperature across sloping blocks can induce variation in crop phenology, e.g. the time of budbreak and flowering, which, in turn, leads to problems with timing key management tasks for the whole block (e.g. pollination and thinning) and final fruit quality (Richardson *et al.*, 2019). A further consideration is to ensure that cold air can drain from low points in sloping orchard blocks in order to reduce the likelihood of frost damage to fruit and vines (Sale and Lyford, 1990) (Fig. 10.5).

Fig. 10.5. Variation in slope within kiwifruit blocks can increase variation in vine phenology and the risk of frost damage if cold air cannot drain from the lowest points of blocks. (Photo: The New Zealand Institute for Plant and Food Research Ltd)

10.5 Future Directions

The optimization of many environmental factors is needed to develop a successful kiwifruit orchard. Climate, particularly temperature, has a significant impact on all aspects of vine and fruit development. Wind protection and water availability are also key factors. As the climate warms, changes in key parameters as well as the frequency of extreme weather events will affect the suitability of current cultivars in existing growing locations. However, this will also provide new opportunities for a wider diversity of kiwifruit cultivars and new growing regions. It is likely that growers will need to adopt new or additional methods of microclimate modification, for example frost control, wind breaks, irrigation, drainage and growing crops indoors or under protection. This may be coupled with more intelligent sensing of plant status as well as the breeding of climate-resilient scions and rootstocks.

Kiwifruit is grown on soils with a wide range of physical, chemical and biological properties. As competition for land increases and some sites become degraded through current usage and extreme climate events, increasing mitigation of soil and site characteristics will be required. This may include changing to contoured, terraced or raised-bed planting systems. Improved management of soil physical characteristics through modification of soil texture, structure, depth and porosity is likely to be required. New understanding of soil biological properties, including the soil biome and how it may be managed using organic principles, will be essential to support optimal vine growth. Modifying soil chemical properties throughout the soil profile as well as tailored application of vine nutrients to each vine will be needed to ensure vine productivity, while limiting losses of nutrients into waterways and the wider environment.

References

Abid, M., Zhang, Y.J., Li, Z., Bai, D.F., Zhong, Y.P. *et al.* (2020) Effect of salt stress on growth, physiological and biochemical characters of four kiwifruit genotypes. *Scientia Horticulturae* 271, 109473. DOI: 10.1016/j.scienta.2020.109473

Allan, P. (2004) Winter chilling in areas with mild winters: its measurement and supplementation. *Acta Horticulturae* 662, 47–52. DOI: 10.17660/ActaHortic.2004.662.3

Asher, C.J., Smith, G.S., Clark, C.J. and Brown, N.S. (1984) Manganese deficiency of kiwifruit (*Actinidia chinensis* Planch.). *Journal of Plant Nutrition* 7, 1497–1509. DOI: 10.1080/01904168409363296

Barber, R.F., Clark, C.J., Ferguson, A.R., Sale, P.R. and Smith, G.S. (1986) Subtropicals. In: Clark, C.J., Smith, G.S., Prasad, M. and Cornforth, I.S. (eds) *Fertiliser Recommendations for Horticultural Crops*. Ministry of Agriculture and Fisheries, Wellington, pp. 17–31.

Barden, J.A. and Neilsen, G.H. (2003) Selecting the orchard site, site preparation and orchard planning and establishment. In: Ferree, D.C. and Warrington, I.J. (eds) *Apples: Botany, Production and Uses*. CAB International, Wallingford, UK, pp. 237–265.

Basile, B., Giaccone, M., Cirillo, C., Ritieni, A., Graziani, G. *et al.* (2012) Photo-selective hail nets affect fruit size and quality in Hayward kiwifruit. *Scientia Horticulturae* 141, 91–97. DOI: 10.1016/j.scienta.2012.04.022

Basile, B., Giaccone, M., Shahak, Y., Forlani, M. and Cirillo, C. (2014) Regulation of the vegetative growth of kiwifruit vines by photo-selective anti-hail netting. *Scientia Horticulturae* 172, 300–307. DOI: 10.1016/j.scienta.2014.04.011

Benge, J., Manhire, J., Pearson, A., Reid, J. and Moller, H. (2007) Differences in soil quality between and within organic and integrated management kiwifruit orchards in New Zealand. *Acta Horticulturae* 753, 599–607. DOI: 10.17660/ActaHortic.2007.753.79

Bird, P.R., Bicknell, D., Bulman, P.A., Burke, S.J.A., Leys, J.F. *et al.* (1993) The role of shelter in Australia for protecting soils, plants and livestock. In: Prinsley, R.T. (ed.) *The Role of Trees in Sustainable Agriculture*. Forestry Sciences series, Vol. 43. Springer, Dordrecht, The Netherlands, pp. 59–86. DOI: 10.1007/978-94-011-1832-3_3

Black, M.Z., Patterson, K.J., Gould, K.S. and Clearwater, M.J. (2012) Physiological responses of kiwifruit vines (*Actinidia chinensis* Planch. var. *chinensis*) to trunk girdling and root pruning. *New Zealand Journal of Crop and Horticultural Science* 40, 31–41. DOI: 10.1080/01140671.2011.603343

Burdon, J., Lallu, N., Francis, K. and Boldingh, H. (2007) The susceptibility of kiwifruit to low temperature breakdown is associated with pre-harvest temperatures and at-harvest soluble solids content. *Postharvest Biology and Technology* 43, 283–290. DOI: 10.1016/j.postharvbio.2006.09.011

Burdon, J., Pidakala, P., Martin, P., Billing, D. and Boldingh, H. (2016) Fruit maturation and the soluble solids harvest index for 'Hayward' kiwifruit. *Scientia Horticulturae* 213, 193–198. DOI: 10.1016/j.scienta.2016.10.027

Carey, P.L., Benge, J.R. and Haynes, R.J. (2009) Comparison of soil quality and nutrient budgets between organic and conventional kiwifruit orchards. *Agriculture, Ecosystems & Environment* 132, 7–15. DOI: 10.1016/j.agee.2009.02.017

Castro, J., Lago, M.C.F., Briones, M.J.I., Gallego, P.P. and Barreal, M.E. (2015) Effects of agricultural practices on soil fauna communities in kiwifruit plantations. *Acta Horticulturae* 1096, 267–273. DOI: 10.17660/ActaHortic.2015.1096.29

Castro, J., Barreal, M.E., Briones, M.J.I. and Gallego, P.P. (2018) Effects of farming practices on the structure of earthworm communities in kiwifruit orchards. *Acta Horticulturae* 1218, 419–425. DOI: 10.17660/ActaHortic.2018.1218.58

Castro, J., Barreal, M.E., Briones, M.J.I. and Gallego, P.P. (2019) Earthworm communities in conventional and organic fruit orchards under two different climates. *Applied Soil Ecology* 144, 83–91. DOI: 10.1016/j.apsoil.2019.07.013

Chartzoulakis, K.S., Therios, I.N., Misopolinos, N.D. and Noitsakis, B.I. (1995) Growth, ion content and photosynthetic performance of salt-stressed kiwifruit plants. *Irrigation Science* 16, 23–28. DOI: 10.1007/BF00208392

Cho, Y., Cho, H., Ma, K., Park, M. and Kim, B. (2013) Effect of subsoiling on soil physical properties and fruit quality in organic conversion kiwifruit (*Actinidia chinensis* 'Haegeum'). *Acta Horticulturae* 1001, 347–352. DOI: 10.17660/ActaHortic.2013.1001.40

Costa, G., Kukuriannis, B. and Monet, R. (1992) Kiwifruit production in Europe. *Acta Horticulturae* 297, 141–149. DOI: 10.17660/ActaHortic.1992.297.17

Cradock-Henry, N.A. (2017) New Zealand kiwifruit growers' vulnerability to climate and other stressors. *Regional Environmental Change* 17, 245–259. DOI: 10.1007/s10113-016-1000-9

Currie, M.B., Patterson, K.J., Snelgar, W.P. and Blattmann, P. (2018) Girdling kiwifruit vines for commercial advantage: opportunities and risks. *Acta Horticulturae* 1218, 405–412. DOI: 10.17660/ActaHortic.2018.1218.56

Deurer, M., Grinev, D., Young, I., Clothier, B.E. and Müller, K. (2009) The impact of soil carbon management on soil macropore structure: a comparison of two apple orchard systems in New Zealand. *European Journal of Soil Science* 60, 945–955. DOI: 10.1111/j.1365-2389.2009.01164.x

Fan, L., Zhou, X., Li, Y., Ji, L., Wu, G. *et al.* (2016) The influence of effective microorganisms on microbes and nutrients in kiwifruit planting soil. *Applied Sciences* 6, 168. DOI: 10.3390/app6060168

Ferrante, P. and Scortichini, M. (2014) Frost promotes the pathogenicity of *Pseudomonas syringae* pv. *actinidiae* in *Actinidia chinensis* and *A. deliciosa* plants. *Plant Pathology* 63, 12–19. DOI: 10.1111/ppa.12070

Gentile, R.M., Malepfane, N.M., van den Dijssel, C., Arnold, N., Liu, J. *et al.* (2021) Comparing deep soil organic carbon stocks under kiwifruit and pasture land uses in New Zealand. *Agriculture, Ecosystems & Environment* 306, 107190. DOI: 10.1016/j.agee.2020.107190

Gibb, H. and Cunningham, S.A. (2013) Restoration of trophic structure in an assemblage of omnivores, considering a revegetation chronosequence. *Journal of Applied Ecology* 50, 449–458. DOI: 10.1111/1365-2664.12054

Gloyne, R. (1955) Some effects of shelter-belts and wind-breaks. *Meteorology Magazine* 84, 272–281.

Greaves, A.J. (1985) Root distribution of kiwifruit (*Actinidia deliciosa*) in a deep sandy loam soil of the Te Puke district, New Zealand. *New Zealand Journal of Agricultural Research* 28, 433–436. DOI: 10.1080/00288233.1985.10430449

Greer, D.H. and Jeffares, D. (1998) Temperature-dependence of carbon acquisition and demand in relation to shoot growth of kiwifruit (*Actinidia deliciosa*) vines grown in controlled environments. *Australian Journal of Plant Physiology* 25, 843–850. DOI: 10.1071/PP98055

Guyot, G. and Seguin, B. (1975) Modification of land roughness and resulting microclimate effects: a field study in Brittany. In: de Vries, D.A. and Afgan, N.H. (eds) *Heat and Mass Transfer in the Biosphere. Part 1: Transfer Processes in the Plant Environment.* Scripta Book Co., Halsted Press, New York, pp. 467–478.

Hathaway, R.L. (1990) Shelter for kiwifruit production. In: Warrington, I.J. and Weston, G.C. (eds) *Kiwifruit: Science and Management*. Ray Richards Publisher in association with the New Zealand Society for Horticultural Science, Auckland, New Zealand, pp. 322–346.

Hewett, E.W. and Young, K. (1981) Critical freeze damage temperatures of flower buds of kiwifruit (*Actinidia chinensis* Planch.). *New Zealand Journal of Agricultural Research* 24, 73–75. DOI: 10.1080/00288233.1981.10420873

Hughes, K.A. and Wilde, R.H. (1989) The effect of poor drainage on the root distribution of kiwifruit vines. *New Zealand Journal of Crop and Horticultural Science* 17, 239–244. DOI: 10.1080/01140671.1989.10428038

Hughes, K.A., de Willigen, P., Gandar, P.W. and Clothier, B.E. (1992) Kiwifruit root systems: structure and function. *Acta Horticulturae* 297, 383–389. DOI: 10.17660/ActaHortic.1992.297.51

Judd, M.J., McAneney, K.J. and Trought, M.C.T. (1986) Water-use by sheltered kiwifruit under advective conditions. *New Zealand Journal of Agricultural Research* 29, 83–92. DOI: 10.1080/00288233.1986.10417978

Klages, K., Boldingh, H. and Smith, G.S. (1999) Accumulation of *myo*-inositol in *Actinidia* seedlings subjected to salt stress. *Annals of Botany* 84, 521–527. DOI: 10.1006/anbo.1999.0946

Koh, Y.J., Lim, M.T., Jeong, I.H., Kim, G.H., Han, T.W. *et al.* (2007) Survey on the occurrence of abiotic diseases on kiwifruit in Korea. *The Plant Pathology Journal* 23, 308–313. DOI: 10.5423/PPJ.2007.23.4.308

Lago, M.C.F., Castro, J., Briones, M.J.I., Gallego, P.P. and Barreal, M.E. (2015) Effect of agricultural management on kiwifruit nutritional plant status, fruit quality and yield. *Acta Horticulturae* 1096, 79–86. DOI: 10.17660/ActaHortic.2015.1096.5

Lago, M.C.F., Gallego, P.P. and Briones, M.J.I. (2019) Intensive cultivation of kiwifruit alters the detrital foodweb and accelerates soil C and N losses. *Frontiers in Microbiology* 10, 686. DOI: 10.3389/fmicb.2019.00686

Lema, M.J., Rodríguez, P., Salinero, C. and Veloso, F. (2010) Seasonal pattern of accumulation of nutrients by kiwifruit 'Hayward' in acid soils of northern Portugal. *Acta Horticulturae* 868, 231–236. DOI: 10.17660/ActaHortic.2010.868.29

Liu, Z., Guo, Q., Feng, Z., Liu, Z., Li, H. *et al.* (2020a) Long-term organic fertilization improves the productivity of kiwifruit (*Actinidia chinensis* Planch.) through increasing rhizosphere microbial diversity and network complexity. *Applied Soil Ecology* 147, 103426. DOI: 10.1016/j.apsoil.2019.103426

Liu, Z., Ma, X., He, N., Zhang, J., Wu, J. *et al.* (2020b) Shifts in microbial communities and networks are correlated with the soil ionome in a kiwifruit orchard under different fertilization regimes. *Applied Soil Ecology* 149, 103517. DOI: 10.1016/j.apsoil.2020.103517

Magliulo, V., d'Andria, R., Giorio, P., Martorella, A. and Quaglietta Chiarandà, F. (1991) Cumulative effects of winter shallow water-table conditions on stem growth and yield of kiwifruit. *New Zealand Journal of Crop and Horticultural Science* 19, 375–380. DOI: 10.1080/01140671.1991.10422879

Man, Y.P., Wang, Y.C., Li, Z.Z., Jiang, Z.W., Yang, H.L. *et al.* (2015) High-temperature inhibition of biosynthesis and transportation of anthocyanins results in the poor red coloration in red-fleshed *Actinidia chinensis*. *Physiologia Plantarum* 153, 565–583. DOI: 10.1111/ppl.12263

McAneney, K.J. and Judd, M.J. (1983) Observations on kiwifruit (*Actinidia chinensis* Planch.) root exploration, root pressure, hydraulic conductivity, and water uptake. *New Zealand Journal of Agricultural Research* 26, 507–510. DOI: 10.1080/00288233.1983.10427029

McAneney, K.J. and Judd, M.J. (1987) Comparative shelter strategies for kiwifruit: a mechanistic interpretation of wind damage measurements. *Agricultural and Forest Meteorology* 39, 225–240. DOI: 10.1016/0168-1923(87)90040-2

McAneney, K.J., Judd, M.J. and Trought, M.C.T. (1984) Wind damage to kiwifruit (*Actinidia chinensis* Planch.) in relation to windbreak performance. *New Zealand Journal of Agricultural Research* 27, 255–263. DOI: 10.1080/00288233.1984.10430427

McAneney, K.J., Salinger, M.J., Porteous, A.S. and Barber, R.F. (1990) Modification of an orchard climate with increasing shelter-belt height. *Agricultural and Forest Meteorology* 49, 177–189. DOI: 10.1016/0168-1923(90)90031-Z

McPherson, H.G., Hall, A.J. and Stanley, C.J. (1992) The influence of current temperature on the time from bud break to flowering in kiwifruit (*Actinidia deliciosa*). *Journal of Horticultural Science* 67, 509–519. DOI: 10.1080/00221589.1992.11516277

McPherson, H.G., Hall, A.J. and Stanley, C.J. (1994) Seasonal and regional variation in budbreak and flowering of kiwifruit vines in New Zealand. *New Zealand Journal of Crop and Horticultural Science* 22, 263–276. DOI: 10.1080/01140671.1994.9513835

McPherson, H.G., Richardson, A.C., Snelgar, W.P. and Currie, M.B. (2001) Effects of hydrogen cyanamide on budbreak and flowering in kiwifruit (*Actinidia deliciosa* 'Hayward'). *New Zealand Journal of Crop and Horticultural Science* 29, 277–285. DOI: 10.1080/01140671.2001.9514189

Morgan, D.C., Warrington, I.J. and Halligan, E.A. (1985) Effect of temperature and photosynthetic photon flux density on vegetative growth of kiwifruit (*Actinidia chinensis*). *New Zealand Journal of Agricultural Research* 28, 109–116. DOI: 10.1080/00288233.1985.10427003

Müller, K., Duwig, C., Tinet, A.-J., Strozzi, A.G., Spadini, L. *et al.* (2019) Orchard management and preferential flow in Andosols – comparing two kiwifruit orchards in New Zealand. *Soil Research* 57, 615–628. DOI: 10.1071/sr18293

Otero, V., Barreal, M.E., Martínez-Núñez, L. and Gallego, P.P. (2010) Nutritional status of kiwifruit in organic and conventional farming systems. *Acta Horticulturae* 868, 155–160. DOI: 10.17660/ActaHortic.2010.868.17

Perry, K.B. (1998) Basics of frost and freeze protection for horticultural crops. *Horttechnology* 8, 10–15. DOI: 10.21273/HORTTECH.8.1.10

Prendergast, P.T., McAneney, K.J., Astill, M.S., Wilson, A.D. and Barber, R.F. (1987) Water extraction and fruit expansion by kiwifruit. *New Zealand Journal of Experimental Agriculture* 15, 345–350. DOI: 10.1080/03015521.1987.10425580

Pyke, N.B., Stanley, C.J. and Warrington, I.J. (1986) Kiwifruit: frost tolerance of plants in controlled frost conditions. *New Zealand Journal of Experimental Agriculture* 14, 443–447. DOI: 10.1080/03015521.1986.10423063

Rahman, M.H., Holmes, A.W. and Saunders, S.J. (2014) Spatio-temporal variation in soil organic carbon under kiwifruit production systems of New Zealand. *Acta Horticulturae* 1018, 279–286. DOI: 10.17660/ActaHortic.2014.1018.29

Řeháček, D., Khel, T., Kučera, J., Vopravil, J. and Petera, M. (2017) Effect of windbreaks on wind speed reduction and soil protection against wind erosion. *Soil and Water Research* 12, 128–135. DOI: 10.17221/45/2016-SWR

Reid, J.B. and Petrie, R.A. (1991) Effects of soil aeration on root demography in kiwifruit. *New Zealand Journal of Crop and Horticultural Science* 19, 423–431. DOI: 10.1080/01140671.1991.10422888

Reid, J.B., Tate, K.G., Brown, N.S. and Cheah, L.H. (1991) Effects of flooding and alluvium deposition on kiwifruit (*Actinidia deliciosa*). 1. Early vine decline. *New Zealand Journal of Crop and Horticultural Science* 19, 247–257. DOI: 10.1080/01140671.1991.10421808

Reid, J.B., Tate, K.G. and Brown, N.S. (1992) Effects of flooding and alluvium deposition on kiwifruit (*Actinidia deliciosa*). 2. Vine performance the following season. *New Zealand Journal of Crop and Horticultural Science* 20, 283–288. DOI: 10.1080/01140671.1992.10421769

Richards, S., Hewson, K., Moller, H., Wharton, D., Campbell, H. *et al.* (2007) Soil biota as indicators of soil quality in organic and integrated management kiwifruit orchards in New Zealand. *Acta Horticulturae* 753, 627–632. DOI: 10.17660/ActaHortic.2007.753.82

Richardson, A.C., Marsh, K.B., Boldingh, H.L., Pickering, A.H., Bulley, S.M. *et al.* (2004) High growing temperatures reduce fruit carbohydrate and vitamin C in kiwifruit. *Plant, Cell & Environment* 27, 423–435. DOI: 10.1111/j.1365-3040.2003.01161.x

Richardson, A.C., Boldingh, H., Kashuba, P., Knight, G. and Ellingham, D. (2019) Flowering time determines the weight and composition of *Actinidia chinensis* var. *chinensis* 'Zesy002' kiwifruit. *Scientia Horticulturae* 246, 741–748. DOI: 10.1016/j.scienta.2018.11.043

Roussos, P.A., Tsafouros, A., Ntanos, E., Denaxa, N.K., Kosta, A. *et al.* (2022) Could black anti-hail net have an extra role as an amelioration agent against heat stress in kiwifruit? *Journal of Berry Research* 12, 131–147. DOI: 10.3233/jbr-211512

Rowland, J., Maschmedt, D. and Liddicoat, C. (2016) Land use potential for agricultural crops in southern South Australia: summary of assessment and mapping methodology. DEWNR Technical note 2016/29, Government of South Australia, Department of Environment, Water and Natural Resources, Adelaide. Available at: at: https://data.environment.sa.gov.au/Content/Publications/LandUsePotential_Descriptions_MappingAndSpatialData.pdf (accessed 29 May 2019).

Sale, P.R. and Lyford, P.B. (1990) Cultural, management and harvesting practices for kiwifruit in New Zealand. In: Warrington, I.J. and Weston, G.C. (eds) *Kiwifruit: Science and Management*. Ray Richards Publisher in association with the New Zealand Society for Horticultural Science, Auckland, New Zealand, pp. 247–296.

Salinger, M.J. and Kenny, G.J. (1995) Climate and kiwifruit cv. 'Hayward'. 2. Regions in New Zealand suited for production. *New Zealand Journal of Crop and Horticultural Science* 23, 173–184. DOI: 10.1080/01140671.1995.9513884

Savé, R. and Serrano, L. (1986) Some physiological and growth responses of kiwi fruit (*Actinidia chinensis*) to flooding. *Physiologia Plantarum* 66, 75–78. DOI: 10.1111/j.1399-3054.1986.tb01236.x

Savian, F., Ginaldi, F., Musetti, R., Sandrin, N., Tarquini, G. *et al.* (2020) Studies on the aetiology of kiwifruit decline: interaction between soil-borne pathogens and waterlogging. *Plant and Soil* 456, 113–128. DOI: 10.1007/s11104-020-04671-5

Smith, G.S., Asher, C.J. and Clark, C.J. (1987a) *Kiwifruit Nutrition: Diagnosis of Nutritional Disorders*, 2nd edn. Agpress Communications, Wellington.

Smith, G.S., Clark, C.J. and Henderson, H.V. (1987b) Seasonal accumulation of mineral nutrients by kiwifruit. *New Phytologist* 106, 81–100. DOI: 10.1111/j.1469-8137.1987.tb04793.x

Smith, G.S., Buwalda, J.G., Green, T.G.A. and Clark, C.J. (1989) Effect of oxygen supply and temperature at the root on the physiology of kiwifruit vines. *New Phytologist* 113, 431–437. DOI: 10.1111/j.1469-8137.1989.tb00354.x

Smith, G.S., Judd, M.J., Miller, S.A. and Buwalda, J.G. (1990) Recovery of kiwifruit vines from transient waterlogging of the root system. *New Phytologist* 115, 325–333. DOI: 10.1111/j.1469-8137.1990.tb00459.x

Snelgar, W.P., Hall, A.J., Ferguson, A.R. and Blattmann, P. (2005) Temperature influences growth and maturation of fruit on 'Hayward' kiwifruit vines. *Functional Plant Biology* 32, 631–642. DOI: 10.1071/FP05062

Snelgar, W.P., Hall, A.J. and McPherson, H.G. (2008) Modelling flower production of kiwifruit (*Actinidia deliciosa*) from winter chilling. *New Zealand Journal of Crop and Horticultural Science* 36, 273–284. DOI: 10.1080/01140670809510244

Sotiropoulos, T.E., Therios, I.N. and Dimassi, K.N. (2005) Uptake of boron by kiwifruit plants under various levels of shading and salinity. *Journal of Plant Nutrition* 27, 1979–1989. DOI: 10.1081/PLN-200030091

Sotiropoulos, T.E., Therios, I.N. and Dimassi, K.N. (2006) Seasonal accumulation and distribution of nutrient elements in fruit of kiwifruit vines affected by boron toxicity. *Australian Journal of Experimental Agriculture* 46, 1639–1644. DOI: 10.1071/EA05081

Strik, B.C. and Cahn, H. (1998) *Growing Kiwifruit*. PNW 507. Pacific Northwest Cooperative Extension, Corvallis, Oregon.

Strik, B.C. and Davis, A.J. (2021) *Growing Kiwifruit. A Guide to Kiwiberries and Fuzzy Kiwifruit for Pacific Northwest Producers.* PNW 507 revision. Pacific Northwest Extension Publishing, Corvallis, Oregon.

Tagliavini, M. and Marangoni, B. (2002) Major nutritional issues in deciduous fruit orchards of northern Italy. *HortTechnology* 12, 26–31. DOI: 10.21273/horttech.12.1.26

Tait, A., Paul, V., Sood, A. and Mowat, A. (2018) Potential impact of climate change on Hayward kiwifruit production viability in New Zealand. *New Zealand Journal of Crop and Horticultural Science* 46, 175–197. DOI: 10.1080/01140671.2017.1368672

Testolin, R. and Messina, R. (1987) Winter cold tolerance of kiwifruit: a survey after winter frost injury in northern Italy. *New Zealand Journal of Experimental Agriculture* 15, 501–504. DOI: 10.1080/03015521.1987.10425604

Tian, N., Wang, J. and Xu, Z.Q. (2011) Overexpression of Na^+/H^+ antiporter gene *AtNHX1* from *Arabidopsis thaliana* improves the salt tolerance of kiwifruit (*Actinidia deliciosa*). *South African Journal of Botany* 77, 160–169. DOI: 10.1016/j.sajb.2010.07.010

Tsadilas, C.D., Dimoyiannis, D. and Samaras, V. (1997) Methods of assessing boron availability to kiwifruit plants growing on high boron soils. *Communications in Soil Science and Plant Analysis* 28, 973–987. DOI: 10.1080/00103629709369847

Vetharaniam, I., Timar, L., Stanley, C.J., Müller, K., van den Dijssel, C. *et al.* (2022a) Modelling climate change impacts on location suitability and spatial footprint of apple and kiwifruit. *Land* 11, 1639. DOI: 10.3390/land11101639

Vetharaniam, I., van den Dijssel, C., Stanley, J., Richardson, A. and Müller, K. (2022b) Projecting suitability of kiwifruit by chill requirements for different locations in New Zealand under climate change. *Acta Horticulturae* 1332, 455–462. DOI: 10.17660/ActaHortic.2022.1332.60

Wall, C., Dozier, W., Ebel, R.C., Wilkins, B., Woods, F. *et al.* (2008) Vegetative and floral chilling requirements of four new kiwi cultivars of *Actinidia chinensis* and *A. deliciosa*. *HortScience* 43, 644–647. DOI: 10.21273/HORTSCI.43.3.644

Wang, N., He, H., Lacroix, C., Morris, C., Liu, Z. *et al.* (2019) Soil fertility, leaf nutrients and their relationship in kiwifruit orchards of China's central Shaanxi province. *Soil Science and Plant Nutrition* 65, 369–376. DOI: 10.1080/00380768.2019.1624481

Wang, S., Huang, C., Tao, J., Zhong, M., Qu, X. *et al.* (2017) Evaluation of chilling requirement of kiwifruit (*Actinidia* spp.) in south China. *New Zealand Journal of Crop and Horticultural Science* 45, 289–298. DOI: 10.1080/01140671.2017.1334670

Wardle, D.A., Yeates, G.W., Bonner, K.I., Nicholson, K.S. and Watson, R.N. (2001) Impacts of ground vegetation management strategies in a kiwifruit orchard on the composition and functioning of the soil biota. *Soil Biology and Biochemistry* 33, 893–905. DOI: 10.1016/S0038-0717(00)00235-2

Warrington, I.J. and Stanley, C.J. (1986) The influence of pre- and post-budbreak temperatures on flowering in kiwifruit. *Acta Horticulturae* 175, 103–107. DOI: 10.17660/ActaHortic.1986.175.15

Weet, C. (1979) Frost damage to kiwifruit vines (in southern California). *Avocado Grower* 3, 26–29, 59.

Zhang, L., Yao, C., Liang, J., Wu, C. and Zhang, Y. (2003) Leaf and soil nutritional status of the 'Qinmei' kiwifruit orchards in Shaanxi Province. *Acta Horticulturae* 610, 153–155. DOI: 10.17660/ActaHortic.2003.610.19

Zhao, T., Li, D., Li, L., Han, F., Liu, X. *et al.* (2017) The differentiation of chilling requirements of kiwifruit cultivars related to ploidy variation. *HortScience* 52, 1676–1679. DOI: 10.21273/hortsci12410-17

Zhong, Y.-P., Qi, X.-J., Chen, J.-Y., Li, Z., Bai, D.-F. *et al.* (2019) Growth and physiological responses of four kiwifruit genotypes to salt stress and resistance evaluation. *Journal of Integrative Agriculture* 18, 83–95. DOI: 10.1016/S2095-3119(18)62011-8

11 Vine Nutrition and Water Requirements

Cristos Xiloyannis, Bartolomeo Dichio* and Alba N. Mininni
University of Basilicata, Matera, Italy

11.1 Introduction

Modern agricultural production systems face challenges from climate change and poorly applied nutritional and irrigation practices. Soil erosion, degradation and salinization, resulting in loss of fertility, along with the pollution of soils, groundwater and air, and emission of greenhouse gases, are increasingly compromising agricultural production and environmental health.

Balanced crop irrigation and nutrition can enhance crop productivity and quality. However, when poorly applied, these practices can severely compromise production, make resource use inefficient, increase waste and threaten environmental health. For these reasons, careful, well-planned irrigation and fertilizer regimes are necessary for sustainable orchard management.

Plants acquire nutrients through water absorption by their roots, so optimal soil-water availability and plant transpiration rates are important to ensure soil nutrients reach all plant organs. Soil nutrient availability, in the quantities and at appropriate times to satisfy vine nutritional requirements, ensures regular plant development and fruit growth. Management of mineral nutrition and irrigation is key to maximizing crop yield and quality and contributes to environmental protection.

The aim of this chapter is to provide information to promote sustainable management of irrigation and nutrition in kiwifruit orchards. Much of the research is related to the Mediterranean environment and its effects on the ecophysiology of plants through high light, temperature and vapour pressure deficit (VPD) and low rainfall conditions. The basic principles and appropriate criteria for the optimal management of nutrition and irrigation in kiwifruit orchards presented in this chapter can, however, be applied at a global level.

11.2 Nutrition

11.2.1 Nutrient uptake, transport and translocation

The application of fertilizers is among the most common activities used to increase agricultural production but can have a high impact on the environment (Medici *et al.*, 2020). Sustainable fertilizer management should provide the nutrients required by the plant by considering their availability in the soil, limiting losses and improving nutrient use efficiency (Bryla, 2019). Understanding nutrient dynamics, including uptake,

*Corresponding author: bartolomeo.dichio@unibas.it

© CAB International 2024. *Kiwifruit: Botany, Production and Uses* (eds A.C. Richardson, J.N. Burdon and A.R. Ferguson)

DOI: 10.1079/9781800620933.0011

translocation and utilization within kiwifruit vines, is essential for the development of appropriate nutrient management practices (Buwalda and Smith, 1987).

Some mineral elements are absorbed from soil solution by roots through diffusion. Diffusion begins when a concentration gradient is established in the soil around the root system. This process involves mainly ions such as phosphorus (P), potassium (K), boron (B), iron (Fe), zinc (Zn) and manganese (Mn), which move slowly and over short distances; thus, root development, structure and density are important for their regular uptake.

Other mineral elements, such as calcium (Ca) and magnesium (Mg), move within the soil towards the roots mainly by mass flow. For such elements, readily available water (RAW) and plant transpiration are important for water and, consequently, nutrient uptake.

Often, mineral elements may be present in the soil, but not available to plants. Incorrect fertilizer inputs can cause imbalances between mineral elements in the soil, increasing competition and availability for root uptake. To define an adequate nutritional plan and ensure correct nutrient uptake and utilization by plants, potential antagonistic relationships between mineral elements should be considered. Adequate K concentrations allow uptake of Fe and Mn, but if K concentrations are too high, they can hinder uptake of Mg, B, nitrogen (N), P and Ca (Hashmatt, 2021). The usual once-a-season application of K can result in plant nutrient deficiencies owing to the temporary high concentrations of K in the soil. Similarly, excess Ca can compete with other elements such as K and Mg for uptake sites on the root. Some forms of Ca (lime) can change soil chemistry by raising the pH until Fe and B become unavailable to plants. Despite soil analysis indicating that the concentration of mineral elements is higher than required, it is necessary also to consider the relative amounts of nutrients. For example, to ensure adequate plant uptake, the Mg:K ratio should be greater than 2, and the Ca:Mg ratio should be between 8 and 12. A promising fertilization strategy to reduce antagonistic relationships between Ca and K involves spatial (application on different soil portions) and temporal (different timing) separation of soil-applied Ca and K fertilizers (Hashmatt, 2021).

All mineral nutrients taken up from the soil by the roots are passively transported upward to the transpiring organs (i.e. leaves and fruit) by the transpiration stream in the xylem. Some elements, such as N, P and K (high mobility) and, to some extent, Mg (lower mobility) can also move within the plant via the phloem stream, which is involved in assimilate allocation, supplying fruit even when they are not transpiring. Elements such as Ca, B, Mn, Zn and Fe have low mobility within the plant and move mainly via the xylem stream, following the transpiration flow towards transpiring plant organs (Fig. 11.1).

11.2.2 Seasonal mineral element accumulation in plant organs and nutrient requirements

Crop demand is estimated from a nutrient balance that considers all orchard components during

Fig. 11.1. Xylem flow is driven by transpiration (blue arrows), directing the flow towards transpiring organs. Multidirectional phloem sap flow is also a pathway for mineral element allocation and recycling within the plant (red arrows). However, phloem sap does not transport some elements such as calcium.

the life of the crop, including nutrient removal from orchards (Ferguson and Eiseman, 1983). A nutrient balance should consider the dynamics of nutrient circulation in the orchard and within the vine, including nutrients returned to the soil via fallen leaves, thinned fruit or prunings; remobilized and translocated within the vine; and removed from the orchard system, mainly in harvested fruit or prunings.

Evaluation of vine nutrient requirements focuses on plant biomass production during the year (leaves, fruit, root turnover) and the growth of permanent structures (trunk, leaders and permanent roots). In actively growing young vines (first 4–5 years from planting), mineral elements absorbed from soil are accumulated in permanent plant structures. In these years, the amount of nutrients removed from soil is low. After reaching full crop production, structural root, trunk and leader growth is limited and the amounts of mineral elements in these permanent plant structures change little over the years. Then, vine nutrients are required mainly for the annual growth of shoots, leaves and fruit as well as root turnover. Nutrient removal and uptake rates change according to the growth increment and annual relative growth rate of these plant organs and is known as the 'plant annual nutrient demand'.

Since the growth of permanent structures is minimal in a mature orchard, the macronutrients accumulated in the annual plant organs reflect the nutrients absorbed from soil during the different vine phenological growth stages. The quantities of mineral elements taken up from the soil and accumulated in different plant organs and eventually returned to the soil, as fallen leaves and prunings, must be considered for correct nutrition management. The macronutrient contents of the annual plant parts of a mature kiwifruit orchard at the end of the growing season are shown in Table 11.1. The nutrients in the harvested fruit are removed from the orchard, while those in the leaves, thinned fruit and prunings are returned to the orchard soil. Some mobile nutrients accumulated in leaves are partially remobilized within the plant during leaf senescence and recycled in the soil after leaf fall. Nutrients that are not as mobile in the soil (P, K, Ca, Mg) are fully recovered (100%), while highly mobile N is only partially recovered (50–90%) depending on soil management (the highest recovery rates are achieved in covered soils and under optimal irrigation management).

The development of annual kiwifruit vine biomass (shoots, leaves and fruit) during the growing season is shown in Fig. 11.2. During the first phase, from the beginning of shoot

Table 11.1. Dry matter and mineral element contents (three-year average) in different plant organs in a mature kiwifruit orchard (*Actinidia chinensis* var. *deliciosa* 'Hayward' on a pergola training system (494 plants ha^{-1}) with an average yield of 35 t fresh weight ha^{-1}). To calculate the nutrients removed from the orchard, harvested fruit were assumed to be removed from the orchard. Sixty-five per cent of the nitrogen in recycled biomass (leaves, prunings, thinned fruit, dead roots and orchard grass cover) was considered to be recovered with the remainder lost through leaching. All other mineral elements in recycled biomass were considered to be returned to the soil. (Adapted from Montanaro *et al*., 2006a)

	Dry matter	N	P	K	Ca	Mg
Vine component	t ha^{-1}			kg ha–1		
Fruit at harvest	5.9	51.8	6.6	95.0	10.1	10.0
Prunings	2.0	13.6	2.4	10.0	7.9	3.7
Leaves	4.3	42.9	3.4	87.0	104.2	16.9
Thinned fruit	0.2	2.1	0.2	9.0	0.3	0.7
Dead roots	2.5	22.1	2.5	38.4	24.5	6.3
Orchard grass cover	2.0	6.7	2.3	68.3	6.3	15.1
Absorbed from soil		139.2	17.4	309.5	153.3	52.7
Removed		82.4	6.6	95.0	10.1	10.0

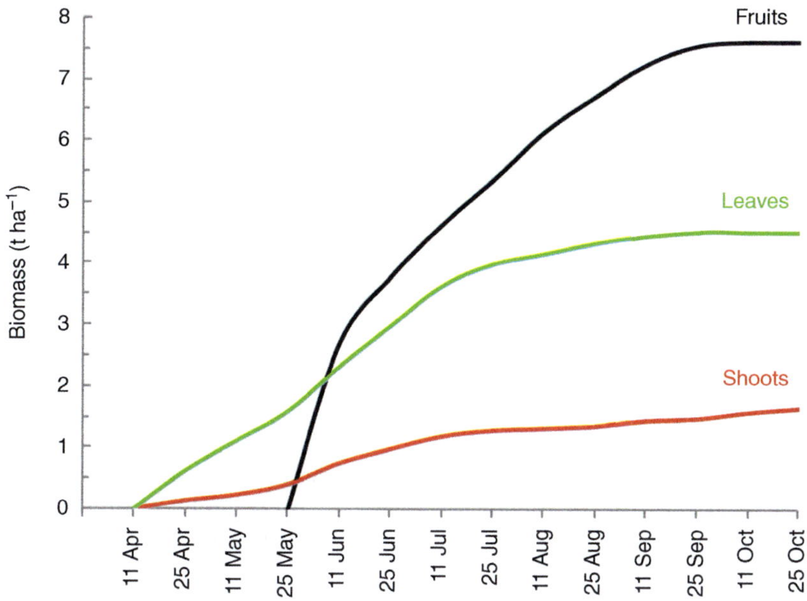

Fig. 11.2. Leaves, shoots and fruit biomass accumulation (t dry weight ha^{-1}) for an *Actinidia chinensis* var. *deliciosa* 'Hayward' orchard on a pergola support system with an annual fruit yield of about 55 t fresh weight ha^{-1} in southern Italy. (From Montanaro, G., Dichio, B. and Xiloyannis, C., University of Basilicata, unpublished)

development after budbreak until the end of fruit set (end of May in the northern hemisphere), nutrient uptake is quite gradual, equal to about 20% of total vine nutritional requirements. In the second phase, between June and mid-July, total nutrient uptake reaches 80% from vegetative biomass development (leaves + shoots) and rapid fruit growth. In the third and final phase (mid-July to the end of September), during which fruit accumulate starch (dry matter) and mature, vegetative growth slows, nutrient uptake is reduced to a further 15% of total vine nutritional requirements. At the end of this phase, kiwifruit vines have absorbed ~95% of their total annual nutrient requirements; thus, nutrient uptake in the subsequent period is limited.

Later in the growing season, when the nutrient uptake rate is lower due to reduced shoot and fruit growth rates, reserves such as starch are accumulated in the permanent plant structures. Reserves need to be restored to optimal levels to support vegetative growth at the start of the following growing season and ensure a full crop yield (see Chapter 14).

Vine nutrient requirements should be estimated by measurement of nutrients in plant tissue samples (leaves, fruit and shoots) during the annual development cycle. A nutrient balance, which considers both inputs and outputs, should be used to calculate nutrients needed to fill deficits in vine nutrient requirements. Outputs consist of fruit at harvest and prunings if they are removed from the orchard. Inputs come from fallen leaves, thinned fruit, grass-cover mowing, dead roots and mulched pruning residues where nutrients are recycled by biomass mineralization (Table 11.1).

11.2.3 Nutrition management

Fertilizer applications including those carried out before orchard planting and during the life of the orchard are usually applied in solid form, through irrigation water (fertigation) or foliar application. Conventional solid fertilizer distribution in the soil is inefficient, particularly in young orchards, tilled soils, hot arid climates and orchards with localized irrigation methods.

Pre-planting fertilization

Soil analyses can provide general information on soil nutritional status and pH levels, which can influence nutrient solubility or cation exchange capacity, affecting plant nutrient uptake. Inputs of organic manure or compost (~30–40 t ha^{-1}) can be used to maintain and improve soil properties and fertility. Generally, mineral elements, such as K, P, Ca and Mg are provided before planting to correct deficiencies and increase levels in the deeper soil layers. For kiwifruit, pre-planting addition of these elements over the soil surface is not recommended where development of the root system is slow and nutrients would not be absorbed until 5–6 years after planting. Any requirements for these minerals can be met by fertigation.

Foliar fertilization

Foliar fertilization involves application of mineral element solutions directly to the leaves, flowers, fruit and branches. This cannot replace soil applications but can be a useful tool to promote prompt nutrient uptake by leaves and fruit in situations where soil nutrient fertilization may trigger nutrient antagonisms (Mia, 2015). Foliar application is a target-specific strategy that facilitates fast nutrient absorption and assimilation (Peuke *et al.*, 1998), and can reduce nutrient competition between vegetative and reproductive sinks. For some trace elements in unavailable forms in the soil, foliar fertilizers can replace soil fertilization and remedy minor deficiencies (Hashmatt *et al.*, 2019).

The effectiveness of foliar fertilization depends on plant characteristics, environmental conditions, type of fertilizer and mode of distribution. The kiwifruit orchard leaf area receiving nutrient solutions varies throughout the year and between years. The anatomical and morphological characteristics of leaves and fruit (cuticle thickness, quantity and quality of waxy substances, characteristics and functionality of stomata, presence of trichomes) influence nutrient uptake. Kiwifruit leaves have many irregular-shaped (anomocytic) stomata on the lower (abaxial) leaf surface that facilitate uptake of foliar nutrients (Ferguson, 1990). The dense covering of hairs on leaves and fruit may influence spray coverage and retention. Moreover, in kiwifruit, the thin fruit epidermis, with high rates of surface conductance compared with other fruit such as apples, allows for efficient nutrient absorption, particularly during the early stages of fruit development (Smith *et al.*, 1995).

Surfactants that can reduce the surface tension of spray droplets may assist foliar nutrient uptake, particularly early in the season (Pathan and Gaskin, 2008). The youngest leaves and those in the inner part of the canopy are generally the most efficient in absorbing mineral elements. Considering the leaf:fruit surface area ratio is important when the fruit is the 'target' of foliar fertilization. The fruit absorbing surface area is very low compared with that of leaves and the leaf:fruit surface area ratio varies throughout the annual cycle. Leaf area reaches almost its maximum at fruit set in kiwifruit. Foliar fertilizations, aiming to increase the concentration in the fruit of mineral elements that are poorly mobile in the phloem (Ca, B, Fe), are very inefficient because almost all the distributed solution is intercepted by the leaves rather than the fruit.

Environmental parameters involved in determining the efficiency of foliar fertilization include temperature, light, relative humidity and wind. These factors influence the evaporation of water from the nutrient solution and uptake by the leaf and fruit.

Fertigation

Supplying nutrients by fertigation, rather than as traditional solid fertilizers, can maintain the soil nutrient balance with less cost while benefiting crop quality and environmental health. Optimal fertigation management aims to make soil nutrients available at the right times and in the right quantities to satisfy plant requirements. Proper fertigation scheduling considers the dynamics of nutrient demand during leaf canopy and fruit development, as well as nutrient inputs and outputs, which are influenced by orchard management. Since nutrient uptake is concentrated in the period between flowering and harvest, as described above, fertilizer inputs should be calculated and applied weekly to meet plant nutrient demand (Table 11.2).

Soil and plant diagnostic tools, such as mobile and quick-assessment kits, should be integral with fine-tuning fertilizer recommendations for different soil types, and should be used as a guide to match the most appropriate fertilizer type for each production system (Bindraban *et al.*, 2015).

Fertigation is considered the most effective method for supplying nutrients in arid climates where plants are dependent on the continuous supply of water. It is used less in humid climates where solid fertilizers are also applied (Marsh and Stowell, 1993). For N, slow-release solid fertilizers are used to decrease nitrate leaching and the risk of environmental pollution (Monastra *et al.*, 1997). Slow-release fertilizers release N over a few months rather than immediately after application.

For correct nutrition management it is important to consider the mobility of the mineral elements in the soil because leaching, e.g. from excessive irrigation or intense rainfall (Gentile *et al.*, 2022), can lead to environmental pollution and unavailability of leached elements to the plant roots (Deurer *et al.*, 2011). N is highly mobile compared with other mineral elements and can be readily leached in sandy soils. To ensure sustainable kiwifruit orchard nutrition, it is important to verify N availability in the root zone from organic N mineralization. The optimal soil nitrate (NO_3^-) concentration is between 15 ppm and 20 ppm (G. Celano, University of Salerno, personal communication). Fertigation ensures relatively steady soil nitrate availability throughout the growing season compared with traditional application of solid fertilizers, where nitrate levels can reach peaks of 90–110 ppm, which increases the risk of leaching and groundwater pollution (Fig. 11.3). Optimized

Table 11.2. Fertigation plan to provide the necessary doses of N, P and K as a percentage of the total mineral distributed, defined according to phenological stages for kiwifruit, as derived from the nutrient balance analysis. (Adapted from Montanaro *et al.*, 2006a)

		Mineral element distribution (%)		
Vine and fruit development stage	Duration (weeks)	N	P	K
Stage 1 – Budbreak until fruit set	6	18	33	18
Stage 2 – Rapid fruitlet growth	6	61	67	62
Stage 3 – Fruit development	7	21	0	20
Total	19	100	100	100

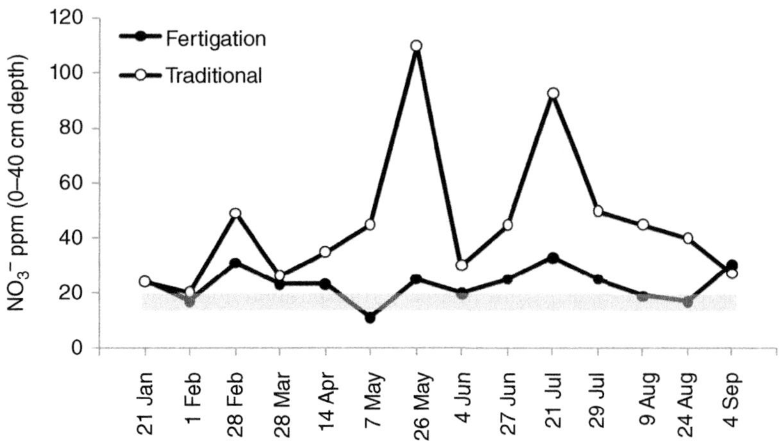

Fig. 11.3. Seasonal pattern of nitrate (NO_3^-) concentration in soil at 0–40 cm depth with fertigation or traditional solid fertilizer applications during the growing season (northern hemisphere). Optimal soil nitrate concentrations (between 15 ppm and 20 ppm) are indicated by the grey bar. (Adapted from Montanaro *et al.*, 2010)

N nutrition management contributes to improved fruit quality and reduced environmental impact (Mills *et al.*, 2008).

Organic fertilization

Sustainable kiwifruit orchard management has positive effects on soil physical, chemical and biological properties, increasing soil quality. Loss of soil organic matter is considered the main cause of decreased soil fertility and soil degradation. Agronomic practices that protect soils, such as cover crops, no tillage, adequate fertilization, compost application, mulching of pruning residues and return of crop residues to the soil, ensure high soil fertility, biodiversity and, consequently, crop yields (Montanaro *et al.*, 2010). The increase of soil organic matter improves soil structure and physico-hydrological properties, increasing aggregate stability, porosity, water infiltration, soil drainage, water retention capacity and soil resistance to compaction, thus reducing water losses and soil erosion and increasing the availability of water for plants (Xiloyannis *et al.*, 2015). (See Chapter 10.)

Higher organic matter content and optimal moisture levels in the upper soil layers, compared with the deeper layers, preserve and promote root activity and turnover. Available Ca concentrations are higher in the first 10 cm of soil depth and decrease rapidly through the top 30 cm of soil. Tillage can damage young roots located in the upper soil layers that are very important for water and nutrient, particularly Ca, uptake. Therefore, preserving root integrity in the upper soil layers ensures greater uptake of water and nutrients, particularly Ca.

Organic amendments such as compost are an excellent replacement for synthetic fertilizers, although typically the nutrients from organic fertilizers do not become immediately available for root uptake (Baldi *et al.*, 2021).The main aim of amendments is to increase soil organic matter and the release of nutrients after mineralization to satisfy vine nutritional requirements. Sustainable management with organic fertilizers is characterized by a better balance between release and use of N and other nutrients compared with conventional fertilization procedures.

Protecting soils by retention of crop residues, application of organic amendments, guided fertilization and reduced soil disturbance (reduced tillage), in combination with proper canopy management including higher bud numbers, increased yields in a kiwifruit orchard by about 50% compared with conventional orchard management practices (i.e. poor canopy management, soil tillage, removal of pruning residues and use of mineral fertilizers) (Montanaro *et al.*, 2010). Management practices that protect soils can positively affect crop yield, mainly by improving the reservoir of soil mineral elements despite a lower mineral fertilization application rate and increasing the annual soil carbon inputs to enhance soil organic carbon pools.

11.2.4 Fruit mineral composition and postharvest quality

The mineral composition of kiwifruit influences fruit quality during pre- and postharvest phases (Ferguson *et al.*, 2003). Fruit nutritional status is significantly affected by differential transport of each nutrient within the plant, which is related to xylem and phloem sap flows. A low Ca concentration and unbalanced nutrient concentration ratios (Ca:N, Ca:K and Ca:Mg) in fruit are generally considered to increase the incidence of fruit disorders (Hocking *et al.*, 2016).

High Ca concentrations are associated with a reduction of physiological disorders in kiwifruit (Thorp *et al.*, 2003), with consequent improvements in storage quality and shelf life (Dichio *et al.*, 2003; Ferguson *et al.*, 2003; Thorp *et al.*, 2003; Montanaro *et al.*, 2006b, 2012). Increased fruit Ca concentration from 0.32 to 0.42 mg g^{-1} dry weight was shown to improve fruit quality in kiwifruit at harvest (Koutinas *et al.*, 2010).

Ca is mainly transported to the leaves owing to their high rates of transpiration; leaves contain ~85–93% of the total Ca taken up. Almost all water absorbed from soil is lost through leaf transpiration and only small amounts through the transpiration of fruit (Table 11.3). This explains the reduced capacity of fruit to acquire Ca.

Kiwifruit berries are only able to transpire during the first phase of their development, when they receive xylem sap containing mineral elements, especially Ca, and fruit physiological and morphological conditions favour mineral accumulation. The seasonal pattern of Ca accumulation mirrors the growth curve of kiwifruit with a higher water demand and xylem sap inflow during the early stage of fruit development and, consequently, a great accumulation of Ca in young kiwifruit.

Ca is supplied to the fruit throughout the 7–8 weeks after fruit set, when the total fruit Ca content is about 80–90% of that present at harvest time. Fruit growth is particularly sensitive to water supply during this period of active cellular division. Water reaches fruit only through the xylem during this short period after fruit set; thus, optimal soil-water conditions are critical during this time.

After this first phase of fruit development, the reduction in fruit transpiration rates and morphological changes to the fruit exocarp prevent further accumulation of mineral elements, so that soil fertilization and foliar applications have little benefit. Moreover, the fruit Ca concentration decreases owing to dilution, because fruit size continues to increase while Ca import slows (Fig. 11.4).

Water and nutrient transport to fruit via the xylem are related to the fruit transpiration rate and vascular xylem efficiency. Ca transport towards fruit is exclusively via the xylem, and a previous study showed the linkage between xylem functionality and well-known patterns of fruit Ca accumulation (Dichio *et al.*, 2003). Xylem functionality decreases during the growing season, so that vessel dysfunction, together with other factors, may cause a sharp reduction in the fruit transpiration rate 8–9 weeks after flowering and, consequently, a decline in Ca accumulation in fruit.

Fruit transpiration changes with daily fluctuations in environmental parameters and is correlated with the VPD (Fig. 11.5). Fruit growing under higher relative humidity conditions (i.e. low VPD) show a lower transpiration rate than when fruit are exposed to higher VPD conditions, which results in a higher rate of transpiration, allowing accumulation of more Ca (Fig. 11.6). Ca accumulation is linearly related to cumulative transpiration under high VPD environmental conditions, as cumulative transpiration is equal to the xylem flow into fruit, allowing fruit to receive all its nutrients via the xylem (Montanaro *et al.*, 2015).

Table 11.3. Transpiration of fully expanded leaves and fruit in an *Actinidia chinensis* var. *deliciosa* 'Hayward' orchard during July in southern Italy. (From Xiloyannis *et al.*, 1999)

Plant organ	Transpired water (m^3 ha^{-1} day^{-1})
Exposed leaves	31.6
Shaded leaves	20.6
Fruit	0.3
Total	52.5

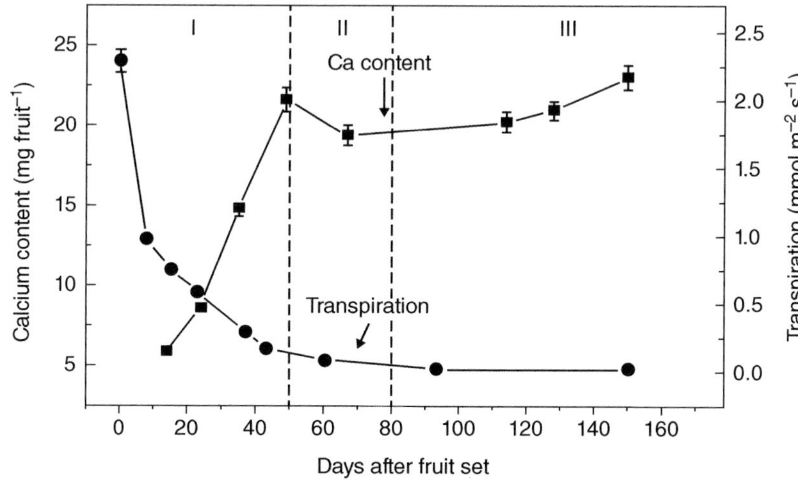

Fig. 11.4. Seasonal pattern of calcium (Ca) accumulation and fruit transpiration in kiwifruit berries (*Actinidia chinensis* var. *deliciosa* 'Hayward'). Stage I, from anthesis to approx. 50 days after fruit set; stage II, 50–80 days after fruit set; stage III, 80–180 days after fruit set. (Adapted from Montanaro *et al.*, 2014)

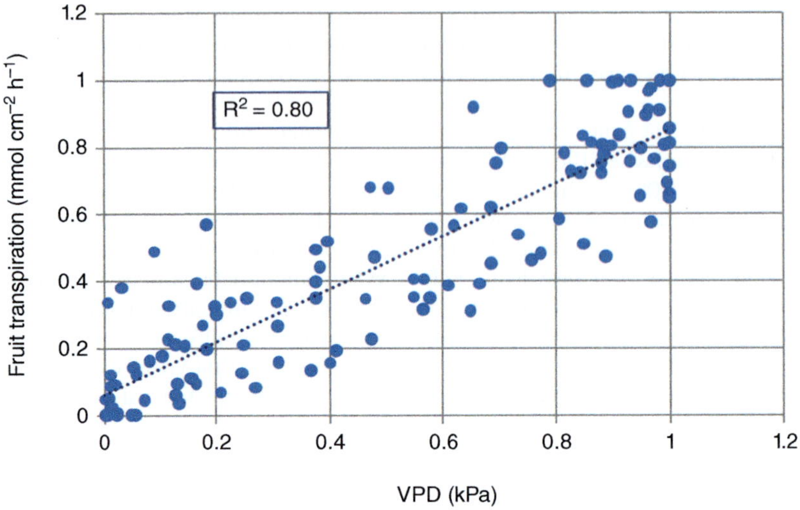

Fig. 11.5. Relationship between (normalized) vapour pressure deficit (VPD) and (normalized) fruit transpiration in mature *Actinidia chinensis* var. *deliciosa* 'Hayward' vines measured hourly for 24 h at 23, 35, 49, 65, 94 and 140 days after full bloom. (Data scaled [min–max normalization, (0,1) range] before plotting.) (Adapted from Montanaro *et al.*, 2014)

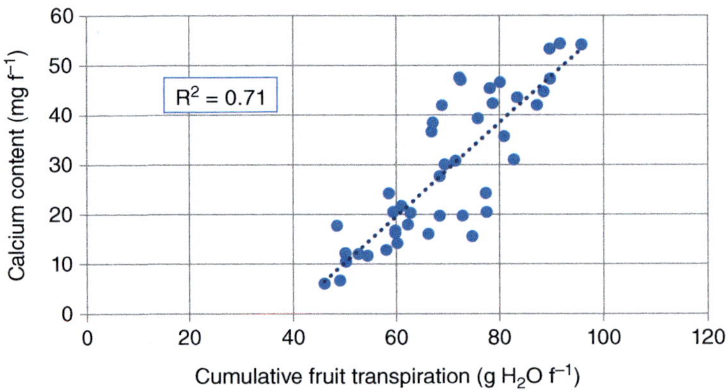

Fig. 11.6. Relationship between calcium accumulation and cumulative fruit transpiration on a per fruit (f) basis for *Actinidia chinensis* var. *deliciosa* 'Hayward' kiwifruit under high vapour pressure deficit conditions. (Adapted from Montanaro *et al.*, 2015)

Several other nutrients may influence fruit quality. For example, K is involved in fruit growth and development during cell expansion; thus, high levels of K could enhance the dilution of fruit Ca. In fact, while the concentrations of both these elements decrease during later fruit development phases, the K:Ca ratio increases. Then, the Ca concentration in leaves increases, whereas the K concentration and the K:Ca ratio decreases. Moreover, K, as well as Mg, are potential competitors of Ca for binding sites at the plasma membrane, potentially reducing membrane stability and increasing the incidence of Ca deficiency (Xiloyannis *et al.*, 2001).

N is also important in directly and indirectly determining fruit quality and storage life (Mills *et al.*, 2008; Gentile *et al.*, 2022). A high N supply increases mean N concentration in the fruit, which reduces fruit storage potential owing to rapid fruit softening (Johnson *et al.*, 1997; Vizzotto *et al.*,

1999; Pacheco *et al.*, 2008; Montanaro *et al.*, 2014). Furthermore, N enhances vegetative shoot growth, with a consequent increase in the leaf transpiration rate, resulting in a greater Ca quantity delivered to the leaves rather than the fruit.

11.3 Irrigation

11.3.1 Water relations and hydraulic features

Root system characterization and development

Choosing appropriate irrigation, water management and distribution requires an understanding of the influence of kiwifruit vine morpho-anatomical characteristics on physiological responses to soil-water availability, root system structure and distribution and root water uptake. Kiwifruit vines have a different root distribution and structure from tree crops such as peach and olive, influencing the soil volume explored by roots (Table 11.4). Kiwifruit root systems have numerous tiny fibrous (feeder) roots that develop from thicker structural and secondary roots, with a high root density and intense development in the upper soil layers (about 50 cm) (Fig. 11.7). In a well-structured (high porosity) soil, a 16-year-old kiwifruit vine can develop a root system of up 20 t ha^{-1} of dry weight with roots occupying up to 80 m^3 of soil and extending below 1.6 m depth (Hughes *et al.*, 1992). However, root

Table 11.4. Soil volume explored by root systems of peach (*Prunus persica* 'Vega', 4.5 × 1.25 m), kiwifruit (*Actinidia chinensis* var. *deliciosa* 'Hayward', 4.5 × 3 m) and olive (*Olea europaea* 'Coratina', 6 × 3 m) plants during the first four years after planting. Data presented per plant (p) and per ha. (From Massai *et al.*, 1993; Xiloyannis *et al.*, 1993, 1999; Dichio *et al.*, 2002)

	Soil volume explored							
	Year 1		Year 2		Year 3		Year 4	
Crop	m^3 p^{-1}	m^3 ha^{-1}	m^{3-} p^{-1}	m^3 ha^{-1}	m^3 p^{-1}	m^3 ha^{-1}	m^3 p^{-1}	m^3 ha^{-1}
Peach 'Vega'	1.22	2,168	3.39	6,024	3.60	6,575	3.60	6,575
Kiwifruit 'Hayward'	0.11	84	0.64	474	1.35	999	1.41	1,043
Olive 'Coratina'	0.50	138	2.9	803	8.6	2,382	12.5	3,475

Fig. 11.7. Kiwifruit root system structure. (**A**) *Actinidia chinensis* var. *deliciosa* 'Hayward' on its own roots. (Photo: C. Xiloyannis); (**B**) *Actinidia chinensis* var. *chinensis* 'Zesy002' grafted onto 'Hayward' roots. (Photo: M. Calabritto, 2021)

growth rates decline with increasing soil depth mainly due to reduced soil porosity (Buwalda and Hutton, 1988).

Soil management can influence root development. For instance, soil compaction resulting from frequent machine traffic and poor drainage can limit root growth and the total soil volume explored by the roots (Hughes and Wilde, 1989) and decrease fruit quality (Cho *et al.*, 2013). Kiwifruit root systems tend to colonize the available soil volume more slowly than other species, but more efficiently exploit water and mineral elements within the limited soil volume explored. Root system structure and spatial distribution are important for optimizing water absorption to prevent water deficit or, conversely, waterlogging. Waterlogging is a major cause of decreased root turnover and a resulting reduction in the development of new feeder roots.

Water relations and physiological traits

Kiwifruit vines have large vessels (200–350 μm diameter) in the xylem, which permits the movement of large volumes of water, about 40–60 L d^{-1}, from soil to leaves with little resistance. Plants with an efficient water transport system, such as kiwifruit, have very low daytime fluctuations in the leaf water potential (isohydric behaviour) and a very limited potential gradient between leaves and roots (Dichio *et al.*, 2013). Under optimal conditions, they quickly transfer large amounts of water from the soil to leaves and almost completely replace water lost by transpiration (Nuzzo *et al.*, 1997).

This system of water transport is very efficient in optimal soil-water conditions, but vulnerable to soil-water deficits. Unlike tree crops, kiwifruit vines have low leaf stomatal density and mild water stress causes stomatal closure, resulting in a reduction in transpiration rates and increases in leaf temperature, making vines extremely sensitive to mild water stress.

Kiwifruit vines can have high water requirements depending on environmental conditions but are also extremely sensitive to waterlogging. Thus, any type of water stress (water deficit or waterlogging), even of short duration, negatively affects plant growth and development, e.g. through a decrease in root turnover capacity and a lack of white root formation, thus compromising plant production.

11.3.2 Water requirements

The benefits of irrigation and water availability on yield and quality production of kiwifruit orchards are well known (Judd and McAneney, 1987; Prendergast *et al.*, 1987; Judd *et al.*, 1989; Gucci *et al.*, 1997; Boini *et al.*, 2022). Irrigation enhances fruit fresh and dry weight, soluble solids content and mineral element content. Generally, an increase in available soil water improves vine nutritional status and influences mineral element dynamics within the plant. Proper irrigation management ensures soil-water status meets the specific needs of kiwifruit orchards according to phenological phases and environmental and soil variables.

To manage water and ensure high production of quality fruit, it is important to use a soil-water balance and proper irrigation scheduling. Irrigation should be carefully managed according to the site-specific evaporative demand and the water-holding capacity of the soil in the root volume to avoid excessively low or high soil-water contents.

Irrigation volumes should be calculated using the reference evapotranspiration (ET_0) values for each site. This is very important considering that kiwifruit cultivation has spread to sites around the world with very low to very high environmental water demand. For example, in southern Italy, which has a Mediterranean climate with a high summer environmental water demand and low rainfall concentrated mainly in the autumn and winter months, vines are exposed to a high environmental water deficit. Conversely, in the Latina area in central Italy, the average annual rainfall is ~90% higher than that of the Metaponto area (southern Italy) and the average annual evapotranspiration is lower than that of Metaponto, with a significantly lower water deficit (Fig. 11.8).

The vine irrigation requirement (i.e. crop evapotranspiration, ET_c) depends on ET_0 and vine characteristics (cultivar, rootstock, number of fruit, leaf area development) that determine the crop coefficient value (K_c). K_c values specific for *Actinidia chinensis* var. *deliciosa* 'Hayward' vines have been reported to vary from 0.5 (April, initial growth stage) to 1.1 (July, rapid growth stage in the northern hemisphere) over the course of the growing season (Xiloyannis *et al.*, 2012). Due to the lack of information on yellow-fleshed

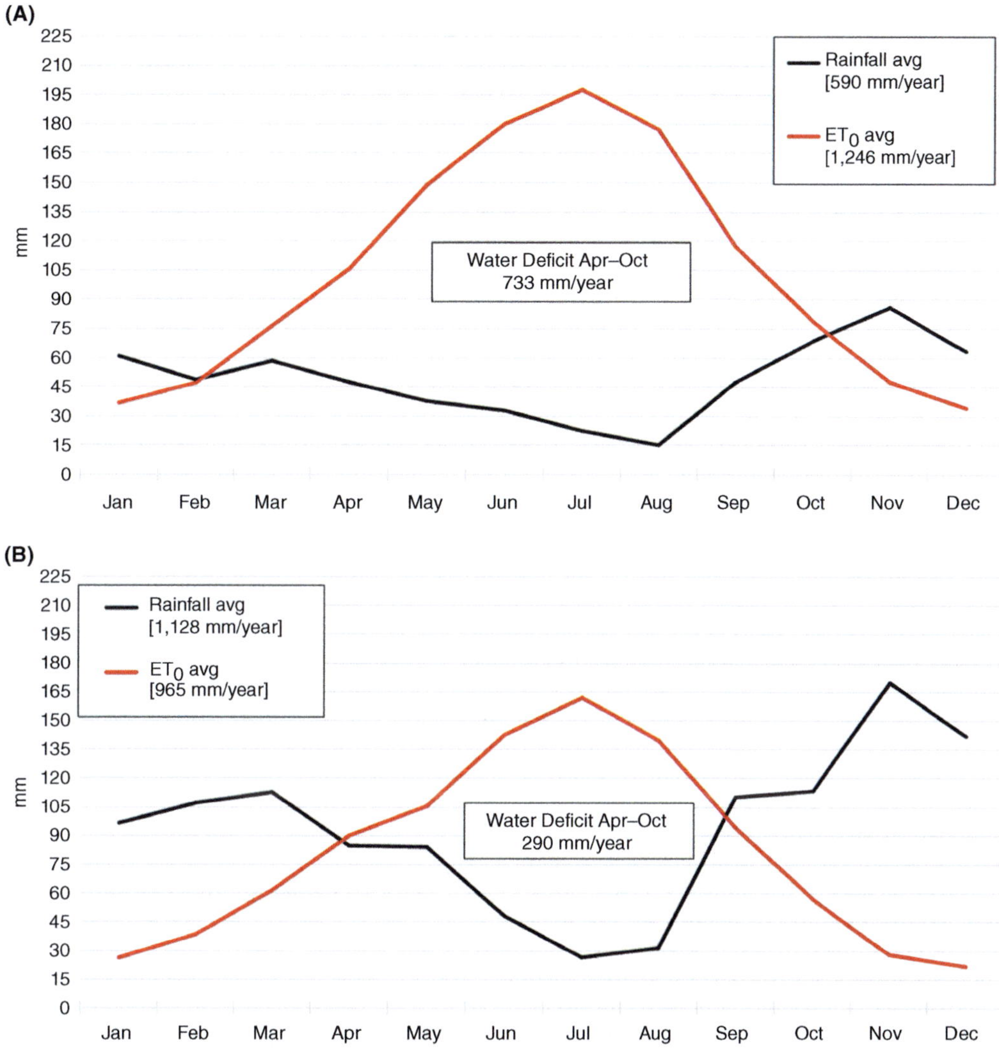

Fig. 11.8. Annual evapotranspiration (ET_0) and rainfall trends (monthly averages). (**A**) 2001–2020 data for the Metaponto regional weather station (Pantanello, southern Italy). (**B**) 2004–2020 data for the Latina regional weather station (Cisterna-Doganella di Ninfa, central Italy).

cultivars, the published K_c values from Xiloyannis *et al.* (2012) are commonly used for the estimation of crop evapotranspiration of these cultivars, although the validity of this should be confirmed.

As kiwifruit vine water requirements depend on the total transpiration area (leaf and fruit) as well as environmental conditions, the K_c could be adjusted by considering the vegetation index. Kiwifruit vine leaf area changes according to orchard age, planting density, training system, planting material type and quality, vegetative phase and cultural practices. The maximum leaf area of a fully productive kiwifruit orchard may range from 25,000 to 50,000 $m^2 ha^{-1}$ depending on the training system. Generally, leaf area significantly increases in the first two months of shoot growth (after budbreak), then slows, reaching a maximum four months after budbreak and remaining almost constant until leaf fall (Fig. 11.9A). Leaf area index values during the first period after budbreak are lower and

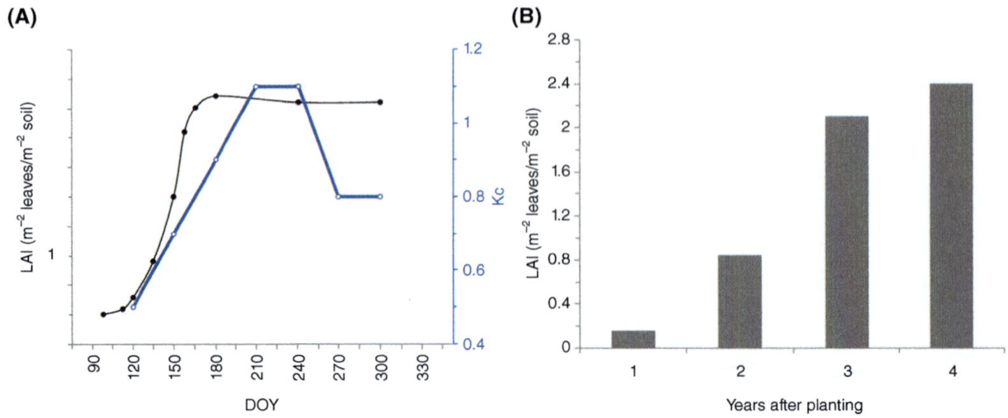

Fig. 11.9. (**A**) Leaf area index (LAI) and crop coefficient value (K_c) trends over one growing season (DOY = day of year) in a pergola-trained *Actinidia chinensis* var. *deliciosa* 'Hayward' orchard, 625 plants ha^{-1}. (**B**) Changes in LAI in a pergola-trained 'Hayward' orchard during the first four years after planting, 741 plants ha^{-1}. (Adapted from Xiloyannis *et al*., 1993, 1999, 2012)

environmental conditions favourable, thus transpiration rate and, consequently, water consumption are limited. In this period, less canopy cover leads to an increase in soil-water loss by evaporation, although this depends on the soil-water content. Increased leaf area and high environmental evaporative demand (i.e. high VPD) enhance vine water requirements, but less water is lost by evaporation from soil, especially for plants trained on a pergola system. Variation in kiwifruit leaf area is considerable during the first 4–5 years after planting. During these years kiwifruit leaf area increases progressively, defining different water requirements for each year. Once orchards reach full production, variations in leaf area only occur during the growing season and not between years (Fig. 11.9B).

Leaf area variations in the first years and during each growing season are important for appropriate nutrition and water management planning. Canopy structure and spatial distribution, defined by training systems, affect water-use efficiency (WUE). WUE is defined as the amount of carbon assimilated per unit water transpired. Exposed leaves receiving sufficient solar radiation to reach maximum photosynthesis rates transpire at about twice the rate of shaded leaves, but their WUE is about ten times that of shaded leaves. Shaded leaves have a transpiration rate of about 50% and a photosynthetic rate only about 10% that of leaves fully exposed to sunlight. The part of the leaf canopy that receives less than 20% of available solar radiation represents a sink rather than a source of assimilates, and uses high quantities of water (about 30% of the total water requirement in the case of a pergola-trained vine) (Xiloyannis *et al*., 1999). Different training systems can modify the ratio of exposed to shaded leaves and thus contribute to water saving and better WUE.

11.3.3 Irrigation management

The irrigation volume (Vi) to be supplied to vines is calculated through a simplified water budget method by considering the vine water requirement (ET_c), the irrigation method efficiency (Eff) and effective rainfall (R):

$$Vi = \left(ET_c - R\right) / Eff$$

This simplified water budget refers to the wetted soil volume since, under conditions of optimal irrigation management, plant water uptake occurs almost entirely from this soil volume. The soil volume affected by irrigation varies considerably according to the irrigation system, the physico-mechanical soil properties and the irrigation volume delivered. The most commonly used irrigation systems in kiwifruit orchards are the drip, sprinkler and a combination of both (Chartzoulakis *et al*., 1991). The total volume of soil explored by the root system can be referred to as two containers: the first is affected by irrigation (container 1), while the remaining volume (container 2) does not receive irrigation but plays an important role as a water reservoir during rainy periods.

In the case of drip irrigation systems, with 30–50 cm between drippers, the soil volume wetted by irrigation is continuous and uniform and can be calculated by considering the width, depth and length of the soil strip, which represents container 1. In the case of irrigation systems that do not uniformly wet a continuous soil area, such as sprinkler irrigation systems, the soil area wetted by each sprinkler, the number of sprinklers per ha and the established wetted depth should be considered. For irrigation purposes, considering kiwifruit root structure and development, the depth of the wetted volume is considered to be 50–60 cm.

The amount of water lost daily from the wetted soil volume and, consequently, the irrigation volumes required to restore the water level, are indirectly calculated using the simplified water budget. Daily irrigation volumes can be added and applied once RAW is depleted from the wetted soil volume. Following this criterion, the maximum irrigation volume is equal to the RAW of container 1 and the irrigation interval is automatically defined as the time interval required to use the RAW.

Knowledge of soil intrinsic characteristics is pivotal to understanding the amount of water that can be stored in soil containers 1 and 2. Soil characteristics (texture, structure, hydrological properties and the soil profile) give information about the soil water-holding capacity and must be considered to define field capacity (FC) and wilting point values. Soil-water retention curves, which relate soil matric potential to soil-water contents, show different trends according to soil texture. Soil-water content between the values of the FC and the wilting point represent soil total available water (AW) which could be absorbed by plants. RAW is a fraction ranging between 30% and 40% of the AW, defined according to cultivar, soil type and the soil volume affected by irrigation. Both the soil-water reservoir and RAW depend on the soil volume explored by the root system and the soil's characteristics and hydrological properties, such as porosity, which affect water infiltration and retention capacity.

To improve soil-water retention capacity and promote maximum accumulation of water, it is important to adopt a sustainable and rational soil management regime. This will improve not only soil organic matter content and soil chemical and biological fertility, but also soil structure and hydrological properties. Soil carbon is a key factor in determining soil water-holding capacity. However, the applied irrigation volume should not exceed the soil water-storage capacity.

The RAW, expressed as a fraction of soil AW, is used as a threshold to define when irrigation should be applied. For kiwifruit vines, this value is about 40% of AW as kiwifruit is highly sensitive to water deficits. Identifying the most appropriate time to start irrigation is of great importance, especially for localized irrigation systems. Currently, a wide range of irrigation systems is used for kiwifruit orchards, depending on the soil physico-hydrological properties and microclimate of the area.

Kiwifruit vines should not be subjected to sustained or regulated deficit-irrigation strategies because all phenological phases are sensitive to small soil-water status variations. Water stress throughout the growing season has negative effects on fruit quality and on the quantity and distribution of assimilates between different plant organs. Adequate quantities of assimilates are important to ensure consistently high production of quality fruit. After harvest, it is important to maintain optimal soil-water content to maintain leaf photosynthetic activity, allowing the storage of greater quantities of reserves in permanent organs. High reserves sustain vegetative growth the following season and increase tissue resistance to low winter temperatures. Shoot development and leaf and fruit growth are sensitive to water stress. Vegetative growth, expressed as increased shoot development and leaf area, is reduced when about 40% of AW has been used and growth stops when the soil-water content is below 50% of AW.

During summer, in most kiwifruit orchards in a semi-arid environment, daily standard evapotranspiration often exceeds 6 mm (Green *et al.*, 2022) and the RAW reserve can only meet vine water requirements for about a day. In Clothier and Green (1994), the primary uptake role of near-surface roots was highlighted, supporting the claim that more efficient and sustainable irrigation management could result from higher frequency applications of smaller amounts of irrigation. During summer, irrigation should be scheduled daily, while in spring and autumn, irrigation is required every 3–4 days. Daily irrigation scheduling ensures root growth in the upper soil layers where, in addition to water uptake, nutrient uptake is also promoted.

Optimal irrigation management in kiwifruit orchards promotes root turnover and encourages the uptake of essential mineral elements from the soil, which enhances fruit development and quality. The water footprint should be assessed to define the efficiency of different orchard systems (depending on plant density, pruning operations, plant size, soil management, irrigation schedule etc.) in managing water resources.

11.3.4 Precision irrigation approaches

Adoption of irrigation technologies and systems that ensure efficient and effective irrigation combines irrigation scheduling management with water-saving practices. Precision irrigation monitors soil–plant–environment variables following a multidisciplinary approach. Innovative precision irrigation and decision-support systems allow careful and constant monitoring of soil-water content and environmental parameters through the interconnection of remote-sensing systems and specific proximal sensors (Fig. 11.10). This makes it possible to modulate irrigation scheduling at any time according to the specific needs of the kiwifruit vines. The difficulty in calculating an accurate K_c underlines a weakness of the traditional water balance method, which is not representative of the actual water needs and therefore not a sustainable water management technique.

Data obtained from sensors are used to define vine water requirements more precisely and thereby refine irrigation scheduling management. A sustainable irrigation strategy should be based on the integration of a daily water balance of the soil volume affected by irrigation with soil

Fig. 11.10. Soil moisture monitoring probes: (**A–C**) Single profile sensors (frequency domain reflectometry (FDR) technology); (**D**) Single profile sensors installed at 20, 30 and 60 cm soil depth (time domain transmissometry technology); (**E, F**) Multi-profile probe sensors (FDR technology), used in conjunction with a data logger for data acquisition and remote control. (Photos: D. Laterza and A. Tuzio)

moisture measurements by probes to ensure the soil moisture content remains within the RAW and FC range throughout the irrigation season. Such a scheduled irrigation volume should be adjusted during the growing season to maintain the soil moisture level in this range and above the critical threshold value, defined as 40% of AW for a kiwifruit orchard (Mininni *et al.*, 2022).

Application of a sustainable irrigation strategy shows that through the integration of a soil-water budget with soil moisture measurements it may be possible to save more than 1,500 m^3 ha^{-1} of irrigation water compared with the total volume suggested using only the water balance method (Table 11.5). Thus, the monitoring of soil-water status and environmental conditions is pivotal in promoting sustainable irrigation strategies and meeting vine water requirements.

Soil moisture probes allow the measurement and control of soil-water status at different depths, including deeper soil layers (50–70 cm). Through the analysis of these data it is possible to evaluate if irrigation water supply is sufficient to satisfy plant water needs and whether the irrigation volume needs to be adjusted to keep the soil-water content between the FC and RAW threshold (Fig. 11.11).

Irrigation management is critical as kiwifruit is highly susceptible to both water deficit and excess from poor irrigation management, and this compromises orchard productive potential. Ensuring a water reservoir in soil layers explored by roots, but not directly irrigated, is important to support root activity away from the irrigated area and prevent water stress caused by poor irrigation volume and timing. Excess irrigation leads to waterlogging and anaerobic conditions of the root system. Root hypoxia has several negative effects on the vine's physiological state and growth, substantially compromising root growth and stomatal activity and causing necrotic leaf spots (Smith *et al.*, 1990). The accumulation of water through the soil profile, the alteration of soil physico-hydrological, chemical and biological properties and the establishment of anoxic conditions all affect the root system and represent factors closely related to the increasingly widespread kiwifruit vine decline syndrome (KVDS) (Sofo *et al.*, 2021; Mininni *et al.*, 2022).

Developing a sustainable irrigation strategy based on the increase in efficiency of irrigation systems and optimization of irrigation scheduling according to soil-water content leads to improved plant WUE and productivity at the orchard level.

11.4 Future Directions

Irrigation and nutrition currently require careful, optimal management to increase resource-use efficiency (water and nutrients) and reduce

Table 11.5. Monthly averages (over 3 years) of evapotranspiration (ET_0), rainfall and irrigation volumes obtained by adjusting the soil-water balance method with the soil-water content monitoring and theoretical irrigation volume calculated using the water balance method in central and southern Italy. (B. Dichio, University of Basilicata, unpublished data)

	Central Italy			Southern Italy		
Month	ET_0 (mm)	Rainfall (mm)	Irrigation (mm)	ET_0 (mm)	Rainfall (mm)	Irrigation (mm)
April	88	53	10	112	0	90
May	128	8	106	169	8	104
June	158	13	97	209	17	127
July	170	16	74	219	1	171
August	135	79	53	179	29	130
September	85	137	30	123	13	69
October	61	5	18	84	45	14
Total	825	311	388	1,095	113	705
Theoretical irrigation volume			585			870

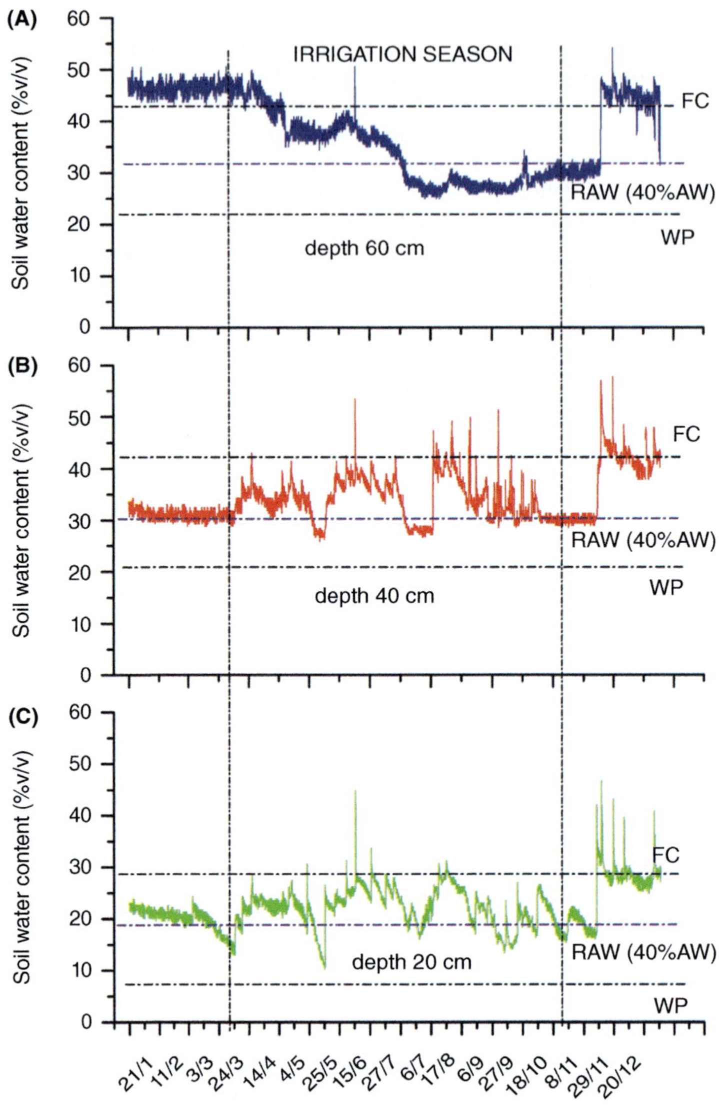

Fig. 11.11. Seasonal trend in soil moisture, detected by SM100 WATERSCOUT SPECTRUM probes, at soil depths of (**A**) 60cm; (**B**) 40cm; and (**C**) 20cm (FC = field capacity; RAW = readily available water; AW = available water; WP = wilting point). Monitoring of soil-water content at different soil depths allows the evaluation of irrigation to supply vine water requirements. In the case of soil-water content in the deeper layers being higher than 80–90% field capacity (FC), the irrigation volumes were reduced by 15–20% to avoid excess water. When the soil moisture values were below 40% AW, an increase of 15–20% of volume was applied. The irrigation strategy applied allowed soil moisture to be kept between the FC and the RAW threshold in the upper layers and to have a good water depletion of deeper layers (where absorption by roots is very low), avoiding waterlogging which is one of the main causes of kiwifruit root death. (From Mininni *et al.*, 2022)

environmental impacts (degradation of soil fertility and underground/surface waters). Kiwifruit orchards, which are very complex systems, require special attention and continuous monitoring of environmental and soil conditions to define or adjust optimal management strategies. Investigating plant physiological processes and plant-water relations, as well as increasing knowledge about crop characteristics, vine susceptibility to water stress and seasonal water requirements, are of primary importance to develop and improve precision irrigation strategies. Proper irrigation and nutrition management contribute to the development of guidelines for growers on irrigation and nutrition best practices that will help to improve kiwifruit orchard productivity, fruit quality and the sustainability of orchard management.

References

Baldi, E., Cavani, L., Mazzon, M., Marzadori, C., Quartieri, M. *et al.* (2021) Fourteen years of compost application in a commercial nectarine orchard: effect on microelements and potential harmful elements in soil and plants. *Science of the Total Environment* 752, 141894. DOI: 10.1016/j.scitotenv.2020.141894

Bindraban, P.S., Dimkpa, C., Nagarajan, L., Roy, A. and Rabbinge, R. (2015) Revisiting fertilisers and fertilisation strategies for improved nutrient uptake by plants. *Biology and Fertility of Soils* 51, 897–911. DOI: 10.1007/s00374-015-1039-7

Boini, A., Cavallina, L., Perulli, G., Bresilla, K., Bortolotti, G. *et al.* (2022) *Actinidia chinensis*: physiological and productive performance under water stress condition. *Acta Horticulturae* 1346, 43–50. DOI: 10.17660/ActaHortic.2022.1346.6

Bryla, D.R. (2019) 4R nutrient stewardship in fruit crops. In: Srivastava, A.K. and Hu, C. (eds) *Fruit Crops: Diagnosis and Management of Nutrient Constraints*. Elsevier, Amsterdam, pp. 509–519. DOI: 10.1016/B978-0-12-818732-6.00036-8

Buwalda, J.G. and Hutton, R.C. (1988) Seasonal changes in root growth of kiwifruit. *Scientia Horticulturae* 36(3-4), 251–260. DOI: 10.1016/0304-4238(88)90059-3

Buwalda, J.G. and Smith, G.S. (1987) Accumulation and partitioning of dry matter and mineral nutrients in developing kiwifruit vines. *Tree Physiology* 3, 295–307. DOI: 10.1093/treephys/3.3.295

Chartzoulakis, K., Michelakis, N. and Vougioukalou, E. (1991) Growth and production of kiwi under different irrigation systems. *Fruits* 46(1), 75–81.

Cho, Y., Cho, H., Ma, K., Park, M. and Kim, B. (2013) Effect of subsoiling on soil physical properties and fruit quality in organic conversion kiwifruit (*Actinidia chinensis* 'Haegeum'). *Acta Horticulturae* 1001, 347–352. DOI: 10.17660/ActaHortic.2013.1001.40

Clothier, B.E. and Green, S.R. (1994) Rootzone processes and the efficient use of irrigation water. *Agricultural Water Management* 25(1), 1–12. DOI: 10.1016/0378-3774(94)90048-5

Deurer, M., Green, S.R., Clothier, B.E. and Mowat, A. (2011) Can product water footprints indicate the hydrological impact of primary production? A case study of New Zealand kiwifruit. *Journal of Hydrology* 408(3-4), 246–256. DOI: 10.1016/j.jhydrol.2011.08.007

Dichio, B., Romano, M., Nuzzo, V. and Xiloyannis, C. (2002) Soil water availability and relationship between canopy and roots in young olive trees (cv. Coratina). *Acta Horticulturae* 586, 255-258. DOI: 10.17660/ActaHortic.2002.586.48

Dichio, B., Remorini, D. and Lang, S. (2003) Developmental changes in xylem functionality in kiwifruit fruit: implications for fruit calcium accumulation. *Acta Horticulturae* 610, 191–195. DOI: 10.17660/ActaHortic.2003.610.25

Dichio, B., Montanaro, G., Sofo, A. and Xiloyannis, C. (2013) Stem and whole-plant hydraulics in olive (*Olea europaea*) and kiwifruit (*Actinidia deliciosa*). *Trees* 27(1), 183–191. DOI: 10.1007/s00468-012-0787-3

Ferguson, A.R. (1990) Stem, branches, leaves and roots of the kiwifruit vine. In: Warrington, I.J. and Weston, G.C. (eds) *Kiwifruit: Science and Management*. Ray Richards Publisher in association with the New Zealand Society for Horticultural Science, Auckland, New Zealand, pp. 58–70.

Ferguson, A.R. and Eiseman, J.A. (1983) Estimated annual removal of macronutrients in fruit and prunings from a kiwifruit orchard. *New Zealand Journal of Agricultural Research* 26, 115–117. DOI: 10.1080/00288233.1983.10420960

Ferguson, I.B., Thorp, T.G., Barnett, A.M., Boyd, L.M. and Triggs, C.M. (2003) Inorganic nutrient concentrations and physiological pitting in 'Hayward' kiwifruit. *The Journal of Horticultural Science and Biotechnology* 78(4), 497–504. DOI: 10.1080/14620316.2003.11511654

Gentile, R.M., Boldingh, H.L., Campbell, R.E., Gee, M., Gould, N. *et al.* (2022) System nutrient dynamics in orchards: a research roadmap for nutrient management in apple and kiwifruit: a review. *Agronomy for Sustainable Development* 42(4), 64. DOI: 10.1007/s13593-022-00798-0

Green, S., Reyes, F., Dichio, B., Mastroleo, M. and Xylogiannis, E. (2022) Water use of yellow-fleshed kiwifruit during an annual cycle. *Acta Horticulturae* 1332, 187–194. DOI: 10.17660/ActaHortic.2022.1332.25
Gucci, R., Massai, R. and Xiloyannis, C. (1997) Mechanisms of drought avoidance in kiwifruit vines. *Acta Horticulturae* 444, 311–316. DOI: 10.17660/ActaHortic.1997.444.48
Hashmatt, M. (2021) Effects of soil & foliar nutrient application strategies for improving fruit quality for 'Zesy002' kiwifruit. PhD thesis. Massey University, Palmerston North, New Zealand.
Hashmatt, M., Morton, A.R., Heyes, J.A., Armour, D., Lowe, T. *et al.* (2019) Effect of pre-harvest foliar calcium application on fruit quality in Gold3 kiwifruit. *Acta Horticulturae* 1253, 327–333. DOI: 10.17660/ActaHortic.2019.1253.43
Hocking, B., Tyerman, S.D., Burton, R.A. and Gilliham, M. (2016) Fruit calcium: transport and physiology. *Frontiers in Plant Science* 7, 569. DOI: 10.3389/fpls.2016.00569
Hughes, K.A. and Wilde, R.J. (1989) The effect of poor drainage on the root distribution of kiwifruit vines. *New Zealand Journal of Crop and Horticultural Science* 17, 239–244. DOI: 10.1080/01140671.1989.10428038
Hughes, K.A., de Willigen, P., Gandar, P.W. and Clothier, B.E. (1992) Kiwifruit root systems: structure & function. *Acta Horticulturae* 297, 383–389. DOI: 10.17660/ActaHortic.1992.297.51
Johnson, R.S., Mitchell, F.G., Crisosto, C.H., Olson, W.H. and Costa, G. (1997) Nitrogen influences kiwifruit storage life. *Acta Horticulturae* 444, 285–289. DOI: 10.17660/ActaHortic.1997.444.44
Judd, M.J. and McAneney, K.J. (1987) Economic analysis of kiwifruit irrigation in a humid climate. *Advances in Irrigation* 4, 307–330. DOI: 10.1016/B978-0-12-024304-4.50011-0
Judd, M.J., McAneney, K.J. and Wilson, K.S. (1989) Influence of water stress on kiwifruit growth. *Irrigation Science* 10, 303–311. DOI: 10.1007/BF00257495
Koutinas, N., Sotiropoulos, T., Petridis, A., Almaliotis, D., Deligeorgis, E. *et al.* (2010) Effects of preharvest calcium foliar sprays on several fruit quality attributes and nutritional status of the kiwifruit cultivar Tsechelidis. *Horticultural Science* 45(6), 984–987. DOI: 10.21273/HORTSCI.45.6.984
Marsh, K.B. and Stowell, B.M. (1993) Effect of fertigation and hydrogen cyanamide on fruit production, nutrient uptake, and fruit quality in kiwifruit. *New Zealand Journal of Crop and Horticultural Science* 21(3), 247–252. DOI: 10.1080/01140671.1993.9513776
Massai, R., Xiloyannis, C., Piccotino, D. and Baroni, G. (1993) Root system growth and conformation of peach grafted on two rootstocks in high density orchards. *Acta Horticulturae* 349, 163-166. DOI: 10.17660/ActaHortic.1993.349.24
Medici, M., Canavari, M. and Toselli, M. (2020) Interpreting environmental impacts resulting from fruit cultivation in a business innovation perspective. *Sustainability* 12, 9793. DOI: 10.3390/su12239793
Mia, M.A.B. (2015) *Nutrition of Crop Plants.* Nova Science Publishers, Inc., Hauppauge, New York.
Mills, T., Boldingh, H., Blattmann, P., Green, S. and Meekings, J. (2008) Nitrogen application rate and the concentration of other macronutrients in the fruit and leaves of gold kiwifruit. *Journal of Plant Nutrition* 31(9), 1656–1675. DOI: 10.1080/01904160802244894
Mininni, A.N., Laterza, D., Tuzio, A.C., Di Biase, R. and Dichio, B. (2022) Soil water content monitoring as a tool for sustainable irrigation strategy in a kiwifruit orchard under semi-arid conditions. *Acta Horticulturae* 1332, 203–209. DOI: 10.17660/ActaHortic.2022.1332.27
Monastra, F., Strabbioli, G., Raparelli, E., Turci, E., Cappelloni, M. *et al.* (1997) Slow-release fertiliser effects on kiwifruit orchard. *Acta Horticulturae* 444, 273–278. DOI: 10.17660/ActaHortic.1997.444.42
Montanaro, G., Dichio, B. and Palese, A.M. (2006a) Influence of environmental factors on mineral elements transport and accumulation in leaves and fruits of kiwifruit plants [*Actinidia deliciosa* (A. Chev.) CF Liang et AR Ferguson]; Basilicata. *Italus Hortus* 13(3), 65–68.
Montanaro, G., Dichio, B., Xiloyannis, C. and Celano, G. (2006b) Light influences transpiration and calcium accumulation in fruit of kiwifruit plants (*Actinidia deliciosa* var. *deliciosa*). *Plant Science* 170(3), 520–527. DOI: 10.1016/j.plantsci.2005.10.004
Montanaro, G., Celano, G., Dichio, B. and Xiloyannis, C. (2010) Effects of soil-protecting agricultural practices on soil organic carbon and productivity in fruit tree orchards. *Land Degradation & Development* 21(2), 132–138. DOI: 10.1002/ldr.917
Montanaro, G., Dichio, B., Xiloyannis, C. and Lang, A. (2012) Fruit transpiration in kiwifruit: environmental drivers and predictive model. *AoB PLANTS* 2012, pls036. DOI: 10.1093/aobpla/pls036
Montanaro, G., Dichio, B., Lang, A., Mininni, A.N., Nuzzo, V. *et al.* (2014) Internal versus external control of calcium nutrition in kiwifruit. *Journal of Plant Nutrition and Soil Science* 177, 819–830. DOI: 10.1002/jpln.201400396
Montanaro, G., Dichio, B., Lang, A., Mininni, A.N. and Xiloyannis, C. (2015) Fruit calcium accumulation coupled and uncoupled from its transpiration in kiwifruit. *Journal of Plant Physiology* 181, 67–74. DOI: 10.1016/j.jplph.2015.04.004

Nuzzo, V., Dichio, B., Xiloyannis, C., Piccotino, D. and Massai, R. (1997) Contribution to transpiration of different tissues of kiwifruit vines from their water reserves. *Acta Horticulturae* 444, 329–334. DOI: 10.17660/ActaHortic.1997.444.51

Pacheco, C., Calouro, F., Vieira, S., Santos, F., Neves, N. *et al.* (2008) Influence of nitrogen and potassium on yield, fruit quality and mineral composition of kiwifruit. *International Journal of Energy and Environment* 1, 9–15.

Pathan, A. and Gaskin, R.E. (2008) Seasonal characterisation of kiwifruit plant surfaces. Plant Protection Chemistry New Zealand Report to NZ Kiwifruit Growers Inc. SFF Project Number 06/090. Available at: https://www.yumpu.com/en/document/view/36183875/seasonal-characterisation-of-kiwifruit-plant-surfaces (accessed 20 April 2023).

Peuke, A.D., Jeschke, W.D., Dietz, K.-J., Schreiber, L. and Hartung, W. (1998) Foliar application of nitrate or ammonium as sole nitrogen supply in *Ricinus communis*. I. Carbon and nitrogen uptake and inflows. *New Phytologist* 138, 675–687. DOI: 10.1046/j.1469-8137.1998. 00158.x

Prendergast, P., McAneney, K.J., Astill, M.S., Wilson, A.D. and Barber, R.F. (1987) Water extraction and fruit expansion by kiwifruit. *New Zealand Journal of Experimental Agriculture* 15(3), 345–350. DOI: 10.1080/03015521.1987.10425580

Smith, G.S., Judd, M.J., Miller, S.A. and Buwalda, J.G. (1990) Recovery of kiwifruit vines from transient waterlogging of the root system. *New Phytologist* 115(2), 325–333. DOI: 10.1111/j.1469-8137.1990.tb00459.x

Smith, G.S., Klages, K.U., Green, T.G.A. and Walton, E.F. (1995) Changes in abscisic acid concentration, surface conductance, and water content of developing kiwifruit. *Scientia Horticulturae* 61(1-2), 13–27. DOI: 10.1016/0304-4238(94)00733-V

Sofo, A., Mininni, A.N., Dichio, B., Mastroleo, M. and Xylogiannis, E. (2022) Physical structure and chemical quality of waterlogged soils in an Italian kiwifruit orchard. *Acta Horticulturae* 1332, 195–201. DOI:10.17660/ActaHortic.2022.1332.26

Thorp, T.G., Ferguson, I.B., Boyd, L.M. and Barnett, A.M. (2003) Fruiting position, mineral concentration and incidence of physiological pitting in 'Hayward' kiwifruit. *Journal of Horticultural Science and Biotechnology* 78(4), 505–511. DOI: 10.1080/14620316.2003.11511655

Vizzotto, G., Lain, O. and Costa, G. (1999) Relationship between nitrogen and fruit quality in kiwifruit. *Acta Horticulturae* 498, 165–172. DOI: 10.17660/ActaHortic.1999.498.19

Xiloyannis, C., Massai, R., Piccotino, D., Baroni, G. and Bovo, M. (1993) Method and technique of irrigation in relation to root system characteristics in fruit growing. *Acta Horticulturae* 335, 505–510. DOI:10.17660/ActaHortic.1993.335.62

Xiloyannis, C., Dichio, B., Montanaro, G., Biasi, R. and Nuzzo, V. (1999) Water use efficiency of pergola-trained kiwifruit plants. *Acta Horticulturae* 498, 151–158. DOI: 10.17660/ActaHortic.1999.498.17

Xiloyannis, C., Celano, G., Montanaro, G., Dichio, B., Sebastiani, L. *et al.* (2001) Water relations, calcium and potassium concentration in fruits and leaves during annual growth in mature kiwifruit plants. *Acta Horticulturae* 564, 129–134. DOI: 10.17660/ActaHortic.2001.564.14

Xiloyannis, C., Montanaro, G. and Dichio, B. (2012) Kiwifruit. In: Steduto, P., Hsiao, T.C., Fereres, F. and Raes, D. (eds) *Crop Yield Response to Water*. FAO Irrigation and Drainage Paper 66, FAO, Rome, pp. 488–497. Available at: https://www.fao.org/3/i2800e/i2800e.pdf (accessed 20 April 2023).

Xiloyannis, C., Montanaro, G., Mininni, A.N. and Dichio, B. (2015) Sustainable production systems in fruit tree orchards. *Acta Horticulturae* 1099, 319–324. DOI: 10.17660/ActaHortic.2015.1099.36.

12 Vine Management Systems

Kevin J. Patterson[1]* and Grant Thorp[2]
[1]*The New Zealand Institute for Plant and Food Research Ltd, Auckland, New Zealand;* [2]*The New Zealand Institute for Plant and Food Research Ltd, Te Puke, New Zealand*

12.1 Introduction

Orchard and vine management systems used for kiwifruit (*Actinidia* spp.) have developed and diversified substantially over the last two decades. 'Hayward' kiwifruit (*A. chinensis* var. *deliciosa*) dominated world trade prior to 2000, and orchard management systems were tailored accordingly. Since ~2000, the development of many new cultivars of *A. chinensis* var. *deliciosa* and *A. chinensis* var. *chinensis* with red-, yellow- and green-coloured flesh, and kiwiberry (*A. arguta*; see Chapter 7), has significantly increased choices for kiwifruit consumers.

Techniques to optimize productivity of new cultivars differing in plant form, production potential and fruit characteristics have been developed. Consumer demand for fruit of consistent flavour and nutritional content produced according to sound sustainability principles have been major drivers for new vine management systems. The devastating effects of bacterial canker, caused by *Pseudomonas syringae* pv. *actinidiae* (Psa) biovar 3, in kiwifruit production areas have also led to new methods of vine management to minimize risks associated with bacterial canker infection.

Productivity of 'Hayward' orchards increased considerably as growers in New Zealand and Italy became more familiar with the vine and its growing requirements. In New Zealand, 'Hayward' yields increased from 25 t ha^{-1} in the 1980s to over 50 t ha^{-1} in 2010 (Sale and Williams, 1985; Thorp *et al.*, 2011). Even higher yields have been reported for 'Hayward' (73.5 t ha^{-1}) (Thorp *et al.*, 2011) and the yellow-fleshed cultivars of *A. chinensis* var. *chinensis*, such as 'Hort16A' (65.7 t ha^{-1}) (Thorp and Barnett, 2011) and 'Zesy002' (91.4 t ha^{-1}) (Thorp *et al.*, 2018). These changes have come about from our greater understanding of growing systems and fruit/vine responses to new growing techniques.

12.2 Canopy Management

Kiwifruit training and pruning systems have undergone significant change within a relatively short period since commercialization of the crop outside of China, where until recently much of the crop was harvested directly from the wild (Li, 1990). Initially, kiwifruit vines were trained on vertical fences or overhead pergolas, with various arrangements of leaders and trunks and spur pruning of shoots to manage the fruiting canopy (Sale and Lyford, 1990) (Fig. 12.1). Significant improvements in yield were not realized until growers identified the benefits of cane

*Corresponding author: kevin.patterson@plantandfood.co.nz

© CAB International 2024. *Kiwifruit: Botany, Production and Uses* (eds A.C. Richardson, J.N. Burdon and A.R. Ferguson)

DOI: 10.1079/9781800620933.0012

Fig. 12.1. During the early stages of industry development in New Zealand, kiwifruit vines were trained on vertical fences and/or overhead pergolas, with various arrangements of leaders and trunks, and spur pruning. (Photo (taken in 1964): C.A. Schroeder)

replacement pruning (Sale and Lyford, 1990), better utilizing the natural growth habit or architecture of the kiwifruit vine.

12.2.1 Vine architecture

Actinidia vine architecture is dominated by the formation of long upright shoots or canes that become pendulous under their own weight (Seleznyova *et al.*, 2002). These long shoots produce leaves continuously, from both pre-formed and neo-formed nodes, until growth ceases at the end of the season. New replacement canes or relay axes (dominant branches or leaders) form near the base of the parent canes, in the transition zone between pre-formed and neo-formed nodes. Other shoots may self-terminate in spring by shoot-tip necrosis, either before a period of internode extension to form short shoots or after internode extension to form medium shoots (see Chapter 14). Branching of short and medium shoots is mainly acrotonic with the strongest shoots developing from buds below the necrotic shoot tip. All branching is sympodial with new shoot growth produced from quiescent or resting axillary meristems and thus always formed by prolepsis (i.e. development of lateral branches from a lateral meristem, after the formation of a bud or following a period of dormancy, when the lateral meristem is split from a terminal meristem). Secondary shoots may form during summer on vigorously growing long shoots (Thorp *et al.*, 2021) or from young plants in the nursery or glasshouse (Iliescu *et al.*, 2021); these shoots are also formed by prolepsis as they arise from quiescent buds from the current season's growth. When pruned, all shoot types produce new shoot growth from buds immediately below the pruning cut.

In horticulture, this natural architecture of kiwifruit vines is utilized by training vines onto pergola (Fig. 12.2) or T-bar support structures and fostering the appropriate mix of short, medium and long shoots to form the fruiting canopy. A single trunk is usually maintained and, initially, two leaders (cordons) are trained horizontally in opposite directions along the support structures. New shoots form along these leaders and are eventually tied down to the support structures in winter as replacement canes (Sale and Lyford, 1990). In the following season, a combination of short, medium and long axillary shoots arising from these replacement canes produces the fruiting canopy. Fruiting canes are generally replaced during winter pruning, so that vines have a similar structure at the beginning of each season. How and when the replacement canes are grown and selected is critical to the success of training systems.

12.2.2 Vine physiology

Improvements in orchard productivity have resulted from a better understanding of the basic principles of whole-vine physiology, advances in light management and control of vegetative growth and adapting these measures to the region and/or cultivar. Annual pruning of kiwifruit vines is undertaken to ensure fruiting canopies can efficiently convert the sun's energy into carbohydrate to support high yields of large, high dry matter fruit with the required taste profile. For 'Hayward' and 'Hort16A', a leaf area index of 3–4 has been suggested to maximize allocation of carbon to fruit growth and regeneration of reserves to support growth in the following season (Buwalda 1991; Tombesi *et al.*, 1994; Snelgar and Martin, 1997; Clearwater

Fig. 12.2. *Actinidia chinensis* var. *deliciosa* 'Hayward' kiwifruit vines in New Zealand trained on overhead pergola support structures with a single trunk, two horizontal leaders (cordons) and replacement canes to form the fruiting canopy. (Photo: The New Zealand Institute for Plant and Food Research Ltd)

et al., 2004). Today's highly productive growing systems are producing more and better fruit from vines with fewer leaves and less vegetative growth than previously (Thorp *et al.*, 2003a, 2011).

12.2.3 Training and pruning systems

Manipulation of kiwifruit vines, involving winter and summer pruning and training of vines onto appropriate support structures, is one of the most important aspects of vine management. The choice of support structure, planting density and pruning system is determined by both tradition and functionality, and differs between countries.

New Zealand

While early plantings in New Zealand were established on vertical fences, by the 1970s, T-bar and pergola support structures with vines trained to a single trunk and two leaders were predominant (Sale, 1992; Berry, 1997). Choice between the two support structures was often determined by cost, as pergolas were more expensive to install than T-bars (Sale and Lyford, 1990). Many growers started with T-bars and then converted them to pergolas to attain better yields of quality fruit.

The New Zealand industry used either total cane replacement pruning or mixed-age, spur pruning systems that retained fruiting canes for multiple seasons to form a semi-permanent fruiting canopy (Fig. 12.3). Cane replacement pruning incurs higher costs during winter and lower costs during summer while mixed-age pruning focuses on spring and summer pruning with less effort during winter. Both systems produce equivalent yields and fruit quality.

Vine planting distances for pergola systems vary from 4.5–6.0 m within rows and from 3.0–5.0 m between rows, with a canopy footprint or allocated surface area per vine of approximately 30 m^2 depending on vine vigour (Sale and Lyford, 1990). Increasing fruit returns in the 1980s encouraged higher planting densities, by double- or triple-planting vines, to increase yields especially from young vines (Berry, 1997; Massai *et al.*, 1997). Unfortunately, this promoted strong vegetative growth from new canes near the trunk which increased pruning costs. Increased vine vigour and regrowth from pruning cuts had a negative effect on fruit quality.

Declining market returns in the late 1980s and poor fruit quality forced growers to look at systems that reduced costs (Sale, 1992; Berry, 1997). Growers moved away from detailed pruning systems towards less precise systems, including mechanical pruning of T-bar trained vines. This promoted the growth of large, high-vigour canes and dense fruiting canopies that produced large fruit with high dry matter (Lai *et al.*, 1990; Sale, 1992; Volz *et al.*, 1992; Smith *et al.*, 1997). However, dense canopies led to increased leaf necrosis and *Botrytis* grey mould (*Botrytis cinerea*) on fruit and higher fruit losses during

Fig. 12.3. *Actinidia chinensis* var. *chinensis* 'Hort16A' kiwifruit vines in New Zealand pruned using total cane replacement (**A**) or mixed-age pruning (**B**) to produce the desired fruiting canopy. (Photos: The New Zealand Institute for Plant and Food Research Ltd)

storage (Sale, 1992; Snelgar *et al.*, 1998; Manning *et al.*, 2010). Excessive pruning during summer to control vigour stimulated more canopy growth, more leaf necrosis and fruit with low dry matter (Minchin *et al.*, 2010). During the 1990s, a return to less vigorous canopies, greater knowledge of the disease epidemiology and adjustments to postharvest handling procedures largely eliminated *Botrytis* fruit rot in the New Zealand industry (Michailides and Elmer, 2000; Manning *et al.*, 2010) (see Chapters 17 and 20).

Key summer pruning decisions are made in spring and include the number and size of medium and long shoots to be retained in the leader zone during summer for use as replacement canes in the following season, hence the term leader pruning (Miller *et al.*, 2001; Thorp *et al.*, 2003a). Preference is given to retention of sufficient flowering, terminated medium shoots rather than non-terminated long shoots, to reduce shading of the next season's fruiting wood and of the current season's fruiting canopy (Grant and Ryugo, 1984). The fruiting canopy is managed by removing non-fruiting shoots (blind shoots) and shoots growing from downward facing buds on the underside of canes. Terminated short and medium shoots are left unpruned to prevent regrowth. The time of pruning is also important as delayed or excessive pruning can increase the production of wasteful vegetative growth that will need to be removed with further summer pruning, an action which may damage the fruit. If the optimum number of canes and buds are retained in winter, leader pruning and blind shoot removal will ensure adequate leaf area is present to supply carbohydrate and minerals to the developing fruit (Thorp *et al.*, 2003b; Buxton, 2005).

Leader pruning, which is based on the concept of a divided canopy, creates a zone of open canopy along the leaders of vines to allow high rates of light transmission to the lower canopy and orchard floor (Buxton, 2005). Leader pruning is important for both conventional cane replacement and mixed-age pruning systems, but the ratio of wood types retained during summer and winter pruning differs (Fig. 12.3). Cane replacement pruning relies on total replacement of fruiting canes with new canes (long shoots) produced from the transition zone near the base of the previous year's cane. For mixed-age pruning, only a proportion of previous fruiting canes are renewed each year and terminated shoot growth on the remaining canes is thinned to retain the required number of winter buds for the following year's canopy and crop.

During winter, pruning is guided by wood type decisions made during the previous summer that ensure the next season's fruiting wood is grown under high light conditions. Further wood type selection during winter focuses on retaining larger diameter canes as they are more fruitful than small diameter canes (Volz *et al.*, 1992; Thorp *et al.*, 2003a). However, larger canes come with the risk of more shoot growth and more summer pruning (Thorp *et al.*, 2003a). With new cultivars, such as 'Zesy002', to obtain high yields of 95 t ha^{-1} it is necessary to retain larger basal diameter canes than is optimal for 'Hayward' (Thorp *et al.*, 2018).

Cane spacing and bud density (number of buds per m^2 of canopy) not only determine canopy density but also the fruiting density (number of fruit per m^2 of canopy), which affects fruit size and ultimately yield (Richardson and McAneney, 1990; Buxton, 2005). An optimum number of winter buds, determined by cultivar, location and management system, evenly spread across the canopy will potentially produce the size, quality and yield of fruit required.

The principles and practices of leader and mixed-age pruning systems apply equally to pergola- and T-bar-trained vines and to all cultivar/rootstock combinations. However, adjustments are required to account for differences in plant vigour resulting from rootstock, soil type and environment. Characteristics of new genotypes also need to be considered. For example, genotypes with very strong vertical shoot growth that is difficult to bend over onto a horizontal pergola or T-bar canopy present a challenge.

The strip male system, in which male and female vines are planted in alternating rows with the males grown in a narrow strip between the opposing female canopies (Fig. 12.4), is becoming more common in kiwifruit orchards (Costa *et al.*, 2018). Like leader pruning, this divided canopy method creates a gap between the overhead canopy of pergola-trained vines to allow more light transmission to the lower canopy (Thorp *et al.*, 2011). Costa *et al.* (2018) reported that most top-performing growers in New Zealand are now using pergola training systems with strip males.

Fig. 12.4. Strip male kiwifruit vines in New Zealand growing as a narrow strip between opposing *Actinidia chinensis* var. *deliciosa* 'Hayward' female canopies. (Photo: The New Zealand Institute for Plant and Food Research Ltd)

Pruning of strip male vines involves the selection of shorter shoots and the development of spur systems along the leaders to facilitate flowering. Well-spaced spur systems that have high exposure to light enable high flower initiation among the shorter shoots present. Shortening of shoots after flowering is carried out to improve light and spray penetration. During summer (late January) in New Zealand, vigorous unwanted growth is typically removed by ripping (breaking out).

Male vines planted in conventional matrix layouts ('box' males) are generally pruned back following flowering to allow light penetration into the canopy of adjacent female vines. In winter, some replacement of canes that have flowered is carried out and the canopy is generally kept open to ensure adequate flower initiation can occur.

A key consideration with the introduction of newer male cultivars into kiwifruit blocks is the compatibility of male and female cultivars (see Chapter 5) and the timing of male flowering to achieve coincidence with adjacent female vines. The increasing range of scion and rootstock cultivars used in modern orchards makes it imperative that appropriate combinations, proven to have flowering coincidence under local conditions, are used.

Italy and France

Pruning and training systems in Italy and France have utilized the T-bar or pergola support structures and cane replacement pruning methods used in New Zealand, with some regional differences (Blanchet and Ellis, 1988; Costa, 1990; Zuccherelli, 1994; Hennion, 2003; Testolin, 2015). In northern Italy and France, growers have mainly used the T-bar training system (Costa, 1990; Costa *et al.*, 1992a; Hennion, 2003; Testolin, 2015). The added advantage of this system, especially in the Emilia Romagna region of Italy, is its compatibility with the use of overhead nets to protect against hailstorms which can cause significant damage to vines (Fig. 12.5).

Fig. 12.5. Overhead netting providing hail protection during summer for kiwifruit vines in the Emilia Romagna region of northern Italy. A gap is maintained between vines so that nets can be unclipped and folded away in winter to avoid damage from build-up of snow. (Photo: The New Zealand Institute for Plant and Food Research Ltd)

The gaps down the centre of the rows of vines on T-bars enable growers to open and close the nets as required. Nets are folded away (closed) during winter to prevent snow collapsing them and opened in summer when there is a risk of hail. Growers tend to use short-cane rather than the long-cane pruning systems used in New Zealand (Costa *et al.*, 1990, 1992b). However, with short canes, fruiting wood is concentrated near the base of canes where growth is most vigorous (Seleznyova *et al.*, 2002), making it difficult to select suitable renewal wood for fruiting in the subsequent season (Costa *et al.*, 1990, 1992b). A long-cane option has been introduced for T-bars with canes bent downwards to increase the canopy area per vine and enable closer row spacing. The two systems produce similar yields, but the modified T-bar with long shoots produces the desired renewal wood types. Other support structures that have been evaluated but not adopted in Italy include a V-trellis and free-spindle training system similar to a system used in temperate tree crops (Costa *et al.*, 1992b, 2018).

In Piedmont, where growers need to harvest earlier than other regions to avoid periods of heavy rainfall and orchard flooding, growers have successfully adopted a long-cane, mixed-age pruning system (Peyracchia system) with partial cane replacement, similar to the leader pruning and mixed-age system used in New Zealand (Vittone *et al.*, 1996; Costa *et al.*, 1997; Anon., 2010; Vittone, 2010). Growers are able to select and grow less vigorous, mostly previously fruited wood by removing excess and late-developing canes from the leader zone of vines during spring. This maintains a balance between old previously fruited wood and new shoot growth, with no pruning of terminated wood types. Comparisons between conventional T-bar-trained vines and vines pruned using the Peyracchia system over three years revealed significant savings in time and increased yields with the Peyracchia system (Vittone *et al.*, 1996; Costa *et al.*, 1997). In the third year, total pruning time was 266 h and yield 30–32 t ha^{-1} with the Peyracchia system compared with 365 h and 22–25 t ha^{-1} for the traditional system (Vittone *et al.*, 1996).

To avoid damaging frosts, some growers in the Piedmont region have also tested tall T-bar systems called Pergoletta Alta, with structures up to 3.0 m high and fruiting canopies extending down to approximately 1 m above the ground (CReSO, 2010). Yields from these tall T-bar vines are reputed to be 20% higher than from adjacent vines on conventional T-bars, but the tall T-bar systems incur additional pruning and harvesting costs as elevated platforms are required. Using the Peyracchia pruning system would reduce pruning costs and is well suited to this type of structure.

An alternative growing system mainly used in Emilia Romagna has been developed from the Geneva Double Curtain (GDC) system used for grapes (Fig. 12.6). In this divided-canopy system, T-bar structures are separated into two leaders running in parallel along the planted row, with one leader having a single fruiting 'curtain' down one side and the other leader having a single fruiting 'curtain' down the opposite side (Costa *et al.*, 1992b; Testolin, 2015). The two leaders are separated by approximately 0.75 m so there is a clear gap between the two during summer, resulting in a horizontally divided canopy separated into two distinct canopies or 'curtains'. This is effectively a modification of the Peyracchia and leader pruning styles where the most important pruning task is to remove excess growth from the leader zone and create a shaft of light to help illuminate the lower canopy zones. 'Hayward' yields were similar for GDC and T-bar-trained vines with 42.4 and 40.6 t ha^{-1}, respectively (Costa *et al.*, 1992b).

In the Latina region, south of Rome, several table grape orchards using a tendone support structure, similar to an overhead pergola

Fig. 12.6. *Actinidia chinensis* var. *deliciosa* 'Hayward' kiwifruit vines in Italy trained with a Geneva Double Curtain system with two horizontal leaders running in parallel along the row. (Photo: The New Zealand Institute for Plant and Food Research Ltd)

structure, were converted to kiwifruit in the 1980s, with growers using the same training and pruning methods on kiwifruit that they had used for their grapevines (Costa, 1990; Testolin, 2015). At that time, the method involved a single trunk with several leaders (long shoots) arranged in a radial pattern from the top of the trunk, often described as a wagon-wheel layout (Fig. 12.7). The wagon-wheel arrangement created a divided canopy with a large area of open canopy at the centre of the vine providing for a shaft of light to help illuminate the lower canopy zones. However, the central origin for replacement cane growth would become very congested with vigorously growing shoots in summer. The solution was to reduce the height of the main trunk, so that the replacement canes originate well below the main canopy. Growers would remove all new shoot growth from near the base of parent canes and retain less vigorous wood types that developed further along the parent canes to form the horizontal fruiting canopy. This became the predominant training system in Latina, until the introduction of the 'Hort16A' cultivar and adoption of New Zealand-style growing systems.

In France, kiwifruit growers have persisted with complete cane renewal pruning with vines on T-bar structures, as previously used in New Zealand (Hennion, 2003). Many growers are now incorporating leader pruning and mixed-age pruning into their orchards.

China

Kiwifruit production systems in China involve many green-, yellow- and red-fleshed cultivars grown across a wide range of environments, soil types and topographies (Huang, 2016). As a consequence, growth characteristics of individual cultivars may vary significantly depending on regional differences in soil type, climate or local differences in Psa infection. The types of systems used to produce kiwifruit vary greatly across the main production areas in the provinces of Shaanxi, Sichuan and Hubei. There is a marked

Fig. 12.7. *Actinidia chinensis* var. *deliciosa* 'Hayward' vines in the Latina region of Italy with a single trunk and several leaders arranged in a radial pattern from near the top of the trunk. (Photo: The New Zealand Institute for Plant and Food Research Ltd)

contrast between methods used in traditional family holdings and newer commercial orchards, where vine management incorporates best practice methodology used in Italy and New Zealand. While kiwifruit has been grown commercially in China for ~40 years, there has been a much greater focus on fruit quality rather than simply on fruit yield since the late 2000s (Huang, 2016).

Early plantings were based on simple structures similar to a conical tepee or umbrella design. Use of small hand-held agricultural equipment rather than tractors enabled close planting, particularly in small family holdings where all available land is utilized. However, with limited space (as low as 6–8 m^2) allocated per vine, growers needed to make hard pruning cuts which stimulated unwanted vigorous shoot growth. Canopy structure is often sub-optimal in traditional orchards, with excessive leaf layering and limited opportunity for development of highly fruitful replacement canes. Newer commercial orchards now utilize pergola structures, planting densities and cane replacement pruning typical of systems used in Europe and New Zealand (Huang, 2016). These systems have had a major influence on yield and fruit quality expectations for kiwifruit growers in China, with techniques such as girdling and optimized pruning systems now being incorporated into traditional production systems.

12.2.4. Innovative growing systems

Further examples of how divided canopies can create zones of open canopy within vines to allow higher rates of light transmission to lower canopy zones include stringing systems for training replacement canes, use of secondary canopies separate from the main canopy and alternate year cropping systems.

Stringing systems for training canopies

The management technique called 'stringing' or string-assisted growing directs the vigorous growth of long shoots into vine frameworks that encourage strong and rapid development of leaders and the fruiting canopy on young vines (Patterson and Currie, 2011; Thorp and Barnett, 2011). The technique involves temporary placement of strings at an angle of 40–60° to the horizontal plane of the pergola for growing shoots to latch onto and twine along (Fig. 12.8; Patterson and Currie, 2011). This enables shoot growth to occur in a high-light environment which encourages subsequent flower production (Richardson *et al.*, 2022). Although expensive to install each year, stringing systems are promoted as a less complicated management system for workers, especially with the more vigorous *A. chinensis* var. *chinensis* cultivars (Patterson and Currie, 2011).

Secondary canopies

Top-performing 'Hayward' orchards in New Zealand produce 36 t ha^{-1}, with some blocks producing 54 t ha^{-1}. Research on the use of secondary canopies, as another form of divided canopy, with reflective ground covers demonstrated that yields could be increased to 72 t ha^{-1} (Fig. 12.9; Thorp *et al.*, 2011).

Within conventional T-bar or pergola training systems, significantly increasing the number of winter buds increases canopy density, resulting in small, poor-tasting fruit with a higher incidence of botrytis rot and fewer flowers being produced in the following season (Snelgar *et al.*, 1998; Manning *et al.*, 2010). New divided-canopy configurations, e.g. carrying fruit on secondary canopies trained as vertical curtains below the main pergola fruiting canopies, were tested to solve these limitations (Thorp *et al.*, 2011). Yields on 'Hayward' vines with secondary canopies and a matrix male layout were increased by 19% to 73.5 t ha^{-1} (Thorp *et al.*, 2011). However, fruit were smaller and had lower dry matter concentration and vines showed leaf yellowing and necrosis at harvest, suggesting vines were at the limits of economic productivity. Secondary canopies were also tested on 'Hayward' vines in a strip male configuration, doubling the canopy surface area and crop load per vine without changing the fruit density. In this case there was a weak relationship between high fruit number and reduced fruit size or dry matter, but, more importantly, there was no effect on leaf quality, suggesting vines were not at their limit of productivity.

Fig. 12.8. String-assisted growing system for kiwifruit vines in New Zealand is particularly useful to promote rapid canopy development with young vines. (Photo: The New Zealand Institute for Plant and Food Research Ltd)

Fig. 12.9. *Actinidia chinensis* var. *deliciosa* 'Hayward' kiwifruit in New Zealand with reflective ground covers and secondary fruiting canopies growing as a hanging curtain beneath the leaders of conventional leader pruned, pergola trained, vines. (Photo: The New Zealand Institute for Plant and Food Research Ltd)

Alternate year cropping

Alternate year cropping (AYC) systems are also an example of divided canopies, separating fruiting and replacement cane canopies (Thorp and Barnett, 2011). AYC is a string-assisted pergola growing system used to crop a single row of vines over the combined area of two rows, every second year (Fig. 12.10). Non-cropping vines grow replacement canes on strings above the fruiting canopy of adjacent cropping vines. In the following year, the roles are reversed: the fruiting canopy of the previously cropped vines is removed, and the replacement cane canopy of the previously non-cropping vines is tied down across two rows. With reflective ground covers, the yields of 'Hort16A' vines were increased by 13% to 65.7 t ha^{-1} (Thorp and Barnett, 2011). Fruit weight was not affected, and fruit also had high dry matter. Increasing cane numbers in AYC did not increase yields as the fruit size declined, suggesting that 'Hort16A' AYC vines with reflective ground covers were close to their limit of economic productivity.

12.2.5 Controlling vegetative regrowth

Tip-squeezing

In the mid-2000s, a simple technique commonly known as 'tip-squeezing' or 'crush tipping' proved very effective at limiting shoot extension

Fig. 12.10. *Actinidia chinensis* var. *chinensis* 'Hort16A' vines in New Zealand trained with an alternate year cropping system. Image shows the pruned leaders of Year 1 cropping vines after complete removal of the previous fruiting canopy. Replacement canes for cropping in Year 2 are shown growing on strings on adjacent non-cropping vines. (Photo: The New Zealand Institute for Plant and Food Research Ltd)

growth of non-terminating, long-fruiting shoots (Patterson and Currie, 2011). Tip-squeezing simply crushes the shoot tip with no stimulatory effect on regrowth from that shoot, in contrast to pruning which stimulates shoot regrowth. This technique alters competition between fruit and shoots for available carbohydrates to favour fruit development (Minchin *et al.*, 2010).

Zero-leaf pruning

Zero-leaf pruning is also used for shoot control, especially at the ends of canes in systems with opposing female canopies. The technique involves cutting the fruiting shoot just above the last (distal) fruit on the lateral so that there are no axillary buds from which secondary regrowths can develop (Patterson and Currie, 2011). This technique is reserved for extreme cases, as if applied too liberally it will negatively impact fruit quality and the productive potential of vines through reduced leaf area.

Stubbing

Stubbing, or the cutting back of new shoots to a few buds, is commonly practised to maintain growing points in the leader zone of vines (Snelgar *et al.*, 1992). Applied to vigorous, upright-growing shoots, the technique usually promotes the formation of two new shoots growing at a flatter angle than the parent shoot. The technique is also useful with stringed growing systems to synchronize the timing of shoot attachment to the strings.

12.3 Crop Management

12.3.1 Setting crop load

Setting bud numbers during winter pruning establishes the basis for crop load and average fruit size in the subsequent season, while flower and fruit thinning is undertaken in spring to ensure uniformity in fruit size and quality across the canopy. Setting bud numbers during winter pruning is determined by the flowering potential of the cultivar being grown, the cane type used (total cane replacement or mixed-age pruning) and seasonal temperature influences on floral budbreak (see Chapter 14).

For 'Hayward' production in New Zealand in the 1980s, fruit yields of ~25 t h^{-1} were achieved with between 30 and 50 fruit m^{-2} of canopy (Sale and Lyford, 1990), and between three and five leaves per fruit on whole vines (Snelgar *et al.*, 1986). There was a significant inverse relationship between fruit number and mean fruit weight at harvest that was relatively consistent across site and season (Richardson and McAneney, 1990; Cooper and Marshall, 1992). While crop load had a strong effect on average fruit size, the effect on fruit quality, e.g. dry matter, was less significant (Richardson *et al.*, 1997). Similar effects occur in other cultivars (Patterson and Currie, 2011).

In most high-producing orchards, removal of excess flowers and fruitlets, particularly lateral fruit and poorly pollinated or mis-shapen fruit, is carried out as soon as fruit shape and potential size can be identified. Both the date of flower opening and flower type within an inflorescence (terminal vs. lateral) may be related to final fruit weight and fruit dry matter concentration at harvest (Richardson *et al.*, 2019). Typically, fruit thinning is based on a guide of 1–2 fruit on spurs (~five leaves), 2–6 fruit on medium-terminated shoots (~7–21 leaves) and 4–6 fruit on long non-terminated shoots.

The plant growth regulator hydrogen cyanamide not only increases and synchronizes flowering on kiwifruit vines (see Chapter 14) but can also be an important flower-thinning agent causing the abortion of lateral flowers (Henzell *et al.*, 1992; Walton and Fowke, 1993; Richardson *et al.*, 2022).

12.3.2 Girdling practices for productivity and fruit quality

Trunk girdling, or cincturing, was originally introduced to promote earlier and increased fruiting on young vines, but any hastening of fruit maturation on mature vines was considered undesirable (Davison, 1990). Development of girdling techniques for 'Hayward' and 'Hort16A' occurred in New Zealand from ~2000 onwards (Patterson and Currie, 2011; Currie *et al.*, 2018). Of all the manipulative techniques introduced into kiwifruit growing systems over the last ten years, trunk girdling has had the greatest effect on increasing flowering and improving fruit growth and development (Fig. 12.11) (Currie *et al.*, 2018). Trunk girdling is now used widely on cultivars in New Zealand and is gaining wide acceptance in Europe and Chile.

Girdling involves the severance and disruption of the periderm, phloem and cambial tissue between the phloem and xylem (Goren *et al.*, 2004). Girdling is thought to prevent roots of girdled vines from competing for carbohydrate; however, changes in hormone flows, e.g. cytokinins in xylem sap, may also be involved (Currie, 1997). Trunk girdling has also been shown to cause a significant decrease in photosynthesis until the girdle heals and a reduction in root growth throughout the season (Black *et al.*, 2012). Multiple girdles can be applied per season. Trunk girdling is preferred over cane girdling as it is much quicker to carry out. In New Zealand, kiwifruit girdles applied in spring and early summer heal rapidly (3–6 weeks) but girdles applied in autumn, when temperatures are cooler, may be slower to heal.

The timing of girdling has been extensively studied in relation to fruit growth and development (Currie *et al.*, 2018). Trunk girdling in spring can increase fruit size while later girdles increase carbohydrate accumulation by fruit (Currie *et al.*, 2018). The increased soluble solids content in fruit from girdled vines

Fig. 12.11. *Actinidia chinensis* var. *deliciosa* 'Hayward' vines in New Zealand with multiple trunk girdles to promote fruit size and dry matter. (Photo: The New Zealand Institute for Plant and Food Research Ltd)

may be mistaken for earlier maturation (Burdon *et al.*, 2011) (see Chapters 16 and 20, for discussion of maturation and harvest indices, respectively).

12.3.3 Root pruning

Root pruning has long been used to dwarf fruiting trees. For kiwifruit, root pruning was developed in New Zealand in the mid-1990s (Black *et al.*, 2011, 2012) to stimulate fruit size and dry matter accumulation rather than reduce shoot growth and flower initiation, which were the focus in apple trials (Schupp and Ferree, 1990). In a study on mature 'Hort16A' vines, severance of ~80% of the major roots rapidly reduced vine water use and leaf water status but did not reduce vegetative growth of shoots (Black *et al.*, 2011). Effects of root pruning on vine water status and fruit growth appear to be quite different to those induced by limited water supply (Black *et al.*, 2011). Typically, kiwifruit root pruning is applied when fruit are actively accumulating carbohydrate (Richardson *et al.*, 1997; Nar dozza *et al.*, 2010; Hall *et al.*, 2013); effects on fruit size and dry matter have been observed in 'Hayward', 'Hort16A' and 'Zesy002' vines. Increases in dry matter and fruit size (up to 15 g) were achieved when a significant proportion of the root biomass was removed (Black *et al.*, 2012). While the technique is used commercially, it is not widely used, partly because of the difficulty in reliably severing a known proportion of the root biomass.

12.4 Modified Environments

Modifications of the kiwifruit vine or fruit growing environment are extensively used in Italy, New Zealand, South Korea and China to improve fruit quality and minimize the impacts of hail, wind and extreme weather events, such as typhoons, on vines.

12.4.1 Overhead netting

Netting is used to protect vines from the damaging effects of hail and wind and to minimize the use of productive orchard land for live shelter belts (McAneney *et al.*, 1984; Zhang *et al.*, 2004). Use of anti-hail nets, shade nets and overhead shelters has become much more common in kiwifruit orchards in Italy and New Zealand over the last ten years (Fig. 12.5). Although the costs of installation and maintenance are significant, overhead netting systems are considered economically viable for high-value cultivars.

Netting systems modify the light intensity, light quality, wind speed, canopy temperature, relative humidity and soil temperature in orchards (Basile *et al.*, 2008, 2014; Mupambi *et al.*, 2018). Practical experience with netting systems in Italy and New Zealand indicates positive benefits on fruit quality (appearance, size and dry matter) and vine growth. Growth of shoots, leaf size and petiole length may be affected along with the ratio of shoot types produced according to netting colour. In research trials of different coloured (photoselective) netting, blue net appeared to reduce vine vigour compared with un-netted vines while red and grey nets stimulated shoot and fruit growth and improved fruit composition (Basile *et al.*, 2008, 2014). Netting may adversely affect insect pollination (see Chapter 13).

12.4.2 Rain-out shelters

Kiwifruit cultivation under plastic has been used to combat Psa infection (Chiabrano and Giacalone, 2018; Costa *et al.*, 2018). Modifications to light, temperature and shelter under plastic increased shoot extension on vines. Plastic coverings are used extensively in Jeju Province, South Korea, to protect vines from excessive rain and wind from typhoons and thereby reduce the incidence of Psa (Shim and Ha, 1999; Kim *et al.*, 2017). Plastic rain-out shelters are more commonly used in Italy than in New Zealand, where growers found them to be prone to severe damage from strong winds. In China, new systems combining the beneficial attributes of netting and plastic coverings are under development to reduce the incidence of Psa in susceptible, red-fleshed cultivars (Fig. 12.12) (Z. Liu, Shaanxi, October 2019, personal communication).

12.4.3 Reflective ground covers

Reflective ground covers (Fig. 12.9) provide a means of increasing light utilization by fruit canopies where leaf shading within canopies or

Fig. 12.12. Use of combined netting and plastic coverings in China to reduce the incidence of bacterial canker *Pseudomonas syringae* pv. *actinidiae* biovar 3 (Psa) for susceptible kiwifruit cultivars. (Photo: The New Zealand Institute for Plant and Food Research Ltd)

overhead netting limits productivity (Thorp *et al.*, 2001; Meinhold *et al.*, 2011; Overbeck *et al.*, 2014; Schmidt *et al.*, 2014). The large leaves and layered canopies of kiwifruit vines typically create dense, shaded areas, particularly in the lower canopy layers (Smith and Buwalda, 1994; Kramer *et al.*, 2022). Reflective covers increased light intensity in the shaded section of the kiwifruit canopy (Kramer *et al.*, 2022). In this study, reflective covers in the inter-row increased fruit fresh and dry weight accumulation compared with untreated vines. Use of reflective ground covers also increased dry matter accumulation by 'Hayward' fruit in the first year of implementation, and increased flowering and crop loads in the second year (Thorp *et al.*, 2001).

12.4.4 Fruit bagging

Preharvest fruit bagging is widely used in China to improve the quality of kiwifruit (Huang, 2013, 2016). While the technique requires significant labour inputs, benefits include increased fruit size, advanced maturity (including flesh colour) and reduced fruit disorders (Sharma *et al.*, 2014). Variations in bagging techniques used on kiwifruit include the type of bag material (cloth, coloured or white paper with or without black liner) and the timing of bag placement and removal from the fruit (Fig. 12.13). The timing of bagging affects chlorophyll development in the outer pericarp, the degree of scuffing damage during rapid fruit expansion and the amount of chemical residues on fruit (Zhao and Xie, 2018). Kiwifruit in Shaanxi Province, China, are bagged for ~4–5 months from early June until harvest in September/October. Fruit bagging may increase air temperature within bags by 0.5–1.5°C above ambient, particularly when temperatures are above ~30°C for much of the day as in the Meixian region of Shaanxi Province (Huang, 2013). For 'Hayward' in this region, bagging increased titratable acid-

Fig. 12.13. Bagging of kiwifruit in China is used to reduce skin blemishes and improve fruit quality. (Photo: The New Zealand Institute for Plant and Food Research Ltd)

ity and lowered total soluble sugar content of ripe fruit.

12.5 Plant Growth Bioregulators

Plant-based biostimulants and synthetic plant growth regulators are collectively termed plant bioregulators and are widely used in kiwifruit production (Costa *et al.*, 2018) to stimulate budbreak, flowering and fruit growth, control vine growth and enhance vine health.

12.5.1 Budbreak and flowering

The growth regulator hydrogen cyanamide has been widely used in the kiwifruit industries in Europe, Chile and New Zealand to enhance budbreak, increase flower numbers per shoot and synchronize flowering. Several alternatives to hydrogen cyanamide are under investigation as budbreak agents (see Chapter 14).

12.5.2 CPPU effects on fruit development

Forchlorfenuron, commonly called CPPU (N-(2-chloro-4-pyridyl)-N'-phenylurea), is a synthetic cytokinin of the phenylurea group that was used to enhance fruit size of 'Hayward' in Italy, Japan, Chile and New Zealand. It is now only used in China to increase fruit size of many green-, yellow- and red-fleshed cultivars. The response of *Actinidia* spp. to CPPU varies: *A. chinensis* var. *chinensis* and *A. chinensis* var. *deliciosa* were generally quite sensitive to CPPU whereas *A. melanandra* and *A. macrosperma* were much less sensitive (Jo *et al.*, 2003). The *A. arguta* cultivar 'Mitsuko' was also very responsive to CPPU application (Kim *et al.*, 2006).

CPPU enhances fruit growth in kiwifruit by stimulating endogenous hormones (Ohara *et al.*, 1997; Brown and Woolley, 2010), which promotes cell division or expansion (Woolley *et al.*, 1992; Antognozzi *et al.*, 1997), and accumulation of water and carbohydrate into fruit

(Antognozzi *et al.*, 1996; Li *et al.*, 2011). Changes in fruit morphology at the cellular level were evident in CPPU-treated fruit, in which the average size of the small parenchyma cells was increased whereas the average size of the large parenchyma cells was unaffected (Patterson *et al.*, 1993). Childerhouse (2009) also showed that these changes altered the ratio of inner and outer pericarp tissue fractions in 'Hort16A'.

Both 'Hayward' and 'Hort16A' fruit respond to CPPU applied during the early stages of fruit growth (Iwahori *et al.*, 1988; Patterson *et al.*, 1993; Cruz-Castillo *et al.*, 1999; Woolley and Cruz-Castillo, 2006; Childerhouse, 2009), with fruit size increases of ~10–30%. Early studies of CPPU on 'Hayward' fruit indicated that positive effects on fruit size of ~33% were associated with negative effects on fruit dry matter of ~9%. Optimum fruit sizing responses are obtained with applications 2–3 weeks after full bloom, with later applications promoting the accumulation of water more than dry matter (Famiani *et al.*, 1998; Cruz-Castillo *et al.*, 2014). CPPU applications before flowering (~20 days before anthesis) have been shown to cause the development of very unevenly shaped fruit (Woolley *et al.*, 1992). Use of CPPU on 'Hayward' may also increase growth of the outer pericarp at the stylar end of the fruit, resulting in small protuberances (Patterson *et al.*, 1993).

CPPU may also alter fruit maturity of 'Hayward', as treated fruit reached 6.2% soluble solids content one week earlier than untreated fruit (Patterson *et al.*, 1993). Influences of CPPU on fruit flesh colour have been observed as a darker green appearance of the outer pericarp tissues just below the exocarp in 'Hayward' (Lawes *et al.*, 1992; Patterson *et al.*, 1993) but reduced chlorophyll content in 'Monty' kiwifruit flesh (Kurosaki and Mochizuki, 1990).

12.5.3 Shoot growth retardants

Kiwifruit pruning traditionally included cutting back all fruiting shoots to retain a fixed number of leaves past the last fruit. Regrowth after pruning was problematic and chemical pruning techniques were developed to minimize this growth (Henzell *et al.*, 1986, 1992; Currie *et al.*, 2011). The application of a gel containing naphthalene acetic acid (NAA) to the pruning cut effectively inhibited new shoot growth. However, the NAA was readily translocated to fruit, substantially increasing fruit size but not dry matter accumulation, resulting in poor-quality fruit (Currie *et al.*, 2011). Use of NAA pruning gels to control excess vegetative growth was replaced with tip-squeezing and, to a lesser extent, zero-leaf pruning.

12.5.4 Benefit® treatments to increase fruit size

The biostimulant product Benefit® was extensively used to enhance fruit size in 'Hort16A' from 2000 to 2010 in New Zealand. Research with 'Hort16A' found that fruit size could be increased by up to 26 g per fruit; however, the dry matter concentration of treated fruit was significantly reduced, and fruit did not reach the required taste profile (Childerhouse, 2009).

12.6 Future Directions

The introduction of new cultivars with high profit margins has stimulated innovations in kiwifruit production, leading to raised expectations of both yield and fruit quality. The Psa epidemic has also significantly changed the way kiwifruit are managed, requiring growers to pay more attention to canopy management.

New green-, yellow- and, more recently, red-fleshed cultivars have created a renewed focus on vine and fruit growth to tailor management systems to individual cultivars. Some new *A. chinensis* var. *chinensis* cultivars are more vigorous than 'Hayward' and require systems to control vigour in order to achieve/maintain their potential for higher productivity and fruit quality. It is essential that growers and researchers work together to identify and mitigate the idiosyncrasies of each new cultivar. The call by Costa and Testolin (2022) for completely different and innovative growing systems is fundamental to the challenge of continuous and sustainable improvement in orchard yields and fruit quality to which the kiwifruit industry must continue to aspire. Further description of the three-dimensional aspects of modern high-performing kiwifruit canopies is warranted, including their study in the various divided canopy systems associated with leader pruning, alternate year cropping, secondary canopy and strip male systems.

There are several opportunities to improve understanding of the basic physiology behind different vine and fruit management techniques. Successful growers and researchers understand the whole-orchard system. This requires combining information from the plethora of mechanistic studies into a more holistic approach using tools such as functional structural plant modelling (Cieslak *et al.*, 2011), with further refinements to incorporate attributes of new cultivars and growing systems, and the biophysical fruit model developed for kiwifruit by Hall *et al.* (2013). This approach would hopefully avoid past issues where mechanistic studies, without interpretation at the whole-plant/orchard system level, have misled industry. For example, by failing to take a whole-vine interpretation, early studies that suggested more and bigger canes would produce higher yields did not allow for the consequences of stimulating more vegetative growth, which ultimately decreased fruit quantity and quality.

References

Anon. (2010) Actinidieto a 'pergoletta alta': caratteristiche degli impianti, la fase di allevamento e indicazioni per la potatura di produzione. In: *Potatura dei Fruittiferi. Supplemento a Coldiretti Informa* n. 125, 22 December, 11–14. Available at: https://docplayer.it/22001466-Potatura-dei-fruttiferi.html (accessed March 2022).

Antognozzi, E., Battistelli, A., Famiani, F., Moscatello, S., Stanica, F. *et al.* (1996) Influence of CPPU on carbohydrate accumulation and metabolism in fruits of *Actinidia deliciosa* (A. Chev.). *Scientia Horticulturae* 65, 37–47. DOI: 10.1016/0304-4238(95)00852-7

Antognozzi, E., Famiani, F., Proietti, F., Tombesi, A., Ferranti, F. *et al.* (1997) Effect of CPPU (cytokinin) treatments on fruit anatomical structure and quality in *Actinidia deliciosa*. *Acta Horticulturae* 444, 459–465. DOI: 10.17660/ActaHortic.1997.444.70

Basile, B., Romano, R., Giaccone, M., Barlotti, E., Colonna, V. *et al.* (2008) Use of photo-selective nets for hail protection of kiwifruit vines in southern Italy. *Acta Horticulturae* 770, 185–192. DOI: 10.17660/ActaHortic.2008.770.21

Basile, B., Giaccone, M., Shahak, Y., Forlani, M. and Cirillo, C. (2014) Regulation of the vegetative growth of kiwifruit vines by photo-selective anti-hail netting. *Scientia Horticulturae* 172, 300–307. DOI: 10.1016/j.scienta.2014.04.011

Berry, P. (1997) The establishment of kiwifruit growing in New Zealand and the development of a research programme. *Acta Horticulturae* 444, 33–36. DOI: 10.17660/ActaHortic.1997.444.1

Black, M.Z., Patterson, K.J., Minchin, P.E.H., Gould, K.S. and Clearwater, M.J. (2011) Hydraulic responses of whole vines and individual roots of kiwifruit (*Actinidia chinensis*) following root severance. *Tree Physiology* 31, 508–518. DOI: 10.1093/treephys/tpr045

Black, M.Z., Patterson, K.J., Gould, K.S. and Clearwater, M.J. (2012) Physiological responses of kiwifruit vines (*Actinidia chinensis* Planch. var. *chinensis*) to trunk girdling and root pruning. *New Zealand Journal of Crop and Horticultural Science* 40(1), 31–41. DOI: 10.1080/01140671.2011.603343

Blanchet, P. and Ellis, R. (1988) *Kiwi: Adaptation et Techniques.* SARL Publications Agricoles, Agen, France.

Brown, E. and Woolley, D.J. (2010) Timing of application and growth regulator interaction effects on fruit growth of two species of *Actinidia*. *Acta Horticulturae* 884, 107–113. DOI: 10.17660/ActaHortic.2010.884.11

Burdon, J., Lallu, N., Pidakala, P. and Barnett, A.M. (2011) Is the 6.2°Brix soluble solids harvest index suitable for 'Hayward' kiwifruit from high productivity orchard management systems? *Acta Horticulturae* 913, 539–546. DOI: 10.17660/ActaHortic.2011.913.73

Buwalda, J.G. (1991) A mathematical model of carbon acquisition and utilisation by kiwifruit vines. *Ecological Modelling* 57, 43–64. DOI: 10.1016/0304-3800(91)90054-5

Buxton, K.N. (2005) Preharvest practices affecting postharvest quality of 'Hayward' kiwifruit. PhD thesis, Massey University, Palmerston North, New Zealand.

Chiabrando, V. and Giacalone, G. (2018) Kiwifruit under plastic covering: impact on fruit quality and on orchard microclimate. *Journal of Food, Nutrition & Agriculture* 1(1), 1–6. DOI: 10.21839/jfna.v1i1.113

Childerhouse, E. (2009) The effect of a natural plant extract and synthetic growth regulators on growth, quality and endogenous hormones of *Actinidia chinensis* and *Actinidia deliciosa* fruit. MSc thesis, Massey University, Palmerston North, New Zealand.

Cieslak, M., Seleznyova, A.N. and Hanan, J. (2011) A functional-structural kiwifruit vine model integrating architecture, carbon dynamics and effects of the environment. *Annals of Botany* 107(5), 747–764. DOI: 10.1093/aob/mcq180

Clearwater, M.J., Lowe, R.G., Hofstee, B.J., Barclay, C., Mandemaker, A.J. *et al.* (2004) Hydraulic conductance and rootstock effects in grafted vines of kiwifruit. *Journal of Experimental Botany* 55, 1371–1382. DOI: 10.1093/jxb/erh137

Cooper, K.M. and Marshall, R.R. (1992) Crop loading and canopy management. *Acta Horticulturae* 297, 501–508. DOI: 10.17660/ActaHortic.1992.297.65

Costa, G. (1990) Kiwi orchard establishment: main factors in planning and management. *Acta Horticulturae* 282, 87–96. DOI: 10.17660/ActaHortic.1990.282.8

Costa, G. and Testolin, R. (2022) Recent changes in the kiwifruit (*Actinidia* spp.) industry since its introduction: Italy as a case study. *Acta Horticulturae* 1332, 1–10. DOI: 10.17660/ActaHortic.2022.1332.1

Costa, G., Succi, F., Biasi, R. and Miserocchi, O. (1990) Vegetative and cropping performance of kiwifruit (cv. Hayward) as related to pruning length and bud number. *Acta Horticulturae* 282, 113–118. DOI: 10.17660/ActaHortic.1990.282.13

Costa, G., Kukuriannis, B. and Monet, R. (1992a) Kiwifruit production in Europe. *Acta Horticulturae* 297, 141–149. DOI: 10.17660/ActaHortic.1992.297.17

Costa, G., Biasi, R., Giuliani, R. and Succi, F. (1992b) Comparison of kiwifruit training systems. *Acta Horticulturae* 297, 427–434. DOI: 10.17660/ActaHortic.1992.297.56

Costa, G., Quadretti, R., Galliano, A. and Vittone, F. (1997) Three-year trial of 2 pergola training systems for kiwifruits: traditional and Peyracchia. Confronto triennale tra due sistemi di potatura e di gestione delle pergolette di actinidia tradizionale e 'Peyracchia'. *Rivista di Frutticoltura e di Ortofloricoltura* 61(5), 67–72.

Costa, G., Ferguson, R., Huang, H. and Testolin, R. (2018) Main changes in the kiwifruit industry since its introduction: present situation and future. *Acta Horticulturae* 1218, 1–16. DOI: 10.17660/ActaHortic.2018.1218.1

CReSO (2010) Actinidieto a 'pergoletta alta': caratteritiche degli impianti, la fase di allevamento e indicazioni per la potatura di produzione. In *Potatura dei Fruittiferi. Supplemento a Coldiretti Informa* n. 125 del 22 December, 11–14. Available at: https://docplayer.it/22001466-Potatura-dei-fruttiferi.html (accessed March 2022).

Cruz-Castillo, J.G., Woolley, D.J. and Lawes, G.S. (1999) Effects of CPPU and other plant growth regulators on fruit development in kiwifruit. *Acta Horticulturae* 498, 173–178. DOI: 10.17660/ActaHortic.1999.498.20

Cruz-Castillo, J.G., Baldicchi, A., Frioni, T., Marocchi, F., Moscatello, S. *et al.* (2014) Pre-anthesis CPPU low dosage application increases 'Hayward' kiwifruit weight without affecting the other qualitative and nutritional characteristics. *Food Chemistry* 158, 224–228. DOI: 10.1016/j.foodchem.2014.01.131

Currie, M.B. (1997) Source-sink relations in kiwifruit: carbohydrate and hormone effects on fruit growth at the cell, organ and whole plant level. PhD thesis, Massey University, Palmerston North, New Zealand.

Currie, M.B., Patterson, K.J., Dawson, T., Ramankutty, P. and Blattmann, P. (2011) Careful use of NAA pruning gels is needed to avoid compromising 'Hort16A' fruit quality. *Acta Horticulturae* 913, 327–335. DOI: 10.17660/ActaHortic.2011.913.43

Currie, M.B., Patterson, K.J., Snelgar, W.P. and Blattmann, P. (2018) Girdling kiwifruit vines for commercial advantage: opportunities and risks. *Acta Horticulturae* 1218, 405–412. DOI: 10.17660/ActaHortic.2018.1218.56

Davison, R.M. (1990) The physiology of the kiwifruit vine. In: Warrington, I.J. and Weston, G.C. (eds) *Kiwifruit: Science and Management*. Ray Richards Publisher in association with the New Zealand Society for Horticultural Science, Auckland, New Zealand, pp.127–154.

Famiani, F., Antognozzi, E., Tombesi, A., Moscatello, S. and Battistelli, A. (1998) CPPU-induced alterations in source-sink relationships in *Actinidia deliciosa*. *Acta Horticulturae* 463, 303–310. DOI: 10.17660/ActaHortic.1998.463.38

Goren, R., Huberman, M. and Goldschmidt, E.E. (2004) Girdling: physiological and horticultural aspects. *Horticultural Reviews* 30, 1–36. DOI: 10.1002/9780470650837.ch1

Grant, J.A. and Ryugo, K. (1984) Influence of within-canopy shading on fruit size, shoot growth, and return bloom in kiwifruit. *Journal of the American Society for Horticultural Science* 109(6), 799–802. DOI: 10.21273/JASHS.109.6.799

Hall, A.J., Minchin, P.E.H., Clearwater, M.J. and Génard, M. (2013) A biophysical model of kiwifruit (*Actinidia deliciosa*) berry development. *Journal of Experimental Botany* 64(18), 5473–5483. DOI: 10.1093/jxb/ert317

Hennion, B. (2003) *Le Kiwi*. Ctifl Monographie, Paris. Available at: https://lafontaine.groupe-esa.com/index.php?lvl=notice_display&id=59374 (accessed 14 December 2022).

Henzell, R.F., Briscoe, M.R. and Lauren, D.R. (1986) Evaluation of two plant growth regulators as chemical pruning agents for kiwifruit vines in summer. *New Zealand Journal of Experimental Agriculture* 14(2), 199–203. DOI: 10.1080/03015521.1986.10426143

Henzell, R.F., Briscoe, M.R. and Gravett, I. (1992) Improving kiwifruit vine productivity with plant growth regulators. *Acta Horticulturae* 297, 345–350. DOI: 10.17660/ActaHortic.1992.297.46

Huang, F.W. (2013) Effect of bagging on growth characteristics, quality and storage of Hayward kiwifruit. MSc thesis, Northwest Agriculture and Forestry University, Xian, China.

Huang, H. (2016) *Kiwifruit: The Genus ACTINIDIA*. Elsevier, London.

Iliescu, L.M., Alexandru, C.M. and Stănică, F. (2021) 'Z1 Vitroplant' – valuable rootstock for kiwifruit cultivars – grafting results. *Scientific Papers Series B, Horticulture* 65(1), 99–110.

Iwahori, S., Tominaga, S. and Yamasaki, T. (1988) Stimulation of fruit growth of kiwifruit, *Actinidia chinensis* Planch., by N-(2-chloro-4-pyridyl)-N'-phenylurea, a diphenylurea-derivative cytokinin. *Scientia Horticulturae* 35, 109–115. DOI: 10.1016/0304-4238(88)90042-8

Jo, Y.S., Cho, H.S., Park, M.Y., Park, J.O., Park, T.D. *et al.* (2003) Comparison of CPPU effect on fruit development in several *Actinidia* species. *Acta Horticulturae* 610, 539–543. DOI: 10.17660/ActaHortic.2003.610.72

Kim, G.H., Jung, J.S. and Koh, Y.J. (2017) Occurrence and epidemics of bacterial canker of kiwifruit in Korea. *Plant Pathology Journal* 33(4), 351–361. DOI: 10.5423/PPJ.RW.01.2017.0021

Kim, J.G., Takami, Y., Mizugami, T., Beppu, K., Fukuda, T. *et al.* (2006) CPPU application on size and quality of hardy kiwifruit. *Scientia Horticulturae* 110, 219–222. DOI: 10.1016/j.scienta.2006.06.017

Kramer, M., Snelgar, P., Kramer-Walter, K., Blattmann, P. and McKenzie, C. (2022) The effect of reflective cloth on light distribution and fruit quality in a kiwifruit orchard. *Acta Horticulturae* 1332, 415–422. DOI: 10.17660/ActaHortic.2022.1332.55

Kurosaki, T. and Mochizuki, T. (1990) Effect of KT-30 treatment on fruit growth and some components of 'Monty' kiwifruit. *Journal of the Japanese Society for Horticultural Science* 59(1), 43–50. DOI: 10.2503/jjshs.59.43

Lai, R., Woolley, D.J. and Lawes, G.S. (1990) The effect of inter-fruit competition, type of fruiting lateral and time of anthesis on the fruit growth of kiwifruit (*Actinidia deliciosa*). *Journal of Horticultural Science* 65(1), 87–96. DOI: 10.1080/00221589.1990.11516034

Lawes, G.S., Woolley, D.J. and Cruz-Castillo, J.G. (1992) Field responses of kiwifruit to CPPU (cytokinin) application. *Acta Horticulturae* 297, 351–356. DOI: 10.17660/ActaHortic.1992.297.47

Li, S. (1990) Kiwi fruit resources and production in China. *Acta Horticulturae* 282, 57–64. DOI: 10.17660/ActaHortic.1990.282.5

Li, X.-X., Kobayashi, F., Ikeura, H. and Hayata, Y. (2011) Chlorophenoxyacetic acid and chloropyridylphenylurea accelerate translocation of photoassimilates to parthenocarpic and seeded fruits of muskmelon (*Cucumis melo*). *Journal of Plant Physiology* 168, 920–926. DOI: 10.1016/j.jplph.2010.11.005

Manning, M.A., Pak, H.A. and Beresford, R.M. (2010) Non-fungicidal control of botrytis storage rots in New Zealand kiwifruit through pre- and postharvest crop management. In: Prusky, D. and Gullino, M.L. (eds) *Post-harvest Pathology. Plant Pathology in the 21st Century*, Vol. 2. Springer Science, Dordrecht, The Netherlands, pp. 183–196. DOI: 10.1007/978-1-4020-8930-5_13

Massai, R., Piccotino, D., Raffaelli, M., Xiloyannis, C. and Baroni, G. (1997) Growth and yield of kiwifruit vines planted at different distances along the row. *Acta Horticulturae* 444, 169–174. DOI: 10.17660/ActaHortic.1997.444.24

McAneney, K.J., Judd, M.J. and Trought, M.C.T. (1984) Wind damage to kiwifruit (*Actinidia chinensis* Planch.) in relation to windbreak performance. *New Zealand Journal of Agricultural Research* 27, 255–263. DOI: 10.1080/00288233.1984.10430427

Meinhold, T., Damerow, L. and Blanke, M. (2011) Reflective materials under hailnet improve orchard light utilisation, fruit quality and particularly fruit colouration. *Scientia Horticulturae* 127, 447–451. DOI: 10.1016/j.scienta.2010.09.006

Michailides, T.J. and Elmer, P.A.G. (2000) Botrytis gray mold of kiwifruit caused by *Botrytis cinerea* in the United States and New Zealand. *Plant Disease* 84(3), 208–223. DOI: 10.1094/PDIS.2000.84.3.208

Miller, S.A., Broom, F.D., Thorp, T.G. and Barnett, A.M. (2001) Effects of leader pruning on vine architecture, productivity and fruit quality in kiwifruit (*Actinidia deliciosa* cv. Hayward). *Scientia Horticulturae* 91(3-4), 189–199. DOI: 10.1016/S0304-4238(01)00259-X

Minchin, P.E.H., Snelgar, W.P., Blattmann, P. and Hall, A.J. (2010) Competition between fruit and vegetative growth in Hayward kiwifruit. *New Zealand Journal of Crop and Horticultural Science* 38(2), 101–112. DOI: 10.1080/01140671003781728

Mupambi, G., Anthony, B.M., Layne, D.R., Musacchi, S., Serra, S. *et al.* (2018) The influence of protective netting on tree physiology and fruit quality of apple: a review. *Scientia Horticulturae* 236, 60–72. DOI: 10.1016/j.scienta.2018.03.014

Nardozza, S., Boldingh, H.L., Richardson, A.C., Costa, G., Marsh, H. *et al.* (2010) Variation in carbon content and size in developing fruit of *Actinidia deliciosa* genotypes. *Functional Plant Biology* 37, 545–554. DOI: 10.1071/fp09301

Ohara, H., Kato, M., Matsui, H., Hirata, N. and Takahashi, E. (1997) Comparison of the levels of endogenous plant growth substances in CPPU-treated and -untreated kiwifruit. *Journal of the Japanese Society for Horticultural Science* 65(4), 693–705. DOI: 10.2503/jjshs.65.693

Overbeck, V., Blanke, M.M. and Schmitz-Eiberger, M.A. (2014) Healthier fruit with Extenday®? *Acta Horticulturae* 1058, 381–388. DOI: 10.17660/actahortic.2014.1058.46

Patterson, K.J. and Currie, M.B. (2011) Optimizing kiwifruit vine performance for high productivity and superior fruit taste. *Acta Horticulturae* 913, 257–268. DOI: 10.17660/actahortic.2011.913.33

Patterson, K.J., Mason, K.A. and Gould, K.S. (1993) Effects of CPPU (N-(2-chloro-4-pyridyl)-N'-phenylurea) on fruit growth, maturity, and storage quality of kiwifruit. *New Zealand Journal of Crop and Horticultural Science* 21(3), 253–261. DOI: 10.1080/01140671.1993.9513777

Richardson, A.C. and McAneney, K.J. (1990) Influence of fruit number on fruit weight and yield of kiwifruit. *Scientia Horticulturae* 42, 233–241. DOI: 10.1016/0304-4238(90)90085-s

Richardson, A.C., McAneney, K.J. and Dawson, T.E. (1997) Carbohydrate dynamics in kiwifruit. *Journal of Horticultural Science* 72, 907–917. DOI: 10.1080/14620316.1997.11515583

Richardson, A.C., Boldingh, H., Kashuba, P., Knight, G. and Ellingham, D. (2019) Flowering time determines the weight and composition *of Actinidia chinensis* var. *chinensis* 'Zesy002' kiwifruit. *Scientia Horticulturae* 246, 741–748. DOI: 10.1016/j.scienta.2018.11.043

Richardson, A.C., Eyre, V., Rebstock, R., Popowski, E. and Nardozza, S. (2022) Factors influencing flower development in kiwifruit vines. *Acta Horticulturae* 1332, 141–153 DOI: 10.17660/ActaHort.2022.1332.20

Sale, P.R. (1992) The evolution of kiwifruit production methods. *Acta Horticulturae* 297, 43–50. DOI: 10.17660/ActaHortic.1992.297.2

Sale, P.R. and Lyford, P.B. (1990) Cultural, management and harvesting practices for kiwifruit in New Zealand. In: Warrington, I.J. and Weston, G.C. (eds) *Kiwifruit: Science and Management*. Ray Richards Publisher in association with the New Zealand Society for Horticultural Science, Auckland, New Zealand, pp. 247–296.

Sale, P.R. and Williams, D.A. (1985) *Kiwifruit Culture*, 2nd edn. Government Printer, Wellington.

Schmidt, T., Hanrahan, I., Castillo, F. and McFerson, J. (2014) Reflective ground covers increase yields of fruit trees. *Acta Horticulturae* 1058, 313–319. DOI: 10.17660/actahortic.2014.1058.37

Schupp, J.R. and Ferree, D.C. (1990) Influence of time of root pruning on growth, mineral nutrition, net photosynthesis and transpiration of young apple trees. *Scientia Horticulturae* 42, 299–306. DOI: 10.1016/0304-4238(90)90053-h

Seleznyova, A.N., Thorp, T.G., Barnett, A.M. and Costes, E. (2002) Quantitative analysis of shoot development and branching patterns in *Actinidia*. *Annals of Botany* 89, 471–482. DOI: 10.1093/aob/mcf069

Sharma, R.R., Reddy, S.V.R. and Jhalegar, M.J. (2014) Pre-harvest fruit bagging: a useful approach for plant protection and improved post-harvest fruit quality – a review. *Journal of Horticultural Science and Biotechnology* 89(2), 101–113. DOI: 10.1080/14620316.2014.11513055

Shim, K.-K. and Ha, Y.-M. (1999) Kiwifruit production and research in Korea. *Acta Horticulturae* 498, 127–131. DOI: 10.17660/actahortic.1999.498.14

Smith, G.S. and Buwalda, J.G. (1994) Kiwifruit. In: Schaffer, B. and Anderson, P.C. (eds) *Handbook of Environmental Physiology of Fruit Crops*. CRC Press, Boca Raton, Florida, pp. 135–163. DOI: 10.1201/9780203719299

Smith, G.S., Miller, S.A. and Curtis, J.P. (1997) Location of fruit with superior characteristics in the canopy of kiwifruit vines. *Acta Horticulturae* 444, 193–198. DOI: 10.17660/ActaHortic.1997.444.28

Snelgar, W.P. and Martin, P.J. (1997) Relationship between leaf area index and fruit size in kiwifruit. *Acta Horticulturae* 444, 199–204. DOI: 10.17660/actahortic.1997.444.29

Snelgar, W.P., Thorp, T.G. and Patterson, K.J. (1986) Optimal leaf:fruit ratios for fruit growth in kiwifruit. *Acta Horticulturae* 175, 115–120. DOI: 10.17660/actahortic.1986.175.17

Snelgar, W.P., Manson, P.J. and Stowell, B.M. (1992) Relationship between time of shoot growth, shoot stubbing, and return bloom of kiwifruit vines. *New Zealand Journal of Crop and Horticultural Science* 20(3), 345–350. DOI: 10.1080/01140671.1992.10421777

Snelgar, W.P., Hopkirk, G., Seeyle, R.J., Martin, P.J. and Manson, P.J. (1998) Relationship between canopy density and fruit quality of kiwifruit. *New Zealand Journal of Crop and Horticultural Science* 26, 223–232. DOI: 10.1080/01140671.1998.9514058

Testolin, R. (2015) Kiwifruit (*Actinidia* spp.) in Italy: the history of the industry, international scientific cooperation and recent advances in genetics and breeding. *Acta Horticulturae* 1096, 47–61. DOI: 10.17660/ActaHortic.2015.1096.2

Thorp, T.G. and Barnett, A.M. (2011) Increasing productivity of alternate year cropping systems for Zespri™ Gold kiwifruit ('Hort16A'). *Acta Horticulturae* 913, 321–326. DOI: 10.17660/ActaHortic.2011.913.42

Thorp, T.G., Barnett, A.B. and Toye, J.D. (2001) Harvesting light in persimmon and kiwifruit orchards with reflective ground covers. *Acta Horticulturae* 557, 363–368. DOI: 10.17660/actahortic.2001.557.48

Thorp, T.G., Barnett, A.M. and Miller, S.A. (2003a) Effects of cane size and pruning system on shoot growth, flowering and productivity of 'Hayward' kiwifruit vines. *Journal of Horticultural Science and Biotechnology* 78(2), 219–224. DOI: 10.1080/14620316.2003.11511609

Thorp, T.G., Ferguson, I.B., Boyd, L.M. and Barnett, A.M. (2003b) Fruiting position, mineral concentration and incidence of physiological pitting in 'Hayward' kiwifruit. *Journal of Horticultural Science and Biotechnology* 78(4), 505–511. DOI: 10.1080/14620316.2003.11511655

Thorp, T.G., Barnett, A.M. and Blattmann, P. (2011) What are the productivity limits for 'Hayward' kiwifruit? *Acta Horticulturae* 913, 419–424. DOI: 10.17660/ActaHortic.2011.913.56

Thorp, T.G., Barnett, A.M., Blattmann, M., Hedderley, D. and Mayston, R. (2018) Optimising vine management to increase yields and improve quality of Zespri™ SunGold kiwifruit (*Actinidia chinensis* var. *chinensis* 'Zesy002'). *Acta Horticulturae* 1218, 397–404.DOI: 10.17660/ActaHortic.2018.1218.55

Thorp, T.G., Barnett, A.M., Civolani, C., Spinelli, R., Lallu, N. *et al.* (2021) Iron chelate applications improve productivity and fruit quality at harvest and during cold storage of *Actinidia chinensis* var. *chinensis* ('Hort16A') kiwifruit growing in high-pH calcareous soils. *European Journal of Horticultural Science* 86(6), 609–619. DOI: 10.17660/eJHS.2021/86.6.4

Tombesi, A., Antognozzi, E. and Palliotti, A. (1994) Optimum leaf area index in T-bar trained kiwifruit vines. *Journal of Horticultural Science* 69, 339–350. DOI: 10.1080/14620316.1994.11516463

Vittone, G. (2010) La qualità inizia dalla potatura: tecniche e suggerimenti per l`actinidia. *Rivista di Frutticoltura e di Ortofloricoltura* 74(3), 64–66.

Vittone, G., Franco, E., Astesano, B., Peyracchia, G. and Peyracchia, B. (1996) Pruning kiwifruits by the Peyracchia method. La potatura dell'actinidia con il metodo 'Peyracchia'. *Informatore Agrario* 52(43), 61–66.

Volz, R.K., Gibbs, H.M. and Lupton, G.B. (1992) Variation in fruitfulness among kiwifruit replacement canes. *Acta Horticulturae* 297, 443–449. DOI: 10.17660/ActaHortic.1992.297.58

Walton, E.F. and Fowke, P.J. (1993) Effect of hydrogen cyanamide on kiwifruit shoot flower number and position. *Journal of Horticultural Science* 68(4), 529–534. DOI: 10.1080/00221589.1993.11516381

Woolley, D.J. and Cruz-Castillo, J.G. (2006) Stimulation of fruit growth of green and gold kiwifruit. *Acta Horticulturae* 727, 291–293. DOI: 10.17660/actahortic.2006.727.34

Woolley, D.J., Lawes, G.S. and Cruz-Castillo, J.G. (1992) The growth and competitive ability of *Actinidia deliciosa* 'Hayward' fruit: carbohydrate availability and response to the cytokinin-active compound CPPU. *Acta Horticulturae* 297, 467–473. DOI: 10.17660/actahortic.1992.297.61

Zhang, J., Scarrow, S. and Barnes, H.T. (2004) Model of the effects of a range of shelter orientations on shading patterns in kiwifruit (*Actinidia* sp.) blocks. *New Zealand Journal of Crop and Horticultural Science* 32, 407–417. DOI: 10.1080/01140671.2004.9514322

Zhao, X.-Y. and Xie, D.-F. (2018) Review on the influences of bagging treatment on pesticide residue in fruits. *Journal of Agricultural Resources and Environment* 35(2), 104–110.

Zuccherelli, G. (1994) *L'Actinidia e i Nuovi Kiwi*. Fruitticoltura Moderna. Edaricole – Edizioni Agricole, Bologna, Italy.

13 Kiwifruit Pollination

Melissa A. Broussard* and David E. Pattemore
The New Zealand Institute for Plant and Food Research Ltd, Hamilton, New Zealand

13.1 Introduction

Pollination is critical for achieving fruit set for members of the genus *Actinidia*, and the amount of pollination a flower receives is linked to the marketable quality of the resulting fruit. There are several key features of this genus that influence the likelihood of successful pollination: the relatively small number of flowers per plant; the need for pollen to be moved between plants as a result of dioecy (see Chapter 3); the lack of nectar production; and the high number of ovules in many species (over 1000 for some *A. chinensis* cultivars), which means that many pollen grains have to be transferred to achieve full fruit size.

13.2 Pollination Biology in *Actinidia*

Kiwifruit (and the genus *Actinidia* by extension) was long considered to have a 'deceptive' pollination system, tricking bees into collecting pollen grains from the pistillate flowers (Fig. 13.1). 'Pistillate pollen' from female plants, lacking a protoplast, was assumed to be of little nutritional value (Schmid, 1978) being lower in carbohydrates and protein (1 g kg^{-1} and 14%, respectively) than 'staminate pollen' from male vines (42–60 g kg^{-1} and 34–41%). However, pistillate pollen is significantly higher than staminate pollen in calcium (2.5–3.5×), manganese (5.0–6.9×) and iron (2.0–6.7×), as well as lipids (3.5–4.1×) (Clark and Lintas, 1992). Lipids are a preferred source of energy for honey bees (Manning, 2001). The ratio between protein and lipids may drive pollinator foraging behaviour (Vaudo *et al.*, 2020) and there may be advantages in offering two classes of pollen with different macronutrient profiles.

13.2.1 Floral production and phenology

The number of flowers is affected by cultivar, climate and management (see Chapter 14) (Ferguson, 1991). Male vines of *A. chinensis* and *A. arguta* produce 2–10 times the number of flowers of female vines owing to a combination of genetics and pruning strategies (Palmer-Jones *et al.*, 1976; Ferguson, 1991; Testolin, 1991).

Flowering duration in *Actinidia* genotypes varies between 4 and 35 days (Brundell, 1975; González *et al.*, 1994; Howpage, 1999; McPherson *et al.*, 2001a), depending on local climate and the use of budbreak enhancers. Commercial cultivars with shorter flowering windows (5–12 days) are generally preferred as fruit maturation

*Corresponding author: melissa.broussard@plantandfood.co.nz

© CAB International 2024. *Kiwifruit: Botany, Production and Uses* (eds A.C. Richardson, J.N. Burdon and A.R. Ferguson)
DOI: 10.1079/9781800620933.0013

Fig. 13.1. (*Left*) Pistillate ('female') and (*right*) staminate ('male') flowers of *Actinidia chinensis* var. *chinensis*. (Photos: H. McBrydie, The New Zealand Institute for Plant and Food Research Ltd)

is then more uniform. However, shorter flowering windows increase the chance of a mismatch between the flowering periods of male and female cultivars.

Flowers of both sexes tend to open between midnight and noon, depending on the climate and weather (Goodwin, 1987; Thakur and Rathore, 1991; Goodwin *et al.*, 2013; Mo *et al.*, 2016). In warmer climates, such as India, flowers are observed to open earlier – between midnight and 06:00 (Thakur and Rathore, 1991) – whereas in cooler climates, such as in New Zealand, flowers open between 05:00 and noon (Goodwin, 1987; Goodwin *et al.*, 2013). Rain has been shown to delay both flower opening and pollen release in several *Actinidia* species (Goodwin, 1987; Mo *et al.*, 2016).

13.2.2 Pollen production and viability

Male kiwifruit flowers produce large numbers of pollen grains which are binucleate, tricolporate and between 15 μm and 25 μm in diameter (Huang, 2016). Pollen production is widely variable between *Actinidia* species and cultivars – ranging from ~500,000 pollen grains per flower to over 3 million in *A. chinensis* var. *deliciosa* (Hopping, 1986) and from 1.4 million to 4 million in *A. arguta* (Stasiak *et al.*, 2020).

The maximum amount of pollen is available at the popcorn (balloon) stage (Fig. 13.2), just before anthesis (Oh *et al.*, 2020). Not all pollen is immediately available when the flower opens. For *A. chinensis* var. *deliciosa* (male and female cultivars), ~50% of pollen is released within 24 h of the petals beginning to unfold (Goodwin, 1987; Hopping, 1990; Chen, Y. *et al.*, 1996), with the remainder released over the subsequent 1–3 days, following an exponential decay pattern (Goodwin, 1987). Daily pollen release can be measured by airborne pollen and is highest between 09:00 and 13:00, roughly corresponding to the period of highest insect activity (McKay, 1978; Malabœuf *et al.*, 1997). Release continues throughout the day at low levels. Pollen can be released earlier through explosive dehiscence induced by vibration at frequencies of 150–900 Hz, such as from the buzzing of bumble bees, e.g. *Bombus terrestris* (179–374 Hz) and carpenter bees, *Xylocopa* spp. (117–346 Hz) (Corbet *et al.*, 1988; King and Ferguson, 1994; King and Buchmann, 2003).

Kiwifruit pollen from staminate plants is generally 80–100% viable (Hopping, 1986; Chen, Y. *et al.*, 1996; Stasiak *et al.*, 2017). Maximum viability occurs at flower opening (Oh *et al.*, 2020) and is only slightly lower several days before anthesis. Environmental factors and inclement weather can reduce kiwifruit pollen viability. It decreases sharply above 30°C (Jansson and Warrington, 1988; Polito and Weinbaum, 1988) and under high humidity (Hopping, 1990; Pyke and Alspach, 1991). Vine

Fig. 13.2. Popcorn-stage *Actinidia chinensis* var. *deliciosa* flower bud. (Photo: H. McBrydie, The New Zealand Institute for Plant and Food Research Ltd)

age can also affect maximum pollen viability, with pollen from young vines having lower viability and a decreased range of temperatures at which pollen is viable compared with older vines (Polito and Weinbaum, 1988).

13.2.3 Effective pollination period

Pistillate kiwifruit flowers can be successfully pollinated throughout the day (Broussard *et al.*, 2021); pollination is not limited to the morning period of high pollen availability. After landing on the stigma, pollen tubes penetrate the style within 7 h (Hopping and Jerram, 1979) and reach the ovary within 74–96 h (Hopping and Jerram, 1979; González *et al.*, 1997).

Stigma viability decreases in the days following anthesis (González *et al.*, 1995a, b). For *A. chinensis* var. *deliciosa* 'Hayward', higher temperatures reduce the window during which pollination results in reasonable (> 50%) fruit set. This window varies from eight days after opening in New Zealand (Hopping, 1990) to four days in Spain and Italy (González *et al.*, 1995b; Gianni and Vania, 2018) and three days in India (Thakur and Rathore, 1991).

13.2.4 Xenia

The source of pollen and its genetic characteristics can affect fruit traits such as fruit set, seed number, size/shape, flesh colour, time to maturity and fruit quality in *Actinidia*, a phenomenon referred to as xenia (Chen, Q. *et al.*, 1996; Qi *et al.*, 2007; Seal *et al.*, 2013a, b, 2016, 2017, 2018; Li *et al.*, 2016; Chen *et al.*, 2019; Stasiak *et al.*, 2019; Oh *et al.*, 2021).

Xenia effects are often variable, particularly when they affect attributes influenced by environment and management factors. For example, fruit soluble solids and dry matter contents respond in some cases (Chen, Q. *et al.*, 1996; Qi *et al.*, 2007; Chen *et al.*, 2019; Stasiak *et al.*, 2019), but not in others (Seal *et al.*, 2017; Oh *et al.*, 2021). Vitamin C content and fruit acidity varied with pollen parent in *A. chinensis* var. *chinensis*, *A. chinensis* var. *deliciosa* and *A. chinensis* × *A. eriantha* hybrids (Chen, Q. *et al.*, 1996; Chen *et al.*, 2019), but not in every case (Oh *et al.*, 2021), and not in *A. arguta* (Stasiak *et al.*, 2019). Some studies have found fruit firmness to change with different pollen sources (Qi *et al.*, 2007; Seal *et al.*, 2013a; Oh *et al.*, 2021), as can fruit storage life (Qi *et al.*, 2007) and date to maturity (Seal *et al.*, 2013b). Fruit colour is often affected, particularly with red-fleshed cultivars of *A. chinensis* var. *chinensis* (Seal *et al.*, 2013b) and *A. arguta* (Stasiak *et al.*, 2019).

13.3 Dependency of Kiwifruit on Pollination

Kiwifruit was classified by Klein *et al.* (2007) as being one of a handful of agricultural crops for which pollination is essential. Because of its dioecious nature, and the complete non-viability of pollen from female vines, pollen must travel between plants in order for pollination to be successful.

13.3.1 Pollen requirement (dose-response relationship)

Kiwifruit has a high fruit-setting rate – often above 90% – and, under the right conditions, can achieve 100% fruit set. High fruit set has negative implications for commercial fruit production as vines, particularly of *A. chinensis*, set 'under-pollinated', unmarketable fruit which are half the size of fully pollinated fruit (González *et al.*, 1998; Gardi *et al.*, 2003; Goodwin *et al.*, 2013).

In kiwifruit, fruit weight is not as good an indicator of commercial pollination success as in other crops. Seed numbers are a more reliable indicator (Fig. 13.3) as environmental conditions, crop load, vine age and many other factors can affect fruit weight and seed set is not always limiting (Hopping, 1986; Lawes *et al.*, 1990; Richardson and McAneney, 1990; Costa *et al.*, 1993; Naik, 2011). Assessments of the pollination requirement of plant genotypes are thus ideally measured as the number of pollen grains required for full seed set (González *et al.*, 1994).

A. chinensis var. *chinensis* 'Zesy002' requires fewer pollen grains than 'Hayward' to achieve full seed set and fruit size (3500–4000 vs. 10,000–13,250) (Broussard *et al.*, 2021). This difference may be due to 'Hayward' having more ovules, as indicated by 'Hayward' fruit producing more seed (1500–2000; Hopping, 1986; Costa *et al.*, 1993) than 'Zesy002' (~650; Seal *et al.*, 2018).

Fig. 13.3. Seed set is a better measure of pollination in *Actinidia* spp. than fruit weight, as illustrated here with two *A. chinensis* var. *chinensis* fruit (*top and bottom*) that differ markedly in seed set (indicated by the presence of black seeds) but are similar in size. (Photo: H. McBrydie, The New Zealand Institute for Plant and Food Research Ltd)

The pollen requirement of flowers of different *Actinidia* taxa is probably highly variable: fruit of cultivars of *A. chinensis* var. *deliciosa* have anywhere from 500–700 seeds (Goodwin *et al.*, 2013; Abbate *et al.*, 2021; Broussard *et al.*, 2021); *A. arguta* fruit from 110–195 seeds (Stasiak *et al.*, 2021), though up to 260 have been reported for some tetraploid selections (Mizugami *et al.*, 2007); *A. eriantha* from 210–978 seeds (Ke *et al.*, 1991); and *A. chrysantha* from 98–314 seeds (Mo *et al.*, 2016).

Flowers can receive too much pollen. Too many bees visiting a flower reduces the seed number, but not fruit weight (Palmer-Jones *et al.*, 1976). Repeated hand pollination has the same effect (Hopping, 1990), suggesting that pollen tube competition may reduce seed set at high pollen loads.

13.3.2 Syncarpy

For a kiwifruit flower to be fully pollinated, every ovule must be reached by a pollen tube. Even if pollen deposition is restricted to only some stigmata, pollen tubes are evenly distributed around the ovary resulting in an even distribution of seeds (Habart,1974; McKay, 1978; Howpage *et al.*, 1998). In 'Hayward', it is possible to fully pollinate the flower by depositing pollen on a few stigmata instead (Howpage *et al.*, 1998). Kiwifruit flowers appear to be perfectly syncarpous. In some fruit crops, mis-shapen fruit have been attributed to poor pollination, which could occur if pollen was not uniformly applied across stigmata, and each stigma was linked only to a discrete set of carpels. This does not appear to be the case with kiwifruit: dropped shoulders (uneven expansion of the fruit at the stem end) seem to be due to infection by *Pseudomonas viridiflava* (blossom rot) (Haslam *et al.*, 1988).

13.3.3 Pollination, fruit size and quality

Earlier, higher-quality flowers have a greater capacity to become larger fruit (McKay,1978; Lai *et al.*, 1990; McPherson *et al.*, 2001b). In general, seed number is correlated with fruit size (linearly or logarithmically depending on the range of seed numbers) for *A. chinensis* (Clinch, 1984; Hopping, 1986; Testolin, 1991), and similar trends have been observed for *A. arguta* (Tiyayon and Strik, 2002). A number of other factors can also affect fruit size and quality independently of pollination (irrigation, weather, disease, crop load, etc.), which is why seed numbers remain the gold standard for assessing pollination success.

Fruit quality may also be improved by better pollination, with sugar concentration increasing and acidity decreasing with increasing seed number (Howpage *et al.*, 2001; Buxton, 2005; Seal *et al.*, 2018; Lee *et al.*, 2019). Increasing seed number also increases fruit calcium content (Ainalidou *et al.*, 2015).

13.3.4 Wind versus insect pollination

Insect vectors are more important than wind in achieving full pollination. Early studies considered that perhaps *Actinidia* flowers were adapted to wind pollination due to a number of floral characteristics including lack of nectar (Ford, 1971; Jones, 1975; Craig *et al.*, 1988), pollen lacking pollenkitt (a sticky covering on pollen grains; Jones, 1975; Craig *et al.*, 1988) and high pollen-to-ovule ratios (Craig *et al.*, 1988). However, subsequent investigations have consistently found that seed numbers are higher and fruit is larger when *A. chinensis* flowers receive insect pollination (Testolin, 1991; González *et al.*, 1998; Abbate *et al.*, 2021), leading these authors to conclude that insect vectors are more important than wind in achieving full pollination.

Insect pollination is also important for wild *Actinidia* species; wind pollination alone results in less fruit set for *A. polygama* (Kawagoe and Suzuki, 2004) and smaller fruit for *A. chrysantha* (Mo *et al.*, 2016) versus insect pollination. For wild plants, wind pollination may act as a 'back-up' in case of low insect pollination.

13.4 Insect Pollinators of Kiwifruit

Over 200 species of insects have been recorded as visiting *Actinidia* flowers. Wild plants have been surveyed only occasionally, but there have been many studies of pollinators in cultivated kiwifruit orchards.

As *Actinidia* flowers do not produce nectar, the viable pollen from male plants and cytoplasm-less pollen from female plants are the only rewards for insect visitors. Many *Actinidia* species have floral odours, with *A. polygama* having particularly strongly scented flowers (Ferguson, 1991). Flower odours have been characterized for *A. chinensis* var. *chinensis* (Tatsuka *et al.*, 1990; Twidle *et al.*, 2018), *A. chinensis* var. *deliciosa* (Nieuwenhuizen *et al.*, 2010; Twidle *et al.*, 2015, 2017) and *A. arguta* (Matich *et al.*, 2003) and include numerous compounds attractive to bees (Twidle *et al.*, 2015, 2017, 2018). Insects, particularly honey bees, can differentiate between staminate and pistillate flowers without landing (Goodwin and Steven, 1993), presumably because of these volatiles.

Flower visitation occurs primarily in the morning, when pollen is available, with most insect activity between 07:00 and 14:00 for *A. chinensis* (McKay, 1978; Macfarlane and Ferguson, 1983; Miñarro and Twizell, 2015; Broussard *et al.*, 2022). Visits to *A. chrysantha* also align with pollen availability, primarily between 09:00 and 12:00, but in rainy weather these can be delayed to 10.00–14.00 (Mo *et al.*, 2016).

13.4.1 Pollination of wild *Actinidia*

Wild *Actinidia* species occur over a wide geographical area – from the far east of Russia down through Korea, Japan, China and southeast Asia (Huang, 2016), alongside a great diversity of insect species. While pollinators of *Actinidia* were examined in the USA and New Zealand in the 1970s and 1980s (McKay, 1978; Macfarlane and Ferguson, 1983), it was not until 1988 that the first known survey of pollinators of *Actinidia chinensis* var. *chinensis*, *A. chinensis* var. *deliciosa* and *A. arguta* was conducted in their native range in Hunan. In this survey, D. Steven found more than 45 species visiting flowers, with *Apis cerana* and *Apis mellifera* the most common (Yang and Wu, 1990). Other visitors included *Xylocopa appendiculata*, *Bombus atripes*, *B. ignitus*, *Ceratina hieroglyphica*, *Osmia pedicornis* and *Lasioglossum* spp.

Other data on wild pollinators are limited. In Japan, *A. polygama* was visited primarily by syrphid flies, followed by bumble bees and solitary bees (Kawagoe and Suzuki, 2004). In China, *A. chrysantha* was visited by bees, syrphid flies, rhiniid flies, planthoppers, crane flies and mosquitoes (Mo *et al.*, 2016).

13.4.2 Managed pollination

Managed kiwifruit pollination is largely achieved with the western honey bee (*Apis mellifera*), ironically the only *Apis* species whose native range does not overlap with *Actinidia* (Han *et al.*, 2012). Managed bumble bees (*Bombus* spp.) have also been used to successfully pollinate kiwifruit (Cutting *et al.*, 2018). Apart from social Apidae, there has been limited use of managed solitary bees. A bee from the family Megachilidae, *Megachile rotundata* (alfalfa leafcutting bee), was unsuccessfully trialled for pollination in New Zealand. Out of 24,000 bees released in kiwifruit orchards, only three males visited kiwifruit flowers, and no kiwifruit pollen was collected in nest cells (Donovan and Read, 1988). Even when caged with kiwifruit vines, *M. rotundata* foraged exclusively on understory weeds (Asteraceae, Fabaceae). An unspecified megachilid bee in California was observed to visit flowers, but did not contact stigmatic surfaces (McKay, 1978). In contrast, several *Osmia* species (particularly *O. cornifrons*) were found to perform adequate pollination services in Japanese orchards (Maeta *et al.*, 2009a).

Honey bee pollination

Honey bees (*Apis* spp.) are the primary flower visitors of kiwifruit in China (Yang and Wu, 1990; Zhu *et al.*, 2003), Japan (Maeta *et al.*, 2009b), India (Sharma *et al.*, 2013), Portugal (Castro *et al.*, 2021), Spain (Miñarro and Twizell, 2015), Italy (Testolin, 1991), USA (McKay, 1978; Strik and Cahn,1998; Jackson, 2001), Chile (Jorquera Araya, 2004), Argentina (Sáez *et al.*, 2019; David *et al.*, 2022), Australia (Howpage *et al.*, 2001) and New Zealand (Macfarlane and Ferguson, 1983; Broussard *et al.*, 2022) (Fig. 13.4). While several other *Apis* species have been reported to forage on kiwifruit, including *Apis cerana* (Yang and Wu, 1990; Zhu *et al.*, 2003; Sharma *et al.*, 2013) and *Apis dorsata* (Sharma *et al.*, 2013), *Apis mellifera* is the primary species managed for kiwifruit pollination, with limited reports of the use of managed *Apis*

Fig. 13.4. Western honey bee (*Apis mellifera*) collecting pollen from a pistillate *Actinidia chinensis* var. *deliciosa* flower. (Photo: B. Cutting, The New Zealand Institute for Plant and Food Research Ltd)

cerana colonies, sometimes alongside *A. mellifera* (Zhu *et al.*, 2003).

Stocking rates of 7–10 *A. mellifera* hives ha^{-1} have often been recommended for *A. chinensis* var. *deliciosa* based on a single study (Palmer-Jones and Clinch, 1974). Subsequent studies found a poor correlation between the number of hives per ha and the number of bees per 1000 flowers in most years (Clinch, 1984). This discrepancy is probably due to the variable strength of bee colonies brought into orchards and resulted in Goodwin (2012) developing a standard for 'pollination strength' hives: 12 frames of bees.

Regardless of colony strengths or hive numbers, a visitation rate of six honey bees per 1000 *A. chinensis* var. *deliciosa* flowers during peak foraging results in optimal pollination (Palmer-Jones *et al.*, 1976; Clinch, 1984; Goodwin, 1987). However, many factors can influence honey bee visitation rates.

Overall honey bee abundance in orchards is correlated with pollen availability, with both declining in the afternoon (Goodwin, 1995). This afternoon decline in activity has been observed worldwide (Clinch, 1984; Vaissière *et al.*, 1996; Zhu *et al.*, 2003; de Piano *et al.*, 2021). Most honey bees visit only staminate or pistillate flowers, with as few as 5% visiting both floral sexes in one foraging trip and authors reporting a preference for male flowers (Palmer-Jones and Clinch, 1974; McKay, 1978; Clinch, 1984; Sano, 1987; Vaissière *et al.*, 1996) or female flowers (Goodwin *et al.*, 2013). If there are equal numbers of staminate and pistillate flowers, honey bees are evenly distributed across the sexes, indicating that relative flower abundance may be the underlying cause of much of the observed preference for staminate flowers (Zhu *et al.*, 2003).

The degree of flower constancy exhibited by honey bees is a critical factor in kiwifruit pollination. For fruit set to occur, an individual honey bee has to start by foraging on male flowers, which carry viable pollen, and then switch to female flowers, which have non-viable pollen but have the receptive stigmata. As bees sequentially visit one female flower after the other, the amount of male pollen they carry decreases, leading to a decrease in the amount of male pollen carried by bees with increasing distance from male vines (Goodwin *et al.*, 2013).

Nectar-rich plants attract honey bees away from kiwifruit, with colonies in kiwifruit orchards collecting as little as 2–4% kiwifruit pollen (Blanchet *et al.*, 1991; Barbattini and Greatti, 2008; de Piano *et al.*, 2021). Similar problems have been reported for *A. arguta*, with honey bees sharply reducing pollen collection from vines within 24 h of introduction (Jackson, 2001). Mass flowering of citrus, legumes (particularly clover), Brassicaceae, honeysuckle (*Lonicera* spp.), poppy and Rosaceae have been noted as strong competitors for pollinators during kiwifruit flowering (McKay, 1978; Macfarlane and Ferguson, 1983; Clinch, 1984; Donovan and Read, 1988; Blanchet *et al.*, 1991; Pomeroy and Fisher, 2002; Gardi *et al.*, 2003; Barbattini and Greatti, 2008).

Attempts have been made to encourage honey bees to forage on kiwifruit flowers. Successful strategies include removal of competing bloom from the surrounding area (Ford, 1971) and in-hive feeding of sugar syrup (Goodwin *et al.*, 1991, 1994; Goodwin, 1997; de Piano *et al.*, 2021). Other attempts to attract honey bees to kiwifruit include use of queen mandibular pheromone (Winston and Slessor, 1998; Howpage, 1999), lavender fragrance (Meroi Arcerito *et al.*, 2021), pre-feeding colonies with sugar syrup and kiwifruit pollen (Gardi *et al.*, 2003; Barbattini and Greatti, 2008), and providing bees with sugar syrup containing kiwifruit flowers (Zhu *et al.*, 2003).

To improve pollination of *A. arguta* vines, Jackson (2001) trialled honey bee colonies fitted with pollen dispensers (with *A. chinensis* var. *deliciosa* pollen). Four out of five *A. arguta* cultivars achieved full fruit set despite almost total lack of staminate *A. arguta* flowers and very low honey bee activity on vines.

Bumble bee pollination

Even without management, bumble bees often visit kiwifruit flowers (Clinch, 1984; Yang and Wu, 1990; Jorquera Araya, 2004; Maeta *et al.*, 2009b; Sharma *et al.*, 2013; Miñarro and Twizell, 2015). Compared with honey bees, bumble bees have a higher proportion of legitimate (i.e. pollen-transferring) visits to flowers, deposit more pollen per visit (Miñarro and Twizell, 2015; Broussard *et al.*, 2022), visit more flowers per minute (Maeta *et al.*, 2009b) and switch between staminate and pistillate kiwifruit flowers more often (Pomeroy and Fisher, 2002; Lee *et al.*, 2019). On an individual basis they are more efficient pollinators of kiwifruit than honey bees, achieving full pollination of *A. chinensis* var. *deliciosa* at 0.9 bees per 1000 flowers (Pomeroy and Fisher, 2002); less than one sixth of the number of honey bees required (Clinch, 1984; Goodwin, 1987). Unlike honey bees, they are able to buzz-pollinate flowers, releasing additional pollen from the anthers (Fig. 13.5).

Use of bumble bees for commercial kiwifruit orchards was first suggested by Wearing (1983) but the first commercial trials in Europe (Biobest, 2016), New Zealand (Cutting *et al.*, 2018), South Korea (Lee *et al.*, 2019) and the US (Abbate *et al.*, 2021) are more recent. Bumble bee colonies are significantly smaller than honey bee colonies, therefore a higher stocking rate of 22–48 hives ha^{-1} is recommended (Cutting *et al.*, 2018).

The recent interest in open-field bumble bees, despite their higher cost per ha, is in part due to the poor performance of honey bees under hail netting, which is increasingly employed to protect crops (Evans *et al.*, 2019; Kendall *et al.*, 2021). Bumble bees appear to be less affected by hail netting and other covered cropping techniques (Kendall *et al.*, 2021), making them an attractive pollinator in these situations despite their higher cost per ha.

Bumble bees can be attracted by competing flowering sources like clover, dandelions, buttercups and honeysuckle, but the attraction is weaker than for co-located honey bees (Pomeroy and Fisher, 2002). They are also less affected by inclement conditions, continuing to forage through rainy periods (Pomeroy and Fisher, 2002) and low temperatures (Miñarro and Twizell, 2015), as well as earlier in the morning and later in the evening than honey bees (Macfarlane and Ferguson, 1983; Maeta *et al.*, 2009b; Broussard *et al.*, 2022).

13.4.3 Other pollinators

Although often less abundant than honey bees, unmanaged pollinators have been reported to be as efficient as honey bees (Macfarlane and Ferguson, 1983; Miñarro and Twizell, 2015; Broussard *et al.*, 2022) and have been noted in multiple cropping systems to make honey bees more efficient pollinators (Brittain *et al.*, 2013).

Fig. 13.5. Buff-tailed bumble bee (*Bombus terrestris*) pollinating a pistillate *Actinidia chinensis* var. *chinensis* flower. (Photo: B. Cutting, The New Zealand Institute for Plant and Food Research Ltd)

Bees

Unmanaged bees visit kiwifruit flowers and have been reported, in rare instances, to deliver a substantial fraction of the crop's pollination (Ford, 1971). However, unmanaged bees are typically observed at lower densities than managed bees. The most commonly reported wild bee visitors to kiwifruit flowers are sweat bees in the genus *Lasioglossum* (Macfarlane and Ferguson, 1983; Maeta *et al.*, 2009b; Sharma *et al.*, 2013; Abbate *et al.*, 2021; Broussard *et al.*, 2022). Other members of the family Halictidae have been recorded as flower visitors, including *Halictus* spp. (Barbattini and Greatti, 2008; Maeta *et al.*, 2009b) and *Corynura chloris* (Jorquera Araya, 2004). Other families of unmanaged bees reported to visit kiwifruit include Apidae: *Xylocopa* spp. (Maeta *et al.*, 2009b; Abbate *et al.*, 2021), Andrenidae: *Andrena* spp. (Barbattini and Greatti, 2008; Maeta *et al.*, 2009b), Colletidae: *Leioproctus* spp. (Macfarlane and Ferguson, 1983; Clinch, 1984; Broussard *et al.*, 2022) and *Hylaeus* spp. (Macfarlane and Ferguson, 1983; Maeta *et al.*, 2009b), and Megachilidae (McKay, 1978).

Of the above, *Leioproctus* spp. were found to have a single-visit pollen deposition similar to honey bees, while *Lasioglossum* spp. had lower deposition rates (Macfarlane and Ferguson, 1983; Broussard *et al.*, 2022). *Xylocopa* spp.

were noted by Yang and Wu (1990) to visit almost exclusively male flowers, so are unlikely to be good pollinators in this system.

Flies

A number of fly species have been reported in commercial kiwifruit orchards, with syrphid flies (hover flies) amongst the most common (Jorquera Araya, 2004; Barbattini and Greatti, 2008; Maeta *et al.*, 2009b; Sharma *et al.*, 2013; Miñarro and Twizell, 2015; Howlett *et al.*, 2017; Abbate *et al.*, 2021; Broussard *et al.*, 2022). Indeed, Barbattini and Greatti (2008) noted that, in their Italian orchards, syrphid flies actually outnumbered both managed and wild bees. The pollination efficiency of several syrphid species has been investigated, finding that these insects deposit a similar amount of pollen as honey bees on a per-visit basis (Miñarro and Twizell, 2015; Howlett *et al.*, 2017; Broussard *et al.*, 2022). However, syrphid flies are also noted to have a high proportion of illegitimate visits to kiwifruit, where they deposit no pollen or do not touch reproductive structures (Testolin, 1991; Miñarro and Twizell, 2015; Howlett *et al.*, 2017; Broussard *et al.*, 2022). Despite this shortcoming, their abundance and high pollen deposition means that syrphid flies likely provide a small but significant amount of supplemental pollination; species with predatory larvae that may control pests in the orchard provide this secondary service in addition to pollination. Additionally, syrphid flies persist at similar densities under hail netting as in the open (Costa *et al.*, 1993), meaning that they may be increasingly important as more orchards go under covers.

Very high numbers of moth flies (Psychodidae) and mosquitoes (Culicidae) visit commercial kiwifruit flowers at night (Broussard *et al.*, 2022). Mosquitoes have also been recorded on wild *Actinidia chrysantha* (Mo *et al.*, 2016) but it is not known if they transferred pollen between flowers.

Several other dipteran families have been noted to visit commercial kiwifruit at low abundances, including Tipulidae, Bibionidae, Stratiomyidae, Anthomyiidae, Calliphoridae and Muscidae (Macfarlane and Ferguson, 1983; Maeta *et al.*, 2009b; Sharma *et al.*, 2013; Abbate *et al.*, 2021; Broussard *et al.*, 2022).

Others

Beetles found on kiwifruit flowers include members of Cerambycidae, Coccinellidae and Chrysomelidae (Macfarlane and Ferguson, 1983; Maeta *et al.*, 2009b; Broussard *et al.*, 2022). *Zorion guttigerum*, a cerambycid, is not a particularly effective pollinator owing to its low single-visit deposition and long flower handling time (Broussard *et al.*, 2022). The damage caused by chafer beetles (Scarabaeidae) eating flower parts probably outweighs any benefit provided as pollinators (Yang and Wu, 1990; Maeta *et al.*, 2009b).

Butterflies and moths occasionally visit kiwifruit flowers (Macfarlane and Ferguson, 1983; Sharma *et al.*, 2013; Miñarro and Twizell, 2015; Abbate *et al.*, 2021; Broussard *et al.*, 2022), but the rarity of these visits means they are unlikely to provide significant pollination services.

13.5 Artificial Pollination

Asynchronous flowering of male and female vines (González *et al.*, 1994; Yao *et al.*, 2010), failure of insect pollination (Costa *et al.*, 1993; Holcroft and Allan, 1994; Yao *et al.*, 2010) and inclement weather during flowering (Ford, 1971; Goodwin, 1987; Hopping, 1990; Holcroft and Allan, 1994) often result in low kiwifruit yields. Artificial pollination, first investigated by Hopping and Jerram (1979), is now the main method of kiwifruit pollination in Italy (Testolin and Ferguson, 2009), southern China (Zhu *et al.*, 2003; Huang, 2016) and South Korea (Lee *et al.*, 2019), and is used for supplemental pollination elsewhere. Many kiwifruit growers in New Zealand use artificial pollination and supply kiwifruit pollen globally (KiwiPollen, 2022; PollenPlus, 2022).

In good conditions, pollination by bees is equivalent to or better than artificial pollination for *A. chinensis* (Jorquera Araya, 2004; Sáez *et al.*, 2019; David *et al.*, 2022). However, when conditions are not optimal and limit pollinator activity, correctly applied supplemental pollination can increase seed number and fruit size (Costa *et al.*, 1993; González *et al.*, 1998; Testolin and Ferguson, 2009; Gianni and Vania, 2018; Lee *et al.*, 2019; Castro *et al.*, 2021).

Pollen is typically applied in two or more passes through the orchard (Hii, 2004), but a single pass at petal fall was shown to be effective in Italian orchards (Gianni and Vania, 2018).

Artificial pollination requires a readily available source of viable pollen. This can be difficult in countries lacking an existing pollen-harvesting infrastructure, although many small-scale collection technologies exist. Pollen can be efficiently applied to kiwifruit dry or in a liquid suspension media (wet pollination) – see Section 13.5.3. The advantage of dry application is that it utilizes the natural electrostatic forces which help transfer pollen from the bee to the flower even before the bee lands (Clarke *et al.*, 2017). Dry pollen can also be redistributed by bees from non-targeted areas (i.e. petals, leaves) to stigmata (Thomson and Eisenhart, 2003). Wet pollination can be used in damp conditions when pollinators may be scarce (King and Ferguson, 1991). The two methods of application are generally equivalent for kiwifruit (King and Ferguson, 1991; Pyke and Alspach, 1991; Gardi *et al.*, 2003), but in some cases wet pollination can result in lower fruit weights than dry pollination (Holcroft and Allan, 1994).

13.5.1 Pollen harvesting

Pollen yields are maximized by collecting flowers at the popcorn stage (King and Ferguson, 1991; Yao *et al.*, 2010; Gianni and Vania, 2018). Flowers are collected into a basket or sack and allowed to dehisce under controlled conditions. Anthers can be manually removed using tweezers (Hopping and Jerram, 1980a; Yao *et al.*, 2010), but this is labour-intensive and only economic in small-scale applications, e.g. breeding programmes. Commercial pollen collection generally involves milling whole flowers to separate anthers from other floral structures, and the use of filter- or cyclone-type machines to extract the anthers from flower debris. Anthers are then dried and pollen extracted using a vacuum and sieve (Hopping, 1990; King and Ferguson, 1991; Gianni and Vania, 2018). Drying the anthers can be rate-limiting, but increasing the temperature or the air flow reduces pollen viability (Hopping, 1990).

Bee-collected kiwifruit pollen can also be used (Pyke and Alspach, 1991, 1992) but can vary in viability and purity. Vacuuming pollen directly from dehiscent flowers in the field has been more successful, producing higher-viability pollen than milling (King and Ferguson, 1991; Sáez *et al.*, 2019). Different processes are shown in Fig. 13.6.

Once collected, pollen can be stored at temperatures below −18°C for 1–6 years depending on collection and storage conditions (see Section 13.5.2), allowing producers to apply pollen from one year to the next.

13.5.2 Pollen storage

Unlike that of many other crops, *Actinidia* pollen can remain viable at room temperature for many days. For example, *A. chinensis* var. *deliciosa* pollen can be stored for 7–14 days at 20°C with little loss of viability (McKay, 1978; Chen, Y. *et al.*, 1996; Yao *et al.*, 2010), becoming non-viable only after 4–8 weeks (Abreu and Oliveira, 2004; Borghezan *et al.*, 2011). With minimal chilling (0–4°C), sufficient pollen is still viable after 6–12 months' storage to achieve moderate fruit set (Naik *et al.*, 2013). Early-season pollen can be used to pollinate pistillate flowers later in the same season but in order to apply pollen collected late the previous season to early flowering cultivars, low temperature storage is required.

Relative humidity affects the success of low-temperature storage of binucleate pollen, such as that of *Actinidia* (Gill, 2014). Depending on the moisture content of the pollen, temperatures below 0°C are suitable for long-term (12+ months) storage and a range of optimal temperatures have been recommended: −20°C (Abreu and Oliveira, 2004); −80°C (Bomben *et al.*, 1999); and −196°C (Bomben *et al.*, 1999). Kiwifruit pollen can be stored for up to six years with modest reductions in pollen viability (Gianni and Vania, 2018).

Pollen storage life may vary with genotype: there was a ten-fold difference in pollen viability among male accessions of *A. eriantha* stored at −20°C (Seyrek *et al.*, 2016). Differences in pollen storability may also occur between species. In a preliminary study by Chen, Y. *et al.* (1996), *A. arguta* pollen could be stored twice as long at 5°C as *A. chinensis* var. *deliciosa* pollen before becoming non-viable. The same study found

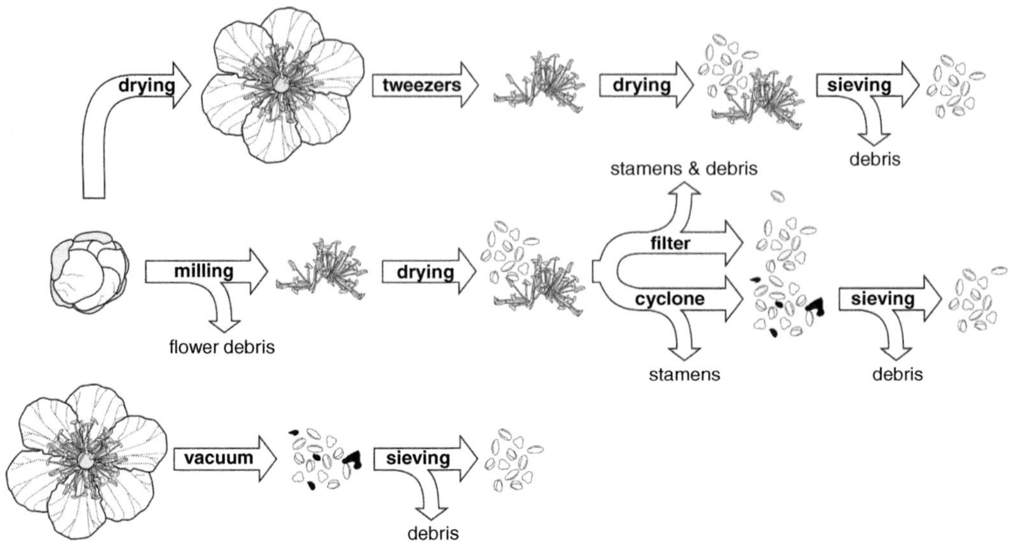

Fig. 13.6. Different processes for extracting kiwifruit pollen. (Illustration M. Broussard, The New Zealand Institute for Plant and Food Research Ltd)

that some pollen of *A. arguta*, *A. chinensis* var. *chinensis*, *A. chinensis* var. *deliciosa* and *A. eriantha* remained viable after being stored at −15°C for one year, but *A. polygama* pollen had become completely non-viable. A study of pollen viability at 18–25°C showed *A. kolomikta* pollen lost all viability after storage for 48 h, much faster than loss of pollen viability in *A. chinensis* (Česonienė, 2000). This difference is possibly due to the high temperature, as *A. kolomikta* pollen stored at 3–5°C maintained good viability for 21 days (Česonienė, 2000).

13.5.3 Artificial pollination technologies

Numerous artificial pollen application technologies have been trialled in kiwifruit over the years, including application with brushes, hand-held purpose-built applicators, tractor-mounted sprayers and robotic pollinator prototypes. The two broad classes of pollen applicators use dry pollen or wet pollen with very different physical constraints in particle/droplet size and application dynamics.

Dry pollination

Dry pollen can be taken directly from a staminate flower and applied to a pistillate flower. However, as the pollen is highly concentrated, an excess can easily be applied. The most promising pollen diluents to avoid this are *Lycopodium* spores (Jackson, 2001), pine pollen (Hopping and Jerram, 1979), and charcoal (Jo *et al.*, 2003). Talc is best avoided as pollen diluted with talc rapidly lost viability (Hopping and Hacking, 1983). The efficiency of dry pollen application can be increased by using an electrostatic charge, which causes more grains to adhere to the stigmata (Gan-Mor *et al.*, 2003). Although this is effective in other crops (Wearing, 1983; Gan-Mor *et al.*, 2003), results with kiwifruit have been inconclusive to date.

Wet pollination

Application of kiwifruit pollen in a liquid suspension was first developed in New Zealand (Hopping and Jerram, 1980a, b; Hopping and Simpson, 1982). The suspension medium, a solution of water, sucrose, $Ca(NO_3)_2{\cdot}4H_2O$ + H_3BO_3, and sodium carboxymethyl cellulose, was taken up by industry and further refined into commercially available pollen suspension aids (KiwiPollen, 2022; PollenPlus, 2022). The quality of the water used for suspension media is critical. Many orchard water supplies include galvanized tanks or pipes that release zinc into the water, significantly reducing pollen viability

(Hopping and Simpson, 1982). Distilled or deionized water and food-grade plastic reservoirs should be used for wet pollination. Typical commercial formulations use pollen rates between 1 g L^{-1} (Hopping and Hacking, 1983; Holcroft and Allan, 1994) and 3 g L^{-1} (Sáez *et al.*, 2019), although up to 8 g L^{-1} has been used in trialling autonomous robotic pollinators (Williams *et al.*, 2020). Numerous application systems have been devised, with different nozzle types to achieve better droplet adhesion to stigmata. Electrostatic charging has also been used successfully with wet pollination systems in other crops, but preliminary results in kiwifruit were equivocal (Hopping, 1990).

Application systems

Targeted manual systems, from brushes to backpack sprayers, are popular for small-scale pollination. However, they are relatively slow, relying on one or more people walking through orchards, often for more than one pass. Pollen use is highly efficient with manual application but as applications can take 2–10 times longer than tractor applications, the high labour costs can make manual systems uneconomic for growers (Hii, 2004; Gianni and Vania, 2018), though they are still used extensively in China, South Korea and Italy and by some orchardists in New Zealand.

Tractor-mounted and tractor-pulled non-targeted broadcast systems have had significant uptake, being used in New Zealand for supplementing insect pollinators, and replacing insect pollination in Italy, South Korea and parts of China. However, non-targeted systems are less efficient with pollen. Many broadcast systems fail to deliver sufficient pollen for full pollination at rates lower than 2 kg pollen ha^{-1} (Hopping, 1990). Owing to the expense of relying solely on artificial pollination, many growers use pollen rates of 160–600 g ha^{-1} to supplement honey bee pollination (Hii, 2004; Gianni and Vania, 2018). To address this, researchers have investigated using electrostatic charge to increase the amount of pollen reaching the active regions of the flower (stigmata), but initial trials for kiwifruit were inconclusive (Gan-Mor *et al.*, 2003), but it has been successful for other crops (Eyles *et al.*, 2022).

Like targeted manual application of pollen, the goal for targeted autonomous pollination is to place exactly the right amount of pollen on the stigmata of individual flowers. Current commercial systems consist of vehicles which broadcast spray throughout the orchard (semi-targeted), although autonomous systems utilizing machine vision to identify and then target flowers individually (targeted) are being developed. An autonomous system developed in New Zealand fully pollinated > 670 export-quality 'Hayward' kiwifruit without contribution from insect pollinators, with key metrics (fruit set, seed count, fruit weight) comparable to insect-pollinated control samples (Williams *et al.*, 2020). A summary of different targeted and untargeted systems is in Table 13.1.

13.5.4 Risks and benefits of artificial pollination

Artificial pollination is not without risks and these should be weighed against the benefits. Pollen can transmit the disease caused by *Pseudomonas syringae* pv. *actinidiae* (Scortichini *et al.*, 2012) (see Chapter 18), causing a range of symptoms, from leaf spotting through to vine death. The global outbreak of this disease around 2010 is thought to have been exacerbated by the importation of infected pollen (Scortichini *et al.*, 2012). Future disease outbreaks may be caused in similar ways, but the use of pollen milled on-orchard reduces the risk. If risks are managed, the benefits can be substantial as artificial pollination can mitigate poor flowering synchrony and lack of pollinators, both of which may happen more frequently because of climate change.

13.6 Summary

Kiwifruit, as a dioecious plant in which neither sex produces nectar, faces significant challenges for successful insect pollination. While visited by over 200 species of invertebrates, only honey bees and bumble bees have been used to pollinate commercial orchards. Because of these challenges, artificial pollination has been relied on instead of bees for the pollination of commercial *A. chinensis* in Asia and parts of Europe and is

Table 13.1. Methods employed in field-scale artificial pollination of kiwifruit.

Precision	Mode	Wet/ dry	General type	Examples
Targeted	Manual	Dry	Pole tipped with a feather brush or velvet	Polli, Polli, New Zealand (Hii, 2004) 'pon-pon' (Gianni and Vania, 2018)
			Hand-held blower device with pollen reservoir	Kiwi Pollen Miniduster, KiwiPollen, New Zealand (KiwiPollen, 2021) KiwiPollen Duster, KiwiPollen, New Zealand (KiwiPollen, 2021) Soffi@PollineZ, Biotac, Verona, Italy (Gianni and Vania, 2018)
			Backpack-mounted hand-held dry applicator	'Speedy' Dall'Agata, Forlì, Italy (Gianni and Vania, 2018) Backpack AirShear, KiwiPollen, New Zealand (Hii, 2004)
		Wet	Backpack-mounted hand-held wet applicator	Cambrian, KiwiPollen, New Zealand (Hii, 2004) ElettroEASY, Volpi, Mantova, Italy (Gianni and Vania, 2018)
Broadcast	Tractor-mounted	Dry	Tractor-mounted dry pollen booms/sprayers	QuadDuster, KiwiPollen, New Zealand (KiwiPollen, 2021) AirShear, KiwiPollen, New Zealand (Hii, 2004) PollenSmart, PollenSmart, New Zealand (PollenSmart, 2021) Scumby, Firman Pollen, USA (Firman Pollen, 2020)
		Wet	Tractor-mounted fogger/fine mist sprayers	Kiwi Pollen Boom Sprayer, KiwiPollen, New Zealand (King and Ferguson, 1991) Spruzz@Polline TR, Gerbaudo, Cuneo, Italy (Gianni and Vania, 2018)
			Tractor-mounted electrostatic sprayers	OnTarget, On Target Spray Systems, USA (OnTarget, 2021) Fruit Tower, LectroBlast, USA (Courtney, 2021)
Targeted broadcast	Autonomous, semi-targeted	Wet	Semi-targeted autonomous robotic system	Drone delivery systems (e.g. Dropcopter) (Courtney, 2021)
	Autonomous, targeted	Wet	Targeted autonomous robotic system	Autonomous pollination robot (Williams *et al.*, 2020)

being trialled for new kiwiberry (*A. arguta*) orchards in the USA.

Challenges currently facing kiwifruit pollination are hail netting, poor bloom synchrony and lack of pollinators. Future work to ensure pollination security should include managed bumble bees for open and covered orchards, investigations into other managed insects (potentially including syrphid flies) and further exploring how different pollen parents affect fruit set and quality of key cultivars, both for insect and artificial pollination systems. Finally, continued investigation into artificial pollen application will provide much-needed insurance against poor bloom synchrony and lack of insect pollinators as we enter an era of extreme weather events.

References

Abbate, A.P., Campbell, J.W., Vinson, E.L. and Williams, G.R. (2021) The pollination and fruit quality of two kiwifruit cultivars (*Actinidia chinensis* var. *chinensis* 'AU Golden Sunshine' and 'AU Gulf Coast Gold') (Ericales: Actinidiaceae) grown in the southeastern United States. *Journal of Economic Entomology* 114(3), 1234–1241. DOI: 10.1093/jee/toab075

Abreu, I. and Oliveira, M. (2004) Fruit production in kiwifruit (*Actinidia deliciosa*) using preserved pollen. *Australian Journal of Agricultural Research* 55(5), 565–569. DOI: 10.1071/AR03211

Ainalidou, A., Karamanoli, K., Menkissoglu-Spiroudi, U., Diamantidis, G. and Matsi, T. (2015) CPPU treatment and pollination: their combined effect on kiwifruit growth and quality. *Scientia Horticulturae* 193, 147–154. DOI: 10.1016/j.scienta.2015.07.011

Barbattini, R. and Greatti, M. (2008) Attività di *Apis mellifera* e di pronubi selvatici su *Actinidia chinensis* e *A. deliciosa* [Activity of *Apis mellifera* and wild pollinators on *Actinidia chinensis* and *A. deliciosa*]. *Notizario ERSA* 4, 35–40.

Biobest (2016) Bumblebees, the better alternative for kiwi pollination Available at: https://www.biobestgroup.com/en/news/bumblebees%2C-the-better-alternative-for-kiwi-pollination (accessed January 2023).

Blanchet, P., Douault, P. and Pouvreau, A. (1991) Kiwifruit (*Actinidia deliciosa* Chev.) pollination: honey bee behaviour and its influence on the fruit. *Acta Horticulturae* 288, 376–380. DOI: 10.17660/ActaHortic.1991.288.61

Bomben, C., Malossini, C., Cipriani, G. and Testolin, R. (1999) Long term storage of kiwifruit pollen. *Acta Horticulturae* 498, 105–110. DOI: 10.17660/ActaHortic.1999.498.12

Borghezan, M., Clauman, A.D., Steinmacher, D.A., Guerra, M.P. and Orth, A.I. (2011) *In vitro* viability and preservation of pollen grain of kiwi (*Actinidia chinensis* var. *deliciosa* (A. Chev.) A. Chev.). *Crop Breeding and Applied Biotechnology* 11, 338–344. DOI: 10.1590/S1984-70332011000400007

Brittain, C., Williams, N., Kremen, C. and Klein, A.-M. (2013) Synergistic effects of non-*Apis* bees and honey bees for pollination services. *Proceedings of the Royal Society B: Biological Sciences* 280(1754): 20122767. DOI: 10.1098/rspb.2012.2767

Broussard, M.A., Goodwin, M., McBrydie, H.M., Evans, L.J. and Pattemore, D.E. (2021) Pollination requirements of kiwifruit (*Actinidia chinensis* Planch.) differ between cultivars 'Hayward' and 'Zesy002'. *New Zealand Journal of Crop and Horticultural Science* 49(1), 30–40. DOI: 10.1080/01140671.2020.1861032

Broussard, M.A., Howlett, B.G., Evans, L.J., McBrydie, H., Cutting, B.T. *et al.* (2022) Pollinator identity and behavior affect pollination in kiwifruit (*Actinidia chinensis* Planch.). *PeerJ* 10, e12963. DOI: 10.7717/peerj.12963

Brundell, D.J. (1975) Quantitative aspects of flowering in the Chinese gooseberry (*Actinidia chinensis* Planch.). *New Zealand Journal of Agricultural Research* 18, 371–374. DOI: 10.1080/00288233.1975.10421061

Buxton, K.N. (2005) Preharvest practices affecting postharvest quality of 'Hayward' kiwifruit. PhD thesis, Massey University, Palmerston North, New Zealand.

Castro, H., Siopa, C., Casais, V., Castro, M., Loureiro, J. *et al.* (2021) Spatiotemporal variation in pollination deficits in an insect-pollinated dioecious crop. *Plants* 10, 1273. DOI: 10.3390/plants10071273

Česonienė, L. (2000) Evaluation and characterisation of male clones of *Actinidia kolomikta*. In: Kaufmane, E. and Libek, A. (eds) *Proceedings of the International Conference Fruit Production and Fruit Breeding*. Estonian Agricultural University, Tartu, Estonia, pp. 236–239.

Chen, L., Wang, S.-Y., Zhong, M., Huang, C.-H., Liao, G.-L. *et al.* (2019) Effects of pollens from the 10 selected *Actinidia* male genotypes on 4 commercial planting kiwifruit female cultivars in Southern China. *New Zealand Journal of Crop and Horticultural Science* 47(3), 155–169. DOI: 10.1080/01140671.2019.1616569

Chen, Q., Zhang, Z., Qin, Z. and Jiang, Y. (1996) Studies on male species selection and pollen induction of 'Jinkui' kiwifruit. *China Fruits* 2, 23–24.

Chen, Y., Li, H., Zhu, D. and Zha, Y. (1996) A study on the viability and storage time of the pollen for *Actinidia*. *Acta Agriculturae Universitatis Henanensis* 30(2), 175–177.

Clark, C.J. and Lintas, C. (1992) Chemical composition of pollen from kiwifruit vines. *New Zealand Journal of Crop and Horticultural Science* 20, 337–344. DOI: 10.1080/01140671.1992.10421776

Clarke, D., Morley, E. and Robert, D. (2017) The bee, the flower, and the electric field: electric ecology and aerial electroreception. *Journal of Comparative Physiology* A 203(9), 737–748. DOI: 10.1007/s00359-017-1176-6

Clinch, P.G. (1984) Kiwifruit pollination by honey bees 1. Tauranga observations, 1978–81. *New Zealand Journal of Experimental Agriculture* 12(1), 29–38. DOI: 10.1080/03015521.1984.10427785

Corbet, S.A., Chapman, H. and Saville, N. (1988) Vibratory pollen collection and flower form: bumble-bees on *Actinidia*, *Symphytum*, *Borago* and *Polygonatum*. *Functional Ecology* 2(2), 147–155. DOI: 10.2307/2389689

Costa, G., Testolin, R. and Vizzotto, G. (1993) Kiwifruit pollination: an unbiased estimate of wind and bee contribution. *New Zealand Journal of Crop and Horticultural Science* 21(2), 189–195. DOI: 10.1080/01140671.1993.9513767

Courtney, R. (2021) High-tech bee boosters: cherry growers turn to tech companies to supplement honey bee pollination. *Good Fruit Grower*. Available at: https://www.goodfruit.com/high-tech-bee-boosters/ (accessed May 2021).

Craig, J.L., Stewart, A.M., Pomeroy, N., Heath, A.C.G. and Goodwin, R.M. (1988) A review of kiwifruit pollination: where to next? *New Zealand Journal of Experimental Agriculture* 16(4), 385–399. DOI: 10.1080/03015521.1988.10425667

Cutting, B.T., Evans, L.J., Paugam, L.I., McBrydie, H.M., Jesson, L.K. *et al.* (2018) Managed bumble bees are viable as pollinators in netted kiwifruit orchards. *New Zealand Plant Protection* 71, 214–220. DOI: 10.30843/nzpp.2018.71.178

David, M.A., Yommi, A., Sánchez, E., Martinez, A., Murillo, N. *et al.* (2022) Strategic use of honey bees (*Apis mellifera* L.) to increase the number and size of fruits in kiwifruit (*Actinidia chinensis* var. *deliciosa*). *European Journal of Agronomy* 133, 126420. DOI: 10.1016/j.eja.2021.126420

de Piano, F.G., Meroi Arcerito, F.R., de Feudis, L., Basilio, A.M., Galetto, L. *et al.* (2021) Food supply in honeybee colonies improved kiwifruit (*Actinidia deliciosa* Liang and Ferguson) (Actinidiaceae: Theales) pollination services. *Revista de la Sociedad Entomológica Argentina* 80(3), 31–36. DOI: 10.25085/rsea.800304

Donovan, B.J. and Read, P.E.C. (1988) The alfalfa leafcutting bee, *Megachile rotundata* (Megachilidae), does not pollinate kiwifruit, *Actinidia deliciosa* var. *deliciosa* (Actinidiaceae). *Journal of Apicultural Research* 27(1), 9–12. DOI: 10.1080/00218839.1988.11100774

Evans, L.J., Cutting, B.T., Jochym, M., Janke, M.A., Felman, C. *et al.* (2019) Netted crop covers reduce honeybee foraging activity and colony strength in a mass flowering crop. *Ecology and Evolution* 9(10), 5708–5719. DOI: 10.1002/ece3.5154

Eyles, A., Close, D.C., Quarrell, S.R., Allen, G.R., Spurr, C.J. *et al.* (2022) Feasibility of mechanical pollination in tree fruit and nut crops: a review. *Agronomy* 12(5), 1113. DOI: 10.3390/agronomy12051113

Ferguson, A.R. (1991) Kiwifruit (*Actinidia*). *Acta Horticulturae* 290, 603–656. DOI: 10.17660/ActaHortic.1991.290.14

Firman Pollen (2020) Scumby applicator instructions. Available at: https://firmanpollen.com/wp-content/uploads/ScumbyApplication.pdf (accessed 25 January 2023).

Ford, I. (1971) Chinese gooseberry pollination. *New Zealand Journal of Agriculture* 122(2), 34–35.

Gan-Mor, S., Bechar, A., Ronen, B., Eisikowitch, D. and Vaknin, Y. (2003) Electrostatic pollen applicator development and tests for almond, kiwi, date, and pistachio – an overview. *Applied Engineering in Agriculture* 19(2), 119–124. DOI: 10.13031/2013.13099

Gardi, T., Famiani, F., Micheli, M. and Moschini, M. (2003). 38th Apimondia International Apicultural Congress, Ljubljana, 24–29 August.

Gianni, T. and Vania, M. (2018) Artificial pollination in kiwifruit and olive trees. In: Mokwaia, P.W. (ed.) *Pollination in Plants*. IntechOpen Limited, London. Available at: https://www.intechopen.com/chapters/59984 (accessed February 2023).

Gill, M. (2014) Pollen storage and viability. *International Journal of Botany and Research* 4(5), 1–18.

González, M.V., Coque, M. and Herrero, M. (1994) Pollinator selection in kiwifruit (*Actinidia deliciosa*). *Journal of Horticultural Science* 69(4), 697–702. DOI: 10.1080/14620316.1994.11516502

González, M.V., Coque, M. and Herrero, M. (1995a) Papillar integrity as an indicator of stigmatic receptivity in kiwifruit (*Actinidia deliciosa*). *Journal of Experimental Botany* 46(2), 263–269. DOI: 10.1093/jxb/46.2.263

González, M.V., Coque, M. and Herrero, M. (1995b) Stigmatic receptivity limits the effective pollination period in kiwifruit. *Journal of the American Society for Horticultural Science* 120(2), 199–202. DOI: 10.21273/JASHS.120.2.199

González, M.V., Coque, M. and Herrero, M. (1997) Kiwifruit flower biology and its implications on fruit set. *Acta Horticulturae* 444, 425–429. DOI: 10.17660/ActaHortic.1997.444.66

González, M.V., Coque, M. and Herrero, M. (1998) Influence of pollination systems on fruit set and fruit quality in kiwifruit (*Actinidia deliciosa*). *Annals of Applied Biology* 132, 349–355. DOI: 10.1111/j.1744–7348.1998.tb05210.x
Goodwin, M. (2012) *Pollination of Crops in Australia and New Zealand*. Publication No. 12/059. Rural Industries Research and Development Corporation, Ruakura, New Zealand.
Goodwin, R.M. (1987) Biology of honeybee (*Apis mellifera* L.) pollination of kiwifruit (*Actinidia deliciosa* (A. Chev.). PhD thesis, University of Auckland, New Zealand.
Goodwin, R.M. (1995) Afternoon decline in kiwifruit pollen collection. *New Zealand Journal of Crop and Horticultural Science* 23(2), 163–171. DOI: 10.1080/01140671.1995.9513883
Goodwin, R.M. (1997) Feeding sugar syrup to honey bee colonies to improve pollination: a review. *Bee World* 78(2), 56–62. DOI: 10.1080/0005772X.1997.11099335
Goodwin, R.M. and Steven, D. (1993) Behaviour of honey bees visiting kiwifruit flowers. *New Zealand Journal of Crop and Horticultural Science* 21, 17–24. DOI: 10.1080/01140671.1993.9513741
Goodwin, R.M., Ten Houten, A. and Perry, J.H. (1991) Effect of variations in sugar presentation to honey bees (*Apis mellifera*) on their collection of kiwifruit (*Actinidia deliciosa*) pollen. *New Zealand Journal of Crop and Horticultural Science* 19(3), 259–262. DOI: 10.1080/01140671.1991.10421809
Goodwin, R.M., Ten Houten, A. and Perry, J.H. (1994) Effect of feeding pollen substitutes to honey bee colonies used for kiwifruit pollination and honey production. *New Zealand Journal of Crop and Horticultural Science* 22(4), 459–462. DOI: 10.1080/01140671.1994.9513858
Goodwin, R.M., McBrydie, H.M. and Taylor, M.A. (2013) Wind and honey bee pollination of kiwifruit (*Actinidia chinensis* 'HORT16A'). *New Zealand Journal of Botany* 51(3), 229–240. DOI: 10.1080/0028825X.2013.806934
Habart, J.L. (1974) La baie de l'*Actinidia chinensis* Planch. var. *chinensis* [The berry of *Actinidia chinensis* Planch. var. *chinensis*]. *Fruits* 29(3), 191–207.
Han, F., Wallberg, A. and Webster, M.T. (2012) From where did the Western honeybee (*Apis mellifera*) originate? *Ecology and Evolution* 2(8), 1949–1957. DOI: 10.1002/ece3.312
Haslam, S.K., Hanson, J.S. and Hanson, A.J. (1988) Preliminary investigation of a small fruit problem in kiwifruit, *Actinidia deliciosa*. *New Zealand Journal of Experimental Agriculture* 16(4), 379–383. DOI: 10.1080/03015521.1988.10425666
Hii, M.J.W. (2004) Kiwifruit flower pollination: wind pollination efficiencies and sprayer jet applications. PhD thesis, University of Canterbury, Christchurch, New Zealand.
Holcroft, D.M. and Allan, P. (1994) Applied research note: Artificial pollination of kiwifruit. *Journal of the South African Society for Horticultural Science* 4, 24–26.
Hopping, M.E. (1986) Kiwifruit. In: Monselise, S.P. (ed.) *Handbook of Fruit Set and Development*. CRC Press, Boca Raton, Florida, pp. 217–232. DOI: 10.1201/9781351073042
Hopping, M.E. (1990) Floral biology, pollination, and fruit set. In: Warrington, I.J. and Weston, G.C. (eds) *Kiwifruit: Science and Management*. Ray Richards Publisher in association with the New Zealand Society for Horticultural Science, Auckland, New Zealand, pp. 71–96.
Hopping, M.E. and Hacking, N.J.A. (1983) A comparison of pollen application methods for the artificial pollination of kiwifruit. *Acta Horticulturae* 139, 41–50. DOI: 10.17660/ActaHortic.1983.139.5
Hopping, M.E. and Jerram, E.M. (1979) Pollination of kiwifruit (*Actinidia chinensis* Planch.): stigma-style structure and pollen tube growth. *New Zealand Journal of Botany* 17(3), 233–240. DOI: 10.1080/0028825X.1979.10426897
Hopping, M.E. and Jerram, E.M. (1980a) Supplementary pollination of tree fruits. II. Field trials on kiwifruit and Japanese plums. *New Zealand Journal of Agricultural Research* 23(4), 517–521. DOI: 10.1080/00288233.1980.10417876
Hopping, M.E. and Jerram, E.M. (1980b) Supplementary pollination of tree fruits. I. Development of suspension media. *New Zealand Journal of Agricultural Research* 23(4), 509–515. DOI: 10.1080/00288233.1980.10417875
Hopping, M.E. and Simpson, L.M. (1982) Supplementary pollination of tree fruits. III. Suspension media for kiwifruit pollen. *New Zealand Journal of Agricultural Research* 25(2), 245–250. DOI: 10.1080/00288233.1982.10420919
Howlett, B.G., Read, S.F.J., Jesson, L.K., Benoist, A., Evans, L.E. *et al.* (2017) Diurnal insect visitation patterns to 'Hayward' kiwifruit flowers in New Zealand. *New Zealand Plant Protection* 70, 52–57. DOI: 10.30843/nzpp.2017.70.27
Howpage, D. (1999) Pollination biology of kiwifruit: influence of honey bees, *Apis mellifera* L, pollen parents and pistil structure. PhD thesis, University of Western Sydney, Hawkesbury, Australia.

Howpage, D., Vithanage, V. and Spooner-Hart, R.N. (1998) Pollen tube distribution in the kiwifruit (*Actinidia deliciosa* A. Chev. C. F. Liang) pistil in relation to its reproductive process. *Annals of Botany* 81, 697–703. DOI: 10.1006/anbo.1998.0615
Howpage, D., Spooner-Hart, R.N. and Vithanage, V. (2001) Influence of honey bee (*Apis mellifera*) on kiwifruit pollination and fruit quality under Australian conditions. *New Zealand Journal of Crop and Horticultural Science* 29(1), 51–59. DOI: 10.1080/01140671.2001.9514160
Huang, H. (2016) *Kiwifruit: The Genus ACTINIDIA*. Elsevier, London.
Jackson, D. (2001) Sustainable Agriculture Research and Education (SARE) Project FNE00-307: Hardy kiwi pollination and production final report. Available at: https://projects.sare.org/project-reports/fne00-307/ (accessed 25 January 2023).
Jansson, D.M. and Warrington, I.J. (1988) The influence of temperature during floral development and germination *in vitro* on the germinability of kiwifruit pollen. *New Zealand Journal of Experimental Agriculture* 16(3), 225–230. DOI: 10.1080/03015521.1988.10425644
Jo, Y.S., Cho, H.S., Ma, K.C., Kim, B.H., Kim, S.C. *et al.* (2003) Utilization of various charcoal powders as new economical pollen diluent for kiwifruit pollination. *Acta Horticulturae* 610, 165–169. DOI: 10.17660/ActaHortic.2003.610.22
Jones, D.L. (1975) Pollinaton is essential for Chinese gooseberries. *Victorian Horticulture Digest* 66, 28–29.
Jorquera Araya, R.E. (2004) Efecto de la polinización entomófila y por aire forzado en el peso, número de semillas y cuaja en frutos de kiwi (*Actinidia deliciosa* (Chev.) Liang et Ferguson nogales, quinta region. [Effect of entomophilic and forced air pollination on weight, number of seeds and fruit set in kiwi (*Actinidia deliciosa* (Chev.) Liang et Ferguson). Nogales fifth region]. MSc thesis, Pontificia Universidad Católica de Valparaíso, Chile.
Kawagoe, T. and Suzuki, N. (2004) Cryptic dioecy in *Actinidia polygama*: a test of the pollinator attraction hypothesis. *Canadian Journal of Botany* 82(2), 214–218. DOI: 10.1139/b03-150
Ke, S.-Q., Huang, R.-H., Wang, S.-M., Xion, Z.-T. and Wu, Z.-W. (1991) Studies on interspecific hybrids of *Actinidia*. *Acta Horticulturae* 297, 133–139. DOI: 10.17660/ActaHortic.1992.297.15
Kendall, L.K., Evans, L.J., Gee, M., Smith, T.J., Gagic, V. *et al.* (2021) The effect of protective covers on pollinator health and pollination service delivery. *Agriculture, Ecosystems and Environment* 319, 107556. DOI: 10.1016/j.agee.2021.107556
King, M.J. and Buchmann, S.L. (2003) Floral sonication by bees: mesosomal vibration by *Bombus* and *Xylocopa*, but not *Apis* (Hymenoptera: Apidae), ejects pollen from poricidal anthers. *Journal of the Kansas Entomological Society* 76(23), 295–305.
King, M.J. and Ferguson, A.M. (1991) Collection and use of dry pollen for pollination of kiwifruit. *New Zealand Journal of Crop and Horticultural Science* 19(4), 385–389. DOI: 10.1080/01140671.1991.10422881
King, M.J. and Ferguson, A.M. (1994) Vibratory collection of *Actinidia deliciosa* (kiwifruit) pollen. *Annals of Botany* 74(5), 479–482. DOI: 10.1006/anbo.1994.1144
KiwiPollen (2021) Dry application products. Available at: https://www.kiwipollen.com/dry-application-products/ (accessed June 2021).
KiwiPollen (2022) Products. Available at: https://www.kiwipollen.com/ (accessed January 2022).
Klein, A.-M., Vaissière, B.E., Cane, J.H., Steffan-Dewenter, I., Cunningham, S.A. *et al.* (2007) Importance of pollinators in changing landscapes for world crops. *Proceedings of the Royal Society B: Biological Sciences* 274(1608), 303–313. DOI: 10.1098/rspb.2006.3721
Lai, R., Woolley, D.J. and Lawes, G.S. (1990) The effect of inter-fruit competition, type of fruiting lateral and time of anthesis on the fruit growth of kiwifruit (*Actinidia deliciosa*). *Journal of Horticultural Science* 65(1), 87–96. DOI: 10.1080/00221589.1990.11516034
Lawes, G.S., Woolley, D.J. and Lai, R. (1990) Seeds and other factors affecting fruit size in kiwifruit. *Acta Horticulturae* 282, 257–264. DOI: 10.17660/ActaHortic.1990.282.33
Lee, K.Y., Cho, H., Yoon, H.J., Lee, S. and Ko, H. (2019) Comparison of the foraging activity and pollination effects of honeybees (*Apis mellifera* L.) and bumblebees (*Bombus terrestris* L.) in gold kiwi (*Actinidia chinensis* P. 'Haegeum'). *Journal of Apiculture* 34(4), 295–304. DOI: 10.17519/apiculture.2019.11.34.4.295
Li, Z., Fang, J.B., Qi, X.J., Lin, M.M., Chen, J.Y. *et al.* (2016) Effects of male plants with different ploidy on the fruit set and fruit characteristics in *Actinidia arguta* kiwifruit. *Journal of Fruit Science* 33(6), 658–663.
Macfarlane, R.P. and Ferguson, A.M. (1983) *The 1980 to 1982 Kiwifruit Survey of Pollination*. DSIR, Wellington.
Maeta, Y., Kawakami, C. and Shimizu, H. (2009a) Preliminary report on the use of three species of Japanese mason bees, the genus *Osmia*, as pollinators of kiwifruit (Hymenoptera, Megachilidae). *Bulletin of the Hoshizaki Green Foundation* 12, 1–17.

Maeta, Y., Kawakami, C., Shimizu, H. and Kitamura, K. (2009b) Flower-visiting insect fauna of kiwifruit and pollination efficacy of some promising bee species. *Bulletin of the Hoshizaki Green Foundation* 12, 19–34.

Malabœuf, F., Vaissière, B.E. and Cour, P. (1997) Pollen flow in the atmosphere of kiwifruit (*Actinidia deliciosa*) orchards. *Acta Horticulturae* 444, 413–418. DOI: 10.17660/ActaHortic.1997.444.64

Manning, R. (2001) Fatty acids in pollen: a review of their importance for honey bees. *Bee World* 82(2), 60–75. DOI: 10.1080/0005772X.2001.11099504

Matich, A.J., Young, H., Allen, J.M., Wang, M.Y., Fielder, S. *et al.* (2003) *Actinidia arguta*: volatile compounds in fruit and flowers. *Phytochemistry* 63(3), 285–301. DOI: 10.1016/S0031-9422(03)00142-0

McKay, S.A. (1978) Pollination and other factors affecting fruit-set and size of kiwifruit. MSc thesis, University of California, Davis, California.

McPherson, H.G., Richardson, A.C., Snelgar, W.P. and Currie, M.B. (2001a) Effects of hydrogen cyanamide on budbreak and flowering in kiwifruit (*Actinidia deliciosa* 'Hayward'). *New Zealand Journal of Crop and Horticultural Science* 29(4), 277–285. DOI: 10.1080/01140671.2001.9514189

McPherson, H.G., Richardson, A.C., Snelgar, W.P., Patterson, K.J. and Currie, M.B. (2001b) Flower quality and fruit size in kiwifruit (*Actinidia deliciosa*). *New Zealand Journal of Crop and Horticultural Science* 29(2), 93–101. DOI: 10.1080/01140671.2001.9514167

Meroi Arcerito, F.R., De Feudis, L.L., Amarilla, L.D., Galetto, L., Mitton, G. *et al.* (2021) Fragrance addition improves visitation by honeybees and fruit quality in kiwifruit (*Actinidia deliciosa*). *Journal of the Science of Food and Agriculture* 101, 5082–5088. DOI: 10.1002/jsfa.11153

Miñarro, M. and Twizell, K.W. (2015) Pollination services provided by wild insects to kiwifruit (*Actinidia deliciosa*). *Apidologie* 46(3), 276–285. DOI: 10.1007/s13592-014-0321-2

Mizugami, T., Kim, J.G., Beppu, K., Fukuda, T. and Kataoka, I. (2007) Observation of parthenocarpy in *Actinidia arguta* selection 'Issai'. *Acta Horticulturae* 753, 199–203. DOI: 10.17660/ActaHortic.2007.753.22

Mo, Q.-H., Li, J.-W., Gong, H.-J., Ye, K.-Y., Jiang, Q.-S. *et al.* (2016) A preliminary study on the reproduction biology of the endangered plant *Actinidia chrysantha*. *Guihaia* 36(6), 640–645.

Naik, S. (2011) Effect of pollination methods on the fruit set, yield and quality of kiwifruit (*Actinidia deliciosa* Liang and Ferguson). MSc thesis, Dr Yashwant Singh Parmar University of Horticulture and Forestry, Nauni, Solan, India.

Naik, S., Rana, P. and Rana, V. (2013) Pollen storage and use for enhancing fruit production in kiwifruit (*Actinidia deliciosa* A. Chev.). *Journal of Applied Horticulture* 15(2), 128–132.

Nieuwenhuizen, N.J., Green, S. and Atkinson, R.G. (2010) Floral sesquiterpenes and their synthesis in dioecious kiwifruit. *Plant Signaling and Behavior* 5(1), 61–63. DOI: 10.4161/psb.5.1.10161

Oh, E.U., Jeong, S.Y., Kang, H.H. and Song, K.J. (2020) Characterization of pollen development in staminate kiwifruit (*Actinidia* sp.) cultivars. *Horticultural Science and Technology* 38(1), 1–8. DOI: 10.7235/HORT.20200001

Oh, E.U., Jeong, S.Y., Kim, J.Y. and Song, K.J. (2021) Response of fruit set and quality and seed formation to ploidy levels of pollen donor in yellow-fleshed kiwifruits. *Horticulture, Environment, and Biotechnology* 62, 9–15. DOI: 10.1007/s13580-020-00293-z

OnTarget (2021) Pollen application sprayers. On Target Spray Systems Ltd. Available at: https://ontarget-spray.com/pollen/ (accessed February 2023).

Palmer-Jones, T. and Clinch, P.G. (1974) Observations on the pollination of Chinese gooseberries variety 'Hayward'. *New Zealand Journal of Experimental Agriculture* 2(4), 455–458. DOI: 10.1080/03015521.1974.10427714

Palmer-Jones, T., Clinch, P.G. and Briscoe, D.A. (1976) Effect of honey bee saturation on the pollination of Chinese gooseberries variety 'Hayward'. *New Zealand Journal of Experimental Agriculture* 4(2), 255–256. DOI: 10.1080/03015521.1976.10425879

Polito, V.S. and Weinbaum, S.A. (1988) Intraclonal variation in pollen germinability in kiwifruit, pistachio and walnut as influenced by tree age. *Scientia Horticulturae* 36(1-2), 97–102. DOI: 10.1016/0304-4238(88)90011-8

PollenPlus (2022) Solutions. Available at: http://www.pollenplus.co.nz/solutions (accessed 25 January 2023).

PollenSmart (2021) PollenSmart models. Available at: https://www.pollensmart.co.nz/models (accessed 30 Jun 2022).

Pomeroy, N. and Fisher, R.M. (2002) Pollination of kiwifruit (*Actinidia deliciosa*) by bumble bees (*Bombus terrestris*): effects of bee density and patterns of flower visitation. *New Zealand Entomologist* 25(1), 41–49. DOI: 10.1080/00779962.2002.9722093

Pyke, N.B. and Alspach, P.A. (1991) Variability in the performance of bee collected kiwifruit pollen. *Acta Horticulturae* 297, 291–297. DOI: 10.17660/ActaHortic.1992.297.38

Pyke, N.B. and Alspach, P.A. (1992) The collection and quality of bee-collected pollen for supplemental pollination in kiwifruit. *The New Zealand Beekeeper* 216, 16–17.

Qi, X.-J., Han, L.-X., Li, M., Xu, S.-K., Zhu, Y.-S. *et al.* (2007) Studies on pollen xenia of kiwifruit. *Journal of Fruit Science* 24, 774–777.

Richardson, A.C. and McAneney, K.J. (1990) Influence of fruit number on fruit weight and yield of kiwifruit. *Scientia Horticulturae* 42(3), 233–241. DOI: 10.1016/0304-4238(90)90085-S

Sáez, A., Negri, P., Viel, M. and Aizen, M.A. (2019) Pollination efficiency of artificial and bee pollination practices in kiwifruit. *Scientia Horticulturae* 246, 1017–1021. DOI: 10.1016/j.scienta.2018.11.072

Sano, E. (1987) Utilization of honeybees on the pollination of kiwis. *Honeybee Science* 8, 172–175.

Schmid, R. (1978) Reproductive anatomy of *Actinidia chinensis* (Actinidiaceae). *Botanische Jahrbücher für Systematik, Planzengesichte und Pflanzengeographie* 100, 149–195.

Scortichini, M., Marcelletti, S., Ferrante, P., Petriccione, M. and Firrao, G. (2012) *Pseudomonas syringae* pv. *actinidiae*: a re-emerging, multi-faceted, pandemic pathogen. *Molecular Plant Pathology* 13(7), 631–640. DOI: 10.1111/j.1364-3703.2012.00788.x

Seal, A.G., Dunn, J.K. and Jia, Y.L. (2013a) Pollen parent effects on fruit attributes of diploid *Actinidia chinensis* 'Hort16A' kiwifruit. *New Zealand Journal of Crop and Horticultural Science* 41(4), 219–229. DOI: 10.1080/01140671.2013.803130

Seal, A.G., Dunn, J.K., De Silva, H.N., McGhie, T.K. and Lunken, R.C.M. (2013b) Choice of pollen parent affects red flesh colour in seedlings of diploid *Actinidia chinensis* (kiwifruit). *New Zealand Journal of Crop and Horticultural Science* 41(4), 207–218. DOI: 10.1080/01140671.2013.803129

Seal, A.G., McGhie, T., Boldingh, H., Rees, J., Blackmore, A. *et al.* (2016) The effect of pollen donor on fruit weight, seed weight and red colour development in *Actinidia chinensis* 'Hort22D'. *New Zealand Journal of Crop and Horticultural Science* 44(1), 1–12. DOI: 10.1080/01140671.2015.1100127

Seal, A.G., Blackmore, A., Jaksons, P., McAtee, P. and Cheng, C.H. (2017) Pollen donor effects on the expression of red flesh colour in kiwifruit. *Acta Horticulturae* 1218, 373–379. DOI: 10.17660/ActaHortic.2018.1218.52

Seal, A.G., Clark, C.J., Sharrock, K.R., de Silva, H.N., Jaksons, P. *et al.* (2018) Choice of pollen donor affects weight but not composition of *Actinidia chinensis* var. *chinensis* 'Zesy002' (Gold3) kiwifruit. *New Zealand Journal of Crop and Horticultural Science* 46(2), 133–143. DOI: 10.1080/01140671.2017.1365732

Seyrek, U.A., Qu, X.Y., Huang, C.H., Tao, J.J., Zhong, M. *et al.* (2016) Influence of storage time on pollen traits in *Actinidia eriantha*. *Agricultural Sciences* 7(6), 373–382. DOI: 10.4236/as.2016.76039

Sharma, M., Mattu, V.K. and Thakur, M.L. (2013) Pollination studies on kiwi crop (*Actinidia deliciosa* Chev.) in Himachal Pradesh, India. *International Journal of Advanced Biological Research* 3(4), 545–548.

Stasiak, A., Stefaniak, J., Łata, B. and Latocha, P. (2017) Efficiency of *A. arguta* (Siebold et Zucc.) Planch. ex. Miq. pollination using *A. deliciosa* pollen. *Journal of International Scientific Publications: Agriculture and Food* 5, 178–187.

Stasiak, A., Latocha, P., Drzewiecki, J., Hallmann, E., Najman, K. *et al.*, (2019) The choice of female or male parent affects some biochemical characteristics of fruit or seed of kiwiberry (*Actinidia arguta*). *Euphytica* 215, 52. DOI: 10.1007/s10681-019-2375-8

Stasiak, A., Łata, B., Bieniasz, M. and Latocha, P. (2020) Morphological variation of male *A. arguta* plants affects their flowering potential and pollen efficiency. *Horticultural Science* 47(2), 100–109. DOI: 10.17221/37/2019-HORTSCI

Stasiak, A., Latocha, P. and Bieniasz, M. (2021) Effect of genetically diverse pollen on pollination, pollen tube overgrow, fruit set and morphology of kiwiberry (*Actinidia arguta*). *Agronomy* 11(9), 1814. DOI: 10.3390/agronomy11091814

Strik, B.C. and Cahn, H. (1998) Growing kiwifruit. Oregon State University Extension Service. PNW 507.

Tatsuka, K., Suekane, S., Sakai, Y. and Sumitani, H. (1990) Volatile constituents of kiwi fruit flowers: simultaneous distillation and extraction versus headspace sampling. *Journal of Agricultural and Food Chemistry* 38(12), 2176–2180. DOI: 10.1021/jf00102a015

Testolin, R. (1991) Male density and arrangement in kiwifruit orchards. *Scientia Horticulturae* 48, 41–50. DOI: 10.1016/0304-4238(91)90151-N

Testolin, R. and Ferguson, A.R. (2009) Kiwifruit (*Actinidia* spp.) production and marketing in Italy. *New Zealand Journal of Crop and Horticultural Science* 37(1), 1–32. 10.1080/01140670909510246

Thakur, B.S. and Rathore, D.S. (1991) Effective pollination period in Chinese gooseberry (*Actinidia chinensis* Planch.). *Gartenbauwissenschaft* 56(5), 230–232.

Thomson, J.D. and Eisenhart, K.S. (2003) Rescue of stranded pollen grains by secondary transfer. *Plant Species Biology* 18(2-3), 67–74. DOI: 10.1111/j.1442-1984.2003.00089.x

Tiyayon, C. and Strik, B. (2002) Flowering and fruiting morphology of hardy kiwifruit, *Actinidia arguta*. *Acta Horticulturae* 610, 171–176. DOI: 10.17660/ActaHortic.2003.610.23

Twidle, A.M., Mas, F., Harper, A.R., Horner, R.M., Welsh, T.J. *et al.* (2015) Kiwifruit flower odor perception and recognition by honey bees, *Apis mellifera*. *Journal of Agricultural and Food Chemistry* 63(23), 5597–5602. DOI: 10.1021/acs.jafc.5b01165

Twidle, A.M., Suckling, D.M., Seal, A.G., Fedrizzi, B., Pilkington, L.I. *et al.* (2017) Identification of *in situ* flower volatiles from kiwifruit (*Actinidia chinensis* var. *deliciosa*) cultivars and their male pollenisers in a New Zealand orchard. *Phytochemistry* 141, 61–69. DOI: 10.1016/j.phytochem.2017.05.011

Twidle, A.M., Barker, D., Seal, A.G., Fedrizzi, B. and Suckling, D.M. (2018) Identification of floral volatiles and pollinator responses in kiwifruit cultivars, *Actinidia chinensis* var. *chinensis*. *Journal of Chemical Ecology* 44(4), 406–415. DOI: 10.1007/s10886-018-0936-2

Vaissière, B.E., Rodet, G., Cousin, M., Botella, L. and Torré Grossa, J.-P. (1996) Pollination effectiveness of honey bees (Hymenoptera: Apidae) in a kiwifruit orchard. *Journal of Economic Entomology* 89(2), 453–461. DOI: 10.1093/jee/89.2.453

Vaudo, A.D., Tooker, J.F., Patch, H.M., Biddinger, D.J., Coccia, M. *et al.* (2020) Pollen protein: lipid macronutrient ratios may guide broad patterns of bee species floral preferences. *Insects* 11(2), 132. DOI: 10.3390/insects11020132

Wearing, C.H. (1983) Expert reviews bee research. *Horticultural News* (New Zealand) 5(3), 5.

Williams, H., Nejati, M., Hussein, S., Penhall, N., Lim, J.Y. *et al.* (2020) Autonomous pollination of individual kiwifruit flowers: toward a robotic kiwifruit pollinator. *Journal of Field Robotics* 37(2), 246–262. DOI: 10.1002/rob.21861

Winston, M.L. and Slessor, K.N. (1998) Honey bee primer pheromones and colony organization: gaps in our knowledge. *Apidologie* 29(1–2), 81–95. DOI: 10.1051/apido:19980105

Yang, L. and Wu, Y. (1990) Preliminary study of kiwifruit pollinating bees in China. *Apiculture of China* 1, 2–4.

Yao, C., Long, Z., Liu, X. and Wang, X. (2010) 不同干及贮藏方法对猕猴桃花粉活力的影响 Effects of different drying and storage methods on kiwifruit pollen viability. *Northern Horticulture* 20, 37–39.

Zhu, Y., Zhou, Z., Mao, J., Hu, H., Jiang, A. *et al.* (2003) Preliminary report on bee pollination technology research in kiwifruit. *South China Fruits* 32(2), 45–46+60.

14 Axillary Bud, Shoot and Flower Development

Annette C. Richardson[1]* and Erika Varkonyi-Gasic[2]
[1]*The New Zealand Institute for Plant and Food Research Ltd, Kerikeri, New Zealand;* [2]*The New Zealand Institute for Plant and Food Research Ltd, Auckland, New Zealand*

14.1 Introduction

The development of axillary buds, shoots and flowers in kiwifruit (*Actinidia* spp.) underpins the phenology of vines. *Actinidia* spp. buds develop over two seasons, starting with their formation in the leaf axils of expanding shoots and floral evocation during spring and summer of the first season. Then buds become quiescent, dormant in winter, with shoot growth and differentiation of flowers occurring during spring of the second season (Fig. 14.1). These processes determine the numbers of flowers and fruit produced by vines, and hence the yield and economic performance of orchards.

Kiwifruit vines are woody perennials that produce few flowers compared with highly floriferous fruit tree crops like apple, peach and citrus. In *Actinidia* spp. buds, inflorescence meristems remain undeveloped during winter and flower differentiation occurs during an extended period from budswell through until anthesis in spring (Table 14.1). This differs from bud and flower development in many fruit tree crops where the flowers are formed before winter and emerge in early spring. This chapter reviews current understanding of the physiological and molecular processes involved in bud, shoot and flower development in *Actinidia* spp. as well as effects of environmental and management factors on their development.

14.2 Axillary Bud and Shoot Development

14.2.1 Budbreak

Dormant buds on one-year-old wood (longer canes and shorter spurs) begin to swell and break in early spring. The timing of events during spring bud, shoot and flower development has been described for both *A. chinensis* var. *deliciosa* 'Hayward' and Alpha (a male selection) by Brundell (1975a) and compared in Hopping (1990) (Table 14.1). Buds begin to swell ten days prior to budbreak, and the leaves have developed and begun to open by 10–15 days after budbreak. Both the timing and spread of budbreak depend on scion and rootstock genetics (Wang *et al.*, 1994; Snowball, 1997b), environmental factors, particularly temperature, and vine management techniques (McPherson *et al.*, 2001). Across the genus *Actinidia*, *A. eriantha* and *A. chinensis* var. *chinensis* genotypes are

*Corresponding author: annette.richardson@plantandfood.co.nz

© CAB International 2024. *Kiwifruit: Botany, Production and Uses* (eds A.C. Richardson, J.N. Burdon and A.R. Ferguson)

DOI: 10.1079/9781800620933.0014

Fig. 14.1. Development of *Actinidia* spp. buds occurs over two seasons from their formation in leaf axils in (**A**) spring and (**B**) summer of the first season, through (**C**) dormancy in winter, then (**D**) budswell, (**E**) budbreak, (**F–I**) shoot growth and flower development in spring of the second season. (Photos: The New Zealand Institute for Plant and Food Research Ltd)

generally the first to break bud, followed by *A. chinensis* var. *deliciosa* 'Hayward' and, finally, *A. arguta* and *A. polygama* genotypes.

Over a range of environments, budbreak occurs earlier in cooler regions that receive higher winter chilling during bud dormancy and later in areas with warm winters (Hall and McPherson, 1997). For example, across geographic regions in New Zealand, the mid-point of budbreak for 'Hayward' vines varied by 32 days, with the earliest budbreak on vines in southern regions with cool winters and the latest in northern growing regions with warmer winters (McPherson *et al.*, 1994). Similarly, the duration of budbreak on 'Hayward' vines varied from 14 days at sites with cooler winters to over 50 days at warmer sites (McPherson *et al.*, 2001). As the duration of budbreak increased, the floral potential and vigour of later-developing buds decreased in a range of *Actinidia* taxa (McPherson *et al.*, 1994; Seleznyova *et al.*, 2002; Foster *et al.*, 2007).

In addition to cool winters, warm temperatures immediately prior to budbreak are important, as warmer conditions advance bud development of 'Hayward' vines and can contribute to variation in vine development between sites and seasons (Hall and McPherson, 1997). Considerable variation in the timing of budbreak also occurred both within and between one-year-old canes and spurs on an *A. chinensis* var. *chinensis* selection (Seleznyova *et al.*, 2002). Buds at the apical end of canes/spurs and upward-facing buds generally break first, while downward-facing buds and those towards the base of 'Hayward' canes are less likely to break (Austin *et al.*, 2002; Friend *et al.*, 2015).

14.2.2 Shoot development

Shoot outgrowth from previously established (pre-formed) embryonic shoots begins at budswell in spring. Embryonic 'Hayward' and Alpha shoots can have as many as 22 pre-formed nodes and leaf primordia (Brundell, 1975a) and may expand into self-terminating short shoots or continue to grow until midsummer through the formation of new leaves, nodes and internodes (neo-formed) after budbreak. Competition for resources between neighbouring shoots causes the apices of some shoots to terminate and stop growing in early spring, resulting in shorter shoots, for example in *A. chinensis* var. *chinensis* 'Hort16A' vines (Clearwater *et al.*, 2006). Shoot

Table 14.1. Key stages of floral morphogenesis in *Actinidia chinensis* var. *deliciosa* 'Hayward' and Alpha (a male selection) buds. (From Brundell, 1975b; Hopping, 1990)

Days from budbreak	Axillary bud development	Flower bud development	
		Pistillate ('Hayward')	Staminate (Alpha)
Dormant		Uninitiated flower primordium	
−10	Budswell	Swollen flower primordium	
−6	Advanced budswell	Initiation of bracts Initiation of sepals	
0	Budbreak	Initiation of petals	
5	Advanced budbreak	Initiation of stamens	
9			Initiation of stigmata
10	Open cluster		
11		Initiation of stigmata	
15	Advanced open cluster		
16			Gynoecial development stops
18		Formation of gynoecial plateau	
27			Initiation of anthers and filaments
34		Initiation of anthers and filaments	
38			Formation of pollen grains
40			Calyx split
45		Initiation of ovules	
50		Calyx split	
53			Open flower
60		Open flower	

termination usually occurs when internode extension ceases in the pre-formed zone of the shoot (Brundell, 1975a) and the apex abscises (Foster *et al.*, 2007). This results in a population of shoot types within canopies that can be defined as short (≤ 9 nodes and terminated), medium (> 9 nodes and terminated) and long shoots (pre-formed and neo-formed nodes (19–90) and not terminated) (Seleznyova *et al.*, 2002) (Fig. 14.2). Cessation of growth of long shoots in summer depends on genotype (Snowball, 1997b), environmental conditions (Richardson *et al.*, 2004) and vine management techniques (Patterson and Currie, 2011).

A developing 'Hayward' shoot contains four basal nodes that produce bud scales, nodes 5–7 that produce transitional leaves, and normal leaves form thereafter (Brundell, 1975a; Snowball, 1995; Walton *et al.*, 1997). There is limited development of bud scales and transitional leaves, which often abscise soon after budbreak (Brundell, 1975b). The bud scales may subtend second-order axillary buds, but these rarely develop (Walton *et al.*, 1997). Inflorescences can develop from meristems at nodes 5–16 but they are most likely to appear at nodes 6–12 where the floral meristems are largest (Walton and Fowke, 1993). Thereafter, both pre-formed (up to node 22) and neo-formed leaves appear on growing shoots with associated development of new axillary buds. There is some variation in the development of shoots of different taxa, e.g. *A. arguta*, *A. polygama* and *A. valvata* shoots do not produce basal buds and the number and position of floral nodes may vary (Snowball, 1997a; Walton and Wu, 1999).

14.2.3 Axillary bud development

'Hayward' buds develop in the leaf axils of growing shoots in spring and summer of the current season, forming embryonic shoots containing leaf primordia and undifferentiated axillary meristems (Walton *et al.*, 1997) (Fig. 14.3). New

Fig. 14.2. Three shoot types in *Actinidia* spp.: (**A**) short (≤ 9 nodes and terminated); (**B**) medium (> 9 nodes and terminated); (**C**) long shoots (pre-formed and neo-formed nodes (19–90) and not terminated) as defined by Seleznyova *et al*. (2002). (Photos: The New Zealand Institute for Plant and Food Research Ltd)

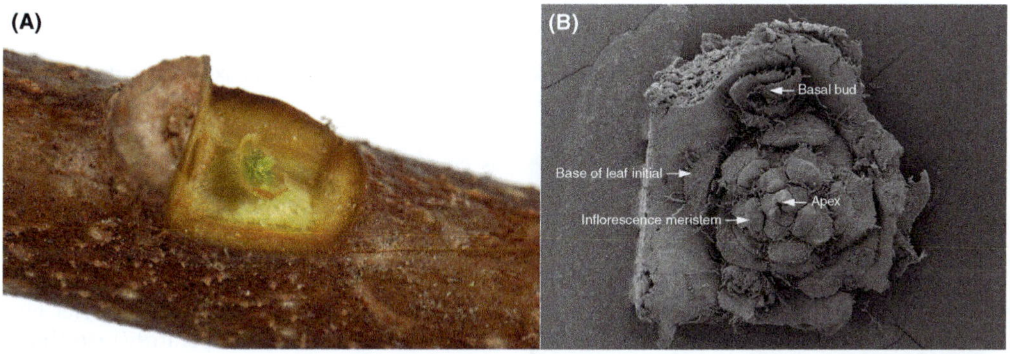

Fig. 14.3. (**A**) An *Actinidia* spp. bud contains an embryonic shoot with (**B**) a spiral phyllotaxy of second-order basal buds, leaf initials with inflorescence meristems in their axils and an apex. (Photos: The New Zealand Institute for Plant and Food Research Ltd)

buds form in the axils of very young leaves along developing shoots. Forty days after a leaf first appears on a new shoot, the associated axillary bud can contain 10–15 leaf initials. At this stage in late spring/early summer, structures that will become inflorescences or second-order basal shoots begin to form within the axils of leaf initials in the new buds of ‘Hayward’ and other taxa (Snowball, 1997a; Walton *et al*., 1997) (Fig. 14.4). The size and floral productivity of axillary buds can vary along shoots and are correlated with the size of subtending leaves (Watson and Gould, 1994; Snowball, 1997a). Both the distribution of leaf initials, inflorescence primordia and second-order axillary buds and the external appearance of buds can vary between *Actinidia* taxa (Snowball, 1997a).

Once the axillary buds are fully formed, they become quiescent and signals from the apex and other organs outside the bud generally prevent them from growing until the following spring (Brundell, 1975a; Walton *et al*., 1997). However, these buds may break in the same season in which they first develop if the shoot apex is removed, or shoots are defoliated. Shoots that develop in this way are rarely floral; however, removal of the apex of shoots of *A. eriantha* genotypes (Fig. 14.5) or defoliation of *A. chinensis* var. *chinensis* ‘Jecy Gold’ shoots in summer have been reported to occasionally result in flowers on shoots (Walton, 1995; Srisook *et al*., 2015, 2016). ‘Chaohong’, a male cultivar that is a hybrid between *A. eriantha* and *A. chinensis* var. *chinensis*, is reported to flower several times during the growing season (Zhang *et al*., 2010).

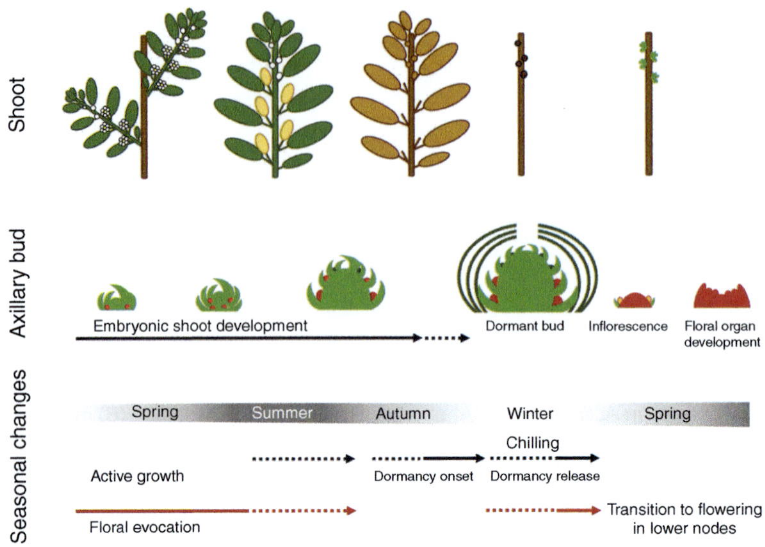

Fig. 14.4. A schematic diagram representing seasonal development of shoot and axillary bud development over two seasons. (Adapted from Varkonyi-Gasic, 2018)

Fig. 14.5. Removal of the apex of *Actinidia eriantha* shoots in summer may result in the production of flowers on the resulting apical shoot. (Photo: The New Zealand Institute for Plant and Food Research Ltd)

14.3 Bud Endodormancy

Actinidia vines, like most deciduous, woody, perennial plants, cycle between growth during favourable conditions and dormancy (lack of visible growth) during unfavourable conditions. Plant growth ceases in response to declining daylength and temperatures and buds become dormant. This deepest state of dormancy is often referred to as endodormancy, when lack of growth is controlled by factors within the bud (Lang, 1987). There are limited data on the environmental cues that induce endodormancy in *Actinidia* buds. However, early studies with *A. chinensis* var. *deliciosa* 'Bruno' suggest that reducing daylength in autumn is the primary driver but this can be modified by temperature, with high temperatures advancing endodormancy and low temperatures delaying it (Lionakis and Schwabe, 1984). Cold tolerance of buds also increases in association with bud endodormancy. In 'Hayward' vines, cold tolerance increased from −6.5°C in autumn to −17°C in midwinter, before decreasing again to −5°C in early spring (Genitsariotis *et al.*, 1999). The accumulation of chilling during winter gradually resets the ability of the bud meristem to resume growth. Then the bud enters ecodormancy, and growth is regulated by the environment (Lang, 1987) and a period of warmer temperature is needed for budbreak to begin.

14.3.1 Winter chilling

Winter chilling of perennial plants, including kiwifruit, is complex and understanding has been hampered by experimental use of many different temperature models, plant units (cuttings, canes, plants), environmental conditions and outcomes (budbreak and flowering responses) to describe results. Maintenance of bud capacity is dependent on chilling to ensure full subsequent development of vegetative and floral meristems (Brundell, 1975b). However, *A. chinensis* var. *deliciosa* does not have an obligate cold temperature requirement for flower development, as cuttings of 'Hayward', Alpha and *A. chinensis* var. *deliciosa* 'Monty' collected before leaf fall were able to flower without cold treatment (Brundell, 1976).

Several temperature models have been used to describe the winter chilling requirements of kiwifruit including chilling hours below 7.2°C (Weinberger, 1950), the Utah or Richardson model (Richardson *et al.*, 1974), the dynamic chill model (Fishman *et al.*, 1987; Erez *et al.*, 1990), the daily positive chill unit model (Linsley-Noakes *et al.*, 1994) and mean air temperatures (Hall and McPherson, 1997). As these models are based on temperature, they are closely related with high correlations reported between the dynamic and Utah models ($r = 0.99$), chilling hours below 7.2°C and the dynamic model ($r = 0.97$) and chilling hours below 7.2°C and the Utah model ($r = 0.95$) (Zhao *et al.*, 2017).

Controlled-environment studies conducted at constant temperatures suggested that the chilling requirement of *A. chinensis* var. *deliciosa* buds is met at 4–11°C. However, studies conducted at varying temperatures, typical of field conditions, indicated that interactions between the negative effects of high temperatures and positive effects of low temperatures affected the accumulation of winter chilling in *A. chinensis* var. *deliciosa* 'Bruno', 'Hayward' and 'Tomuri' buds, suggesting a more complex relationship (Nee, 1991; Salinger and Kenny, 1995; Therios and Porlingis, 1997).

It is not clear when endodormancy starts in *Actinidia* buds and winter chill accumulation should begin. Many studies have used a calendar date (e.g. the beginning of winter) as the start of endodormancy (Brundell, 1976) while others have used the start of positive chill accumulation (Zhao *et al.*, 2017) or the date of the autumn equinox (Hall and McPherson, 1997). New studies using consistent methodologies are required to define the continuum of dormancy, as well as the relationship between winter temperatures and vine physiology. This understanding of winter chilling and endodormancy could then be used to compare different cultivars, growing regions and the effects of various management techniques.

The depth and duration of endodormancy, as well as winter chilling requirements, vary greatly between *Actinidia* taxa. In a cutting experiment of *A. chinensis* var. *deliciosa* genotypes, Alpha and 'Monty' required a shorter period of winter chilling at 4°C than 'Hayward' for budbreak and flowering to occur (Brundell, 1976). Cuttings of *A. chinensis* var. *chinensis* 'Golden Sunshine' and 'Golden Dragon' required 700 h and 800 h (Utah chill units), respectively, while within *A. chinensis* var. *deliciosa*, 'AU Fitzgerald' needed 800 h and 'Hayward' and the male cultivars 'AU Authur' and 'Matua' required 900 h of chilling to reach maximum budbreak (Wall *et al.*, 2008) (Table 14.2). Zhao *et al.* (2017) suggested that the winter chilling required to reach 50% budbreak (Table 14.2) was positively correlated with the ploidy of the genotype; however, we suggest that these differences may reflect the environment that the different genotypes originated from rather than their ploidy level. Rootstocks can also affect the chilling requirement of scions, e.g. *A. chinensis* var. *deliciosa* 'Allison' on an unnamed *A. chinensis* var. *deliciosa* rootstock required more chilling compared with 'Allison' on its own roots (Allan *et al.*, 1997).

14.3.2 Bud endodormancy mechanisms

Establishment of endodormancy in kiwifruit is accompanied by significant physiological, metabolic and transcriptomic changes in buds. Early metabolic studies showed that during the cessation of growth and onset of endodormancy, respiration of 'Hayward' buds declined and remained low from late autumn until the end of winter (McPherson *et al.*, 1997). Sucrose and hexose concentrations in 'Hayward' bud meristems increased rapidly in late autumn as leaves abscised and buds became dormant, and concentrations of these sugars remained constant until the end of winter (Richardson *et al.*, 2010). A rapid increase in bud respiration and a marked decline in sucrose as well as an increase in hexose concentrations in 'Hayward' bud meristems

Table 14.2. The Utah (Richardson) chill units required to reach 50% budbreak (Zhao *et al.*, 2017) or maximum budbreak (Wall *et al.*, 2008) for a range of *Actinidia* cultivars.

Cultivar	Taxon	Ploidy	Sex	Chill units	Reference
'Donghong'	*A. chinensis* var. *chinensis*	2*x*	female	316	Zhao *et al.* (2017)
'Wuzhi-7'	*A. chinensis* var. *chinensis*	2*x*	female	673	Zhao *et al.* (2017)
'Jinyu'	*A. chinensis* var. *chinensis*	2*x*	female	693	Zhao *et al.* (2017)
'Golden Dragon'	*A. chinensis* var. *chinensis*	2*x*	female	800	Wall *et al.* (2008)
'Golden Sunshine'	*A. chinensis* var. *chinensis*	4*x*	female	700	Wall *et al.* (2008)
'Jinxia'	*A. chinensis* var. *chinensis*	4*x*	female	728	Zhao *et al.* (2017)
'Jintao'	*A. chinensis* var. *chinensis*	4*x*	female	962	Zhao *et al.* (2017)
'H-15'	*A. chinensis* var. *deliciosa*	4*x*	female	991	Zhao *et al.* (2017)
'AU Fitzgerald'	*A. chinensis* var. *deliciosa*	6*x*	female	800	Wall *et al.* (2008)
'Hayward'	*A. chinensis* var. *deliciosa*	6*x*	female	900	Wall *et al.* (2008)
'Matua'	*A. chinensis* var. *deliciosa*	6*x*	male	900	Wall *et al.* (2008)
'AU Authur'	*A. chinensis* var. *deliciosa*	6*x*	male	900	Wall *et al.* (2008)
'Chunmi-1'	*A. chinensis* var. *deliciosa*	6*x*	female	913	Zhao *et al.* (2017)
'Bruno'	*A. chinensis* var. *deliciosa*	6*x*	female	956	Zhao *et al.* (2017)
'Jinkui'	*A. chinensis* var. *deliciosa*	6*x*	female	961	Zhao *et al.* (2017)

3–4 weeks before budswell suggest that this is when the transition between endo- and ecodormancy occurs (McPherson *et al.*, 1997; Richardson *et al.*, 2010).

More recently, widespread changes in transcriptome profiles have been identified during both the onset and release of buds from endodormancy. As *A. chinensis* var. *chinensis* 'Hort16A' and 'Zesy002' buds entered endodormancy, the expression of developmental regulators was reduced and the expression of genes implicated in oxidative, temperature, osmotic or water stress, synthesis of heat-shock proteins, accumulation of osmoprotectants and sequestering of iron and nitrogen were upregulated (Brian *et al.*, 2021; Voogd *et al.*, 2022). Abscisic acid (ABA) levels in buds increased during endodormancy and there was an increase in the expression of ABA and drought response genes in 'Hayward' buds (Wu *et al.*, 2017; Wu, 2018).

Homologs of the *Arabidopsis* flowering repressor *SHORT VEGETATIVE PHASE* (*SVP*), often called *DORMANCY-ASSOCIATED MADS-box* genes (*DAM*), have been identified as central regulators of dormancy (Yang *et al.*, 2021). The expression pattern of *SVP*-like genes has been correlated with dormancy of many perennial plants, including 'Hayward' (Wu *et al.*, 2012) in which four *SVP* genes have been characterized. Expression of *SVP1*, *SVP2* and *SVP4* is elevated in buds during winter (Wu *et al.*, 2012) and their roles in bud dormancy have been confirmed by overexpression in kiwifruit (Wu *et al.*, 2017; Wu, 2018). Analysis has suggested a synergistic mode of action between the *SVP* genes to prevent premature budbreak before the winter chilling requirement has been fulfilled (Wu, 2018; Wu *et al.*, 2018). A direct interaction between *SVP2* and the regulatory elements of genes involved in the ABA pathway and dehydration-mediated growth repression suggests that *SVP2* modulates developmental and osmotic stress pathways in 'Hayward' (Wu *et al.*, 2017; Wu, 2018).

Expression of the phosphatidylethanolamine-binding protein (PEBP) genes *FLOWERING LOCUS T* (*FT*), *TERMINAL FLOWER1* (*TFL1*)/*CENTRORADIALIS* (*CEN*) and *BROTHER OF FT AND TFL1* (*BFT*) in buds is associated with various stages of dormancy, winter chilling, resumption of growth and flowering in 'Hayward' and 'Hort16A' vines (Varkonyi-Gasic *et al.*, 2013; Voogd *et al.*, 2017; Varkonyi-Gasic, 2018) (Fig. 14.6). Although these genes are better known for their role in regulation of flowering across angiosperms (Wickland and Hanazawa, 2015; Putterill and Varkonyi-Gasic, 2016), they have been implicated in regulation of growth in trees (Hsu *et al.*, 2011). In kiwifruit, editing of *BFT* using CRISPR-Cas9 did not affect flowering but delayed the onset of dormancy, giving rise to an ever-growing 'Hort16A' vine (Herath *et al.*, 2022) (Fig. 14.7).

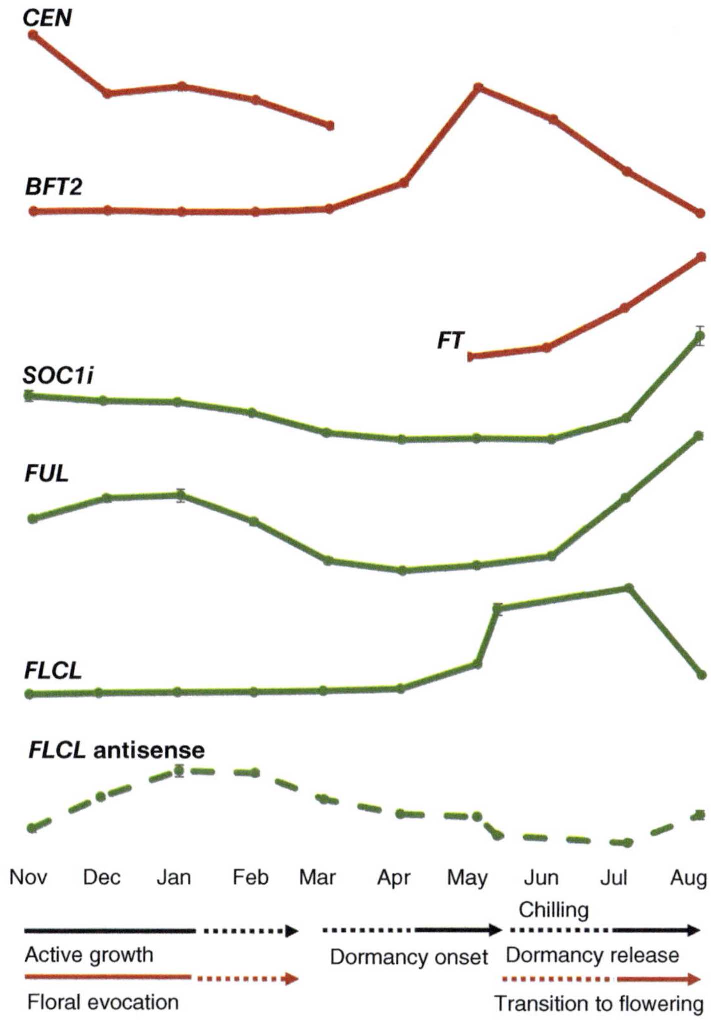

Fig. 14.6. A schematic diagram of expression of key phosphatidylethanolamine-binding protein (PEBP; red lines) and MADS-box genes (green lines) in axillary buds over a season in New Zealand. (Data from Brian *et al*., 2021 and Voogd *et al*., 2022)

Accumulation of winter chilling and dormancy release in 'Hort16A' and 'Zesy002' buds are characterized by cell wall modification and a reduction in stress response factors, followed by accumulation of growth- and photosynthesis related transcripts; and, finally, expression of cell division and expansion genes at budbreak (Brian *et al*., 2021; Voogd *et al*., 2022). Genes with homology to *Arabidopsis* floral integrator gene *SUPPRESSOR OF OVEREXPRESSION OF CONSTANS1* (*SOC1*) displayed distinctive expression patterns in buds during dormancy, and overexpression of one (*SOC1i*) was able to promote budbreak in 'Hort16A' (Voogd *et al*., 2015). Multiple homologs of *APETALA1* (*AP1*) and *FRUITFUL* (*FUL*) are associated with budbreak and resumption of shoot growth in 'Hort16A' and 'Zesy002' (Brian *et al*., 2021; Voogd *et al*., 2022). Surprisingly, a gene with homology to *FLOWERING LOCUS C* (*FLC*), a major floral repressor in *Arabidopsis*, has been shown to promote bud growth in 'Zesy002' and is gradually upregulated by cold (Voogd *et al*., 2022).

Fig. 14.7. CRISPR-Cas9-mediated mutagenesis of kiwifruit *BFT2* results in an ever-growing but not early-flowering *Actinidia chinensis* var. *chinensis* 'Hort16A' phenotype. (Photo: The New Zealand Institute for Plant and Food Research Ltd)

14.4 Flower Development

In many woody perennials, including *Actinidia*, flower development takes place over two seasons, but many of the processes involved are yet to be fully understood. It is possible that flower development is started by environmental changes inducing signals in the leaves that travel to the shoot apical meristem (SAM) to induce floral evocation (commitment to flowering) and floral initiation (the appearance of differentiated cells in the SAM). It is completed with the differentiation of inflorescences and anthesis in spring of the following season (Walton *et al.*, 1997; Noyce *et al.*, 2019). Morphological, physiological and molecular aspects of floral development during these stages have been studied in several *Actinidia* spp. and are described below.

14.4.1 Floral evocation

Timing

Considerable effort has been devoted to establishing the timing of floral evocation and initiation in *A. chinensis* var. *deliciosa* buds. Generally, studies have shown that inflorescence meristems appear in developing axillary buds during spring and summer of the first season of flower development, concurrent with anthesis and early fruit development of the current season's crop (Brundell, 1975b; Fabbri *et al.*, 1992; Snelgar and Manson, 1992; Snowball 1997a; Walton *et al.*, 1997, 2001). As shoots grow during spring and early summer of the current season, buds are formed in new leaf axils and undergo floral evocation sequentially along developing shoots, over several weeks (Snowball, 1995; Walton *et al.*, 1997). It has been suggested that later-developing 'Hayward' buds contain fewer inflorescence primordia (Snowball, 1996) and later-developing shoots also produce fewer floral buds (Walton *et al.*, 2000).

The timing of floral evocation in *Actinidia* buds has been inferred from the appearance of inflorescence meristem primordia. Early studies, in which developing 'Hayward' shoots were defoliated to manipulate floral development, indicated that floral evocation was influenced by leaf metabolism, but it was unclear whether carbohydrate supply or flowering time signals were involved in the process (Snelgar and Manson, 1992; Snowball, 1996). A combination of shoot girdling with either defoliation or shading of developing shoots showed that removal of leaf signals (defoliation) prevented floral evocation, while shading (lower carbohydrate supply) reduced, but

Fig. 14.8. (**A**) An *Actinidia chinensis* var. *chinensis* 'Hort16A' inflorescence contains a single flower while an *A. latifolia* inflorescence (**B**) may contain up to 198 flowers. (Photos: The New Zealand Institute for Plant and Food Research Ltd)

did not eliminate, subsequent flower production by 'Hayward' buds (Richardson, 2014).

Genetic regulation of transition to flowering

Flowering is regulated by the activation of independent flowering pathways in response to endogenous and environmental stimuli which converge on pathway integrator genes, most notably *FT* (Amasino, 2010). The FT protein is transported through the phloem as 'florigen' to the SAM to determine floral fate (Turck *et al.*, 2008; Putterill and Varkonyi-Gasic, 2016) by outcompeting the flowering repressor TFL1/CEN proteins (Bradley *et al.*, 1996; Périlleux *et al.*, 2019), which then leads to activation of floral meristem identity genes.

Early molecular studies showed that transcripts with homology to the *Arabidopsis* floral meristem identity genes *LEAFY* (*LFY*) and *AP1* accumulated in newly developed 'Hayward' buds when floral evocation is believed to occur (Walton *et al.*, 2001). Similarly, a *FUL*-like gene associated with bract (early inflorescence) development was also expressed in newly formed axillary buds in spring (Varkonyi-Gasic *et al.*, 2011). However, unlike in *Arabidopsis*, *AP1/FUL* and *LFY* were not strictly confined to floral meristems but were also associated with active vegetative development in 'Hort16A' and 'Zesy002' vines (Varkonyi-Gasic *et al.*, 2011; Brian *et al.*, 2021). Therefore, it remains unclear if expression of these genes reflects floral evocation (i.e. inflorescence fate) or vegetative development.

Diploid *A. chinensis* var. *chinensis* 'Hort16A' and Red5 genotypes have three *FT* and multiple *CEN* and *BFT* genes (Varkonyi-Gasic *et al.*, 2013; Voogd *et al.*, 2017; Pilkington *et al.*, 2018) that are differentially expressed at specific phenological stages (Fig. 14.6). All three *Actinidia FT* (*AcFT*) genes can promote flowering and restore normal flowering time in the late-flowering *Arabidopsis ft-1* mutant, confirming that *Actinidia* FT proteins act as flowering signals (Varkonyi-Gasic *et al.*, 2013; Voogd *et al.*, 2017). They all promoted flowering strongly in 'Hort16A', giving rise to *in vitro* flower development in tissue culture. Expression of *AcFT* and *AcFT2* was elevated in basal, source leaves, leading to speculation that they may act as 'florigen' in kiwifruit (Varkonyi-Gasic *et al.*, 2013; Voogd *et al.*, 2017; Moss *et al.*, 2018). However, *FT* genes were expressed in different tissues, implying they have other roles in addition to promotion of flowering (Varkonyi-Gasic *et al.*, 2013; Voogd *et al.*, 2017).

In *A. eriantha*, *Actinidia CEN (AcCEN)* fully repressed flowering and all *CEN* genes delayed flowering in *Arabidopsis*, confirming their roles in repression of flowering (Varkonyi-Gasic *et al.*, 2013; Voogd *et al.*, 2017). *AcCEN and AcCEN4* were the most abundant *CEN* genes, and their expression was confined to the shoot tip and axillary bud. CRISPR-Cas9 editing of both *AcCEN* and *AcCEN4* in 'Hort16A' produced a compact, precocious (reduced juvenile stage) vine, and editing both alleles of either gene produced a rapidly maturing, small vine (Varkonyi-Gasic *et al.*, 2019). This proves that *CEN* and *CEN4* are central repressors of the onset of flowering in kiwifruit. In the absence of *CEN* or *CEN4*, flowering was confined to a single terminal flower in very compact vines, or a terminal flower and axillary flowers in the upper nodes of small vines. These compact, rapid-cycling plants could easily be adopted for indoor growth and present opportunities for fast breeding and testing gene function (Akagi *et al.*, 2019; Varkonyi-Gasic *et al.*, 2021).

14.4.2 Floral differentiation

Timing and stages of floral differentiation

Inflorescence development in *Actinidia* buds starts at budswell in spring (about ten days before budbreak) in the second season of flower development and continues for 7–12 weeks until anthesis (Table 14.1) (Brundell, 1975b). Differentiation of a 'Hayward' inflorescence begins during budswell with the appearance of a bract primordium generally followed by initiation of a large terminal flower and smaller lateral flower meristems (Brundell, 1975b; Polito and Grant, 1984). By budbreak, sepals have initiated in terminal flowers and soon after the petals develop between and inside the sepals. At the advanced budbreak stage, five days after budbreak (DAB), the petals form, and the stamens are initiated at the base of the petals. Soon after, the leaves on the newly developing shoot begin to open (open cluster 10 DAB), and a whorl of stigma initials develops inside the stamens. This whorl expands to form the ovary topped by styles and stigmata (18 DAB). The anthers and filaments develop at

34 DAB, ovules form in the ovary at 45 DAB and flowers reach anthesis at 60 DAB. The development of inflorescences occurs in a similar manner on both female 'Hayward' and male Alpha vines, with only small differences in the timing of events and the formation of the gynoecium and stamens (Brundell, 1975b). Lateral flowers develop in a similar way to terminal flowers, but the timing of events is delayed, and their development is often halted, and flowers aborted before petal formation (Brundell, 1975b). For details on flower morphology see Chapter 3 and on anthesis see Chapter 13.

Typically, genotypes of *A. chinensis* var. *chinensis* and *A. chinensis* var. *deliciosa* produce inflorescences that are each a dichasium, with a terminal or king flower and two lateral flowers (Brundell, 1975b; Richardson *et al.*, 2001). In several commercial female *A. chinensis* var. *chinensis* and *A. chinensis* var. *deliciosa* cultivars, only the terminal flower may develop (Ferguson, 1984). However, many species can produce compound dichasia, with those of *A. latifolia* containing up to 198 flowers (Snowball, 1997a) (Fig. 14.8). Male *Actinidia* vines flower more profusely than females, producing more floral shoots and inflorescences with more flowers (Snowball, 1997a; Akagi *et al.*, 2019).

Variation in the shape and development of floral primordia leads to differences and aberrations in the size and shape of the ovary in 'Hayward' flowers (Brundell, 1975b). The size and shape of ovaries was related to fruit size at harvest in 'Hayward' and 'Zesy002' fruit (Patterson *et al.*, 1999; Richardson *et al.*, 2019). In 'Hayward', flattened ovaries led to flat fruit (Watson and Gould, 1994) and fan-shaped fruit developed from flattened ovaries in which the ovary of the terminal flower fused with that of one or more lateral flower meristems. In 'Hayward', these aberrant flowers are generally found at the basal end of shoots (Watson and Gould, 1993), although this may not be true for all genotypes.

Genetic regulation of floral differentiation

Floral organ identity in angiosperms is defined by classes of conserved homeotic genes, according to the ABC model (Coen and Meyerowitz, 1991). The letters A, B and C, together with D and E, represent the overlapping activities of five classes of floral genes involved in organ morphogenesis. Flower development in *Actinidia* is consistent with this model, with the potential exception of a discrete A class activity for sepal identity. The MADS-box genes *FUL*-like, *FUL*, *APETALA3* (*AP3*: *AP3*-1 and *AP3*-2), *PISTILLATA* (*PI*), *AGAMOUS* (*AG*), and *SEPALLATA* (*SEP*, *SEP1*, *SEP3* and *SEP4*) were identified from the *Actinidia* EST database (Crowhurst *et al.*, 2008; Varkonyi-Gasic *et al.*, 2011) and additional members of each of these gene classes have been found in the 'Hort16A' genome (Brian *et al.*, 2021). Apart from *FUL*-like and *FUL*, gene expression was confined to reproductive organs. *PI* was specifically expressed in petal and stamen tissues whilst *AG* was expressed in stamen, pistil and developing fruit tissues. *AP3* and *SEP* genes were more broadly expressed in flowers and developing fruit (Varkonyi-Gasic *et al.*, 2011; Brian *et al.*, 2021). Expression of these genes is clearly associated with bud and flower developmental stages (Fig. 14.9).

Expression of *AP2* (the only non-MADS-box A-class gene) in flowers is consistent with a role in petal formation (Bowman *et al.*, 1989, 1991). The flower phenotype of a natural *A. chinensis* var. *deliciosa* mutant, Pukekohe dwarf, in which multiple petal whorls developed instead of stamens, was associated with increased expression of *AP2* and the absence of microRNA miR172 (Fig. 14.10) (Varkonyi-Gasic *et al.*, 2012).

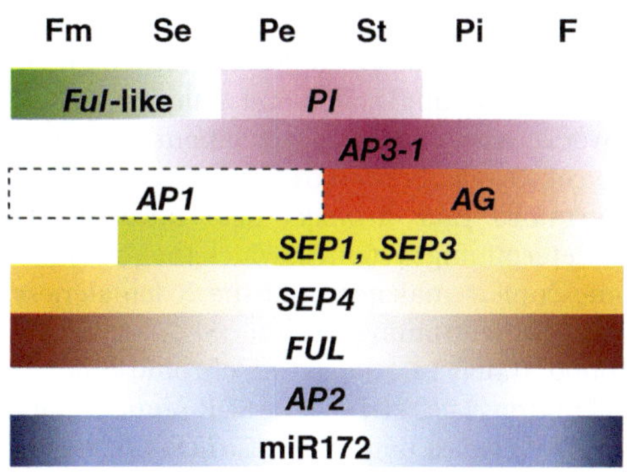

Fig. 14.9. A model for floral meristem and floral organ specification in kiwifruit. Colour intensity reflects gene expression levels. Genes: *AG*, *AGAMOUS*; *AP*, *APETALA*; *FUL*, *FRUITFUL*; *PI*, *PISTILLATA*; *SEP*, *SEPALLATA*. F, fruit; Fm, floral meristem; miR172, microRNA R172; Pe, petal; Pi, pistil; Se, sepal; St, stamen. (Modified from Varkonyi-Gasic *et al.*, 2011)

Fig. 14.10. A natural *Actinidia chinensis* var. *deliciosa* Pukekohe dwarf mutant with multipetallate flowers (*AP2*-mediated promotion of petal over stamen fate in the absence of microRNA R172). (Photos: The New Zealand Institute for Plant and Food Research Ltd)

This is consistent with the antagonistic action of *AP2* and *AG* in specifying outer (sepal and petal) and inner (reproductive organ) flower whorls and the *AP2*/miRNA172 module defining the number of petals in other species, such as roses (Wollmann *et al.*, 2010; François *et al.*, 2018).

14.4.3 Floral sex expression

All *Actinidia* are functionally dioecious, with separate male and female plants bearing unisexual flowers (see Chapter 3). Female (pistillate) flowers undergo androecial development but lack functional pollen and male (staminate) flowers cease gynoecial development upon initiation of stigmas. Previous research (see Chapter 9) determined that *Actinidia* taxa have an XY sex-determining system, in which the Y chromosome confers maleness and the X femaleness. The sex-determining region (SDR) in Red5, around 6 Mbp in size, is on chromosome 25 (Pilkington *et al.*, 2019). The SDR sequence and candidate genes named *Shy Girl* (*SyGl*), encoding a cytokinin response regulator, and *Friendly Boy* (*FrBy*), a fasciclin-like gene essential for male gametogenesis, have been identified (Akagi *et al.*, 2018, 2019). The *SyGl* gene represses gynoecial development through altered cytokinin metabolism and signalling. This was demonstrated in a naturally occurring hermaphrodite with a large deletion including the *SyGl* locus, and CRISPR-Cas9 editing of a male *A. chinensis* var. *chinensis* cultivar 'Bruce', where loss of *SyGl* restored gynoecial development and allowed self-fertility (Akagi *et al.*, 2018, 2019; Varkonyi-Gasic *et al.*, 2021). Expression of the male-promoting gene *FrBy* in a female cultivar ('Hort16A') gave rise to hermaphrodite plants (Akagi *et al.*, 2019) (Fig. 14.11).

14.4.4 Floral abortion

Actinidia spp. can produce up to 14 inflorescences per shoot, but many floral meristems fail to develop or abort partway through development (Fig. 14.12), and this has a significant effect on flower number (Brundell, 1975c; Snowball, 1997b). Inflorescence and floral organ development occur from budswell to anthesis in spring during peak demand for scarce resources. Competition for carbohydrate between leaves, the shoot apex and inflorescences can limit organ development (Brundell, 1975c; Piller *et al.*, 1998; Foster *et al.*, 2007). In 'Hayward' vines, floral abortion occurs during the early stages of floral differentiation in inflorescences that form above the four basal bracts at the bottom of shoots (Brundell, 1975b; Grant and Ryugo, 1982; Walton and Fowke, 1993). Inflorescence abortion just above the basal bracts in 'Hayward' shoots has been associated

Fig. 14.11. An early-flowering hermaphrodite kiwifruit, obtained via CRISPR-Cas9-mediated mutagenesis of *CENTRORADIALIS* genes (*CEN* and *CEN4*) and expression of a sex-determining gene *FRIENDLY BOY* in *Actinidia chinensis* var. *chinensis* 'Hort16A'. (Photo: The New Zealand Institute for Plant and Food Research Ltd)

with the application of hydrogen cyanamide to vines (Walton and Fowke, 1993). Competition for resources during sex organ development, in the later stages of floral differentiation (Hopping and Jerram, 1979) and ovary growth (Lai *et al.*, 1990), can lead to floral abortion further along the shoot with smaller ovaries leading to smaller fruit (Lai *et al.*, 1990; Piller and Meekings, 1997; Piller *et al.*, 1998; Richardson *et al.*, 2019).

Fig. 14.12. A developing *Actinidia chinensis* var. *chinensis* 'Hort16A' shoot with an aborted inflorescence (indicated by the arrow pointing to the lower leaf axil) and four developing inflorescences. (Photo: The New Zealand Institute for Plant and Food Research Ltd)

14.5 Managing Shoot and Bud Development

All *Actinidia* taxa are lianas. In their natural environment, vines grow rapidly from the ground to well-lit positions in the forest canopy, supported by other plants, especially trees (Rowe and Speck, 2005). The growth habit of vines determines their seasonal development pattern and how growth in one season impacts growth in subsequent seasons. Shoots grow rapidly from budbreak in early spring until midsummer, with 'Hayward' shoots reaching growth rates of 60 mm day^{-1} (Greer and Jeffares, 1998), after which shoots become lignified woody canes that store carbohydrate (Richardson *et al.*, 2018). The dynamics of shoot growth, and the ability of shoots to store carbohydrate, influence the characteristics and composition of one-year-old wood and its ability to support budbreak and rapid shoot growth during spring of the following season. This ability is related to factors including shoot vigour, sun exposure, temperature, time of development and carbohydrate accumulation in shoots during the previous growing season.

14.5.1 Shoot growth and vigour

In general, the size of one-year-old wood (canes and spurs) has been shown to have a positive impact on its performance in the following spring. For example, thicker one-year-old 'Hayward' canes produced 20% more shoots and 40% more flowers than thinner canes (Thorp *et al.*, 2003). In *A. arguta* vines, canes had a slightly higher budbreak and produced more floral shoots than smaller terminated spurs, but this did not affect productivity per unit length (Tiyayon and Strik, 2003). Mapping whole vines of two *A. chinensis* var. *chinensis* genotypes showed that although flower production was three-fold higher per unit length on spurs (average length 0.2 m) than on canes (average length 1.6 m), the floral productivity per bud was similar (Richardson *et al.*, 2001). The size of one-year-old wood and its productivity is influenced by several factors including shoot orientation, time of development and environmental conditions during the previous growing season.

The orientation of shoots during their development has a large effect on shoot vigour and the subsequent floral productivity of one-year-old wood. When 'Hayward' shoots grew at 60° above the horizontal on a Y-trellis, the number of flowers they subsequently produced was reduced by 58% compared with that of vines grown on a T-bar trellis (i.e. 0° above horizontal) (Snelgar and Manson, 1990). It has been suggested that training shoots of the vigorous *A. chinensis* var. *chinensis* cultivars 'Hort16A' and 'Zesy002' at 40–60° above a horizontal pergola canopy results in greater productivity compared with those growing within the canopy, owing to higher light exposure (Patterson and Currie, 2011). However, a study of three cultivars ('Zesy002', 'Zes008' (*A. chinensis* var. *chinensis*) and 'Zes007' (*A. chinensis* var. *deliciosa*)) showed that when shoots were all well lit, those growing at 45°, 60° or 90° produced more vigorous canes (increased basal diameter and cane length) and these canes produced 50% fewer flowers than well-lit shoots that grew at their natural orientation (Richardson *et al.*, 2022). Therefore, flower numbers in shoots trained at an angle above a pergola canopy are determined by a trade-off between cane vigour and high light exposure (Fig. 14.13).

The vigour of growing shoots is also influenced by the growing environment, particularly temperature. Increasing the day/night temperatures in a growth chamber from 22/13°C to 30/25°C during summer and early autumn resulted in a ten-fold increase in shoot growth of 'Hayward' vines and a failure to flower in the following season (Richardson *et al.*, 2004). An increase in mean minimum summer temperature (from 12°C to 15°C) over two seasons was also associated with an increase in the duration of shoot growth, a doubling of the amount of shoot growth and a 50% reduction in subsequent flowering of 'Hayward' vines (Richardson, 2014). However, reducing the light available to shoots, by shading during the very early stages of growth, did not consistently affect shoot growth or subsequent floral productivity of 'Hayward' canes (Snelgar and Hopkirk, 1988; Snelgar *et al.*, 1991; Richardson, 2014).

The productivity of one-year-old wood can be influenced by the timing of development in the previous season. In a study of 'Hayward' vines, shoots that developed during late summer (February in New Zealand) produced 42–56% fewer flowers in the following spring compared with those that developed in late spring (November) (Snelgar *et al.*, 1992a). Another study of 'Hayward' shoots showed that those that developed in early spring were larger and subsequently produced 80–100% more flowers per shoot than shoots that developed later in the season (Walton *et al.*, 2000). In that study, early shoots that carried fruit were smaller than early shoots that did not fruit; however, their subsequent floral productivity per unit length or per bud was similar. A recent study of several cultivars showed that late-developing shoots

Fig. 14.13. *Actinidia chinensis* var. *chinensis* 'Zesy002' shoots trained at an angle above a pergola canopy are well lit but vigorous. (Photo: The New Zealand Institute for Plant and Food Research Ltd)

on less-floriferous cultivars ('Hayward' and 'Zesy002') were less productive than early shoots, but there was no effect of time of shoot development on a highly floriferous cultivar *A. chinensis* var. *deliciosa* × *A. chinensis* var. *chinensis* ('Zesh004') (Richardson *et al.*, 2022).

14.5.2 Shoot reserve storage

Carbohydrate stored in vines during summer and autumn supports rapid development of flowers and shoots in the following spring. Carbohydrate reserves accumulate in 'Hayward' vines from late summer until leaf fall in winter (Richardson, 2014). Root and cane reserves are the main source of carbohydrate for budbreak and growth of new shoots, as the carbon balance of young 'Hayward' shoots remains negative for at least four weeks after budbreak (Piller and Meekings, 1997; Greer *et al.*, 2003; Richardson *et al.*, 2018). There was a positive correlation between carbohydrate concentrations in internodes and buds of 'Hayward' canes in winter and the amount of budbreak, floral budbreak and the number of flowers produced in the following spring (Richardson, 2014). Competition for carbohydrate between leaves, the shoot apex and inflorescences in spring can affect shoot development by limiting leaf growth and causing shoot tip and flower abortion (Brundell, 1975b; Piller and Meekings, 1997; Foster *et al.*, 2007). Several factors, including light, thinning, girdling, defoliation and root pruning, can affect the accumulation of vine reserves in woody tissues and hence shoot and flower development in the following season.

The amount of light to which leaves are exposed has a significant effect on vine reserves. Floral production of 'Hayward' vines was reduced by growth under reduced amounts of photosynthetically active radiation (PAR) but not under altered light quality (low or high red:far-red light ratios) (Morgan *et al.*, 1985). Shading 'Hayward' vines in the orchard throughout the growing season was associated with a 20–50% reduction in their subsequent floral production (Snelgar *et al.*, 1991). Later studies found that shading vines during midsummer or autumn, when vines accumulate reserves, had the most effect on their subsequent floral production.

Shading shoots from midsummer to autumn reduced floral production of 'Hayward' by 15–23% (Snelgar *et al.*, 1992b), *A. chinensis* var. *chinensis* 'Cuiyu' by 36% and *A. chinensis* var. *chinensis* 'Fengyue' by 90%, but had no effect on *A. chinensis* var. *deliciosa* 'Miliang No. 1' in the following spring (Wang *et al.*, 2007). Shading *A. arguta* vines for two months prior to harvest reduced the number of flowers produced in the following spring by 48% (Tiyayon and Strik, 2004). Early defoliation of canes in autumn has also been associated with reduced flower numbers on 'Hayward' canes in the following season (Snelgar *et al.*, 1997). Shading 'Hayward' vines in late summer and autumn over two growing seasons reduced the total root carbohydrate by 40–50% and the cane carbohydrate by 17% (Richardson *et al.*, 2018) and has also been associated with reduced sucrose in bud meristems in winter (Richardson *et al.*, 2010).

There is competition between the fruit and woody storage organs for carbohydrate during summer and autumn. Altering the partitioning of carbohydrate between these sinks can affect their growth and storage capacity and hence the quantity of reserves to support bud growth in the following spring. For example, girdling 'Hayward' canes or vine trunks from late summer to autumn altered carbohydrate partitioning between the root system, the canopy and fruit (Richardson *et al.*, 2018). Girdling 'Hayward' canes at this time increased cane carbohydrate by 19% (Richardson, 2014), boosted sucrose concentrations in meristems during winter (Richardson *et al.*, 2010) and increased flower production by 34–40% in the following season (Snelgar and Manson, 1992) and 22–140% over three consecutive seasons (Richardson *et al.*, 2018). In contrast, when carbohydrate supply to 'Hort16A' vines was limited by high crop load and low leaf number, the vines produced fewer flowers over four seasons than those that had high carbohydrate supply owing to a low crop load and high leaf number (Boyd and Barnett, 2011).

14.6 Winter Dormancy

Actinidia vines cease growth and become more cold-tolerant during autumn, and buds become endodormant as the environment becomes less favourable. Once buds become endodormant, they require a period of winter chilling to restore their ability to grow, followed by a period of warmth to stimulate bud growth (ecodormancy). During winter, key management techniques that influence budbreak and flower development include winter pruning (see Chapter 12) and the application of budbreak enhancers. It is also useful to be able to predict both the onset of budbreak and floral productivity of vines in the following season (McPherson *et al.*, 1994).

14.6.1 Predicting the time of budbreak and flower number

The success of temperature-based models to predict *Actinidia* vine phenology is highly site-specific. For example, in China, Zhao *et al.* (2017) found that the Dynamic and Utah models both gave good predictions of the date that a range of cultivars would reach 50% budbreak. In New Zealand, Hall and McPherson (1997) found a weighted model based on mean temperature during autumn and winter and a smooth transition from winter chilling to spring warming gave the best predictions of mean budbreak date for 'Hayward' vines. A similar response was reported for 'Hayward' by Salinger *et al.* (1993) if there was sufficient winter chilling; however, if there was insufficient winter chilling, the relationship between the time of budbreak and temperature was dependent on cold temperatures right up to budbreak. Predicting the start of budbreak for 'Zesy002' is more difficult than for 'Hayward'. Recently a molecular test has been used to indicate the application date for specific dormancy regulators including hydrogen cyanamide (Collen, 2016) and Erger® (Hoeberichts *et al.*, 2017) to achieve optimum responses from 'Hayward' and 'Zesy002' vines.

Bud respiration rates and autumn and winter temperatures have also been used to predict the number of flowers on kiwifruit vines. The respiration rate of 'Hayward' buds in autumn (17 weeks before budbreak) predicted flower production of vines in the following spring (McPherson *et al.*, 1997). Snelgar *et al.* (2008) found a relationship between mean temperatures during winter (May, June and July in New Zealand) and flower numbers produced by 'Hayward' buds, with an increase of 0.45 flowers per winter bud per 1°C decrease in mean winter temperature.

14.6.2 Techniques to improve budbreak and flowering

Several techniques, including the application of budbreak enhancers such as hydrogen cyanamide and management techniques such as pruning and cane tipping, have been used to manage budbreak and flower production in kiwifruit. Hydrogen cyanamide is a highly effective budbreak enhancer that has been widely used on 'Hayward' vines since the 1980s (Linsley-Noakes, 1989) in many countries including New Zealand (Richardson *et al.*, 1994; McPherson *et al.*, 2001), Italy (Costa *et al.*, 1997; Inglese *et al.*, 1998), Portugal (Veloso *et al.*, 2003), Spain (Salinero *et al.*, 2007), Brazil (Schuck and Petri, 1995) and Thailand (Pichakum *et al.*, 2018). It has also been used on *A. chinensis* var. *deliciosa* 'Bruno' (Nee, 1991; Pichakum *et al.*, 2018) and 'Allison' vines (Allan, 2004) as well as *A. chinensis* var. *chinensis* 'Hort16A' and 'Zesy002' vines (Hernández and Craig, 2015a, 2016a). Hydrogen cyanamide has multiple benefits, including increasing budbreak, the proportion of flowering shoots and the number of flowers produced per shoot (collectively measured by the number of flowers per winter bud (FWB)), as well as thinning unwanted lateral flowers from inflorescences and advancing and reducing the duration of budbreak and flowering (Linsley-Noakes, 1989; McPherson *et al.*, 2001) (Fig. 14.14).

Hydrogen cyanamide is generally applied to vines 4–5 weeks before budbreak (Schuck and Petri, 1995; Costa *et al.*, 1997), although applications from 2–8 weeks before budbreak have been reported to be effective. In general, 3% and 4% active ingredient (ai) gave the best results in terms of balancing optimal floral productivity and minimal phytotoxicity on 'Hayward' vines over three seasons (Richardson *et al.*, 1994). Applications of 3% ai to 'Hayward' vines across four regions in New Zealand over four seasons advanced the mid-point of budbreak by 40 days and flowering by 20 days, reduced the duration of both budbreak and flowering, increased FWB by up to 1.0 and reduced lateral flower numbers by up to 10% (McPherson *et al.*, 2001).

Several studies have examined the mechanisms underlying responses of 'Hayward' to hydrogen cyanamide. A microarray study showed that hydrogen cyanamide altered the expression of many genes/proteins associated with stress and signalling, including glutathione S-transferase, and those associated with the reactivation of growth (Walton *et al.*, 2009). Wu *et al.* (2012) showed that both hydrogen cyanamide and winter chilling were able to overcome the maintenance of bud endodormancy by *SVP* expression. Both the accumulation of proline (up to 6.5-fold higher) and increased utilization of sucrose have been shown in buds treated with hydrogen cyanamide compared with untreated buds (Walton *et al.*, 1998; Richardson *et al.*, 2010). Morphological studies suggest that floral bud differentiation may be advanced by hydrogen cyanamide application (Engin *et al.*, 2010)

Fig. 14.14. Buds on (**A**) untreated and (**B**) hydrogen cyanamide treated *Actinidia chinensis* var. *chinensis* 'Zesy002' vines. (Photos: The New Zealand Institute for Plant and Food Research Ltd)

and this may explain why increases in ovary fresh weight (6–20%) and fruit size (11%) can occur in treated vines (Patterson *et al.*, 1999).

More recently, several new budbreak enhancers have been tested on kiwifruit vines, although data have only been published for Armobreak®, BluPrins® and Erger® (Table 14.3). Erger® applications to 'Hayward', 'Hort16A' and 'Zesy002' vines (Di Tommaso *et al.*, 2011; Hernández and Craig, 2015a, 2016a, b; Hoeberichts *et al.*, 2017) resulted in increases in FWB compared with untreated vines. BluPrins® applications to 'Hayward' and 'Zesy002' and Armobreak® applications to 'Hayward' and 'Hort16A' vines also increased FWB compared with untreated vines (Hernández and Craig, 2015a, b, 2016a). A combination of 10% garlic oil and 2% mineral oil also increased total budbreak in single node 'Bruno' cuttings but effects on flowering were not determined (Santos and Gilreath, 2006). A gene expression study of the mode of action of Erger® on 'Hayward' buds implicated genes associated with sugar, nitrogen, hormone and stress metabolism (Hoeberichts *et al.*, 2017).

In very warm growing regions such as South Africa, flower production was increased threefold with a combination of management factors including 30–40% shade, evaporative cooling and the application of hydrogen cyanamide (Allan, 2004). In New Zealand, tipping or removing the apical bud on 'Hayward' canes after the onset of xylem sap movement in spring increased flower numbers by 20–51% in 'Hayward' vines (Manson *et al.*, 1991), particularly where the amount of winter chilling was low (Manson and Snelgar, 1995). Several pruning techniques have also been used to improve the floral productivity of vines including wood selection and vine management systems (see Chapter 12).

14.7 Future Perspectives

Much focus has been placed on specific phases of bud and flower development in key commercial *Actinidia* cultivars. Developing technologies, particularly genomic and genetic tools, and new understanding from other plants has expanded our knowledge of bud and flower development in *Actinidia*. However, it has also exposed large gaps in our knowledge. A more comprehensive overview is required to define and fill these gaps, using studies with consistent terminology and methodologies across a range of genetic material and environmental conditions.

Climate change will be a key factor determining future vine productivity and profitability across current and new growing regions of the world. In the past, winter chilling has been a key focus for flower production in many deciduous crops, including kiwifruit. However, given that bud and flower development take place over two seasons, a much wider seasonal view is required to develop understanding of the effects of environmental conditions on vine productivity. It is also important that such studies link the biological responses of the plant closely with environmental factors. In the past, many studies of winter chilling focused heavily on temperature models with limited correlations to vine biology or investigation of the mechanisms involved. Being able to understand and predict the effects of seasonal variation and extreme weather events

Table 14.3. The increase in number of flowers per winter bud (FWB) in vines treated with budbreak enhancers compared with untreated vines of several *Actinidia* cultivars.

Budbreak enhancer	Cultivar	Increase in FWB	References
Armobreak®	'Hayward'	0.38	Hernández and Craig (2016b)
Armobreak®	'Hort16A'	0.87	Hernández and Craig (2015a)
Bluprins®	'Hayward'	0.15	Hernández and Craig (2015b)
Bluprins®	'Zesy002'	0.85	Hernández and Craig (2015b)
Erger®	'Hayward'	0.33	Di Tommaso *et al.* (2011)
Erger®	'Hayward'	0.35	Hernández and Craig (2016b)
Erger®	'Hayward'	0.85	Hoeberichts *et al.* (2017)
Erger®	'Hort16A'	0.63	Di Tommaso *et al.* (2011)
Erger®	'Hort16A'	0.31	Hernández and Craig (2015a)
Erger®	'Zesy002'	0.86	Hernández and Craig (2016a)

on vines will be important to match cultivars to growing environments in seasons to come.

Kiwifruit production and associated research have been largely based on a limited number of cultivars, often with similar phenology and biology, which are managed using a limited repertoire of management techniques. In recent years there has been a rapid increase in the number of *Actinidia* cultivars grown commercially, including kiwiberry (*A. arguta*), *A. chinensis* var. *chinensis* and *A. chinensis* var. *deliciosa* cultivars, with a further suite of male, female and hermaphrodite vines under development. The diverse range of *Actinidia* germplasm has a much wider range of behaviours that is yet to be explored. The sustainability of some management practices and increasing labour scarcity is likely to influence how vine bud and flower development is managed in future. With these changes, highly integrated and sustainable systems that consider the seasonal cycle of development of individual cultivars and growing regions will be required. Understanding genetic differences and how their effects are influenced by environmental conditions and management manipulations will be essential for the economic viability of future kiwifruit production.

References

Akagi, T., Henry, I.M., Ohtani, H., Morimoto, T., Beppu, K. *et al.* (2018) A Y-encoded suppressor of feminization arose via lineage-specific duplication of a cytokinin response regulator in kiwifruit. *Plant Cell* 30, 780–795. DOI: 10.1105/tpc.17.00787

Akagi, T., Pilkington, S.M., Varkonyi-Gasic, E., Henry, I.M., Sugano, S.S. *et al.* (2019) Two Y-chromosome-encoded genes determine sex in kiwifruit. *Nature Plants* 5, 801–809. DOI: 10.1038/s41477-019-0489-6

Allan, P. (2004) Winter chilling in areas with mild winters: its measurement and supplementation. *Acta Horticulturae* 662, 47–52. DOI: 10.17660/ActaHortic.2004.662.3

Allan, P., Linsley-Noakes, G.C., Holcroft, D.M., Brunette, S.A., Burnett, M.J. *et al.* (1997) Kiwifruit research in a subtropical area. *Acta Horticulturae* 444, 37–42. DOI: 10.17660/ActaHortic.1997.444.2

Amasino, R. (2010) Seasonal and developmental timing of flowering. *Plant Journal* 61, 1001–1013. DOI: 10.1111/j.1365-313X.2010.04148.x

Austin, P.T., Hall, A.J., Snelgar, W.P. and Currie, M.J. (2002) Modelling kiwifruit budbreak as a function of temperature and bud interactions. *Annals of Botany* 89, 695–706. DOI: 10.1093/aob/mcf113

Bowman, J.L., Smyth, D.R. and Meyerowitz, E.M. (1989) Genes directing flower development in *Arabidopsis*. *The Plant Cell* 1, 37–52. DOI: 10.1105/tpc.1.1.37

Bowman, J.L., Smyth, D.R. and Meyerowitz, E.M. (1991) Genetic interactions among floral homeotic genes of *Arabidopsis*. *Development* 112, 1–20. DOI: 10.1242/dev.112.1.1

Boyd, L.M. and Barnett, A.M. (2011) Manipulation of whole-vine carbon allocation using girdling, pruning, and fruit thinning affects fruit numbers and quality in kiwifruit. *HortScience* 46, 590–595. DOI: 10.21273/HORTSCI.46.4.590

Bradley, D., Vincent, C., Carpenter, R. and Coen, E. (1996) Pathways for inflorescence and floral induction in *Antirrhinum*. *Development* 122, 1535–1544. DOI: 10.1242/dev.122.5.1535

Brian, L., Warren, B., McAtee, P., Rodrigues, J., Nieuwenhuizen, N. *et al.* (2021) A gene expression atlas for kiwifruit (*Actinidia chinensis*) and network analysis of transcription factors. *BMC Plant Biology* 21, 121. DOI: 10.1186/s12870-021-02894-x

Brundell, D.J. (1975a) Flower development of the Chinese gooseberry (*Actinidia chinensis* Planch.) I. Development of the flowering shoot. *New Zealand Journal of Botany* 13, 473–483. DOI: 10.1080/0028825X.1975.10430338

Brundell, D.J. (1975b) Flower development of the Chinese gooseberry (*Actinidia chinensis* Planch.) II. Development of the flower bud. *New Zealand Journal of Botany* 13, 485–496. DOI: 10.1080/0028825X.1975.10430339

Brundell, D.J. (1975c) Effect of defoliation, shading, and tipping of shoots on flower bud development and vegetative growth of Chinese gooseberry (*Actinidia chinensis* Planch.). *New Zealand Journal of Agricultural Research* 18, 365–370. DOI: 10.1080/00288233.1975.10421060

Brundell, D.J. (1976) The effect of chilling on the termination of rest and flower bud development of the Chinese gooseberry. *Scientia Horticulturae* 4, 175–182. DOI: 10.1016/S0304-4238(76)80010-6

Clearwater, M.J., Seleznyova, A.N., Thorp, T.G., Blattmann, P., Barnett, A.M. *et al.* (2006) Vigor-controlling rootstocks affect early shoot growth and leaf area development of kiwifruit. *Tree Physiology* 26, 505–515. DOI: 10.1093/treephys/26.4.505

Coen, E.S. and Meyerowitz, E.M. (1991) The war of the whorls – genetic interactions controlling flower development. *Nature* 353, 31–37. DOI: 10.1038/353031a0

Collen, C. (2016) NSure launches BreakNSure test. Available at: www.fruitnet.com/asiafruit/nsure-launches-breaknsure-test/169654.article (accessed 30 September 2022).

Costa, G., Vizzotto, G. and Lain, O. (1997) Fruiting performance of kiwifruit cv Hayward affected by use of cyanamide. *Acta Horticulturae* 444, 473–478. DOI: 10.17660/ActaHortic.1997.444.72

Crowhurst, R.N., Gleave, A.P., MacRae, E.A., Ampomah-Dwamena, C., Atkinson, R.G. *et al.* (2008) Analysis of expressed sequence tags from *Actinidia*: applications of a cross species EST database for gene discovery in the areas of flavor, health, color and ripening. *BMC Genomics* 9, 351. DOI: 10.1186/1471-2164-9-351

Di Tommaso, G., Piaggesi, A. and Mills, R. (2011) Bud break: safe and efficient method of overcoming winter dormancy – New Zealand experiences with 'Hort16A' and 'Hayward' kiwifruit. *Acta Horticulturae* 913, 275–282. DOI: 10.17660/ActaHortic.2011.913.35

Engin, H., Gökbayrak, Z. and Dardeniz, A. (2010) Effects of hydrogen cyanamide on the floral morphogenesis of kiwifruit buds. *Chilean Journal of Agricultural Research* 70, 503–509. DOI: 10.4067/S0718-58392010000300018

Erez, A., Fishman, S., Linsley-Noakes, G.C. and Allan, P. (1990) The dynamic model for rest completion in peach buds. *Acta Horticulturae* 276, 165–174. DOI: 10.17660/ActaHortic.1990.276.18

Fabbri, A., Lisetti, M. and Benelli, C. (1992) Studies on flower induction in kiwifruit. *Acta Horticulturae* 297, 217–222. DOI: 10.17660/ActaHortic.1992.297.28

Ferguson, A.R. (1984) Kiwifruit: a botanical review. *Horticultural Reviews* 6, 1–64. DOI: 10.1002/9781118060797.ch1

Fishman, S., Erez, A. and Couvillon, G.A. (1987) The temperature dependence of dormancy breaking in plants: mathematical analysis of a 2-step model involving a cooperative transition. *Journal of Theoretical Biology* 124, 473–483. DOI: 10.1016/s0022-5193(87)80221-7

Foster, T.M., Seleznyova, A.N. and Barnett, A.M. (2007) Independent control of organogenesis and shoot tip abortion are key factors to developmental plasticity in kiwifruit (*Actinidia*). *Annals of Botany* 100, 471–481. DOI: 10.1093/aob/mcm140

François, L., Verdenaud, M., Fu, X., Ruleman, D., Dubois, A. *et al.* (2018) A miR172 target-deficient AP2-like gene correlates with the double flower phenotype in roses. *Scientific Reports* 8, 12912. DOI: 10.1038/s41598-018-30918-4

Friend, A.P., Alspach, P., Diack, R.N., Seymour, S.M., Shirtliff, A.M. *et al.* (2015) Bud orientation on horizontally trained canes modifies budbreak and shoot fruitfulness of *Actinidia deliciosa* 'Hayward' vines. *Acta Horticulturae* 1096, 87–92. DOI: 10.17660/ActaHortic.2015.1096.6

Genitsariotis, M., Sfakiotakis, E. and Diamandidis, G. (1999) Frost hardiness of kiwifruit buds in relation to seasonal changes of fatty acids. *Acta Horticulturae* 498, 179–184. DOI: 10.17660/ActaHortic.1999.498.21

Grant, J.A. and Ryugo, K. (1982) Influence of developing shoots on flowering potential of dormant buds of *Actinidia chinensis*. *HortScience* 17, 977–978.

Greer, D.H. and Jeffares, D. (1998) Temperature-dependence of carbon acquisition and demand in relation to shoot growth of kiwifruit (*Actinidia deliciosa*) vines grown in controlled environments. *Australian Journal of Plant Physiology* 25, 843–850. DOI: 10.1071/PP98055

Greer, D.H., Cirillo, C. and Norling, C.L. (2003) Temperature-dependence of carbon acquisition and demand in relation to shoot and fruit growth of fruiting kiwifruit (*Actinidia deliciosa*) vines grown in controlled environments. *Functional Plant Biology* 30, 927–937. DOI: 10.1071/FP03111

Hall, A.J. and McPherson, H.G. (1997) Modelling the influence of temperature on the timing of bud break in kiwifruit. *Acta Horticulturae* 444, 401–406. DOI: 10.17660/ActaHortic.1997.444.62

Herath, D., Voogd, C., Mayo-Smith, M., Yang, B., Allan, A.C. *et al.* (2022) CRISPR-Cas9-mediated mutagenesis of kiwifruit *BFT* genes results in an evergrowing but not early flowering phenotype. *Plant Biotechnology Journal* 20, 2064–2076. DOI: 10.1111/pbi.13888

Hernández, G. and Craig, R. (2015a) Effects of alternatives to hydrogen cyanamide on commercial 'Hort16A' kiwifruit production. *Acta Horticulturae* 1096, 109–115. DOI: 10.17660/ActaHortic.2015.1096.9

Hernández, G. and Craig, R. (2015b) Exploring alternative bud break enhancing product in 'Hayward' and 'Zesy002' (Gold3) by painting application. *Acta Horticulturae* 1096, 275–285. DOI: 10.17660/ActaHortic.2015.1096.30

Hernández, G. and Craig, R.L. (2016a) Identifying agrichemical alternatives to hydrogen cyanamide in 'Zesy002' ('Gold3') kiwifruit. *Acta Horticulturae* 1130, 123–129. DOI: 10.17660/ActaHortic.2016.1130.18

Hernández, G. and Craig, R.L. (2016b) Effects of alternatives to hydrogen cyanamide on commercial 'Hayward' kiwifruit production. *Acta Horticulturae* 1130, 131–138. DOI: 10.17660/ActaHortic.2016.1130.19

Hoeberichts, F.A., Povero, G., Ibañez, M., Strijker, A., Pezzolato, D. *et al.* (2017) Next Generation Sequencing to characterise the breaking of bud dormancy using a natural biostimulant in kiwifruit (*Actinidia deliciosa*). *Scientia Horticulturae* 225, 252–263. DOI: 10.1016/j.scienta.2017.07.011

Hopping, M.E. (1990) Floral biology, pollination, and fruit set. In: Warrington, I.J. and Weston, G.C. (eds) *Kiwifruit: Science and Management*. Ray Richards Publisher in association with the New Zealand Society for Horticultural Science, Auckland, New Zealand, pp. 71–96.

Hopping, M.E. and Jerram, E.M. (1979) Pollination of kiwifruit (*Actinidia chinensis* Planch.): stigma-style structure and pollen tube growth. *New Zealand Journal of Botany* 17, 233–240. DOI: 10.1080/0028825X.1979.10426897

Hsu, C.-Y., Adams, J.P., Kim, H., No, K., Ma, C. *et al.* (2011). *FLOWERING LOCUS T* duplication coordinates reproductive and vegetative growth in perennial poplar. *Proceedings of the National Academy of Sciences USA* 108, 10756–10761 DOI 10.1073/pnas.1104713108

Inglese, P., Gullo, G. and Pace, L.S. (1998) Effect of cyanamide on budbreak and cane fruitfulness for 'Hayward' kiwifruit in relation to cane length and time of application. *New Zealand Journal of Crop and Horticultural Science* 26, 45–53. DOI: 10.1080/01140671.1998.9514039

Lai, R., Woolley, D.J. and Lawes, G.S. (1990) The effect of inter-fruit competition, type of fruiting lateral and time of anthesis on the fruit growth of kiwifruit (*Actinidia deliciosa*). *Journal of Horticultural Science* 65, 87–96. DOI: 10.1080/00221589.1990.11516034

Lang, G.A. (1987) Dormancy: a new universal terminology. *HortScience* 22, 817–820. DOI: 10.21273/HORTSCI.22.5.817

Linsley-Noakes, G.C. (1989) Improving flowering of kiwifruit in climatically marginal areas using hydrogen cyanamide. *Scientia Horticulturae* 38, 247–259. DOI: 10.1016/0304-4238(89)90072-1

Linsley-Noakes, G.C., Allan, P. and Matthee, G. (1994) Modification of rest completion prediction models for improved accuracy in South African stone fruit orchards. *Journal of the Southern African Society for Horticultural Sciences* 4, 13–15.

Lionakis, S.M. and Schwabe, W.W. (1984) Some effects of daylength, temperature and exogenous growth regulator application on the growth of *Actinidia chinensis* Planch. *Annals of Botany* 54, 485–501. DOI: 10.1093/oxfordjournals.aob.a086819

Manson, P.J. and Snelgar, W.P. (1995) Regional variations in the response of kiwifruit vines to time of cane tipping. *New Zealand Journal of Crop and Horticultural Science* 23, 67–71. DOI: 10.1080/01140671.1995.9513869

Manson, P.J., Snelgar, W.P. and Seelye, R.J. (1991) Relationships between timing of cane tipping and yield of kiwifruit (*Actinidia deliciosa*). *Journal of Horticultural Science* 66, 415–422. DOI: 10.1080/00221589.1991.11516170

McPherson, H.G., Hall, A.J. and Stanley, C.J. (1994) Seasonal and regional variation in budbreak and flowering of kiwifruit vines in New Zealand. *New Zealand Journal of Crop and Horticultural Science* 22, 263–276. DOI: 10.1080/01140671.1994.9513835

McPherson, H.G., Snelgar, W.P., Manson, P.J. and Snowball, A.M. (1997) Bud respiration and dormancy of kiwifruit (*Actinidia deliciosa*). *Annals of Botany* 80, 411–418. DOI: 10.1006/anbo.1997.0455

McPherson, H.G., Richardson, A.C., Snelgar, W.P. and Currie, M.B. (2001) Effects of hydrogen cyanamide on budbreak and flowering in kiwifruit (*Actinidia deliciosa* 'Hayward'). *New Zealand Journal of Crop and Horticultural Science* 29, 277–285. DOI: 10.1080/01140671.2001.9514189

Morgan, D.C., Stanley, C.J. and Warrington, I.J. (1985) The effects of simulated daylight and shade-light on vegetative and reproductive growth in kiwifruit and grapevine. *Journal of Horticultural Science* 60, 473–484. DOI: 10.1080/14620316.1985.11515654

Moss, S.M.A., Wang, T., Voogd, C., Brian, L.A., Wu, R. *et al.* (2018) *AcFT* promotes kiwifruit in vitro flowering when overexpressed and *Arabidopsis* flowering when expressed in the vasculature under its own promoter. *Plant Direct* 2, e00068. DOI: 10.1002/pld3.68

Nee, C.C. (1991) Breaking dormancy in kiwifruit with chilling or chemical treatments. *New Zealand Journal of Crop and Horticultural Science* 19, 419–421. DOI: 10.1080/01140671.1991.10422887

Noyce, P.W., Offler, C.E., Steel, C.C. and Grof, C.P.L. (2019) Timing of floral evocation in the grapevine (*Vitis vinifera* L. cv. Chardonnay) is identified by cyto-histological changes in the vegetative shoot apical meristem. *Australian Journal of Grape and Wine Research* 25, 252–265. DOI: 10.1111/ajgw.12391

Patterson, K.J. and Currie, M.B. (2011) Optimising kiwifruit vine performance for high productivity and superior fruit taste. *Acta Horticulturae* 913, 257–268. DOI: 10.17660/ActaHortic.2011.913.33

Patterson, K.J., Snelgar, W.P., Richardson, A.C. and McPherson, H.G. (1999) Flower quality and fruit size of Hayward kiwifruit. *Acta Horticulturae* 498, 143–150. DOI: 10.17660/ActaHortic.1999.498.16

Périlleux, C., Bouché, F., Randoux, M. and Orman-Ligeza, B. (2019) Turning meristems into fortresses. *Trends in Plant Science* 24, 431–442. DOI: 10.1016/j.tplants.2019.02.004

Pichakum, A., Chaiwimol, W., Meetam, M. and Songnuan, W. (2018) Responses of green kiwifruit grown in low-chill area to hydrogen cyanamide application. *Acta Horticulturae* 1206, 97–103. DOI: 10.17660/ActaHortic.2018.1206.14

Pilkington, S.M., Crowhurst, R., Hilario, E., Nardozza, S., Fraser, L. *et al.* (2018) A manually annotated *Actinidia chinensis* var. *chinensis* (kiwifruit) genome highlights the challenges associated with draft genomes and gene prediction in plants. *BMC Genomics* 19, 257. DOI: 10.1186/s12864-018-4656-3

Pilkington, S.M., Tahir, J., Hilario, E., Gardiner, S.E., Chagné, D. *et al.* (2019) Genetic and cytological analyses reveal the recombination landscape of a partially differentiated plant sex chromosome in kiwifruit. *BMC Plant Biology* 19, 172. DOI: 10.1186/s12870-019-1766-2

Piller, G.J. and Meekings, J.S. (1997) The acquisition and utilization of carbon in early spring by kiwifruit shoots. *Annals of Botany* 79, 573–581. DOI: 10.1006/anbo.1996.0382

Piller, G.J., Greaves, A.J. and Meekings, J.S. (1998) Sensitivity of floral shoot growth, fruit set and early fruit size in *Actinidia deliciosa* to local carbon supply. *Annals of Botany* 81, 723–728. DOI: 10.1006/anbo.1998.0624

Polito, V.S. and Grant, J.A. (1984) Initiation and development of pistillate flowers in *Actinidia chinensis*. *Scientia Horticulturae* 22, 365–371. DOI: 10.1016/S0304-4238(84)80008-4

Putterill, J. and Varkonyi-Gasic, E. (2016) FT and florigen long-distance flowering control in plants. *Current Opinion in Plant Biology* 33, 77–82. DOI: 10.1016/j.pbi.2016.06.008

Richardson, A. (2014) How does carbohydrate supply limit flower development in grape and kiwifruit vines? PhD thesis. Charles Sturt University, Wagga Wagga, Australia.

Richardson, A., Dawson, T.E., Hampton, R.E. and Blank, R.H. (1994) The effects of hydrogen cyanamide and mineral oil on kiwifruit performance. *Proceedings of the 47th New Zealand Plant Protection Conference*, 314–319. DOI: 10.30843/nzpp.1994.47.11045

Richardson, A., Snelgar, W.P., De Silva, H.N. and Paul, H.R. (2001) Variation in flowering within *Actinidia chinensis* vines. *New Zealand Journal of Crop and Horticultural Science* 29, 103–110. DOI: 10.1080/01140671.2001.9514168

Richardson, A., Marsh, K.B., Boldingh, H.L., Pickering, A.H., Bulley, S.M. *et al.* (2004) High growing temperatures reduce fruit carbohydrate and vitamin C in kiwifruit. *Plant, Cell & Environment* 27, 423–435. DOI: 10.1111/j.1365-3040.2003.01161.x

Richardson, A., Walton, E.F., Meekings, J.S. and Boldingh, H.L. (2010) Carbohydrate changes in kiwifruit buds during the onset and release from dormancy. *Scientia Horticulturae* 124, 463–468. DOI: 10.1016/j.scienta.2010.02.010

Richardson, A., Boldingh, H.L., Kashuba, M.P., Nardozza, S. and Greer, D.H. (2018) Kiwifruit reserves: balancing vine growth and fruit productivity. *Acta Horticulturae* 1218, 163–169. DOI: 10.17660/ActaHortic.2018.1218.22

Richardson, A., Boldingh, H., Kashuba, P., Knight, G. and Ellingham, D. (2019) Flowering time determines the weight and composition of *Actinidia chinensis* var. *chinensis* 'Zesy002' kiwifruit. *Scientia Horticulturae* 246, 741–748. DOI: 10.1016/j.scienta.2018.11.043

Richardson, A., Eyre, V., Rebstock, R., Popowski, E. and Nardozza, S. (2022) Factors influencing flower development in kiwifruit vines. *Acta Horticulturae* 1332, 141–153. DOI: 10.17660/ActaHortic.2022.1332.20

Richardson, E.A., Seeley, S.D. and Walker, D.R. (1974) A model for estimating the completion of rest for 'Redhaven' and 'Elberta' peach trees. *HortScience* 9, 331–332. DOI: 10.21273/HORTSCI.9.4.331

Rowe, N. and Speck, T. (2005) Plant growth forms: an ecological and evolutionary perspective. *New Phytologist* 166, 61–72. DOI: 10.1111/j.1469-8137.2004.01309.x

Salinero, M.C., Piñón, M.P. and Sainz, M.J. (2007) Field response to CPPU application in kiwifruit (*Actinidia deliciosa* 'Hayward') orchards treated with hydrogen cyanamide in northwest Spain. *Acta Horticulturae* 753, 555–559. DOI: 10.17660/ActaHortic.2007.753.73

Salinger, M.J. and Kenny, G.J. (1995) Climate and kiwifruit cv 'Hayward'. 2. Regions in New Zealand suited for production. *New Zealand Journal of Crop and Horticultural Science* 23, 173–184. DOI: 10.1080/01140671.1995.9513884

Salinger, M.J., Kenny, G.J. and Morleybunker, M.J. (1993) Climate and kiwifruit cv Hayward. 1. Influences on development and growth. *New Zealand Journal of Crop and Horticultural Science* 21, 235–245. DOI: 10.1080/01140671.1993.9513775

Santos, B.M. and Gilreath, J.P. (2006) Influence of nitrogen fertilization and support systems on passion fruit yield and economic feasibility. *HortTechnology* 16, 43–45. DOI: 10.21273/HORTTECH.16.1.0043

Schuck, E. and Petri, J.L. (1995) The effect of concentrations and application of hydrogen cyanamide on kiwifruit dormancy breaking. *Acta Horticulturae* 395, 177–184. DOI: 10.17660/ActaHortic.1995.395.19

Seleznyova, A.N., Thorp, T.G., Barnett, A.M. and Costes, E. (2002) Quantitative analysis of shoot development and branching patterns in *Actinidia*. *Annals of Botany* 89, 471–482. DOI: 10.1093/aob/mcf069

Snelgar, W.P. and Hopkirk, G. (1988) Effect of overhead shading on yield and fruit quality of kiwifruit (*Actinidia deliciosa*). *Journal of Horticultural Science* 63, 731–742. DOI: 10.1080/14620316.1988.11515918

Snelgar, W.P. and Manson, P.J. (1990) Influence of cane angle on flower evocation, flower numbers, and productivity of kiwifruit vines (*Actinidia deliciosa*). *New Zealand Journal of Crop and Horticultural Science* 18, 225–232. DOI: 10.1080/01140671.1990.10428099

Snelgar, W.P. and Manson, P.J. (1992) Determination of the time of flower evocation in kiwifruit vines. *New Zealand Journal of Crop and Horticultural Science* 20, 439–447. DOI: 10.1080/01140671.1992.10418062

Snelgar, W.P., Manson, P.J. and Hopkirk, G. (1991) Effect of overhead shading on fruit size and yield potential of kiwifruit (*Actinidia deliciosa*). *Journal of Horticultural Science* 66, 261–273. DOI: 10.1080/00221589.1991.11516153

Snelgar, W.P., Manson, P.J. and Stowell, B.M. (1992a) Relationship between time of shoot growth, shoot stubbing, and return bloom of kiwifruit vines. *New Zealand Journal of Crop and Horticultural Science* 20, 345–350. DOI: 10.1080/01140671.1992.10421777

Snelgar, W.P., Manson, P.J. and Martin, P.J. (1992b) Influence of time of shading on flowering and yield of kiwifruit vines. *Journal of Horticultural Science* 67, 481–487. DOI: 10.1080/00221589.1992.11516273

Snelgar, W.P., Manson, P.J. and McPherson, H.G. (1997) Evaluating winter chilling of kiwifruit using excised canes. *Journal of Horticultural Science* 72, 305–315. DOI: 10.1080/14620316.1997.11515517

Snelgar, W.P., Hall, A.J. and McPherson, H.G. (2008) Modelling flower production of kiwifruit (*Actinidia deliciosa*) from winter chilling. *New Zealand Journal of Crop and Horticultural Science* 36, 273–284. DOI: 10.1080/01140670809510244

Snowball, A.M. (1995) The seasonal cycle of leaf, shoot and bud development in kiwifruit. *Journal of Horticultural Science* 70, 787–797. DOI: 10.1080/14620316.1995.11515353

Snowball, A.M. (1996) The timing of flower evocation in kiwifruit. *Journal of Horticultural Science* 71, 335–347. DOI: 10.1080/14620316.1996.11515413

Snowball, A.M. (1997a) Axillary shoot bud development in selected *Actinidia* species. *New Zealand Journal of Crop and Horticultural Science* 25, 233–242. DOI: 10.1080/01140671.1997.9514012

Snowball, A.M. (1997b) Seasonal cycle of shoot development in selected *Actinidia* species. *New Zealand Journal of Crop and Horticultural Science* 25, 221–231. DOI: 10.1080/01140671.1997.9514011

Srisook, W., Lim, C.-K., Oh, E.U., Yi, K., Kim, S.-C. *et al.* (2015) Effect of artificial defoliation on cane regrowth and fruit development in 'Jecy Gold' kiwifruit. *Horticulture, Environment, and Biotechnology* 56, 22–26. DOI: 10.1007/s13580-015-0119-9

Srisook, W., Lim, C.-K., Oh, E.U., Yi, K., Prathibhana, H.M. *et al.* (2016) Defoliation time influences vine regrowth, off-season flowering, and fruit quality in 'Jecy Gold' kiwifruit vines. *Horticulture, Environment, and Biotechnology* 57, 219–224. DOI: 10.1007/s13580-016-0029-5

Therios, I.N. and Porlingis, I.C. (1997) Effect of low temperatures and their alternation with higher ones on breaking bud dormancy of *Actinidia deliciosa* (cvs. Hayward and Bruno). *Acta Horticulturae* 444, 389–394. DOI: 10.17660/ActaHortic.1997.444.60

Thorp, T.G., Barnett, A.M. and Miller, S.A. (2003) Effects of cane size and pruning system on shoot growth, flowering and productivity of 'Hayward' kiwifruit vines. *Journal of Horticultural Science & Biotechnology* 78, 219–224. DOI: 10.1080/14620316.2003.11511609

Tiyayon, C. and Strik, B. (2003) Effect of fruiting cane origin on fruitfulness of hardy kiwifruit, *Actinidia arguta*. *New Zealand Journal of Crop and Horticultural Science* 31, 179–186. DOI: 10.1080/01140671.2003.9514250

Tiyayon, C. and Strik, B. (2004) Influence of time of overhead shading on yield, fruit quality, and subsequent flowering of hardy kiwifruit, *Actinidia arguta*. *New Zealand Journal of Crop and Horticultural Science* 32, 235–241. DOI: 10.1080/01140671.2004.9514301

Turck, F., Fornara, F. and Coupland, G. (2008) Regulation and identity of florigen: FLOWERING LOCUS T moves center stage. *Annual Review of Plant Biology* 59, 573–594. DOI: 10.1146/annurev.arplant.59.032607.092755

Varkonyi-Gasic, E. (2018) Regulation of flowering in kiwifruit. *Acta Horticulturae* 1218, 129–134. DOI: 10.17660/ActaHortic.2018.1218.16

Varkonyi-Gasic, E., Moss, S.M., Voogd, C., Wu, R., Lough, R.H. *et al.* (2011) Identification and characterization of flowering genes in kiwifruit: sequence conservation and role in kiwifruit flower development. *BMC Plant Biology* 11, 72. DOI: 10.1186/1471-2229-11-72

Varkonyi-Gasic, E., Lough, R.H., Moss, S.M.A., Wu, R. and Hellens, R.P. (2012) Kiwifruit floral gene *APETALA2* is alternatively spliced and accumulates in aberrant indeterminate flowers in the absence of miR172. *Plant Molecular Biology* 78, 417–429. DOI: 10.1007/s11103-012-9877-2

Varkonyi-Gasic, E., Moss, S.M.A., Voogd, C., Wang, T., Putterill, J. *et al.* (2013) Homologs of *FT*, *CEN*, and *FD* respond to developmental and environmental signals affecting growth and flowering in the perennial vine kiwifruit. *New Phytologist* 198, 732–746. DOI: 10.1111/nph.12162

Varkonyi-Gasic, E., Wang, T.C., Voogd, C., Jeon, S., Drummond, R.S.M. *et al.* (2019) Mutagenesis of kiwifruit *CENTRORADIALIS*-like genes transforms a climbing woody perennial with long juvenility and axillary flowering into a compact plant with rapid terminal flowering. *Plant Biotechnology Journal* 17, 869–880. DOI: 10.1111/pbi.13021

Varkonyi-Gasic, E., Wang, T., Cooney, J., Jeon, S., Voogd, C. *et al.* (2021) *Shy Girl*, a kiwifruit suppressor of feminization, restricts gynoecium development via regulation of cytokinin metabolism and signalling. *New Phytologist* 230, 1461–1475. DOI: 10.1111/nph.17234

Veloso, A., Oliveira, M. and Antunes, M.D.C. (2003) The effect of hydrogen cyanamide on bud break and yield of kiwifruit in northwest Portugal. *Acta Horticulturae* 610, 161–164. DOI: 10.17660/ActaHortic.2003.610.21

Voogd, C., Wang, T. and Varkonyi-Gasic, E. (2015) Functional and expression analyses of kiwifruit *SOC1*-like genes suggest that they may not have a role in the transition to flowering but may affect the duration of dormancy. *Journal of Experimental Botany* 66, 4699–4710. DOI: 10.1093/jxb/erv234

Voogd, C., Brian, L.A., Wang, T., Allan, A.C. and Varkonyi-Gasic, E. (2017) Three *FT* and multiple *CEN* and *BFT* genes regulate maturity, flowering, and vegetative phenology in kiwifruit. *Journal of Experimental Botany* 68, 1539–1553. DOI: 10.1093/jxb/erx044

Voogd, C., Brian, L.A., Wu, R., Wang, T., Allan, A.C. *et al.* (2022) A MADS-box gene with similarity to *FLC* is induced by cold and correlated with epigenetic changes to control budbreak in kiwifruit. *New Phytologist* 233, 2111–2126. DOI: 10.1111/nph.17916

Wall, C., Dozier, W., Ebel, R.C., Wilkins, B., Woods, F. *et al.* (2008) Vegetative and floral chilling requirements of four new kiwi cultivars of *Actinidia chinensis* and *A. deliciosa*. *HortScience* 43, 644–647. DOI: 10.21273/HORTSCI.43.3.644

Walton, E.F. (1995) Occurence of secondary floral shoots in summer on *Actinidia eriantha* Benth. *New Zealand Journal of Crop and Horticultural Science* 23, 341–343. DOI: 10.1080/01140671.1995.9513907

Walton, E.F. and Fowke, P.J. (1993) Effect of hydrogen cyanamide on kiwifruit shoot flower number and position. *Journal of Horticultural Science* 68, 529–534. DOI: 10.1080/00221589.1993.11516381

Walton, E.F. and Wu, R.-M. (1999) Buds on *Actinidia arguta* shoots do contain axillary meristems. *New Zealand Journal of Crop and Horticultural Science* 27, 181–185. DOI: 10.1080/01140671.1999.9514094

Walton, E.F., Fowke, P.J., Weis, K. and McLeay, P.L. (1997) Shoot axillary bud morphogenesis in kiwifruit (*Actinidia deliciosa*). *Annals of Botany* 80, 13–21. DOI: 10.1006/anbo.1996.0381

Walton, E.F., Podivinsky, E., Wu, R.-M., Reynolds, P.H.S. and Young, L.W. (1998) Regulation of proline biosynthesis in kiwifruit buds with and without hydrogen cyanamide treatment. *Physiologia Plantarum* 102, 171–178. DOI: 10.1034/j.1399-3054.1998.1020203.x

Walton, E.F., Richardson, A.C., Waller, J.E. and Dow, B.W. (2000) Effect of time of cane initiation on subsequent fruitfulness in kiwifruit. *New Zealand Journal of Crop and Horticultural Science* 28, 271–275. DOI: 10.1080/01140671.1997.9514148

Walton, E.F., Podivinsky, E. and Wu, R.-M. (2001) Bimodal patterns of floral gene expression over the two seasons that kiwifruit flowers develop. *Physiologia Plantarum* 111, 396–404. DOI: 10.1034/j.1399–3054.2001.1110318.x

Walton, E.F., Wu, R.-M., Richardson, A.C., Davy, M., Hellens, R.P. *et al.* (2009) A rapid transcriptional activation is induced by the dormancy-breaking chemical hydrogen cyanamide in kiwifruit (*Actinidia deliciosa*) buds. *Journal of Experimental Botany* 60, 3835–3848. DOI: 10.1093/jxb/erp231

Wang, Z.-Y., Patterson, K.J., Gould, K.S. and Lowe, R.G. (1994) Rootstock effects on budburst and flowering in kiwifruit. *Scientia Horticulturae* 57, 187–199. DOI: 10.1016/0304-4238(94)90140-6

Wang, Z.-Y., Yuan, F.-R., He, K.-J. and Bu, F.-W. (2007) Effects of overhead shading on yield and fruit quality of kiwifruit in regions with high temperatures in summer. *Acta Horticulturae* 753, 399–405. DOI: 10.17660/ActaHortic.2007.753.51

Watson, M. and Gould, K.S. (1993) The development of fruit shape in kiwifruit: growth characteristics and positional differences. *Journal of Horticultural Science* 68, 185–194. DOI: 10.1080/00221589.1993.11516342

Watson, M. and Gould, K.S. (1994) Development of flat and fan-shaped fruit in *Actinidia chinensis* var. *chinensis* and *Actinidia deliciosa*. *Annals of Botany* 74, 59–68. DOI: 10.1006/anbo.1994.1094

Weinberger, J.H. (1950) Chilling requirements of peach varieties. *Proceedings of the American Society for Horticultural Science* 56, 122–128.

Wickland, D.P. and Hanzawa, Y. (2015) The *FLOWERING LOCUS T/TERMINAL FLOWER 1* gene family: Functional evolution and molecular mechanisms. *Molecular Plant* 8, 983–997. DOI: 10.1016/j.molp.2015.01.007

Wollmann, H., Mica, E., Todesco, M., Long, J.A. and Weigel, D. (2010) On reconciling the interactions between *APETALA2*, miR172 and *AGAMOUS* with the ABC model of flower development. *Development* 137, 3633–3642. DOI: 10.1242/dev.036673

Wu, R.-M. (2018) Kiwifruit *SVP2*, a dormancy regulator. *European Journal of Horticultural Science* 83, 231–235. DOI: 10.17660/eJHS.2018/83.4.3

Wu, R.-M., Walton, E.F., Richardson, A.C., Wood, M., Hellens, R.P. *et al.* (2012) Conservation and divergence of four kiwifruit *SVP*-like MADS-box genes suggest distinct roles in kiwifruit bud dormancy and flowering. *Journal of Experimental Botany* 63, 797–807. DOI: 10.1093/jxb/err304

Wu, R.-M., Wang, T., Warren, B.A.W., Allan, A.C., Macknight, R.C. *et al.* (2017) Kiwifruit *SVP2* gene prevents premature budbreak during dormancy. *Journal of Experimental Botany* 68, 1071–1082. DOI: 10.1093/jxb/erx014

Wu, R.-M., Wang, T., Allan, A.C., Macknight, R.C. and Varkonyi-Gasic, E. (2018) Overexpression of both *AcSVP1* and *AcSVP4* delays budbreak in kiwifruit *A. chinensis* var. *deliciosa*, but only *AcSVP1* delays flowering in model plants. *Environmental and Experimental Botany* 153, 262–270 DOI: 10.1016/j.envexpbot.2018.06.010

Yang, Q., Gao, Y., Wu, X., Moriguchi, T., Bai, S. *et al.* (2021) Bud endodormancy in deciduous fruit trees: advances and prospects. *Horticulture Research* 8, 139. DOI: 10.1038/s41438-021-00575-2

Zhang, L., Li, Z.-Z., Wang, Y., Jiang, Z., Wang, S. *et al.* (2010) Vitamin C, flower color and ploidy variation of hybrids from a ploidy-unbalanced *Actinidia* interspecific cross and SSR characterization. *Euphytica* 175, 133–143. DOI: 10.1007/s10681-010-0194-z

Zhao, T., Li, D., Li, L., Han, F., Liu, X. *et al.* (2017) The differentiation of chilling requirements of kiwifruit cultivars related to ploidy variation. *HortScience* 52, 1676–1679. DOI: 10.21273/hortsci12410-17

15 Vine Canopy and Root Physiology

Nick Gould[1]* and Michael J. Clearwater[2]
[1]*The New Zealand Institute for Plant and Food Research Ltd, Te Puke, New Zealand;* [2]*The University of Waikato, Hamilton, New Zealand*

15.1 Introduction

With the rapid growth in international kiwifruit production over the last 30 years, understanding of kiwifruit vine physiology has also increased. This new knowledge has contributed to the development and release of new scion and rootstock cultivars, increasing the diversity and quality of the fruit available, and has improved management tools to increase vine productivity and the range of environments in which kiwifruit can be grown. Understanding canopy development and leaf function has led to the development of efficient low-vigour canopy systems that optimize leaf area index, crop load and the reproductive cycle over excessive root and shoot growth. Significant advances have been made in understanding the kiwifruit root system, its roles in storage of reserves, water and nutrient uptake and generating xylem pressure in spring. This chapter reviews advances in the understanding of kiwifruit shoot and root physiology. Unless stated otherwise, the chapter deals with the commercial cultivars of *Actinidia chinensis* var. *chinensis* and *A. chinensis* var. *deliciosa* (syn. *A. deliciosa*).

15.2 Phenological Growth Stages

Knowledge of the annual growth cycle of vine structures is important for understanding vine metabolism and the uptake and partitioning of resources throughout the season. Early in winter, kiwifruit vines are dormant. Although the absolute dates of each phenological stage vary slightly between cultivars and with climate (Thorp *et al.*, 1990; McPherson *et al.*, 1994), generally the season's growth starts in winter with the production of new roots (Fig. 15.1). After about 50–100 days of root growth (depending on when root growth started), new shoots emerge in spring from buds that formed the previous summer, and a period of rapid canopy development occurs. Initially, shoots extend and new leaves emerge as the flowers develop and open about 40–60 days after budbreak (DAB), although the time is variable depending on cultivar (Brundell, 1975; McPherson *et al.*, 1994; Snowball, 1995; Richardson, 2015) (see Chapter 14). Following fruit set, there is a period of about 60–70 days of rapid fruit growth driven by cell division and cell expansion (McPherson *et al.*, 2001; Richardson *et al.*, 2011; Le Lievre

*Corresponding author: nick.gould@plantandfood.co.nz

© CAB International 2024. *Kiwifruit: Botany, Production and Uses* (eds A.C. Richardson, J.N. Burdon and A.R. Ferguson)
DOI: 10.1079/9781800620933.0015

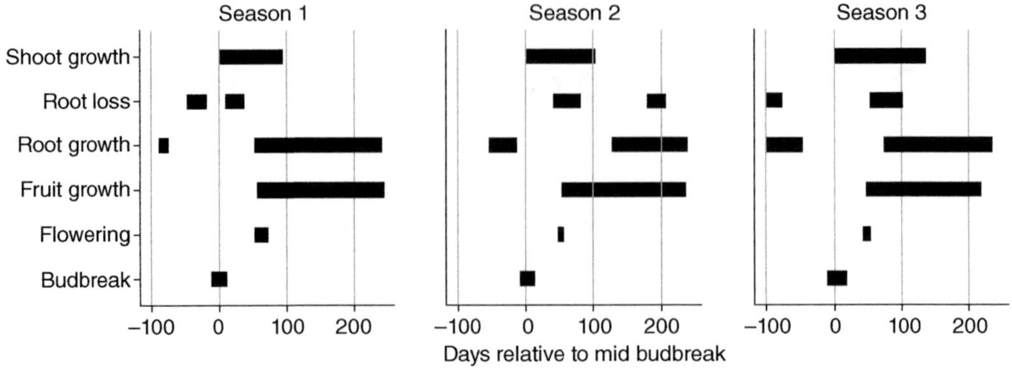

Fig. 15.1. Timing of key growth stages of *Actinidia chinensis* var. *deliciosa* 'Hayward' vines grown in Kerikeri, New Zealand over three years. (From Richardson, 2015)

et al., 2021), followed by a slower growth phase where the fruit accumulates carbohydrate and matures (Minchin *et al.*, 2003; Nardozza *et al.*, 2013; Le Lievre *et al.*, 2021) (see Chapters 16 and 20).

After budbreak in spring, the development of new roots stops and there is a period of root death concurrent with the period of rapid shoot growth which continues for about 60–90 days. After this interval of rapid growth, shoot growth continues at a slower rate up to about 120 DAB. During this period new root growth starts again, continuing until after fruit harvest, sometimes with a late summer/autumn flush (Buwalda and Hutton, 1988; Black, 2012; Richardson, 2015; Bennewitz *et al.*, 2019; Le Lievre *et al.*, 2021). The timing and extent of root growth and death depends on the season and may occur throughout spring until autumn with a continuous turnover of roots during this period, (Fig. 15.1) (Reid *et al.*, 1993; Richardson, 2015).

15.3 The Canopy

15.3.1 Canopy establishment

Rapid canopy establishment in spring is important to provide photosynthate for fruit growth and to replenish vine reserves. Increasing the availability of carbohydrate during the rapid fruit growth phase has been shown to increase fruit size and quality. Manipulation of the leaf:fruit ratio and spring trunk girdling can increase carbohydrate availability by reducing competition for carbohydrate or increasing the photosynthetic rate (Minchin *et al.*, 2010; Currie *et al.*, 2018) (see Chapter 12). The rate of canopy development in spring and early summer is temperature-dependent. In a three-year study on unpruned *Actinidia chinensis* var. *deliciosa* 'Hayward' shoots, rapid canopy development continued for around 100 DAB in the first two seasons, but in the third season, where the early summer minimum temperatures were 2.0–2.6°C higher on average compared with the first two seasons (15.0°C vs. 12.5°C), rapid growth continued for 145 DAB (Fig. 15.1) (Richardson, 2015). Increased shoot vigour is not necessarily at the expense of fruit growth. Trials carried out on potted 'Hayward' vines grown in controlled environment rooms have also demonstrated increased shoot growth with increasing temperatures (Greer and Jeffares, 1998; Richardson *et al.*, 2004). The increased shoot growth enhanced surplus carbon acquisition (i.e. carbon not required for the observed increase in shoot growth), making more carbohydrate available for the fruit during its development (Greer and Jeffares, 1998).

15.3.2 Leaf function

Kiwifruit plants utilize the C3 photosynthetic pathway, with inorganic carbon fixed and metabolized into sugar within the chloroplasts of the leaf mesophyll cells (Zheng *et al.*, 2020). Photosynthesis in kiwifruit, like other C3 crops, is highly responsive to a range of environmental factors, including light, temperature, air humidity, soil water supply, CO_2 concentration and vine mineral nutrient status. For individual kiwifruit leaves grown in full sun and optimal conditions, light-saturated rates of photosynthesis (Pmax) are usually between 10 and 20 μmol CO_2 m^{-2} s^{-1} for selections of *A. chinensis* var. *chinensis*, *A. chinensis* var. *deliciosa*, *A. eriantha* and *A. arguta*, with the highest reported rates reaching ~30 μmol CO_2 m^{-2} s^{-1} (Greer, 1995; Jo *et al.*, 2008a; Richardson *et al.*, 2021). Scion and rootstock affect Pmax, with *A. arguta* (unstated genotype) leaves having a lower Pmax than selections of *A. chinensis* var. *chinensis*, *A. chinensis* var. *deliciosa* and *A. eriantha* (Jo *et al.*, 2008b), while leaves of *A. chinensis* var. *chinensis* 'Zes008' grown on *A. chinensis* var. *deliciosa* 'Bruno' rootstock have a higher photosynthetic rate than 'Zes008' grown on *A. macrosperma* (unstated genotype) rootstock (Richardson *et al.*, 2021). Lower Pmax values are observed for younger leaves or those on plants grown indoors, in partial shade or subject to environmental stresses like high altitude (Greer, 1995, 1996; Gong *et al.*, 2015).

As leaves mature, photosynthetic rate per unit leaf area increases; early-developing leaves reach maximum rates just before anthesis and rates remain stable throughout the season (Greer, 1996; Richardson, 2015). However, Pmax increases along the shoot with the basal, early-maturing leaves having lower rates compared with later-maturing leaves in the middle of the shoot. A decrease in Pmax at the apical end of shoots at 80–95 DAB in selections of *A. chinensis* var. *chinensis*, *A. chinensis* var. *deliciosa*, *A. eriantha* and *A. arguta* may be explained by leaves not being mature at this time (Greer, 1996; Jo *et al.*, 2008b; Richardson, 2015). Shoot type also has an effect, with Pmax of leaves on short shoots (less than ten leaves) declining from a maximum of 16 to ~8 μmol CO_2 m^{-2} s^{-1} over the season but Pmax of leaves on longer shoots remaining at their higher rate of ~16 μmol CO_2 m^{-2} s^{-1}. It is unclear why the rate reduces in leaves of short shoots; however, leaf senescence, proximity to fruit and shading by longer shoots may be involved (Richardson *et al.*, 2021).

15.3.3 Environmental effects on photosynthesis

Actinidia are characteristic of C3 plants in that photosynthesis gradually saturates with increasing light and intercellular leaf CO_2 concentration (Ci) (Fig. 15.2a, b). CO_2 release rates in the dark are usually between 1 and 2 μmol m^{-2} s^{-1}, and the light compensation point (light required to achieve net positive photosynthesis) occurs between 20 and 30 μmol m^{-2} s^{-1} photosynthetically active radiation (PAR). Depending on the growth environment, the light-saturation point of photosynthesis for individual leaves can vary between ~750 and ~1500 μmol m^{-2} s^{-1} PAR. In contrast to the light response, the photosynthetic response of *Actinidia* leaves to CO_2 has not been well studied, and there are no published estimates of leaf mesophyll conductance to CO_2. Maximum rates of ribulose-1,5-bisphosphate carboxylation vary from 37 to 59 μmol CO_2 m^{-2} s^{-1} for *A. chinensis* var. *chinensis* and *A. chinensis* var. *deliciosa*, and maximum rates of electron transport vary from 112 to 157 μmol m^{-2} s^{-1} for these varieties (Clearwater *et al.*, 2004; Woodward, 2007). These rates are comparable with those reported for other relatively fast-growing, woody, perennial C3 species (Wullschleger, 1993), but are likely to vary between cultivars and growth environments (Jo *et al.*, 2008b) and in response to stress (see below). Similarly, the stomatal behaviour of *Actinidia* leaves is comparable with other perennial C3 species adapted to mesic environments that have relatively high rates of photosynthesis and transpiration. In mature and unstressed *Actinidia* species in full sun, stomatal conductance values between 0.1 and 0.4 mol m^{-2} s^{-1} are typically reported (Clearwater *et al.*, 2004; Montanaro *et al.*, 2007). However, both photosynthesis and stomatal conductance decline as humidity falls and the vapour pressure deficit rises (Fig. 15.2c), and during periods of drought stress. Stomatal limitation of photosynthesis can therefore contribute to the reductions in photosynthesis and vine productivity observed during periods of environmental stress (see below).

15.3.4 Whole canopy carbon acquisition

To examine the effects of treatments or cultivar differences on photosynthetic performance,

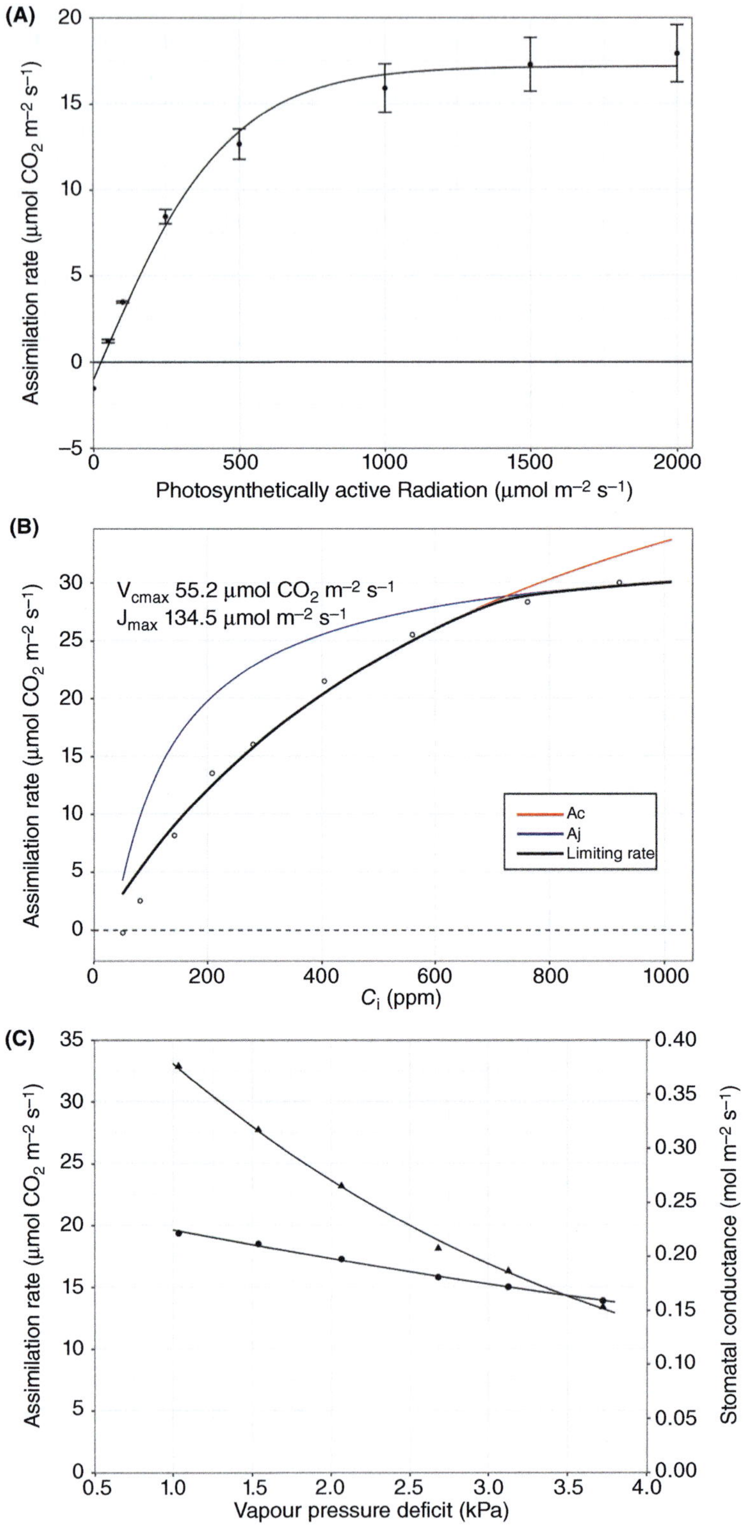

Fig. 15.2. Examples of leaf photosynthetic responses to light, CO_2 and vapour pressure deficit for cropping *Actinidia chinensis* var. *chinensis* 'Hort16A' vines growing in Latina, Italy. (**A**) The response of the

most researchers have relied on the use of portable photosynthesis systems to measure the photosynthetic properties of individual leaves. A few studies have measured the photosynthetic performance of entire kiwifruit canopies (Ac) or used simulation models to examine how leaf-level performance scales to that of the whole plant. Ac measurements are important as they provide information on the combined photosynthetic rate of leaves throughout the canopy layers, which all contribute to carbon fixation. The information can guide pruning systems and optimization of light penetration through the canopy. 'Hayward' vines grown within large 26 m^3 ethyl vinyl acetate cuvettes throughout a season had an Ac over 30 μmol CO_2 m^{-2} s^{-1} midsummer, saturating at around 1000 μmol m^{-2} s^{-1} PAR (Buwalda and Greaves, 1997). The depth of the canopy (e.g. a leaf area index of 3) and the lower leaves absorbing scattered and reflected light likely account for the higher photosynthetic rates per unit area and higher saturation points in whole-canopy measurements compared with single leaf measurements.

Several studies have attempted to model total rates of carbon acquisition by 'Hayward' vines growing on pergola or T-bar systems. The models suggest a range of 1049–1773 g of carbon fixed per m^2 of ground cover for the growing season, with the variation in estimated values likely related to differences in the methods used to model canopy development, photosynthetic rates, and growing systems (pergola compared with T-bar). These values align with known rates of biomass accumulation, including shoot, root and fruit growth, and carbon required for maintenance. Net carbon acquisition of the vine follows a steep linear increase from budbreak until canopy establishment (~60–90 DAB), slowing until ~150 DAB before declining until leaf fall (Buwalda, 1991a; Greer *et al.*, 2004). Carbon required for the new growth of fruit and shoots represented large carbon sinks, accounting for 20% and 15%, respectively, of the total carbon fixed in the season (Buwalda, 1991a). However, maintenance of the new shoot biomass (seasonal shoot growth plus leaves) was the largest carbon sink over the season, accounting for 25% of total carbon fixed, and was only superseded by fruit maintenance from about 160 DAB.

Patterns of carbon use and mobilization from storage in spring depend upon the time taken for the leaves to become autotrophic (i.e. able to fix enough carbon to support their own growth and respiration) and contribute to spring canopy and fruit development. Whole-canopy measurements of vine photosynthesis suggest there are at least 25 DAB when respiration exceeds leaf canopy photosynthetic rate until the canopy becomes a net producer of fixed carbon (Buwalda and Greaves, 1997). In contrast, calculation of shoot carbon acquisition and use from leaf photosynthesis measurements suggests that the shoot becomes autotrophic around 35–60 DAB (Greer and Jeffares, 1998; Greer *et al.*, 2003; Richardson, 2015). Whole-vine modelling shows that total vine carbon requirements appear to be greater than the amount of carbon fixed by the shoots during the first 30 DAB and again 60–110 DAB, mainly due to shoot and fruit growth and total vine biomass maintenance, highlighting the importance of stored carbon during this period (Buwalda, 1991a). Modelled carbon storage and mobilization are consistent with empirical studies that have shown significant (three-fold) reductions in root and wood carbohydrate concentrations and root death during the period of canopy development (up to ~100 DAB) (Richardson *et al.*, 2018). Thus, a surplus of photosynthate and its movement into storage later in the season is important for the following season's productivity.

15.3.5 Source/sink relationships

Once carbon is fixed, any not used locally in the leaf is exported in the phloem. In most plants,

Fig. 15.2. Continued.

CO_2 assimilation rate to photosynthetically active radiation. (**B**) The response of the CO_2 assimilation rate to intercellular CO_2 concentration (*Ci*). (**C**) The response of the CO_2 assimilation rate (circles) and stomatal conductance (triangles) to the vapour pressure deficit at the leaf surface. Data recorded with a portable photosynthesis system, n=5 leaves, error bars ± SE. Ac = net assimilation rate limited by activity of ribulose-1,5-bisphosphate carboxylase/oxygenase (Rubisco); Aj = net assimilation rate limited by ribulose-1,5-bisphosphate concentration; Jmax (μmol m^{-2} s^{-1}) = maximum rate of electron transport; Vcmax (μmol CO_2 m^{-2} s^{-1}) = maximum rate of ribulose-1,5-bisphosphate carboxylation. (From Clearwater, M., University of Waikato, unpublished) sucrose is the common form of carbon transported in the phloem and it is transported in an *A. arguta* selection and 'Hayward' (Klages *et al.*, 1998). However, in *A. arguta*, planteose, a galactose + sucrose trisaccharide, has been found to be exported from leaves to fruit in equal or greater amounts than sucrose. Planteose is an isomer from the raffinose family of oligosaccharides and is known to have a role in seed storage in sesame (Dey, 1980) and short-term storage in the vasculature of cyclamen (Rothe *et al.*, 1999). However, the role of planteose in long-distance transport is still unclear, but it has been suggested that the extra carbon per osmoticum compared with sucrose may increase the efficiency of transporting carbon against high osmotic gradients (Boldingh *et al.*, 2015).

Competition for carbohydrate between growing and storage sink tissues influences carbon partitioning. This was clearly illustrated in trials where sink tissue was manipulated or removed. For example, in a hydroponic trial, removing roots resulted in an increase in radiolabelled photosynthate to the fruit (Black *et al.*, 2012a), whilst in a field trial, the removal of 80% of the roots at 80 days after mid-flowering (coinciding with starch accumulation in the fruit) increased fruit size, dry matter and soluble solids (Black *et al.*, 2011). Removing all fruit from 'Hayward' vines increased shoot growth compared with vines where all the fruit were left on, including reducing the percentage of self-terminating shoots as resources were made available for shoot growth (Xiloyannis *et al.*, 1997). Removing the sink tissues, such as fruit, also enabled the vine to develop greater starch and soluble carbohydrate storage pools. Trunk girdling is a common management technique to isolate the root system from the shoot and thus stop transport of photosynthate to the root. Removing competition for carbohydrate by the roots increases both fruit size and dry matter, depending on the timing of girdling (Currie *et al.*, 2018).

Removing sink carbon demand by removing or isolating sink tissues has direct effects on the regulation of the photosynthetic system. Severe root pruning reduces photosynthesis and stomatal conductance, with a reduction in leaf water potential demonstrating a degree of water stress. However, the absence of any change in Ci or $\delta^{13}C$ concentration in the leaf suggests that the photosynthetic response is due to the downregulation of the photosynthetic system rather than just a reduction in gas exchange through reduced stomatal conductance. These data are suggestive of feedback from sink carbon requirements, possibly through photosynthate accumulation in the leaves (Black *et al.*, 2011). Likewise, trunk girdling reduced photosynthetic rate and stomatal conductance with no immediate effect on Ci or $\delta^{13}C$ concentration (Black *et al.*, 2012b).

15.4 The Root System

15.4.1 Root growth and distribution

Root distribution throughout the soil is important for anchorage as well as uptake of water and nutrients. The roots are also an important carbohydrate storage organ: accumulation of starch grains in the roots occurs initially in the cortex in younger roots but progresses to phloem parenchyma and xylem ray cells in older roots (Bardi, 2020). Root systems consist of larger structural roots (up to 5 cm in diameter) and smaller secondary roots and fine roots roughly 0.2 cm or smaller in diameter (McAneney and Judd, 1983). The three-dimensional distribution of roots in soil is dependent upon soil type, soil bulk density and, in a commercial setting, competition from other vines impacting on lateral spread (Hughes and Wilde, 1988). The lateral extension of the root system has been shown to extend up to 2.5 m from the base of the trunk (McAneney and Judd, 1983). In the sandy loam soils of the Bay of Plenty in New Zealand, rooting depths of 3 m have been recorded for *A. chinensis* var. *deliciosa* 'Matua' and at least 4 m for 'Hayward' (McAneney and Judd, 1983; Greaves, 1985). However, compact clay layers in Northland, New Zealand soils may limit rooting depths to just 50 cm (McAneney and Judd, 1983).

When the roots reached 4 m depth in sandy loam soils, almost 50% of the root system was found at 2–4 m; the same study mentions an anecdotal report of kiwifruit roots reaching 10 m in sandy loam soil (Greaves, 1985). However, the majority of the active 'feeder' roots are found in the top 50 cm and in well-watered vines they preferentially take up water from the top layer of the soil (Green and Clothier, 1995). The genetics of the rootstock are likely to influence root growth and distribution (see Chapter 6); however, most early studies were carried out using seedling rootstocks or unnamed rootstocks. The interaction between rootstock and scion affects the growth and behaviour of both the root and shoot systems, although data on root systems are scarce. In young plants, the root system is generally slower than the canopy to fill its available area, with the canopy filling its allocated area by four years after grafting while the root system can take up to ten years to achieve this (Hughes *et al.*, 1992). Periods of root growth are generally associated with availability of photosynthate and sink priority. Carbohydrate required for fruit and shoot growth appears to take priority over that for root growth (Buwalda and Hutton, 1988). Reducing carbon fixation through leaf removal resulted in a greater decrease in the root growth rate than in the fruit growth rate (Buwalda and Smith, 1990; Buwalda, 1991b). Removal of canopy also appeared to affect the rate of root turnover, with a reduction in root growth rate in vines that had a low leaf:fruit ratio (Buwalda, 1991b).

15.4.2 Root-to-shoot transport and the xylem

The root system is the primary site for water and mineral nutrient uptake. Long-distance transport of water and solutes to the shoots predominantly occurs in the xylem, with retranslocation occurring in the phloem. Solution flow in the xylem is driven by transpiration or hydrostatic pressure (referred to here as root pressure) depending on the environmental conditions. During the day, negative water potentials in the leaf tissue (following water loss through open stomata) drive water flow from roots to shoots. In the absence of transpiration, the movement of solutes into the xylem in the roots and along the xylem pathway can generate osmotic flow of water into the xylem, creating positive root pressure and driving solutes to the shoot.

15.4.3 Root hydraulic conductance and water flow

Actinidia vines have a high requirement for water as indicated by their high transpiration rates (Judd *et al.*, 1989). To supply adequate water, the xylem vessel diameter and hydraulic conductance of kiwifruit are high compared with other fruiting tree crops (Dichio *et al.*, 1999; Muramatsu and Hiraoka, 2008). Kiwifruit root xylem vessel diameters range from 0.12–0.5 mm with interconnecting radial conductance that allows water uptake over their surface area (McAneney and Judd, 1983). Consequently, the leaf specific hydraulic conductance (the hydraulic conductivity per unit pressure gradient, taking into account the leaf area) of *Actinidia* root systems is relatively high, ranging from 0.7 to 1.7 kg MPa^{-1} m^{-2} s^{-1} × 10^4, depending on species (Clearwater *et al.*, 2004; Black *et al.*, 2011). Although having large xylem vessel diameters and associated high hydraulic conductivity may allow kiwifruit to maximize water uptake from brief periods of rain, it also makes vines more vulnerable to xylem air embolisms and cavitation of the water column, thus reducing hydraulic conductance in periods of drought (Tyree and Ewers, 1991).

Sap flow from kiwifruit roots differs depending on the availability and distribution of water throughout the root zone. The water demands of the shoot system can be supplied by a partially wetted root system as water uptake rates are similar to when the entire root zone is supplied with water (Green and Clothier, 1995). Short-term fluctuations in sap flow have been shown to be independent of changes in root hydraulic conductance, e.g. the excision of 80% of the root system increased sap flow in the remaining roots without any increase in their root hydraulic conductance (Black *et al.*, 2011). Longer term, changes in root growth, including the production of root hairs, are likely to change water uptake rates (Green and Clothier, 1995). The presence of cortex in larger roots suggests symplastic continuity between the cortex and the vasculature and the possibility of water absorption in the secondary thickened roots (Bardi, 2020). The ability of the root system to absorb water over its entire surface area is consistent with calculations made from saturated soil hydraulic conductivity and peak vine water-flow measurements (McAneney and Judd, 1983).

15.4.4 Generation of root pressure

During winter, transpiration ceases, and xylem vessels may fill with air. Before leaf budbreak, in early spring, the xylem vessels refill by movement of water down a water potential gradient into the apoplastic space. The water potential gradient is generated by remobilization of solutes from storage tissues along the vascular pathway, and uptake of mineral ions by the roots. Water reserves mobilized from root storage can make up 8% of all water transported in commercially grown 'Hayward' (Nuzzo *et al.*, 1996). However, most water is taken up via the roots, which creates positive hydrostatic pressure from the base of the vine to the shoots, resulting in the filling of vessels from the ground up. The time of initiation of root pressure and the amount of pressure generated depends on the size of the vine, amount of solute stored in the vine, root system size and the genetics of the rootstock. Pressures as high as 0.15 MPa have been measured in the xylem of *A. macrosperma* (unstated genotype) during spring and flow rates up to 415 ml per 24 h have been measured from bleeding sap collected from cut 'Hayward' canes (Peterlunger *et al.*, 1990; Clearwater *et al.*, 2007).

The amount of pressure generated is correlated with the osmotic potential of the xylem sap and appears to be related to the vigour of the vine during spring growth. A comparison of seven *Actinidia* spp. rootstocks (*A. chrysantha*, *A. chinensis* var. *deliciosa*, *A. eriantha*, *A. hemsleyana*, *A. kolomikta*, *A. macrosperma* and *A. polygama*) showed that higher-vigour species generated higher root pressure in spring (Clearwater *et al.*, 2007). During early spring, root pressure remains positive until transpiration from new leaves generates negative xylem pressure during the day, with positive pressure returning after stomatal closure at night, under humid conditions. However, kiwifruit leaves can lose water at night under less humid, windy conditions, resulting in negative xylem pressure throughout the 24-hour cycle (Green *et al.*, 1989; Black *et al.*, 2011).

In addition to water, the xylem is important for the transportation of amino acids, hormones, certain soluble carbohydrates, potassium, magnesium, phosphate, zinc, sulphur, iron, copper, boron, calcium, malate, glutamine, nitrate, ammonium, arginine, beta-galactosidase, glucose, fructose, sucrose, sorbitol, *myo*-inositol and malic acid. The composition of xylem sap varies with year, time of year, cultivar, nutrient supply, water availability, size of root system and location within the vine, making direct comparisons between studies difficult (Ferguson *et al.*, 1981, 1983; Clark *et al.*, 1986; Peterlunger *et al.*, 1990; Sotiropoulos *et al.*, 2002; Buxton *et al.*, 2007; Nardozza *et al.*, 2015). During periods of positive root pressure, xylem sap composition reflects the solutes available to generate the water potential gradient, e.g. high concentrations prior to budbreak drive osmotic flow in the xylem. High soluble sugar concentrations, particularly hexose sugars, are found in the xylem at this time, likely due to starch reserves in the root and trunk being remobilized as sugars (Nardozza *et al.*, 2015).

Solute concentrations within the xylem decrease after budbreak and xylem pressure is dissipated in the transpiration stream (Ferguson *et al.*, 1983; Peterlunger *et al.*, 1990; Nardozza *et al.*, 2015). Prior to budbreak, position on the vine and tissue type affect xylem sap composition, with the highest solute concentrations generally in one-year-old shoots, decreasing through the plant to the roots (Ferguson *et al.*, 1981) owing to the availability of stored compounds for remobilization. Once vines start to transpire, solute concentrations in one-year-old shoots reduce. However, nitrate concentrations follow the opposite pattern, with higher concentrations in the trunk and roots than in the shoot (Ferguson *et al.*, 1981), likely due to the reduction of nitrate in tissues along the pathway as well as greater storage pools in the root and trunk. Nitrogen is predominately taken up as nitrate by kiwifruit roots where approximately half is reduced to ammonium and the remainder is transported to leaves before assimilation (Ledgard and Brier, 1991).

Calcium (Ca) is particularly important in cell wall and membrane stability, cell extension and plant signalling (Marschner, 1995). Uptake and transport of Ca in the xylem is important for plant and fruit development as it is one of the few nutrients that are immobile in the phloem and hence Ca flow and distribution are mainly determined by the transpiration stream. As a result, fruit from short shoots at the distil end of canes with a low leaf:fruit ratio had low Ca concentrations at harvest compared with fruit from long shoots at the proximal end of canes that had a high leaf:fruit ratio (Thorp *et al.*, 2003). About 80% of total fruit Ca content is accumulated within the first few weeks after fruit set when the higher transpiration-facilitated flow to the young fruit makes them a stronger sink for Ca than more mature fruit (Montanaro *et al.*, 2012, 2014).

15.5 Environmental Stress Responses

15.5.1 Water deficit stress

Water stress reduces the starch content of leaves (Miller *et al.*, 1998), reduces vine growth and increases the root:shoot dry biomass ratio as the growth of the root system takes priority over the shoot (Chartzoulakis *et al.*, 1993). A reduction in soil water availability during extended dry periods in summer results in decreased water supply to the leaves, and although stomatal conductance decreases below leaf water potentials of −1.0 MPa, the vine continues to lose water (Judd *et al.*, 1989; Chartzoulakis *et al.*, 1997), causing wilting and ultimately necrosis of leaves (Nuzzo *et al.*, 1996). Drought also causes a reduction in xylem pressure to more negative values, increasing the probability of cavitation and embolism formation within the xylem vessels. Although there have been no detailed studies of the vulnerability of *Actinidia* species to drought-induced embolism in xylem, their liana growth habit, large-diameter xylem vessels and adaptation to high-rainfall environments suggest they are likely to be highly vulnerable to drought-induced hydraulic failure (Dichio *et al.*, 2013). Using magnetic resonance imaging, increased embolism formation was observed as xylem tensions decreased below a modest threshold of −0.5 MPa in *A. chinensis* var. *chinensis* 'Hort16A' and 'Hayward' stems (Clearwater and Clark, 2003). It is difficult to easily identify the onset of water stress as clear visual symptoms such as leaf wilt occur after water stress has already reduced fruit growth rate and dry matter accumulation. Recovery of vines from short-term water stress can be rapid. 'Hayward' vines stopped fruit expansion and exhibited leaf wilt when midday leaf water potentials were below −1.0 MPa (Judd *et al.*, 1989) but both leaf and fruit turgor recovered within 24 h of watering.

A reduction in stomatal conductance from water deficits also affects photosynthetic rate and the partitioning of carbohydrate within vines. When the soil moisture level was reduced to 65% of field capacity for one-year-old potted 'Hayward' plants, their photosynthetic rate was reduced by 69%, concomitant with a 55% reduction in stomatal conductance and a midday leaf water potential of −1.63 MPa (Chartzoulakis *et al.*, 1993). However, the Ci in leaves remained high, indicating that a reduction in carbon fixation in the chloroplasts, not the drop in stomatal conductance, was the primary driving force for the reduction in photosynthetic rate. Further investigations demonstrated that photosystem II (PSII) stress under drought conditions was linked to a midday decline in photosynthesis (Montanaro *et al.*, 2007).

Tolerance to drought stress differs between *Actinidia* species. One-year-old rooted cuttings of *A. arguta* and *A. eriantha* had higher drought tolerance than *A. chinensis* var. *chinensis* and *A. chinensis* var. *deliciosa* cuttings (unstated genotypes) (Jo *et al.*, 2008c). It was suggested that a well-developed cuticle layer, low stomatal density and a small stomatal aperture contributed to water stress tolerance in *A. arguta* and high trichome density contributed to water stress tolerance in *A. eriantha* (Jo *et al.*, 2008c).

Melatonin, which appears to act as a plant growth regulator and antioxidant in higher plants, may have a role in mitigating drought stress (Arnao and Hernández-Ruiz, 2014). *A. chinensis* var. *deliciosa* 'Qinmei' seedlings that were treated with melatonin prior to water stress were found to be more tolerant to drought than non-melatonin-treated plants (Liang *et al.*, 2019). Melatonin alleviated the inhibitory effects of drought on photosynthesis, leading to stronger plants with greater biomass including increased root mass. The antioxidant properties of melatonin appeared to reduce membrane injury of water-stressed kiwifruit plants (Liang *et al.*, 2019).

15.5.2 Waterlogging

Kiwifruit root systems are sensitive to waterlogging and associated anoxia. Vines growing on heavy soils exposed to high rainfall can exhibit death of white feeder roots, wilting, leaf necrosis, leaf drop and may even die following a waterlogging event. In severe cases, vines can decline and die in the year following a flooding event (Smith *et al.*, 1989; Reid *et al.*, 1991, 1992).

The response of the vine depends on the duration of the waterlogging event. Within days of the start of a waterlogging event, new root growth ceases, stomatal conductance is reduced, and xylem water potential and osmotic potential increase. The reduction in stomatal conductance during these first few days does not appear to be caused by a leaf water deficit as leaf turgor is not affected, suggesting direct signalling from the roots rather than a secondary response to water loss (Savé and Serrano, 1986). Several plant hormones that contribute to stress signalling under waterlogged conditions have been identified. Leaf and root abscisic acid and indole-acetic acid contents have been shown to increase with root zone hypoxia, with a concomitant reduction in leaf and root zeatin riboside and gibberellin, the roles of which are consistent with a reduction in root and shoot growth in the stressed plants (Mi *et al.*, 2009).

As the waterlogging period extends over several weeks, more severe impacts are observed in the shoot with a reduction in leaf turgor, water potential, transpiration rate and shoot growth rate, as well as the appearance of leaf chlorosis and necrosis (Savé and Serrano, 1986).

The ability of the vine to withstand waterlogging may be affected by several factors: the size of the root mass, the nitrogen and carbohydrate content of the roots and the genotype. Vines waterlogged for seven days showed no sign of stress above ground despite losing 65% of their initial root length (Reid and Petrie, 1991; Li *et al.*, 2020). 'Hayward' vines with low nitrogen status were more susceptible to waterlogging damage than high nitrogen status vines (Smith and Miller, 1992). Root carbohydrate content has been associated with a vine's ability to recover from waterlogging stress, with higher sucrose in roots of *A. valvata* thought to play a role through modulating fermentative enzymes to reduce damaging accumulation of lactic acid and ethanol (Li *et al.*, 2020).

Genetics affect the ability of a vine to respond to waterlogging stress. A comparison of waterlogging tolerance among genotypes of *A. chinensis* var. *chinensis*, *A. chinensis* var. *deliciosa*, *A. eriantha* and *A. arguta*, grown on their own roots, showed all species had reduced photosynthetic rates after nine days of waterlogging. However, *A. arguta* and *A. eriantha* maintained pre-treatment photosynthesis levels for six days whereas *A. chinensis* var. *chinensis and A. chinensis* var. *deliciosa* had reduced photosynthetic rates after three days (Jo *et al.*, 2008a). *A. arguta* outperformed the other genotypes in root activity and tolerance to leaf necrosis following waterlogging. *A. chinensis* var. *chinensis* was the most susceptible to leaf necrosis, followed by *A. chinensis* var. *deliciosa*, *A. eriantha* and then *A. arguta*. This is consistent with other studies in which *A. chinensis* var. *deliciosa* performed better than *A. chinensis* var. *chinensis* (Mi *et al.*, 2009; Zhang *et al.*, 2013). *A. valvata* is more tolerant to waterlogging stress than *A. chinensis* var. *deliciosa*, but there is variability among *A. valvata* genotypes (Li *et al.*, 2020). *A. macrosperma* genotypes have also shown promise, with *A. macrosperma* and *A. arguta* outperforming *A. chinensis* var. *deliciosa* and *A. polygama* under challenging conditions associated with kiwifruit vine decline syndrome in Italy (Mian *et al.*, 2022).

15.5.3 High temperatures

Actinidia chinensis is described as being adapted to the cool, humid, partially shaded environments from its original forest margin habitat (Ferguson, 1984), with reduced growth and photosynthetic performance in hotter, drier sun-exposed environments. Although shoot growth rates of 'Hayward' appear optimal between 20°C and 25°C (Morgan *et al.*, 1985), the optimal photosynthetic rate per leaf area appears to occur around 16°C for 'Hayward' and *A. arguta* (unstated genotype) (Laing, 1985; Shim *et al.*, 1985).

Response curves demonstrate that the relationship between photosynthetic rate and temperature of 'Hayward' leaves is dependent on the temperature the vines were grown at prior to a sudden change in temperature. Vines grown at continuous temperatures of 20°C or 25°C were more adaptable and had higher photosynthetic rates when the temperatures were suddenly reduced to 15°C or raised to 30°C, compared with plants that were previously grown at 10°C, 15°C or 30°C (Laing, 1985).

Early work on the effects of temperature and light on photosynthesis was carried out on kiwifruit in New Zealand in a moderate climate, with adequate water supply and temperatures generally under 30°C. Photosynthetic rate can fluctuate during the day and photoinhibition has been recorded in *A. chinensis* var. *deliciosa* selections at 10°C (Greer, 1990; Greer and Laing, 1992). Shading has been shown to limit Pmax, with leaves from the sub-canopy having a much lower photosynthetic light response curve with a Pmax of about 20 μmol CO_2 m^{-2} s^{-1} compared with 30 μmol CO_2 m^{-2} s^{-1} for leaves in full light at the top of the canopy. Light saturation occurs at about 600 μmol m^{-2} s^{-1} in the shaded leaves compared with 800 μmol m^{-2} s^{-1} in the exposed leaves (Greer, 1995). Shaded leaves were able to acclimate to photoinhibition within days by increasing total xanthophyll concentration to increase non-photochemical quenching (Greer, 1995).

In kiwifruit-producing regions such as China and Italy, maximum temperatures often reach 30°C and can exceed 40°C. High temperatures heat the leaf and fruit surface, causing high rates of water loss and a decline in photosynthetic rates during the day. Under these conditions, shading vines can improve photosynthetic efficiency by reducing leaf temperature, increasing humidity and conserving leaf water potential (He *et al.*, 2007; Montanaro *et al.*, 2009). Temperatures over 30°C increase the susceptibility of PSII to stress (Montanaro *et al.*, 2007; Li *et al.*, 2021). Artificial shading has been used to reduce photoinhibition and improve water use efficiency (WUE). Reducing photosynthetic photon flux density by 50% to leaves that were acclimatized to full light and had been water-stressed to levels typical for an Italian summer (pre-dawn water potentials of −1.0 MPa) reduced photoinhibition by maintaining PSII efficiency at 0.8, compared with leaves that were left exposed to full sunlight (PSII efficiency of 0.65), and increased WUE by 15–20% (Montanaro *et al.*, 2009). WUE peaked at ~7.5 μmol (CO_2) mmol (H_2O)$^{-1}$ in irrigated shaded vines in the morning in comparison with ~1 μmol (CO_2) mmol (H_2O)$^{-1}$ in droughted sunlit plants at 3 pm. Gas exchange of CO_2 and H_2O through stomata across the leaf surface is important for regulating both photosynthesis and vine water use. Under extreme temperature conditions, a balance is required between maintaining a high *Ci*, regulating leaf temperature and maintaining leaf water potential (Lawson and Vialet-Chabrand, 2019). Water potential of kiwifruit leaves decreases during the day, demonstrating behaviour typical of anisohydric plants that keep stomatal conductance high throughout the day to maintain photosynthetic rates even as water potential drops (Montanaro *et al.*, 2009).

15.5.4 Frost and tolerance to low temperatures

Kiwifruit plants are susceptible to frost damage, although their ability to tolerate freezing temperatures varies with the time of year. Frost damage causes splitting of the trunk and woody tissue and death of vegetative and flower buds (Pyke *et al.*, 1986; Hartmann *et al.*, 2021). Temperatures as low as −1.5°C to −2°C for less than 60 minutes can damage vegetative and flower buds in spring (Hewett and Young, 1981; Nanos *et al.*, 1997). Dormant buds are more resilient: a reduction in subsequent budbreak occurred in plants exposed to temperatures of −7°C or below, and plant damage/death occurred at or below −9°C, although some plants survived −13°C (Pyke *et al.*, 1986). In Italy, 'Matua' and 'Hayward' plants survived a winter frost of −23°C, with regrowth from adventitious trunk and leader buds and reduced flowering in the following spring (Testolin and Messina, 1987). Frost damage is also associated with increased infection and spread of *Pseudomonas syringae* pv. *actinidiae* through water, nutrients and soluble sugars released by frost-ruptured cells, and cane and trunk cracking providing an entry point for the pathogen (Ferrante and Scortichini, 2014).

Year-to-year and site-to-site variability in the ability of plants to survive frost is likely to be due to differing acclimation conditions of the plants. Differences in frost tolerance have also been observed with plant age and between cultivars. Young plants have been shown to be more susceptible than mature plants (Ferrante and Scortichini, 2014). The response of two-year-old *A. chinensis* var. *chinensis* and *A. chinensis* var. *deliciosa* plants to a −4.1°C frost in autumn differed, with plants of *A. chinensis* var. *chinensis* displaying more root and trunk damage than those of *A. chinensis* var. *deliciosa* genotypes (Hartmann *et al.*, 2021). At more extreme temperatures, 7 h of cold treatment to −8°C during winter killed ~40% of one-year-old *A. chinensis* var. *chinensis* 'Soreli' plants but no 'Hayward' plants (Ferrante and Scortichini, 2014). A series of trials looking at cold tolerance of dormant buds and shoots, young shoots and young seedlings indicated that, overall, *A. arguta* was more tolerant to low (freezing) temperatures than *A. eriantha*, *A. chinensis* var. *chinensis* and *A. chinensis* var. *deliciosa* (Jo *et al.*, 2008d).

15.5.5 Salinity

Kiwifruit vines are susceptible to high levels of sodium chloride (NaCl) in the soil, with salinity tolerance varying between genotypes of the same species and across species; tolerance decreased from *A. valvata*, *A. longicarpa*, *A. chinensis* var. *deliciosa* to *A. hemsleyana* (Zhong *et al.*, 2019; Abid *et al.*, 2020). Salinity can reduce shoot and root biomass accumulation, leaf number and leaf size, photosynthetic rate and leaf chlorophyll content and, if severe enough, can cause the death of roots, shoots and leaves (Chartzoulakis *et al.*, 1995; Abid *et al.*, 2020). At high external NaCl concentrations, Na^+ and Cl^- accumulate in the roots with an associated reduction in potassium (K^+) ions, presumably to regulate the ionic balance. There is no effect of salinity on K^+ concentration in leaves since kiwifruit can exclude Na^+ from accumulating in leaves; however, leaf Cl^- increases at high external NaCl concentrations. Increased leaf Cl^- concentration is thought to contribute to reduced photosynthetic rates and stomatal conductance (Chartzoulakis *et al.*, 1995).

15.5.6 High pH soils

Plants grown in high pH soils are often deficient in iron, phosphorus and zinc through the neutralization of protons and other acidic compounds exuded from roots during nutrient uptake. This is manifested through leaf chlorosis, particularly in 'Hort16A', and sometimes fruit chlorosis as well as reduced leaf area, leaf number and shoot biomass (Thorp *et al.*, 2021). In *A. chinensis* var. *chinensis* 'Qihong' plants grown at pH 7.8, roots were initially able to maintain proton efflux and maintain or even increase ferric chelate reductase (FCR; the enzyme required to reduce iron(III) for cell uptake) activity for ~3 weeks from the start of treatment, and then both FCR activity and proton efflux declined (Wang *et al.*, 2019a). Sustained periods of high pH are also associated with increased accumulation of succinic acid in roots, suggesting a possible link to the Krebs cycle (Wang *et al.*, 2019a).

Leaf chlorosis induced by high pH soils can be reversed by foliar application of a combination of malic acid, citric acid, sorbitol and Fe^{III}-sulphate (Rombolà *et al.*, 2002). There is some evidence that the response to high pH soils differs between kiwifruit genotypes, as leaf number and leaf biomass were more affected in *A. chinensis* var. *chinensis* 'Nongda Jinmi' scion than *A. chinensis* var. *deliciosa* 'Xuxiang' scion when both were grown on 'Qinmei' rootstock. The difference may be due to the origins of the two scions, with 'Xuxiang' originating from an area of higher pH soils in southern China and 'Nongda Jinmi' genotype from lower pH soils at higher latitudes (Wang *et al.*, 2019b).

15.6 Concluding Remarks

A. chinensis shares many of the physiological traits and growth habits typical of liana plants. In its typically cool, humid, partially shaded forest-margin natural environment, vigorous vertical shoot growth of up to 7 m occurs. In a cultivated environment, vines are normally grown on a horizontal pergola system with rapid annual canopy replacement every spring. The vigorous nature of shoot growth can affect vine productivity and careful management is required to balance resource allocation between competing shoots, fruit and roots throughout the season. Healthy kiwifruit plants have high photosynthetic rates and water requirements. As the commercial cultivation of kiwifruit spreads further from its natural habit, the environment begins to have an impact on vine resource capture and utilization, and a range of management techniques, e.g. girdling, pruning, covered cropping and irrigation, are needed to optimize productivity. Ultimately, as global productivity increases and consumers seek new cultivars, *Actinidia* genetic diversity will play a key role in overcoming the physiological constraints of many current cultivars through successful rootstock and scion breeding programmes.

References

Abid, M., Zhang, Y.J., Li, Z., Bai, D.F., Zhong, Y.P. *et al.* (2020) Effect of salt stress on growth, physiological and biochemical characters of four kiwifruit genotypes. *Scientia Horticulturae* 271, 109473. DOI: 10.1016/j.scienta.2020.109473

Arnao, M.B. and Hernández-Ruiz, J. (2014) Melatonin: plant growth regulator and/or biostimulator during stress? *Trends in Plant Science* 19, 789–797. DOI: 10.1016/j.tplants.2014.07.006

Bardi, L. (2020) Early kiwifruit decline: a soil-borne disease syndrome or a climate change effect on plant–soil relations? *Frontiers in Agronomy* 2, 3. DOI: 10.3389/fagro.2020.00003

Bennewitz, E. von, Suazo, S., Keutgen, A.J., Lošák, T. and Carrasco-Benavides, M. (2019) Seasonal root, shoot, and fruit growth patterns in kiwifruit (*Actinidia deliciosa* A. Chev.) in central Chile. *Erwerbs-Obstbau* 61, 283–292. DOI: 10.1007/s10341-019-00429-1

Black, M.Z. (2012) The kiwifruit (*Actinidia* sp.) vine root system: responses to vine manipulations. PhD thesis, University of Waikato, Hamilton, New Zealand.

Black, M.Z., Patterson, K.J., Minchin, P.E.H., Gould, K.S. and Clearwater, M.J. (2011) Hydraulic responses of whole vines and individual roots of kiwifruit (*Actinidia chinensis*) following root severance. *Tree Physiology* 31, 508–518. DOI: 10.1093/treephys/tpr045

Black, M.Z., Minchin, P.E.H., Gould, N., Patterson, K.J. and Clearwater, M.J. (2012a) Measurement of Bremsstrahlung radiation for in vivo monitoring of ^{14}C tracer distribution between fruit and roots of kiwifruit (*Actinidia arguta*) cuttings. *Planta* 236, 1327–1337. DOI: 10.1007/s00425-012-1685-z

Black, M.Z., Patterson, K.J., Gould, K.S. and Clearwater, M.J. (2012b) Physiological responses of kiwifruit vines (*Actinidia chinensis* Planch. var. *chinensis*) to trunk girdling and root pruning. *New Zealand Journal of Crop and Horticultural Science* 40, 31–41. DOI: 10.1080/01140671.2011.603343

Boldingh, H., Richardson, A., Minchin, P. and MacRae, E. (2015) Planteose is a major sugar translocated in *Actinidia arguta* 'Hortgem Tahi'. *Scientia Horticulturae* 193, 261–268. DOI: 10.1016/j.scienta.2015.07.009

Brundell, D.J. (1975) Flower development of the Chinese gooseberry (*Actinidia chinensis* Planch.). I. Development of the flowering shoot. *New Zealand Journal of Botany* 13, 473–483. DOI: 10.1080/0028825X.1975.10430338

Buwalda, J.G. (1991a) A mathematical model of carbon acquisition and utilisation by kiwifruit vines. *Ecological Modelling* 57, 43–64. DOI: 10.1016/0304-3800(91)90054-5

Buwalda, J.G. (1991b) Root growth of kiwifruit vines and the impact of canopy manipulations. In: McMichael, B.L. and Persson, H. (eds) *Plant Roots and Their Environment.* Developments in Agricultural and Managed Forest Ecology Vol. 24. Elsevier, Amsterdam, pp. 431–441. DOI: 10.1016/B978-0-444-89104-4.50059-1

Buwalda, J.G. and Greaves, A.J. (1997) Whole-plant photosynthesis in kiwifruit vines – temporal changes on diurnal and seasonal scales. *Acta Horticulturae* 444, 343–348. DOI: 10.17660/ActaHortic.1997.444.53

Buwalda, J.G. and Hutton, R.C. (1988) Seasonal changes in root growth of kiwifruit. *Scientia Horticulturae* 36, 251–260. DOI: 10.1016/0304-4238(88)90059-3

Buwalda, J.G. and Smith, G.S. (1990) Effects of partial defoliation at various stages of the growing season on fruit yields, root growth and return bloom of kiwifruit vines. *Scientia Horticulturae* 42, 29–44. DOI: 10.1016/0304-4238(90)90145-5

Buxton, K.N., Clearwater, M.J., Giles-Hansen, K., Hewett, E.W. and Ferguson, I.B. (2007) Comparison of xylem sap mineral concentrations between kiwifruit shoot types using spittlebugs for non-destructive sampling of sap. *Functional Plant Biology* 34, 1029–1037. DOI: 10.1071/fp07091

Chartzoulakis, K.S., Noitsakis, B.I. and Therios, I.N. (1993) Photosynthesis, plant growth and carbon allocation in kiwi, cv. *Hayward*, as influenced by water deficits. *Acta Horticulturae* 335, 227–234. DOI: 10.17660/ActaHortic.1993.335.26

Chartzoulakis, K.S., Therios, I.N., Misopolinos, N.D. and Noitsakis, B.I. (1995) Growth, ion content and photosynthetic performance of salt-stressed kiwifruit plants. *Irrigation Science* 16, 23–28. DOI: 10.1007/bf00208392

Chartzoulakis, K.S., Therios, I. and Noitsakis, B. (1997) The effect of soil water deficit on internal water relations of kiwifruit. *Acta Horticulturae* 444, 305–310. DOI: 10.17660/ActaHortic.1997.444.47

Clark, C.J., Holland, P.T. and Smith, G.S. (1986) Chemical composition of bleeding xylem sap from kiwifruit vines. *Annals of Botany* 58, 353–362. DOI: 10.1093/oxfordjournals.aob.a087213

Clearwater, M.J. and Clark, C.J. (2003) *In vivo* magnetic resonance imaging of xylem vessel contents in woody lianas. *Plant, Cell & Environment* 26, 1205–1214. DOI: 10.1046/j.1365-3040.2003.01042.x

Clearwater, M.J., Lowe, R.G., Hofstee, B.J., Barclay, C., Mandemaker, A.J. *et al.* (2004) Hydraulic conductance and rootstock effects in grafted vines of kiwifruit. *Journal of Experimental Botany* 55, 1371–1382. DOI: 10.1093/jxb/erh137

Clearwater, M.J., Blattmann, P., Luo, Z. and Lowe, R.G. (2007) Control of scion vigour by kiwifruit rootstocks is correlated with spring root pressure phenology. *Journal of Experimental Botany* 58, 1741–1751. DOI: 10.1093/jxb/erm029

Currie, M.B., Patterson, K.J., Snelgar, W.P. and Blattmann, P. (2018) Girdling kiwifruit vines for commercial advantage: opportunities and risks. *Acta Horticulturae* 1218, 405–412. DOI: 10.17660/ActaHortic.2018.1218.56

Dey, P.M. (1980) Biosynthesis of planteose in *Sesamum indicum*. *FEBS Letters* 114, 153–156. DOI: 10.1016/0014-5793(80)80882-9

Dichio, B., Baldassarre, R., Nuzzo, V., Biasi, R. and Xiloyannis, C. (1999) Hydraulic conductivity and xylem structure in young kiwifruit vines. *Acta Horticulturae* 498, 159–164. DOI: 10.17660/ActaHortic.1999.498.18

Dichio, B., Montanaro, G., Sofo, A. and Xiloyannis, C. (2013) Stem and whole-plant hydraulics in olive (*Olea europaea*) and kiwifruit (*Actinidia deliciosa*). *Trees* 27, 183–191. DOI: 10.1007/s00468-012-0787-3

Ferguson, A.R. (1984) Kiwifruit: a botanical review. *Horticultural Reviews* 6, 1–64. DOI: 10.1002/9781118060797.ch1

Ferguson, A.R., Eiseman, J.A. and Dale, J.R. (1981) Xylem sap from *Actinidia chinensis*: gradients in sap composition. *Annals of Botany* 48, 75–80. DOI: 10.1093/oxfordjournals.aob.a086099

Ferguson, A.R., Eiseman, J.A. and Leonard, J.A. (1983) Xylem sap from *Actinidia chinensis*: seasonal changes in composition. *Annals of Botany* 51, 823–833. DOI: 10.1093/oxfordjournals.aob.a086533

Ferrante, P. and Scortichini, M. (2014) Frost promotes the pathogenicity of *Pseudomonas syringae* pv. *actinidiae* in *Actinidia chinensis* and *A. deliciosa* plants. *Plant Pathology* 63, 12–19. DOI: 10.1111/ppa.12070

Gong, H.J., Li, J.W., Ye, K.Y., Jiang, Q.S. and Liu, P.P. (2015) Effect of altitudes on leaf photosynthetic characteristics and fruit nutrition of 'Hongyang' kiwifruit (*Actinidia chinensis*). *Acta Horticulturae* 1096, 295–301. DOI: 10.17660/ActaHortic.2015.1096.32

Greaves, A.J. (1985) Root distribution of kiwifruit (*Actinidia deliciosa*) in a deep sandy loam soil of the Te Puke district, New Zealand. *New Zealand Journal of Agricultural Research* 28, 433–436. DOI: 10.1080/00288233.1985.10430449

Green, S.R. and Clothier, B.E. (1995) Root water uptake by kiwifruit vines following partial wetting of the root zone. *Plant and Soil* 173, 317–328. DOI: 10.1007/bf00011470

Green, S.R., McNaughton, K.G. and Clothier, B.E. (1989) Observations of night-time water use in kiwifruit vines and apple trees. *Agricultural and Forest Meteorology* 48, 251–261. DOI: 10.1016/0168-1923(89)90072-5

Greer, D.H. (1990) The combined effects of chilling and light stress on photoinhibition of photosynthesis and its subsequent recovery. *Plant Physiology and Biochemistry* (Paris) 28, 447–455.

Greer, D.H. (1995) Effect of canopy position on the susceptibility of kiwifruit (*Actinidia deliciosa*) leaves on vines in an orchard environment to photoinhibition throughout the growing season. *Australian Journal of Plant Physiology* 22, 299–309. DOI: 10.1071/pp9950299

Greer, D.H. (1996) Photosynthetic development in relation to leaf expansion in kiwifruit (*Actinidia deliciosa*) vines during growth in a controlled environment. *Australian Journal of Plant Physiology* 23, 541–549. DOI: 10.1071/pp9960541

Greer, D.H. and Jeffares, D. (1998) Temperature-dependence of carbon acquisition and demand in relation to shoot growth of kiwifruit (*Actinidia deliciosa*) vines grown in controlled environments. *Australian Journal of Plant Physiology* 25, 843–850. DOI: 10.1071/pp98055

Greer, D.H. and Laing, W.A. (1992) Low-temperature and bright-light induced photoinhibition of photosynthesis in kiwifruit leaves. *Acta Horticulturae* 297, 315–321. DOI: 10.17660/ActaHortic.1992.297.42

Greer, D.H., Cirillo, C. and Norling, C.L. (2003) Temperature-dependence of carbon acquisition and demand in relation to shoot and fruit growth of fruiting kiwifruit (*Actinidia deliciosa*) vines grown in controlled environments. *Functional Plant Biology* 30, 927–937. DOI: 10.1071/fp03111

Greer, D.H., Seleznyova, A.N. and Green, S.R. (2004) From controlled environments to field simulations: leaf area dynamics and photosynthesis of kiwifruit vines (*Actinidia deliciosa*). *Functional Plant Biology* 31, 169–179. DOI: 10.1071/fp03151

Hartmann, T.P., Scheiner, J.J., Stein, L.A., King, A.R. and Feagely, S.E. (2021) Young field-grown kiwifruit plants' response to early autumn frost injury in Texas. *HortTechnology* 31, 125–135. DOI: 10.21273/HORTTECH04670-20

He, K., Wang, Z. and Wang, R. (2007) Effects of overhead shading in summer on ecological factors and photosynthesis of kiwifruit orchard. *Journal of Fruit Science* 24, 616–619.

Hewett, E.W. and Young, K. (1981) Critical freeze damage temperatures of flower buds of kiwifruit (*Actinidia chinensis* Planch.). *New Zealand Journal of Agricultural Research* 24, 73–75. DOI: 10.1080/00288233.1981.10420873

Hughes, K.A. and Wilde, R.H. (1988) Root distributions and their interactions with the soil. *Proceedings Agronomy Society of New Zealand* 18, 33–36.

Hughes, K.A., de Willigen, P., Gandar, P.W. and Clothier, B.E. (1992) Kiwifruit root systems: structure & function. *Acta Horticulturae* 297, 383–390. DOI: 10.17660/ActaHortic.1992.297.51

Jo, Y.-S., Cho, H.-S., Liu, I.-S. and Kim, W.-S. (2008a) Assessment of waterlogging tolerance and change in net photosynthetic rate of major *Actinidia species* during flooding treatment. *Acta Horticulturae* 773, 289–293. DOI: 10.17660/ActaHortic.2008.773.43

Jo, Y.-S., Cho, H.-S., Park, M.-Y. and Kim, W.-S. (2008b) Comparative photosynthetic characteristics among major *Actinidia* species. *Acta Horticulturae* 773, 271–276. DOI: 10.17660/ActaHortic.2008.773.40

Jo, Y.-S., Cho, H.-S., Park, M.-Y., Bang, G.-P. and Kim, W.-S. (2008c) Drought stress tolerance of *Actinidia arguta* and *Actinidia eriantha*. *Acta Horticulturae* 773, 283–287. DOI: 10.17660/ActaHortic.2008.773.42

Jo, Y.-S., Cho, H.-S., Park, J.-O., Kim, T.-C. and Kim, W.-S. (2008d) Assay of cold tolerance of *Actinidia eriantha*. *Acta Horticulturae* 773, 277–282. DOI: 10.17660/ActaHortic.2008.773.41

Judd, M.J., McAneney, K.J. and Wilson, K.S. (1989) Influence of water stress on kiwifruit growth. *Irrigation Science* 10, 303–311. DOI: 10.1007/bf00257495

Klages, K., Donnison, H., Boldingh, H. and MacRae, E. (1998) *myo*-Inositol is the major sugar in *Actinidia arguta* during early fruit development. *Australian Journal of Plant Physiology* 25, 61–68. DOI: 10.1071/PP97052

Laing, W.A. (1985) Temperature and light response curves for photosynthesis in kiwifruit (*Actinidia chinensis*) cv. Hayward. *New Zealand Journal of Agricultural Research* 28, 117–124. DOI: 10.1080/00288233.1985.10427004

Lawson, T. and Vialet-Chabrand, S. (2019) Speedy stomata, photosynthesis and plant water use efficiency. *New Phytologist* 221, 93–98. DOI: 10.1111/nph.15330

Le Lievre, D., Anderson, R., Boldingh, H., Cooney, J., Seelye, R. *et al.* (2021) Modifying carbohydrate supply to fruit during development changes the composition and flavour of *Actinidia chinensis* var. *chinensis* 'Zesy002' kiwifruit. *Plants* (Basel) 10, 1328. DOI: 10.3390/plants10071328

Ledgard, S.F. and Brier, G.J. (1991) A ^{15}N method to quantify the sources of nitrogen for new growth of non-legumes, examined using kiwifruit. *Annals of Botany* 68, 75–78. DOI: 10.1093/oxfordjournals.aob.a088221

Li, H., Wang, J., Zhang, W., Bai, Q., Zhang, T. *et al.* (2021) Effects of high temperature stress on leaf chlorophyll fluorescence characteristics of kiwifruit. *Journal of Applied Meteorological Science* 32, 468–478. DOI: 10.11898/1001-7313.20210408

Li, Z., Zhong, Y., Bai, D., Lin, M., Qi, X. *et al.* (2020) Comparative analysis of physiological traits of three *Actinidia valvata* Dunn genotypes during waterlogging and post-waterlogging recovery. *Horticulture, Environment, and Biotechnology* 61, 825–836. DOI: 10.1007/s13580-020-00276-0

Liang, D., Ni, Z., Xia, H., Xie, Y., Lv, X. *et al.* (2019) Exogenous melatonin promotes biomass accumulation and photosynthesis of kiwifruit seedlings under drought stress. *Scientia Horticulturae* 246, 34–43. DOI: 10.1016/j.scienta.2018.10.058

Marschner, H. (1995) Functions of mineral nutrients: macronutrients. In: Marschner, H. (ed.) *Mineral Nutrition of Higher Plants*, 2nd edn. Academic Press, San Diego, California, pp. 285–299.

McAneney, K.J. and Judd, M.J. (1983) Observations on kiwifruit (*Actinidia chinensis* Planch.) root exploration, root pressure, hydraulic conductivity, and water uptake. *New Zealand Journal of Agricultural Research* 26, 507–510. DOI: 10.1080/00288233.1983.10427029

McPherson, H.G., Hall, A.J. and Stanley, C.J. (1994) Seasonal and regional variation in budbreak and flowering of kiwifruit vines in New Zealand. *New Zealand Journal of Crop and Horticultural Science* 22, 263–276. DOI: 10.1080/01140671.1994.9513835

McPherson, H.G., Richardson, A.C., Snelgar, W.P., Patterson, K.J. and Currie, M.B. (2001) Flower quality and fruit size in kiwifruit (*Actinidia deliciosa*). *New Zealand Journal of Crop and Horticultural Science* 29, 93–101. DOI: 10.1080/01140671.2001.9514167

Mi, Y.-F., Ma, F.-W. and Ma, X.-W. (2009) Effects of root-zone hypoxia stress on growth and endogenous hormones in young kiwifruit seedlings with different resistant ability. *Acta Horticulturae Sinica* 36, 163–170.

Mian, G., Cipriani, G., Saro, S., Martini, M. and Ermacora, P. (2022) Potential of different *Actinidia* genotypes as resistant rootstocks for preventing kiwifruit vine decline syndrome. *Horticulturae* 8, 627. DOI: 10.3390/horticulturae8070627

Miller, S.A., Smith, G.S., Boldingh, H.L. and Johansson, A. (1998) Effects of water stress on fruit quality attributes of kiwifruit. *Annals of Botany* 81, 73–81. DOI: 10.1006/anbo.1997.0537

Minchin, P.E.H., Richardson, A.C., Patterson, K.J. and Martin, P.J. (2003) Prediction of final weight for *Actinidia chinensis* 'Hort16A' fruit. *New Zealand Journal of Crop and Horticultural Science* 31, 147–157. DOI: 10.1080/01140671.2003.9514247

Minchin, P.E.H., Snelgar, W.P., Blattmann, P. and Hall, A.J. (2010) Competition between fruit and vegetative growth in Hayward kiwifruit. *New Zealand Journal of Crop and Horticultural Science* 38, 101–112. DOI: 10.1080/01140671003781728

Montanaro, G., Dichio, B. and Xiloyannis, C. (2007) Response of photosynthetic machinery of field-grown kiwifruit under Mediterranean conditions during drought and re-watering. *Photosynthetica* 45, 533–540. DOI: 10.1007/s11099-007-0091-4

Montanaro, G., Dichio, B. and Xiloyannis, C. (2009) Shade mitigates photoinhibition and enhances water use efficiency in kiwifruit under drought. *Photosynthetica* 47, 363–371. DOI: 10.1007/s11099-009-0057-9

Montanaro, G., Dichio, B., Xiloyannis, C. and Lang, A. (2012) Fruit transpiration in kiwifruit: environmental drivers and predictive model. *AoB Plants* 2012, pls036. DOI: 10.1093/aobpla/pls036

Montanaro, G., Dichio, B., Lang, A., Mininni, A.N., Nuzzo, V. *et al.* (2014) Internal versus external control of calcium nutrition in kiwifruit. *Journal of Plant Nutrition and Soil Science* 177, 819–830. DOI: 10.1002/jpln.201400396

Morgan, D.C., Warrington, I.J. and Halligan, E.A. (1985) Effect of temperature and photosynthetic photon flux density on vegetative growth of kiwifruit (*Actinidia chinensis*). *New Zealand Journal of Agricultural Research* 28, 109–116. DOI: 10.1080/00288233.1985.10427003

Muramatsu, N. and Hiraoka, K. (2008) Hydraulic conductance and xylem anatomy in fruit tree shoots. *Journal of the Japanese Society for Horticultural Science* 77, 122–127. DOI: 10.2503/jjshs1.77.122

Nanos, G.D., Sfakiotakis, E.M., Boufidis, C. and Tsitouridis, K. (1997) Frost damage and protection of kiwi vines. *Acta Horticulturae* 444, 233–236. DOI: 10.17660/ActaHortic.1997.444.35

Nardozza, S., Boldingh, H.L., Osorio, S., Höhne, M., Wohlers, M. *et al.* (2013) Metabolic analysis of kiwifruit (*Actinidia deliciosa*) berries from extreme genotypes reveals hallmarks for fruit starch metabolism. *Journal of Experimental Botany* 64, 5049–5063. DOI: 10.1093/jxb/ert293

Nardozza, S., Boldingh, H., Richardson, A., Walter, M., Kashuba, P. *et al.* (2015) Kiwifruit xylem sap: composition and in vitro growth of a virulent strain of *Pseudomonas syringae* pv. *actinidiae*. *Acta Horticulturae* 1095, 123–128. DOI: 10.17660/ActaHortic.2015.1095.15

Nuzzo, V., Dichio, B., Montanaro, G., Celano, G. and Xiloyannis, C. (1996) Risposta di piante di actinidia in piena produzione alle limitate disponibilità idriche del suolo (Response of outdoor kiwifruits to soil water stress). *Atti del Convegno Nazionale.* Societa Orticola Italiana, Florence, Italy, pp. 209–220.

Peterlunger, E., Marangoni, B., Testolin, R., Vizzotto, G. and Costa, G. (1990) Carbohydrates, organic acids and mineral elements in xylem sap bleeding from kiwifruit canes. *Acta Horticulturae* 282, 273–282. DOI: 10.17660/ActaHortic.1990.282.35

Pyke, N.B., Stanley, C.J. and Warrington, I.J. (1986) Kiwifruit: frost tolerance of plants in controlled frost conditions. *New Zealand Journal of Experimental Agriculture* 14, 443–447. DOI: 10.1080/03015521.1986.10423063

Reid, J.B. and Petrie, R.A. (1991) Effects of soil aeration on root demography in kiwifruit. *New Zealand Journal of Crop and Horticultural Science* 19, 423–431. DOI: 10.1080/01140671.1991.10422888

Reid, J.B., Tate, K.G., Brown, N.S. and Cheah, L.H. (1991) Effects of flooding and alluvium deposition on kiwifruit (*Actinidia deliciosa*): 1. Early vine decline. *New Zealand Journal of Crop and Horticultural Science* 19, 247–257. DOI: 10.1080/01140671.1991.10421808

Reid, J.B., Tate, K.G. and Brown, N.S. (1992) Effects of flooding and alluvium deposition on kiwifruit (*Actinidia deliciosa*). 2. Vine performance the following season. *New Zealand Journal of Crop and Horticultural Science* 20, 283–288. DOI: 10.1080/01140671.1992.10421769

Reid, J.B., Sorensen, I. and Petrie, R.A. (1993) Root demography in kiwifruit (*Actinidia deliciosa*). *Plant, Cell & Environment* 16, 949–957. DOI: 10.1111/j.1365-3040.1993.tb00518.x

Richardson, A.C. (2015) How does carbohydrate supply limit flower development in grape and kiwifruit vines? PhD thesis, Charles Sturt University, Wagga Wagga, Australia.

Richardson, A.C., Marsh, K.B., Boldingh, H.L., Pickering, A.H., Bulley, S.M. *et al.* (2004) High growing temperatures reduce fruit carbohydrate and vitamin C in kiwifruit. *Plant, Cell & Environment* 27, 423–435. DOI: 10.1111/j.1365-3040.2003.01161.x

Richardson, A.C., Boldingh, H.L., McAtee, P.A., Gunaseelan, K., Luo, Z. *et al.* (2011) Fruit development of the diploid kiwifruit, *Actinidia chinensis* 'Hort16A'. *BMC Plant Biology* 11, 182. DOI: 10.1186/1471-2229-11-182

Richardson, A.C., Boldingh, H.L., Kashuba, M.P., Nardozza, S. and Greer, D.H. (2018) Kiwifruit reserves: balancing vine growth and fruit productivity. *Acta Horticulturae* 1218, 163–169. DOI: 10.17660/ActaHortic.2018.1218.22

Richardson, A.C., Eyre, V., Kashuba, P., Ellingham, D., Jenkins, H. *et al.* (2021) Early shoot development affects carbohydrate supply and fruit quality of red-fleshed *Actinidia chinensis* var. *chinensis* 'Zes008'. *Agronomy* 11, 66. DOI: 10.3390/agronomy11010066

Rombolà, A.D., Dallari, S., Quartieri, M. and Scudellari, D. (2002) Effect of foliar-applied Fe sources, organic acids and sorbitol on the re-greening of kiwifruit leaves affected by lime-induced iron chlorosis. *Acta Horticulturae* 594, 349–355. DOI: 10.17660/ActaHortic.2002.594.43

Rothe, K., Porzel, A., Neumann, S. and Grimm, E. (1999) Characteristics of the phloem path: analysis and distribution of carbohydrates in the petiole of *Cyclamen*. *Journal of Experimental Botany* 50, 1807–1816. DOI: 10.1093/jxb/50.341.1807

Savé, R. and Serrano, L. (1986) Some physiological and growth responses of kiwi fruit (*Actinidia chinensis*) to flooding. *Physiologia Plantarum* 66, 75–78. DOI: 10.1111/j.1399-3054.1986.tb01236.x

Shim, K.K., Ahn, Y.H. and Yoo, M.S. (1985) Studies on photosynthesis of perennial vines (*Actinidia arguta*, *Celastrus orbiculatus*, *Lonicera japonica*, *Lonicera japonica* var. *aureoreticulata*, *Lonicera sempervirens*). *Journal of the Korean Society for Horticultural Science* 26, 44–50.

Smith, G.S. and Miller, S.A. (1992) Effects of root anoxia on the physiology of kiwifruit vines. *Acta Horticulturae* 297, 401–407. DOI: 10.17660/ActaHortic.1992.297.53

Smith, G.S., Buwalda, J.G., Green, T.G.A. and Clark, C.J. (1989) Effect of oxygen supply and temperature at the root on the physiology of kiwifruit vines. *New Phytologist* 113, 431–437. DOI: 10.1111/j.1469-8137.1989.tb00354.x

Snowball, A.M. (1995) The seasonal cycle of leaf, shoot and bud development in kiwifruit. *Journal of Horticultural Science* 70, 787–795.

Sotiropoulos, T.E., Therios, I.N. and Dimassi, K.N. (2002) Seasonal variation and chemical composition of bleeding xylem sap of kiwifruit vines irrigated with high boron water. *Journal of Plant Nutrition* 25, 1239–1248. DOI: 10.1081/pln-120004385

Testolin, R. and Messina, R. (1987) Winter cold tolerance of kiwifruit: a survey after winter frost injury in Northern Italy. *New Zealand Journal of Experimental Agriculture* 15, 501–504. DOI: 10.1080/03015521.1987.10425604

Thorp, T.G., Zhang, J. and Lay-Yee, M. (1990) Horticultural characteristics of seven pistillate and three staminate New Zealand cultivars of kiwifruit. *New Zealand Journal of Crop and Horticultural Science* 18, 233–240. DOI: 10.1080/01140671.1990.10428100

Thorp, T.G., Ferguson, I.B., Boyd, L.M. and Barnett, A.M. (2003) Fruiting position, mineral concentration and incidence of physiological pitting in 'Hayward' kiwifruit. *The Journal of Horticultural Science and Biotechnology* 78, 505–511. DOI: 10.1080/14620316.2003.11511655

Thorp, T.G., Barnett, A.M., Civolani, C., Spinelli, R., Lallu, N. *et al.* (2021) Iron chelate applications improve productivity and fruit quality at harvest and during cold storage of *Actinidia chinensis* var. *chinensis* ('Hort16A') kiwifruit growing in high-pH calcareous soils. *European Journal of Horticultural Science* 86, 609–619. DOI: 10.17660/eJHS.2021/86.6.4

Tyree, M.T. and Ewers, F.W. (1991) Tansley Review No. 34. The hydraulic architecture of trees and other woody plants. *The New Phytologist* 119, 345–360. DOI: 10.1111/j.1469-8137.1991.tb00035.x

Wang, N., Jiao, X., Guo, T., Li, C., Liu, Z. *et al.* (2019a) Time course of physiological responses in kiwifruit induced by bicarbonate. *Trees: Structure and Function* 33, 1711–1722. DOI: 10.1007/s00468-019-01892-9

Wang, N., Yao, C., Li, M., Li, C., Liu, Z. *et al.* (2019b) Anatomical and physiological responses of two kiwifruit cultivars to bicarbonate. *Scientia Horticulturae* 243, 528–536. DOI: 10.1016/j.scienta.2018.09.011

Woodward, T.J. (2007) Variation in 'Hayward' kiwifruit quality characteristics. PhD thesis, University of Waikato, Hamilton, New Zealand.

Wullschleger, S.D. (1993) Biochemical limitations to carbon assimilation in C_3 plants: a retrospective analysis of the A/C_i curves from 109 species. *Journal of Experimental Botany* 44, 907–920. DOI: 10.1093/jxb/44.5.907

Xiloyannis, C., Massai, R., Piccotino, D., Raffaelli, M. and Bovo, M. (1997) Effects of cropload on annual growth and dry matter partitioning in young kiwi vines. *Acta Horticulturae* 444, 213–218. DOI: 10.17660/ActaHortic.1997.444.31

Zhang, C., Zhang, H., Xiao, J., Ma, C. and Xie, M. (2013) Comparison on seedling waterlogging tolerance of three cultivars of *Actinidia chinensis* Planch. *Acta Agriculturae Zhejiangensis* 25, 1007–1012.

Zheng, S., Cui, N., Gong, D., Wang, Y., Hu, X. *et al.* (2020) Relationship between stable carbon isotope discrimination and water use efficiency under deficit drip irrigation of kiwifruit in the humid areas of South China. *Agricultural Water Management* 240, 106300. DOI: 10.1016/j.agwat.2020.106300

Zhong, Y.-P., Qi, X.-J., Chen, J.-Y., Li, Z., Bai, D.-F. *et al.* (2019) Growth and physiological responses of four kiwifruit genotypes to salt stress and resistance evaluation. *Journal of Integrative Agriculture* 18, 83–95. DOI: 10.1016/s2095-3119(18)62011-8

16 Fruit Development: Growth, Maturation and Ripening

Simona Nardozza* and Jeremy N. Burdon
The New Zealand Institute for Plant and Food Research Ltd, Auckland, New Zealand

16.1 Introduction

Fruit morphology is fairly well conserved within the *Actinidia* genus, although there is variation in fruit size, shape, stylar end shape, skin type and flesh colour (see Chapter 3). Ripe fruit composition, including sugars, organic acids, vitamin C and fibre, also differs within the genus. Some metabolic features make studied species of the *Actinidia* genus unique, including planteose as a key translocation sugar, the importance of cytokinin in fruit growth, starch as the main storage carbohydrate and autocatalytic ethylene production only occurring after the main ripening changes.

Fruit development is the sequence of physical and biochemical changes that brings about growth, maturation, ripening and senescence. This chapter describes kiwifruit growth and maturation with a focus on metabolic control.

16.2 Fruit Structure and Composition

In botanical terms, fruit from *Actinidia* species are berries that develop from superior multicarpellate ovaries (Hopping, 1976a). The fruit comprise four tissue zones (Fig. 16.1A): epicarp, outer pericarp, inner pericarp and core (Habart, 1974; Hopping, 1976a; Schmid, 1978; Ferguson, 1984) (see Chapter 3).

16.2.1 Tissue zones

The nature of the epicarp of mature kiwifruit varies among species and may consist of dead tissue with a suberized periderm and a coating of stiff uniseriate or multiseriate hairs (e.g. *A. chinensis* var. *deliciosa*) or a hairless epidermis protected by a thick cuticle (e.g. *A. arguta*) (Hallett and Sutherland, 2005). The periderm originates from three layers: the phellem (cork), the phellogen (cork meristem) and the phelloderm (Macnee *et al.*, 2020). The epidermal skin trait is semi-dominant over the peridermal skin trait. It is controlled by four quantitative trait loci (QTL) and the skin phenotype is suggested to be the result of a balance between cuticle strength and propensity to russet (Macnee *et al.*, 2021). Exploration of the genomic regions associated to the QTLs identified candidates for the trait, including cutin and suberin biosynthetic genes and MYB transcription factors (Macnee *et al.*, 2021).

*Corresponding author: simona.nardozza@plantandfood.co.nz

© CAB International 2024. *Kiwifruit: Botany, Production and Uses* (eds A.C. Richardson, J.N. Burdon and A.R. Ferguson)
DOI: 10.1079/9781800620933.0016

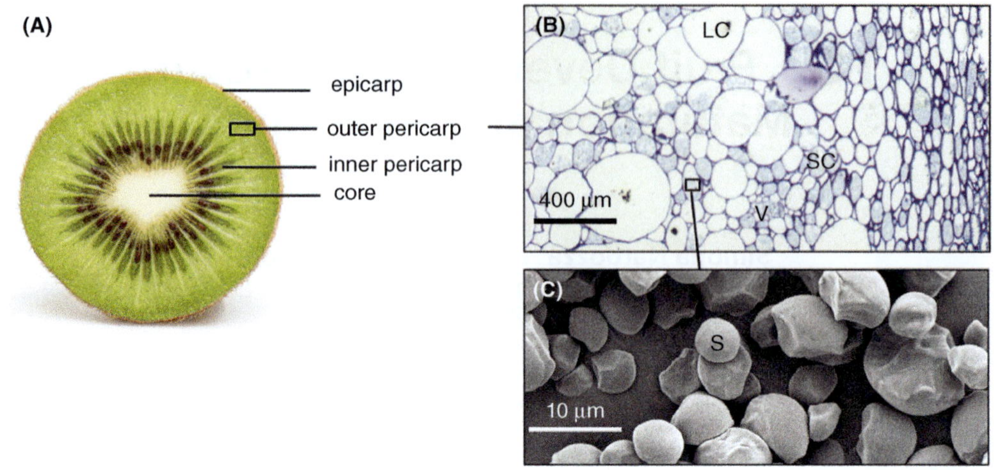

Fig. 16.1. (**A**) *Actinidia chinensis* var. *deliciosa* 'Hayward' fruit tissues. (**B**) Cross-section of fruit outer pericarp. (**C**) Scanning electron microscope image of starch granules. LC, large cell; S, starch; SC, small cell; V, vasculature. (Photos: The New Zealand Institute for Plant and Food Research Ltd)

Thin-walled parenchyma cells make up the fleshy outer pericarp (Fig. 16.1B): either large ovoid cells (diameter 0.5–0.8 mm) with low starch concentrations or small round cells (diameter 0.1–0.2 mm) packed with starch granules (Fig. 16.1C) resulting in high starch concentrations (Hallett *et al.*, 1992; Nardozza *et al.*, 2010a). The inner pericarp is made up of distended septum cells enclosing locules with multiple seeds (Hopping, 1976a; Hallett *et al.*, 1992). Chloroplasts in the inner and outer pericarp persist late into development. Chlorophyl degradation causing degreening occurs at different rates in different genotypes (Hopping, 1976b; Pilkington *et al.*, 2012). The core is also fleshy, composed of small isodiametric parenchyma cells containing starch (Hopping, 1976a; Guo *et al.*, 2013).

Vascular bundles are essential for fruit development and growth, as they transport water, sugars, minerals and hormones from the pedicel to the various fruit tissues. The pericarp and core tissues are rich in vasculature (Habart, 1974; Beever and Hopkirk, 1990; Guo *et al.*, 2013) (Fig. 16.2), with ventral and dorsal vascular bundles running from the pedicel end to the stylar end of the fruit. Ventral vascular bundles are present along the length of the core and together with dorsal vascular bundles define the boundaries of the inner pericarp. Ovule traces connect the ovules to the ventral vascular bundles, providing support to the growing seeds. Lateral vascular bundles branch from the dorsal vascular bundle to the outer pericarp. In *A. chinensis* var. *deliciosa* 'Qinmei' fruit, vasculature within the core and between alternate septa has been described (Guo *et al.*, 2013).

The contribution each tissue makes to the final fruit fresh weight (FW) depends on the cultivar. In *A. chinensis* var. *deliciosa* 'Hayward', the core is the smallest tissue and accounts for 7% of FW, the inner pericarp contributes 36% and the outer pericarp is the largest tissue, accounting for 57% (MacRae *et al.*, 1989). Similar to 'Hayward', the outer pericarp is the largest tissue in the fruit of most kiwifruit genotypes (Nardozza *et al.*, 2010a; Richardson *et al.*, 2021; Table 16.1), however, by the end of fruit growth the inner pericarp tissue is the largest in *A. chinensis* var. *chinensis* 'Hort16A' fruit (Richardson *et al.*, 2011). The core tissue in *A. chinensis* var. *deliciosa* genotypes is usually larger than in *A. chinensis* var. *chinensis* genotypes (Table 16.1).

16.2.2 Composition

The fruit of *A. chinensis* var. *chinensis* and *A. chinensis* var. *deliciosa* are 75–85% water and 15–25% dry matter (DM) (Beever and Hopkirk,

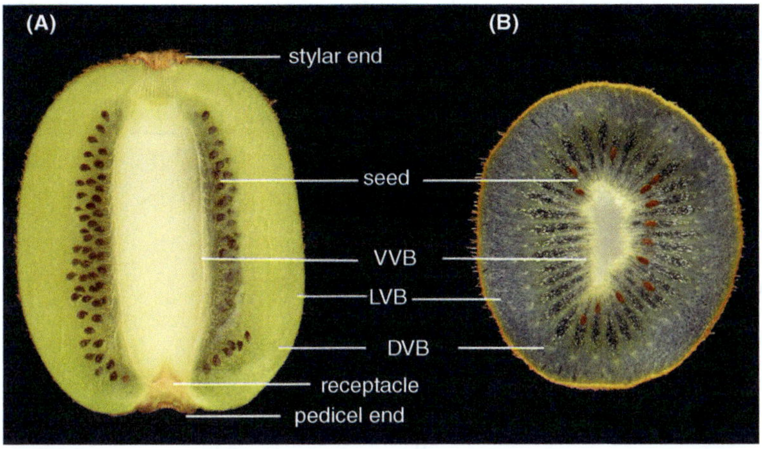

Fig. 16.2. Vasculature in fruit of *Actinidia chinensis* var. *deliciosa* 'Hayward'. Major vascular bundles in (**A**) longitudinal and (**B**) transverse sections. The transverse section has been cleared in lactate. DVB, dorsal vascular bundle; LVB, lateral vascular bundle; VVB, ventral vascular bundle. (Photos: The New Zealand Institute for Plant and Food Research Ltd)

Table 16.1. The proportion of outer pericarp, inner pericarp and core in the fruit of different kiwifruit genotypes.

Taxon and cultivar/ genotype	Outer pericarp	Inner pericarp	Core	References
Actinidia chinensis var. *deliciosa* 'Hayward'	57% FW	36% FW	7% FW	MacRae *et al*. (1989)
Actinidia chinensis var. *deliciosa* various genotypes	50–55% ECSA	40% ECSA	5–10% ECSA	Nardozza *et al.* (2010a)
Actinidia chinensis var. *chinensis* 'Hort16A'	40% ECSA	58% ECSA	2% ECSA	Richardson *et al*. (2011)
Actinidia chinensis var. *chinensis* 'Zesy002'	60–70% ECSA	20–30% ECSA	2–3% ECSA	Richardson *et al*. (2021)

FW, fresh weight; ECSA, equatorial cross-sectional area

1990). DM is highest in the central core (24–25%), lowest in the outer pericarp (15–16%) and intermediate in the inner pericarp (17–18%) (MacRae *et al.*, 1989). DM mostly comprises primary metabolites including structural and non-structural carbohydrates, organic acids and proteins. Also present, albeit in low concentrations, are secondary metabolites such as phenolic acids, flavonoids, chlorophylls, carotenoids and vitamins, as well as minerals and phytohormones (Drummond, 2013; Nardozza *et al.*, 2013, 2020b) (Table 16.2). Fruit composition changes during development.

Kiwifruit fruit development is similar in many aspects to the model fruit, tomato (Nardozza *et al.*, 2010b). However, there are some fundamental differences: organic acids, not soluble sugars, drive fruit osmotic pressure; starch is the major non-structural carbohydrate stored during fruit development (Nardozza *et al.*, 2010b); and despite kiwifruit being commonly referred to as a climacteric fruit (Kader, 2002), the main ripening events occur in the absence of an ethylene burst or respiratory climacteric (see Section 16.8). These differences show that there is no single fruit model. Furthermore, kiwifruit is a good model for starch metabolism during fruit growth and development, as at its peak prior to ripening, starch may represent 40% of the total DM (Nardozza *et al.*, 2010b). On ripening, the

Table 16.2. Ripe kiwifruit (*Actinidia chinensis*) composition per 100 g fruit. Source: USDA National Nutrient Database for Standard Reference. (From Haytowitz *et al.*, 2019)

Name	Unit	'Hayward' (entry 09148)	'Zesy002' (entry 09520)
Water	g	83.9	82.4
Energy	kcal	64.0	63.0
Protein	g	1.1	1.0
Total lipid (fat)	g	0.4	0.3
Ash	g	0.6	0.5
Carbohydrate			
Carbohydrate, by difference	g	14.0	15.8
Fibre, total dietary	g	3.0	1.4
Sugars, total	g	9.0	12.3
Sucrose	g	0.2	1.2
Glucose	g	4.1	5.3
Fructose	g	4.4	5.8
Minerals			
Calcium	mg	35.0	17.0
Iron	mg	0.2	0.2
Magnesium	mg	15.7	12.0
Phosphorus	mg	34.0	25.0
Potassium	mg	198.0	315.0
Sodium	mg	5.0	3.0
Zinc	mg	0.1	0.1
Copper	mg	0.1	0.2
Manganese	mg	0.1	0.0
Selenium	µg	0.2	0.4
Vitamins and other components			
Vitamin C, total ascorbic acid	mg	74.7	161.0
Thiamin	mg	0.0	0.0
Riboflavin	mg	0.0	0.1
Niacin	mg	0.4	0.2
Pantothenic acid	mg	0.2	0.0
Vitamin B-6	mg	0.1	0.1
Folate, total	µg	26.0	310.0
Vitamin A, RAE	µg	4.0	1.0
Carotene, beta	µg	52.0	14.0
Lutein + zeaxanthin	µg	122.0	24.0
Vitamin E (alpha-tocopherol)	mg	1.3	1.4
Vitamin K (phylloquinone)	µg	40.3	6.1

RAE, retinol activity equivalents

starch is converted to soluble sugars (MacRae *et al.*, 1992), commonly estimated as the soluble solids content (SSC). DM at harvest can be used to predict ripe fruit soluble solids content (rSSC) derived from the starch degradation (Burdon *et al.*, 2004). The fruit sugar:acid balance is an important consumer quality trait (Klee and Tieman, 2018) owing to its effect on perceived fruit sweetness, overall flavour perception and consumer liking in many fruit crops, e.g. apple, grape, citrus and kiwifruit (Richardson *et al.*, 1997; Crisosto and Crisosto, 2001, 2002; Harker *et al.*, 2002; Burdon *et al.*, 2004; Nardozza *et al.*, 2011) (see Chapter 21).

16.3 Growth

Fruit growth is described by the increase in fruit size over time. Fruit dimensions, fruit FW and fruit dry weight (DW) have all been used to plot changes in fruit size. The LDD (product of the three linear fruit dimensions: length, smallest diameter and largest diameter) is a reliable non-destructive method for estimating fruit size that is highly correlated with fruit FW in 'Hayward' (Snelgar *et al.*, 1992; Yuan *et al.*, 2020), 'Hort16A' (Minchin *et al.*, 2003) and 'Zesy002' (Yuan *et al.*, 2020). This high correlation with fruit FW reflects a proportional growth of fruit in its three dimensions (Watson and Gould, 1993). Fruit volume as measured by immersion is not always a good predictor of fruit size owing to inaccuracies created by the hair on the epicarp (Minchin *et al.*, 2003).

Kiwifruit berry DW growth increases linearly for most of the period from fruit set to harvest (Hopping, 1976a). FW growth follows a sigmoidal pattern (single, double or triple) in line with fluctuations in growth rates due to genetic, physiological and environmental factors (Pratt and Reid, 1974; Hopping, 1976a; Walton and De Jong, 1990; Richardson *et al.*, 2011).

16.3.1 Fruit growth stages

Kiwifruit growth has been subdivided into different stages according to physical or metabolic changes and the focus of the researchers (e.g. early growth, overall growth or maturation). Various classification systems have been reported in the kiwifruit literature to describe fruit growth stages (Fig. 16.3).

The basic model of fruit growth defines phases of cell division, cell enlargement and slow cell enlargement (Hopping, 1976a). In the work by Hopping (1976a), 'Hayward' growth was separated into stage I (0–58 days after

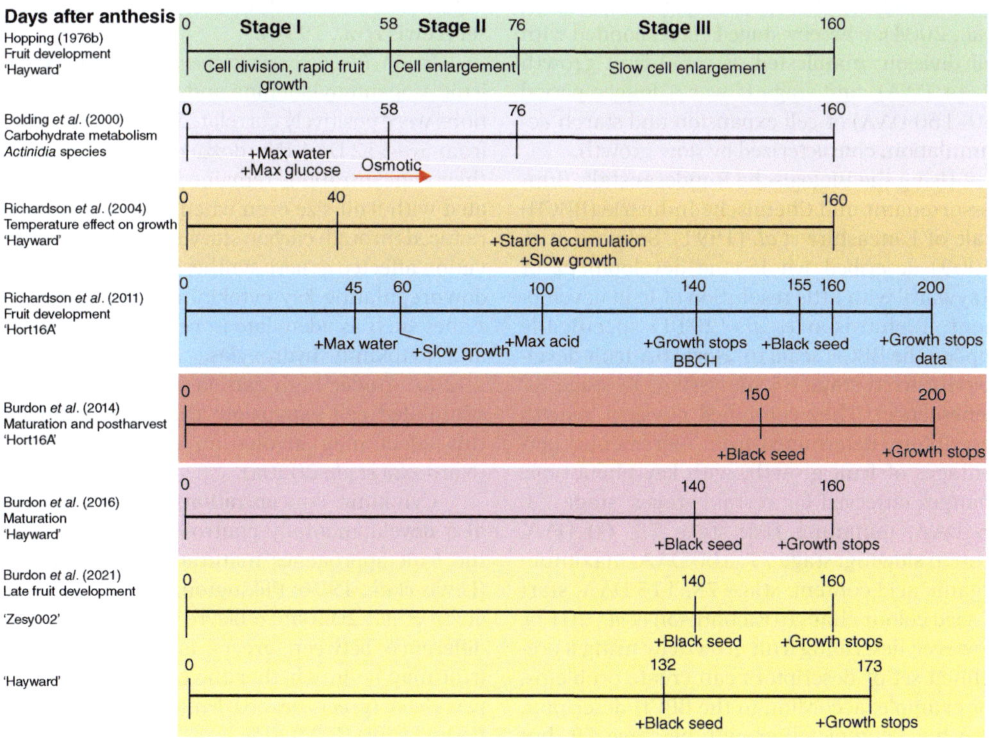

Fig. 16.3. Kiwifruit growth stages. The various studies looked at development for different reasons and a range of aspects of fruit growth have been defined. The + sign indicates features added by the different studies to the Hopping (1976a) model. BBCH, Biologische Bundesanstalt, Bundessortenamt und Chemische Industrie scale. Note: for Richardson *et al.* (2011), 'Growth stops BBCH' refers to when growth should stop according the BBCH scale; 'Growth stops data' refers to when growth stops according to experimental data.

anthesis, DAA), stage II (58–76 DAA) and stage III (76–160 DAA). Rapid growth in stage I involved cell division of all tissues and cell enlargement. Slowing of this cell enlargement in the inner pericarp and core, resulting in reduced growth, characterized stage II. A second growth period, with cell enlargement in the inner pericarp and core, marked stage III. Changes in metabolites may be observed across these fruit stages. In a study of five *Actinidia* taxa by Boldingh *et al.* (2000), the rapid fruit growth in stage I corresponded, in most taxa, with a spike in glucose concentration and high water concentration (~90% FW), indicating an osmotic drive to cell enlargement. Measurement of organic acids in later studies revealed that organic acids make much higher contributions to the fruit osmotic potential early in fruit development than soluble sugars (Nardozza *et al.*, 2010b). Fruit growth stages have also been described in terms of fruit growth and starch accumulation (Richardson *et al.*, 2004), whereby stage I corresponded with cell division, manifested as rapid fruit growth (0–44 DAA), and stage II was a longer period (40–160 DAA) of cell expansion and starch accumulation, characterized by slow growth.

Using the Biologische Bundesanstalt, Bundessortenamt und Chemische Industrie (BBCH) scale of Lancashire *et al.* (1991), Salinero *et al.* (2009) described whole-vine development of 'Hayward' with little resolution of fruit development. Richardson *et al.* (2011) specifically applied the BBCH scale to 'Hort16A' fruit development from stage 65 (flowering) to stage 90 (senescence). They combined physical growth descriptors, determined from incremental percentages of fruit growth, with key phenotypic changes observed for certain stages: stage 73, 45 DAA, minimum DM; stage 75, 60 DAA, growth slowing; stage 77, 100 DAA, maximum organic acid content; stage 78, 115 DAA, start of seed colour change (Richardson *et al.*, 2011). However, describing fruit growth by using a predefined set of descriptors can create problems. For example, according to the BBCH descriptor, fruit have completed growth by stage 79, but kiwifruit continued to grow until 210 DAA, before the onset of rapid fruit softening (Richardson *et al.*, 2011). By BBCH stage 80, most seeds have turned colour; however, the fruit is still growing. While the BBCH scale is useful for most of fruit growth, it has low resolution during the final stages of fruit maturation and ripening (Richardson *et al.*, 2011). However, this 'Hort16A' BBCH scale was useful for evaluating fruit development of 'Hort16A' continuous-flowering lines produced by editing a flowering repressor gene (Varkonyi-Gasic *et al.*, 2019), as a true control was not available (Wang *et al.*, 2021).

16.3.2 Phytohormones and fruit growth

Fruit endogenous cytokinin concentrations have been correlated with kiwifruit cell division (Lewis *et al.*, 1996), cell expansion (Nardozza *et al.*, 2020a), maturity and ripening (Lewis *et al.*, 1996; Pilkington *et al.*, 2013; Le Lievre *et al.*, 2021). Cytokinin concentrations (i.e. of isopentenyladenine and zeatin) were high from flowering to 30 DAA in 'Hayward' fruit, suggesting an involvement in cell division around fruit set (Lewis *et al.*, 1996).

In *A. chinensis* var. *chinensis* 'Zes006' kiwifruit, isopentenyladenine and zeatin concentrations were positively correlated with fruit weight from 56–112 DAA (Nardozza *et al.*, 2020a), and their concentrations remained positively correlated with fruit size even when fruit growth was reduced through carbon starvation. Carbon starvation affected *de novo* synthesis of cytokinins by downregulating key cytokinin biosynthetic enzymes such as adenylate isopentenyltransferase and cytokinin hydroxylase (Nardozza *et al.*, 2020a). Under high carbon supply, cytokinins stimulated cell expansion; the mechanism for this effect may involve an expansin protein (Nardozza *et al.*, 2020a).

Cytokinin concentrations in kiwifruit are also developmentally controlled, increasing as the fruit approaches maturation and ripening (Lewis *et al.*, 1996; Pilkington *et al.*, 2013; Nardozza *et al.*, 2020a; Le Lievre *et al.*, 2021). The difference between green- and yellow-fleshed fruit may lie in whether the cytokinin signal is perceived (green-fleshed fruit) or not (yellow-fleshed fruit) (Pilkington *et al.*, 2013).

The roles of other phytohormones in kiwifruit growth are less known. Whilst auxins are known to be crucial in the cell expansion phase of some fruits (de Jong *et al.*, 2010; Csukasi *et al.*, 2011; Devoghalaere *et al.*, 2012), there was no correlation between free auxins and fruit weight in 'Zes006' kiwifruit (Nardozza *et al.*, 2020a). Gibberellins and jasmonic acid changes during development were negatively correlated with kiwifruit weight (Nardozza *et al.*, 2020a), suggesting the role of gibberellins in kiwifruit contrasts with that in other fleshy fruits, where they drive cell expansion (Kumar *et al.*, 2014). Abscisic acid is involved in control of ripening in 'Hayward' kiwifruit (Huang *et al.*, 2021), but in 'Zes006' kiwifruit, increasing abscisic acid concentration during fruit development suggests it may have a role in fruit maturation (Nardozza *et al.*, 2020a).

16.4 Carbohydrate Metabolism during Fruit Growth

Carbohydrate composition changes during fruit development and varies among genotypes (Table 16.3).

16.4.1 Starch

Starch is an insoluble non-structural carbohydrate that can consist of two glucan polymers, the linear amylose and the branched amylopectin. Amylose concentration during development in kiwifruit ranges from 10% to12% of total starch in 'Hayward' and from 15% to 19% of total starch in 'Zesy002' (Li and Zhu, 2018a). Kiwifruit has a type-B polymorph starch, where amylopectin is less densely packed and higher in water retention than type-A polymorphs (e.g. cereals). Kiwifruit amylopectin belongs to group 4 (Bertoft *et al.*, 2008), with a low proportion of short-branching chains and high proportion of long-branching chains (Li and Zhu, 2018b).

Kiwifruit starch is stored predominantly in chloroplasts in the outer pericarp (Hallett *et al.*, 1995). Starch accumulation allows large amounts of carbon to be stored during fruit development without changing the osmotic pressure (Hall

Table 16.3. Carbohydrate composition of developing fruit from different kiwifruit genotypes.

Species and cultivar/genotype	Carbohydrate	Content (mg g^{-1} FW) at ~50 DAA	~100 DAA	~150 DAA	References
Actinidia chinensis var. *deliciosa* various genotypes	Starch	4–5	30–50	50–80	Nardozza *et al.* (2010b)
	Sucrose	1	1–2	3–7	
	Fructose	1	1	8	
	Glucose	5	2	6	
	myo-Inositol	1	1	1	
Actinidia chinensis var. *deliciosa* 'Hayward'	Starch	5	50	85	Nardozza *et al.* (2017)
	Sucrose	1.5	2	5	
	Fructose	1.5	2	9	
	Glucose	7	3	5	
	myo-Inositol	2	2	2	
Actinidia chinensis var. *chinensis* 'Zesy002'	Starch	4.5	56	90	Le Lievre *et al.* (2021)
	Sucrose	0.6	2.3	1.2	
	Fructose	4.6	6.4	9.0	
	Glucose	11.4	8.1	11.1	
	myo-Inositol	1.3	1.3	1.4	
Actinidia chinensis var. *chinensis* 'Zes006'	Starch	5	60	60	Nardozza *et al.* (2020b)
	Sucrose	1	1	5	
	Fructose	1	4	15	
	Glucose	4	5	12	
	myo-Inositol	1	1	1	
Actinidia chinensis var. *chinensis* 'Hort16A'	Starch	2	27.1	58.8	Richardson *et al.* (2011)
	Sucrose	4.3	1.8	4.2	
	Fructose	2.0	1.6	4.9	
	Glucose	0.6	0.7	3.4	
	myo-Inositol	2.2	2.0	1.3	

DAA, days after anthesis; FW, fresh weight

et al., 2013). A dynamic process of synthesis and degradation determines starch accumulation in kiwifruit (Wegrzyn and MacRae, 1995). The increase in transcripts of beta-amylase 9 (*BAM9*) and alpha-amylase2 (*AMY2*) genes early in fruit development has been associated with cytosolic starch/carbon turnover (Nardozza *et al.*, 2013). Net starch accumulation begins 50 DAA, following completion of cell division and the glucose peak (Richardson *et al.*, 2004; Nardozza *et al.*, 2013). ADP-glucose pyrophosphorylase (AGPase) may be a key enzyme in starch synthesis as higher AGPase activity and transcription of *APL4* (AGPase large subunit 4) were associated with genotypes with higher starch concentrations (Nardozza *et al.*, 2013).

16.4.2 Soluble carbohydrates

At 20–40 DAA, glucose concentration peaks in *A. chinensis* var. *deliciosa* genotypes (Klages *et al.*, 1998; Boldingh *et al.*, 2000; Nardozza *et al.*, 2013) and *myo*-inositol concentrations peak in *A. arguta* genotypes (Klages *et al.*, 1998; Boldingh *et al.*, 2000). In both cases, these peaks coincided with maximum fruit water content. While glucose may contribute in part to rapid fruit growth through an osmotic role (Nardozza *et al.*, 2010b, 2013), it has been suggested, but not proved, that *myo*-inositol may act as an osmo-protectant in *A. arguta* (Klages *et al.*, 1998).

The glucose peak has been associated with neutral invertase enzyme activity (Nardozza *et al.*, 2013), resulting in equal amounts of glucose and fructose. The involvement of a fructokinase has been suggested to metabolize fructose quickly, so only glucose increases (Klages *et al.*, 1998). Fructokinase and hexokinase enzymes showed similar patterns of activity during fruit development (Nardozza *et al.*, 2013), but *FK4* (fructokinase 4) transcripts were very high early in fruit development. The decline in glucose (as well as in gamma-aminobutyric acid and trehalose) may act as a signal for cell expansion and net starch accumulation (Nardozza *et al.*, 2013). The *myo*-inositol peak in *A. arguta* has not yet been associated with any particular enzyme or gene in kiwifruit.

The kiwifruit translocated sugars, sucrose and planteose, appear not to have a significant role during fruit development, with sucrose concentrations remaining low until maturity is reached and planteose concentrations being low in fruit, suggesting they are quickly metabolized after unloading (Nardozza *et al.*, 2013).

16.4.3 Structural carbohydrates

Metabolism of fruit structural carbohydrates and cell walls during growth has been overlooked compared with the changes occurring during ripening (Schröder and Atkinson, 2006). The main cell wall structural carbohydrates in whole ripening fruit are cellulose (insoluble), galactose and uronic acid. Early work on 'Hayward' showed a steady increase in cell wall carbohydrates with increasing fruit weight up to 90 DAA and then a plateau (Gallego and Zarra, 1997). Solubilization of cell wall-bound galactose started 90 DAA whilst polyuronide degradation occurred later, from 150 DAA. The main cell wall structural carbohydrates in the outer pericarp of 'Hayward' fruit during growth are pectin, hemicelluloses (e.g. xyloglucans) and cellulose (Li *et al.*, 2006). Pectins are degraded during early fruit expansion (56–84 DAA), whilst xyloglucans are more stable and their turnover is associated with cell enlargement (Li *et al.*, 2006).

16.5 Organic Acids during Fruit Growth

Citric acid, quinic acid, malic acid and ascorbic acid are present in kiwifruit (Bulley *et al.*, 2009; Marsh *et al.*, 2009). Quinic acid accounts for 40–60%, citric acid 40–60% and malic acid 10% of total acids in mature, unripe *Actinidia* fruit (Cheng *et al.*, 2004; Nishiyama *et al.*, 2008). Quinic acid peaks at 30–50 mg g^{-1} FW at 50 DAA, making it the most abundant organic acid during rapid fruit growth, with levels gradually decreasing later in fruit development (Nardozza *et al.*, 2010b). It has the strongest impact of the acids on taste (Marsh *et al.*, 2006). Quinic acid is a metabolite of the shikimic acid pathway, a key biosynthetic pathway for many secondary metabolites. Key enzymes in this pathway include quinate dehydrogenase (QDH) and shikimate dehydrogenase (SDH). Both SDH and QDH activities have been measured in fruit from different *Actinidia* species (Marsh *et al.*, 2009); it remains unknown if a QDH primary enzyme or dehydroquinate dehydratase (DQH)/SDH enzyme is responsible for the QDH activity. Marsh *et al.* (2009) proposed that the quinic acid/shikimic acid branching point is regulated by a bifunctional DQH/SDH enzyme.

Ascorbic acid (vitamin C) is a key health metabolite in kiwifruit. Ascorbic acid concentration is ~75 mg 100 g^{-1} FW in 'Hayward', ~160 mg 100 g^{-1} FW in 'Zesy002' (Table 16.2) and can reach up to 800 mg 100 g^{-1} FW in *A. eriantha* fruit (Bulley *et al.*, 2009). Most ascorbic acid is synthesized early in fruit development (Bulley *et al.*, 2009), and concentrations decline through dilution as fruit grow. In kiwifruit, ascorbic acid biosynthesis is dominated by the L-galactose pathway (Bulley *et al.*, 2009), where GDP-galactose phosphorylase and GDP-mannose epimerase are the rate-limiting enzymes (Bulley and Laing, 2016) of a tightly controlled and transcriptionally regulated system (Laing *et al.*, 2015).

16.6 Flesh Colour

The genus *Actinidia* exhibits a wide range of skin and flesh colours (Montefiori *et al.*, 2005, 2009) (Fig. 16.4). Skin colour is largely shades of brown for fruit with a peridermal skin, with the more vivid reds and yellows being restricted to the epidermal skin types of kiwiberries (*A. arguta, A. melanandra, A. polygama, A. macrosperma* and *A. valvata*) and hybrids.

Flesh colour of the major commercially available kiwifruit (*A. chinensis* var. *deliciosa* and *A. chinensis* var. *chinensis*) is green, yellow or red (see Chapter 4).

Green-fleshed kiwifruit retain chlorophyll in chloroplasts during fruit development and ripening (Possingham *et al.*, 1980). The retention of chlorophyll (green flesh colour) is due to slow chlorophyll degradation (degreening) during fruit development and ripening (Ben-Arie *et al.*, 1982; Gross, 1982).

Yellow-fleshed cultivars, including 'Hort16A', 'Jintao' and 'Zesy002', degreen late in development revealing the underlying pigmentation from xanthophylls and carotenoids (Nishiyama, 2007; Montefiori *et al.*, 2009). Rates of degreening vary among commercial cultivars (Gambi *et al.*, 2018; Ghasemnezhad *et al.*, 2022). Chlorophyll degradation in yellow-fleshed cultivars is temperature-dependent both on- and off-vine (De Silva *et al.*, 2007; Burdon and Lallu, 2022).

Contrary to some suggestions (McGhie and Ainge, 2002), degreening in yellow-fleshed cultivars is not ripening related (Burdon *et al.*, 2014). The timing of degreening is affected by the climatic conditions of the growing season, so that it is possible to have either fully degreened, firm, low SSC, unripe fruit or green-fleshed, soft, high SSC, ripe fruit (Burdon *et al.*, 2014). This may explain the variable descriptions of genotypes as having a yellow or greenish-yellow flesh.

Red-fleshed cultivars can be classified according to the localization of the red pigments (i.e. anthocyanin) into starburst fruit, where anthocyanin is only present in the locules of the inner pericarp (Fig. 16.4C), or block-red, where anthocyanin accumulates in both the inner and outer pericarp (Fig. 16.4D) (Montefiori *et al.*,

Fig. 16.4. Flesh colour examples of commercial kiwifruit. (**A**) A green-fleshed kiwifruit, *Actinidia chinensis* var. *deliciosa* 'Hayward'. (**B**) A yellow-fleshed kiwifruit, *A. chinensis* var. *chinensis* 'Zesy002'. (**C**) A starburst red-fleshed kiwifruit, *A. chinensis* var. *chinensis* 'Hort22D'. (**D**) A block red-fleshed kiwifruit, *A. chinensis* var. *chinensis* 'Zes008'. (Photos: The New Zealand Institute for Plant and Food Research Ltd)

2005, 2009). Starburst fruit commercialized so far have a yellow to greenish-yellow outer pericarp. The amount of anthocyanin accumulated during fruit development varies substantially, even within the same orchard and/or vine. Factors contributing to the variability include rootstock, cane type, shoot type, leaf numbers and carbon supply (Nardozza *et al.*, 2020b; Richardson *et al.*, 2021). The final appearance of the fruit flesh when eating ripe will depend on the fruit degreening to give the ideal bright red-on-yellow appearance of a starburst fruit (Nardozza *et al.*, 2015; McAtee *et al.*, 2022).

16.6.1 Chlorophyll metabolism

Degreening starts with the degeneration of grana and thylakoid membranes in chloroplasts, which marks their transition to chromoplasts (Possingham *et al.*, 1980; Montefiori *et al.*, 2009). Chlorophyll degradation is a multi-step process (Barry, 2009; Pilkington *et al.*, 2012) (see Chapter 9) leading to the formation of a colourless compound. Chlorophyll breakdown coincides with increased cytokinin concentration during kiwifruit maturation and ripening, but cytokinin metabolic control differs between green- and yellow-fleshed cultivars (Pilkington *et al.*, 2013; Le Lievre *et al.*, 2021). Cytokinin's role in kiwifruit flesh chlorophyll degradation, and thus degreening, is yet to be established (Pilkington *et al.*, 2013).

16.6.2 Carotenoid metabolism

Carotenoids in kiwifruit fruit flesh are largely xanthophylls (e.g. lutein, neoxanthin and violaxanthin) and β-carotene (McGhie and Ainge, 2002; Nishiyama *et al.*, 2005; Montefiori *et al.*, 2009). Green- and yellow-fleshed kiwifruit have similar carotenoid composition and there is no increase in carotenoids with ripening. In yellow-fleshed cultivars, the chlorophyll loss reveals the underlying colour (Nishiyama *et al.*, 2005; Montefiori *et al.*, 2009). However, it has been reported that carotenoids may increase after harvest if fruit are held at 20°C (Xia *et al.*, 2020).

16.6.3 Anthocyanin metabolism

The anthocyanins responsible for pigmentation of red-fleshed kiwifruit cultivars are flavonoids (Montefiori *et al.*, 2005, 2009; Comeskey *et al.*, 2009). Localization of anthocyanin in the inner and/or outer pericarp fruit tissues is genetically determined (Fig. 16.4c, d), although the loci responsible have not been identified. All the genes in the anthocyanin biosynthetic pathway have been identified and mostly characterized (Montefiori *et al.*, 2011; Wang *et al.*, 2019). Whilst MYB10 can switch on anthocyanin synthesis in kiwifruit, it is not the sole regulator and its gene expression is not linked to the presence or absence of anthocyanin (Nardozza *et al.*, 2020b). Early in fruit development, the flesh of red-fleshed kiwifruit cultivars is mostly green, although some anthocyanin might accumulate around flowering and disappear soon after (Montefiori *et al.*, 2011; Nardozza *et al.*, 2020b; Richardson *et al.*, 2021). Anthocyanin starts accumulating from 80 DAA, when starch accumulation is at its maximum rate (Nardozza *et al.*, 2020b). Carbon starvation has a negative effect on anthocyanin synthesis via a mechanism involving trehalose 6-phosphate sensing and the transcriptional repressor MYB27 (Nardozza *et al.*, 2020b).

The ploidy level of the pollen donor affects anthocyanin accumulation in kiwifruit, with higher ploidies yielding lower pigmentation (Seal *et al.*, 2018). Anthocyanin accumulation in kiwifruit is also affected by heat, with temperatures above 25°C reducing red pigmentation in 'Hongyang' (Man *et al.*, 2015). Storage at low temperature stimulates anthocyanin accumulation in 'Hongyang' (Li B. *et al.*, 2017) by upregulating the anthocyanin biosynthetic genes. It would be interesting to understand if this upregulation is stimulated by the increase in sugars from starch degradation (Naing *et al.*, 2021).

16.7 Maturation

Maturation is the developmental stage associated with the cessation of growth that precedes fruit ripening. There is sometimes confusion between maturation and ripening, but they are two very distinct physiological stages. A mature fruit is unripe but has the competence to ripen if detached from the plant. Commercial fruit must also ripen to a good flavour and withstand storage. Assessment and definition of maturation in kiwifruit has largely been based on fruit composition using easily measured indices, particularly SSC (Harman, 1981; Burdon, 2019). Understanding maturation is important for selecting the correct harvesting criteria (see Chapter 20). 'Hayward' fruit develop the capacity to be stored at low temperatures over a period of ~6–8 weeks (Burdon *et al.*, 2016). Understanding this process as well as the physiological stages of fruit development is therefore relevant to predicting the postharvest behaviour of the fruit (Burdon, 2022).

16.7.1 Soluble solids accumulation

The sigmoidal SSC accumulation pattern associated with the completion of fruit growth and fruit ripening is common to all cultivars. It was initially believed that the increased rate of SSC accumulation was because of starch breakdown (Harman, 1981) induced by low night temperatures (Seager *et al.*, 1992). However, the same pattern occurs irrespective of temperature, e.g. in newer cultivars which mature earlier than 'Hayward' when it is warmer (Burdon *et al.*, 2021). The original link between increased SSC accumulation rate and starch breakdown being dependent on cold temperatures is now considered coincidental, related to the late maturation of 'Hayward' fruit and the prevailing cool autumn temperatures in the 1970s (Burdon, 2015, 2019).

It now appears that at the transition between fruit growth and maturation, soluble sugars are still unloaded into the fruit but are not stored as starch, causing the initial increase in SSC accumulation rate (Burdon, 2022) (see Chapter 20). The increased accumulation rate continues as net starch degradation occurs during fruit ripening. SSC reaches a plateau when the fruit is fully ripe and all starch has been converted to soluble sugars. Thus, the sigmoidal pattern of SSC change is under genetic control unless environmental conditions (i.e. cold) occur when the fruit are responsive.

Starch degradation and increased SSC can also be triggered by preharvest treatment with Ethrel® (which releases ethylene) (Bowen *et al.*, 1988). This was originally considered to speed up maturation, as indicated by the increased SSC. However, the small increase in softening also recorded may indicate the triggering of ripening (Burdon *et al.*, 2017). The capacity of the fruit to respond to the environment (i.e. cold temperatures) and start ripening (starch breakdown, increase in SSC and softening) has been shown to develop over time during maturation (McAtee *et al.*, 2015; Burdon, 2015; Burdon *et al.*, 2016).

16.7.2 Seed development

Knowledge on kiwifruit seed development biology is fragmented. Hopping (1976a) showed that the nucellus and endosperm developed during fruit growth stages I and II. Embryo growth only occurred from stage II and continued to mid-stage III. Later work, summarized in Table 16.4, showed that the zygote did not start dividing until 56–70 days post-pollination and that the embryo took 112 days to form (Harvey *et al.*, 1987; Harvey and Fraser, 1988).

Hopping (1976b) suggested the involvement of an unknown 'seed factor' in fruit development. Although this has not yet been identified, cytokinins could be a likely candidate for a role in fruit size determination (Lewis *et al.*, 1996; Nardozza *et al.*, 2020a). Seed number has been positively correlated with fruit size in 'Hayward', 'Monty' and 'Bruno' (Hopping and Hacking, 1983), but not in other cultivars, e.g. 'Hort16A' (Seal *et al.*, 2013).

As in other fruit species, a xenia effect of pollen genes on seed and fruit development has been observed in kiwifruit (Qi *et al.*, 2007; Seal *et al.*, 2013, 2018). These studies showed that the origin of the pollen affected fruit traits including fruit set, fruit weight, SSC, DM concentration, weight and number of seeds, as well as the expression of red pigmentation in red-fleshed cultivars. 'Hort16A' pollinated by higher ploidy males produced heavier fruit (Seal *et al.*, 2013).

The seed coat colour change to dark brown/black and coat hardening are signs of seed maturity and are considered markers of fruit growth progression (Richardson *et al.*, 2011; Burdon *et al.*, 2021).

Table 16.4. Kiwifruit seed developmental stages from ovule initiation prior to anthesis to starch accumulation in cotyledons. (From Harvey and Fraser, 1988)

Stage	Time (days post-pollination)
Ovule initiation	−45[1]; −27[2]
Epidermal differentiation of cells of ovule (will become seed coat)	−45[1]; −27[2]
Meiosis	−12[1]; −7[2]
Flowering	0
Vacuolated uninucleate zygote	2
Endosperm dividing embryo sac into two sub-chambers	4–7
Dividing zygote, start	56–70
Endosperm storage products	56–70
Dividing zygote, to heart-shaped	84
Epidermal layer differentiated into seed coat	84
Endosperm cell disintegrating around embryo	98
Dividing zygote, to fully formed straight embryo	112
Starch in cotyledons	112

[1]*Actinidia chinensis* var. *deliciosa;* [2]*Actinidia chinensis* var. *chinensis*

16.7.3 Cessation of fruit growth

Towards the end of maturation, fruit growth gradually slows and ceases. Shrinkage in fruit volume has been reported in 'Hort16A' (Minchin *et al.*, 2003) using LDD measurements; however, this appears to be an artefact as possible changes in fruit shape during growth caused by relative differences among the three dimensions were not considered (Yuan *et al.*, 2020).

16.8 Ripening

In kiwifruit, ripening is judged as the change to rapid fruit softening, the onset of net starch breakdown and a subsequent rapid increase in soluble sugars (Burdon, 2019; Burdon *et al.*, 2021). Detailed monitoring of fruit development on the vine allows the change into ripening to be clearly identified. This occurs after seed maturation and the cessation of growth (Richardson *et al.*, 2011; Burdon *et al.*, 2014, 2016, 2021).

16.8.1 Control of ripening

The rates of softening and starch breakdown increase as development progresses or are triggered prematurely by ethylene or low temperature.

A key element in understanding the control of kiwifruit ripening is the status of the fruit in the climacteric–non-climacteric continuum. Kiwifruit is often categorized as climacteric (Kader, 2002), despite evidence that many ripening events are not associated with autocatalytic ethylene production (Fig. 16.5), which occurs once fruit are already soft or even senescent (Kim *et al.*, 1999, Richardson *et al.*, 2011). As a consequence, kiwifruit has been categorized as atypically climacteric (Antunes, 2007).

Kiwifruit ripening was initially described as climacteric based on its ethylene production, whilst noting that the respiratory pattern was atypical, but not considering the firmness of the fruit (Pratt and Reid, 1974). By considering the ripeness and softness of the fruit, the endogenous ethylene production in kiwifruit tends to increase only as the fruit approaches eating ripeness, prior to senescence (Richardson *et al.*, 2011). In 'Hayward' and 'Hort16A' fruit, this occurs when firmness reaches ~1 kgf or lower (Kim *et al.*, 1999; Ritenour *et al.*, 1999; Richardson *et al.*, 2011) (Fig. 16.5). Exceptions to this seem to be associated with some form of physical, physiological or pathological damage (Feng *et al.*, 2003). Thus, it appears that the early interpretation was hampered by overlooking the actual process of ripening and focusing on the change in respiration or ethylene production with time after harvest. Further confusion may arise as kiwifruit are very responsive to trace concentrations of exogenous ethylene. Kiwifruit respond to ethylene with an increase in respiratory activity (as seen in both climacteric and non-climacteric fruit).

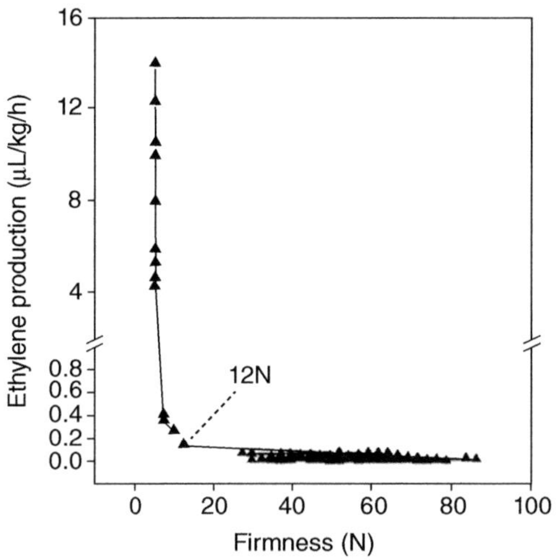

Fig. 16.5. Relationship between ethylene production and firmness of individual *Actinidia chinensis* var. *deliciosa* 'Hayward' kiwifruit. N.B. 9.81 N = 1 kgf. (From Kim *et al.*, 1999)

Softening may also be induced in kiwifruit by low temperatures, a process not mediated by ethylene (Mworia *et al.*, 2012; McAtee *et al.*, 2015; Minas *et al.*, 2016; Asiche *et al.*, 2018; Gunaseelan *et al.*, 2019; Mitalo *et al.*, 2019). The maximal temperature-induced softening rate is at ~8°C (Burdon *et al.*, 2017). Low storage temperature (4°C) prior to ethylene treatment has been reported to increase the ethylene sensitivity of fruit compared with fruit held at a high temperature of 25°C (Murakami *et al.*, 2014).

The main enzymes of ethylene biosynthesis are 1-aminocyclopropane-1-carboxylic acid synthase (ACS) and 1-aminocyclopropane-1-carboxylic acid oxidase (ACO) (Adams and Yang, 1979). Transcriptomic studies confirmed that changes in ACS and ACO gene transcripts are not linked to softening changes during ripening, reinforcing the view that softening is not naturally controlled by ethylene (Burdon *et al.*, 2021). Of more interest is the role of ethylene receptors and transcription factors. Ethylene receptors are negative regulators of ethylene signalling, meaning they inhibit the ethylene cascade reactions in the absence of ethylene (Hall *et al.*, 1999). The receptors are inactivated upon detection of ethylene, thereby stimulating ethylene response genes (*CTR1*, *EIN2*, *ERF2* and finally *EIN3/EIL*) (Yin *et al.*, 2008). However, the studies on the role of these genes were performed when fruit were already soft and ripe (Yin *et al.*, 2008; McAtee *et al.*, 2015). Fruit of 'Zesy002' on the vine progressed through to fruit softening with no significant changes in the transcription of the ethylene signalling genes (Burdon *et al.*, 2021).

Transcription factors associated with fruit ripening are the NAC class *NON-RIPENING* (*NOR*) genes that can inhibit ripening in tomato (Barry and Giovannoni, 2007) and the *MADS* class *RIPENING INHIBITOR*-like genes (*RIN*) associated with autocatalytic ethylene ripening (McAtee *et al.*, 2015). Kiwifruit has five NOR-like genes: *NAC1* to *NAC4* (Nieuwenhuizen *et al.*, 2015) and *NAC6* (Asiche *et al.*, 2018). In 'Zesy002' there was strong *NAC2* transcription associated with the progression of on-vine ripening; the *RIN/SEP4* transcript abundance decreased whilst *MADS2* transcripts increased (Burdon *et al.*, 2021).

16.8.2 Softening

Fruit softening is a complex co-ordinated chain of biochemical events (Schröder and Atkinson, 2006; Atkinson and Schröder, 2016). Kiwifruit softening is routinely measured as the force required to puncture the outer pericarp (epicarp removed) with a penetrometer (Watkins and Harman, 1981). The pattern of softening is sigmoidal with three clear phases (Fig. 16.6; MacRae and Redgwell, 1992), both on and off the vine (Kempler *et al.*, 1992; Burdon *et al.*, 2017). The pattern of softening in storage is a continuation of the pattern based on the progress that had occurred prior to harvest (see Chapter 20).

Little attention has been paid to Phase 1 softening, which occurs largely on the vine during fruit maturation; yet there is softening at a slow rate during this phase that can be assessed by the standard penetrometer measurement. The change from Phase 1 to Phase 2 softening appears to be the transition into ripening, which is developmentally driven but can also be induced by cold or ethylene. Kiwifruit Phase 2 softening is slowed slightly by 1-methylcyclopropene (1-MCP, an ethylene action inhibitor), suggesting some role for ethylene in this softening phase. The slow Phase 3 softening is characterized by a

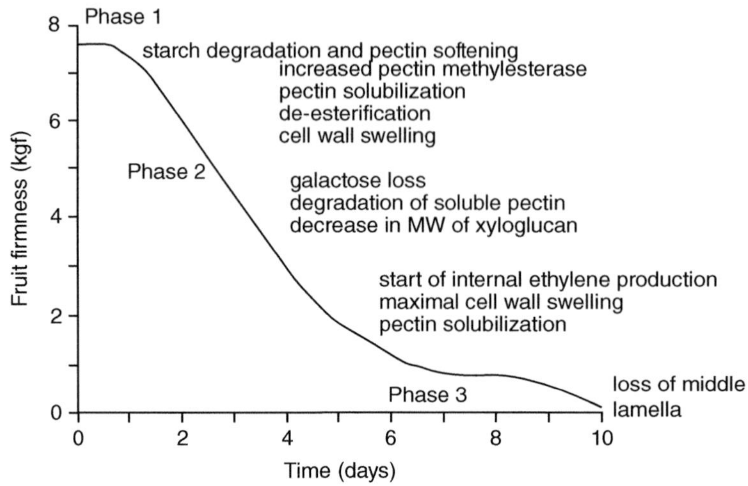

Fig. 16.6. Key events in kiwifruit softening of ethylene-treated *Actinidia chinensis* var. *deliciosa* 'Hayward' fruit (Phases 1–3 are softening phases). MW, molecular weight. (From Schröder, R., and Atkinson, R.G., 2006. ©Scion 2006. Available at: https://www.scionresearch.com/services/science-publications/new-zealand-journal-of-forestry-science [accessed 29 July 2023]; modified from MacRae and Redgwell, 1992)

burst of ethylene production, whilst there is a possible additional Phase 4 (Atkinson *et al.*, 2011) described as senescence.

Generally, the softening of the outer and inner pericarp and core shows some degree of co-ordination, although this can differ among genotypes or within a genotype dependent on storage conditions (Li, H. *et al.*, 2017). Softening of the outer pericarp and core may differ in the timing of the change to rapid softening and the pattern of softening during Phase 2 (Burdon *et al.*, 2017). Ethylene treatment can help to co-ordinate softening of the different tissues by stimulating all to soften simultaneously.

Most of the early work looking at cell wall polysaccharide changes used ethylene-treated fruit, where ethylene triggers or accelerates Phase 2 softening (Fig. 16.6). Recently, cell wall polysaccharide composition and softening of kiwifruit have been compared without ethylene treatment in genotypes with contrasting fruit softening behaviours (Fullerton *et al.*, 2020). Whilst a difference in xyloglucanase activity was found between fast- and slow softening genotypes, structural cell wall polysaccharide differences (e.g. polyuronides and xyloglucans) suggested different accessibility of degradative enzymes to specific cell wall polysaccharides during softening.

16.8.3 Starch metabolism

Starch concentration peaks as fruit mature owing to net starch accumulation (Burdon *et al.*, 2021). At the end of fruit maturation, there is a change from net starch accumulation to no starch change and finally to net starch degradation (Burdon, 2022). Alpha-amylase (Wegrzyn and MacRae, 1995) and beta-amylases (Gunaseelan *et al.*, 2019) have key roles in kiwifruit starch degradation. Ripening-associated starch breakdown is triggered either by programmed genetic control (Burdon *et al.*, 2021) or, if the fruit is competent to respond, by cold temperatures or ethylene (Burdon *et al.*, 2016; Gunaseelan *et al.*, 2019).

16.8.4 Volatile production

When ripe, kiwifruit are not regarded as highly aromatic fruit but can produce large amounts of volatiles when over-ripe (Young *et al.*, 1983). The characteristic 'Hayward' flavour is based on the grassy aroma of the C-6 aldehydes, n-hexanal and 2-hexenal when unripe, and the fruity aroma of the ester, ethyl butanoate when ripe (Young and Paterson, 1985, 1995; Gilbert *et al.*, 1996). Volatile production declines with storage

(Young and Paterson, 1985), particularly the aldehydes (Burdon *et al.*, 2005).

Ester production in kiwifruit tends to occur only once the fruit are ripe or over-ripe (Stec *et al.*, 1989; Burdon *et al.*, 2005). Ethylene plays a key role in volatile production by regulating lipoxygenase gene expression, which is associated with aroma volatile production (Zhang *et al.*, 2009, 2020). Volatiles may therefore depend on when ethylene is produced, possibly associated with fruit firmness, suggesting that fruit need to be softer later in storage for ethylene and volatile production (Paterson *et al.*, 1991). In 'Hayward', the key ester, ethyl butanoate, may not be produced until after the fruit have passed eating softness of 0.6–0.8 kgf (Burdon *et al.*, 2005). Production of esters in 'Hort16A' fruit has been associated with ethylene production and alcohol acyltransferase activity (Günther *et al.*, 2015). More recently, a fatty acid desaturase gene (*AdFAD1*), an aldehyde dehydrogenase gene (*AdALDH2*) and an alcohol acyltransferase gene (*AdAT17*) have been associated with ripening-related ester biosynthesis in kiwifruit (Zhang *et al.*, 2020).

The basic aldehyde/ester mix found in 'Hayward' fruit also occurs in other cultivars, but there may be other specific aroma volatiles. The terpenes, cineole and limonene, have been identified in 'Hort16A' fruit (Burdon *et al.*, 2007; Friel *et al.*, 2007; Wang *et al.*, 2011; Nieuwenhuizen *et al.*, 2015; Zeng *et al.*, 2020). Terpenes are more commonly seen in *A. arguta* genotypes, especially in the flowers (Matich *et al.*, 2003).

16.9 Summary and Outlook

Our understanding of kiwifruit growth, maturation and ripening has increased significantly in the last 15–20 years, with a particular focus on the metabolic control of development.

Future research is likely to increasingly use new tools such as gene editing to study how metabolite accumulation is controlled by signalling networks, genes or transcription factors during fruit development. For example, continuously flowering kiwifruit lines, with a significantly reduced time to obtain fruit, are a good model system to study fruit developmental physiology and its metabolic control. The metabolic role of seed development in fruit growth has not yet been explored, and the interaction between seed and flesh needs to be clarified. A greater focus on the interaction of the genotype with the environment will ensure that the increasing numbers of new cultivars are grown under optimal conditions for fruit growth, maturation and ripening to ensure the desired fruit phenotype is obtained (e.g. red-fleshed cultivars that are consistently red-fleshed; fruit of expected size and quality).

References

Adams, D.O. and Yang, S.F. (1979) Ethylene biosynthesis: identification of 1-aminocyclopropane-1-carboxylic acid as an intermediate in the conversion of methionine to ethylene. *Proceedings of the National Academy of Sciences* 76, 170–174. DOI: 10.1073/pnas.76.1.170

Antunes, M.D.C. (2007) The role of ethylene in kiwifruit ripening and senescence. *Stewart Postharvest Review* 3(2), 1–8. DOI: 10.2212/spr.2007.2.9

Asiche, W.O., Mitalo, O.W., Kasahara, Y., Tosa, Y., Mworia, E.G. *et al.* (2018) Comparative transcriptome analysis reveals distinct ethylene–independent regulation of ripening in response to low temperature in kiwifruit. *BMC Plant Biology* 18, 47. DOI: 10.1186/s12870-018-1264-y

Atkinson, R.G. and Schröder, R. (2016) Genetics of fruit softening. In: Testolin, R., Huang, H.-W. and Ferguson, A.R. (eds) *The Kiwifruit Genome*. Springer International Publishing, Cham, Switzerland, pp. 205–217. DOI: 10.1007/978-3-319-32274-2_16

Atkinson, R.G., Gunaseelan, K., Wang, M.Y., Luo, L., Wang, T. *et al.* (2011) Dissecting the role of climacteric ethylene in kiwifruit (*Actinidia chinensis*) ripening using a 1-aminocyclopropane-1-carboxylic acid oxidase knockdown line. *Journal of Experimental Botany* 62, 3821–3835. DOI: 10.1093/jxb/err063

Barry, C.S. (2009) The stay-green revolution: recent progress in deciphering the mechanisms of chlorophyll degradation in higher plants. *Plant Science* 176, 325–333. DOI: 10.1016/j.plantsci.2008.12.013

Barry, C.S. and Giovannoni, J.J. (2007) Ethylene and fruit ripening. *Journal of Plant Growth Regulation* 26, 143–159. DOI: 10.1007/s00344-007-9002-y

Beever, D.J. and Hopkirk, G. (1990) Fruit development and fruit physiology. In: Warrington, I.J. and Weston, G.C. (eds) *Kiwifruit: Science and Management*. Ray Richards Publisher in association with the New Zealand Society for Horticultural Science, Auckland, New Zealand, pp. 97–126.

Ben-Arie, R., Gross, J. and Sonego, L. (1982) Changes in ripening-parameters and pigments of the Chinese gooseberry (kiwi) during ripening and storage. *Scientia Horticulturae* 18, 65–70. DOI: 10.1016/0304-4238(82)90104-2

Bertoft, E., Piyachomkwan, K., Chatakanonda, P. and Sriroth, K. (2008) Internal unit chain composition in amylopectins. *Carbohydrate Polymers* 74, 527–543. DOI: 10.1016/j.carbpol.2008.04.011

Boldingh, H., Smith, G.S. and Klages, K. (2000) Seasonal concentrations of non-structural carbohydrates of five *Actinidia* species in fruit, leaf and fine root tissue. *Annals of Botany* 85, 469–476. DOI: 10.1006/anbo.1999.1094

Bowen, J.H., Lowe, R.G. and MacRae, E.A. (1988) The effect of a pre-harvest treatment with Ethrel on the starch content of kiwifruit. *Scientia Horticulturae* 35, 251–258. DOI: 10.1016/0304-4238(88)90118-5

Bulley, S. and Laing, W. (2016) The regulation of ascorbate biosynthesis. *Current Opinion in Plant Biology* 33, 15–22. DOI: 10.1016/j.pbi.2016.04.010

Bulley, S.M., Rassam, M., Hoser, D., Otto, W., Schünemann, N. *et al.* (2009) Gene expression studies in kiwifruit and gene over-expression in *Arabidopsis* indicates that GDP-L-galactose guanyltransferase is a major control point of vitamin C biosynthesis. *Journal of Experimental Botany* 60, 765–778. DOI: 10.1093/Jxb/Ern327

Burdon, J. (2015) Soluble solids revisited: a maturity or harvest index for kiwifruit. *Acta Horticulturae* 1096, 257–266. DOI: 10.17660/ActaHortic.2015.1096.28

Burdon, J. (2019) Kiwifruit biology: the commercial implications of fruit maturation. *Horticultural Reviews* 46, 385–421. DOI: 10.1002/9781119521082.ch8

Burdon, J. (2022) Understanding kiwifruit physiology: why we can predict aspects of postharvest performance. *Acta Horticulturae* 1332, 253–260. DOI: 10.17660/ActaHortic.2022.1332.34

Burdon, J. and Lallu, N. (2022) Minimising chilling injury in Italian-grown 'Hort16A' kiwifruit. *Acta Horticulturae* 1332, 343–349. DOI: 10.17660/ActaHortic.2022.1332.45

Burdon, J., McLeod, D., Lallu, N., Gamble, J., Petley, M. *et al.* (2004) Consumer evaluation of 'Hayward' kiwifruit of different at-harvest dry matter contents. *Postharvest Biology and Technology* 34, 245–255. DOI: 10.1016/j.postharvbio.2004.04.009

Burdon, J., Lallu, N., Billing, D., Burmeister, D., Yearsley, C. *et al.* (2005) Carbon dioxide scrubbing systems alter the ripe fruit volatile profiles in controlled-atmosphere stored 'Hayward' kiwifruit. *Postharvest Biology and Technology* 35, 133–141. DOI: 10.1016/j.postharvbio.2004.07.010

Burdon, J., Wang, M., Friel, E. and Lallu, N. (2007) Aroma volatiles of SmartFresh^{SM} treated 'Hayward' and 'Hort16A' kiwifruit. *Acta Horticulturae* 753, 763–768. DOI: 10.17660/ActaHortic.2007.753.100

Burdon, J., Pidakala, P., Martin, P., McAtee, P.A., Boldingh, H.L. *et al.* (2014) Postharvest performance of the yellow-fleshed 'Hort16A' kiwifruit in relation to fruit maturation. *Postharvest Biology and Technology* 92, 98–106. DOI: 10.1016/j.postharvbio.2014.01.004

Burdon, J., Pidakala, P., Martin, P., Billing, D. and Boldingh, H. (2016) Fruit maturation and the soluble solids harvest index for 'Hayward' kiwifruit. *Scientia Horticulturae* 213, 193–198. DOI: 10.1016/j.scienta.2016.10.027

Burdon, J., Pidakala, P., Martin, P. and Billing, D. (2017) Softening of 'Hayward' kiwifruit on the vine and in storage: the effects of temperature. *Scientia Horticulturae* 220, 176–182. DOI: 10.1016/j.scienta.2017.04.004

Burdon, J., Martin, P., Ireland, H., Schaffer, R., McAtee, P. *et al.* (2021) Transcriptomic analysis reveals differences in fruit maturation between two kiwifruit cultivars. *Scientia Horticulturae* 286, 110207. DOI: 10.1016/j.scienta.2021.110207

Cheng, C.H., Seal, A.G., Boldingh, H.L., Marsh, K.B., MacRae, E.A. *et al.* (2004) Inheritance of taste characters and fruit size and number in a diploid *Actinidia chinensis* (kiwifruit) population. *Euphytica* 138, 185–195. DOI: 10.1023/B:EUPH.0000046802.28347.41

Comeskey, D.J., Montefiori, M., Edwards, P.J.B. and McGhie, T.K. (2009) Isolation and structural identification of the anthocyanin components of red kiwifruit. *Journal of Agricultural and Food Chemistry* 57, 2035–2039. DOI: 10.1021/jf803287d

Crisosto, C.H. and Crisosto, G.M. (2001) Understanding consumer acceptance of early harvested 'Hayward' kiwifruit. *Postharvest Biology and Technology* 22, 205–213. DOI: 10.1016/S0925-5214(01)00097-7

Crisosto, C.H. and Crisosto, G.M. (2002) Understanding American and Chinese consumer acceptance of 'Redglobe' table grapes. *Postharvest Biology and Technology* 24, 155–162. DOI: 10.1016/S0925-5214(01)00189-2

Csukasi, F., Osorio, S., Gutierrez, J.R., Kitamura, J., Giavalisco, P. *et al.* (2011) Gibberellin biosynthesis and signalling during development of the strawberry receptacle. *New Phytologist* 191, 376–390. DOI: 10.1111/j.1469-8137.2011.03700.x

de Jong, M., Wolters-Arts, M., García-Martínez, J.L., Mariani, C. and Vriezen, W.H. (2010) The *Solanum lycopersicum* AUXIN RESPONSE FACTOR 7 (*SlARF7*) mediates cross-talk between auxin and gibberellin signalling during tomato fruit set and development. *Journal of Experimental Botany* 62, 617–626. DOI: 10.1093/jxb/erq293

De Silva, H.N., Hall, A.J., Burdon, J., Lallu, N., Connolly, P. *et al.* (2007) Modelling the effect of holding temperature on flesh de-greening of 'Hort16A' (ZESPRI™ GOLD) kiwifruit. *Acta Horticulturae* 753, 769–775. DOI: 10.17660/ActaHortic.2007.753.101

Devoghalaere, F., Doucen, T., Guitton, B., Keeling, J., Payne, W. *et al.* (2012) A genomics approach to understanding the role of auxin in apple (*Malus* x *domestica*) fruit size control. *BMC Plant Biology* 12, 7. DOI: 10.1186/1471-2229-12-7

Drummond, L. (2013) The composition and nutritional value of kiwifruit. *Advances in Food and Nutrition Research* 68, 33–57. DOI: 10.1016/B978-0-12-394294-4.00003-1

Feng, J., Maguire, K.M. and MacKay, B.R. (2003) Factors affecting ethylene production of Hayward kiwifruit. *Acta Horticulturae* 610, 203–209. DOI: 10.17660/ActaHortic.2003.610.27

Ferguson, A.R. (1984) Kiwifruit: a botanical review. *Horticultural Reviews* 6, 1–64. DOI: 10.1002/9781118060797.ch1

Friel, E.N., Wang, M., Taylor, A.J. and MacRae, E.A. (2007) *In vitro* and *in vivo* release of aroma compounds from yellow-fleshed kiwifruit. *Journal of Agricultural and Food Chemistry* 55, 6664–6673. DOI: 10.1021/jf063733x

Fullerton, C.G., Prakash, R., Ninan, A.S., Atkinson, R.G., Schaffer, R.J. *et al.* (2020) Fruit from two kiwifruit genotypes with contrasting softening rates show differences in the xyloglucan and pectin domains of the cell wall. *Frontiers in Plant Science* 11, 964. DOI: 10.3389/fpls.2020.00964

Gallego, P.P. and Zarra, I. (1997) Changes in cell wall composition and water-soluble polysaccharides during kiwifruit development. *Annals of Botany* 79, 695–701. DOI: 10.1006/anbo.1997.0399

Gambi, F., Pilkington, S.M., McAtee, P.A., Donati, I., Schaffer, R.J. *et al.* (2018) Fruit of three kiwifruit (*Actinidia chinensis*) cultivars differ in their degreening response to temperature after harvest. *Postharvest Biology and Technology* 141, 16–23. DOI: 10.1016/j.postharvbio.2018.03.009

Ghasemnezhad, M., Fullerton, C., Billing, D., Ampomah-Dwamena, C., Burdon, J. *et al.* (2022) Fruit degreening in different cultivars of kiwifruit. *Acta Horticulturae* 1332, 285–293. DOI: 10.17660/ActaHortic.2022.1332.38

Gilbert, J.M., Young, H., Ball, R.D. and Murray, S.H. (1996) Volatile flavour compounds affecting consumer acceptability of kiwifruit. *Journal of Sensory Studies* 11, 247–259. DOI: 10.1111/j.1745-459X.1996.tb00044.x

Gross, J. (1982) Pigment changes in the pericarp of the Chinese gooseberry or kiwi fruit (*Actinidia chinensis*) cv. Bruno during ripening. *Gartenbauwissenschaft* 47, 162–167.

Gunaseelan, K., McAtee, P.A., Nardozza, S., Pidakala, P., Wang, R. *et al.* (2019) Copy number variants in kiwifruit *ETHYLENE RESPONSE FACTOR/APETALA2* (*ERF/AP2*)-like genes show divergence in fruit ripening associated cold and ethylene responses in *C-REPEAT/DRE BINDING FACTOR*-like genes. *PLOS ONE* 14, e0216120. DOI: 10.1371/journal.pone.0216120

Günther, C.S., Marsh, K.B., Winz, R.A., Harker, R.F., Wohlers, M.W. *et al.* (2015) The impact of cold storage and ethylene on volatile ester production and aroma perception in 'Hort16A' kiwifruit. *Food Chemistry* 169, 5–12. DOI: 10.1016/j.foodchem.2014.07.070

Guo, X.-M., Xiao, X., Wang, G.-X. and Gao, R.-F. (2013) Vascular anatomy of kiwi fruit and its implications for the origin of carpels. *Frontiers in Plant Science* 4, 391. DOI: 10.3389/fpls.2013.00391

Habart, J.L. (1974) La baie de l'*Actinidia chinensis* Planch. var. *chinensis*. *Fruits* 29, 191–207.

Hall, A.E., Chen, Q.G., Findell, J.L., Schaller, G.E. and Bleecker, A.B. (1999) The relationship between ethylene binding and dominant insensitivity conferred by mutant forms of the ETR1 ethylene receptor1. *Plant Physiology* 121, 291–300. DOI: 10.1104/pp.121.1.291

Hall, A.J., Minchin, P.E.H., Clearwater, M.J. and Génard, M. (2013) A biophysical model of kiwifruit (*Actinidia deliciosa*) berry development. *Journal of Experimental Botany* 64, 5473–5483. DOI: 10.1093/jxb/ert317

Hallett, I.C. and Sutherland, P.W. (2005) Structure and development of kiwifruit skins. *International Journal of Plant Sciences* 166, 693–704. DOI: 10.1086/431232

Hallett, I.C., MacRae, E.A. and Wegrzyn, T.F. (1992) Changes in kiwifruit cell wall ultrastructure and cell packing during postharvest ripening. *International Journal of Plant Sciences* 153, 49–60. DOI: 10.1086/297006

Hallett, I.C., Wegrzyn, T.F. and MacRae, E.A. (1995) Starch degradation in kiwifruit: in vivo and in vitro ultrastructural studies. *International Journal of Plant Sciences* 156, 471–480. DOI: 10.1086/297269

Harker, F.R., Marsh, K.B., Young, H., Murray, S.H., Gunson, F.A. *et al.* (2002) Sensory interpretation of instrumental measurements 2: sweet and acid taste of apple fruit. *Postharvest Biology and Technology* 24, 241–250. DOI: 10.1016/S0925-5214(01)00157-0

Harman, J. (1981) Kiwifruit maturity. *The Orchardist of New Zealand* 54, 126–127, 130.

Harvey, C.F. and Fraser, L.G. (1988) Floral biology of two species of *Actinidia* (Actinidiaceae). II. Early embryology. *Botanical Gazette* 149, 37–44. DOI: 10.1086/337689

Harvey, C.F., Fraser, L.G., Pavis, S.E. and Considine, J.A. (1987) Floral biology of two species of *Actinidia* (Actinidiaceae). I. The stigma, pollination, and fertilization. *Botanical Gazette* 148, 426–432. DOI: 10.1086/337673

Haytowitz, D.B., Ahuja, J.K.C., Wu, X., Somanchi, M., Nickle, M. *et al.* (2019) USDA National Nutrient Database for Standard Reference, Legacy Release. Available at: https://catalog.data.gov/dataset/usda-national-nutrient-database-for-standard-reference-legacy-release (accessed 3 April 2023).

Hopping, M.E. (1976a) Structure and development of fruit and seeds in Chinese gooseberry (*Actinidia chinensis* Planch.). *New Zealand Journal of Botany* 14, 63–68. DOI: 10.1080/0028825X.1976.10428651

Hopping, M.E. (1976b) Effect of exogenous auxins, gibberellins, and cytokinins on fruit development in Chinese gooseberry (*Actinidia chinensis* Planch.). *New Zealand Journal of Botany* 14, 69–75. DOI: 10.1080/0028825X.1976.10428652

Hopping, M.E. and Hacking, N.J.A. (1983) A comparison of pollen application methods for the artificial pollination of kiwifruit. *Acta Horticulturae* 139, 41–50. DOI: 10.17660/ActaHortic.1983.139.5

Huang, W., Billing, D., Cooney, J., Wang, R. and Burdon, J. (2021) The role of ethylene and abscisic acid in kiwifruit ripening during postharvest dehydration. *Postharvest Biology and Technology* 178, 111559. DOI: 10.1016/j.postharvbio.2021.111559

Kader, A.A. (2002) *Postharvest Technology of Horticultural Crops*, 3rd edn. University of California Agriculture and Natural Resources Publication 3311, Oakland, California.

Kempler, C., Kabaluk, J.T. and Toivonen, P.M.A. (1992) Effect of environment and harvest date on maturation and ripening of kiwifruit in British Columbia. *Canadian Journal of Plant Science* 72, 863–869. DOI: 10.4141/cjps92-108

Kim, H.O., Hewett, E.W. and Lallu, N. (1999) The role of ethylene in kiwifruit softening. *Acta Horticulturae* 498, 255–261. DOI: 10.17660/ActaHortic.1999.498.29

Klages, K., Donnison, H., Boldingh, H. and MacRae, E. (1998) *myo*-Inositol is the major sugar in *Actinidia arguta* during early fruit development. *Australian Journal of Plant Physiology* 25, 61–67. DOI: 10.1071/PP97052

Klee, H.J. and Tieman, D.M. (2018) The genetics of fruit flavour preferences. *Nature Reviews Genetics* 19, 347–356. DOI: 10.1038/s41576-018-0002-5

Kumar, R., Khurana, A. and Sharma, A.K. (2014) Role of plant hormones and their interplay in development and ripening of fleshy fruits. *Journal of Experimental Botany* 65, 4561–4575. DOI: 10.1093/jxb/eru277

Laing, W.A., Martínez-Sánchez, M., Wright, M.A., Bulley, S.M., Brewster, D. *et al.* (2015) An upstream open reading frame is essential for feedback regulation of ascorbate biosynthesis in Arabidopsis. *Plant Cell* 27, 772–786. DOI: 10.1105/tpc.114.133777

Lancashire, P.D., Bleiholder, H., van den Boom, T., Langelüddeke, P., Stauss, R. *et al.* (1991) A uniform decimal code for growth stages of crops and weeds. *Annals of Applied Biology* 119, 561–601. DOI: 10.1111/j.1744-7348.1991.tb04895.x

Le Lievre, D., Anderson, R., Boldingh, H., Cooney, J., Seelye, R. *et al.* (2021) Modifying carbohydrate supply to fruit during development changes the composition and flavour of *Actinidia chinensis* var. *chinensis* 'Zesy002' kiwifruit. *Plants* 10(7), 1328. DOI: 10.3390/plants10071328

Lewis, D.H., Burge, G.K., Schmierer, D.M. and Jameson, P.E. (1996) Cytokinins and fruit development in the kiwifruit (*Actinidia deliciosa*). 1. Changes during fruit development. *Physiologia Plantarum* 98, 179–186. DOI: 10.1111/j.1399-3054.1996.tb00690.x

Li, B., Xia, Y., Wang, Y., Qin, G. and Tian, S. (2017) Characterization of genes encoding key enzymes involved in anthocyanin metabolism of kiwifruit during storage period. *Frontiers in Plant Science* 8, 341. DOI: 10.3389/fpls.2017.00341

Li, D. and Zhu, F. (2018a) Starch structure in developing kiwifruit. *International Journal of Biological Macromolecules* 120A, 1306–1314. DOI: 10.1016/j.ijbiomac.2018.08.128

Li, D. and Zhu, F. (2018b) Characterization of polymer chain fractions of kiwifruit starch. *Food Chemistry* 240, 579–587. DOI: 10.1016/j.foodchem.2017.07.079

Li, H., Billing, D., Pidakala, P. and Burdon, J. (2017) Textural changes in 'Hayward' kiwifruit during and after storage in controlled atmospheres. *Scientia Horticulturae* 222, 40–45. DOI: 10.1016/j.scienta.2017.05.013

Li, X., Nakagawa, N., Nevins, D.J. and Sakurai, N. (2006) Changes in the cell-wall polysaccharides of outer pericarp tissues of kiwifruit during development. *Plant Physiology and Biochemistry* 44, 115–124. DOI: 10.1016/j.plaphy.2006.03.005

Macnee, N.C., Rebstock, R., Hallett, I.C., Schaffer, R.J. and Bulley, S.M. (2020) A review of current knowledge about the formation of native peridermal exocarp in fruit. *Functional Plant Biology* 47, 1019–1031. DOI: 10.1071/FP19135

Macnee, N.C., Hilario, E., Tahir, J., Currie, A., Warren, B. *et al.* (2021) Peridermal fruit skin formation in *Actinidia* sp. (kiwifruit) is associated with genetic *loci* controlling russeting and cuticle formation. *BMC Plant Biology* 21, 334. DOI: 10.1186/s12870-021-03025-2

MacRae, E. and Redgwell, R. (1992) Softening in kiwifruit. *Postharvest News and Information* 3(3), 49N–52N.

MacRae, E.A., Bowen, J.H. and Stec, M.G.H. (1989) Maturation of kiwifruit (*Actinidia deliciosa* cv. Hayward) from two orchards: differences in composition of the tissue zones. *Journal of the Science of Food and Agriculture* 47, 401–416. DOI: 10.1002/jsfa.2740470403

MacRae, E.A., Quick, W.P., Benker, C. and Stitt, M. (1992) Carbohydrate metabolism during postharvest ripening in kiwifruit. *Planta* 188, 314–323. DOI: 10.1007/BF00192797

Man, Y.-P., Wang, Y.-C., Li, Z.-Z., Jiang, Z.-W., Yang, H.-L. *et al.* (2015) High-temperature inhibition of biosynthesis and transportation of anthocyanins results in the poor red coloration in red-fleshed *Actinidia chinensis*. *Physiologia Plantarum* 153, 565–583. DOI: 10.1111/ppl.12263

Marsh, K.B., Friel, E.N., Gunson, A., Lund, C. and MacRae, E. (2006) Perception of flavour in standardised fruit pulps with additions of acids or sugars. *Food Quality and Preference* 17, 376–386. DOI: 10.1016/j.foodqual.2005.04.011

Marsh, K.B., Boldingh, H.L., Shilton, R.S. and Laing, W.A. (2009) Changes in quinic acid metabolism during fruit development in three kiwifruit species. *Functional Plant Biology* 36, 463–470. DOI: 10.1071/FP08240

Matich, A.J., Young, H., Allen, J.M., Wang, M.Y., Fielder, S. *et al.* (2003) *Actinidia arguta*: volatile compounds in fruit and flowers. *Phytochemistry* 63, 285–301. DOI: 10.1016/S0031-9422(03)00142-0

McAtee, P.A., Richardson, A.C., Nieuwenhuizen, N.J., Gunaseelan, K., Hoong, L. *et al.* (2015) The hybrid non-ethylene and ethylene ripening response in kiwifruit (*Actinidia chinensis*) is associated with differential regulation of *MADS*-box transcription factors. *BMC Plant Biology* 15, 304. DOI: 10.1186/s12870-015-0697-9

McAtee, P.A., Nardozza, S., Richardson, A., Wohlers, M. and Schaffer, R.J. (2022) A data driven approach to assess complex colour profiles in plant tissues. *Frontiers in Plant Science* 12, 808138. DOI: 10.3389/fpls.2021.808138

McGhie, T.K. and Ainge, G.D. (2002) Color in fruit of the genus *Actinidia*: carotenoid and chlorophyll compositions. *Journal of Agricultural and Food Chemistry* 50, 117–121. DOI: 10.1021/jf010677l

Minas, I.S., Tanou, G., Karagiannis, E., Belghazi, M. and Molassiotis, A. (2016) Coupling of physiological and proteomic analysis to understand the ethylene- and chilling-induced kiwifruit ripening syndrome. *Frontiers in Plant Science* 7, 120. DOI: 10.3389/fpls.2016.00120

Minchin, P.E.H., Richardson, A.C., Patterson, K.J. and Martin, P.J. (2003) Prediction of final weight for *Actinidia chinensis* 'Hort16A' fruit. *New Zealand Journal of Crop and Horticultural Science* 31, 147–157. DOI: 10.1080/01140671.2003.9514247

Mitalo, O.W., Tokiwa, S., Kondo, Y., Otsuki, T., Galis, I. *et al.* (2019) Low temperature storage stimulates fruit softening and sugar accumulation without ethylene and aroma volatile production in kiwifruit. *Frontiers in Plant Science* 10, 888. DOI: 10.3389/fpls.2019.00888

Montefiori, M., McGhie, T.K., Costa, G. and Ferguson, A.R. (2005) Pigments in the fruit of red-fleshed kiwifruit (*Actinidia chinensis* and *Actinidia deliciosa*). *Journal of Agricultural and Food Chemistry* 53, 9526–9530. DOI: 10.1021/jf051629u

Montefiori, M., McGhie, T.K., Hallett, I.C. and Costa, G. (2009) Changes in pigments and plastid ultrastructure during ripening of green-fleshed and yellow-fleshed kiwifruit. *Scientia Horticulturae* 119, 377–387. DOI: 10.1016/j.scienta.2008.08.022

Montefiori, M., Espley, R.V., Stevenson, D., Cooney, J., Datson, P.M. *et al.* (2011) Identification and characterisation of F3GT1 and F3GGT1, two glycosyltransferases responsible for anthocyanin biosynthesis in red-fleshed kiwifruit (*Actinidia chinensis*). *Plant Journal* 65, 106–118. DOI: 10.1111/j.1365-313X.2010.04409.x

Murakami, S., Ikoma, Y. and Yano, M. (2014) Low temperature increases ethylene sensitivity in *Actinidia chinensis* 'Rainbow Red' kiwifruit. *Journal of the Japanese Society for Horticultural Science* 83, 322–326. DOI: 10.2503/jjshs1.CH-104

Mworia, E.G., Yoshikawa, T., Salikon, N., Oda, C., Asiche, W.O. *et al.* (2012) Low-temperature-modulated fruit ripening is independent of ethylene in 'Sanuki Gold' kiwifruit. *Journal of Experimental Botany* 63, 963–971. DOI: 10.1093/jxb/err324

Naing, A.H., Xu, J., Park, K.I., Chung, M.Y. and Kim, C.K. (2021) Sucrose enhances anthocyanin accumulation in Torenia by promoting expression of anthocyanin biosynthesis genes. *Horticulturae* 7, 219. DOI: 10.3390/horticulturae7080219

Nardozza, S., Hallett, I.C., McCartney, R., Richardson, A.C., MacRae, E.A. *et al.* (2010a) Is fruit anatomy involved in variation in fruit starch concentration between *Actinidia deliciosa* genotypes? *Functional Plant Biology* 38, 63–74. DOI: 10.1071/FP10158

Nardozza, S., Boldingh, H.L., Richardson, A.C., Costa, G., Marsh, H. *et al.* (2010b) Variation in carbon content and size in developing fruit of *Actinidia deliciosa* genotypes. *Functional Plant Biology* 37, 545–554. DOI: 10.1071/FP09301

Nardozza, S., Gamble, J., Axten, L.G., Wohlers, M.W., Clearwater, M.J. *et al.* (2011) Dry matter content and fruit size affect flavour and texture of novel *Actinidia deliciosa* genotypes. *Journal of the Science of Food and Agriculture* 91, 742–748. DOI: 10.1002/jsfa.4245

Nardozza, S., Boldingh, H.L., Osorio, S., Höhne, M., Wohlers, M. *et al.* (2013) Metabolic analysis of kiwifruit (*Actinidia deliciosa*) berries from extreme genotypes reveals hallmarks for fruit starch metabolism. *Journal of Experimental Botany* 64, 5049–5063. DOI: 10.1093/jxb/ert293

Nardozza, S., Kashuba, P., McCaughan, L., Philippe, M., Wohlers, M. *et al.* (2015) Leaves are important to obtain consistent red flesh pigmentation in *Actinidia chinensis* fruit. *Scientia Horticulturae* 197, 496–503. DOI: 10.1016/j.scienta.2015.10.008

Nardozza, S., Boldingh, H.B., Wohlers, M.W., Gleave, A.P., Luo, Z. *et al.* (2017) Exogenous cytokinin application to *Actinidia chinensis* var. *deliciosa* 'Hayward' fruit promotes fruit expansion through water uptake. *Horticulture Research* 4, 17043. DOI: 10.1038/hortres.2017.43

Nardozza, S., Cooney, J., Boldingh, H.L., Hewitt, K.G., Trower, T. *et al.* (2020a) Phytohormone and transcriptomic analysis reveals endogenous cytokinins affect kiwifruit growth under restricted carbon supply. *Metabolites* 10, 23. DOI: 10.3390/metabo10010023

Nardozza, S., Boldingh, H.L., Kashuba, M.P., Feil, R., Jones, D. *et al.* (2020b) Carbon starvation reduces carbohydrate and anthocyanin accumulation in red-fleshed fruit via trehalose 6-phosphate and MYB27. *Plant, Cell & Environment* 43, 819–835. DOI: 10.1111/PCE.13699

Nieuwenhuizen, N.J., Chen, X., Wang, M.Y., Matich, A.J., Perez, R.L. *et al.* (2015) Natural variation in monoterpene synthesis in kiwifruit: transcriptional regulation of terpene synthases by NAC and ETHYLENE-INSENSITIVE3-like transcription factors. *Plant Physiology* 167, 1243–1258. DOI: 10.1104/pp.114.254367

Nishiyama, I. (2007) Fruits of the *Actinidia* genus. *Advances in Food Nutrition Research* 52, 293–324. DOI: 10.1016/s1043-4526(06)52006-6

Nishiyama, I., Fukuda, T. and Oota, T. (2005) Genotypic differences in chlorophyll, lutein, and β-carotene contents in the fruits of *Actinidia* species. *Journal of Agricultural and Food Chemistry* 53, 6403–6407. DOI: 10.1021/jf050785y

Nishiyama, I., Fukuda, T., Shimohashi, A. and Oota, T. (2008) Sugar and organic acid composition in the fruit juice of different *Actinidia* varieties. *Food Science and Technology Research* 14, 67–73. DOI: 10.3136/fstr.14.67

Paterson, V.J., MacRae, E.A. and Young, H. (1991) Relationships between sensory properties and chemical composition of kiwifruit (*Actinidia deliciosa*). *Journal of the Science of Food and Agriculture* 57, 235–251. DOI: 10.1002/jsfa.2740570208

Pilkington, S.M., Montefiori, M., Jameson, P.E. and Allan, A.C. (2012) The control of chlorophyll levels in maturing kiwifruit. *Planta* 236, 1615–1628. DOI: 10.1007/s00425-012-1723-x

Pilkington, S.M., Montefiori, M., Galer, A.L., Emery, R.J.N., Allan, A.C. *et al.* (2013) Endogenous cytokinin in developing kiwifruit is implicated in maintaining fruit flesh chlorophyll levels. *Annals of Botany* 112, 57–68. DOI: 10.1093/aob/mct093

Possingham, J.V., Coote, M. and Hawker, J.S. (1980) The plastids and pigments of fresh and dried Chinese gooseberries (*Actinidia chinensis*). *Annals of Botany* 45, 529–533. DOI: 10.1093/oxfordjournals.aob.a085856

Pratt, H.K. and Reid, M.S. (1974) Chinese gooseberry: seasonal patterns in fruit growth and maturation, ripening, respiration and the role of ethylene. *Journal of the Science of Food and Agriculture* 25, 747–757. DOI: 10.1002/jsfa.2740250702

Qi, X.J., Han, L.X., Li, M., Xu, S.K., Zhu, Y.S. *et al.* (2007) Studies on pollen xenia of kiwifruit. *Journal of Fruit Science* 24, 774–777.

Richardson, A.C., Marsh, K.B. and MacRae, E.A. (1997) Temperature effects on satsuma mandarin fruit development. *Journal of Horticultural Science* 72, 919–929. DOI: 10.1080/14620316.1997.11515584

Richardson, A.C., Marsh, K.B., Boldingh, H.L., Pickering, A.H., Bulley, S.M. *et al.* (2004) High growing temperatures reduce fruit carbohydrate and vitamin C in kiwifruit. *Plant, Cell & Environment* 27, 423–435. DOI: 10.1111/j.1365-3040.2003.01161.x

Richardson, A.C., Boldingh, H.L., McAtee, P.A., Gunaseelan, K., Luo, Z. *et al.* (2011) Fruit development of the diploid kiwifruit, *Actinidia chinensis* 'Hort16A'. *BMC Plant Biology* 11, 182. DOI: 10.1186/1471-2229-11-182

Richardson, A.C., Eyre, V., Kashuba, P., Ellingham, D., Jenkins, H. *et al.* (2021) Early shoot development affects carbohydrate supply and fruit quality of red-fleshed *Actinidia chinensis* var. *chinensis* 'Zes008'. *Agronomy* 11, 66. DOI: 10.3390/agronomy11010066

Ritenour, M.A., Crisosto, C.H., Garner, D.T., Cheng, G.W. and Zoffoli, J.P. (1999) Temperature, length of cold storage and maturity influence the ripening rate of ethylene-preconditioned kiwifruit. *Postharvest Biology and Technology* 15, 107–115. DOI: 10.1016/S0925-5214(98)00074-X

Salinero, M.C., Vela, P. and Sainz, M.J. (2009) Phenological growth stages of kiwifruit (*Actinidia deliciosa* 'Hayward'). *Scientia Horticulturae* 121, 27–31. DOI: 10.1016/j.scienta.2009.01.013

Schmid, R. (1978) Reproductive anatomy of *Actinidia chinensis* (Actinidiaceae). *Botanischer Jahrbücher für Systematik, Pflanzengeschichte und Pflanzegeographie* 100, 149–195.

Schröder, R. and Atkinson, R.G. (2006) Kiwifruit cell walls: towards an understanding of softening? *New Zealand Journal of Forestry Science* 36, 112–129.

Seager, N.G., Hewett, E.W., Warrington, I.J. and MacRae, E.A. (1992) The effect of temperature on the rate of kiwifruit maturation using controlled environments. *Acta Horticulturae* 297, 247–253. DOI: 10.17660/ActaHortic.1992.297.32

Seal, A.G., Dunn, J.K. and Jia, Y.L. (2013) Pollen parent effects on fruit attributes of diploid *Actinidia chinensis* 'Hort16A' kiwifruit. *New Zealand Journal of Crop and Horticultural Science* 41, 219–229. DOI: 10.1080/01140671.2013.803130

Seal, A.G., Blackmore, A., Jaksons, P., McAtee, P. and Cheng, C.H. (2018) Pollen donor effects on the expression of red flesh colour in kiwifruit. *Acta Horticulturae* 1218, 373–379. DOI: 10.17660/ActaHortic.2018.1218.52

Snelgar, W.P., Manson, P.J. and Martin, P.J. (1992) Influence of time of shading on flowering and yield of kiwifruit vines. *Journal of Horticultural Science* 67, 481–487. DOI: 10.1080/00221589.1992.11516273

Stec, M.G.H., Hodgson, J.A., MacRae, E.A. and Triggs, C.M. (1989) Role of fruit firmness in the sensory evaluation of kiwifruit (*Actinidia deliciosa* cv. Hayward). *Journal of the Science of Food and Agriculture* 47, 417–433. DOI: 10.1002/jsfa.2740470404

Varkonyi-Gasic, E., Wang, T., Voogd, C., Jeon, S., Drummond, R.S.M. *et al.* (2019) Mutagenesis of kiwifruit CENTRORADIALIS-like genes transforms a climbing woody perennial with long juvenility and axillary flowering into a compact plant with rapid terminal flowering. *Plant Biotechnology Journal* 17, 869–880. DOI: 10.1111/pbi.13021

Walton, E.F. and De Jong, T.M. (1990) Growth and compositional changes in kiwifruit berries from three Californian locations. *Annals of Botany* 66, 285–298. DOI: 10.1093/oxfordjournals.aob.a088027

Wang, L., Tang, W., Hu, Y., Zhang, Y., Sun, J. *et al.* (2019) A MYB/bHLH complex regulates tissue-specific anthocyanin biosynthesis in the inner pericarp of red-centered kiwifruit *Actinidia chinensis* cv. Hongyang. *Plant Journal* 99, 359–378. DOI: 10.1111/tpj.14330

Wang, M.Y., MacRae, E., Wohlers, M. and Marsh, K. (2011) Changes in volatile production and sensory quality of kiwifruit during fruit maturation in *Actinidia deliciosa* 'Hayward' and *A. chinensis* 'Hort16A'. *Postharvest Biology and Technology* 59, 16–24. DOI: 10.1016/j.postharvbio.2010.08.010

Wang, R., Nardozza, S., Nieuwenhuizen, N.J., Wang, T., Wang, M.Y. *et al.* (2021) Kiwifruit maturation, ripening and environmental response is not affected by *CENTRORADIALIS (CEN)* gene-editing. *New Zealand Journal of Crop and Horticultural Science* 49, 277–293. DOI: 10.1080/01140671.2021.1915342

Watkins, C. and Harman, J. (1981) Use of penetrometer to measure flesh firmness of fruit. *The Orchardist of New Zealand* 54, 14, 16.

Watson, M. and Gould, K.S. (1993) The development of fruit shape in kiwifruit: growth characteristics and positional differences. *Journal of Horticultural Science* 68, 185–194. DOI: 10.1080/00221589.1993.11516342

Wegrzyn, T. and MacRae, E. (1995) Alpha-amylase and starch degradation in kiwifruit. *Journal of Plant Physiology* 147, 19–28. DOI: 10.1016/S0176-1617(11)81407-0

Xia, H., Wang, X., Su, W., Jiang, L., Lin, L. *et al.* (2020) Changes in the carotenoids profile of two yellow-fleshed kiwifruit cultivars during storage. *Postharvest Biology and Technology* 164, 111162. DOI: 10.1016/j.postharvbio.2020.111162

Yin, X.-R., Chen, K.-S., Allan, A.C., Wu, R.-M., Zhang, B. *et al.* (2008) Ethylene-induced modulation of genes associated with the ethylene signalling pathway in ripening kiwifruit. *Journal of Experimental Botany* 59, 2097–2108. DOI: 10.1093/jxb/ern067

Young, H. and Paterson, V.J. (1985) The effects of harvest maturity, ripeness and storage on kiwifruit aroma. *Journal of the Science of Food and Agriculture* 36, 352–358. DOI: 10.1002/jsfa.2740360507

Young, H. and Paterson, V.J. (1995) Characterisation of bound flavour components in kiwifruit. *Journal of the Science of Food and Agriculture* 68, 257–260. DOI: 10.1002/jsfa.2740680218

Young, H., Paterson, V.J. and Burns, D.J.W. (1983) Volatile aroma constituents of kiwifruit. *Journal of the Science of Food and Agriculture* 34, 81–85. DOI: 10.1016/S0176-1617(11)81407-0

Yuan, H., Martin, P., Fullerton, C., Gould, N., Hall, A. *et al.* (2020) Refining the estimation of kiwifruit size from linear fruit dimensions. *Scientia Horticulturae* 262, 108878. DOI: 10.1016/j.scienta.2019.108878

Zeng, Y., Wang, M.Y., Hunter, D.C., Matich, A.J., McAtee, P.A. *et al.* (2020) Sensory-directed genetic and biochemical characterization of volatile terpene production in kiwifruit. *Plant Physiology* 183, 51–66. DOI: 10.1104/pp.20.00186

Zhang, A., Zhang, Q., Li, J., Gong, H., Fan, X. *et al.* (2020) Transcriptome co-expression network analysis identifies key genes and regulators of ripening kiwifruit ester biosynthesis. *BMC Plant Biology* 20, 103. DOI: 10.1186/s12870-020-2314-9

Zhang, B., Yin, X.-R., Li, X., Yang, S.-I., Ferguson, I.B. *et al.* (2009) Lipoxygenase gene expression in ripening kiwifruit in relation to ethylene and aroma production. *Journal of Agricultural and Food Chemistry* 57, 2875–2881. DOI: 10.1021/jf9000378

17 Vine and Fruit Diseases

Joy L. Tyson[1]*, Irene Donati[2] and Kerry R. Everett[1]
[1]*The New Zealand Institute for Plant and Food Research Ltd, Auckland, New Zealand;* [2]*Zespri Fresh Produce, Bologna, Italy*

17.1 Introduction

Kiwifruit were considered essentially disease-free when first grown in New Zealand (Schroeder and Fletcher, 1967). As more commercial orchards were developed and cropping intensified, fruit and vine diseases began to be reported. By 1969, seven organisms had been recorded causing disease on *Actinidia chinensis* in New Zealand: '*Armillaria mellea*' (now identified as *A. novae-zelandiae*), *Botrytis cinerea*, *Glomerella cingulata*, *Phytophthora cactorum*, *P. cinnamomi*, *Sclerotinia sclerotiorum* and *Agrobacterium tumefaciens* (Dingley, 1969).

Fifty years later, the number of known diseases of kiwifruit and their associated pathogens has increased to include many new fungi, bacteria and viruses, although kiwifruit diseases are still most commonly caused by fungal pathogens. The most important bacterial pathogen is *Pseudomonas syringae* pv. *actinidiae* (Psa), which causes kiwifruit bacterial canker (see Chapter 18). Since first recorded in 2003 (Clover *et al.*, 2003), the number of viruses reported in kiwifruit has continued to increase as more specimens are examined. Some kiwifruit diseases of unknown aetiology, such as kiwifruit vine decline syndrome (KVDS) in Italy, have also been reported.

Diseases of kiwifruit can be classified broadly into diseases of vine or fruit, with many of the causative pathogens present on both. The impact of each disease is likely to be cultivar-dependent, and some diseases may only be significant in certain countries or regions. This chapter gives an overview of the main diseases of kiwifruit and updates the previous work by Brook (1990).

17.2 Vine Diseases

17.2.1 Root and crown pathogens

Armillaria *species (white root rot)*

SYNONYMY. *Armillariella mellea*

DISTRIBUTION. *Armillaria mellea* is recorded on kiwifruit in Spain (Aguin *et al.*, 2004), Italy (Ciccarese *et al.*, 1992), China (Lin *et al.*, 1994) and Greece (Thomidis and Exadaktylou, 2012). In New Zealand, only *A. novae-zelandiae* is present (Brook, 1990; Horner, 1992).

SYMPTOMS. *Armillaria* spp. cause root rot. Initial symptoms are lack of thrift with yellowing leaves and stunted fruit, progressing to vine death after 1–2 years (Brook, 1990). Thick rhizomorphs (dense, aggregated strands of mycelium) develop

*Corresponding author: joy.tyson@plantandfood.co.nz

© CAB International 2024. *Kiwifruit: Botany, Production and Uses* (eds A.C. Richardson, J.N. Burdon and A.R. Ferguson)
DOI: 10.1079/9781800620933.0017

on the roots, while dense white mycelial mats form beneath the bark and up the trunk. Cracking and swelling of the trunk base may be accompanied by red exudates (Horner, 1992). Often, mushrooms form at the base of the vines.

DISEASE CYCLE. Inoculum is from previously infected woody debris in the soil or airborne basidiospores from nearby forested areas. The stumps of felled shelterbelt trees, particularly those treated with herbicides, can be infected by basidiospores (Horner, 1992). The fungus spreads via soilborne rhizomorphs or root contact (Brook, 1990). Discrete areas of infected vines are found in an orchard, often adjacent to shelterbelts and particularly the stumps of felled willows (Hood, 1989).

CONTROL. Control is largely preventative, such as root pruning of adjacent shelterbelt trees (particularly willow) and complete stump removal of felled trees (Horner, 1992). Cultural control includes 'sluicing', in which the soil around the crown and upper roots of the infected vine is removed, and waterblasting and 'surgery' used to remove diseased tissue and fungal mycelium (Hood and Shaw, 2022).

Phytophthora *species (collar rot, root rot)*

DISTRIBUTION. Numerous *Phytophthora* species affect kiwifruit, with different countries reporting different combinations of species. The most common species are *P. cactorum*, *P. cinnamomi*, *P. citrophthora*, *P. cryptogea* and *P. megasperma* (French, 1987; Shivas, 1989; Stewart and McCarrison, 1991; Förster and Coffey, 1993; Erwin and Ribeiro, 1996; Cooke *et al.*, 2000; Akilli *et al.*, 2011; Rahman *et al.*, 2014).

SYMPTOMS. Symptoms within the canopy are often non-specific and inconspicuous whilst the infection is spreading. These may include reduced terminal shoot growth, stunted foliage, wilting and discolouration of the leaves, and eventual vine collapse and death (Latorre *et al.*, 1991; Lee Y.-H. *et al.*, 2001). Infected roots and crowns develop a typical red-brown rot with a distinctive margin between healthy white tissue and the discoloured area, and cankers may form.

DISEASE CYCLE. *Phytophthora* diseases are multicyclic. They are exacerbated by waterlogging and soil temperatures > 10°C, when the sporangia release motile zoospores that infect young roots. Some species can also produce chlamydospores, while others are homothallic and can form oospores in the absence of the opposite mating type, assuring long-term survival (Gallegly and Hong, 2008).

CONTROL. Prevention of *Phytophthora* disease requires planting disease-free vines and avoiding waterlogging and soil compaction. Once present, control strategies are aimed at improving soil drainage and managing irrigation. Organic amendments, mulching, cover cropping and other agronomic practices can reduce disease impact. Copper-based protectant/curative fungicides can be used. Effective contact fungicides include ethylene bis-dithiocarbamate fungicides such as mancozeb and zineb. Systemic fungicides such as the acylanilide, metalaxyl, provide efficient control. Phosphonates such as fosetyl-Al are commonly used; these act directly on the pathogen, as well as indirectly through host defence mechanisms.

Fusarium solani *species complex (root rot, collar rot)*

SYNONYMY. *Fusarium solani* (syn. *Nectria haematococca*) has recently been found to be a complex of morphologically similar species. Although no research has been undertaken on the members of the complex associated with *Actinidia*, Savian *et al.* (2020a) recovered many *F. solani* sequence types from flood-challenged kiwifruit vines in Italy. In New Zealand, sequencing of isolates has also shown that there are multiple *F. solani* types in the wood of vines with visible trunk disease (J.L. Tyson, 2023, unpublished results).

DISTRIBUTION. *F. solani* has been recorded for many years in association with root and crown rots and trunk disease of kiwifruit with published findings from Italy (Porta-Puglia and Corazza, 1985), Turkey (Yaman and Türkkan, 2017) and China (Huang and Qi, 1998).

SYMPTOMS. In Italy, isolation and re-inoculation of *F. solani* into wounded kiwifruit bark resulted in browning of the wood; *F. solani* was then re-isolated from the diseased tissues, satisfying Koch's postulates (Porta-Puglia and Corazza, 1985). The microorganisms most frequently isolated from 'Hayward' wood decay in Italy in 1997 were *F. solani* and *Phomopsis* sp. (Gianetti *et al.*, 2002). Yaman and Türkkan (2017) listed the many genera of root-rotting fungi found on kiwifruit and noted that *F. solani* caused more severe root rot than *F. oxysporum* and *Rhizoctonia solani*.

DISEASE CYCLE. The disease cycle is unclear; however, it is likely that macro- and microconidia will be rain-splashed and ascospores both rain-splashed and windblown. Red perithecia can develop on or under the bark of the collar or trunk. There is uncertainty over the role of *F. solani* in root and trunk disease and whether it acts as a primary or secondary invader.

In China, where *P. cinnamomi* and *F. solani* were recovered from kiwifruit root rots, *P. cinnamomi* had a higher pathogenicity, and *F. solani* was likely to have been a secondary pathogen (Huang and Qi, 1998). In Italy, species of *Phaeoacremonium* and *F. solani* were isolated from elephantiasis-affected vines (characterized by trunk swelling, defoliation and fruit that fail to mature). *Phaeoacremonium* spp. were suggested as the initial invaders, predisposing the vines to the entry of *F. solani* (Nipoti *et al.*, 2004).

Other fungal root pathogens

Verticillium dahliae is a minor wilt pathogen of kiwifruit in New Zealand (Brook, 1990). In Turkey, it has been associated with wilt accompanied by leaf chlorosis, defoliation and discolouration within the stem vascular tissues (Türkkan *et al.*, 2020). In Chile, a sudden wilt and dieback of vines was initially found to be caused by *V. alboatrum* (Auger *et al.*, 2009, 2011); this pathogen was then re-identified as *V. nonalfalfae* (Kasson *et al.*, 2014) and appears to be significantly more aggressive to kiwifruit vines than *V. dahliae*.

Ilyonectria spp. (syn. *Cylindrocarpon* spp.) have been associated with root and trunk diseases of kiwifruit. The first record was of '*Cylindrocarpon* sp.' associated with kiwifruit wood decay in Italy in 1996 (Gianetti *et al.*, 2002). Published records of the *Ilyonectria* group associated with kiwifruit diseases currently include *I. liriodendri* from the USA (California) and Turkey (Erper *et al.*, 2011, 2013; Lawrence *et al.*, 2019), *I. europaea* from France and Turkey (Cabral *et al.*, 2012; Erper *et al.*, 2013), *I. robusta* from Turkey (Erper *et al.*, 2011, 2013), and *I. radicicola* from Italy (Tacconi *et al.*, 2015). The closely related *Dactylonectria pauciseptata* (as *C. pauciseptatum*) and *D. torresensis* (as *I. torresensis*) have been recorded in Turkey (Erper *et al.*, 2013).

Kiwifruit symptoms associated with *Ilyonectria* and *Dactylonectria* spp. include necrotic lesions on woody roots and the base of the rootstock, loss of feeder roots, leaf wilt and dieback. Pathogenicity tests have shown that *I. liriodendri*, *I. europaea*, *I. robusta* and *D. torresensis* can induce root rot symptoms on young kiwifruit (Erper *et al.*, 2011, 2013).

17.2.2 Kiwifruit trunk diseases

Kiwifruit trunk diseases (KTDs) have been reported in Italy (Di Marco *et al.*, 2000), New Zealand (Johnston *et al.*, 2010) and France (Hennion *et al.*, 2003). KTDs are now a serious impediment to kiwifruit production and are associated with a variety of external and internal wood discolouration and decay symptoms, including collar rots, cracking, swelling or discrete bulging of the trunk and leaders, dieback and sudden death. The principal components of KTDs have yet to be elucidated, including the associated pathogens, order of infection, climate and time for symptoms to develop.

A wide range of phytopathogenic fungi can be associated with KTDs. Few of the pathogens involved evoke unique symptoms that allow them to be distinguished from others, and KTDs may be disease complexes, with many pathogens involved.

Fungi recovered from kiwifruit with vine decline, decay or kiwifruit elephantiasis include *Cadophora* spp. (*C. luteo-olivacea*, *C. malorum*, *C. melinii*, *Lecythophora luteoviridis*, *Phaeoacremonium minimum* (as *Ph. aleophilum*), *Ph. fraxinopennsylvanicum* (as *Ph. mortoniae*), *Ph. parasiticum* and *Fomitiporia punctata*) (Di Marco *et al.*, 2000, 2003, 2004; Prodi *et al.*, 2008). *Botryosphaeria* spp. and *B. dothidea* cause wood decay and dieback, respectively, in Italy and Greece (Abbatecola *et al.*, 2008; Thomidis and Exadaktylou, 2010). *B. dothidea* also causes vine dieback in China

(Zeng and Huang, 2000) and Japan (Kinugawa and Sato, 2003).

In France, isolations from vines with symptoms of an esca-like disease, with wood decay and bulging trunks or leaders, yielded *Ph. aleophilum* and *Ph. viticola*, *Botryosphaeria obtusa*, *Diaporthe* spp., *Cylindrocarpon* sp. and *Fusarium* sp., as well as members of the largely saprophytic genera *Alternaria*, *Cephalosporium*, *Cladosporium*, *Paecilomyces* and *Penicillium* (Hennion *et al.*, 2003). In Italy, *Cadophora luteo-olivacea* was the most commonly isolated fungus from leaders with swelling and dieback from five orchards (Riccioni *et al.*, 2007). More recently, *Fusarium avenaceum* was reported causing a canker disease in China; symptoms included black necrotic lesions and cracked areas in the trunk (Sun *et al.*, 2022). In New Zealand, *Neobulgaria alba* has been associated with swollen trunks and decay or staining within the trunks (Johnston *et al.*, 2010).

Ceratocystis fimbriata *(ceratocystis wilt)*

Ceratocystis fimbriata sensu lato is a complex of cryptic pathogenic fungi affecting a range of hosts including mango, eucalyptus and taro. The first report of *C. fimbriata* on kiwifruit was from Brazil in 2010 (Sônego *et al.*, 2010) and over the last ten years ceratocystis wilt has had a significant economic impact on kiwifruit production there (Sônego *et al.*, 2010; Ferreira *et al.*, 2013; Piveta *et al.*, 2013). Phylogenetic analyses separated isolates from kiwifruit into five haplotypes, with the four most common haplotypes having a similar pathogenicity (Piveta *et al.*, 2016). In New Zealand, *C. fimbriata* has been recorded only on sweet potato (Pennycook, 1989), isolates from which were not pathogenic to kiwifruit (Tyson *et al.*, 2020).

Vines affected by *C. fimbriata* develop wilted and curled-up leaves in the whole canopy or a single cane. Trunk cross-sections reveal radial reddish-brown staining of the xylem parenchyma (Fig. 17.1). Crop load is reduced, and most infected vines eventually die (Ferreira *et al.*, 2013, 2017; Piveta *et al.*, 2016).

Ceratocystis species have several dispersal strategies. The fungus is soilborne and infection may occur through the roots; spread to adjacent vines may also occur through root anastomoses. Pruning wounds can provide infection sites and pruning implements and insects such as ambrosia beetles may vector the disease. In addition, movement of infected whole vines or budwood may transfer the pathogen to new areas (Piveta *et al.*, 2016).

A targeted control strategy for ceratocystis wilt has yet to be identified. Potential management

Fig. 17.1. Stained trunk of a kiwifruit vine infected by *Ceratocystis fimbriata*. (Photo: J.L. Tyson, The New Zealand Institute for Plant and Food Research Ltd)

options include disposal of diseased plants, disinfection of tools, severing of root grafts and the development of resistant cultivars.

Neonectria microconidia (*kiwifruit neonectria canker*)

SYNONYMY. *Cylindrocarpon* sp.

DISTRIBUTION. *N. microconidia* is recorded in China and Japan on various hosts (Zhao *et al.*, 2011; Hirooka *et al.*, 2013), but is currently reported only on kiwifruit in New Zealand (MPI, 2016; KVH, 2020).

SYMPTOMS. Symptoms of kiwifruit neonectria canker include swelling and cracking of trunks and leaders, and cankers of the wood, often with the eventual death of the leader above the canker. Cross-sections of swollen leaders reveal discolouration of the internal tissues. Red perithecia may form on or under the bark, and cankers are often associated with old wounds.

DISEASE CYCLE. The disease cycle of *N. microconidia* within kiwifruit, or any host, is currently unknown. The fungus has two spore types: sexual ascospores formed in red perithecia, and asexual macro- and microconidia that are often formed in cream-coloured sporodochia. It is likely that the ascospores will be windborne and waterborne, while the conidia will be spread by rain splash and windblown rain. Thus, conidia will likely cause new infections on the same or adjacent vines, whilst ascospores may account for longer-distance spread. Although the conditions required for the formation and release of conidia and ascospores are unknown, it is likely that, as for *N. ditissima* (the causal agent of European canker of apples), the entry points for new infections will be wounds (Weber and Børve, 2021) arising from pruning, grafting and girdling, as well as natural wounds.

17.2.3 Kiwifruit vine decline syndrome

DISTRIBUTION. KVDS (also called moria, Moria del kiwi or Verona vine decline) has been observed in Italy since 2012, originally localized in the northern provinces. Over time, KVDS has spread to other regions and since 2019 has been observed in central-south Italy (Tacconi *et al.*, 2015; Ermacora *et al.*, 2020).

SYMPTOMS. The above-ground symptoms are a progressive withering of the leaves and lack of fruit development. The roots appear compromised, with widespread decay of the cortex. New root growth is limited to the top 5–10 cm of soil. Symptoms first appear in late spring although the root health problems probably start during the previous autumn. Symptoms peak during summer, when the vines collapse due to heat stress and lack of functional roots. Vines die within two years of above-ground symptoms appearing (Sorrenti *et al.*, 2016; Donati *et al.*, 2020).

CAUSES. KVDS is a multifactorial disease (Bardi, 2020). It is linked to poor root health, probably from mismanagement of soil quality and suboptimal orchard practices (inappropriate irrigation and drainage systems) coupled with heavy rainfall and floods, resulting in anoxia and associated root diseases. The existence of a 'disease complex' rather than a unique causal agent has been hypothesized. The pathogens found are typical of water-saturated conditions, including oomycetes of the *Phytophthora*, *Pythium* and *Phytopythium* genera and *Desarmillaria tabescens* or bacteria of the *Clostridium* genus (Tacconi *et al.*, 2015; Donati *et al.*, 2020; Savian *et al.*, 2020b; Spigaglia *et al.*, 2020).

CONTROL. Effective KVDS management is likely to be achieved through an integrated approach, with short- and long-term strategies. Improving soil physical characteristics and biological fertility (organic matter, mycorrhizae, *Trichoderma*) are important to reduce or prevent disease progression. Water management to promote root turnover and uptake of essential minerals is key. Correct soil preparation by tillage and fertilization before planting are also important (Sorrenti *et al.*, 2016). The adoption of resistant or tolerant rootstocks seems promising as some *Actinidia* species (e.g. *A. macrosperma*, *A. arguta*) appear less susceptible than *A. chinensis* (Ermacora *et al.*, 2020).

A common trait of multifactorial diseases is the ability of the pathogens to live asymptomatically within their hosts until abiotic stresses trigger the disease. Causal mechanisms for KVDS still need confirmation to identify management strategies.

17.2.4 Diseases of the canopy

Sclerotinia sclerotiorum *(sclerotinia blossom blight, twig blight)*

DISTRIBUTION. *S. sclerotiorum* has been identified in New Zealand (Pennycook, 1985), China (Jiao *et al.*, 2014), South Korea (Lee *et al.*, 2015) and Italy (Testolin and Ferguson, 2009).

SYMPTOMS. Sclerotinia blossom blight results in pale brown withered blossoms and pedicels (Pennycook, 1985). In dry years, the blossoms dry out without secondary spread. In wet years, the blossoms rot and white mycelium often becomes apparent on and within dried blossoms. This mycelium often consolidates into hard black sclerotia, usually within the blossom and occasionally externally. Fruit scarring can occur where infections of the fruit surface establish and then dry out, leaving a superficial, sometimes sunken, scarred appearance (Hoyte, 2001). Watery, whitish fruit rots occur most commonly during midsummer, sometimes with sclerotia on the surface. The fruit rots and scarring tend to originate where flower debris has adhered to the fruit surface, or where a moist microclimate has formed between two adjacent fruits (Pennycook, 1985).

DISEASE CYCLE. The cycle begins during kiwifruit flowering, when apothecia (Fig. 17.2) develop from sclerotia in the soil and litter. In New Zealand, apothecia continue to emerge from mid-spring through to early summer (Hoyte *et al.*, 1992). Ascospores discharged from the apothecia drift to the canopy where they readily infect petals when wet for several hours (Pennycook, 1985). Fruit drop, scarring or rot result where infected petals and other floral debris remain in contact with fruit (Pennycook, 1985; Hoyte, 2000). Fruit damage can occur at any time during the season when weather conditions are suitable for fungal growth. Sclerotia form on the fallen diseased blossoms and fruit during spring and summer, remaining dormant in the soil and leaf litter over winter and germinating to form new apothecia and ascospores the following season.

CONTROL. Control is targeted at the pre-flowering and flowering periods. In New Zealand, the systemic fungicide, fluopyram, or the protectant fungicide, trifloxystrobin (AgriMedia Ltd, 2021), are applied before flowering.

Fig. 17.2. Apothecia of *Sclerotinia sclerotiorum*. (Photo: R.A. Fullerton, The New Zealand Institute for Plant and Food Research Ltd)

Pseudocercospora actinidiae *(black spot, sooty spot, black mould)*

SYNONYMY. *Pseudocercospora hangzhouensis* has also been associated with sooty spot of kiwifruit (Braun *et al.*, 2015).

DISTRIBUTION. *P. actinidiae* was first described from leaf spots of *Actinidia callosa* in China (Deighton, 1976). The recorded distribution now includes China (Jiang *et al.*, 1995), South Korea (Braun *et al.*, 2015), Japan and Taiwan (Kikuhara and Nakashima, 2008) and Brazil (Araujo *et al.*, 2021).

SYMPTOMS. *P. actinidiae* affects the leaves, branches and fruit. The first symptoms are small, grey floccose spots on the undersides of leaves. The spots enlarge, becoming dark grey to black and velvety, and may coalesce. Yellow chlorotic spots then appear on the upper surface of the leaves, corresponding to the lesion site, and expand into a yellow-brown to dark brown necrotic spot. Leaves may wither and abscise (Jiang *et al.*, 1995; Ai, 2007; Kikuhara and Nakashima, 2008).

On leaders and canes, spindle- or oval-shaped lesions with a yellow- or reddish-brown water stain form. The lesions become sunken, then expand; longitudinal cracking appears, and the branches swell. Small black spots (ascomata, the teleomorphic spore-forming structures) or velvety grey layers of mycelium may form on the diseased tissues (Jiang *et al.*, 1995; Ai, 2007).

Initial symptoms shown by fruit on the vine are grey spots of mycelium, which enlarge until the mould layer falls off, leaving a 0.2–1.0 cm circular sunken lesion. Externally, the lesion appears brown to purple-brown, while internally a hard, cone-shaped necrotic rot develops. The diseased fruit abscise easily (Jiang *et al.*, 1995; Ai, 2007). In Japan, the sunken spots on fruit can be up to 2 cm in diameter, sometimes with dark mycelial masses at the centre. Postharvest fruit rot can also occur (Kikuhara and Nakashima, 2008).

DISEASE CYCLE. *P. actinidiae* overwinters in diseased branches and leaf debris; these are the primary sources of infection of new growth in spring (Jiang *et al.*, 1995; Ai, 2007). Ascomata are produced during flowering, releasing spores which are spread with rain-splash and windborne rain. Disease incidence is severe in periods of continuous rain during late spring to early summer (Ai, 2007).

Leaf symptoms are first seen in spring and peak by midsummer. Lower leaves on the vine are the first to develop symptoms, which then spread to upper leaves. Fruit lesions appear from late spring to early summer, peaking in late summer. The branches become latently colonized during summer and autumn, with lesions appearing in autumn (Jiang *et al.*, 1995; Ai, 2007).

CONTROL. Cultural control consists of opening up the canopy during winter and summer pruning to improve ventilation and light and reduce humidity. The inoculum load can be reduced by removal of obviously diseased plant material from the orchard (Jiang *et al.*, 1995; Ai, 2007).

Leaf diseases

A powdery mildew disease of kiwifruit recorded in South Korea and China may be caused by *Phyllactinia* sp., *P. actinidiae*, *P. actinidiae-latifoliae*, *P. imperialis*, *P. salmonii*, *Uncinula actinidiae* or *U. necator* (Lin *et al.*, 1994; Lin and Gao, 1995; Kim *et al.*, 2017; Wang and Huang, 2020). Typical symptoms are circular patches of whitish mycelia and abundant conidia on leaves, which may turn brown, wither and abscise. The disease is more prevalent in dry conditions with high temperatures (Kim *et al.*, 2017; Wang and Huang, 2020).

Several kiwifruit rust fungi have been recorded in China, including *Pucciniastrum actinidiae*, *Uromyces* sp. and *Uredo* sp. (Lin *et al.*, 1994, 1995; Song *et al.*, 1997). Symptoms begin on leaves as small orange lesions, which become grey-brown as they expand (Song *et al.*, 1997). *Uromyces* sp. on kiwifruit overwinters as teliospores in pruning debris on the orchard floor. In spring, basidiospores are carried to the canopy by wind where they infect the leaves, with spots developing several days later (Zhu, 2020).

*Bacterial blossom blight (*Pseudomonas *spp.)*

SYNONYMY. Bacterial blossom blight has been attributed to several *Pseudomonas* species, including *Ps. viridiflava*, *Pseudomonas* sp., *Ps. syringae*,

Ps. savastanoi and *Ps. syringae* pv. *syringae*. The disease may be caused by a complex of pseudomonads; Everett and Henshall (1994) showed a succession during flowering of *Ps. viridiflava* followed by *Ps. syringae* and finally *Ps. fluorescens*. Psa, the cause of kiwifruit bacterial canker, can also cause blossom bud rot (see Chapter 18).

DISTRIBUTION. Blossom blight of 'Hayward' kiwifruit occurs in most kiwifruit-growing countries: Italy (Balestra *et al.*, 2008), Spain (González and Rodicio, 2007), Turkey (Demir, 2000), Portugal (Balestra *et al.*, 2009), New Zealand (Everett and Henshall, 1994), USA (Conn *et al.*, 1993), Japan (Ieki, 1993), South Korea (Koh *et al.*, 2001) and China (Hu *et al.*, 1998).

SYMPTOMS. Blossom blight initially affects kiwifruit flower buds by rotting the anthers, which turn brown (Fig. 17.3). The disease may affect only the anthers or may spread to other floral tissues, which in turn become brown. In severe cases, the bud will drop from the flower stalk. Less severe infection may lead to mis-shapen fruitlets. Dark angular lesions surrounded by yellow haloes may develop on the leaves (Young *et al.*, 1997).

DISEASE CYCLE. The disease cycle and means of overwintering are largely unresolved. The development of blossom blight is thought to be associated with rainfall during flowering (Pennycook and Triggs, 1992; Koh *et al.*, 2003).

CONTROL. Management includes ensuring good ventilation of the canopy during flowering, vine girdling and the use of orchard covers. Chemical control may involve application of antibiotics and/or copper formulations during flowering (Miyoshi and Tachibana, 1996; Koh *et al.*, 2003; Shin *et al.*, 2004a, b; Zhang *et al.*, 2019).

17.2.5 Viruses of kiwifruit

Kiwifruit were once thought to be resistant to viruses (Caciagli and Lovisolo, 1987). While some indicative symptoms were observed, e.g. mosaic disease in China (Lin and Gao, 1995), no kiwifruit viruses were definitively identified until *Apple stem grooving virus* (ASGV) was reported in kiwifruit plants from China (Clover *et al.*, 2003). Various virus species have since been recorded in kiwifruit vines in China (Zheng *et al.*, 2014), India (Bhardwaj *et al.*, 2014), Italy (Biccheri *et al.*,

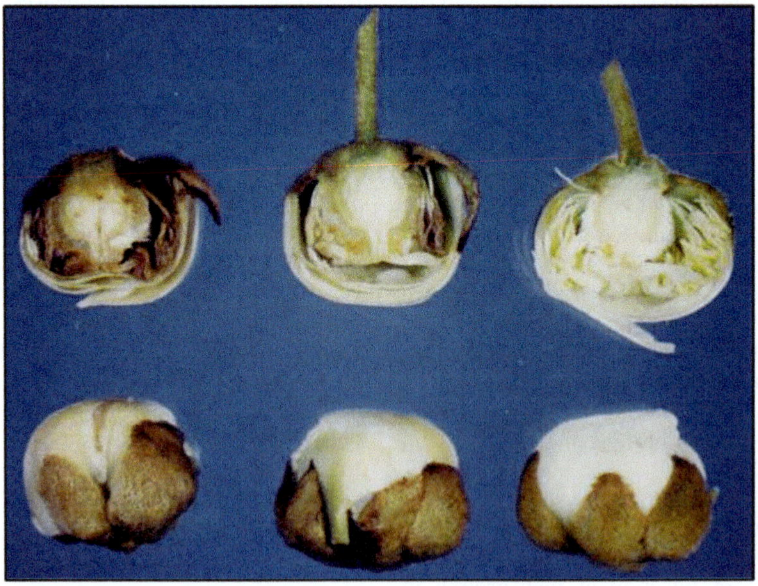

Fig. 17.3. Dissected kiwifruit flower buds affected by bacterial blossom blight: diseased (*left, middle*) and healthy (*right*). Note the chocolate brown (diseased) anthers in the bud on the left. (Photo: K.R. Everett, The New Zealand Institute for Plant and Food Research Ltd)

2012; Blouin *et al.*, 2013), South Korea (Cho *et al.*, 2017) and New Zealand (Blouin *et al.*, 2012). No viroids or phytoplasmas have been reported infecting kiwifruit.

Twenty-four viruses have now been reported from kiwifruit; 19 of these have been categorized into three groups: non-specialist viruses that infect a broad range of host plants (n=7); kiwifruit-adapted viruses (10); and disease-causing viruses (2) (Blouin *et al.*, 2013; Veerakone *et al.*, 2018; Zhao *et al.*, 2019; Wang *et al.*, 2020). In addition to these, a further five viruses have been recently described (Zhao *et al.*, 2021, 2022).

The two viruses that cause disease in kiwifruit are *Cherry leaf roll virus* (CLRV) (Woo *et al.*, 2012) and *Pelargonium zonate spot virus* (PZSV) (Biccheri *et al.*, 2012). CLRV is associated with leaf spots, fruit deformity, yield reduction, bark cracking and cane wilting (Fig. 17.4), although research to demonstrate the cause is lacking (Blouin *et al.*, 2013). PZSV is associated with concentric chlorotic and necrotic ring and line patterns on leaves, and deformation of the fruit (Biccheri *et al.*, 2012).

Infection with most non-specialist viruses is either asymptomatic or causes yellow spots and/or patterns on leaves. Kiwifruit-adapted viruses can be asymptomatic, or cause leaf symptoms such as ringspots, mottles, vein and leaf clearing or interveinal chlorosis.

Control of viruses relies on planting virus-free material and sterilizing pruning tools to prevent sap transmission. Several viruses found in kiwifruit, including CLRV, *Ribgrass mosaic virus*

Fig. 17.4. Kiwifruit infected with *Cherry leaf roll virus*. *Clockwise from top left*: vine cracking and leaf wilting; necrotic patches on *Actinidia chinensis* var. *deliciosa* 'Hayward' leaf; chlorotic areas on *A. chinensis* var. *chinensis* 'Zesy002' leaf; fruit of *A. chinensis* var. *chinensis* 'Hort16A' without a beak. (Photos: K.M. Chooi, The New Zealand Institute for Plant and Food Research Ltd)

and *Turnip vein-clearing virus* (Blouin *et al.*, 2013), are mechanically transmissible, which is of particular concern during pruning.

One kiwifruit-adapted virus, *Actinidia seed-borne latent virus* (ASbLV), is pollen-transmitted (Amponsah *et al.*, 2022). Several viruses found in kiwifruit are vector-transmitted on other hosts including aphids, thrips, nematodes, chytrids, mealy bugs, scale and mites (Bragard *et al.*, 2013). Transmission by arthropods has not been reported for kiwifruit.

17.3 Postharvest Disease

Fruit disease can be seen on orchard (e.g. *Sclerotinia* and *Pseudocercospora* spp.) or at harvest (e.g. *Botryosphaeria* spp.), but most fruit rots develop during or after storage. These can be broadly classified as botrytis rot, ripe rots or wound rots (Figs 17.5, 17.6 and 17.7, respectively).

17.3.1 *Botrytis cinerea* (botrytis rot, stem-end rot, grey mould)

SYNONYMY. *Botryotinia fuckeliana*

DISTRIBUTION. *B. cinerea* is likely to be found wherever kiwifruit is grown, including China (Xue *et al.*, 2017), Italy (Bisiach *et al.*, 1984), France (Hennion *et al.*, 2003), South Korea (Koh *et al.*, 2005), New Zealand (Ford, 1971) and Chile (Pinilla *et al.*, 1994).

SYMPTOMS. *B. cinerea* grows relatively rapidly at low temperatures, developing early in storage (Manning *et al.*, 2003). The rots start at the picking wound, developing a distinctive 'tidemark' as the rot progresses evenly towards the stylar end of the fruit (Fig. 17.5). The affected flesh is glassy and water-soaked (Brook, 1991). As the rot develops, white floccose mycelium, which becomes grey, often emerges, particularly at the picking wound (Manning *et al.*, 2003). Mycelium can aggregate to form dark sclerotia, or spread, developing into a 'nest' of mycelium covering several fruits (Brook, 1990).

INFECTION CYCLE. The picking wound becomes infected at harvest from conidia released from necrotic leaf tissue in the canopy, or in the picking bags (Brook, 1991). Infected senescent leaves, pruning debris and discarded fruit that fall under the canopy are possible sources of inoculum for infection of necrotic, ageing leaves the following summer (Pennycook, 1985).

CONTROL. Botrytis stem-end rots can be controlled by reducing necrotic tissue in the canopy (Manning *et al.*, 2010). Pruning to reduce canopy density, planting shelter to minimize wind damage and avoiding damage from passing tractors will reduce necrotic leaf tissue. In addition, 'curing' picking wounds by holding fruit at ambient temperatures for 2–3 days before cool storage can reduce infection (Pennycook and Manning, 1992) (see Chapter 20).

Fig. 17.5. *Botrytis* stem-end rots. (Photo: J.L. Tyson, The New Zealand Institute for Plant and Food Research Ltd)

7.3.2 Ripe rots

These develop as the fruit ripen in or after storage and are frequently categorized by their position on the fruit. Thus, they are commonly named stem-end rots (SER), which originate at the picking scar, body rots (BR), on the main body of the fruit, or blossom-end rots (BER), from the floral remnants at the stylar end of the fruit. A suite of fungi is associated with ripe rots; most commonly *Diaporthe* spp., *Botryosphaeria* spp., *Neofabraea actinidiae* and '*Cylindrocarpon*' sp., as well as a host of minor rot fungi (Manning *et al.*, 2003).

Diaporthe *spp.*

SYNONYMY. *Phomopsis* spp.
The taxonomic status of *Diaporthe* that infect kiwifruit is problematic due to unstable culture morphology, causing them to be referred to as *Diaporthe* spp. or *Phomopsis* spp. in early publications. To date, 28 species names have been assigned to isolates from kiwifruit, including *Diaporthe actinidiae* (Sommer and Beraha, 1975), *D. perniciosa* (Yi and Lee, 1998), *D. ambigua* (Auger et al., 2013) and *D. hongkongensis* (Erper *et al.*, 2017). These species names may not be useful as the taxonomy of this genus remains unresolved.

DISTRIBUTION. *Diaporthe* species are found on kiwifruit worldwide, including New Zealand (Hawthorne *et al.*, 1982), Italy (Luongo *et al.*, 2011), Chile (Auger *et al.*, 2013), China (Bai *et al.*, 2017) and South Korea (Lee, J.-G. *et al.*, 2001).

SYMPTOMS. *Diaporthe* spp. are the most commonly isolated fungi from fruit rots of all cultivars of kiwifruit. They cause SER and BR on fruit after harvest (Fig. 17.6d–f) (Manning *et al.*, 2003). While it has been suggested that wounding is required for postharvest infection (Luongo *et al.*, 2011; Ledesma *et al.*, 2021), field inoculations with spores on sound fruit result in postharvest rots (Fullerton *et al.*, 2007).

Kiwifruit vine dieback and canker have also been attributed to this genus in Italy (Nipoti *et al.*,

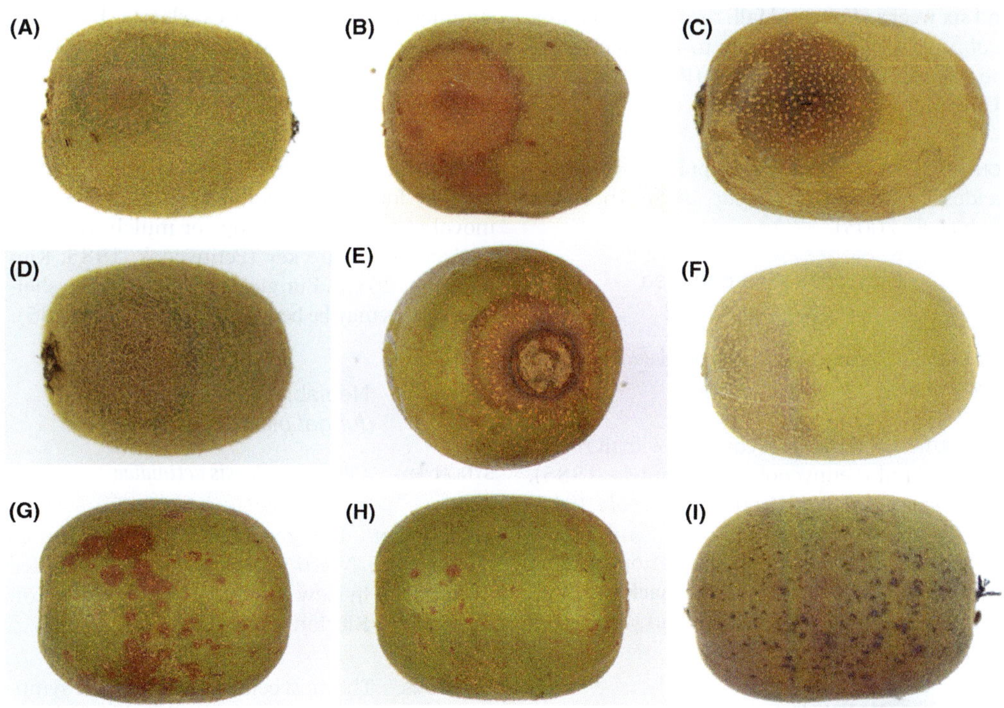

Fig. 17.6. Ripe rots on the fruit of several kiwifruit cultivars. (**A, B**) *Botryosphaeria* sp. (**C**) *Phoma* sp. (**D–F**) *Diaporthe* sp. (**G, H**) *Neofabraea actinidiae*. (**I**) *Colletotrichum* sp. (Photos: J.L. Tyson, The New Zealand Institute for Plant and Food Research Ltd)

2003), Chile (Diaz *et al.*, 2021), Greece (Thomidis, 2009) and China (Du *et al.*, 2021).

DISEASE CYCLE. The anamorph and teleomorph stages of *Diaporthe* species are commonly found on bark or dead wood in kiwifruit orchards, including on pruning stubs and old peduncles in the canopy and on pruning debris on the orchard floor. In New Zealand, they are also found on almost all shelter belt species used in kiwifruit orchards (Fullerton *et al.*, 2007). Both spore-forming structures (pycnidia and long-necked perithecia) sporulate prolifically during wet weather. The conidia are dispersed by rain; the ascospores are extruded in mucilage from perithecia and are likely to be also rain dispersed.

Diaporthe spp. appear to infect fruit throughout the season, surviving as latent infections within the skin. Research using nitrate non-utilizing (nit) mutants failed to define a period of maximum susceptibility, suggesting that infection could occur at any time during fruit development (Fullerton *et al.* 2007). However, other findings suggest initial and maximal infection occur three and six weeks after petal fall, respectively (Wang *et al.*, 2013). Chilling injury to fruit facilitates the growth of *Diaporthe* rots (Fullerton *et al.*, 2007; Manning *et al.*, 2016).

CONTROL. In-orchard spray application of fungicides can be effective, e.g. in South Korea (Koh *et al.*, 2005).

Botryosphaeria dothidea
(side rot, dimple rot)

SYNONYMY. *Botryosphaeria berengeriana, Botryosphaeria mutila, Fusicoccum aesculi*

DISTRIBUTION. *B. dothidea* causes kiwifruit rots in New Zealand (Pennycook and Samuels, 1985), China, Japan and South Korea (Zhou *et al.*, 2015) and Iran (Nazerian *et al.*, 2019), and causes kiwifruit shoot blight in Greece (Thomidis and Exadaktylou, 2010) and vine dieback in China (Zeng and Huang, 2000) and Japan (Kinugawa and Sato, 2003).

SYMPTOMS. *B. dothidea* causes distinctive lesions early in storage. It is isolated less frequently from later lesions and rarely from lesions on over-ripe fruit, in which the closely related *Neofusicoccum parvum* and *N. luteum* are more common (Pennycook and Samuels, 1985).

Initial symptoms are small, shallow brown dimples (2–5 mm diameter) on the fruit surface. Beneath these dimples is a thin layer of dry, yellowish flesh. Some dimples develop into larger lesions as the fruit ripen (Pennycook, 1985). *Botryosphaeria* ripe rots on kiwifruit are often solitary, forming a distinctive oval 'thumbprint' with a narrow, water-soaked margin, commonly on the sides of the fruit (Fig. 17.6A, B) (Pennycook and Samuels, 1985; Kim and Koh, 2018).

DISEASE CYCLE. Asexual spores (conidia) are produced within pycnidia embedded in dead host tissue. Conidia are normally dispersed by rain-splash and windborne rain. The sexual spores (ascospores) are typically ejected forcibly from the fruiting bodies (pseudothecia), which are also embedded in dead host tissue. The ascospores can be airborne or waterborne (Sutton, 1981).

Zhou *et al.* (2015) proposed a disease cycle in kiwifruit whereby *B. dothidea* pycnidia or pseudothecia release spores in spring or early summer which infect leaves, shoots, buds and fruit. New pycnidia develop during the growing season, also infecting the fruit.

CONTROL. Reducing the inoculum pressure in the orchard is suggested as a primary control strategy for *B. dothidea* (Zhou *et al.*, 2015). Removal of kiwifruit prunings or mulching to encourage decay are key (Pennycook, 1985; Kim and Koh, 2018). Fungicide applications to shelterbelt trees may be beneficial (Pennycook, 1985).

Neofabraea actinidiae
(fungal pitting, ripe rot)

SYNONYMY. *Cryptosporiopsis actinidiae*

DISTRIBUTION. First recorded as *Myxosporium* sp. (Ford, 1971), *N. actinidiae* has only been recorded on kiwifruit in New Zealand and Italy (Johnston *et al.*, 2004; Riccioni *et al.*, 2007).

SYMPTOMS. The most common postharvest symptom is 'fungal pitting', i.e. small depressions on the sides of the fruit (Fig. 17.6G, H) (Fullerton *et al.*, 2007). The symptoms begin as small red-brown flecks, often centred on lenticels (Tyson *et al.*,

2019), and progress to dark, sunken pits with dry, yellow flesh beneath the skin (Manning *et al.*, 2003). Some pits later develop into larger rots.

DISEASE CYCLE. The source of *N. actinidiae* inoculum within kiwifruit orchards remains unknown; however, *N. actinidiae* conidia are likely to be waterborne (Tyson *et al.*, 2019); ascospores, should they be produced, will be airborne.

The teleomorph of this species is yet to be found in culture or in the field (Tyson *et al.*, 2019), although DNA sequencing suggests a *Neofabraea* teleomorphic state (Johnston *et al.*, 2014).

The majority of fruit infection is suggested to occur during flowering, with the fungus living endophytically during fruit development, remaining quiescent until the fruit ripen (Fullerton *et al.*, 2007).

In Italy, *N. actinidiae* was isolated from kiwifruit leaders with swelling and dieback and, while not found in most of the vines, its presence suggests its life cycle may be partly within the vine (Riccioni *et al.*, 2007).

CONTROL. Currently, there are no specific control options. Controlled-atmosphere storage with high (> 5%) CO_2 may exacerbate *Neofabraea* expression (Manning *et al.*, 2003).

Cadophora luteo-olivacea *(pitting, side rot)*

SYNONYMY. *Phialophora luteo-olivacea, Phialophora goidanichii*

DISTRIBUTION. *C. luteo-olivacea* is a widely distributed ascomycete, reported as a postharvest kiwifruit disease in Italy (Spadaro *et al.*, 2010) and Chile (Auger *et al.*, 2018). The pathogen is frequently associated with grapevine wood decay and has also been isolated from apple and pear fruit rots (Gramaje *et al.*, 2011; Wenneker *et al.*, 2016; Carneiro *et al.*, 2022).

SYMPTOMS. Symptoms generally appear after 12 weeks of storage, especially when followed by a few days at room temperature (Spadaro *et al.*, 2010). The symptoms are circular or oval sunken spots, ~2–15 mm in diameter, with white spongy flesh under the skin and an obvious margin (Testoni *et al.*, 1997).

DISEASE CYCLE. The disease is recorded sporadically and seems to be related to environmental conditions. The epidemiology remains poorly understood, although in kiwifruit orchards, *C. luteo-olivacea* is normally a soilborne microorganism (Auger *et al.*, 2018).

CONTROL. The incidence of skin pitting is correlated with fruit of low dry matter, high nitrogen content and, to a lesser extent, low calcium content (Spadaro *et al.*, 2010). Postharvest calcium chloride treatments have resulted in a slightly lower incidence of skin pitting (Di Lenarda *et al.*, 2011).

Neonectria microconidia *(stem end rot, cylindrocarpon rot)*

SYNONYMY. *Cylindrocarpon* sp., *Cylindrocarpon* cf. *candidum*

N. microconidia is a novel species recognized within the *N. coccinea* species group in China in 2011 (Zhao *et al.*, 2011). *N. microconidia* was subsequently recorded in Japan (Hirooka *et al.*, 2013), and most recently by this name in New Zealand (MPI, 2016). The fungus causing 'Cylindrocarpon' fruit rots in New Zealand has now been re-identified as *N. microconidia*.

DISTRIBUTION. *N. microconidia* is currently only recorded on kiwifruit in New Zealand (Manning *et al.*, 2003, 2016).

SYMPTOMS. *N. microconidia* has been known to cause fruit rots of kiwifruit in New Zealand since the commercialization of 'Hort16A' in the late 1990s. The pathogen is most often associated with a SER with conspicuous yellowish or white mycelium in the picking wound (Manning *et al.*, 2003, 2016), although it is also able to cause BR.

DISEASE CYCLE. On 'Hort16A', *N. microconidia* was often limited to the area of physical damage at the picking scar and, thus, Manning *et al.* (2003) believed that it required damaged tissue to infect the fruit. The rots generally appear after ~16 weeks in cool storage. This is a relatively minor fruit rot compared with *Diaporthe* spp., *Botryosphaeria* spp. and *Botrytis cinerea*.

N. microconidia has also been associated with a canker disease of kiwifruit in New Zealand (see Section 17.2.2).

Minor fruit rot pathogens

There are numerous minor fruit-rotting fungi, including species of *Alternaria*, *Colletotrichum*, *Fusarium*, *Pestalotiopsis* and *Phoma*.

Alternaria sp. is an uncommon kiwifruit-rot pathogen in New Zealand (Manning *et al.*, 2003). In China, *A. alternata* causes a postharvest rot of kiwifruit, often called black spot (Li *et al.*, 2012). It can also cause a pre-harvest black rot of kiwiberry fruit, which begins at the stylar end and may eventually cause early fruit drop (Kim *et al.*, 2017). *Alternaria* spp. can also cause storage stain of fruit.

Colletotrichum species tend to cause BR, often referred to as anthracnose. The infections start as small dark specks (Fig. 17.6I), which may later expand into obvious rots. Although the species involved vary, *Colletotrichum* rots are found in all kiwifruit-growing countries. For many years the recorded species on kiwifruit were *C. acutatum* and *C. gloeosporioides* (Hawthorne *et al.*, 1982; Manning *et al.*, 2003; Kim, M.J. *et al.*, 2018); with novel species of *Colletotrichum* currently being determined by molecular means, this now includes *C. aenigma* (Wang *et al.*, 2019), *C. boninense* (Zhang *et al.*, 2015), *C. karstii* (Jiang, 2016) and *C. nymphaeae* (Kim, G.H., *et al.*, 2018).

Multiple *Fusarium* spp. are minor kiwifruit rot pathogens; these are most often recorded simply as *Fusarium* sp. (Manning *et al.*, 2003). Those identified to the species level include *F. tricinctum* from South Korea (Kim, M.J. *et al.*, 2018), *F. acuminatum* from New Zealand and China (Pennycook, 1985; Wang *et al.*, 2015a), and *F. avenaceum*, *F. incarnatum*, *F. proliferatum* and *F. solani* from China (Li *et al.*, 2012; Yang *et al.*, 2018; Zhao *et al.*, 2020; Zhang *et al.*, 2021).

Pestalotiopsis spp. are very minor ripe rot-forming pathogens in kiwifruit, which can also be found as wound rots, in New Zealand (Manning *et al.*, 2003) and China (Li *et al.*, 2012).

Fruit lesions caused by *Phoma* spp. are dark brown and hard, with the flesh beneath the skin of a similar colour, often sharply defined from the adjacent healthy tissue (Fig. 17.6C) (Manning *et al.*, 2003).

17.3.3 Wound rots

Rots readily develop on fruit that have been damaged during picking or packing (Fig. 17.7). Most fungi associated with wound rots are not able to penetrate intact fruit: therefore control simply involves care during fruit handling and grading for defects. The most common pathogens infecting wounds on fruit are *Penicillium* spp. (blue mould), *Botrytis cinerea* (grey mould) and *Mucor* sp. (pin mould).

Penicillium spp. are common postharvest wound rot fungi of many fruit, with *P. expansum* recorded on kiwifruit in New Zealand (Manning *et al.*, 2003), China (Wang *et al.*, 2015b) and Greece (Prodromou *et al.*, 2018).

Botrytis cinerea is better known as a storage rot pathogen that enters through picking wounds (see Section 17.3.1) but is also found infecting wounds in the fruit flesh.

Mucor sp. is usually found causing soft rots on fruit in association with other fungi, wounds or over-ripe fruit, often with visible mould arising from the symptomatic area (Manning *et al.*, 2003).

Boeremia exigua (syn. *Phoma exigua*) is also frequently recovered from wound rots of kiwifruit (Hawthorne *et al.*, 1982; Brook, 1991; Manning *et al.*, 2003).

Cladosporium sp. and *Epicoccum* spp. are generally considered to be opportunistic fungi on fruit and are also often associated with wounds. They are isolated at low rates from kiwifruit postharvest rots (Manning *et al.*, 2003). *Alternaria alternata* has also been recorded in China as a wound-rot pathogen of kiwifruit (Li *et al.*, 2017).

Acknowledgements

The authors would like to thank Stephen Hoyte and Robin MacDiarmid for helpful comments on *Sclerotinia sclerotiorum* and kiwifruit viruses.

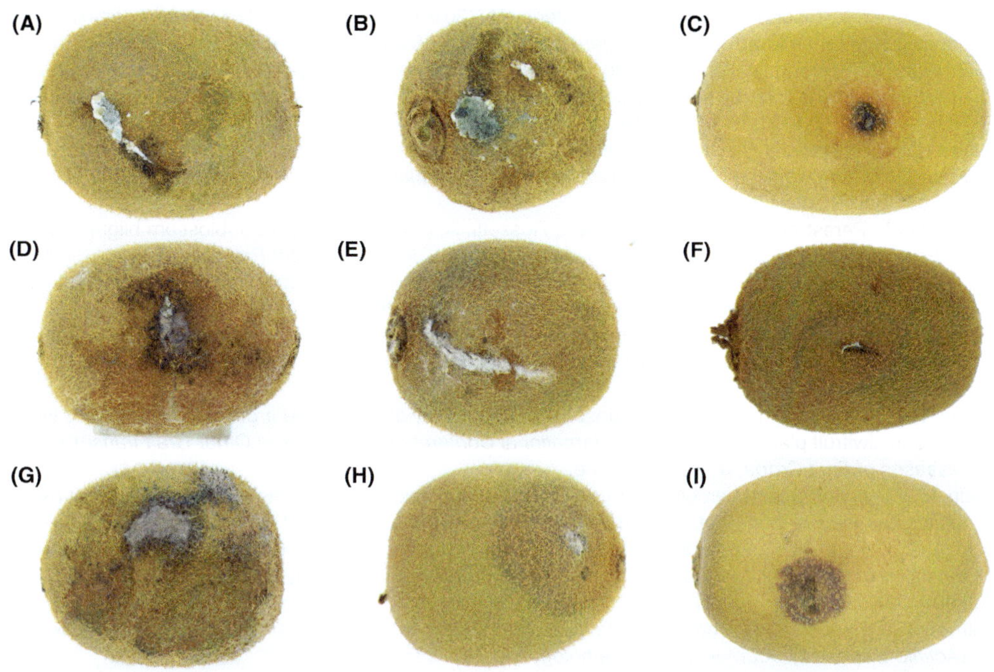

Fig. 17.7. Wound rots on the fruit of several kiwifruit cultivars. (**A, B**) *Penicillium* sp. (**C**) *Fusarium* sp. (**D, E**) *Botrytis cinerea*. (**F**) *Diaporthe* sp. (**G**) *Mucor* sp. (**H**) *Pestalotiopsis* sp. (**I**) *Phoma* sp. (Photo: J.L. Tyson, The New Zealand Institute for Plant and Food Research Ltd)

References

Abbatecola, A., Caponero, A., Camele, I., Xiloyannis, C. and Frisullo, S. (2008) Le alterazioni del legno dell'*Actinidia* in Basilicata: carie o mal dell'esca? *Giornate Fitopatologiche* 2, 219–220.

AgriMedia Ltd (2021) *New Zealand Novachem Agrichemical Manual 2021/22*. AgriMedia Ltd, Christchurch, New Zealand.

Aguin, O., Mansilla, J.P. and Sainz, M.J. (2004) Intraspecific variation in *Armillaria* species from shrubs and trees in north-western Spain. *Phytopathologia Mediterranea* 43, 299–304.

Ai, F. (2007) Occurrence and control of black spot disease of kiwifruit. *South China Fruits* 2007, 79.

Akilli, S., Serçe, Ç.U., Katircioğlu, Y.Z., Karakaya, A. and Maden, S. (2011) Involvement of *Phytophthora citrophthora* in kiwifruit decline in Turkey. *Journal of Phytopathology* 159, 579–581. DOI: 10.1111/j.1439-0434.2011.01798.x

Amponsah, N.T., van den Brink, R., Datson, P.M., Austin, P.T., Horner, M. *et al.* (2022) Actinidia seed-borne latent virus (ASbLV, *Betaflexiviridae*) is transmitted paternally and maternally at high rates. *Australasian Plant Pathology* 51, 399–407. DOI: 10.1007/s13313-022-00867-8

Araujo, L., Pinto, F.A.M.F., de Andrade, C.C.L., Gomes, L.B., Mituti, T. *et al.* (2021) *Pseudocercospora actinidiae* causes sooty spot disease on kiwifruit in Santa Catarina, Brazil. *Australasian Plant Disease Notes* 16, 22. DOI: 10.1007/s13314-021-00436-w

Auger, J., Pérez, I., Fullerton, R.A. and Esterio, M. (2009) First report of *Verticillium* wilt of gold kiwifruit, *Actinidia chinensis* cv. Hort 16A, caused by *Verticillium albo-atrum* in Chile. *Plant Disease* 93, 553. DOI: 10.1094/pdis-93-5-0553b

Auger, J., Pérez, I., Esterio, M. and Fullerton, R.A. (2011) Outbreak of a lethal wilt form of gold kiwifruit, *Actinidia chinensis* 'Hort16A' in Chile. *Acta Horticulturae* 913, 473–479. DOI: 10.17660/ActaHortic.2011.913.63

Auger, J., Pérez, I. and Esterio, M. (2013) *Diaporthe ambigua* associated with post-harvest fruit rot of kiwifruit in Chile. *Plant Disease* 97, 843. DOI: 10.1094/PDIS-10-12-0990-PDN

Auger, J., Pérez, I., Osorio-Navarro, C. and Esterio, M. (2018) First report of *Cadophora luteo-olivacea* causing side rot on kiwifruit in Chile. *Plant Disease* 102, 680. DOI: 10.1094/PDIS-09-17-1349-PDN

Bai, Q., Wang, G.P., Hong, N., Guo, Y.S. and Fu, M. (2017) First report of *Diaporthe tulliensis* and *Diaporthe actinidiae* causing kiwifruit stem canker in Hubei and Anhui Provinces, China. *Plant Disease* 101, 508. DOI: 10.1094/PDIS-10-16-1445-PDN

Balestra, G.M., Mazzaglia, A. and Rossetti, A. (2008) Outbreak of bacterial blossom blight caused by *Pseudomonas viridiflava* on *Actinidia chinensis* kiwifruit plants in Italy. *Plant Disease* 92, 1707. DOI: 10.1094/pdis-92-12-1707a

Balestra, G.M., Perestrelo, L., Mazzaglia, A. and Rossetti, A. (2009) First report of blossom blight caused by *Pseudomonas syringae* on kiwifruit plants in Portugal. *Journal of Plant Pathology* 91, 231.

Bardi, L. (2020) Early kiwifruit decline: a soil-borne disease syndrome or a climate change effect on plant–soil relations? *Frontiers in Agronomy* 2, 3. DOI: 10.3389/fagro.2020.00003

Bhardwaj, P., Ram, R., Zaidi, A.A. and Hallan, V. (2014) Characterization of *Apple stem grooving virus* infecting *Actinidia deliciosa* (Kiwi) in India. *Scientia Horticulturae* 176, 105–111. DOI: 10.1016/j.scienta.2014.06.038

Biccheri, R., Babini, A.R., Blouin, A.G., Lanzoni, C., Pisi, A. *et al.* (2012) *Pelargonium zonate spot virus* infecting kiwifruit plants in Italy. *22nd International Conference on Virus and Other Graft Transmissible Diseases of Fruit Crops*, 3–8 June, Rome.

Bisiach, M., Minervini, G. and Vercesi, A. (1984) Biological and epidemiological aspects of the kiwi fruit (*Actinidia chinensis* Planchon) rot, caused by *Botrytis cinerea* Pers. *Rivista di Patologia Vegetale* 20, 38–55.

Blouin, A.G., Chavan, R.R., Pearson, M.N., MacDiarmid, R.M. and Cohen, D. (2012) Detection and characterisation of two novel vitiviruses infecting *Actinidia*. *Archives of Virology* 157, 713–722. DOI: 10.1007/s00705-011-1219-1

Blouin, A.G., Pearson, M.N., Chavan, R.R., Woo, E.N.Y., Lebas, B.S.M. *et al.* (2013) Viruses of kiwifruit (*Actinidia* species). *Journal of Plant Pathology* 95, 221–235.

Bragard, C., Caciagli, P., Lemaire, O., Lopez-Moya, J.J., MacFarlane, S. *et al.* (2013) Status and prospects of plant virus control through interference with vector transmission. *Annual Review of Phytopathology* 51, 177–201. DOI: 10.1146/annurev-phyto-082712-102346

Braun, U., Crous, P.W. and Nakashima, C. (2015) Cercosporoid fungi (*Mycosphaerellaceae*) 4. Species on dicots (*Acanthaceae* to *Amaranthaceae*). *IMA Fungus* 6, 373–469. DOI: 10.5598/imafungus.2015.06.02.09

Brook, P.J. (1990) Diseases of kiwifruit. In: Warrington, I.J. and Weston, G.C. (eds) *Kiwifruit Science and Management*. Ray Richards Publisher in association with New Zealand Society for Horticultural Science, Auckland, New Zealand, pp. 420–428.

Brook, P.J. (1991) Botrytis stem-end rot and other storage diseases of kiwifruit: a review. *Acta Horticulturae* 297, 545–550. DOI: 10.17660/ActaHortic.1992.297.71

Cabral, A., Rego, C., Nascimento, T., Oliveira, H., Groenewald, J.Z. *et al.* (2012) Multi-gene analysis and morphology reveal novel *Ilyonectria* species associated with black foot disease of grapevines. *Fungal Biology* 116, 62–80. DOI: 10.1016/j.funbio.2011.09.010

Caciagli, P. and Lovisolo, O. (1987) Behaviour of kiwifruit plants (*Actinidia deliciosa*) in regard to virus infections. *Advances in Horticultural Science* 1, 11–14.

Carneiro, G.A., Walcher, M. and Baric, S. (2022) *Cadophora luteo-olivacea* isolated from apple (*Malus domestica*) fruit with post-harvest side rot symptoms in northern Italy. *European Journal of Plant Pathology* 162, 247–255. DOI: 10.1007/s10658-021-02388-4

Cho, S.-Y., Kim, H. and Yi, S.-I. (2017) First report of *Actinidia virus A* and *B* infecting *Actinidia chinensis* in South Korea. *Plant Disease* 101, 1560. DOI: 10.1094/pdis-03-17-0379-pdn

Ciccarese, F., Frisullo, S. and Amenduni, M. (1992) Observations of foot rots of *Actinidia* in southern Italy. *Informatore Fitopatologico* 42(2), 57–58.

Clover, G.R.G., Pearson, M.N., Elliott, D.R., Tang, Z., Smales, T.E. *et al.* (2003) Characterization of a strain of *Apple stem grooving virus* in *Actinidia chinensis* from China. *Plant Pathology* 52, 371–378. DOI: 10.1046/j.1365-3059.2003.00857.x

Conn, K.E., Gubler, W.D. and Hasey, J.K. (1993) Bacterial blight of kiwifruit in California. *Plant Disease* 77, 228–230. DOI: 10.1094/Pd-77-0228

Cooke, D.E.L., Drenth, A., Duncan, J.M., Wagels, G. and Brasier, C.M. (2000) A molecular phylogeny of *Phytophthora* and related Oomycetes. *Fungal Genetics and Biology* 30, 17–32. DOI: 10.1006/fgbi.2000.1202

Deighton, F.C. (1976) Studies on *Cercospora* and allied genera. VI. *Pseudocercospora* Speg., *Pantospora* Cif. and *Cercoseptoria* Petr. *Mycological Papers* 140, 1–168.

Demir, G. (2000) First report of bacterial diseases on kiwifruit (*Actinidia deliciosa*) in Turkey. *Journal of Turkish Phytopathology* 29, 7–18.
Di Lenarda, S., Martini, M., Osler, R., Vizzotto, G., Prete, G. *et al.* (2011) Biological and agronomical strategies for controlling skin-pitting of kiwifruit. *Acta Horticulturae* 913, 501–508. DOI: 10.17660/ActaHortic.2011.913.67
Di Marco, S., Calzarano, F., Gams, W. and Cesari, A. (2000) A new wood decay of kiwifruit in Italy. *New Zealand Journal of Crop and Horticultural Science* 28, 69–72. DOI: 10.1080/01140671.2000.9514124
Di Marco, S., Osti, F. and Spada, G. (2003) The wood decay of kiwifruit and first control measures. *Acta Horticulturae* 610, 291–294. DOI: 10.17660/ActaHortic.2003.610.38
Di Marco, S., Calzarano, F., Osti, F. and Mazzullo, A. (2004) Pathogenicity of fungi associated with a decay of kiwifruit. *Australasian Plant Pathology* 33, 337–342. DOI: 10.1071/ap04024
Diaz, G.A., Zoffoli, J.P., Ferrada, E.E. and Lolas, M. (2021) Identification and pathogenicity of *Diplodia*, *Neofusicoccum*, *Cadophora*, and *Diaporthe* species associated with cordon dieback in kiwifruit cultivar Hayward in Central Chile. *Plant Disease* 105, 1308–1319. DOI: 10.1094/pdis-05-20-0988-re
Dingley, J.M. (1969) *Records of Plant Diseases in New Zealand*. Plant Diseases Division, Department of Scientific and Industrial Research. Bulletin 192. Government Printer, Wellington.
Donati, I., Cellini, A., Sangiorgio, D., Caldera, E., Sorrenti, G. *et al.* (2020) Pathogens associated to kiwifruit vine decline in Italy. *Agriculture* 10, 119. DOI: 10.3390/agriculture10040119
Du, Y., Wang, X., Guo, Y., Xiao, F., Peng, Y. *et al.* (2021) Biological and molecular characterization of seven *Diaporthe* species associated with kiwifruit shoot blight and leaf spot in China. *Phytopathologia Mediterranea* 60, 177–198. DOI: 10.36253/phyto-12013
Ermacora, P., Cipriani, G., Savian, F., Testolin, R., Tosi, L. *et al.* (2020) Moria del kiwi, capire le cause per mettere in atto i rimedi. *Rivista di Frutticoltura e di Ortofloricoltura* 84(7), 12–17.
Erper, I., Tunali, B., Agustí-Brisach, C. and Armengol, J. (2011) First report of *Cylindrocarpon liriodendri* on kiwifruit in Turkey. *Plant Disease* 95, 76. DOI: 10.1094/pdis-09-10-0623
Erper, I., Agustí-Brisach, C., Tunali, B. and Armengol, J. (2013) Characterization of root rot disease of kiwifruit in the Black Sea region of Turkey. *European Journal of Plant Pathology* 136, 291–300. DOI: 10.1007/s10658-012-0163-6
Erper, I., Turkkan, M., Ozcan, M., Luongo, L. and Belisario, A. (2017) Characterization of *Diaporthe hongkongensis* species causing stem-end rot on kiwifruit in Turkey. *Journal of Plant Pathology* 99(3), 779–782.
Erwin, D.C. and Ribeiro, O.K. (1996) *Phytophthora Diseases Worldwide*. American Phytopathological Society Press, St. Paul, Minnesota.
Everett, K.R. and Henshall, W.R. (1994) Epidemiology and population ecology of kiwifruit blossom blight. *Plant Pathology* 43, 824–830. DOI: 10.1111/j.1365-3059.1994.tb01627.x
Ferreira, M.A., Harrington, T.C., Piveta, G. and Alfenas, A.C. (2013) Genetic variability suggests that *Ceratocystis fimbriata* is native to Rio Grande do Sul, Brazil, where it is causing a new wilt disease on kiwifruit. *Phytopathology* 103, 43.
Ferreira, M.A., Harrington, T.C., Piveta, G. and Alfenas, A.C. (2017) Genetic variability suggests that three populations of *Ceratocystis fimbriata* are responsible for the ceratocystis wilt epidemic on kiwifruit in Brazil. *Tropical Plant Pathology* 42, 86–95. DOI: 10.1007/s40858-017-0131-y
Ford, I. (1971) Chinese gooseberry pest and disease control. *New Zealand Journal of Agriculture* 122(3), 86–89.
Förster, H. and Coffey, M.D. (1993) Molecular taxonomy of *Phytophthora megasperma* based on mitochondrial and nuclear DNA polymorphisms. *Mycological Research* 97, 1101–1112. DOI: 10.1016/s0953-7562(09)80511-x
French, A.M. (1987) *California Plant Disease Host Index. Part 1: Fruit and Nuts*. California Department of Food and Agriculture, Division of Plant Industry, Sacramento, California.
Fullerton, R.A., Rheinländer, P.A., Manning, M.A. and Casonato, S.G. (2007) Infection and rot expression by *Cryptosporiopsis actinidiae* and *Phomopsis* sp. in kiwifruit, *Actinidia chinensis* 'Hort16A'. *Acta Horticulturae* 753, 653–660. DOI: 10.17660/ActaHortic.2007.753.86
Gallegly, M.E. and Hong, C. (2008) *Phytophthora: Identifying Species by Morphology and DNA Fingerprints*. American Phytopathological Society Press, St. Paul, Minnesota.
Gianetti, G., Cotroneo, A., Garofalo, M.C., Galliano, A. and Vittone, F. (2002) Acquisizioni preliminari sul deperimento lignicolo dell'actinidia in Piemonte. *Rivista di Frutticoltura e di Ortofloricoltura* 64(9), 65–68.
González, A.J. and Rodicio, M.R. (2007) Disease of floral buds of kiwifruit caused by *Pseudomonas viridiflava* and *Pseudomonas syringae*. *Boletin de Sanidad Vegetal, Plagas* 33, 517–525.

Gramaje, D., Mostert, L. and Armengol, J. (2011) Characterization of *Cadophora luteo-olivacea* and *C. melinii* isolates obtained from grapevines and environmental samples from grapevine nurseries in Spain. *Phytopathologia Mediterranea* 50, S112–S126. DOI: 10.14601/Phytopathol_Mediterr-8723

Hawthorne, B.T., Rees-George, J. and Samuels, G.J. (1982) Fungi associated with leaf spots and post-harvest fruit rots of kiwifruit (*Actinidia chinensis*) in New Zealand. *New Zealand Journal of Botany* 20, 143–150. DOI: 10.1080/0028825x.1982.10428835

Hennion, B., Lecomte, P., Larignon, P., Baudry, A., Durpaire, M.P. *et al.* (2003) First observations of a wood decay (Esca-like disease) on kiwifruit in France. *Acta Horticulturae* 610, 409–413. DOI: 10.17660/ActaHortic.2003.610.54

Hirooka, Y., Rossman, A.Y., Zhuang, W.-Y., Salgado-Salazar, C. and Chaverri, P. (2013) Species delimitation for *Neonectria coccinea* group including the causal agents of beech bark disease in Asia, Europe, and North America. *Mycosystema* 32, 485–517.

Hood, I. (1989) Armillaria root disease in New Zealand forests. *New Zealand Journal of Forestry Science* 19, 180–197.

Hood, I. and Shaw, C. (2022) *Armillaria novae-zelandiae*. In: *Invasive Species Compendium*. Available at: www.cabi.org/isc (accessed February 2022).

Horner, I.J. (1992) Epidemiology of *Armillaria* root-rot of kiwifruit. *Acta Horticulturae* 297, 573–578. DOI: 10.17660/ActaHortic.1992.297.75

Hoyte, S. (2000) The role of adhering floral tissues in sclerotinia disease of kiwifruit. *New Zealand Plant Protection* 53, 138–142. DOI: 10.30843/nzpp.2000.53.3666

Hoyte, S. (2001) Epidemiology and management of *Sclerotinia sclerotiorum* (Lib.) de Bary in kiwifruit (*Actinidia deliciosa* (A. Chev.)). PhD thesis, Massey University, Palmerston North, New Zealand.

Hoyte, S.M., Bedford, R.J. and Gouk, S.C. (1992) Field observations of *Sclerotinia sclerotiorum*: apothecial development and sclerotial production. *Proceedings of the 45th New Zealand Plant Protection Conference* 45, 188–192. DOI: 10.30843/nzpp.1992.45.11223

Hu, F., Fang, D., Young, J. and Xie, L. (1998) Identification of the pathogen caused bacterial blight of kiwifruit in China. *Acta Phytopathologica Sinica* 28, 175–181.

Huang, Y. and Qi, P. (1998) Study on the cause of kiwifruit root rot in Guangdong Province. *Journal of South China Agricultural University* 19(4), 19–22, 35.

Ieki, H. (1993) Kiwifruit [*Actinidia deliciosa*] diseases in Japan. *Japan Pesticide Information* 61, 11–13.

Jiang, F. (2016) Isolation and identification of kiwifruit spoilage fungi and study on the control effect of chitosan. MSc thesis, Guizhou Medical University, China.

Jiang, J.K., Gao, R.X. and Lin, Y.J. (1995) Investigation on *Pseudocercospora actinidiae*. *Journal of Fruit Science* 12, 182–184.

Jiao, H., Wu, Y., Qu, X. and Wang, X. (2014) Occurrence and control of kiwifruit sclerotinia disease in Shaanxi Province. *Deciduous Fruits* 46, 37–38.

Johnston, P.R., Manning, M.A., Meier, X., Park, D. and Fullerton, R.A. (2004) *Cryptosporiopsis actinidiae* sp. nov. *Mycotaxon* 89, 131–136.

Johnston, P.R., Park, D. and Manning, M.A. (2010) *Neobulgaria alba* sp. nov. and its *Phialophora*-like anamorph in native forests and kiwifruit orchards in New Zealand. *Mycotaxon* 113, 385–396. DOI: 10.5248/113.385

Johnston, P.R., Seifert, K.A., Stone, J.K., Rossman, A.Y. and Marvanova, L. (2014) Recommendations on generic names competing for use in *Leotiomycetes* (*Ascomycota*). *IMA Fungus* 5, 91–120. DOI: 10.5598/imafungus.2014.05.01.11

Kasson, M.T., Short, D.P.G., O'Neal, E.S., Subbarao, K.V. and Davis, D.D. (2014) Comparative pathogenicity, biocontrol efficacy, and multilocus sequence typing of *Verticillium nonalfalfae* from the invasive *Ailanthus altissima* and other hosts. *Phytopathology* 104, 282–292. DOI: 10.1094/PHYTO-06-13-0148-R

Kikuhara, K. and Nakashima, C. (2008) Sooty spot of kiwifruit caused by *Pseudocercospora actinidiae* Deighton. *Journal of General Plant Pathology* 74, 185–187. DOI: 10.1007/s10327-008-0077-5

Kim, G.H. and Koh, Y.J. (2018) Diagnosis and integrated management of major fungal fruit rots on kiwifruit in Korea. *Research in Plant Disease* 24, 113–122. DOI: 10.5423/rpd.2018.24.2.113

Kim, G.H., Kim, D.R., Park, S.-Y., Lee, Y.S., Jung, J.S. *et al.* (2017) Incidence rates of major diseases of kiwiberry in 2015 and 2016. *Plant Pathology Journal* 33, 434–439. DOI: 10.5423/PPJ.NT.01.2017.0010

Kim, G.H., Choi, D.H., Park, S.Y. and Koh, Y.J. (2018) First report of anthracnose caused by *Colletotrichum nymphaeae* on kiwiberry in Korea. *Plant Disease* 102, 1455. DOI: 10.1094/PDIS-11-17-1768-PDN

Kim, M.-J., Chae, D.-H., Kwon, Y., Kwack, Y.-B. and Kwak, Y.-S. (2018) Occurrences of major diseases and pests on 'Goldone', 'Redvita', 'Garmrok', new cultivars of kiwifruit. *Research in Plant Disease* 24, 123–131. DOI: 10.5423/rpd.2018.24.2.123

Kinugawa, M. and Sato, T. (2003) Kiwifruit dieback caused by *Diaporthe* sp. and *Botryosphaeria dothidea*, kiwifruit soft rot fungi. *Annals of the Phytopathological Society of Japan* 69, 373–383.

Koh, Y.J., Lee, D.H., Shin, J.S. and Hur, J.-S. (2001) Chemical and cultural control of bacterial blossom blight of kiwifruit caused by *Pseudomonas syringae* in Korea. *New Zealand Journal of Crop and Horticultural Science* 29, 29–34. DOI: 10.1080/01140671.2001.9514157

Koh, Y.J., Jung, J.S. and Hur, J.S. (2003) Current status of occurrence of major diseases on kiwifruits and their control in Korea. *Acta Horticulturae* 610, 437–443. DOI: 10.17660/ActaHortic.2003.610.58

Koh, Y.J., Hur, J.S. and Jung, J.S. (2005) Postharvest fruit rots of kiwifruit (*Actinidia deliciosa*) in Korea. *New Zealand Journal of Crop and Horticultural Science* 33, 303–310. DOI: 10.1080/01140671.2005.9514363

KVH (2020) KVH Information Sheet. *Neonectria*: symptom identification, monitoring and management advice for growers. Available at: https://kvh.org.nz/assets/documents/Growers-tab/Neonectria_Information_Sheet_2020_FINAL.pdf (accessed 2 June 2023).

Latorre, B.A., Alvarez, C. and Rebeiro, O.K. (1991) *Phytophthora* root rot of kiwifruit in Chile. *Plant Disease* 75, 949–952. DOI: 10.1094/pd-75-0949

Lawrence, D.P., Nouri, M.T. and Trouillas, F.P. (2019) Taxonomy and multi-locus phylogeny of cylindrocarpon-like species associated with diseased roots of grapevine and other fruit and nut crops in California. *Fungal Systematics and Evolution* 4, 59–75. DOI: 10.3114/fuse.2019.04.06

Ledesma, M., Shin, Y.H., Whiteman, S., Tyson, J., Kim, G.H. *et al.* (2021) Causal fungus of side rot on 'Zesy002' kiwifruit in Jeju Island, South Korea. *Research in Plant Disease* 27, 24–31. DOI: 10.5423/rpd.2021.27.1.24

Lee, J.-G., Lee, D.-H., Park, S.-Y., Hur, J.-S. and Koh, Y.-J. (2001) First report of *Diaporthe actinidiae*, the causal organism of stem-end rot of kiwifruit in Korea. *The Plant Pathology Journal* 17(2), 110–113.

Lee, J.H., Kwon, Y.H., Kwack, Y.-B. and Kwak, Y.-S. (2015) Report of postharvest rot of kiwifruit in Korea caused by *Sclerotinia sclerotiorum*. *International Journal of Food Microbiology* 206, 81–83. DOI: 10.1016/j.ijfoodmicro.2015.05.001

Lee, Y.-H., Jee, H.-J., Cha, K.-H., Ko, S.-J. and Park, K.B. (2001) Occurrence of *Phytophthora* root rot on kiwifruit in Korea. *The Plant Pathology Journal* 17, 154–158.

Li, C., Jiang, J.X., Leng, J.H., Li, B.M., Yu, Q. *et al.* (2012) Isolation and identification of pathogenic fungi causing fruit rot of kiwifruit in Fengxin county. *Acta Agriculturae Universitatis Jiangxiensis* 34, 259–263.

Li, L., Pan, H., Liu, W., Chen, M.Y. and Zhong, C.H. (2017) First report of *Alternaria alternata* causing postharvest rot of kiwifruit in China. *Plant Disease* 101, 1046. DOI: 10.1094/PDIS-11-16-1611-PDN

Lin, G., Hu, C. and Gao, R. (1994) Research progress on kiwifruit diseases in China. *Fujian Fruits* 1994, 22–25.

Lin, Y. and Gao, R. (1995) Survey and identification of *Actinidia* spp. diseases in Fujian, China. *Journal of Fujian Agricultural University* 24, 49–53.

Lin, Y., Guo, Y., Fang, Z. and Zhang, W. (1995) Diseases of *Actinidia chinensis* and its control 2. *Quarterly of Forest By-product and Speciality in China* 32, 13–15.

Luongo, L., Santori, A., Riccioni, L. and Belisario, A. (2011) *Phomopsis* sp. associated with post-harvest fruit rot of kiwifruit in Italy. *Journal of Plant Pathology* 93, 205–209.

Manning, M., Meier, X., Olsen, T.L. and Johnston, P.R. (2003) Fungi associated with fruit rots of *Actinidia chinensis* 'Hort16A' in New Zealand. *New Zealand Journal of Crop and Horticultural Science* 31, 315–324. DOI: 10.1080/01140671.2003.9514267

Manning, M., Pak, H.A. and Beresford, R.M. (2010) Non-fungicidal control of botrytis storage rot in New Zealand kiwifruit through pre- and postharvest crop management. In: Prusky, D. and Gullino, M.L. (eds) *Postharvest Pathology*. Plant Pathology in the 21st Century, Vol. 2. Springer, Dordrecht, Heidelberg, London, New York, pp. 183–196. DOI: 10.1007/978-1-4020- 8930-5_13

Manning, M., Burdon, J., De Silva, N., Meier, X., Pidakala, P. *et al.* (2016) Maturity and postharvest temperature management affect rot expression in 'Hort16A' kiwifruit. *Postharvest Biology and Technology* 113, 40–47. DOI: 10.1016/j.postharvbio.2015.10.012

Miyoshi, T. and Tachibana, Y. (1996) Epidemiological studies on bacterial blossom blight of kiwifruit 3. Control of bacterial blossom blight of kiwifruit by ringing treatment and its mechanism of disease suppression. *Annals of the Phytopathological Society of Japan* 62, 528–532. DOI: 10.3186/jjphytopath.62.528

MPI (Ministry for Primary Industries) (2016) Pest watch: 24 October–15 February 2016. *Surveillance* 43(1), 31.

Nazerian, E., Mirabolfathy, M., Ashnaei, S.P. and Beiki, F. (2019) Characterization of *Botryosphaeria dothidea* as new pathogen of kiwifruit in Iran. *Journal of Plant Protection Research* 59, 134–137. DOI: 10.24425/jppr.2019.126035

Nipoti, P., Sandalo, S., Prodi, A., Credi, R., Spada, G. *et al.* (2003) An unusual wood disease of kiwifruit in Italy. *Acta Horticulturae* 610, 253–259. DOI: 10.17660/ActaHortic.2003.610.33

Nipoti, P., Sandalo, S., Prodi, A., Credi, R., Spada, G. *et al.* (2004) Una nuova malattia dell'actinidia, l'ipertrofia del tronco o 'elefantiasi'. *Informatore Agrario* 60(4), 79–82.

Pennycook, S.R. (1985) Fungal fruit rots of *Actinidia deliciosa* (kiwifruit). *New Zealand Journal of Experimental Agriculture* 13, 289–299. DOI: 10.1080/03015521.1985.10426097

Pennycook, S.R. (1989) *Plant Diseases Recorded in New Zealand*. Vol. 2: *Fungal Plant Diseases Recorded in New Zealand*. Plant Diseases Division, DSIR, Auckland, New Zealand.

Pennycook, S.R. and Manning, M.A. (1992) Picking wound curing to reduce botrytis storage rot of kiwifruit. *New Zealand Journal of Crop and Horticultural Science* 20, 357–360. DOI: 10.1080/01140671.1992.10421779

Pennycook, S.R. and Samuels, G.J. (1985) *Botryosphaeria* and *Fusicoccum* species associated with ripe fruit rot of *Actinidia deliciosa* (kiwifruit) in New Zealand. *Mycotaxon* 24, 445–458.

Pennycook, S.R. and Triggs, C.M. (1992) Bacterial blossom blight of kiwifruit – a 5-year survey. *Acta Horticulturae* 297, 559–565. DOI: 10.17660/ActaHortic.1992.297.73

Pinilla, C.B., Alvarez, A.M. and Garcia, G.M.A. (1994) Postharvest stem-end rot caused by *Botrytis cinerea* in kiwifruits. *Revista Fruticola* 15(2), 63–66.

Piveta, G., Alfenas, A.C., Muniz, M.d.F.B., Valdebenito-Sanhueza, R.M. and Ferreira, M.A. (2013) Ocorrência de *Ceratocystis fimbriata* em kiwi (*Actinidia deliciosa*) no sul do Brasil. *Revista Brasileira de Fruticultura* 35, 665–669. DOI: 10.1590/s0100-29452013000200040

Piveta, G., Ferreira, M.A., Muniz, M.d.F.B., Valdetaro, D., Valdebenito-Sanhueza, R. *et al.* (2016) *Ceratocystis fimbriata* on kiwifruit (*Actinidia* spp.) in Brazil. *New Zealand Journal of Crop and Horticultural Science* 44, 13–24. DOI: 10.1080/01140671.2016.1143020

Porta-Puglia, A. and Corazza, L. (1985) *Fusarium solani* (Mart.) Sacc. associato ad un marciume del colletto dell'*Actinidia*. *Rivista della Ortoflorofrutticoltura Italiana* 69(1), 81–83.

Prodi, A., Sandalo, S., Tonti, S., Nipoti, P. and Pisi, A. (2008) *Phialophora*-like fungi associated with kiwifruit elephantiasis. *Journal of Plant Pathology* 90, 487–494.

Prodromou, I., Thomidis, T. and Zambounis, A. (2018) First report of *Penicillium expansum* (Link) Thom. causing postharvest fruit rot of kiwifruit in northern Greece. *Plant Disease* 102, 1851. DOI: 10.1094/pdis-11-17-1804-pdn

Rahman, M.Z., Uematsu, S., Coffey, M.D., Uzuhashi, S., Suga, H. *et al.* (2014) Re-evaluation of Japanese *Phytophthora* isolates based on molecular phylogenetic analyses. *Mycoscience* 55, 314–327. DOI: 10.1016/j.myc.2013.11.005

Riccioni, L., Manning, M., Valvassori, M., Haegi, A., Casonato, S. *et al.* (2007) A new disease: leader die-back in *Actinidia chinensis* 'Hort16A' in Italy. *Acta Horticulturae* 753, 669–675. DOI: 10.17660/ActaHortic.2007.753.88

Savian, F., Ginaldi, F., Musetti, R., Sandrin, N., Tarquini, G. *et al.* (2020a) Studies on the aetiology of kiwifruit decline: interaction between soil-borne pathogens and waterlogging. *Plant and Soil* 456, 113–128. DOI: 10.1007/s11104-020-04671-5

Savian, F., Martini, M., Ermacora, P., Paulus, S. and Mahlein, A.-K. (2020b) Prediction of the kiwifruit decline syndrome in diseased orchards by remote sensing. *Remote Sensing* 12, 2194. DOI: 10.3390/rs12142194

Schroeder, C.A. and Fletcher, W.A. (1967) The Chinese gooseberry (*Actinidia chinensis*) in New Zealand. *Economic Botany* 21, 81–92. DOI: 10.1007/BF02897178

Shin, J.-S., Park, J.-K., Kim, G.-H., Jung, J.-S., Hur, J.-S. *et al.* (2004a) Cultural control of bacterial blossom blight using trunk girdling and rainproof installation over kiwifruit trees. *Research in Plant Disease* 10, 304–309. DOI: 10.5423/rpd.2004.10.4.304

Shin, J.-S., Park, J.-K., Kim, G.-H., Jung, J.-S., Hur, J.-S. *et al.* (2004b) Optimum spray program of preventive bactericides for the control of bacterial blossom blight of kiwifruit. *Research in Plant Disease* 10, 297–303. DOI: 10.5423/RPD.2004.10.4.297

Shivas, R.G. (1989) Fungal and bacterial diseases of plants in Western Australia. *Journal of the Royal Society of Western Australia* 72, 1–62.

Sommer, N.F. and Beraha, L. (1975) *Diaporthe actinidiae*, a new species causing stem-end rot of Chinese gooseberry fruits. *Mycologia* 67, 650–653. DOI: 10.1080/00275514.1975.12019787

Sônego, O.R., Ferreira, M.A., Sanhueza, R.M., Gava, R., Garrido, L.R. *et al.* (2010) First report of *Ceratocystis* wilt in kiwi (Primeiro relato da murcha-de-ceratocystis em kiwi). *Tropical Plant Pathology* 35, S233.

Song, X., Zheng, W., Zhang, X. and Ma, S. (1997) An overview of the investigation of kiwifruit diseases and insect pests in Shaanxi Province. *Forest Pest and Disease* 1997, 17–19.

Sorrenti, G., Toselli, M., Reggidori, G., Spinelli, F., Tosi, L. *et al.* (2016) Implicazioni della gestione idrica nella 'Moria del kiwi' del veronese. *Rivista di Frutticoltura e di Ortofloricoltura* 76(3), 45–51.

Spadaro, D., Galliano, A., Pellegrino, C., Gilardi, G., Garibaldi, A. *et al.* (2010) Dry matter, mineral composition, and commercial storage practices influence the development of skin pitting caused by *Cadophora luteo-olivacea* on kiwifruit 'Hayward'. *Journal of Plant Pathology* 92, 349–356.

Spigaglia, P., Barbanti, F., Marocchi, F., Mastroleo, M., Baretta, M. *et al.* (2020) *Clostridium bifermentans* and *C. subterminale* are associated with kiwifruit vine decline, known as *moria*, in Italy. *Plant Pathology* 69, 765–774. DOI: 10.1111/ppa.13161

Stewart, A. and McCarrison, A.M. (1991) Pathogenicity and relative virulence of seven *Phytophthora* species on kiwifruit. *New Zealand Journal of Crop and Horticultural Science* 19, 73–76. DOI: 10.1080/01140671.1991.10418109

Sun, Q., Xie, Y.-Z., Chen, T.-M., Zhang, J.-P., Laborda, P. *et al.* (2022) First report of *Fusarium avenaceum* causing canker disease on kiwi tree in China. *Plant Disease* 106, 1762. DOI: 10.1094/PDIS-11-21-2459-PDN

Sutton, T.B. (1981) Production and dispersal of ascospores and conidia by *Physalospora obtusa* and *Botryosphaeria dothidea* in apple orchards. *Phytopathology* 71, 584–589. DOI: 10.1094/Phyto-71-584

Tacconi, G., Paltrinieri, S., Mejia, J.F., Fuentealba, S.P., Bertaccini, A. *et al.* (2015) Vine decline in kiwifruit: climate change and effect on waterlogging and *Phytophthora* in north Italy. *Acta Horticulturae* 1096, 93–97. DOI: 10.17660/ActaHortic.2015.1096.7

Testolin, R. and Ferguson, A.R. (2009) Kiwifruit (*Actinidia* spp.) production and marketing in Italy. *New Zealand Journal of Crop and Horticultural Science* 37, 1–32. DOI: 10.1080/01140670909510246

Testoni, A., Grassi, B., Quaroni, S., Saracchi, M. and Sardi, P. (1997) Pitting on kiwifruit in storage caused by *Phialophora* sp. *Acta Horticulturae* 444, 751–756. DOI: 10.17660/ActaHortic.1997.444.115

Thomidis, T. (2009) First report of *Phomopsis actinidiae* causing cankers on shoots of kiwifruit in Greece. *Journal of Plant Pathology* 91, 499.

Thomidis, T. and Exadaktylou, E. (2010) First report of *Botryosphaeria dothidea* causing shoot blight of kiwifruit (*Actinidia deliciosa*) in Greece. *Plant Disease* 94, 1503. DOI: 10.1094/pdis-07-10-0528

Thomidis, T. and Exadaktylou, E. (2012) Effectiveness of cyproconazole to control *Armillaria* root rot of apple, walnut and kiwifruit. *Crop Protection* 36, 49–51. DOI: 10.1016/j.cropro.2012.02.003

Türkkan, M., Şahin, N., Özer, G., Evgin, Z., Yaman, M. *et al.* (2020) First report of *Verticillium dahliae* causing Verticillium wilt on kiwifruit in Ordu, Turkey. *Journal of Plant Pathology* 102, 221–222. DOI: 10.1007/s42161-019-00359-7

Tyson, J.L., Manning, M.A., Everett, K.R. and Fullerton, R.A. (2019) *Neofabraea actinidiae* in New Zealand kiwifruit orchards: current status and knowledge gaps. *New Zealand Plant Protection* 72, 75–82. DOI: 10.30843/nzpp.2019.72.268

Tyson, J.L., Manning, M.A. and Wright, P.J. (2020) Pathogenicity of *Ceratocystis fimbriata* from New Zealand kumara on kiwifruit cultivars. *New Zealand Plant Protection* 73, 6–12. DOI: 10.30843/nzpp.2020.73.11717

Veerakone, S., Liefting, L.W., Tang, J. and Ward, L.I. (2018) The complete nucleotide sequence and genome organisation of a novel member of the family *Betaflexiviridae* from *Actinidia chinensis*. *Archives of Virology* 163, 1367–1370. DOI: 10.1007/s00705-017-3701-x

Wang, C.W., Ai, J., Fan, S.T., Lv, H.Y., Qin, H.Y. *et al.* (2015a) *Fusarium acuminatum*: a new pathogen causing postharvest rot on stored kiwifruit in China. *Plant Disease* 99, 1644. DOI: 10.1094/pdis-01-15-0021-pdn

Wang, C.W., Ai, J., Lv, H.Y., Qin, H.Y., Yang, Y.M. *et al.* (2015b) First report of *Penicillium expansum* causing postharvest decay on stored kiwifruit (*Actinidia arguta*) in China. *Plant Disease* 99, 1037. DOI: 10.1094/PDIS-12-14-1274-PDN

Wang, D. and Huang, G. (2020) Comprehensive control technology of main diseases and insect pests of *Actinidia arguta*. *Northern Fruits* 2020, 46–49. DOI: 10.16376/j.cnki.bfgs.2020.03.019

Wang, J., Liu, D., Liu, Y., Zhan, L. and Zhang, Y. (2013) Experiment on infection regularity of fruit rot of kiwifruit and control. *Journal of Zhejiang Forestry Science and Technology* 33, 55–57.

Wang, Y., Qin, H.Y., Liu, Y.X., Fan, S.T., Sun, D. *et al.* (2019) First report of anthracnose caused by *Colletotrichum aenigma* on *Actinidia arguta* in China. *Plant Disease* 103(2), 372. DOI: 10.1094/pdis-07-18-1137-pdn

Wang, Y., Zhai, L., Wen, S., Yang, Z., Wang, G. *et al.* (2020) Molecular characterization of a novel emaravirus infecting *Actinidia* spp. in China. *Virus Research* 275, 197736. DOI: 10.1016/j.virusres.2019.197736

Weber, R.W.S. and Børve, J. (2021) Infection biology as the basis of integrated control of apple canker (*Neonectria ditissima*) in Northern Europe. *CABI Agriculture and Bioscience* 2, 5. DOI: 10.1186/s43170-021-00024-z

Wenneker, M., Pham, K.T.K., Lemmers, M.E.C., de Boer, F.A., van Leeuwen, P.J. *et al.* (2016) First report of *Cadophora luteo-olivacea* causing side rot on 'Conference' pears in the Netherlands. *Plant Disease* 100, 2162. DOI: 10.1094/pdis-02-16-0212-pdn

Woo, E.N.Y., Lebas, B.S.M., Veerakone, S., Tang, J., Ward, L. *et al.* (2012) Molecular detection and characterisation of *Cherry leaf roll virus* in New Zealand. 10th Australasian Plant Virology Workshop, 19–22 November, Hamner Springs, New Zealand.

Xue, L.H., Liu, Y., Liu, H.Y. and Li, C.J. (2017) First report of *Botrytis cinerea* causing fruit rot on kiwifruit (*Actinidia chinensis*) in China. *Plant Disease* 101, 1054. DOI: 10.1094/PDIS-12-16-1819-PDN

Yaman, M. and Türkkan, M. (2017) Determination of the inhibitory effects of organic and inorganic salts against some fungi causing root rot disease on kiwifruit. *Anadolu Tarim Bilimleri Dergisi* 32, 291–302. DOI: 10.7161/omuanajas.297081

Yang, C., Wu, Z.Q., Qi, H., Chen, C.Q., Wang, X. *et al.* (2018) *Fusarium solani*, a new pathogen causing postharvest fruit rot on kiwifruit (*Actinidia deliciosa*) in China. *Plant Disease* 102, 443. DOI: 10.1094/pdis-07-17-0939-pdn

Yi, J.-H. and Lee, D.-H. (1998) Mycological characteristics and pathogenicity of *Phomopsis mali* causing fruit decays of Japanese apricot, apple and kiwifruit. *Korean Journal of Plant Pathology* 14, 109–114.

Young, J.M., Gardan, L., Ren, X.-Z. and Hu, F.-P. (1997) Genomic and phenotypic characterization of the bacterium causing blight of kiwifruit in New Zealand. *Plant Pathology* 46, 857–864. DOI: 10.1046/j.1365-3059.1997.d01-72.x

Zeng, S. and Huang, F. (2000) Preliminary study on the brown rot disease of kiwifruit. *South China Fruits* 29(4), 51.

Zhang, G., Li, R., Li, M., Gu, H., Li, X. *et al.* (2021) Investigation and identification on fungal diseases of *Actinidia chinensis* in Qiandongnan Prefecture. *Forest Pest and Disease* 2021, 11–18.

Zhang, M., He, L., Zhang, M., Yuan, X., Du, F. *et al.* (2015) Identification of pathogenic fungi from Qinmei kiwifruit at stage of post-harvest storage. *Acta Agriculturae Boreali-Occidentalis Sinica* 24, 132–137.

Zhang, Z., Long, Y., Yang, S., Li, L. and Yin, X. (2019) Occurrence and control measures of kiwifruit flower rot. *South China Fruits* 48, 159–164.

Zhao, L., Yang, W., Zhang, Y., Wu, Z., Wang, Q.-C. *et al.* (2019) Occurrence and molecular variability of kiwifruit viruses in *Actinidia deliciosa* 'Xuxiang' in the Shaanxi Province of China. *Plant Disease* 103, 1309–1318. DOI: 10.1094/pdis-09-18-1570-re

Zhao, L., Cao, M., Huang, Q., Wang, Y., Sun, J. *et al.* (2021) Occurrence and distribution of *Actinidia* viruses in Shaanxi Province of China. *Plant Disease* 105, 929–939. DOI: 10.1094/pdis-06-20-1190-re

Zhao, L., Wang, J., Sujata, S., Sun, J., Wang, Y. *et al.* (2022) Viromics reveals the viral diversity in cultivated and wild kiwifruit. *Plant Disease* 106, 3120–3126. DOI: 10.1094/PDIS-03-22-0669-RE

Zhao, P., Luo, J., Zhuang, W., Liu, X. and Wu, B. (2011) DNA barcoding of the fungal genus *Neonectria* and the discovery of two new species. *Science China – Life Sciences* 54, 664–674. DOI: 10.1007/s11427-011-4184-8

Zhao, Z., Zhu, Q., Zhi, T., Li, Q., Long, Y. *et al.* (2020) First report of fruit blotch on kiwifruit caused by *Fusarium avenaceum* in China. *Plant Disease* 104, 1545. DOI: 10.1094/pdis-10-19-2075-pdn

Zheng, Y.Z., Wang, G.P., Hong, N., Zhou, J.F., Yang, Z.K. *et al.* (2014) First report of *Actinidia virus* A and *Actinidia virus* B on kiwifruit in China. *Plant Disease* 98, 1590. DOI: 10.1094/pdis-04-14-0420-pdn

Zhou, Y., Gong, G., Cui, Y., Zhang, D., Chang, X. *et al.* (2015) Identification of Botryosphaeriaceae species causing kiwifruit rot in Sichuan Province, China. *Plant Disease* 99, 699–708. DOI: 10.1094/pdis-07-14-0727-re

Zhu, X.-D. (2020) Occurrence and control of common leaf diseases in kiwifruit. *Journal of Agricultural Catastrophology* 10, 7–8, 14.

18 The Bacterial Canker of Kiwifruit Caused by *Pseudomonas syringae* pv. *actinidiae*

Marco Scortichini[1]*, Francesco Spinelli[2] and Matt D. Templeton[3]
[1]Research Centre for Olive, Fruit and Citrus Crops, Rome, Italy; [2]University of Bologna, Bologna, Italy; [3]The New Zealand Institute for Plant and Food Research Ltd, Auckland, New Zealand

18.1 Introduction

Pseudomonas syringae pv. *actinidiae* (Psa) is the causal agent of bacterial canker of kiwifruit, a disease responsible for major economic losses in the kiwifruit industry worldwide. Psa is composed of different populations named 'biovars' that affect, to different extents, the kiwifruit industries of different countries (Table 18.1). The pathogen was first isolated and described in Japan (Takikawa *et al.*, 1989), and subsequently in Italy (Scortichini, 1994), South Korea (Koh *et al.*, 1994) and China (Wang *et al.*, 1992; Liang *et al.*, 2000) in 1989–1992. Psa biovar 1 was found in Japan, China and Italy, but in Italy had a very limited occurrence. Psa biovar 2 was reported solely in South Korea. These biovars, which affected only *Actinidia chinensis* var. *deliciosa* 'Hayward', caused economic losses in the Asian countries (Kim *et al.*, 2017). A very aggressive population of the pathogen, biovar 3, appeared during 2007–2012 in all major areas of kiwifruit cultivation of the world (i.e. Europe, New Zealand, Asia and Chile), and caused severe symptoms and consequent economic losses everywhere (Balestra *et al.*, 2009; Everett *et al.*, 2011; Vanneste, 2017). Initially, it was yellow-fleshed cultivars (*A. chinensis* var. *chinensis*), 'Hort16A' and 'Jintao', that were mainly affected and, subsequently, green-fleshed cultivars (*A. chinensis* var. *deliciosa*) such as 'Hayward' (Ferrante and Scortichini, 2009; Scortichini *et al.*, 2012). Two years after the first report of the disease, there had been an estimated loss of €60 million and, during 2010–2012, more than 2000 ha of kiwifruit orchards were destroyed or severely damaged by the bacterium in Latina Province, Italy (Donati *et al.*, 2020). In 2012, two years after the first record of the disease in New Zealand, the cost of Psa to the New Zealand kiwifruit industry was estimated at NZ$126 million, with a projected cost of approximately NZ$740–885 million over the subsequent 15 years (Green and Saunders, 2012). In Europe, Psa is currently included in the A2 European and Mediterranean Plant Protection Organization (EPPO) list of pests 'recommended for regulation as quarantine pests'. Therefore, pollen and plant material from third countries must be accompanied by a Psa-free phytosanitary certificate. Psa is a quarantine pathogen for Western Australia (Government of Western Australia, 2007).

*Corresponding author: marco.scortichini@crea.gov.it

© CAB International 2024. *Kiwifruit: Botany, Production and Uses* (eds A.C. Richardson, J.N. Burdon and A.R. Ferguson)
DOI: 10.1079/9781800620933.0018

Table 18.1. Main characteristics of *Pseudomonas syringae* pv. *actinidiae* (Psa) biovars.

Biovar	Occurrence	Virulence	Molecular marker	Detection
Psa 1	Japan, China, Italy	High (Japan, China) Low (Italy)	Phaseolotoxin	Lee *et al.* (2016) Ruinelli *et al.* (2017) Suzaki *et al.* (2022)
Psa 2	South Korea	High	Coronatine	Lee *et al.* (2016) Ruinelli *et al.* (2017)
Psa 3	Worldwide	Very high	Single Nucleotide Polymorphism (SNP)	Gallelli *et al.* (2011, 2014) Ruinelli *et al.* (2017) Andersen *et al.* (2018) Suzaki *et al.* (2022)
Psa 4*	Worldwide	Very low	Effector *hopL1*	Andersen *et al.* (2018)
Psa 5	Japan	Low	Effector *hopL1*	Fujikawa and Sawada (2016)
Psa 6	Japan	Low	Phaseolotoxin Coronatine	Fujikawa and Sawada (2019)

**P. syringae* pv. *actinidifoliorum*

18.2 Symptoms of Bacterial Canker of Kiwifruit

All current *A. chinensis* var. *chinensis* and *A. chinensis* var. *deliciosa* cultivars are considered susceptible, to different extents, to Psa. The symptoms induced by Psa biovar 3 are very similar in all continents and, according to the season, can be summarized as follows:

- spring – bud rot and drop, bud and shoot wilting, leaf spotting, cane shrivel, flower necrosis and drop;
- summer – extensive cane shrivel, initial canker formation on leaders and trunk;
- autumn – reddening of lenticels, extensive canker formation on leaders and trunk;
- winter – extensive oozing from cankers.

However, symptom occurrence, severity and incidence are dependent on climatic conditions and on the timing and frequency of weather events, such as prolonged rain during spring; winter and spring frosts; and hail; all of which favour infection by the pathogen and its spread.

18.3 Origin and Comparative Genomics of *Pseudomonas syringae* pv. *actinidiae*

There is now a comprehensive and ongoing collection of Psa whole-genome sequences isolated from all major kiwifruit-growing regions since the disease was first reported in Japan in 1989 (Takikawa *et al.*, 1989). Included in this collection are the complete sequences of several strains of each of Psa biovars 1, 3 and 5 (Marcelletti *et al.*, 2011; Mazzaglia *et al.*, 2012; Templeton *et al.*, 2015; Poulter *et al.*, 2017, 2018; Firrao *et al.*, 2018; Ho *et al.*, 2019). Collectively, these sequences have given us a wealth of information to investigate the origin, evolution and pathogenicity of Psa. Analysis of single-nucleotide polymorphisms (SNPs) in the core genome of Psa biovars has given insight into the origins of this pathogen. Comparative genomics of 60 isolates of Psa revealed a set of around 4600 core genes common to the four biovars that had been characterized at that time (McCann *et al.*, 2017).

Previous analysis of the topology of a phylogenetic tree generated from SNPs in the core genomes of the different biovars of Psa indicated that each biovar was a distinct sampling from an, as yet, unidentified source population of Psa (McCann *et al.*, 2013). More comprehensive phylogenetic analysis has shown that Korean strains (Psa biovar 2) isolated between 1997 and 2014 and Psa biovar 5 isolated from Japan form the most basal clades of canker-causing Psa. The coexistence of multiple clades in both Japan and South Korea suggests that the source population of all Psa may be in either of those countries rather than in China, where only Psa biovar 3 has so far been found (Ferrante *et al.*, 2015; McCann *et al.*, 2017). The lack of historical isolates makes it difficult to draw more definite conclusions as to the location of the source population and the timing of introductions into other countries.

Within Psa biovar 3, the isolates from China possessed three different versions of an integrative conjugative element (ICE). One of these, PacICE1, was also present in all isolates of the New Zealand strains while a second, PacICE2, was present in all European isolates, and a third, PacICE3, was found in the Chilean isolates. From this, it was deduced that Psa biovar 3 had originated in China (Butler *et al.*, 2013). An extensive sampling of Psa from a wide variety of locations in China followed by an extensive SNP analysis of over 80 isolates provided more conclusive proof that the pandemic Psa biovar 3 originated in China (McCann *et al.*, 2017). The variation in the Psa biovar 3 isolates collected from China indicate that this biovar had been present there for some time and that the pandemic isolates were from a distinct, more recent, sub-clade (McCann *et al.*, 2017).

While Psa biovar 3 originated in China, the epidemiological pathways that led to its introduction via separate routes to Europe, New Zealand and Chile are less well understood. Prior to the Psa pandemic, there was considerable international trade of kiwifruit pollen. This is because high rates of pollination are required for large fruit, and these are more consistently obtained through artificial pollination rather than by use of bees (Hopping, 1986). It has been shown from the analysis of SNPs and other characteristics that pollen, originating from China, Chile and New Zealand, was most probably the source of repeated Psa biovar 3 introductions into South Korea since 2014 (Kim *et al.*, 2020).

In addition to the conserved core genome, Psa possesses a large accessory or flexible genome that varies considerably in size and composition between the five biovars characterized so far and, to a lesser extent, within biovars (McCann *et al.*, 2013, 2017; Fujikawa and Sawada, 2019). The accessory genome contains genes that are not essential for normal growth or primary metabolism, but which might give an isolate a competitive advantage in a particular ecological niche, and hence includes genes involved in pathogenicity. The observed accessory genome variation between Psa biovars and isolates appears to be driven by extensive horizontal transfer of plasmids, ICEs and transposons (McCann *et al.*, 2013; Poulter *et al.*, 2018). More recent on-orchard evolution of Psa biovar 3 by horizontal gene transfer has been highlighted by the rapid assimilation of elements with genes conferring resistance to copper applications (Colombi *et al.*, 2017).

Pathovars (i.e. a set of strains that show restricted pathogenicity to certain host plants) of *P. syringae* cause disease by secreting a collection of proteins, known as effectors, from the bacterial cytoplasm through the host plasma membrane using the Type III secretion system, a specialized dynamic macromolecular structure (Troisfontaines and Cornelis, 2005). The protein effector complements of a *P. syringae* pathogen can be supplemented by various toxic small molecules that often have specific roles in facilitating disease. Psa is a good example of this, where different biovars have various combinations of two well-characterized *P. syringae* toxins: phaseolotoxin and coronatine (Fujikawa and Sawada, 2019).

The specific effectors employed by a particular isolate collectively cause disease and are responsible for the host specificity of a pathovar. Dissection of the role of individual effectors from Psa biovar 3 has commenced. It has been shown that, unlike *P. syringae* pv. *tomato* DC3000 (Pto), only *avrE1* in the conserved effector locus is essential for pathogenicity (Jayaraman *et al.*, 2020). Furthermore, *avrE1* and *hopR1* are additively required for *in planta* growth and lesion development. This is unlike the redundancy described for these effectors in Pto and indicates that these two Psa biovar 3 genes are key determinants for pathogenicity on *A. chinensis* var. *chinensis* (Jayaraman *et al.*, 2020). Unlike diploid cultivars of *A. chinensis* var. *chinensis* that are highly susceptible to Psa biovar 3, other *Actinidia* species, particularly in the *A. arguta* complex, show a high level of resistance to Psa biovar 3 (Datson *et al.*, 2015; Liu *et al.*, 2017). Resistance appears to arise from the hypersensitive response, which implies the induction of effector-triggered immunity (Hemara *et al.*, 2022). Four effectors – *hopAW1a*, *avrRpm1a*, *hopF1c* and *hopZ5a* – could trigger the hypersensitive response in *A. arguta*, implying the presence of at least one major gene for resistance to Psa biovar 3 (Hemara *et al.*, 2022). These genes may have application in breeding resistance to Psa biovar 3 into more susceptible *Actinidia* cultivars. However, a classical gene-to-gene resistance may not account for the qualitative differences in cultivar susceptibility, challenging breeding for resistance.

Indeed, complex characters linked with the plant phenological cycle (e.g. flowering time and duration, bud reactivation after dormancy) and plant morphology (e.g. presence and density of trichomes) may contribute to cultivar susceptibility.

18.4 Isolation and Detection

Psa can be isolated from both symptomatic and asymptomatic plant material using classical or semi-selective culture media. Nutrient sucrose agar and the King's medium B with antibiotics added are commonly used to isolate Psa from symptomatic plant parts as well as from dormant cuttings, nursery plants and pollen (EPPO, 2014; Loreti *et al.*, 2018). In Europe, any first report of pathogen isolation and detection in a new area should carefully follow the official procedures stipulated by EPPO (EPPO, 2014). In parallel, several techniques are currently available for rapid and reliable Psa detection both in symptomatic and asymptomatic plant material. A duplex PCR assay (Gallelli *et al.*, 2011; Loreti *et al.*, 2018) can be used to detect and discriminate Psa from other closely related phytopathogens, namely *P. s.* pv. *theae* and *P. s.* pv. *tomato*, both from plant extracts and pure bacterial cultures. In addition, other techniques can be applied for rapid and reliable Psa detection and discrimination of Psa biovars. Psa biovars 1 and 2 can be detected and distinguished by following the procedures of Lee *et al.* (2016), based on specific markers and PCR analysis. The pandemic Psa biovar 3 can be detected by applying a real-time PCR coupled with a qualitative PCR based on the *hrpW* gene (Gallelli *et al.*, 2014; Loreti *et al.*, 2018). In New Zealand, a real-time quantitative PCR assay based on the *hopA1* gene, which has undergone a rearrangement unique to Psa biovar 3, is used routinely (Andersen *et al.*, 2018). In South Korea, a multiplex PCR assay is used for detecting Psa biovar 3 (Koh *et al.*, 2014). In addition, two loop-mediated isothermal amplification (LAMP) assays can detect and distinguish Psa biovars 1, 2 and 3 from symptomatic and asymptomatic naturally infected material (Ruinelli *et al.*, 2017). Another LAMP method, based on specific biovar effectors, enables the detection and differentiation of Psa biovars 1 and 3 (Suzaki *et al.*, 2022). *P. s.* pv. *actinidifoliorum*, a pseudomonad closely related to the Psa biovars, is often also isolated from kiwifruit specimens. This bacterium, previously named Psa biovar 4, causes leaf spotting that can be confused with that caused by Psa. A qPCR technique, based on whole-genome analyses of the strains and specific pathovar effectors, can be used to detect and differentiate these phytopathogens directly from plant material (Andersen *et al.*, 2018). Psa biovars 5 and 6 can be detected and differentiated from the other biovars by PCR analyses based on primers targeting specific regions of the bacterial genome (Fujikawa and Sawada, 2016, 2019).

18.5 Infection Process

Psa can epiphytically colonize several plant organs including leaves, flowers, anthers, pollen and buds. Once the Psa epiphytic population reaches the infection threshold, which ranges between 10^3 and 2.5×10^4 colony-forming units (CFU) g^{-1} depending on cultivar susceptibility, endophytic colonization occurs (Donati *et al.*, 2020). Among the major commercial cultivars, male cultivars and many *A. chinensis* var. *chinensis* cultivars have the lowest infection threshold (~10^3 CFU), indicating a high susceptibility. In contrast, 'Hayward' and *A. chinensis* var. *chinensis* 'Zesy002' have the lowest susceptibility with infection thresholds of 2.5 and 1.6×10^4 CFU g^{-1}, respectively.

The rate of epiphytic growth is influenced by environmental parameters. The rate peaks at 20°C and 60% relative humidity, conditions often encountered in spring and autumn (Fig. 18.1). Cankers exuding bacterial ooze during the end of winter are considered the primary source of inoculum for the disease cycle (Fig. 18.2). Epiphytic growth occurs both on the nutrient-deficient environment provided by leaves (Fig. 18.3) and the nutrient-rich environment provided by flower stigmata (Fig. 18.4). In the hostile leaf habitat, Psa aggregates in micro-colonies embedded in a biofilm matrix at the base of trichomes and around stomata to better resist environmental stresses such as desiccation and UV radiation (Fig. 18.3E). Trichomes offer optimal growth conditions owing to their ability to retain water and secrete a variety of organic

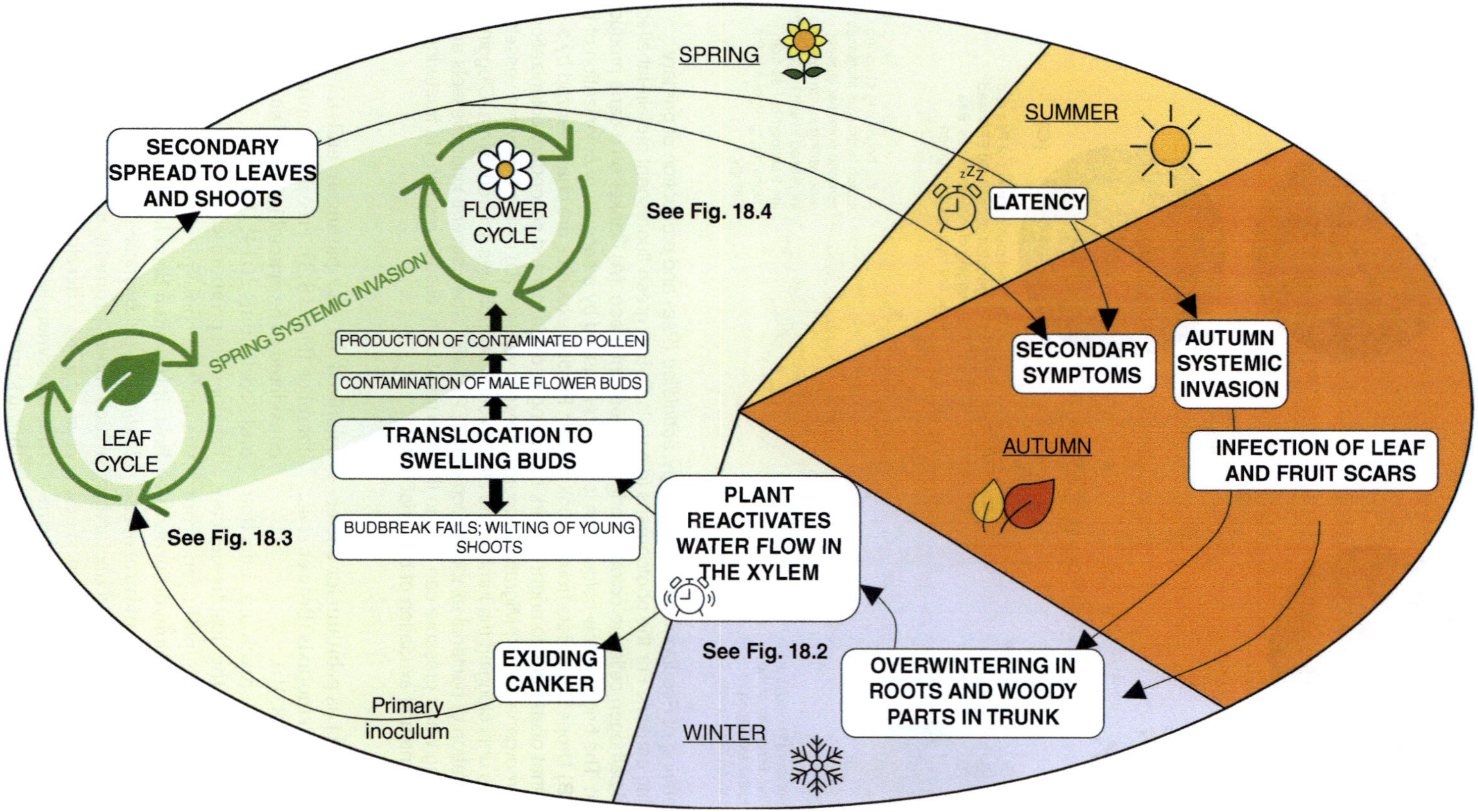

Fig. 18.1. Epidemiological cycle of bacterial canker of kiwifruit caused by *Pseudomonas syringae* pv. *actinidiae* (Psa). Colours refer to winter (blue), spring (green), summer (yellow) and autumn (orange). The production of the primary inoculum via exuding cankers is described in Fig. 18.2. Descriptions of leaf and flower cycles are provided in Figs 18.3 and 18.4, respectively. For further details see Donati *et al.* (2020).

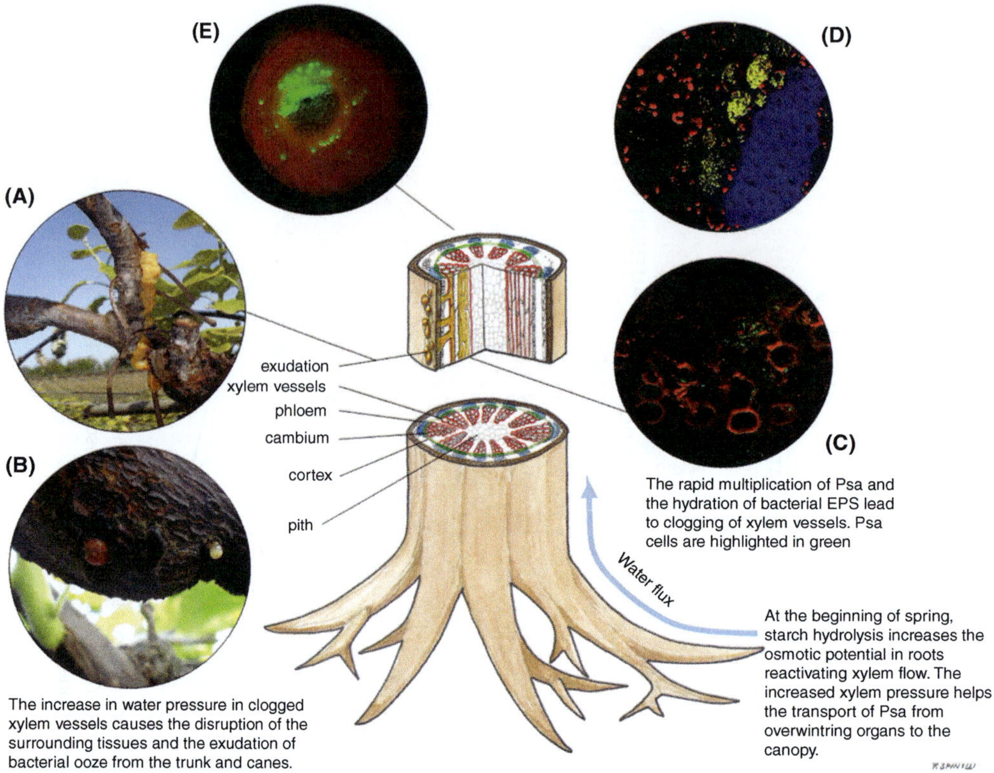

Fig. 18.2. Overwintering of *Pseudomonas syringae* pv. *actinidiae* (Psa) and production of primary inoculum via exuding cankers. For the photographs in panels C–E, a green fluorescent protein-labelled Psa was used, and pathogen cells are coloured bright fluorescent green. (**A**) Canker on a cane producing abundant exudation. The bright yellow colouration is probably caused by a secondary colonization of the exudate by yeast. (**B**) Drops of exudate from a trunk canker. Pristine exudates are characterized by a whitish colouration that often turns to bright or dark red owing to oxidation and secondary colonization by yeast and other microorganisms. (**C**) Migration occurs via xylem vessels. (**D**) Inside the xylem vessels, Psa rapidly multiplies until clogging the lumen. The blue colouration indicates sclerenchyma. Clogging is facilitated by the hydration of bacterial exopolysaccharide (EPS). (**E**) When several xylem vessels are clogged, the increasing water pressure can lead to the rupture of plant tissues resulting in fracturing of the cortex and rhytidome and exudation of bacterial ooze.

compounds that may act as pabulum for bacterial growth or activate chemotaxis (Brewer *et al.*, 1991; Kim, 2019) (Fig. 18.3D).

The need for Psa to reach an infection threshold before invasion of the host tissues can explain the isolation of the pathogen from asymptomatic organs. After epiphytic growth, Psa can invade the host tissues via natural openings such as stomata (Fig. 18.3C) and nectarthodes or wounds such as leaf abscission scars, pedicel end damage after fruit harvest and pruning cuts. Leaf wetness is essential for infection, providing a continuum between the leaf surface colonized by the pathogen and the stomatal chamber (Fig. 18.3). Freezing wounds are also considered to be crucial infection sites (Ferrante and Scortichini, 2014; Gao *et al.*, 2016). While Psa does not promote ice nucleation on plant tissue and the consequent penetration within the host, it has often been found in association with *P. syringae* pv. *syringae* (Purahong *et al.*, 2018), which does promote ice nucleation (Lindow *et al.*, 1978). Therefore, it has been suggested that the synergistic co-presence of the two pathovars may facilitate host invasion in early spring (Donati *et al.*, 2020).

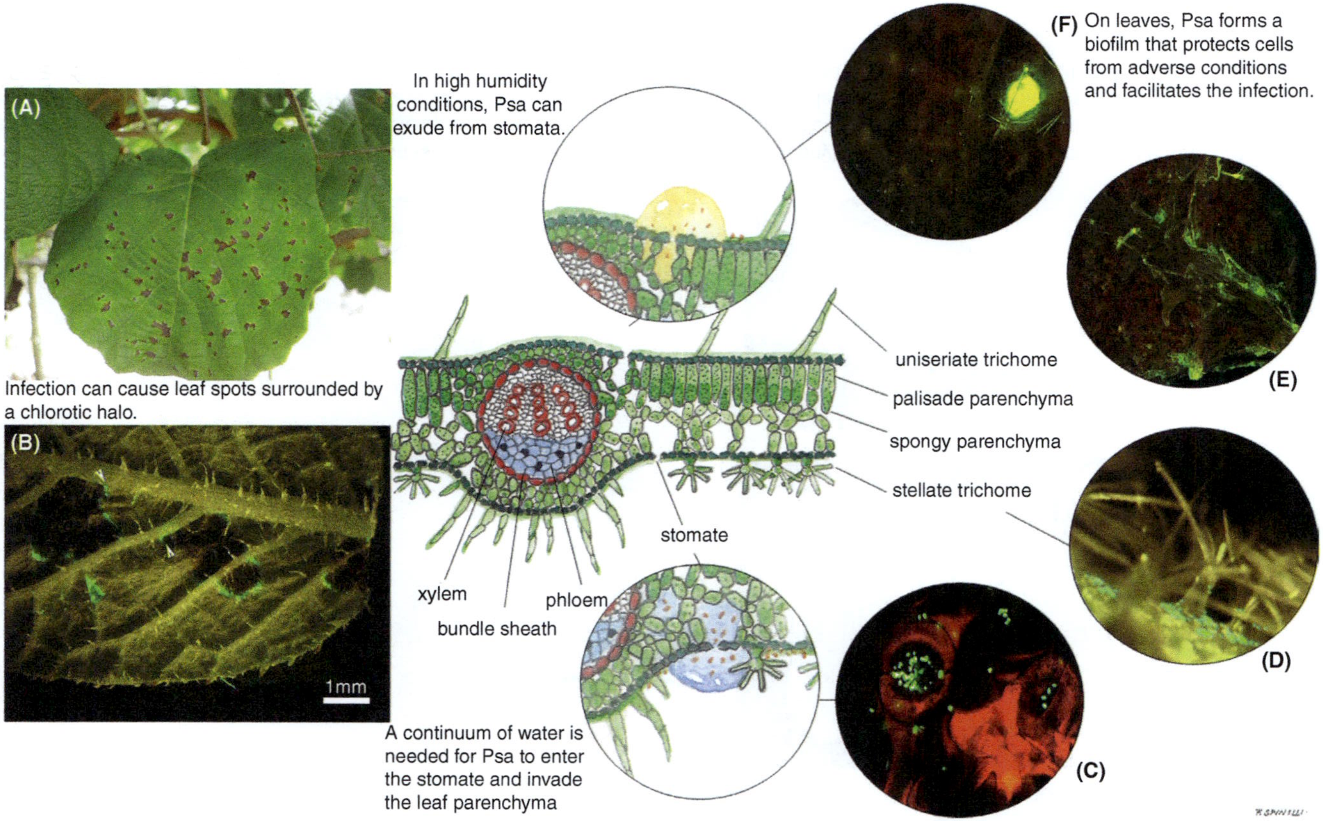

Fig. 18.3. Leaf cycle. During spring, *Pseudomonas syringae* pv. *actinidiae* (Psa) colonizes leaves. For the photographs in panels B–F, a green fluorescent protein (GFP)-labelled Psa was used, and pathogen cells are coloured bright fluorescent green. (**A**) Infection on leaves. (**B**) Photograph of a young *Actinidia chinensis* var. *deliciosa* abaxial leaf lamina taken under UV illumination to detect GFP-labelled Psa. Each spot is surrounded by a fluorescent green border owing to a high Psa population, suggesting that necrosis is caused by the endophytic colonization of the leaf parenchyma. When the colonization reaches a vein (white arrows) infection becomes systemic. Psa is unable to penetrate the leaf cuticle, thus infection occurs via stomata, broken trichomes or wounds. Before infection, Psa multiplies epiphytically. For both multiplication and host invasion, leaf wetness is required. A water film creates a continuum from leaf surface to the stomatal chamber and parenchyma, allowing Psa migration. Psa is able to perceive plant molecules driving the chemiotactic migration toward stomata (Cellini *et al.*, 2022). (**C**) Confocal laser-scanning microscope photograph of stomata colonized by GFP-labelled Psa. (**D**) Trichomes provide a protected environment for epiphytic growth of Psa. (**E**) During epiphytic growth, Psa cells are embedded in a biofilm structure formed by bacterial exopolysaccharide that protects the pathogen from abiotic (UV, desiccation) and biotic (bacteriophages) stress. (**F**) Under high humidity, Psa can exude from stomata producing hyaline drops that may facilitate Psa spread to healthy leaves.

Fig. 18.4. Flower cycle. During spring, *Pseudomonas syringae* pv. *actinidiae* (Psa) can endophytically infect male and female flower buds. Infected male flowers produce infected pollen grains that may facilitate the spread of the pathogen. Infected female flowers do not produce fruit and undergo rapid decay. In all microscopy photographs, a green fluorescent protein-labelled Psa strain was used and Psa cells are coloured bright fluorescent green. (**A**) Symptomatic flower cluster. (**B**) Symptomatic flower buds with browning sepals. (**C**) Infected ovary with Psa colonizing the base of the pistil. (**D**) Infected pollen grains. (**E**) Infected dehiscent anther. (**F, G**) Infected pistil with Psa colonizing the stigma and style. (**H**) Flower infection leading to a systemic invasion of the host plant via the flower pedicel. For further details see Donati *et al*. (2018).

In female flowers, Psa rapidly multiplies on the stigmata (Fig. 18.4F, G). Infection can occur directly in the stigmata or, after migration along the style, in the nectarthodes. Psa can endophytically migrate from the leaves and flowers to shoots and canes via the apoplast (Spinelli *et al.*, 2011; Petriccione *et al.*, 2014). Infection can lead to systemic invasion and to symptoms throughout the vine, including organs distant from the initial infection site, and eventually to plant death. For example, leaf spots can be either the consequence of a localized infection or the result of secondary spread of the pathogen. This phenomenon has led to frequent misinterpretation of the cause–effect relationship of symptom development, thus complicating the dissection of the bacterial canker epidemiological cycle.

Migration occurs preferentially through xylem vessels, but Psa has also been observed in sieve elements and sclerenchyma fibres (Fig. 18.2D). Migration occurs both acropetally and basipetally. In favourable environmental conditions, above-ground infections can result, in a few months, in complete colonization of the plant from shoot apex to the root system. Despite the plant having different mechanisms to restrict migration, such as the production of tyloses (Renzi *et al.*, 2012), Psa can pass the branching nodes and the graft union. Psa shows an active motility-based chemotaxis driven by the recognition of kiwifruit-specific metabolites that may facilitate the migration (Patel *et al.*, 2014; Cellini *et al.*, 2020). Xylem flow towards transpiring leaves can contribute to acropetal Psa movement. Basipetal migration may occur as mass flow owing to the increase in hydrostatic pressure caused by the production of exopolysaccharides and other water-attracting solutes (Fig. 18.2). The plant apoplast is also effectively colonized by Psa and some plant defence proteins are induced upon colonization (Petriccione *et al.*, 2014). The precise mechanisms regulating host colonization have not been fully elucidated.

18.6 Spread and Vectoring

Psa can be spread inside the orchard by rain-splashing and wind-driven droplets of exudate, contaminated pollen and insect and pollinator activity. Agricultural practices also play an important role in Psa dissemination (Donati *et al.*, 2014). In fact, activities such as pruning and thinning create wounds that can be exploited by Psa to systemically invade the host (Gao *et al.*, 2016; Mauri *et al.*, 2016). Contaminated tools may further facilitate the spread of the pathogen. Sucking insects, such as *Metcalfa pruinosa*, have been demonstrated to vector Psa in controlled conditions (Donati *et al.*, 2017). Furthermore, Psa infection induces changes in volatile organic compound (VOC) emissions that attract *M. pruinosa*, suggesting that this pest may also have a role in Psa dissemination in natural conditions (Donati *et al.*, 2017). Since pollen is often contaminated by Psa and infected staminate flowers produce contaminated pollen (Tontou *et al.*, 2014), honey bees and other pollinators may also disperse the pathogen (Pattemore *et al.*, 2014; Donati *et al.*, 2018). Nonetheless, Psa cannot survive in beehives, thus limiting the importance of these pollinators in the long-range movement of the pathogen.

Long-distance spread is primarily caused by human activities such as the importation and movement of contaminated plant material, including flowers and pollen. However, typhoons and the atmospheric water cycle are also suggested as being responsible for long-distance spread of the pathogen (Scortichini, 2019).

18.7 Disease Cycle

The disease cycle of bacterial canker of kiwifruit is complex owing to the ability of the pathogen to epiphytically colonize, and thereafter infect, several organs at different phenological phases of the host (Fig. 18.1). Moreover, despite overwintering cankers being considered the main source of inoculum fuelling the disease cycle, several other organs, including leaves, can produce bacterial exudates that act as inoculum. Leaf and flower cycles are responsible for spring infections and for the rapid dissemination of the pathogen, while autumn infections occur primarily via leaf abscission scars, pedicel end damage after fruit harvest and lenticels. In early spring, reactivation of plant metabolism driven by the hydrolysis of starch in reserve organs leads to an increase in root water pressure. Under these conditions, Psa exploits the xylem water transport to migrate from the roots and trunk, the main overwintering sites, to swelling buds, causing the failure of budbreak and the wilting of young shoots.

The rise in xylem pressure and rapid pathogen multiplication contribute to the reactivation of cankers and the formation of new ones. Root pressure provides both water and nutrients for Psa growth and this pressure, together with hydration and swelling of the bacterial exopolysaccharides inside xylem vessels, facilitates trunk cracking and bacterial exudation. These exuding cankers are the primary source of inoculum for flower and leaf infections. During the leaf cycle, the pathogen multiplies epiphytically, rapidly reaching the infection threshold. Infection occurs primarily via stomata and broken trichomes (Kim, 2019). Penetration via stomata is facilitated by the production of toxins, such as coronatine, or accessorial virulence factors such as ethylene, which may promote stomatal opening (Cellini *et al.*, 2021). Nonetheless, strains lacking toxins or ethylene production retain a high level of virulence suggesting that these factors may help infection but are not essential. Moreover, leaf spots, under relative humidity exceeding 80%, can also be exudation points acting as a source of inoculum for plant-to-plant dissemination of the pathogen (Fig. 18.3F). In less susceptible cultivars, such as 'Hayward', after penetration via stomata has occurred, the plant can limit Psa migration toward leaf veins by producing reactive oxygen species (ROS), leading to the death of colonized tissues and the accumulation of callose in neighbouring tissues (Cellini *et al.*, 2014).

In the flower cycle, Psa can endophytically migrate to anthers, leading to the production of contaminated pollen (Donati *et al.*, 2018). Pollen-mediated dispersal contributes to a rapid dissemination of the pathogen to pistillate flowers where systemic infection occurs. Leaf and flower infections result in primary symptoms, such as leaf spots and flower necrosis, respectively, but they can also lead to systemic infection and secondary symptoms (i.e. cane dieback and canker on leaders) during the following autumn. In summer, once the average daily temperature rises above 22°C and leaf wetness remains below 4 h per day for at least three consecutive days, the disease enters a latency period. Even though older leaves tend to be less susceptible, summer latency seems loosely correlated with leaf aging (Donati *et al.*, 2020). In fact, at the beginning of autumn, when environmental conditions are again favourable for disease development, secondary symptoms appear, providing the source of inoculum for a secondary infection, which occurs at leaf abscission scars and sites of pedicel end damage after fruit harvest. Both spring and autumn systemic infection allow overwintering of the pathogen in the roots and trunk and, in highly infected vines, in dormant buds.

18.8 Control

Agronomic techniques can promote or reduce Psa virulence. An excessive supply of nitrogen appears to promote higher multiplication, whereas a balanced fertilization that supplies nitrogen, phosphorus and potassium appears to reduce disease severity, while depletion of calcium and manganese increases symptom severity (Mauri *et al.*, 2016). Irrigation affects bacterial canker outcomes: symptom development was promoted in kiwifruit orchards with insufficient water, whereas orchards that received sufficient irrigation to maintain soil field capacity were less prone to infection (Mauri *et al.*, 2016).

Rain and wind are key factors in the spread of Psa within and between orchards. Field trials have shown that screens that protect the plant canopy reduce Psa dispersal (Black *et al.*, 2015). However, covering the orchard with protective structures requires substantial modification of the orchard and high investment, including the introduction of ventilation and irrigation systems, and installation of screens to manage light and heat control. While venting promotes air exchange and reduces the temperature and humidity, orchardists should ensure there is no physical damage to the foliage during ventilation. If traditional overhead irrigation systems are used, plants should be watered in the morning to allow the foliage to dry. Modifying the training system to reduce the woody volume of the plant and enhancing air circulation within the canopy may also improve management of kiwifruit bacterial canker.

Copper-based compounds are traditionally used to protect kiwifruit from Psa. They are mainly used during spring and autumn to reduce plant-organ colonization and protect pedicel and leaf scars after harvest and leaf fall. These compounds are also used after frost, hail and pruning. However, copper-resistant Psa biovar 3 strains have frequently been reported in New Zealand (25% of all Psa strains) (Colombi *et al.*, 2017) since copper-based sprays began to be used extensively after the first record of Psa in the country in 2011. The original Psa population introduced into New Zealand was sensitive to copper and has since acquired the genes encoding copper resistance from other bacterial species (McCann *et al.*, 2013). Chitosan-based compounds (Scortichini, 2014) and resistance inducers may also be integrated into control strategies. Resistance inducers that stimulate the salicylic acid pathway, such as acibenzolar-S-methyl, result in significant reduction of the disease (Cellini *et al.*, 2014). Under orchard conditions, the peak of treatment resistance occurs 21 days after acibenzolar-S-methyl application, but treatment efficacy lasts for more than one month (Cellini *et al.*, 2014). Indeed, the activation of salicylic acid-dependent defences counteracts the mechanisms exploited by Psa to facilitate stomatal opening and host colonization. Biological control is another promising method to reduce Psa infection, especially during flowering when other compounds are not allowed or may cause phytotoxicity (Wicaksono *et al.*, 2018). Among biological control agents, strains of *Lactobacillus plantarum* have been shown to prevent infection by producing antimicrobial compounds that inhibit Psa epiphytic growth (Daranas *et al.*, 2019).

18.9 Future Trends

Bacterial populations evolve rapidly following an evolutionary trajectory that may lead to the development of resistance to antimicrobial substances (e.g. copper, antibiotics) and/or changes in virulence and host range. Therefore, to avoid future pandemics or re-emerging outbreaks, constant monitoring of pathogen population structure and genetics is necessary.

The molecular mechanisms of Psa–kiwifruit interactions still need to be fully elucidated. For example, the identification of susceptibility genes may allow innovative strategies for breeding and control, including the silencing of S-genes or CRISPR/Cas9 genome editing of kiwifruit.

Pathogenic bacteria activate virulence based on their perception of the host and the surrounding microbial community. Deciphering the social behaviours of Psa may open new strategies to prevent the activation of virulence and host infection.

Other important aspects of the prevention of future Psa outbreaks include the harmonization of diagnostic protocols and the development of *in situ*, rapid, inexpensive and user-friendly citizen-based diagnostic tools.

Finally, alternatives to chemical pesticides including, but not limited to, phytobiota manipulation, innovative biomolecules (e.g. organic metal chelates, antimicrobial peptides) and the use of bacteriophages should be investigated.

References

Andersen, M.T., Templeton, M.D., Rees-George, J., Vanneste, J.L., Cornish, D.A. *et al.* (2018) Highly specific assays to detect isolates of *Pseudomonas syringae* pv. *actinidiae* biovar 3 and *Pseudomonas syringae* pv. *actinidifoliorum* directly from plant material. *Plant Pathology* 67, 1220–1230. DOI: 10.1111/ppa.12817

Balestra, G.M., Mazzaglia, A., Quattrucci, A., Renzi, M. and Rossetti, A. (2009) Current status of bacterial canker spread on kiwifruit in Italy. *Australasian Plant Disease Notes* 4, 34–36. DOI: 10.1071/DN09014

Black, M.Z., Casonato, S. and Bent, S. (2015) Opportunities for environmental modification to control *Pseudomonas syringae* pv. *actinidiae* in kiwifruit. *Acta Horticulturae* 1105, 253–259. DOI: 10.17660/ActaHortic.2015.1105.36

Brewer, C.A., Smith, W.K. and Volgemann, T.C. (1991) Functional interaction between leaf trichomes, leaf wettability, and the optical properties of water droplets. *Plant, Cell & Environment* 14, 955–962. DOI: 10.1111/j.1365-3040.1991.tb00965.x

Butler, M.I., Stockwell, P.A., Black, M.A., Day, R.C., Lamont, I.L. *et al.* (2013) *Pseudomonas syringae* pv. *actinidiae* from recent outbreaks of kiwifruit bacterial canker belong to different clones that originated in China. *PLOS ONE* 8, e57464. DOI: 10.1371/journal.pone.0057464

Cellini, A., Fiorentini, L., Buriani, G., Yu, J., Donati, I. *et al.* (2014) Elicitors of the salicylic acid pathway reduce incidence of bacterial canker of kiwifruit caused by *Pseudomonas syringae* pv. *actinidiae*. *Annals of Applied Biology* 165, 441–453. DOI: 10.1111/aab.12150

Cellini, A., Donati, I., Fiorentini, L., Vandelle, E., Polverari, A. *et al.* (2020) *N*-Acyl homoserine lactones and Lux solos regulate social behaviour and virulence of *Pseudomonas syringae* pv. *actinidiae*. *Microbial Ecology* 79, 383–396. DOI: 10.1007/s00248-019-01416-5

Cellini, A., Donati, I., Farneti, B., Khomenko, I., Buriani, G. *et al.* (2021) A breach in plant defences: *Pseudomonas syringae* pv. *actinidiae* targets ethylene signalling to overcome *Actinidia chinensis* pathogen responses. *International Journal of Molecular Sciences* 22, 4375. DOI: 10.3390/ijms22094375

Cellini, A., Buriani, G., Correia, C., Fiorentini, L., Vandelle, E. *et al.* (2022) Host-specific signal perception by PsaR2 LuxR solo induces *Pseudomonas syringae* pv. *actinidiae* virulence traits. *Microbiological Research* 260, 127048. DOI: 10.1016/j.micres.2022.127048

Colombi, E., Straub, C., Künzel, S., Templeton, M.D., McCann, H.C. *et al.* (2017) Evolution of copper resistance in the kiwifruit pathogen *Pseudomonas syringae* pv. *actinidiae* through acquisition of integrative conjugative elements and plasmids. *Environmental Microbiology* 19, 819–832. DOI: 10.1111/1462-2920.13662

Daranas, N., Rosselló, G., Cabrefiga, J., Donati, I., Francés, J. *et al.* (2019) Biological control of bacterial plant diseases with *Lactobacillus plantarum* strains selected for their broad-spectrum activity. *Annals of Applied Biology* 174, 92–105. DOI: 10.1111/aab.12476

Datson, P., Nardozza, S., Manako, K., Herrick, J., Martinez-Sanchez, M. *et al.* (2015) Monitoring the *Actinidia* germplasm for resistance to *Pseudomonas syringae* pv. *actinidiae*. *Acta Horticulturae* 1095, 181–184. DOI: 10.17660/ActaHortic.2015.1095.22

Donati, I., Buriani, G., Cellini, A., Mauri, S., Costa, G. *et al.* (2014) New insights on the bacterial canker of kiwifruit (*Pseudomonas syringae* pv. *actinidiae*). *Journal of Berry Research* 4, 53–67. DOI: 10.3233/JBR-140073

Donati, I., Mauri, S., Buriani, G., Cellini, A. and Spinelli, F. (2017) Role of *Metcalfa pruinosa* as a vector for *Pseudomonas syringae* pv. *actinidiae*. *Plant Pathology Journal* 33, 554–560. DOI: 10.5423/PPJ.OA.04.2017.0074

Donati, I., Cellini, A., Buriani, G., Mauri, S., Kay, C. *et al.* (2018) Pathways of flower infection and pollen-mediated dispersion of *Pseudomonas syringae* pv. *actinidiae*, the causal agent of kiwifruit bacterial canker. *Horticulture Research* 5, 56. DOI: 10.1038/s41438-018-0058-6

Donati, I., Cellini, A., Sangiorgio, D., Vanneste, J.L., Scortichini, M. *et al.* (2020) *Pseudomonas syringae* pv. *actinidiae*: ecology, infection dynamics and disease epidemiology. *Microbial Ecology* 80, 81–102. DOI: 10.1007/s00248-019-01459-8

EPPO (European and Mediterranean Plant Protection Organization) (2014) Diagnostics PM 7/120 (1), *Pseudomonas syringae* pv. *actinidiae*. *OEPP/EPPO Bulletin* 44, 360–375. DOI: 10.1111/epp.12171

Everett, K.R., Taylor, R.K., Romberg, M.K., Rees-George, J., Fullerton, R.A. *et al.* (2011) First report of *Pseudomonas syringae* pv. *actinidiae* causing kiwifruit bacterial canker in New Zealand. *Australasian Plant Disease Notes* 6, 67–71. DOI: 10.1007/s13314-011-0023-9

Ferrante, P. and Scortichini, M. (2009) Identification of *Pseudomonas syringae* pv. *actinidiae* as causal agent of bacterial canker of yellow kiwifruit (*Actinidia chinensis* Planchon) in central Italy. *Journal of Phytopathology* 157, 768–770. DOI: 10.1111/j.1439-0434.2009.01550.x

Ferrante, P. and Scortichini, M. (2014) Frost promotes the pathogenicity of *Pseudomonas syringae* pv. *actinidiae* in *Actinidia chinensis* and *A. deliciosa* plants. *Plant Pathology* 63, 12–19. DOI: 10.1111/ppa.12070

Ferrante, P., Takikawa, Y. and Scortichini, M. (2015) *Pseudomonas syringae* pv. *actinidiae* strains isolated from past and current epidemics to *Actinidia* spp. reveal a diverse population structure of the pathogen. *European Journal of Plant Pathology* 142, 677–689. DOI: 10.1007/s10658-015-0643-6

Firrao, G., Torelli, E., Polano, C., Ferrante, P., Ferrini, F. *et al.* (2018). Genomic structural variations affecting virulence during clonal expansion of *Pseudomonas syringae* pv. *actinidiae* biovar 3 in Europe. *Frontiers in Microbiology* 9, 656. DOI: 10.3389/fmicb.2018.00656

Fujikawa, T. and Sawada, H. (2016) Genome analysis of the kiwifruit canker pathogen *Pseudomonas syringae* pv. *actinidiae* biovar 5. *Scientific Reports* 6, 21399. DOI: 10.1038/srep21399

Fujikawa, T. and Sawada, H. (2019) Genome analysis of *Pseudomonas syringae* pv. *actinidiae* biovar 6, which produces the phytotoxins, phaseolotoxin and coronatine. *Scientific Reports* 9, 3836. DOI: 10.1038/s41598-019-40754-9

Gallelli, A., L'Aurora, A. and Loreti, S. (2011) Gene sequence analysis for the molecular detection of *Pseudomonas syringae* pv. *actinidiae*: developing diagnostic protocols. *Journal of Plant Pathology* 93, 425–435.

Gallelli, A., Talocci, S., Pilotti, M. and Loreti, S. (2014) Real-time and qualitative PCR for detecting *Pseudomonas syringae* pv. *actinidiae* isolates causing recent outbreaks of kiwifruit bacterial canker. *Plant Pathology* 63, 264–276. DOI: 10.1111/ppa.12082

Gao, X., Huang, Q., Zhao, Z., Han, Q., Ke, X. *et al.* (2016) Studies on the infection, colonization and movement of *Pseudomonas syringae* pv. *actinidiae* in kiwifruit tissues using a GFPuv-labeled strain. *PLOS ONE* 11, e0151169. DOI: 10.1371/journal.pone.0151169

Government of Western Australia (2007) *Biosecurity and Agriculture Management Act* 2007. Available at: www.legislation.wa.gov.au/legislation/statutes.nsf/main_mrtitle_2736_homepage.html (accessed 4 October 2022).

Green, G. and Saunders, C. (2012) *The Costs of Psa-V to the New Zealand Kiwifruit Industry and the Wider Community*. Research report no. 327. Agribusiness & Economic Research Unit, Lincoln University, New Zealand.

Hemara, L.M., Jayaraman, J., Sutherland, P.W., Montefiori, M., Arshed, S. *et al.* (2022) Effector loss drives adaptation of *Pseudomonas syringae* pv. *actinidiae* to *Actinidia arguta*. *PLOS Pathology* 18(5), e1010542. DOI: 10.1371/journal.ppat.1010542

Ho, J., Taiaroa, G., Butler, M.I. and Poulter, R.T.M. (2019) The genome sequence of M228, a Chinese isolate of *Pseudomonas syringae* pv. *actinidiae*, illustrates insertion sequence element mobility. *Microbiology Resource Announcements* 8, e01427-18. DOI: 10.1128/MRA.01427-18

Hopping, M.E. (1986) Kiwifruit. In: Monselise, S.P. (ed.) *Handbook of Fruit Set and Development*. CRC Press, Boca Raton, Florida, pp. 217–232.

Jayaraman, J., Yoon, M., Applegate, E.R., Stroud, E.A. and Templeton, M.D. (2020) AvrE1 and HopR1 from *Pseudomonas syringae* pv. *actinidiae* are additively required for full virulence on kiwifruit. *Molecular Plant Pathology* 21, 1467–1480. DOI: 10.1111/mpp.12989

Kim, G.H., Jung, J.S. and Koh, Y.J. (2017) Occurrence and epidemics of bacterial canker of kiwifruit in Korea. *Plant Pathology Journal* 33, 351–361. DOI: 10.5423/PPJ.RW.01.2017.0021

Kim, G.H., Lee, Y.-S., Jung, J.S., Koh, Y.-J., Poulter, R.T.M. *et al.* (2020) Genomic analyses of *Pseudomonas syringae* pv. *actinidiae* isolated in Korea suggest the transfer of the bacterial pathogen via kiwifruit pollen. *Journal of Medical Microbiology* 69, 132–138. DOI: 10.1099/jmm.0.001115

Kim, K.W. (2019) Plant trichomes as microbial habitats and infection sites. *European Journal of Plant Pathology* 154, 157–169. DOI: 10.1007/s10658-018-01656-0

Koh, H.S., Cha, B.J., Chung, H.J. and Lee, D.H. (1994) Outbreak and spread of bacterial canker in kiwifruit. *Korean Journal of Plant Pathology* 10, 68– 72.

Koh, H.S., Kim, G.H., Lee, Y.S., Koh, Y.J. and Jung, J.S. (2014) Molecular characteristics of *Pseudomonas syringae* pv. *actinidiae* strains isolated in Korea and a multiplex PCR assay for haplotype differentiation. *Plant Pathology Journal* 30, 96–101. DOI: 10.5423/PPJ.NT.09.2013.0095

Lee, Y.S., Kim, G.H., Koh, Y.J., Zhuang, Q. and Jung, J.S. (2016) Development of specific markers for identification of biovars 1 and 2 strains of *Pseudomonas syringae* pv. *actinidiae*. *Plant Pathology Journal* 32, 162–167. DOI: 10.5423/PPJ.NT.10.2015.0224

Liang, Y., Zhang, X., Tian, C., Gao, A. and Wang, P. (2000) Pathogenic identification of kiwifruit bacterial canker in Shaanxi. *Journal of Northwest Forestry College* 15, 37–39.

Lindow, S.E., Arny, D.C. and Upper, C.D. (1978) Distribution of ice nucleation-active bacteria on plants in nature. *Applied and Environmental Microbiology* 36, 831–838. DOI: 10.1128/aem.36.6.831-838.1978

Liu, Y., Li, D., Zhang, Q., Song, C., Zhong, C. *et al.* (2017) Rapid radiations of both kiwifruit hybrid lineages and their parents shed light on a two-layer mode of species diversification. *New Phytologist* 215, 877–890. DOI: 10.1111/nph.14607

Loreti, S., Cunty, A., Pucci, N., Chabirand, A., Stefani, E. *et al.* (2018) Performance of diagnostic tests for the detection and identification of *Pseudomonas syringae* pv. *actinidiae* (Psa) from woody samples. *European Journal of Plant Pathology* 152, 657–676. DOI: 10.1007/s10658-018-1509-5

Marcelletti, S., Ferrante, P., Petriccione, M., Firrao, G. and Scortichini, M. (2011) *Pseudomonas syringae* pv. *actinidiae* draft genomes comparison reveal strain-specific features involved in adaptation and virulence to *Actinidia* species. *PLOS ONE* 6, e27297. DOI: 10.1371/journal.pone.0027297

Mauri, S., Cellini, A., Buriani, G., Donati, I., Costa, G. *et al.* (2016) Optimization of cultural practices to reduce the development of *Pseudomonas syringae* pv. *actinidiae*, causal agent of the bacterial canker of kiwifruit. *Journal of Berry Research* 6, 355–371. DOI: 10.3233/JBR-160115

Mazzaglia, A., Studholme, D.J., Taratufolo, M.C., Cai, R., Almeida, N.F. *et al.* (2012) *Pseudomonas syringae* pv. *actinidiae* (PSA) isolates from recent bacterial canker of kiwifruit outbreaks belong to the same genetic lineage. *PLOS ONE* 7, e36518. DOI: 10.1371/journal.pone.0036518

McCann, H.C., Rikkerink, E.H.A., Bertels, F., Fiers, M., Lu, A. *et al.* (2013) Genomic analysis of the kiwifruit pathogen *Pseudomonas syringae* pv. *actinidiae* provides insight into the origins of an emergent plant disease. *PLOS Pathogens* 9, e1003503. DOI: 10.1371/journal.ppat.1003503

McCann, H.C., Li, L., Liu, Y., Li, D., Pan, H. *et al.* (2017) Origin and evolution of the kiwifruit canker pandemic. *Genome Biology and Evolution* 9, 932–944. DOI: 10.1093/gbe/evx055

Patel, H.K., Ferrante, P., Covaceuszach, S., Lamba, D., Scortichini, M. *et al.* (2014) The kiwifruit emerging pathogen *Pseudomonas syringae* pv. *actinidiae* does not produce AHLs but possesses three LuxR solos. *PLOS ONE* 9, e87862. DOI: 10.1371/journal.pone.0087862

Pattemore, D.E., Goodwin, R.M., McBrydie, H.M., Hoyte, S.M. and Vanneste, J.L. (2014) Evidence of the role of honey bees (*Apis mellifera*) as vectors of the bacterial plant pathogen *Pseudomonas syringae*. *Australasian Plant Pathology* 43, 571–575. DOI 10.1007/s13313-014-0306-7

Petriccione, M., Salzano, A.M., Di Cecco, I., Scaloni, A. and Scortichini, M. (2014) Proteomic analysis of the *Actinidia deliciosa* leaf apoplast during biotrophic colonization by *Pseudomonas syringae* pv. *actinidiae*. *Journal of Proteomics* 101, 43–62. DOI: 10.1016/j.jprot.2014.01.030

Poulter, R., Taiaroa, G., Sumpter, N., Stockwell, P. and Butler, M. (2017) Complete genome sequence of the kiwifruit pathogen *Pseudomonas syringae* pv. *actinidiae* biovar 5, originating from Japan. *Genome Announcements* 5, e01409-17. DOI: 10.1128/genomeA .01409-17.

Poulter, R.T.M., Ho, J., Handley, T., Taiaroa, G. and Butler, M.I. (2018) Comparison between complete genomes of an isolate of *Pseudomonas syringae* pv. *actinidiae* from Japan and a New Zealand isolate of the pandemic lineage. *Scientific Reports* 8, 10915. DOI: 10.1038/s41598-018-29261-5

Purahong, W., Orrù, L., Donati, I., Perpetuini, G., Cellini, A. *et al.* (2018) Plant microbiome and its link to plant health: host species, organs and *Pseudomonas syringae* pv. *actinidiae* infection shaping bacterial phyllosphere communities of kiwifruit plants. *Frontiers in Plant Science* 9, 1563. DOI: 10.3389/fpls.2018.01563

Renzi, M., Copini, P., Taddei, A.R., Rosetti, A., Gallipoli, L. *et al.* (2012) Bacterial canker on kiwifruit in Italy: anatomical changes in the wood and in the primary infection sites. *Phytopathology* 102, 827–840. DOI: 10.1094/ PHYTO-02-12-0019-R

Ruinelli, M., Schneeberger, P.H.H., Ferrante, P., Bühlmann, A., Scortichini, M. *et al.* (2017) Comparative genomics-informed design of two LAMP assays for detection of the kiwifruit pathogen *Pseudomonas syringae* pv. *actinidiae* and discrimination of isolates belonging to the pandemic biovar 3. *Plant Pathology* 66, 140–149. DOI: 10.1111/ppa.12551

Scortichini, M. (1994) Occurrence of *Pseudomonas syringae* pv. *actinidiae* on kiwifruit in Italy. *Plant Pathology* 43, 1035–1038. DOI: 10.1111/j.1365-3059.1994.tb01654.x

Scortichini, M. (2014) Field efficacy of chitosan to control *Pseudomonas syringae* pv. *actinidiae*, the causal agent of kiwifruit bacterial canker. *European Journal of Plant Pathology* 140, 887-892. DOI: 10.1007/s10658-014-0515-5

Scortichini, M. (2019) The cycle of disease and population structure of *Pseudomonas syringae* pv. *actinidiae*. *Acta Horticulturae* 1243, 55–57. DOI: 10.17660/ActaHortic.2019.1243.9

Scortichini, M., Marcelletti, S., Ferrante, P., Petriccione, M. and Firrao, G. (2012) *Pseudomonas syringae* pv. *actinidiae*: a re-emerging, multi-faceted, pandemic pathogen. *Molecular Plant Pathology* 13, 631–640. DOI: 10.1111/j.1364-3703.2012.00788.x

Spinelli, F., Donati, I., Vanneste, J.L., Costa, M. and Costa, G. (2011) Real time monitoring of the interactions between *Pseudomonas syringae* pv. *actinidiae* and *Actinidia* species. *Acta Horticulturae* 913, 461–465. DOI: 10.17660/ActaHortic.2011.913.61

Suzaki, K., Sawada, H. and Kisaki, G. (2022) Loop-mediated isothermal amplification of bacterial effector genes to detect *Pseudomonas syringae* pv. *actinidiae* biovars 1 and 3. *Journal of General Plant Pathology* 88, 2–9. DOI: 10.1007/s10327-021-01030-9

Takikawa, Y., Serizawa, S., Ichikawa, T., Tsuyumu, S. and Goto, M. (1989) *Pseudomonas syringae* pv. *actinidiae* pv. nov.: the causal bacterium of canker of kiwifruit in Japan. *Annals of the Phytopathological Society of Japan* 55, 437–444.

Templeton, M.D., Warren, B.A., Andersen, M.T., Rikkerink, E.H.A. and Fineran, P.C. (2015) Complete DNA sequence of *Pseudomonas syringae* pv. *actinidiae*, the causal agent of kiwifruit canker disease. *Genome Announcements* 3, e01054-15. DOI: 10.1128/genomeA.01054-15

Tontou, R., Giovanardi, D. and Stefani, E. (2014) Pollen as a possible pathway for the dissemination of *Pseudomonas syringae* pv. *actinidiae* and bacterial canker of kiwifruit. *Phytopathologia Mediterranea* 53, 333–339. DOI: 10.14601/Phytopathol_Mediterr-13757

Troisfontaines, P. and Cornelis, G.R. (2005) Type III secretion: more systems than you think. *Physiology* 20, 326–339. DOI: 10.1152/physiol.00011.2005

Vanneste, J.L. (2017) The scientific, economic, and social impacts of the New Zealand outbreak of bacterial canker of kiwifruit (*Pseudomonas syringae* pv. *actinidiae*). *Annual Review of Phytopathology* 55, 377–399. DOI: 10.1146/annurev-phyto-080516-035530

Wang, Z., Tang, X. and Liu, S. (1992) Identification of the pathogenic bacterium for bacterial canker on *Actinidia* in Sichuan. *Journal of Southwest Agricultural University* 14(06), 500–503.

Wicaksono, W.A., Jones, E.E., Casonato, S., Monk, J. and Ridgway, H.J. (2018) Biological control of *Pseudomonas syringae* pv. *actinidiae* (Psa), the causal agent of bacterial canker of kiwifruit, using endophytic bacteria recovered from a medicinal plant. *Biological Control* 116, 103–112. DOI: 10.1016/j.biocontrol.2017.03.003

19 Pests and Their Management

David P. Logan[1]*, Qiguo Zhuang[2] and Aldo Pollini[3]
[1]*The New Zealand Institute for Plant and Food Research Ltd, Auckland, New Zealand;* [2]*Sichuan Provincial Academy of Natural Resource Sciences, Chengdu, Sichuan, China;* [3]*Imola, Bologna, Italy*

19.1 Introduction

The objective of this chapter is to review published information on the pests of commercial cultivars of kiwifruit (generally *Actinidia chinensis* var. *chinensis* or *A. chinensis* var. *deliciosa*). The commercial cultivation of *A. chinensis* is relatively recent, with the first orchard established in New Zealand in the 1930s and large-scale production occurring from the late 1970s (Ferguson and Bollard, 1990). Kiwifruit germplasm was exported from China as seeds and the risk of spreading endemic pests is likely to have been minimal. Initially, commercial crops in China and elsewhere were relatively free of pests. However, as plantings increased and vines matured, increasing numbers of pests were reported (Song *et al.*, 1998; Huang, 2016). Du *et al.* (2011) found that arthropod communities differed in kiwifruit orchards of different ages. Orchards that were three years old were characterized by lower diversity and greater variability in arthropod numbers than orchards that were 9 and 12 years old. Increased arthropod incidence on old vines may be related to the increased availability of suitable feeding and egg-laying sites as vines age, the lag time for pest populations to develop and the long generation times of some species. The endemic distribution of *A. chinensis* is relatively extensive in China (Huang and Ferguson, 2001) and it is reasonable to assume that pests on cultivated kiwifruit in China originally came from wild *Actinidia*. However, many pests were first reported from other crops before being reported from kiwifruit and can be considered generalists. Examples include endemic leafrollers in New Zealand, Chile and China. Other pest species have been introduced and are often cosmopolitan and polyphagous species such as some armoured scale insects.

In this review, major pests are those species that have been frequently reported in the literature to damage kiwifruit buds and fruit or are regulated in large markets such as China, Japan and South Korea. Minor pests are those that are not regulated or regulated only in minor markets, are reported to feed on leaves, stems and roots, are only rarely reported, or are reported to have limited impact. The division is somewhat arbitrary as the economic impact of pests is typically hard to quantify (Ireland *et al.*, 2020). While gastropods, nematodes and birds have occasionally been reported as pests of kiwifruit (Grandison, 1990; Ferguson, 1991; Michailides and Morgan, 1996; Nakamura and Tsuchiya, 2000), the focus of this chapter is on arthropods, which make up the overwhelming proportion of pests. Pests are considered separately for

*Corresponding author: david.logan@adelaide.edu.au

© CAB International 2024. *Kiwifruit: Botany, Production and Uses* (eds A.C. Richardson, J.N. Burdon and A.R. Ferguson)

DOI: 10.1079/9781800620933.0019

three major production regions: China (including South Korea and Japan); New Zealand; and Italy and its neighbours. Pests of kiwifruit grown in other significant production regions (Iran, Greece, Chile, Turkey, France, USA and Portugal) are considered together in Section 19.5.

19.2 Pests of Kiwifruit in China

19.2.1 Major pests

White peach scale, *Pseudaulacaspis pentagona* (Targioni-Tozzetti) (Hemiptera: Diaspididae) (Figs 19.1, 19.2A), is a polyphagous species native to Asia, which has spread to western Asia, Europe and North America (Hanks and Denno, 1993; Lu *et al.*, 2020). Its presence on fruit can restrict market access due to unacceptable blemish, and heavy infestations can also reduce plant vigour. In Sichuan Province, *P. pentagona* is a pest of 'Hayward', 'Hongyang' and other locally grown kiwifruit cultivars and has three generations per year (Zhuang *et al.*, 2016), starting in spring, summer and autumn. The life cycle is complex. Fertilized adult females are the overwintering stage and are tolerant of cold temperatures. On 'Hayward' kiwifruit stored near 0°C, adult females retained the ability to reproduce after 128 days and some persisted for over 200 days (Stannard *et al.*, 2019). Females lay at least 100 eggs during a period of 2–4 weeks (Hanks and Denno, 1993). Hatching in Sichuan is relatively synchronized in spring, taking, on average, 11 days (Zhuang *et al.*, 2016). The first instar is a mobile dispersal stage ('crawler'). Female crawlers settle on fruit, stems or wood, insert their stylets and spin a circular scale cap, remaining sessile for the rest of their life. They develop through two more instars to reach sexual maturity when they have dull white or yellow circular scale caps 2–2.5 mm in diameter. Males have narrow elongate scale caps and moult four times before becoming midge-like winged adults that only live for a day (Stoetzel, 1976). Adult females release a pheromone to attract males. In Sichuan, adult males fly during a brief period of 5–7 days (Zhuang *et al.*, 2016). Egg laying begins about three weeks after mating.

P. pentagona is managed by broad-spectrum insecticides (Bazrafshan *et al.*, 2010) or mineral oil (Zhuang *et al.*, 2015). Overwintering scale can be removed by brushing before budbreak in spring (Huang, 2016). Models that predict the phenology of crawlers based on day-degree accumulation have been developed as an aid to timing sprays (Takeda, 2004; Zhuang *et al.*, 2016). Other important scale species are *Aspidiotus destructor* Signoret, *P. cockerelli* (Cooley) (Diaspididae) and *Drosicha corpulenta* (Kuwana) (Monophlebidae) (Huang, 2016).

Stinkbug species (Hemiptera: Pentatomidae) are pests of kiwifruit, particularly in Shaanxi Province (Teulon and Xu, 2018). The two most frequently reported pentatomids in a survey of the Chinese literature were the brown marmorated stink bug, *Halyomorpha halys* (Stål), and the yellow-spotted stink bug, *Erthesina fullo* (Thunberg) (Xu and Teulon, 2019) (Figs 19.1, 19.2B, C). Pentatomids use stylets to feed on phloem, which causes bud and flower drop and fruit deformation. Feeding on fruit pulp reduces fruit quality by leaving bruise marks and white, corky areas, and can reduce storage life (Feng, 2007; Francati *et al.*, 2021).

The brown marmorated stink bug is a polyphagous species endemic to eastern Asia. Its native range is China, South Korea and Japan and it has spread to Europe and North and South America (Hoebeke and Carter, 2003; Lee *et al.*, 2013). Adults are 12–17 mm in length with a brown speckled body (Fig. 19.2B). Annually, there are one or two generations in northeastern China, South Korea and Japan and 4–6 generations in southeastern China (Guangdong Province) (Lee *et al.*, 2013). Males and females use vibrational signals in courtship (Polajnar *et al.*, 2016). Adult males also release volatile compounds during courtship, and nymphs and adults of both sexes produce defence compounds when disturbed. Females lay about 200 eggs, on average, in multiple batches of approximately 28 every 4–5 days (Nielsen *et al.*, 2008; Lee *et al.*, 2013). There are five nymphal instars and development is fastest when temperatures are 25–30°C (Nielsen *et al.*, 2008; Kriticos *et al.*, 2017). In autumn (September–November), adults disperse to cool, dry refugia to overwinter. In some populations, a second generation occurs in late summer and early autumn. Feeding by adults and nymphs occurs in summer and autumn, causing damage to up to 30% of fruit (Teulon and Xu, 2018; Chen *et al.*, 2020). The egg parasitoids *Anastatus* spp. and *Trissolcus* spp. (Hymenoptera: Eupelmidae and Platygastridae, respectively) are the most effective natural enemies of brown marmorated stink bug, with parasitism rates of up to 85% (Lee *et al.*, 2013).

Yellow-spotted stink bug is native to eastern and southern Asia, occurring from Japan to Indonesia and west to India (Mi *et al.*, 2020). Adults are approximately 20 mm long and have a black background colour with yellow spots (Fig. 19.2C). Annually, there are three generations in Yunnan Province in the south, two generations in eastern provinces and one or two generations in northern provinces, such as Shaanxi (Mi *et al.*, 2020). Adult females overwinter under bark, fallen leaves and similar refugia. In mid-spring, females start laying eggs in batches of about 12 and adults of the first generation of the cropping season are present in midsummer. The second generation of adults occurs during autumn (August–October) and can cause major damage to fruit (Feng, 2007; Zhang, 2018). Management of pentatomids includes physical destruction of overwintering adults and the application of insecticides to all stages (Teulon and Xu, 2018).

Larvae of the summerfruit tortrix (*Adoxophyes orana* (Fischer von Röslerstamm)) (Lepidoptera: Tortricidae) damage kiwifruit, particularly smooth-skinned 'Hongyang', by grazing on the skin and burrowing into the pulp (Wang, 2005; Huang, 2016) (Figs 19.1, 19.2D, E, F). Adults have wingspans of 15–22 mm, with the yellow-brown males slightly smaller than the grey-brown females. The species occurs from Europe to Japan and is a primary pest of pome and stonefruit (Milonas and Savopoulou-Soultani, 2004; Park *et al.*, 2008). In Sichuan Province, there may be three or four generations per year (Huang, 2016; Ma *et al.*, 2018). Second or third instar larvae overwinter under bark or on dead branches (Wang, 2005). Spring and summer generations damage young fruit and their flight is monitored with pheromone traps to inform spray timing.

19.2.2 Minor pests

Spotted lanternfly, *Lycorma delicatula* (White) (Figs 19.1, 19.2G) is a planthopper (Hemiptera: Fulgoridae) native to China with a wide host range (Hong and Li, 1994; Du *et al.*, 2011). It also occurs on kiwifruit in South Korea (Park *et al.*, 2009). Adults are 21–27 mm long with grey forewings and distinctive red and black patches on their rear wings. Feeding by both adults and nymphs on phloem tissue can lead to shoot deformation, and the production of honeydew, which results in sooty mould (Feng, 2000). Spotted lanternfly has a one-year life cycle (Lee *et al.*, 2019). In autumn, females lay batches of 30–50 eggs, which are protected by an ootheca, on tree trunks, vine trunks and a wide range of inanimate objects (Barringer *et al.*, 2015; Lee *et al.*, 2019). Eggs hatch in spring and there are four nymphal stages, with new adults first occurring in midsummer. In China, management of spotted lanternfly in kiwifruit includes the use of insecticides and cultural practices such as removal and destruction of egg batches (Mi *et al.*, 2007; Wu, 2012).

Many other insects have been reported as locally or occasionally important pests. Feeding by the hawthorn spider mite, *Amphitetranychus viennensis* Zacher (Acarina: Tetranychidae), a major pest of pome and stonefruit from China and Japan to Europe, can cause leaf and shoot deformation of kiwifruit (Du *et al.*, 2011). At least two species of fruit-piercing moths, *Oraesia excavata* Butler and *Adris tyrannus* Guenée (Lepidoptera: Noctuidae), damage kiwifruit in coastal areas (Huang, 2016). Larvae of a cossid moth, *Cossus cossus* L., and a hepialid moth, *Endoclita excrescens* (Butler), tunnel in wood, causing branch death (Song *et al.*, 1998). New-season canes can be damaged by larvae of the grape clearwing moth, *Paranthrene regalis* (Butler) (Lepidoptera: Sesiidae) (Huang, 2016). Many species of leafhopper (Hemiptera: Cicadellidae) feed on leaf vascular tissues and may cause leaf drop when present at high densities (Huang, 2016). Adult scarabs (Coleoptera: Scarabaeidae) can defoliate new plants and will also damage buds and fruit. There are at least 11 species of scarab, with most belonging to the Rutelinae and Melonthinae, which have leaf-feeding adults and subterranean root-feeding larvae (Song *et al.*, 1997; Huang, 2016). Others are flower scarabs (Cetoniinae), which have adults that are attracted to nectar and damaged fruit and larvae that live in decomposing plant material. Adults

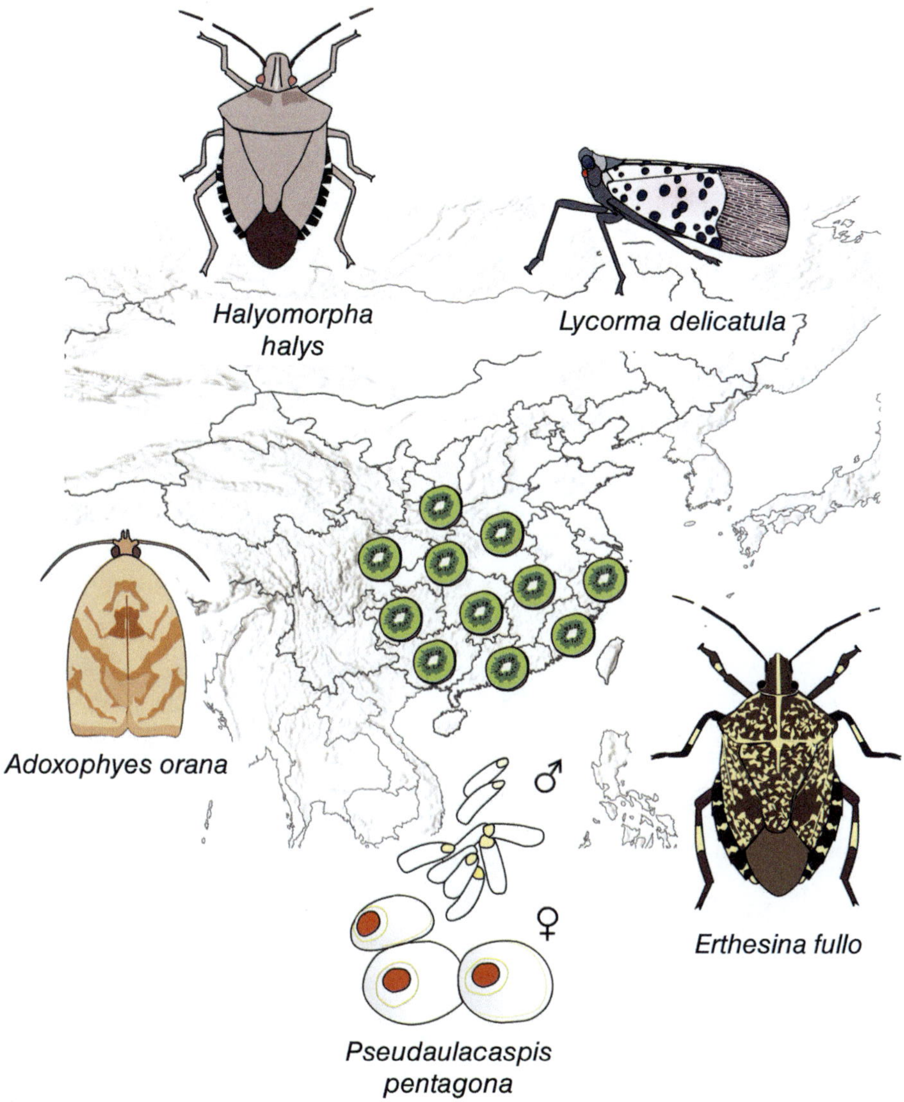

Fig. 19.1. Pests of kiwifruit and areas of major production in China. Images of pests are not drawn in scale to each other. Kiwifruit includes all commercial cultivars, not just those with green flesh. Provincial boundaries are shown for China while other boundaries are national.

and larvae of a minute scavenger beetle, *Cortinicara gibbosa* (Herbst) (Coleoptera: Latridiidae), were reported to burrow into fruit, sometimes causing premature fruit drop (Guo and Shi, 2003; Xu and Teulon, 2019). This is unusual as latridiid beetles generally feed on fungi (Hinton, 1941; Chapman *et al.*, 2016).

19.3 Pests of Kiwifruit in New Zealand

19.3.1 Major pests

The major pests of commercial kiwifruit cultivars grown in New Zealand are an introduced

Fig. 19.2. Pests of kiwifruit grown in China and examples of damage. (**A**) White peach scale, *Pseudaulacaspis pentagona*. (**B**) Brown marmorated stinkbug, *Halyomorpha halys*. (**C**) Yellow-spotted stinkbug, *Erthesina fullo*. (**D**) Adult summerfruit tortrix, *Adoxophyes orana*. (**E**) Dead tortricid larva killed by insecticide, and burrowing damage to the fruit pulp. (**F**) Damage to the fruit epidermis by tortricid larvae. (**G**) Spotted lanternfly, *Lycorma delicatula*. (Photos: A, E, F – Qiguo Zhuang; B – www.shutterstock.com 1863793501; C – New Zealand Institute for Plant and Food Research Ltd.; D – www.shutterstock.com 1157458783; G – www.shutterstock.com 2189428791)

scale insect, endemic leafrollers and an introduced planthopper (Fig. 19.3). These are pests of the green-fleshed cultivar 'Hayward' and yellow-fleshed cultivars 'Zesy002' and 'Hort16A'. The susceptibility of the red-fleshed cultivar 'Zes008' to these pests is not well studied.

Greedy scale, *Hemiberlesia rapax* (Comstock) (Hemiptera: Diaspididae), is a cosmopolitan species with a broad host range (Henderson, 2011) (Fig. 19.4A). In New Zealand kiwifruit, it has two broadly overlapping generations each year, is uniparental and reproduces by parthenogenesis (Steven, 1990; Blank *et al.*, 1996; Hill *et al.*, 2008). All stages feed on parenchyma, and sometimes phloem cells, and occur on fruit, leaves and bark (Mauchline and Hill, 2005). Eggs are laid beneath the scale cap of the mature female scale insect and the mobile first-instar nymphs ('crawlers') escape from under the cap after hatching. Females produced similar numbers of crawlers (94–110) on a range of *Actinidia* genotypes (Hill *et al.*, 2009). Some aerial dispersal occurs (Blank *et al.*, 1990, 1999); however, most crawlers locate new settlement sites by walking along vine trunks and canes (Hill *et al.*, 1986). There they insert their stylets, complete a waxy cap within 24 hours of hatching and remain for the rest of their lives (Hill and Holmes, 2009). Peaks in the number of crawlers occur in November and March (Steven, 1990; Blank *et al.*, 1996; Hill *et al.*, 2008). Crawlers of the second generation settle on fruit, which can result in market access restrictions (Blank *et al.* 1992). First-instar scale moult to second and then third instars, each time adding their cast exuvia to the scale cap and expanding it. There is a pre-reproductive period before third-instar scale insects become reproductively mature and start to lay eggs (Steven, 1990). Generation times were 129–142 days on a range of kiwifruit genotypes including 'Hayward' and 'Hort16A' (Hill *et al.*, 2009). Natural enemies of greedy scale include fungal entomopathogens (*Cosmospora* spp.) (Tyson *et al.*, 2005), parasitoids of early instars (Steven, 1990; Logan and Thomson, 2002), and both specialist and generalist predators (Hill, 1989; Hill *et al.*, 2007a; Logan *et al.*, 2017). The effects of natural enemies are not sufficient to limit scale populations on kiwifruit to the low levels required to meet export requirements (Steven, 1990; Logan *et al.*, 2017).

Management of greedy scale insects is by application of one or two insecticides prior to flowering and one or more applications of mineral oil after flowering (McKenna *et al.*, 2013). Spray application in the post-flowering period depends on the results of monitoring (Stevens *et al.*, 1997) and is limited to a ~1-month period in late summer due to the risk of phytotoxicity to fruit (McKenna *et al.*, 1997).

Endemic leafrollers can damage over half of all fruit when left unmanaged (Steven, 1990). Live caterpillars found on fruit also pose a significant phytosanitary risk for many markets. The most important leafroller pest species is the brown-headed leafroller, *Ctenopseustis obliquana* (Walker) (Lepidoptera: Tortricidae) (Figs 19.3, 19.4B, C). This species is widespread throughout New Zealand and occurs in different habitats from fruit crops to exotic and native forest (Dugdale, 1990; Shaw *et al.*, 1994). Larvae are most common on kiwifruit during summer and can damage fruit from just after fruit set until harvest (Tomkins *et al.*, 1991; Stevens *et al.*, 1995). Eggs are laid in batches of up to 150 on the smooth upper surface of leaves (Steven, 1990). Larvae create secure feeding sites by joining plant parts with silk webbing. They moult four or five times, reaching about 20 mm in length before pupating in the webbing between plant parts or amongst groundcover and leaf litter. The moth is 8–12 mm long with a variable brown wing pattern. There may be between four and six generations per year in northern kiwifruit-growing regions, and three or four generations in the southern-most kiwifruit-growing region of New Zealand (Nelson).

Near harvest time, in late summer and autumn, larvae of a second species, black-lyre leafroller, *Cnephasia jactatana* (Walker) (Lepidoptera: Tortricidae) (Fig. 19.3), may damage fruit, particularly of 'Hayward' (McKenna and Stevens, 2007). There are two generations each year of *C. jactatana* with a large flight of moths in spring and a second, smaller, flight in mid- to late summer. Caterpillars of the summer generation are found on 'Hayward' fruit at harvest. They typically take refuge in the decayed stamen remains and feed on the surrounding skin, sometimes burrowing into the surrounding fruit epidermis (Steven, 1990).

Management of leafrollers is by use of selective insecticides, particularly products

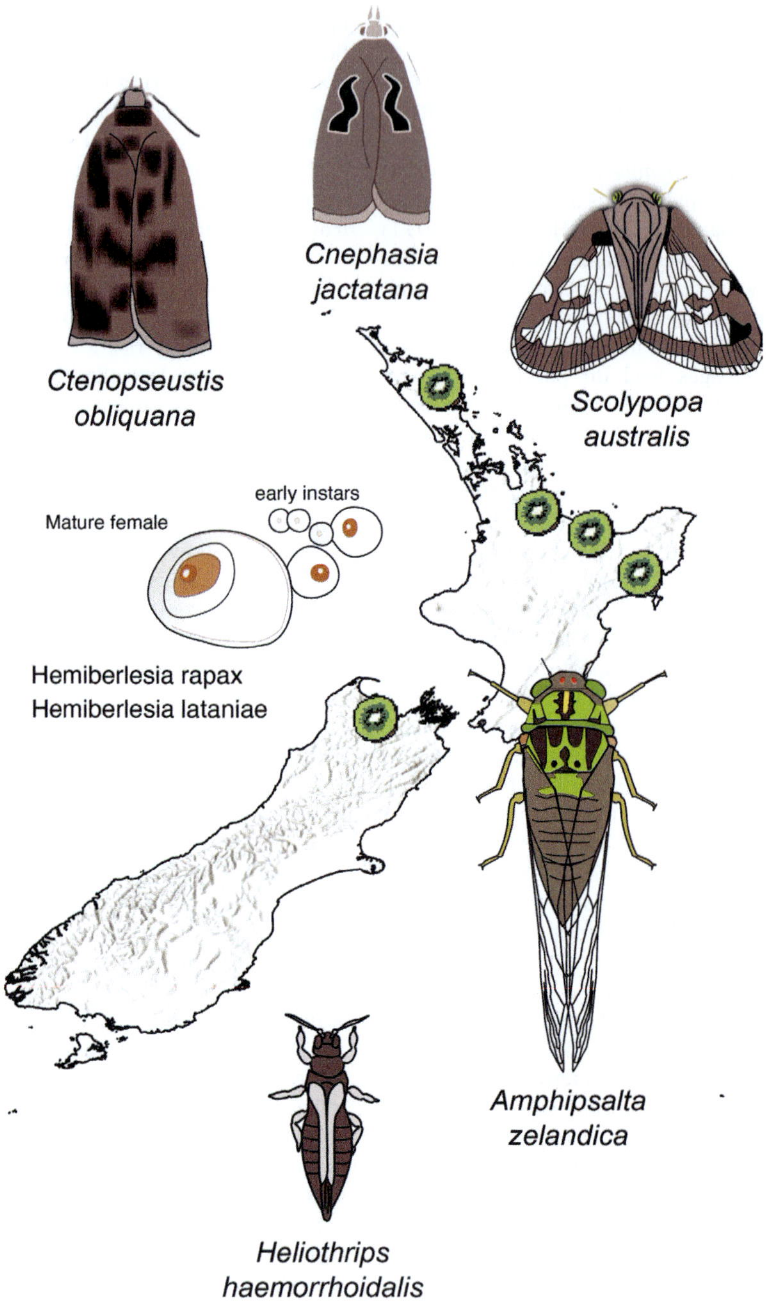

Fig. 19.3. Pests of kiwifruit and areas of major production in New Zealand. Images of pests are not drawn in scale to each other. Kiwifruit includes all commercial cultivars, not just those with green flesh.

Fig. 19.4. Pests of kiwifruit grown in New Zealand. (**A**) Greedy scale, *Hemiberlesia rapax*. (**B**) Brown-headed leafroller moth, *Ctenopseustis obliquana*. (**C**) Mature larva of *C. obliquana*. (**D**) Passionvine hopper, *Scolypopa australis*. (**E**) Sooty moulds growing on fruit. (**F**) Plantain weevil, *Mecinus pascuorum*, on fruit. (**G**) Chorus cicada, *Amphipsalta zelandica*. (Photos: The New Zealand Institute for Plant and Food Research Ltd)

containing *Bacillus thuringiensis* and insect growth regulators. These are applied at the time of fruit set and up to five weeks later, the period of highest risk of damage. Leafroller-related damage is relatively rare on kiwifruit in New Zealand, which may reflect the efficacy of these spray treatments. There may also be some suppression of populations outside kiwifruit orchard blocks by natural enemies such as braconid and ichneumonid parasitoids and tachinids (Steven, 1990; Suckling *et al.*, 1998; Shaw *et al.*, 2001).

Passionvine hopper, *Scolypopa australis* (Walker) (Hemiptera; Ricaniidae) (Figs 19.3, 19.4D), is an introduced pest that was regarded as a minor pest until the 2010s (Steven, 1990; Logan *et al.*, 2020). Both nymphs and adults feed on phloem tissues and excrete honeydew, which is a substrate for sooty moulds (Fig. 19.4E). Fruit with sooty moulds are unsuitable for market and are stripped from vines prior to harvest or removed during postharvest sorting. Passionvine hopper has a broad host range and tends to invade kiwifruit crops from bordering vegetation (Logan *et al.*, 2002). Immigrant adults make up the majority of passionvine hopper in kiwifruit as resident populations tend to be small (Steven, 1990). The life cycle takes one year. Eggs are the longest-lived, and overwintering, stage (Steven, 1990). Females lay at least 100 eggs each, inserting them individually into cavities excavated in thin dead twigs, along the edges of wooden structures and, less often, into growing plant stems and leaf midribs during late summer and autumn (Fletcher, 1979). Nymphs hatch in late spring and aggregate to feed on the tips of actively growing shoots and on senescing leaves during summer. Adults are present from midsummer until late autumn or early winter. Natural enemies, including two egg parasitoids and many generalist predators, are not effective at preventing significant fruit loss (Logan *et al.*, 2020). Management is by a winter spray to the canopy to kill eggs within orchard blocks, and vegetation clearing and insecticide application to kill nymphs on orchard boundaries (Gaskin *et al.*, 2012).

19.3.2 Minor pests

Latania scale, *Hemiberlesia lataniae* (Signoret) (Hemiptera: Diaspididae), is a common species on 'Hayward' and some pollenizer cultivars but not on the yellow-fleshed cultivar 'Hort16A' (Hill *et al.*, 2009). Like greedy scale, latania scale is a cosmopolitan species with a broad host range. The life cycle, feeding behaviour and natural enemy complex of latania scale are very similar to those of greedy scale (Hill *et al.*, 2009). Host plant resistance to latania scale occurs in some yellow-fleshed and green-fleshed selections and the number of quantitative trait loci involved has been investigated (Hill *et al.*, 2015, 2019). Species of scale insects other than *H. lataniae* and *H. rapax* are relatively rare and include other diaspidids such as *Aspidiotus nerii* Bouché and some coccids (Hodgson and Henderson, 2000; Edwards *et al.*, 2008; Keall *et al.*, 2020).

Leafrollers, other than brown-headed leafroller and black-lyre leafroller, occasionally occur on kiwifruit leaves but cause relatively little damage to fruit. They include the endemic green-headed leafrollers, *Planotortrix* spp., a second brown-headed leafroller, *Ctenopseustis herana* (Felder & Rogenhofer), and the introduced light brown apple moth, *Epiphyas postvittana* (Walker) (Lepidoptera: Tortricidae) (Steven, 1990).

Greenhouse thrips, *Heliothrips haemorrhoidalis* (Bouché) (Thysanoptera: Thripidae) (Fig. 19.3), is a cosmopolitan polyphagous species that feeds in colonies on leaves (Martin and Mound, 2004). Damage to kiwifruit leaves can be severe. More importantly, adults and immatures dislodged onto fruit from leaves during harvest can become a market access issue. Densities of greenhouse thrips on kiwifruit are low in summer, increasing in late summer to peak in autumn when extensive damage to leaves can be observed (Logan *et al.*, 2021). As kiwifruit is deciduous, new infestations must establish each spring and summer by invasion from adjacent host plants such as *Cryptomeria japonica* shelter trees. Adults lay eggs individually below the leaf epidermis during a lifetime of up to four months. There are two larval stages that feed on the leaf epidermis, and non-feeding pre-pupal and pupal stages.

Newly emerged adult plantain weevils, *Mecinus pascuorum* (Gyllenhal) (Coleoptera: Curculionidae), are passenger pests on 'Zesy002' kiwifruit (Poulton *et al.*, 2021) (Fig. 19.4F). Plantain weevils are of European origin and have one known host plant, the narrow leaf plantain *Plantago lanceolata* L., which is a component of pastures, waste areas and roadsides (Dickason, 1968). There is one generation each year in New Zealand, with eggs and larvae present during late spring and early summer.

Chorus cicada, *Amphipsalta zelandica* (Boisduval) (Hemiptera: Cicadidae) (Figs 19.3, 19.4G), is an endemic species first regarded as a pest during the early 2000s (Logan and Connolly, 2005). Nymphs and adults feed on the xylem sap of a wide variety of plants. Excreta from feeding adults is associated with sooty mould and adults also damage fruiting canes during egg laying. Kiwifruit orchard blocks adjacent to native forest tend to have the highest densities of chorus cicada (Logan *et al.*, 2011). The entire life cycle takes 3–4 years, with adults living for only two weeks in midsummer (Logan *et al.*, 2014). Batches of, on average, 80 eggs are laid in kiwifruit canes and hatch between late autumn and late spring. The first-instar nymphs fall to the ground where they enter the soil and start feeding on roots. There are five nymphal instars, with the final-instar nymphs emerging from soil to become adults (Logan *et al.*, 2014). Management is based on the application of an insecticide to kill eggs in the bare canopy during winter (Gaskin *et al.*, 2012) and removal of pruned canes from the orchard block to reduce the number of nymphs entering the soil.

Wheat bug, *Nysius huttoni* White (Hemiptera: Lygaeidae), is native to New Zealand. These insects may be incidental passengers on kiwifruit intended for export (Steven, 1990; Giejsztowt *et al.*, 2015). Nymphs and adults are epigeic, feeding mainly on fallen seeds on the ground (Eyles, 1965). Orchard load-out zones, which are open, often gravelled, areas where newly harvested fruit is temporarily stored in bins, are common habitats (Rowe *et al.*, 2014). Good weed control in these areas is recommended to reduce the risk of infestation of bins during harvest.

Newly harvested kiwifruit can host a wide range of passengers including collembola, mites, thrips and small beetles (Stevens and McKenna, 2007; Woolf *et al.*, 2019). Many of these organisms feed on detritus or are predators of detritivores. For example, two species of litter moths, *Stathmopoda horticola* Dugdale and *S. skelloni* (Butler) (Lepidoptera: Oecophoridae), can occasionally occur on fruit. They feed on the dead stamens and styles but do little damage to the fruit itself (Steven, 1990). Postharvest treatment, particularly brushing, removes most of these arthropods from fruit but any that remain may pose market access risks or quality issues for export fruit (Stevens and McKenna, 2007).

19.4 Pests of Kiwifruit in Italy

19.4.1 Major pests

White peach scale is a major pest of kiwifruit in Italy (Hill *et al.*, 2007b) (Figs 19.5, 19.2A). In the past, it was a serious pest of 'Hort16A' in the Latina region of Italy, but it is less problematic for 'Hayward' (Testolin and Ferguson, 2009). Fruit infested with *P. pentagona* are not suitable for sale and can trigger market restrictions on whole fruit lines. Hill *et al.* (2007b) found that an insecticide spray timed to coincide with the release of crawlers from the first generation was as effective as more comprehensive spray programmes that incorporated additional applications against the second and third generations. There are many natural enemies of *P. pentagona*. The parasitoid *Encarsia berlesei* (Howard) (Hymenoptera: Aphelinidae) is widespread in Europe and the Middle East and can achieve up to 35–45% parasitism in kiwifruit orchards (Bianchi *et al.*, 1995; Paloukis *et al.*, 1997).

Brown marmorated stink bug has recently been reported as a pest of kiwifruit in Italy (Francati *et al.*, 2021) (Figs 19.5, 19.2B). There are two generations each year (Costi *et al.*, 2017). Egg batches, nymphs and adults occur on vines of both 'Hayward' and 'Zesy002' kiwifruit from late spring (late May) until early autumn (October) in the Emilia-Romagna region (Francati *et al.*, 2021). In one survey, fruit of unsprayed 'Hayward' and 'Zesy002' vines were damaged from early summer to early autumn (Francati *et al.*, 2021). The incidence of damage averaged 57% and was similar for both cultivars (Francati *et al.*, 2021), consistent with a laboratory study in which adults fed equally on 'Hayward' and 'Zesy002' fruit (Lara *et al.*, 2018). Insecticides are applied during summer and autumn but may not provide effective control (Bariselli *et al.*, 2016), and some orchards use exclusion netting (Candian *et al.*, 2018). Parasitoids native to Europe, the most important of which is considered to be *Anastatus bifasciatus* (Geoffroy) (Hymenoptera: Eupelmidae) (Stahl *et al.* 2019), vary in their effects (Francati *et al.*, 2021). A second species of stink bug, green vegetable bug, *Nezara viridula* (L.) (Hemiptera: Pentatomidae) (Figs 19.5, 19.6A), can damage 'Zesy002' fruit (Ceroni *et al.*, 2004; Francati *et al.*, 2021). It has two

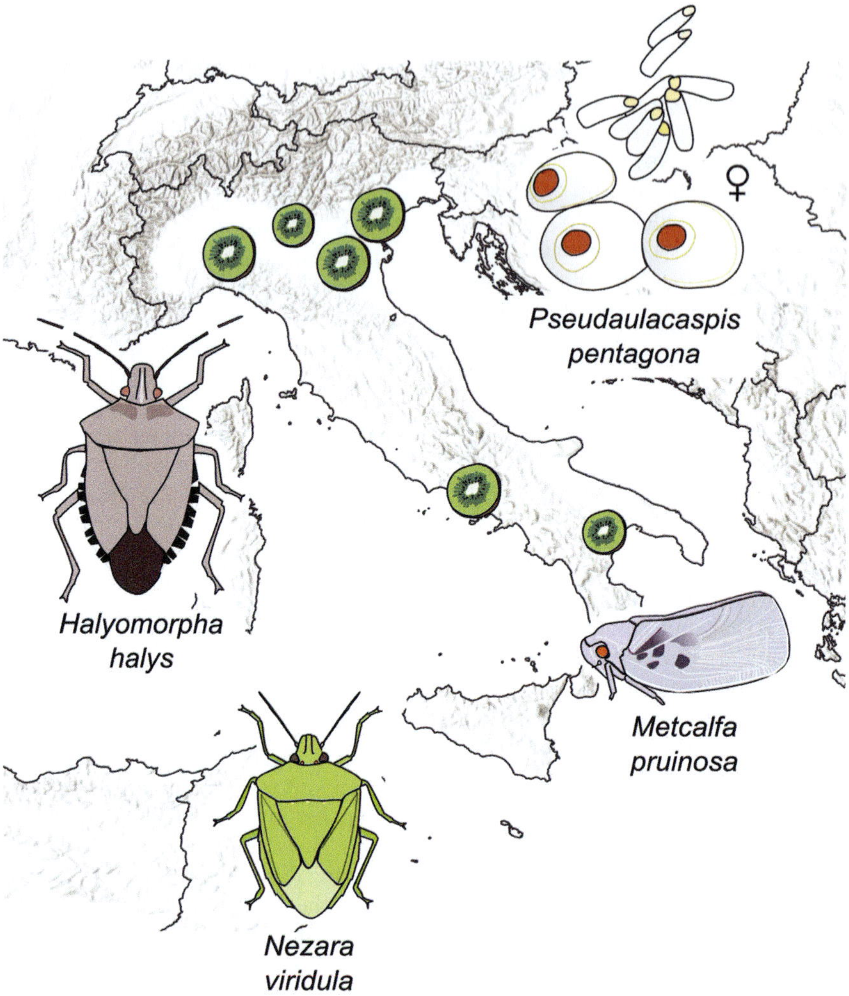

Fig. 19.5. Pests of kiwifruit and areas of major production in Italy. Images of pests are not drawn in scale to each other. Kiwifruit includes all commercial cultivars, not just those with green flesh.

generations each year and may be managed by sprays (Ceroni *et al.*, 2004).

19.4.2 Minor pests

The citrus flatid planthopper, *Metcalfa pruinosa* (Say) (Hemiptera: Flatidae) (Figs 19.5, 19.6B), is a polyphagous phloem-feeding species introduced to Europe from North America (Grozea *et al.*, 2011). It is a potential vector of kiwifruit canker, *Pseudomonas syringae* pv. *actinidiae* (Donati *et al.*, 2017), and honeydew from feeding leads to sooty mould on fruit (Güncan, 2015). In Italy and Greece, there is one generation per year (Tremblay and Priore, 1994; Souliotis *et al.*, 2008). Eggs hatch in spring and early summer, develop through five instars and become adults by mid to late summer (Tremblay and Priore, 1994; Souliotis *et al.*, 2008). Eggs are laid in crevices on tree trunks and are the overwintering stage. A dryinid parasite of nymphs, *Neodryinus typhlocybae* (Ashmead) (Hymenoptera: Dryinidae), provides some control, particularly in

Fig. 19.6. Pests of kiwifruit grown in Italy and other kiwifruit-producing countries. (**A**) Green vegetable bug, *Nezara viridula*. (**B**) Citrus flatid, *Metcalfa pruinosa*. (**C**) Fig wax scale, *Ceroplastes rusci*. (**D**) *Ricania simulans*. (Photos: A, www.shutterstock.com 1853564104; B, www.shutterstock.com 2015084210; C, www.shutterstock.com 1472724908; D, www.shutterstock.com 1677619303)

orchards adjacent to unmanaged areas (Giusti and Rossi, 2002).

The grapevine moth, *Lobesia botrana* Den. et Schiff. (Lepidoptera: Tortricidae), is a polyphagous European species that is a pest of grapevines (Thiéry *et al.*, 2018). It was recorded to damage 1.7% of kiwifruit in a southern Italian orchard adjacent to a vineyard (Moleas, 1988). In the following year there was no damage to kiwifruit but significant (up to 45%) damage to grapes nearby. Greenhouse thrips are present in Italy and can damage kiwifruit leaves (Güncan, 2015; Espinosa *et al.*, 2016). Mediterranean fruit fly, *Ceratitis capitata* (Wiedemann) (Diptera: Tephritidae), damaged 0.6% of mature 'Hayward' kiwifruit in Pisa, Italy, and is also a pest of southern Italian 'Hayward' kiwifruit (Caroli and Loni, 1991). Spotted-wing drosophila, *Drosophila suzukii* (Matsumura) (Diptera: Drosophilidae), is a serious pest of thin-skinned crops such as blueberry, grapes, strawberries and kiwiberry (*Actinidia arguta*) (Walsh *et al.*, 2011). Kiwifruit is attacked but is not a preferred host (Andreazza *et al.*, 2017). When given a choice amongst the fruit of 12 different commercial crops, *D. suzukii* laid few eggs in kiwifruit (Olazcuaga *et al.*, 2019). Survival from egg to adult on a puree made from ripe kiwifruit was relatively high, indicating that traits such as fruit shape and colour, and skin characteristics, are important determinants of susceptibility.

The pest status of other insects is not clear. Fig wax scale, *Ceroplastes rusci* (Linnaeus) (Hemiptera: Coccidae) (Fig. 19.6C), is a widespread, polyphagous species (Pfeiffer, 1997) that can be common in late summer on kiwifruit (Scaltriti and Antonucci, 1982; Testolin, 2015). The leafhopper, *Empoasca vitis* (Göthe) (Hemiptera: Cicadellidae), is a pest of grapevines that can also complete its life cycle on kiwifruit (Moutous and Fos, 1975). Adults and nymphs are phloem feeders and there are three generations per year in northern Italy (Fornasiero *et al.*, 2012). In a pot experiment, feeding by *E. vitis* led to marginal rolling and scorching of leaf margins of kiwifruit (Tavella and Arzone, 1989).

19.5 Other Kiwifruit-producing Countries

White peach scale is recorded on kiwifruit in Greece (Paloukis *et al.*, 1997), Turkey (Güncan, 2015) and Iran (Toorani *et al.*, 2019). The planthoppers *M. pruinosa* and *Ricania simulans* (Walker) (Hemiptera: Ricaniidae) (Fig. 19.6D) are associated with sooty mould on fruit in Turkey. A leafhopper, *Edwardsiana salicicola* (Edwards) (Hemiptera: Cicadellidae), and greenhouse thrips damage kiwifruit leaves in Turkey (Güncan, 2015). Brown marmorated stinkbug is a pest of kiwifruit in Greece (Andreadis *et al.*, 2018) and is present in Turkey (Güncan and Gümüş, 2019).

False spider mite, *Brevipalpus chilensis* (Baker) (Acarina: Tenuipalpidae), is a quarantine pest of kiwifruit exported from Chile (González, 2006). The mites feed on buds, new shoots and leaves and are a primary pest of grapes and some other fruit crops but only occasionally damage kiwifruit (González, 1986, 2006). The leafrollers *Proeulia auraria* (Clarke) and *P. chrysopteris* (Butler) (Lepidoptera: Tortricidae) are endemic to Chile and Argentina where they are pests of kiwifruit and other fruit crops (González, 1986; Flores *et al.*, 2021). Larvae damage the fruit pedicel and epidermis during feeding and all stages are market access pests. Latania and greedy scale are also present on kiwifruit in Chile (Steven *et al.*, 1997).

The South American fruit fly, *Anastrepha fraterculus* (Wiedemann) (Diptera: Tephritidae), occurs from Mexico to Argentina, but not in Chile. There are multiple generations throughout the year and primary hosts include figs, citrus, guava and apple (White and Elson-Harris, 1992). Adults emerged from fallen fruit of 'Hayward' and other green-fleshed cultivars in southern Brazil (Hickel and Schuck, 1993); however, experimental evidence suggested that green-fleshed kiwifruit rarely supported any development of eggs and larvae (Lorscheiter *et al.*, 2012).

19.6 Summary and Challenges for Pest Management

The largest number and diversity of pest species have been reported in China, consistent with this country having the largest area of production and being the origin of *A. chinensis*. The major pests tend to be polyphagous species (Table 19.1) such as diaspidid scale insects, tortricid leafrollers and stinkbugs, and, to a lesser extent, planthoppers.

Meeting regulatory and market requirements for environmental sustainability and food safety while being able to manage current and new pests effectively is a challenge for kiwifruit production. The main tool for managing pests is the application of agrochemicals. For example, in New Zealand, sprays are applied during known high-risk periods or in response to crop monitoring to control scale insects and leafroller caterpillars on kiwifruit under the industry integrated pest management programme, Kiwigreen. However, for many pests there are no registered agrochemicals available and few options for management. There is government regulation which limits the availability of agrochemicals based on agronomic performance, and human and environmental toxicology. Markets impose further restrictions on the use of registered agrochemicals by fixing acceptable maximum residue limits for fruit. It is not economical to register some agrochemicals for use. Kiwifruit is a minor crop on the global scale relative to other crops such as wheat, cotton and soybean and is often not attractive as a market (Lamichhane *et al.*, 2015). These limitations are compounded by a slowing rate of

Table 19.1. Key pests reported from China, New Zealand (NZ), Italy and other significant kiwifruit-growing regions (Iran, Greece, Chile, Turkey, France, USA and Portugal). Pests are included if they were reported to cause production losses through direct damage or fruit blemish, or market access restrictions. Symbols indicate relative importance for each species on kiwifruit: ** major pest; * minor pest; + present but not reported as a pest; – absent.

Higher taxonomy	Species	China	NZ	Italy	Other
Arachnida					
Acarina: Tenuipalpidae	*Brevipalpus chilensis* Baker	–	–	–	**
Insecta					
Hemiptera: Cicadidae	*Amphipsalta zelandica* (Boisduval)	–	*	–	–
Hemiptera: Diaspididae	*Aspidiotus destructor* Signoret	*	–	+	+
	Aspidiotus nerii Bouché	+	*	+	+
	Hemiberlesia lataniae (Signoret)	*	*	*	*
	Hemiberlesia rapax (Comstock)	+	**	*	*
	Pseudaulacaspis cockerelli (Cooley)	*	–	+	+
	Pseudaulacaspis pentagona (Targioni-Tozzetti)	**	–	**	**
Hemiptera: Flatidae	*Metcalfa pruinosa* (Say)	–	–	*	+
Hemiptera: Fulgoridae	*Lycorma delicatula* (White)	*	–	–	+
Hemiptera: Monophlebidae	*Drosicha corpulenta* (Kuwana)	*	–	–	+
Hemiptera: Pentatomidae	*Erthesina fullo* (Thunberg)	**	–	–	–
	Halyomorpha halys (Stål)	**	–	**	*
	Nezara viridula (Linnaeus)	+	*	**	+
Hemiptera: Ricaniidae	*Scolypopa australis* (Walker)	–	**	–	–
Lepidoptera: Tortricidae	*Adoxophyes orana* (Fischer von Röslerstamm)	**	–	+	+
	Cnephasia jactatana (Walker)	–	*	–	–
	Ctenopseustis obliquana (Walker)	–	**	–	–
	Proeulia auraria (Clarke)	–	–	–	*
	Proeulia chrysopteris (Butler)	–	–	–	*
Thysanoptera: Thripidae	*Heliothrips haemorrhoidalis* (Bouché)	+	*	*	*

discovery of novel active ingredients and an increasing development time and cost for agrochemicals (Sparks and Lorsbach, 2017). This is because relatively few new active ingredients can meet the environmental and human health safety requirements while retaining comparable efficacy to existing active ingredients. As a consequence of increasing development costs, there has been a consolidation of companies into a small number of large entities that are more able to sustain an agrochemical development programme (Phillips, 2020). Generic agrochemicals acceptable to regulators are an option (Nishimoto, 2019). Agrochemical companies are likely to continue to supply those that are inexpensive to manufacture and for which key intermediate ingredients are available. Generic agrochemicals may become increasingly important, particularly in developing countries, as the need for food continues to grow. Biopesticides and bioactive natural products remain a strong focus of research but, with a few exceptions, have rarely been able to compete with agrochemicals (Glare and O'Callaghan, 2019). Further investment is needed in understanding pest biology, the impact of pest damage on kiwifruit crops and tools that complement existing agrochemicals. These include cultural approaches such as netting orchard blocks (Candian *et al.*, 2018), selecting for tolerance to pests in breeding programmes (e.g. Hill *et al.*, 2019), and conserving and enhancing existing biological control agents.

A further challenge is the threat posed by invasive pests that can disrupt existing market access and growing systems. Species recorded from kiwifruit that are undergoing range expansion and are potential threats to production in many regions include the brown marmorated stinkbug and the spotted lanternfly. Tephritid fruit flies such as the Mediterranean fruit fly are also of concern, although some cultivars of kiwifruit are relatively poor hosts (Follett *et al.*, 2019).

19.7 Acknowledgements

Thanks to Jessie Bong, Shaun Slattery and Riccardo Spinelli of Zespri International who reviewed the text, and Grace Xu of The New Zealand Institute for Plant and Food Research Ltd who translated papers published in Chinese.

References

Andreadis, S.S., Navrozidis, E.I., Farmakis, A. and Pisalidis, A. (2018) First evidence of *Halyomorpha halys* (Hemiptera: Pentatomidae) infesting kiwi fruit (*Actinidia chinensis*) in Greece 1. *Journal of Entomological Science* 53, 402–405. DOI: 10.18474/JES18-19.1

Andreazza, F., Bernadi, D., Dos Santos, R.S.S., Garcia, F.R.M., Oliveira, E.E. *et al.* (2017) *Drosophila suzukii* in southern neotropical region: current status and future perspectives. *Neotropical Entomology* 46, 591–605. DOI: 10.1007/s13744-017-0554-7

Bariselli, M., Bugiani, R. and Maistrello, L. (2016) Distribution and damage caused by *Halyomorpha halys* in Italy. *OEPP/EPPO Bulletin* 46, 332–334. DOI: 10.1111/epp.12289

Barringer, L.E., Donovall, L.R., Spichiger, S.-E., Lynch, D. and Henry, D. (2015) The first New World record of *Lycorma delicatula* (Insecta: Hemiptera: Fulgoridae). *Entomological News* 125, 20–23. DOI: 10.3157/021.125.0105

Bazrafshan, M., Razmjou, J., Damavandian, M.R. and Dastjerdi, H.R. (2010) Toxicity of several insecticides to white peach scale, *Pseudaulacaspis pentagona* Targioni (Hemiptera: Diaspididae). *Munis Entomology & Zoology* 5(Suppl.), 1020–1024.

Bianchi, A., Guarino, L., Maffeo, E. and Pacchiacucchi, A. (1995) Observations on *Pseudaulacaspis pentagona* and its natural enemies. *L'Informatore Agrario* 51(45), 47–50.

Blank, R.H., Olson, M.H. and Lo, P.L. (1990) Armoured scale (Hemiptera: Diaspididae) aerial invasion into kiwifruit orchards from adjacent host plants. *New Zealand Journal of Crop and Horticultural Science* 18, 81–87. DOI: 10.1080/01140671.1990.10428076

Blank, R.H., Olson, M.H. and Gill, G.S.C. (1992) Armoured scale, *Hemiberlesia lataniae* and *H. rapax* (Hemiptera, Diaspididae), infestation of kiwifruit rejected for export at two packhouses from 1987 to 1991. *New Zealand Journal of Crop and Horticultural Science* 20, 397–405. DOI: 10.1080/01140671.1992.10418056

Blank, R.H., Gill, G.S.C. and Upsdell, M.P. (1996) Greedy scale, *Hemiberlesia rapax* (Hemiptera: Diaspididae), phenology on kiwifruit leaves and wood. *New Zealand Journal of Crop and Horticultural Science* 24, 239–248. DOI: 10.1080/01140671.1996.9513958

Blank, R.H., Gill, G.S.C. and Dow, B.W. (1999) Armoured scale (Hemiptera: Diaspididae) distribution in kiwifruit blocks with reference to shelter. *New Zealand Journal of Crop and Horticultural Science* 27, 1–12. DOI: 10.1080/01140671.1999.9514074

Candian, V., Pansa, M.G., Briano, R., Peano, C., Tedeschi, R. *et al.* (2018) Exclusion nets: a promising tool to prevent *Halyomorpha halys* from damaging nectarines and apples in NW Italy. *Bulletin of Insectology* 71, 21–30.

Caroli, L. and Loni, A. (1991) *Ceratitis capitata* on kiwifruits in Tuscany. *Informatore Fitopatologico* 41(12), 13–16.

Ceroni, M.R., Graziani, S. and Pollini, A. (2004) *Nezara viridula* harmful to *Actinidia*. *Informatore Agrario* 60(44), 37–38.

Chapman, R.B., Marris, J.W.M. and Drummond, J.B. (2016) Survey of insect pests of stored grain in New Zealand. *New Zealand Plant Protection* 69, 285–289. DOI: 10.30843/nzpp.2016.69.5898

Chen, J.-H., Avila, G.A., Zhang, F., Guo, L.F., Sandanayaka, M. *et al.* (2020) Field cage assessment of feeding damage by *Halyomorpha halys* on kiwifruit orchards in China. *Journal of Pest Science* 93, 953–963. DOI: 10.1007/s10340-020-01216-8

Costi, E., Haye, T. and Maistrello, L. (2017) Biological parameters of the invasive brown marmorated stink bug, *Halyomorpha halys*, in southern Europe. *Journal of Pest Science* 90, 1059–1067. DOI: 10.1007/s10340-017-0899-z

Dickason, E.A. (1968) Observations on the biology of *Gymnaetron pascuorum* (Gyll.) (Coleoptera: Curculionidae). *The Coleopterists' Bulletin* 22, 11–15.

Donati, I., Mauri, S., Buriani, G., Cellini, A. and Spinelli, F. (2017) Role of *Metcalfa pruinosa* as a vector for *Pseudomonas syringae* pv. *actinidiae*. *The Plant Pathology Journal* 33, 554–560. DOI: 10.5423/PPJ.OA.04.2017.0074

Du, C., Zhao, H.Y., Gao, H.H., Li, J.F., Luo, K. *et al.* (2011) Arthropod community re-establishment and niche of the main groups in kiwifruit orchards. *Acta Ecologica Sinica* 31, 3246–3254.

Dugdale, J.S. (1990) Reassessment of *Ctenopseustis* Meyrick and *Planotortrix* Dugdale with descriptions of two new genera (Lepidoptera: Tortricidae). *New Zealand Journal of Zoology* 17, 437–465. DOI: 10.1080/03014223.1990.10422943

Edwards, R., Carraher, C., Poulton, J., Sandanayaka, M., Todd, J.H. *et al.* (2008) DNA diagnostics of three armored scale species on kiwifruit in New Zealand. *Journal of Economic Entomology* 101, 1944–1949. DOI: 10.1603/0022-0493-101.6.1944

Espinosa, B., Altieri, A. and Sannino, L. (2016) *Heliothrips haemorrhoidalis* harmful to kiwi in Campania. *Informatore Agrario* 72(45), 58–59.

Eyles, A.C. (1965) Notes on the ecology of *Nysius huttoni* White (Heteroptera: Lygaeidae). *New Zealand Journal of Science* 8, 494–502.

Feng, H. (2000) Anniversary calendar on kiwifruit pest prevention. *Northwest Horticulture* 6, 38–39.

Feng, H. (2007) The occurrence and control of pests in kiwifruit orchards. *Northwest Horticulture* 12, 22.

Ferguson, A.R. (1991) Kiwifruit (*Actinidia*). *Acta Horticulturae* 290, 603–656. DOI: 10.17660/ActaHortic.1991.290.14

Ferguson, A.R. and Bollard, E.G. (1990) Domestication of the kiwifruit. In: Warrington, I.J. and Weston, G.C. (eds) *Kiwifruit: Science and Management*. Ray Richards Publisher in association with the New Zealand Society for Horticultural Science, Auckland, New Zealand, pp. 165–246.

Fletcher, M.J. (1979) Egg types and oviposition behaviour in some Fulgoroid leafhoppers (Homoptera, Fulgoroidea). *The Australian Entomologist* 6, 13–18.

Flores, M.F., Bergmann, J., Ballesteros, C., Arraztio, D. and Curkovic, T. (2021) Development of monitoring and mating disruption against the Chilean leafroller *Proeulia auraria* (Lepidoptera: Tortricidae) in orchards. *Insects* 12, 625. DOI: 10.3390/insects12070625

Follett, P.A., Jamieson, L., Hamilton, L. and Wall, M. (2019) New associations and host status: infestability of kiwifruit by the fruit fly species *Bactrocera dorsalis*, *Zeugodacus cucurbitae*, and *Ceratitis capitata* (Diptera: Tephritidae). *Crop Protection* 115, 113–121. DOI: 10.1016/j.cropo.2018.09.007

Fornasiero, D., Duso, C., Pozzebon, A., Tomasi, D., Gaiotti, F. *et al.* (2012) Effects of irrigation on the seasonal abundance of *Empoasca vitis* in North-Italian vineyards. *Journal of Economic Entomology* 105, 176–185. DOI: 10.1603/EC11218

Francati, S., Masetti, A., Martinelli, R., Mirandola, D., Anteghini, G. *et al.* 2021. *Halyomorpha halys* (Hemiptera: Pentatomidae) on kiwifruit in Northern Italy: phenology, infestation, and natural enemies assessment. *Journal of Economic Entomology* 114, 1733–1742. DOI: 10.1093/jee/toab126

Gaskin, R.E., Logan, D.P., May, W., Rowe, C., Steele, K.D. *et al.* (2012) Control of passionvine hopper and cicada eggs on dormant canes of kiwifruit with bifenthrin sprays containing a new super-penetrant adjuvant. *New Zealand Plant Protection* 65, 100–105. DOI: 10.30843/nzpp.2012.65.5395

Giejsztowt, J., Hill, M.G., Nielsen, M.-C., Davidson, M.M. and Butler, R.C. (2015) *Nysius huttoni* on exported kiwifruit: the role of olfactory cues. *New Zealand Plant Protection* 68, 451. DOI: 10.30843/nzpp.2015.68.5864

Giusti, I. and Rossi, R. (2002) La coltivazione dell'actinidia in agricoltura biologica. *Rivista di Frutticoltura e di Ortofloricoltura* 64(9), 30–35.

Glare, T.R. and O'Callaghan, M. (2019) Microbial biopesticides for control of invertebrates: progress from New Zealand. *Journal of Invertebrate Pathology* 165, 82–88. DOI: 10.1016/j.jip.2017.11.014

González, R.H. (1986) Plagas del kiwi en Chile. *Revista Fruticola* 7(1), 13–27.

González, R.H. (2006) Biology, quarantine risks and control alternatives of the grape flat mite, *Brevipalpus chilensis* Baker (Acarina: Tenuipalpidae). *Revista Fruticola* 27(3), 77–88.

Grandison, G.S. (1990) Nematodes and kiwifruit. In: Warrington, I.J. and Weston, G.C. (eds) *Kiwifruit: Science and Management*. Ray Richards Publisher in association with the New Zealand Society for Horticultural Science, Auckland, New Zealand, pp. 413–419.

Grozea, I., Gogan, A., Virteiu, A-.M., Grozea, A., Stef, R. *et al.* (2011) *Metcalfa pruinosa* Say (Insecta: Homoptera: Flatidae): a new pest in Romania. *African Journal of Agricultural Research* 6, 5870–5877. DOI: 10.5897/AJAR11.478

Güncan, A. (2015) Current status of the kiwifruit pests in Turkey. *Acta Horticulturae* 1096, 371–376. DOI: 10.17660/ActaHortic.2015.1096.43

Güncan, A. and Gümüş, E. (2019) Brown marmorated stink bug, *Halyomorpha halys* (Stål, 1855) (Hemiptera: Heteroptera, Pentatomidae), a new and important pest in Turkey. *Entomological News* 128, 204–210. DOI: 10.3157/021.128.0208

Guo, X. and Shi, X. (2003) The biology and integrated management of important kiwifruit pests in Shaanxi province. *China Fruits* 1, 45–46.

Hanks, L.M. and Denno, R.F. (1993) The white peach scale, *Pseudaulacaspis pentagona* (Targioni-Tozzetti) (Homoptera: Diaspididae): life history in Maryland, host plants, and natural enemies. *Proceedings of the Entomological Society of Washington* 95, 79–98.

Henderson, R.C. (2011) Diaspididae (Insecta: Hemiptera: Coccoidea), *Fauna of New Zealand* 66. Manaaki Whenua Press, Lincoln, Canterbury, New Zealand. DOI: 10.7931/J2/FNZ.66

Hickel, E.R. and Schuck, E. (1993) Ocorrência da mosca-das-frutas, *Anastrepha fraterculus* (Diptera, Tephritidae) em frutos de quivi. *Pesquisa Agropecuária Brasileira* 28, 1345–1347.

Hill, M.G. (1989) Diaspididae, armoured scales (Homoptera). In: Cameron, P.J., Hill, R.L., Bain, J. and Thomas, W.P. (eds) *A Review of Biological Control of Invertebrate Pests and Weeds in New Zealand 1874 to 1987*. CAB International, Institute of Biological Control, Wallingford, UK, pp. 177–182.

Hill, M.G. and Holmes, T. (2009) An analysis of latania scale (*Hemiberlesia lataniae*) crawler settlement behaviour on kiwifruit leaves and bark. *New Zealand Plant Protection* 62, 56–62. DOI: 10.30843/nzpp.2009.62.4785

Hill, M.G., Allan, D.J., McKenna, C. and Steven, D. (1986) Monitoring greedy scale crawler activity on kiwifruit vines. *Proceedings of the New Zealand Weed and Pest Control Conference* 39, 138–142.

Hill, M.G., Charles, J.G., Lupton, T.S., Logan, D.P., Allan, D.J. *et al.* (2007a) Releases from 1994–1997 and recovery of the parasitic mite *Hemisarcoptes coccophagus* Meyer (Acari: Hemisarcoptidae) in Gisborne and Bay of Plenty. *New Zealand Entomologist* 30, 63–70. DOI: 10.1080/00779962.2007.9722152

Hill, M.G., Pollini, A., Ceroni, M.R., Spinelli, R. and Dobson, S.J. (2007b) Controlling white peach scale on 'Hort16A' in Italy. *Acta Horticulturae* 753, 693–702. DOI: 10.17660/ActaHortic.2007.753.91

Hill, M.G., Mauchline, N. and Stannard, K.A. (2008) Predicting armoured scale insect (Homoptera: Diaspididae) phenology on kiwifruit (*Actinidia* sp.). *New Zealand Journal of Crop and Horticultural Science* 36, 253–262. DOI: 10,1080/01140670809510242

Hill, M.G., Mauchline, N.A., Hall, A.J. and Stannard, K.A. (2009) Life table parameters of two armoured scale insect (Hemiptera: Diaspididae) species on resistant and susceptible kiwifruit (*Actinidia* spp.) germplasm. *New Zealand Journal of Crop and Horticultural Science* 37, 335–343. DOI: 10.1080/01140671.2009.9687588

Hill, M.G., Wurms, K.V., Davy, M.W., Gould, E., Allan, A. *et al.* (2015) Transcriptome analysis of kiwifruit (*Actinidia chinensis*) bark in response to armoured scale insect (*Hemiberlesia lataniae*) feeding. *PLOS ONE* 10, e0141664. DOI: 10.1371/journal.pone.0141664

Hill, M.G., Zhuang, Q.G., Mauchline, N.A., Sutherland, P.W., Stannard, K.A. *et al.* (2019) Testing the susceptibility of kiwifruit (*Actinida chinensis* var. *deliciosa*) germplasm to *Hemiberlesia lataniae*. *New Zealand Journal of Crop and Horticultural Science* 47, 1–10. DOI: 10.1080/01140671.2018.1444643

Hinton, H.E. (1941) The Lathridiidae of economic importance. *Bulletin of Entomological Research* 32, 191–247.

Hodgson, C.J. and Henderson, R.C. (2000) Coccidae (Insecta: Hemiptera: Coccoidea). *Fauna of New Zealand* 41. Manaaki Whenua Press, Lincoln, New Zealand. DOI: 10.7931/J2/FNZ.41

Hoebeke, E.R. and Carter, M.E. (2003) *Halyomorpha halys* (Stål) (Heteroptera: Penatatomidae): a polyphagous plant pest from Asia newly detected in North America. *Proceedings of the Entomological Society of Washington* 105, 225–237.

Hong, H.L. and Li, B.B. (1994) Occurrence and control of major insect pests on Chinese kiwi fruit. *Hubei Plant Protection* 3, 17–18.

Huang, H. (2016) *Kiwifruit: The Genus ACTINIDIA*. Elsevier, London.

Huang, H. and Ferguson, A.R. (2001) Review: kiwifruit in China. *New Zealand Journal of Crop and Horticultural Science* 29, 1–14. DOI: 10.1080/01140671.2001.9514154

Ireland, K.B., van Klinken, R., Cook, D.C., Logan, D., Jamieson, L. *et al.* (2020) Plant pest impact metric system (PPIMS): framework and guidelines for a common set of metrics to classify and prioritise plant pests. *Crop Protection* 128, 105003. DOI: 10.1016/j.cropro.2019.105003

Keall, J.B., Hartnett, D.E., McCambridge, J.E., Shaw, N.M., Pearce, B.M. *et al.* (2020) The armoured scale *Oceanaspidiotus spinosus* (Hemiptera: Diaspididae) established in New Zealand. *New Zealand Entomologist* 43, 15–22. DOI: 10.1080/00779962.2020.1736749

Kriticos, D.J., Kean, J.M., Phillips, C.B., Senay, S.D., Acosta, H. *et al.* (2017) The potential global distribution of the brown marmorated stink bug, *Halyomorpha halys*, a critical threat to plant biosecurity. *Journal of Pest Science* 90, 1033–1043. DOI: 10.1007/s10340-017-0869-5

Lamichhane, J.R., Arendse, W., Dachbrodt-Saaydeh, S., Kudsk, P., Roman, J.C. *et al.* (2015) Challenges and opportunities for integrated pest management in Europe: a telling example of minor uses. *Crop Protection* 74, 42–47. DOI: 10.1016/j.cropro.2015.04.005

Lara, J.R., Kamiyama, M., Hernandez, G., Lewis, M. and Hoddle, M.S. (2018) Laboratory assessment of feeding injury and preference of brown marmorated stink bug, *Halyomorpha halys* Stål (Hemiptera: Pentatomidae), for *Actinidia chinensis* var. *deliciosa* 'Hayward' (Zespri® Green) and *Actinidia chinensis* var. *chinensis* 'Zesy002' (Zespri® SunGold). *New Zealand Entomologist* 41, 12–24. DOI: 10.1080/00779962.2018.1438758

Lee, D.H., Short, B.D., Joseph, S.V., Bergh, J.C. and Leskey, T.C. (2013) Review of the biology, ecology, and management of *Halyomorpha halys* (Hemiptera: Pentatomidae) in China, Japan, and the Republic of Korea. *Environmental Entomology* 42, 627–641. DOI: 10.1603/EN13006

Lee, D.H., Park, Y.L. and Leskey, T.C. (2019) A review of biology and management of *Lycorma delicatula* (Hemiptera: Fulgoridae), an emerging global invasive species. *Journal of Asia-Pacific Entomology* 22, 589–596. DOI: 10.1016/j.aspen.2019.03.004

Logan, D. and Connolly, P. (2005) Cicadas from kiwifruit orchards in New Zealand and identification of their final instar exuviae (Cicadidae: Homoptera). *New Zealand Entomologist* 28, 37–48. DOI: 10.1080/00779962.2005.9722684

Logan, D.P. and Thomson, C. (2002) Temperature-dependent development of parasitoids on two species of armoured scale insects. *New Zealand Plant Protection* 55, 361–367. DOI: 10.30843/nzpp.2002.55.3933

Logan, D.P., Allison, P. and Stannard, K. (2002) Selection of wild hosts for feeding by passion vine hopper *Scolypopa australis* (Walker) (Hemiptera: Ricaniidae) in the Bay of Plenty. *New Zealand Plant Protection* 55, 368–373. DOI: 10.30843/nzpp.2002.55.3934

Logan, D.P., Hill, M.G., Connolly, P.G., Maher, B.J. and Dobson, S.J. (2011) Influence of landscape structure on endemic cicadas in New Zealand kiwifruit orchards. *Agricultural and Forest Entomology* 13, 259–271. DOI: 10.1111/j.1461-9563.2010.00517.x

Logan, D.P., Rowe, C.A. and Maher, B.J. (2014) Life-history of chorus cicada, an endemic pest of kiwifruit (Homoptera; Cicadidae). *New Zealand Entomologist* 37, 96–106. DOI: 10.1080/00779962.2014.897302

Logan, D.P., Maher, B. and Rowe, C.A. (2017) Predation of diaspidid scale insects on kiwifruit vines by European earwigs, *Forficula auricularia*, and steel-blue ladybirds, *Halmus chalybeus*. *Biocontrol* 62, 469–479. DOI: 10.1007/s10526-017-9797-9

Logan, D.P., Rowe, C.A. and Connolly, P.G. (2020) Long-term decline in the parasitism rate of passionvine hopper eggs (*Scolypopa australis*). *Biocontrol* 65, 547–558. DOI: 10.1007/s10526-020-10027-w

Logan, D.P., Rowe, C., McKenna, C., Herrick, J. and Rogers, P. (2021) Phenology of greenhouse thrips (*Heliothrips haemorrhoidalis*) on kiwifruit vines, shelter trees and alternative host plants. *New Zealand Plant Protection* 74, 55–61. DOI: 10.30843/nzpp.2021.74.11743

Lorscheiter, R., Redaelli, L.R., Botton, M. and Pimentel, M.Z. (2012) Caracterização de danos causados por *Anastrepha fraterculus* (Wiedemann) (Diptera, Tephritidae) e desenvolvimento larval em frutos de duas cultivares de quivizeiro (*Actinidia* spp.). *Revista Brasileira de Fruticultura* 34, 67–76. DOI: 10.1590/S0100-29452012000100011

Lu, Y., Zhao, Q., Cheng, L., Zhao, L., Zhang, H. *et al.* (2020) The potential global distribution of the white peach scale *Pseudaulacaspis pentagona* (Targioni Tozzetti) under climate change. *Forests* 11(2), 192. DOI: 10.3390/f11020192

Ma, R., Liu, B., Jian, L., Cheng, L., Zhao, J. *et al.* (2018) The important diseases and insect pests of red kiwifruit in different growth stages and their control techniques. *Yunnan Agricultural Science and Technology* 1, 102–103.

Martin, N.A. and Mound, L.A. (2004) Host plants for some New Zealand thrips (Thysanoptera: Terebrantia). *New Zealand Entomologist* 27, 119–123. DOI: 10.1080/00779962.2004.9722133

Mauchline, N.A. and Hill, M.G. (2005) Settlement of armoured scale insects on fruit of commercial *Actinidia* spp. *New Zealand Plant Protection* 58, 294–298. DOI: 10.30843/nzpp.2005.58.4266

McKenna, C. and Stevens, P.S. (2007) A comparison of lepidopteran damage to 'Hort16A' and 'Hayward' kiwifruit. *New Zealand Plant Protection* 60, 254–258. DOI: 10.30843/nzpp.2007.60.4619

McKenna, C., Stevens, P.S. and Steven, D. (1997) Phytotoxicity to kiwifruit from sprays of mineral oil. *Acta Horticulturae* 444, 779–784. DOI: 10.17660/ActaHortic.1997.444.119

McKenna, C., Gaskin, R., Horgan, D., Dobson, S. and Jia, Y. (2013) Efficacy of a postharvest spirotetramat spray against armoured scale insects on kiwifruit vines. *New Zealand Journal of Crop and Horticultural Science* 41, 105–116. DOI: 10.1080/01140671.2013.784710

Mi, H.B., Zhang, G. and Liu, B. (2007) Main technologies of pest management on Shaanxi organic kiwi production. *Academic Periodical of Farm Products Processing* 3, 85–87.

Mi, Q., Zhang, J., Gould, E., Chen, J., Sun, Z. *et al.* (2020) Biology, ecology, and management of *Erthesina fullo* (Hemiptera: Pentatomidae): a review. *Insects* 11, 346. DOI: 10.3390/insects11060346

Michailides, T.J. and Morgan, D.P. (1996) Effect of snail (*Helix aspersa*) damage on Botrytis gray mold caused by *Botrytis cinerea* in kiwifruit. *Plant Disease* 80, 1141–1146. DOI: 10.1094/PD-80-1141

Milonas, P.G. and Savopoulou-Soultani, M. (2004) Diapause termination in overwintering larvae of a Greek strain of *Adoxophyes orana* (Lepidoptera: Tortricidae). *Environmental Entomology* 33, 513–519. DOI: 10.1603/0046-225X-33.3.513

Moleas, T. (1988) *Lobesia botrana* Den. et Schiff. (Tortricidae: Lepidoptera), a potential danger to kiwifruit (*Actinidia chinensis* Planchon). *Informatore Fitopatologico* 38(12), 71–73.

Moutous, G. and Fos, A. (1975) Damage to *Actinidia chinensis* in southwest France by *Empoasca vitis* (Homoptera, Typhlocybidae). *Revue de Zoologie Agricole et de Pathologie Végétale* 74, 43–44.

Nakamura, K. and Tsuchiya, M. (2000) Preventing damage to buds of kiwifruit by brown-eared bulbuls, *Hypsipetes amaurotis*, using distress call. *Japanese Journal of Applied Entomology and Zoology* 44, 27–33. DOI: 10.1303/jjaez.2000.27

Nielsen, A.L., Hamilton, G.C. and Matahda, D. (2008) Developmental rate estimation and life table analysis for *Halyomorpha halys* (Hemiptera: Pentatomidae). *Environmental Entomology* 37, 348–355. DOI: 10.1093/ee/37.2.348

Nishimoto, R. (2019) Global trends in the crop protection industry. *Journal of Pesticide Science* 44, 141–147. DOI: 10.1584/jpestics.D19-101

Olazcuaga, L., Rode, N.O., Foucaud, J., Facon, B., Ravigné, V. *et al.* (2019) Oviposition preference and larval performance of *Drosophila suzukii* (Diptera: Drosophilidae), spotted-wing Drosophila: effects of fruit identity and composition. *Environmental Entomology* 48, 867–881. DOI: 10.1093/ee/nvz062

Paloukis, S.S., Navrozidis, E.I. and Kukuryanis, V.H. (1997) Contribution to the integrated control of *Pseudaulacaspis pentagona* Targ.-Toz. (Homoptera: Diaspididae) on kiwifruit trees (*Actinidia chinensis*). *Acta Horticulturae* 444, 797–802. DOI: 10.17660/ActaHortic.1997.444.122

Park, H., Park, I.J., Lee, S.Y., Han, K.S., Yang, C.Y. *et al.* (2008) Molecular identification of *Adoxophyes orana* complex (Lepidoptera: Tortricidae) in Korea and Japan. *Journal of Asia-Pacific Entomology* 11, 49–52. DOI: 10.1016/j.aspen.2008.04.002

Park, J.-D., Kim, M.-Y., Lee, S.-G., Shin, S.-C., Kim, J.-H. *et al.* (2009) Biological characteristics of *Lycorma delicatula* and the control effects of some insecticides. *Korean Journal of Applied Entomology* 48, 53–57. DOI: 10.5656/KSAE.2009.48.1.053

Pfeiffer, D.G. (1997) Deciduous fruit trees. In: Ben-Dov, Y. and Hodgson, C.J. (eds) *Soft Scale Insects: Their Biology, Natural Enemies and Control*. Elsevier Science Publishers B.V., Amsterdam, pp. 293–322.

Phillips, M.W.A. (2020) Agrochemical industry development, trends in R&D and the impact of regulation. *Pest Management Science* 76, 3348–3356. DOI: 10.1002/ps.5728

Polajnar, J., Maistrello, L., Bertarella, A. and Mazzoni, V. (2016) Vibrational communication of the brown marmorated stink bug (*Halyomorpha halys*). *Physiological Entomology* 41, 249–259. DOI: 10.1111/phen.12150

Poulton, J., Logan, D. and Barraclough, E. (2021) Phenology and ovarian development of the introduced weevil *Mecinus pascuorum*, a passenger pest of kiwifruit in New Zealand. *New Zealand Journal of Crop and Horticultural Science* 49, 294–304. DOI: 10.1080/01140671.2021.1877159

Rowe, C.A., Hill, M.G. and Logan, D.P. (2014) On-orchard habitats for *Nysius huttoni* – an important kiwifruit passenger pest. *New Zealand Plant Protection* 67, 326. DOI: 10.30843/nzpp.2014.67.5769

Scaltriti, G.P. and Antonucci, A. (1982) Note on some insects injurious to *Actinidia* crops. *Informatore Fitopatologico* 32(3), 47–48.

Shaw, P.W., Cruickshank, V.M. and Suckling, D.M. (1994) Geographic changes in leafroller species composition in Nelson orchards. *New Zealand Journal of Zoology* 21, 289–294. DOI: 10.1080/03014223.1994.9517997

Shaw, P.W., Lo, P. and Wallis, D.R. (2001) Recent introduction and establishment of the leafroller parasitoid *Trigonospila brevifacies* (Hardy) (Diptera Tachinidae) in Nelson. *New Zealand Plant Protection* 54, 33–36. DOI: 10.30843/nzpp.2001.54.3735

Song, X., Zheng, W., Zhang, X. and Ma, S. (1997) Investigation of kiwifruit pests and diseases in Shaanxi Province. *Forest Pest and Disease* 4, 17–19.

Song, X., Wang, P., Zhang, X., Ma, S. and Li, Y. (1998) Investigation and analysis on occurrence and damage of diseases and pests on kiwifruit in the Shaanxi province. *Journal of Northwest Forestry College* 13, 79–84.

Souliotis, C., Papanikolaou, N.E., Papachristos, D. and Fatouros, N. (2008) Host plants of the planthopper *Metcalfa pruinosa* (Say) (Hemiptera: Flatidae) and observations on its phenology in Greece. *Hellenic Plant Protection Journal* 1, 39–41.

Sparks, T.C. and Lorsbach, B.A. (2017) Perspectives on the agrochemical industry and agrochemical discovery. *Pest Management Science* 73, 672–677. DOI: 10.1002/ps.4457
Stahl, J.M., Babendreier, D., Marazzi, C., Caruso, S., Costi, E. *et al.* (2019) Can *Anastatus bifasciatus* be used for augmentative biological control of the brown marmorated stink bug in fruit orchards? *Insects* 10, 108. DOI: 10.3390/insects10040108
Stannard, K.A., Hill, M.G., Voscarelli, A., Spinelli, R. and Connolly, P. (2019) Measuring the survival of white peach scale (*Pseudaulacaspis pentagona*) on kiwifruit (*Actinidia chinensis* var. *deliciosa* 'Hayward') in cool storage in Italy. *New Zealand Journal of Crop and Horticultural Science* 47, 198–207. DOI: 10.1080/01140671.2019.1629965
Steven, D. (1990) Entomology and kiwifruit. In: Warrington, I.J. and Weston, G.C. (eds) *Kiwifruit Science and Management*. Ray Richards Publisher in association with the New Zealand Society for Horticultural Science, Auckland, New Zealand, pp. 362–412.
Steven, D., Valenzuela, L. and González, R.H. (1997) Kiwifruit pests in Chile. *Acta Horticulturae* 444, 773–777. DOI: 10.17660/ActaHortic.1997.444.118
Stevens, P.S. and McKenna, C.E. (2007) The effectiveness of two brushing systems for post harvest disinfestation of kiwifruit. *New Zealand Plant Protection* 60, 259–263. DOI: 10.30843/nzpp.2007.60.4620
Stevens, P.S., McKenna, C.E. and Steven, D. (1995) The timing of lepidopteran damage to kiwifruit. *Proceedings of the New Zealand Plant Protection Conference* 48, 130–134.
Stevens, P.S., McKenna, C.E., Blank, R.H., Tomkins, A.R. and Steven, D. (1997) Comparison of armoured scale spray thresholds in kiwifruit. *Proceedings of the New Zealand Plant Protection Conference* 50, 288–292.
Stoetzel, M.B. (1976) Scale-cover formation in the Diaspididae (Homoptera: Coccoidea). *Proceedings of the Entomological Society of Washington* 78, 323–331.
Suckling, D.M., Burnip, G.M., Walker, J.T.S., Shaw, P.W., McLaren, G.F. *et al.* (1998) Abundance of leafrollers and their parasitoids on selected host plants in New Zealand. *New Zealand Journal of Crop and Horticultural Science* 26, 193–203. DOI: 10.1080/01140671.1998.9514055
Takeda, M. (2004) Effects of temperature on oviposition in overwintering females and hatch in first-generation larvae of *Pseudaulacaspis pentagona* (Hemiptera: Diaspididae). *Applied Entomology and Zoology* 39, 15–26. DOI: 10.1303/aez.2004.15
Tavella, L. and Arzone, A. (1989) Feeding punctures of *Empoasca vitis* in *Actinidia*, *Alnus*, *Vitis* (Homoptera, Auchenorrhyncha, Cicadellidae). *Annali della Facoltà di Scienze Agrarie della Università degli Studi di Torino* 15, 69–80.
Testolin, R. (2015) Kiwifruit (*Actinidia* spp.) in Italy: the history of the industry, international scientific cooperation and recent advances in genetics and breeding. *Acta Horticulturae* 1096, 47–61. DOI: 10.17660/ActaHortic.2015.1096.2
Testolin, R. and Ferguson, A.R. (2009) Kiwifruit (*Actinidia* spp.) production and marketing in Italy. *New Zealand Journal of Crop and Horticultural Science* 37, 1–32. DOI: 10.1080/01140670909510246
Teulon, D. and Xu, B. (2018) Biosecurity risks from stink bugs to New Zealand kiwifruit identified in Chinese language literature. *New Zealand Plant Protection* 71, 140–150. DOI: 10.30843/nzpp.2018.71.163
Thiéry, D., Louâpre, P., Muneret, L., Rusch, A., Sentenac, G. *et al.* (2018) Biological protection against grape berry moths: a review. *Agronomy for Sustainable Development* 38, 15. DOI: 10.1007/s13593-018-0493-7
Tomkins, A.R., Steven, D., McKenna, C., Lees, N. and Sale, P. (1991) Monitoring leafroller moths in Bay of Plenty kiwifruit orchards. *Proceedings of the New Zealand Weed and Pest Control Conference* 44, 52–57. DOI: 10.30843/nzpp.1991.44.10803
Toorani, A.H., Abbasipour, H. and Dehghan-Dehnavi, L. (2019) Biodiversity and population fluctuations of parasitoids of the white peach scale, *Pseudaulacaspis pentagona* (Targioni-Tozzetti) (Hemiptera: Diaspididae), in kiwifruit orchards in Northern Iran. *Hellenic Plant Protection Journal* 12, 12–21. DOI: 10.2478/hppj-2019-0003
Tremblay, E. and Priore, R. (1994) *Metcalfa pruinosa* has arrived in Campania. *Informatore Agrario* 50, 69–71.
Tyson, J.L., Henderson, R.C., Fullerton, R.A., Jamieson, L.E. and Froud, K.J. (2005) Distribution and new host records for *Cosmospora aurantiicola* and *Cosmospora flammea*: entomopathogens of Diaspididae in New Zealand. *New Zealand Plant Protection* 58, 283–287. DOI: 10.30843/nzpp.2005.58.4264
Walsh, D.B., Bolda, M.P., Goodhue, R.E., Dreves, A.J., Lee, J. *et al.* (2011) *Drosophila suzukii* (Diptera: Drosophilidae): invasive pest of ripening soft fruit expanding its geographic range and damage potential. *Journal of Integrated Pest Management* 2, 1–7. DOI: 10.1603/IPM10010

Wang, M.-Z. (2005) Study on quality system of Hongyang – pests, diseases and their control techniques. *Resource Development & Market* 21, 443–446.

White, I.M. and Elson-Harris, M.M. (1992) *Fruit Flies of Economic Significance: Their Identification and Bionomics*. CAB International, Wallingford, UK.

Woolf, A.B., Page-Weir, N.E.M., Oliver, R.J., Wood, M., Hawthorne, A.J. *et al.* (2019) Effectiveness of air-knives for pest removal from kiwifruit. *Acta Horticulturae* 1256, 387–392. DOI: 10.17660/ActaHortic.2019.1256.55

Wu, B.G. (2012) Preliminary investigation and control of the pests on kiwi fruit. *Sichuan Agricultural Science & Technology* 7, 45.

Xu, B. and Teulon, D.A.J. (2019) Arthropod pests of kiwifruit identified from Chinese language literature. *New Zealand Plant Protection* 72, 281. DOI: 10.30843/nzpp.2019.72.325

Zhang, Y. (2018) The occurrence of stink bugs on kiwifruits. *Shaanxi Journal of Agricultural Sciences* 64, 27–28.

Zhuang, Q.G., Wang, L.H., Li, M.Z., Hou, T.P. and Xie, Y. (2015) 'EnSpray 99' mineral oils for white peach scale, *Pseudaulacaspis pentagona* and phytotoxicity to 'Hongyang' kiwifruit. *Acta Horticulturae* 1096, 363–369. DOI: 10.17660/ActaHortic.2015.1096.42

Zhuang, Q.G., Hou, T.P. and Hill, M.G. (2016) Predicting white peach scale phenology on kiwifruit in Sichuan. *New Zealand Journal of Crop and Horticultural Science* 44, 218–226. DOI: 10.1080/01140671.2016.1184692

20 Postharvest: Fresh Fruit Harvest, Storage and Supply

Jeremy N. Burdon* **and Ruiling Wang**
The New Zealand Institute for Plant and Food Research Ltd, Auckland, New Zealand

20.1 Introduction

The successful harvest, storage and supply of fresh kiwifruit (*Actinidia* spp.) are dependent on understanding the fruit physiology at harvest and the response of the fruit to handling and storage conditions. The basic premise for commercial kiwifruit postharvest systems is that the fruit is harvested mature (capable of ripening) but unripe and then ripens slowly during storage, or more rapidly at higher temperatures, possibly with the use of ethylene. When in storage for up to six months, fruit is exposed to low temperatures and technologies such as controlled atmospheres (CA) to slow ripening (Burdon and Lallu, 2011). All this must occur without any physical, physiological or pathological damage.

This chapter focuses on the science underpinning the commercial harvest and postharvest management of the main *A. chinensis* var. *deliciosa* and *A. chinensis* var. *chinensis* cultivars. For a discussion of kiwiberries, see Chapter 7.

20.2 Fruit Biology

Key aspects of fruit biology underpinning postharvest behaviour include fruit maturation, softening, sensitivity to ethylene and chilling response. Fruit development, including growth, maturation and ripening, is described in detail in Chapter 16, and commercial implications of maturation have been reviewed by Burdon (2019a). Key elements are only described briefly here.

Kiwifruit is often erroneously described as having climacteric fruit (Kader, 2002), a term which usually means that the fruit ripening is accompanied by elevated respiration and ethylene production. Increased ethylene production in kiwifruit only occurs once fruit is soft (Kim *et al.*, 1999; Ritenour *et al.*, 1999; Richardson *et al.*, 2011), unless damaged in some way (Hyodo and Fukasawa, 1985; Feng *et al.*, 2003a). Thus, kiwifruit ripening (softening, starch breakdown) occurs mostly before an increase in ethylene production. However, despite the lack of ethylene production during the main phase of fruit ripening, the ethylene action inhibitor 1-methylcyclopropene (MCP) can slow softening slightly, suggesting some role for ethylene (Regiroli and Vriends, 2007). Thus, kiwifruit has been described as atypical with reference to a strict climacteric categorization (Antunes, 2007).

While not producing large amounts of ethylene until soft, unripe kiwifruit is sensitive to low concentrations of ethylene (McDonald, 1990; Jabbar and East, 2016). Ethylene acceler-

*Corresponding author: jeremy.burdon@plantandfood.co.nz

© CAB International 2024. *Kiwifruit: Botany, Production and Uses* (eds A.C. Richardson, J.N. Burdon and A.R. Ferguson)
DOI: 10.1079/9781800620933.0020

ates ripening, making ethylene contamination of storage atmospheres a serious commercial risk. Consequently, ethylene monitoring and scrubbing systems are commonly used in kiwifruit stores.

The pattern of kiwifruit softening, on and off the vine, is sigmoidal, with three clearly identifiable phases (Fig. 20.1) (Kempler *et al.*, 1992; MacRae and Redgwell, 1992; Burdon *et al.*, 2017). Phase 1 is slow softening, largely on the vine, phase 2 is rapid softening that occurs as the fruit ripens associated with increased starch breakdown, and phase 3 is slow softening when close to ripe. A fourth softening phase has been identified associated with increased ethylene production and senescence of the fruit (Atkinson *et al.*, 2011). In the postharvest context, the general sigmoidal softening pattern can be altered by the maturity at which fruit is harvested and several aspects of the storage environment (Fig. 20.1).

Kiwifruit is normally stored at ~0°C and at this temperature there is a risk of chilling damage, particularly for fruit harvested at a 'low' maturity (early) when chilling sensitive, and then stored for too long.

Finally, after what may be months in storage, the consumer expects a good eating experience. In addition to all the risks described in this chapter, the basic eating quality of the fruit depends largely on the firmness and sugar/acid balance (see Chapter 21).

20.3 Harvest

20.3.1 Harvest indices

Harvest indices should be simple, reproducible and cheap to measure and should help to ensure the quality of the fruit as far as possible. The New Zealand kiwifruit industry progressed from no harvest indices, to harvesting by date (1960s), to a soluble solids content (SSC)-based index (Harman, 1981) to multiple indices at different stages of development to match specific marketing periods (Burdon, 2019a). The concept of an optimum harvest index can only be applied if the marketing period is defined.

The initial selection of a minimum 6.2°Brix SSC in New Zealand for *A. chinensis* var. *deliciosa* 'Hayward' fruit for export (Harman, 1981) was dependent on cool autumn temperatures promoting early starch breakdown (Seager *et al.*, 1996), although cool autumns are now less common (Burdon, 2015). The current understanding of fruit SSC is that it is under genetic control unless cool temperatures occur at a time when the fruit is responsive (Fig. 20.2). The early

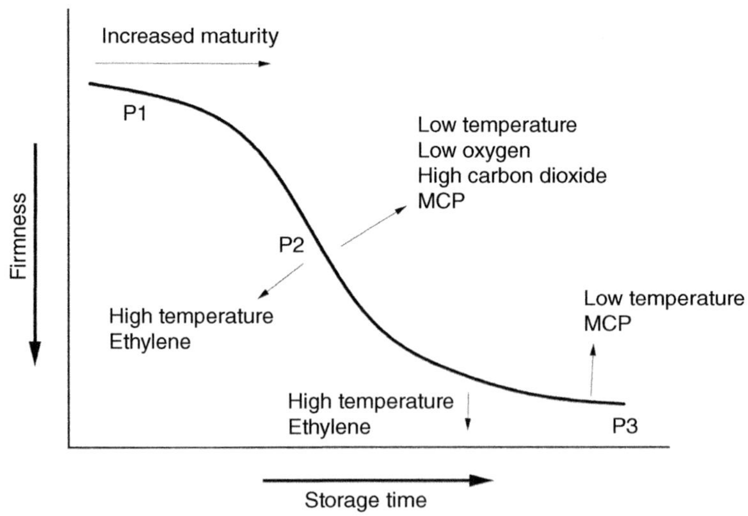

Fig. 20.1. A generalized softening curve for kiwifruit showing the basic sigmoidal pattern with three phases (P1–P3) and how manipulation of kiwifruit physiology can be used to affect softening in storage. MCP, 1-methylcyclopropene. (From Burdon, 2019b)

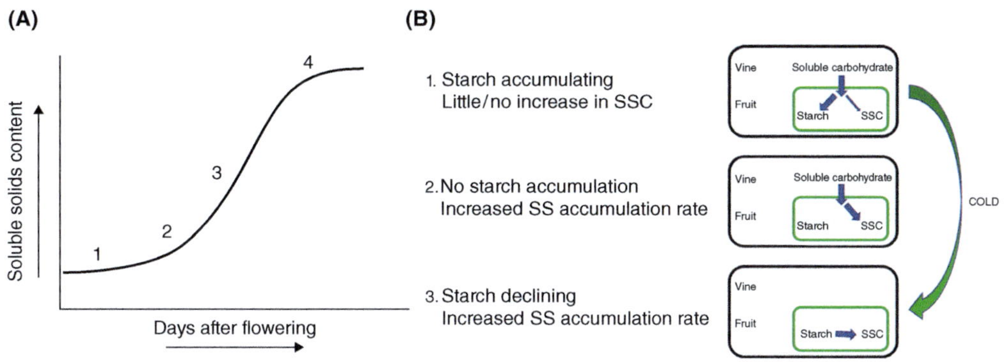

Fig. 20.2. (**A**) The pattern of fruit soluble solids content (SSC) during kiwifruit maturation and ripening (~100–210 days after flowering). The overall sigmoidal pattern of SSC accumulation is shown with suggested phases based on soluble carbohydrate/starch metabolism: 1 – little increase in SSC as soluble carbohydrates entering the growing fruit are converted to starch; 2 – increase in rate of soluble solids (SS) accumulation from either (i) import of soluble carbohydrates not converted to starch or (ii) starch breakdown induced by low temperature; 3 – starch breakdown irrespective of environmental temperatures (or after harvest); 4 – slowing and cessation of SS accumulation rate as all starch has been broken down. (**B**) The changing soluble carbohydrate/starch metabolism. (From Burdon, 2022)

part of the increased rate of soluble solids accumulation is due to a shutdown of starch accumulation at a time when soluble carbohydrates are still being imported into the fruit. Starch breakdown as part of the ripening process occurs later.

Harvest indices have become more complex with the introduction of new cultivars which develop at different rates and have different fruit characteristics to consider. These new cultivars have shown 'Hayward' fruit to be late-maturing (Burdon *et al.*, 2021). Yellow-fleshed cultivars are initially green and must undergo the loss of chlorophyll (degreening) to reveal the desired yellow flesh colour (Burdon, 2018) (see Chapter 16). Flesh colour change may not be tightly associated with other aspects of fruit maturation and ripening, including SSC and firmness (Burdon *et al.*, 2014a) and may be completed before ripening commences.

In general, the colour of a fully green kiwifruit flesh is ~115°h and this reduces to a minimum of ~97°h when fully degreened (Minchin *et al.*, 2003; Gambi *et al.*, 2018), with colour values below ~103°h considered as degreened. The commonest way of measuring colour is with a Minolta Chroma Meter (models CR-300 or CR-400), or similar, after removal of the skin and flesh to a depth of ~2 mm. This method was adopted for 'Hort16A' fruit specifically to quantify degreening, which occurs last just under the fruit skin (Patterson *et al.*, 2003; Ghasemnezhad *et al.*, 2022). Commercially, yellow-fleshed fruit must be either fully degreened before harvest, or at least have degreened sufficiently at harvest to allow completion of degreening off the vine before storage (Burdon and Lallu, 2022). Thus, flesh colour is a major component of a harvest index for yellow-fleshed kiwifruit.

Irrespective of when harvested, fruit must eventually ripen to a good eating quality. Fruit harvested early for immediate marketing must have developed the capacity to ripen, usually in response to ethylene. If harvested too early, fruit does not ripen fully unless given a long, or continuous, ethylene treatment (Lallu *et al.*, 1989; Ritenour *et al.*, 1999; McAtee *et al.*, 2015; Burdon *et al.*, 2016). The capacity to respond fully to ethylene develops over a few weeks while fruit is still on the vine, although there is no easily visible change in the fruit at this time (McAtee *et al.*, 2015; Burdon *et al.*, 2017).

Even if the fruit does ripen fully without disorder, it must also taste good. For kiwifruit, consumer acceptability is strongly influenced by the ripe fruit SSC (rSSC; McMath *et al.*, 1992; Crisosto and Crisosto, 2001; Burdon *et al.*, 2004) (see Chapter 21). Since kiwifruit accumulates a large amount of starch in the unripe fruit, it is possible to predict the rSSC at harvest from the dry matter content of the fruit (Burdon *et al.*, 2004). Initially, this relationship was most significant for early-harvested fruit; however, as orchard productivity has increased (see Chapter 12), it has become more important to ensure that sufficient carbohydrate is accumulated to give an acceptable-tasting fruit. The dry matter content is only of value as an indicator of the ripe fruit eating quality, it is not itself an indicator of maturity. The availability at harvest of a current SSC and a predicted rSSC has led to the concept of 'percent solubilized' (Jordan *et al.*, 2000; Feng *et al.*, 2003b). This concept adds little of value; the point of interest is the change to starch breakdown as a threshold rather than the position on a 0–100 scale for SSC. In addition, once converted to a ratio, individual values are lost. The at-harvest SSC is indicative of the fruit development whilst the dry matter content is indicative of the rSSC and thus consumer liking.

20.3.2 Harvest practice

Generally, all the fruit from an orchard, or a precisely defined maturity area, is harvested at the same time. This allows for speed and scale to minimize costs but may result in damage to the fruit.

Harvesting requires the fruit to be snapped from the pedicel to leave a clean break. When picked without a clean break, a small stub of the pedicel may be left attached to the fruit, which can damage other fruit. The propensity for a clean break seems dependent on both the genotype and fruit maturity.

Fruit is usually harvested into a small container either worn by the picker or hung in the canopy, before being consolidated into a larger crate or bin in the field. In some countries, including China, fruit is bagged early in development to protect from pests and physical damage. The fruit and bag are harvested together and the bag removed at the packhouse before further grading or storage. The crates or bins of fruit are moved to the packhouse, preferably on the day of harvest.

20.4 Postharvest Systems

Systems developed for 'Hayward' fruit deliver high-quality fruit throughout months of storage. With the global commercialization of new cultivars, different biology and storage responses are apparent, despite the similar basic fruit biology. Consequently, the prolonged storage periods obtained, and the physical resilience seen for 'Hayward' fruit may be lacking. Even for a single cultivar such as 'Hayward', the way the fruit is handled from orchard to the retailer can differ (Fig. 20.3).

In many countries, fruit is harvested into plastic or wooden field bins, each containing 250–300 kg of fruit. The fruit may be packed into the final packaging immediately after harvest, or graded, sized and then stored in bins to be packed just in time for dispatch. A further option is for the bins of fruit to be placed directly into CA storage with no prior grading. The use of small plastic crates overcomes infrastructure or orchard conditions that preclude the use of larger bins. Kiwifruit postharvest performance is largely judged by firmness retention and the fruit remaining free of disorders.

20.5 Pre-grading Treatments

20.5.1 Curing

Kiwifruit curing was developed during the 1980s to manage stem end rots (SER) in 'Hayward' fruit caused by *Botrytis cinerea* (Pennycook and Manning, 1992; Manning *et al.*, 2010) (see Chapter 17). The basic principle is

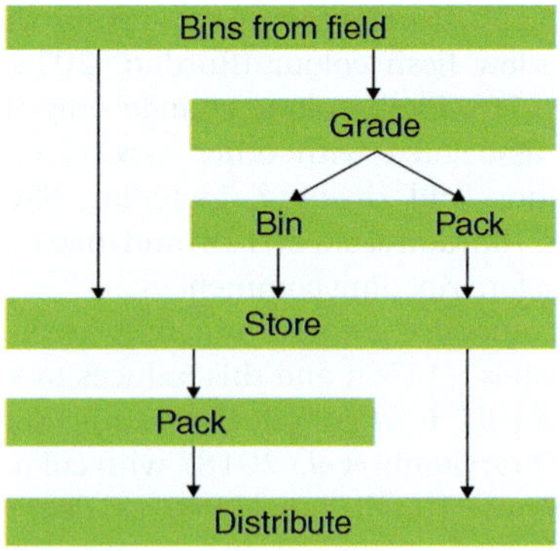

Fig. 20.3. The movement of kiwifruit through a packhouse and coolstore showing some options. Bins holding 250–300 kg of fruit may be replaced by crates holding ~20 kg of fruit.

that picking damages the fruit, creating sites where *Botrytis* can invade. Holding the fruit under cover at ambient conditions for 48–72 h before transferring to cool storage reduces the subsequent incidence of SER in storage. Curing appears to be linked to water loss, but the physiological and pathological interaction has not yet been fully defined. Nowadays, curing tends to be applied almost by default as part of the logistics of handling harvested fruit irrespective of the risk of botrytis SER.

20.5.2 Temperature conditioning

Temperature conditioning refers to a range of postharvest operations to manage specific fruit outcomes. Beyond curing, these outcomes include degreening yellow-fleshed cultivars, reducing the expression of chilling damage, and ripening fruit, in particular low-maturity early-season fruit. The common feature is that the fruit are held above the normal storage temperatures, contradicting the traditional concept of removing field heat as soon as possible after harvest.

Degreening

Degreening of yellow-fleshed kiwifruit cultivars is temperature-sensitive and not responsive to ethylene. Temperatures below ~5°C slow chlorophyll loss (de Silva *et al.*, 2007; Gambi *et al.*, 2018) and, based on the relative changes in flesh colour and softening, the best degreening occurs at 7–10°C. As the fruit is relatively immature, there is usually a short period of little or no softening after harvest, during which time degreening should already be occurring (Fig. 20.4), leaving degreened fruit that are still sufficiently firm for handling. Degreening can be achieved under CA, although the rate of colour change may be slowed (Burdon *et al.*, 2022).

Chilling sensitivity

Temperature conditioning to manage degreening and chilling sensitivity (susceptibility to chilling damage) tend to be required for the same fruit. Generally, fruit of yellow-fleshed cultivars tends to be sensitive to chilling while still green-fleshed, although a fully degreened fruit might still be chilling-sensitive. Holding fruit for several

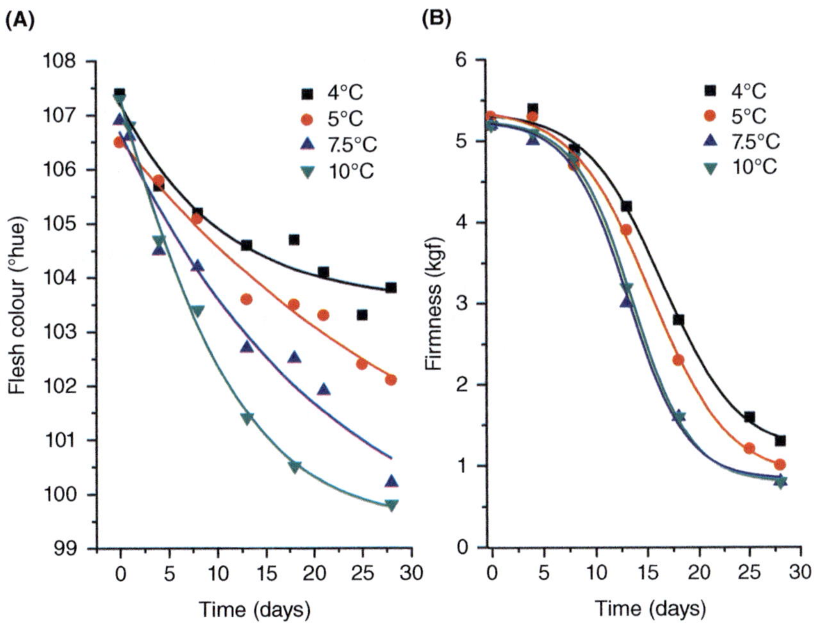

Fig. 20.4. Flesh colour (**A**) and firmness (**B**) of *Actinidia chinensis* var. *chinensis* 'Hort16A' kiwifruit harvested when flesh colour was approximately 107° hue angle and then held at 4°C, 5°C, 7.5°C or 10°C. Values are means of four orchards, 30 fruit per orchard. (From Burdon and Lallu, 2022)

days at 7–10°C before transfer to cool storage reduces chilling sensitivity, as well as favouring degreening (Burdon and Lallu, 2011, 2022; Wang, R. *et al.*, 2022). Holding fruit under CA during degreening slowed the loss of chilling sensitivity (Burdon *et al.*, 2022), possibly associated with the slowed loss of firmness. This agrees with the general concept that a firmer fruit early in storage may indicate chilling sensitivity (Alavi *et al.*, 2021).

20.5.3 Fungicide application

Generally, kiwifruit are cleaned and graded dry. If a postharvest fungicide is applied, it is usually as a dip or drench when fruit reach the packhouse.

20.6 Grading and Packing

20.6.1 Grading

In general, grading creates a uniform product for packing. Orchard debris is removed by brushing, unacceptable fruit are removed, and the fruit sorted by size (weight). Fruit may be deemed unacceptable for a range of reasons, from the minor or cosmetic, such as shape, to major defects such as being damaged or rotten.

Traditionally, identification of unacceptable fruit was by people observing every individual fruit on the grading line. More recently, this role has been automated, with commercially developed vision systems identifying defective fruit. This is beneficial where the availability and cost of labour are limiting. One challenge has been to match the image processing and decision making to the speed at which a commercial grader operates. As the vision systems have developed, it has become possible to identify defects not immediately visible to the human eye, such as bruising or soft patches on fruit.

Beyond vision systems, near infra-red (NIR) spectroscopy has been used for non-destructive evaluation of kiwifruit (McGlone and Kawano, 1998; Walsh *et al.*, 2020). Initial observations generated acceptable estimates of fruit dry matter and SSC, but less so for firmness (McGlone and Kawano, 1998). With the proliferation of low-cost and even hand-held NIR devices, it is a question of what degree of accuracy of estimate is good enough for the desired purpose, which may be maturity monitoring, storage risk or evaluation of small numbers of fruit in breeding programmes.

Kiwifruit with chilling injury are not always visible to the eye and visible (vis)/NIR and dual-laser systems can help to identify affected fruit (Wang, Z. *et al.*, 2021, 2022a). This research has also suggested that it may be possible to discriminate between defect symptoms arising from different origins: chilling, fungal and physical (Wang, Z. *et al.*, 2022b).

Grading to remove disordered fruit after storage is clearly wasteful, although it does maintain consumer confidence. Assessing fruit for risk before harvest or storage would be better. Fruit SSC and firmness can function well as risk indicators at the extremes of maturity but are not precise for fruit of intermediate maturity (Burdon *et al.*, 2014b). While NIR spectroscopy can identify fruit of low maturity within a batch of harvested fruit from factors such as SSC (Clark *et al.*, 2004), thresholds for risk are not the same across all orchards (Burdon *et al.*, 2014b). Assessment at harvest or whilst grading before storage misses the point that chilling injury is strongly affected by the cooling rate, storage temperature and storage duration, all of which must be considered along with the fruit biology (Burdon, 2022).

20.6.2 Packing

Packaging unitizes and protects the fruit during storage and distribution. Unitization may be as a single-layer tray, a pack with two layers or a bulk-fill pack that is loose-filled with fruit. In addition, the package provides a surface on which to display product information. When in layers, an inner pocket pack or tray separates and, to some degree, immobilizes individual fruit. These trays may be surrounded by a polyethylene liner to reduce water loss. In multiple-layer or bulk-fill packs, the liner is replaced by a bag. While this bag normally has ventilation holes and is not sealed, in other circumstances sealed, selectively permeable modified atmosphere (MA) bags may be used. MA bags function by creating a low O_2 and high CO_2 atmosphere around the fruit, slowing fruit ripening in the same way as a CA (see Section 20.8.1). The MA is temperature-dependent, based on fruit respiration, and increased temperature can turn the fruit anaerobic. In addition, any ethylene released inside the bag will accumulate and accelerate fruit softening if not removed by an ethylene absorber.

There is an increased trend towards automation of the packing process, from box assembly, through filling and closure to palletization. The sustainability of all packaging materials is under scrutiny. Compressed fibre-pocket packs have been shown to damage fruit through vibration during transport (Lallu *et al.*, 1999; Fig. 20.7C), leading to a change to plastic, although currently the sustainability of plastic use is being reviewed.

20.7 Storage: Responses to Temperature

Often, discussion of temperature responses in fruit, including kiwifruit, is simply about the storage temperature. This ignores the fact that how fruit reach the storage temperature is critical to storage performance and, for kiwifruit, other temperature responses are significant to postharvest handling.

20.7.1 Temperature management for storage

Parameters to be considered for the cooling of kiwifruit include the cultivar, harvest maturity, timing after harvest, cooling rate and final storage temperature (usually ~0–1°C). The goal is to have fruit at the storage temperature as soon as possible without inducing chilling damage.

Generally, chilling susceptibility of *A. chinensis* var. *chinensis* fruit appears to be greater than that of 'Hayward' fruit (Patterson *et al.*, 2003). While this may be considered a difference between *A. chinensis* var. *chinensis* and *A. chinensis* var. *deliciosa*, it may simply reflect a greater chilling tolerance of 'Hayward' fruit. Other *A. chinensis* var. *deliciosa* cultivars appear less chilling-tolerant than 'Hayward' (Burdon *et al.*, 2007; Wang, R. *et al.*, 2022).

Fruit maturity is a key element of chilling susceptibility. For less mature chilling-sensitive fruit, delaying cooling for curing or degreening may also reduce the risk of chilling damage (Lallu and Webb, 1997; Burdon and Lallu, 2022). However, once fruit starts ripening on the vine there is little risk of chilling damage, and the ability to slow ripening in storage is limited without the application of immediate cooling.

The rate of cooling to storage temperature may influence both the rate of softening and the occurrence of chilling symptoms (Lallu and Webb, 1997; Yang *et al.*, 2013a; Zhao *et al.*, 2015). Extremes of cooling rate may be considered as forced-air cooling directly to storage temperature in a matter of hours, or room cooling, which for packed fruit may take up to two weeks to reach the final storage temperature (Burdon and Lallu, 2011). An alternative may be using forced-air cooling to remove much of the heat from the fruit before allowing slower cooling to the final storage temperature (McDonald, 1990). A further refinement is that the storage temperature may be decreased as the fruit ripens (McDonald, 1990; Jabbar and East, 2016).

20.7.2 Temperature-induced softening

For generalized comments about higher or lower temperatures, it is essential to note the temperature range being considered. An example is given below for temperature-induced softening. At ~0°C, even small changes in temperature can alter the softening pattern of the fruit (Fig. 20.5A). Compared with fruit at 0°C, fruit at 5°C soften more rapidly and are softer when the change to phase 3 slow softening occurs (Fig. 20.5B). In contrast, over a wider range of temperatures, fruit soften quicker at ~8°C than if colder or warmer (Fig. 20.5C).

Temperatures below ~10°C can induce softening in stored kiwifruit (Mworia *et al.*, 2012; Burdon *et al.*, 2017; Asiche *et al.*, 2018), provided that the fruit has developed the capacity to respond before harvest (Burdon *et al.*, 2017). The rate of softening at 4°C and 12°C may be similar (Fig. 20.5C), suggesting an induction of softening as fruit are cooled below

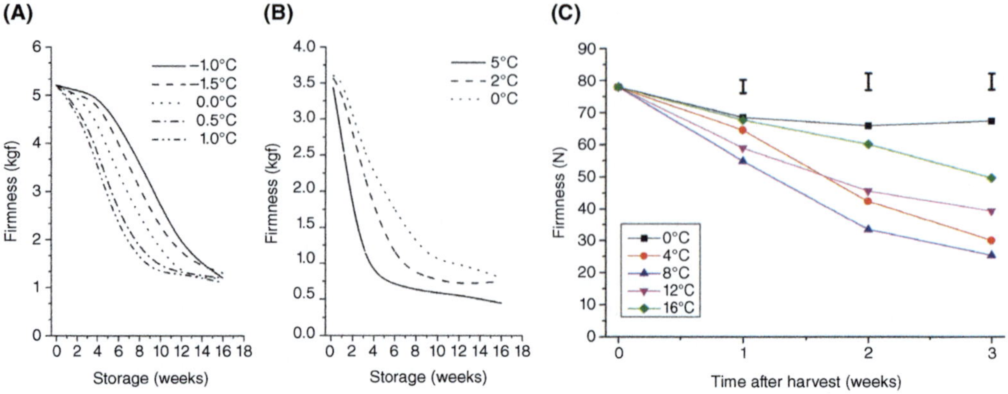

Fig. 20.5. Softening responses of kiwifruit to temperature. (**A, B**) *Actinidia chinensis* var. *chinensis* 'Hort16A' fruit (from Patterson *et al.*, 2003); (**C**) *A. chinensis* var. *deliciosa* 'Hayward' fruit (from Burdon *et al.*, 2017). 1 kgf = 9.81 N.

10°C, and then a slowing of the rate as temperature approaches 0°C. Softening in response to low temperatures has been shown to be independent of ethylene (Mworia *et al.*, 2012), although low temperature may increase the sensitivity of fruit to ethylene (Murakami *et al.*, 2014).

20.8 Storage: Responses to Atmosphere

20.8.1 Controlled and modified atmosphere storage

Generally, CA and MA slow kiwifruit softening (Lallu *et al.*, 2003; Thompson *et al.*, 2018; Burdon, 2020). CA actively maintains the atmosphere around the fruit through monitoring and adjusting both O_2 and CO_2 and removing ethylene. In MA, the final atmosphere is dependent on the fruit's temperature-dependent respiratory activity and the selective permeability of the film used. The softening state of the fruit when CA or MA is applied affects the degree of benefit to firmness retention that can be achieved: Phase 1 softening may be extended, and phase 2 softening slowed. From initial work on 'Hayward' fruit in the late 1970s and 1980s (McDonald and Harman, 1982; Harman and McDonald, 1983; Arpaia *et al.*, 1985), the beneficial effects of CA have been demonstrated for *A. chinensis* var. *chinensis* 'Hort16A' (Lallu *et al.*, 2003), 'Hongyang' (Li *et al.*, 2015) and 'Zesy002' (Burdon *et al.*, 2022). Beneficial atmospheres are usually cited as ranges such as 1–3% O_2 and 2–5% CO_2 (Thompson *et al.*, 2018; Burdon, 2020). The fruit responds to each gas individually and synergistically when combined (McDonald and Harman, 1982; Arpaia *et al.*, 1985). Early kiwifruit CA research tended to emphasize the beneficial effect of high CO_2 (McDonald and Harman, 1982; Arpaia *et al.*, 1985). The current trend in CA research and application is to maximize the benefit from low O_2 by storing at the lowest safe O_2 based on a fruit response (termed dynamic CA; Thompson *et al.*, 2018).

How CA slows kiwifruit ripening, largely the reduction in softening and associated cell-wall changes (Arpaia *et al.*, 1984), has received only limited attention. In general, low O_2 concentrations or elevated CO_2 concentrations in the storage atmosphere may reduce respiratory activity, ethylene production and ethylene action (Kanellis *et al.*, 2009; Saltveit, 2020). Elevated CO_2 in the storage atmosphere may acidify fruit tissues, affecting metabolism.

In 'Hayward' fruit, low O_2 or high CO_2 atmospheres have been linked with anaerobic metabolism: the production of ethanol and acetaldehyde (Antunes and Sfakiotakis, 1997; Yanez Lopez *et al.*, 1999) and the associated enzymes alcohol dehydrogenase, pyruvate decarboxylase and lactate dehydrogenase (Botondi *et al.*, 2011). The presence of anaerobic metabolites in

'Hayward' fruit adversely affects sensory quality (Thomai and Sfakiotakis, 1997). Storage in low O_2 reduced the ethylene-producing capacity of 'Hayward' fruit, through low 1-aminocyclopropane-1-carboxylic acid oxidase activity (Antunes and Sfakiotakis, 2002). Also, in 'Hayward' fruit, 5% CO_2 modulated the expression of individual ethylene response factor transcription factors (Yin *et al.*, 2012).

The slow declines in total flavonoids, phenolics and antioxidant activity during storage of 'Hayward' fruit were unaffected by CA (Yildirim and Bayir, 2013), although in 'Hongyang' fruit, CA was suggested to maintain antioxidative systems (Xia *et al.*, 2016).

Chlorophyll loss from 'Hayward' fruit is slowed by CA (Lallu *et al.*, 2005). Even 1% CO_2 may improve chlorophyll retention, with an atmosphere of 1% O_2/1% CO_2 giving better green colour retention than 1% O_2 (Thomai and Sfakiotakis, 1997). Thus, CA may be beneficial in maintaining the fresh green flesh colour of 'Hayward' fruit. However, CA may slow the off-vine degreening of 'Zesy002' fruit (Burdon *et al.*, 2022).

20.8.2 Ethylene

The responsiveness of kiwifruit to ethylene is significant commercially in limiting storage. It is also important for research, where the presence of ethylene can significantly affect the interpretation of a result. For example, the inhibition of softening by MCP will appear more significant in the presence of exogenous ethylene; likewise, any acceleration of ripening may not be detected if fruit is already being stimulated to ripen by exogenous ethylene.

Avoidance of ethylene in a coolstore can be simply through ventilation with ethylene-free air. Alternatively, ethylene can be removed from store atmospheres by scrubbing with a catalytic convertor or ozone, or by chemical reaction or absorption (Thompson *et al.*, 2018). While the mode of action may differ, the effect on the fruit should be similar if ethylene is removed to the same degree. However, the effect is limited to protecting fruit that are not producing ethylene, i.e. preventing 'bad' fruit from affecting the rest, or protecting from exogenous sources of ethylene.

Another concept is to make fruit unresponsive to ethylene by binding MCP irreversibly to ethylene receptors. MCP has a limited effect on the main (phase 2) softening but leaves fruit a little firmer at phase 3 when fruit are close to the limit for commercial handling or are ready to eat. While the effect on phase 3 softening occurs when the fruit is naturally producing ethylene, the effect during phase 2 is more difficult to explain. Nevertheless, MCP protects kiwifruit from exogenous ethylene, with some degree of this efficacy being relatively long-lasting (Gong *et al.*, 2020). The use of MCP is common throughout the global kiwifruit industry.

20.8.3 Ozone

There has been increased interest in the use of ozone to disinfect horticultural produce (Glowacz, 2020). This has resulted in the availability of systems for use in coolstores and shipping containers. While ozone is commonly recognized as safe, leaving no residues on fresh produce, as a strong oxidizing agent it might damage the produce, damage aspects of the coolstore structure and have detrimental effects on human health.

Generally, ozone is reported to reduce rots caused by *Botrytis* and *Penicillium* and reduce mould build-up in packhouses and coolstores. However, its efficacy is questionable in controlling other species of fungi present as latent infections on fruit at harvest. While *Botrytis* and *Penicillium* rots are found on kiwifruit, they tend to occur only on damaged fruit, botrytis SER being a specific case.

From a physiological perspective, a more interesting finding is that ozone may slow kiwifruit ripening (Minas *et al.*, 2018). Ozone can remove ethylene from storage atmospheres and thereby slow kiwifruit ripening caused by ethylene contamination. A less-ripe kiwifruit is less susceptible to the ripe rots that only develop on ripe fruit.

20.9 Fruit Ripening

It is sometimes desirable to ripen kiwifruit in a controlled way using ethylene (Lallu *et al.*, 1989;

Crisosto, 1997; Ritenour *et al.*, 1999; Burdon, 2019a). This is particularly important for early-harvested fruit that is not close to ripening naturally and will shrivel and remain unripe unless treated. The capacity to respond to ethylene develops gradually over a few weeks while the fruit is still in phase 1 of softening (McAtee *et al.*, 2015; Burdon *et al.*, 2017). During this time, the fruit changes from being unresponsive, through being partially responsive (whilst ethylene is present) to being fully responsive and ripening fully in response to a short ethylene treatment. Therefore, ripening depends on the fruit maturity, temperature (generally > 10°C) and concentration and duration of ethylene application. A key aspect of ripening with ethylene is that it coordinates the ripening of all the fruit being treated; fruit within a tray, trays within a pallet and amongst pallets, resulting in uniformity among the fruit.

20.10 Fruit Quality

Kiwifruit quality is generally assessed by the fruit firmness, a basic assessment of ripening and the presence of any defects. Defects can usefully be divided into categories based on the source of the defect: physical, physiological and pathological.

20.10.1 Physical

Kiwifruit does not have a substantial peel tissue. Most commercial cultivars have a peridermal skin composed of several layers of dead suberized cells, while kiwiberries have a living skin with a cuticle-covered epidermis (Hallett and Sutherland, 2005). Cultivars differ in the sensitivity of their skin to damage, or its ability to protect the underlying tissues. In fully mature fruit, the periderm layers are difficult to dislodge, although in early-harvested 'Hort16A' fruit, the skin could be lifted easily (Fig. 20.6A). Also, in 'Hort16A' fruit, the flesh under the skin dried out to a dark brown/black colour once exposed to air, an aspect not always seen in other cultivars. The presence of phenolic compounds may also account for the unusual circular discolouration around points of impact on this cultivar (Fig. 20.6B, C) and stylar end discolouration associated with chilling injury (Fig. 20.8C).

Physical damage comes from forces generated both among fruit and between fruit and equipment or packaging. The design of fruit handling equipment and packaging should minimize forces applied to the fruit. The most common damage symptom reported in kiwifruit is bruising, which may occur following impact (Mencarelli *et al.*, 1996; Ahmadi, 2012; Du *et al.*, 2019; Gao *et al.*, 2021), compression (Gao *et al.*, 2021) or vibration (Lallu *et al.*, 1999; Tabatabaekoloor *et al.*, 2013). The duration and magnitude of forces that fruit is exposed to in a supply chain can be quantified, allowing commercially relevant forces to be applied to fruit for research purposes.

Impact damage may not be immediately visible externally. Damage in response to an impact depends on fruit texture and temperature, the impact force applied (Ahmadi, 2012) and the impact surface (Mencarelli *et al.*, 1996). In firm fruit, a common sign of impact damage is a white line in the outer pericarp, caused by a lack of starch breakdown in damaged cells (Fig. 20.7A). In softer fruit, the affected tissues may become

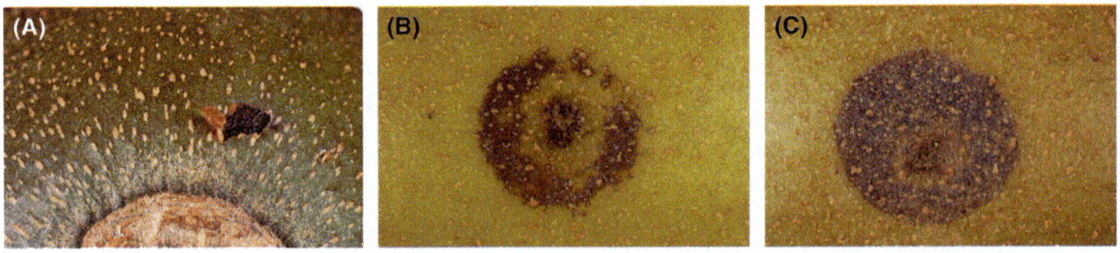

Fig. 20.6. *Actinidia chinensis* var. *chinensis* 'Hort16A' fruit showing lifted skin and discolouration of the flesh underneath in a low-maturity fruit (**A**), and the hypersensitive response to impact damage (**B, C**). (Photos: The New Zealand Institute for Plant and Food Research Ltd)

water-soaked and form a bruise. The water-soaked area tends to be softer than the surrounding tissue and may sometimes be visible through the skin as a darker area. A bruise may not always be visible externally but may be determined by vis/NIR hyperspectral imaging (Lu *et al.*, 2011) or by NIR spectroscopy (Gao *et al.*, 2021).

Fruit compression may flatten the fruit surface, cause an indentation (Fig. 20.7D) or, if severe, cause water soaking of the outer pericarp. There are normally two or more affected areas of the fruit where the compression force has been applied (Fig. 20.7B). As with impact damage, water-soaked areas are generally termed bruises. While it is more difficult to determine when compression loading is excessive, any threshold may depend on the firmness of the fruit and may differ between genotypes. Minor compression marks may recover once the compression force is removed. Compression damage occurs commonly when fruit are in bulk bins or bulk-fill packs, or where packs are over-filled.

Vibration can lead to friction damage caused by rubbing between adjacent fruit or between the fruit and packaging (Lallu *et al.*, 1999; Tabatabaekoloor *et al.*, 2013; Fig. 20.7C). Simulated transport vibration treatments increased the rate of water loss from 'Xuxiang' kiwifruit whilst not leaving visible symptoms (Wei *et al.*, 2019). Where bruises are associated with vibration forces, it may be where vibration causes the fruit to become tightly packed into the constrained space of a bulk-fill pack.

20.10.2 Physiological

Chilling damage

The most frequently reported physiological disorders of stored kiwifruit are associated with chilling temperatures. The best-characterized chilling damage is low-temperature breakdown (LTB; Lallu, 1997). Earlier reports of tissue breakdown (Harman, 1981; Crisosto *et al.*, 1984) and pericarp translucency and granulation

Fig. 20.7. Kiwifruit showing different types of physical damage. (**A**) Impact damage. (**B**) Compression damage. (**C**) Friction damage on *Actinidia chinensis* var. *deliciosa* 'Hayward' fruit. (**D**) Compression damage on *A. chinensis* var. *chinensis* 'Zesy002' fruit. (Photos: The New Zealand Institute for Plant and Food Research Ltd)

(Arpaia *et al.*, 1994a) seen late in storage were probably LTB. Symptoms of LTB start at the stylar end of the fruit and develop more extensively throughout the fruit flesh with increased severity (Fig. 20.8A). The initial symptom is a granular appearance in the outer pericarp, just under the skin, starting with a few small patches, then a full ring around the fruit. This granular appearance may be caused by air bubbles (Bauchot *et al.*, 1999), possibly associated with large cells in the outer pericarp. In more severe LTB, water-soaked tissues may develop in the outer or inner pericarp (Fig. 20.8A).

LTB in kiwifruit develops slowly during storage. Considering LTB as a disruption of the normal ripening process may account for the suggestion that LTB can be reduced by CA or MA (Biswas and Brummell, 2019; Gwanpua *et al.*, 2019), but it may be that symptom expression is simply delayed by the slower ripening rate.

Thus far, the description of LTB symptoms has applied to 'Hayward' fruit, although similar and additional symptoms may also be seen in other cultivars. Most obvious is the discolouration of the skin at the stylar end of 'Zesy002' and 'Hort16A' fruit (Fig. 20.8B, C). In 'Hort16A' fruit, this was limited to fruit harvested very early and usually identifiable as green-fleshed. Dark water-soaked core tissues tend to be seen in cultivars including 'Zesy002' and 'Hort16A' (Fig. 20.8B, C). Random patches of brown, discoloured tissue on the surface of 'Hort16A' fruit were linked to rot development on the disordered tissues (Manning *et al.*, 2016).

A further symptom seen in some cultivars is the outer pericarp remaining firm whilst the inner pericarp becomes soft and water-soaked, resulting in a fruit that has a rubbery feel. In 'Zesy002' fruit, the development of this symptom may be accompanied by some degree of discolouration of the outer pericarp (Fig. 20.8B). The disruption of tissue integrity by the chilling disorder releases phenolic compounds that contribute to discolouration, with lignification suggested as a chilling response (Suo *et al.*, 2018).

The ordered progression of LTB symptoms from the stylar end of the fruit in 'Hayward' contrasts with a more general presence of water-soaked and discoloured tissues more recently described in 'Hongyang' (Yang *et al.*, 2012; Ma *et al.*, 2014; Suo *et al.*, 2018; Wang *et al.*, 2020), 'Xuxiang' (Yang *et al.*, 2013a) and 'Cuixiang' (Jiao *et al.*, 2021) kiwifruit.

In kiwifruit, aspects of antioxidant metabolism have been associated with postharvest changes and altered chilling damage expression. This agrees with general concepts of oxidative stress and membrane fluidity for chilling injury (reviewed by Sevillano *et al.*, 2009; Biswas and Brummell, 2019). In general, for a range of cultivars including 'Hayward', 'Hongyang' and 'Xuxiang', fruit maturity or temperature treatments have been shown to affect both chilling damage and oxidative metabolism, but without demonstrating a strong causal link (Yang *et al.*, 2012, 2013a, b; Wang *et al.*, 2020). In some instances, the timing of chilling symptom expression was changed, whereas the effects on metabolism were largely seen as changes in the magnitude of metabolite concentration or enzyme activity, with no change in the timing. This questions how tightly the changes in metabolism relate to the expression of chilling symptoms or to other changes that occur as the fruit ripens at low temperature.

The difficulty in understanding kiwifruit chilling symptoms may be that the mechanism is often reviewed and investigated as a single biological response, whereas it has long been recognized that the nature of the physiological dysfunction is highly relevant when assessing different symptoms of chilling damage in different fruit species.

The difference between chilling and freezing damage is sometimes questioned. A simple observation is that when a whole fruit has been frozen and thawed, the distinction between inner and outer pericarp is lost (Fig. 20.8D). However, depending on circumstances, the whole fruit may not be frozen and initial symptoms tend to occur at the stem end of the fruit (OECD, 1992; Arpaia *et al.*, 1994b; Gerasopoulos *et al.*, 2006).

Skin markings

A range of skin markings may develop on kiwifruit during storage (Burdon and Lallu, 2011; Gwanpua *et al.*, 2019). Pitting is used as a general term for surface depressions on the fruit with two main causes: physiological and fungal.

Physiological pitting occurs as discrete areas 2–3 mm in diameter where cellular collapse

Fig. 20.8. Symptoms of chilling damage in kiwifruit. (**A**) *Actinidia chinensis* var. *deliciosa* 'Hayward'; (**B**) *A. chinensis* var. *chinensis* 'Zesy002' (**C**) *A. chinensis* var. *chinensis* 'Hort16A'. (**D**) Freezing damage in 'Hayward' kiwifruit. (Photos: The New Zealand Institute for Plant and Food Research Ltd)

under the skin causes a depression which then dries and discolours (Fig. 20.9A, D). The affected fruit tend to come from areas of dense vine canopy (Burdon and Lallu, 2011). The symptoms develop after long-term storage, possibly exacerbated by low temperatures (Lallu and Webb, 1997; Lallu *et al.*, 2003). Pitting has been associated with storage temperatures of −1°C or −1.5°C or cooling with air at −8°C (Gorini, 1991). The orchard-specific nature of physiological pitting was hypothesized to be associated with mineral content, or a vine factor, but extensive research failed to identify a definitive cause (Thorp *et al.*, 2003; Boyd *et al.*, 2006).

One causal agent of fungal pitting is *Neofabraea actinidiae* (syn. *Cryptosporiopsis actinidiae*), which causes orange/brown sunken lesions in which the rotted tissue is dry (Brook, 1992; Manning and Lallu, 1997; Fig. 20.10A). Pitting symptoms have also been attributed to *Cadophora luteo-olivacea* (Spadaro *et al.*, 2010), previously reported as *Phialophora* sp. (Testoni *et al.*, 1997). *N. actinidiae* is regarded as a ripe rot, only growing once the fruit are close to ripe (Tyson *et al.*, 2019). However, under high CO_2 (e.g. ≥ 5%) in CA or MA storage, the rot can grow on firmer fruit (Lallu *et al.*, 2003).

Lenticel spots are 1–2 mm diameter dark patches around lenticels (Fig. 20.9B, E), likely caused by dehydration around the lenticel, possibly in combination with phenolic compounds. In some lenticel spots, fungal growth was found, but there was no fungal growth in most instances (Burdon and Lallu, 2011).

A final category of physiological skin marking is raised spots which appear to be based on lenticels that have not developed normally, but instead have been isolated by further cell division (Fig. 20.9C, F).

A non-physiological discolouration of the fruit surface in storage is referred to as storage stain or simply stain (Hopkirk and Clark, 1992) or juice blotch (Sommer *et al.*, 1994) (Fig. 20.10B). This is caused by the growth of a saprophytic fungus such as *Alternaria*, commonly where juice from damaged fruit has contaminated the surface of an otherwise sound fruit (Sommer *et al.*, 1994). While the fungus does not penetrate the fruit, it is unsightly and difficult to remove.

Shrivel

Shrivel of kiwifruit is associated with water loss. A threshold water loss of ~4% for shrivel

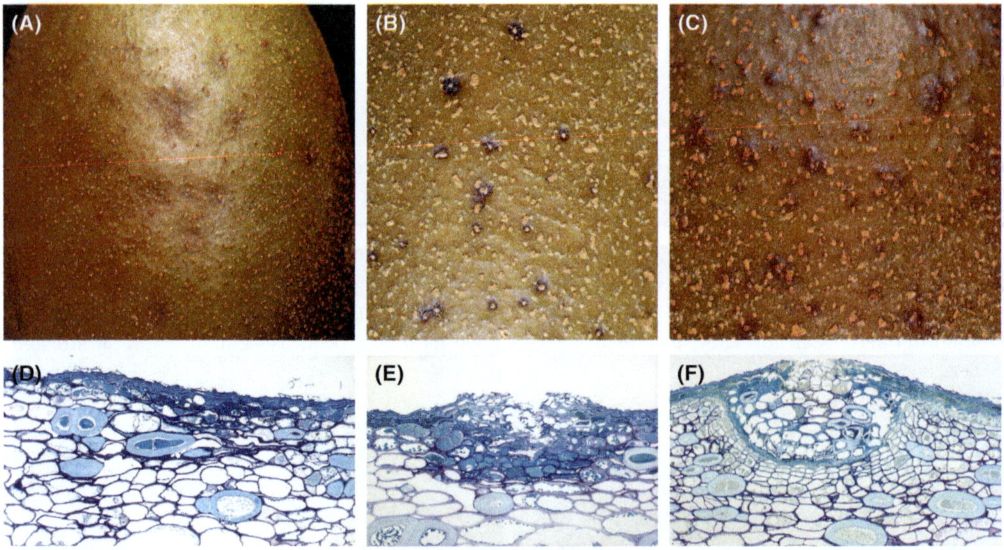

Fig. 20.9. Skin markings on *Actinidia chinensis* var. *chinensis* 'Hort16A' kiwifruit. (**A**) Physiological pitting. (**B**) Lenticel spots. (**C**) Raised spots. Light microscopy showing toluidine blue stained resin sections: (**D**) a physiological pit; (**E**) a lenticel spot; (**F**) a raised spot. (Photos: The New Zealand Institute for Plant and Food Research Ltd)

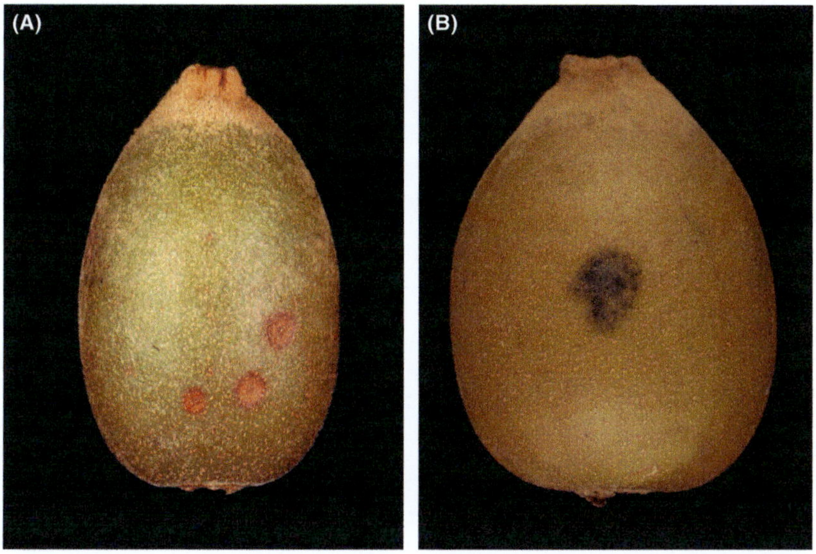

Fig. 20.10. Skin markings on *Actinidia chinensis* var. *chinensis* 'Hort16A' kiwifruit. (**A**) Fungal pitting. (**B**) Storage stain/blotch. (Photos: The New Zealand Institute for Plant and Food Research Ltd)

symptoms is often quoted for 'Hayward' fruit that have softened during long-term low-temperature storage (Burdon and Lallu, 2011). Water loss is largely driven by a high vapour pressure deficit (VPD), exacerbated in storage by a lack of packaging and partially filled coolstores. The VPD of the storage environment is the result of the refrigeration system used in the coolstore, with small evaporators and large temperature differentials increasing water loss. With temperatures of ~0°C, the ice that develops on refrigeration coils and water lost on defrosting are reminders of the water being lost from the fruit.

Research on fruit shrivel is commonly reduced to investigations of the skin permeance and manipulating water loss by inclusion of packaging barriers or applying surface coatings to the fruit. Skin permeance differs among cultivars, and while permeance depends on fruit development, at commercial maturity the permeance is considerably lower than earlier in development (Smith *et al.*, 1995; Montanaro *et al.*, 2012). Permeance is also increased by sun exposure that causes a darker skin, and such areas tend to shrivel earlier than other parts of the fruit or less sun-exposed fruit.

'Zesy003' fruit have shown that fruit biology may play a more complex role in shrivel expression beyond skin permeance (Burdon *et al.*, 2014c, 2015). In 'Zesy003' fruit, shrivel occurred with little water loss once the fruit were soft, possibly the result of an inability to re-equilibrate water within the fruit to replace that lost at the fruit surface.

Other documented disorders

Observations are limited for other documented disorders, but they live on through reviews such as this one. It is to be expected that as new cultivars are commercialized, new disorders, or different symptoms of common disorders, such as chilling injury, may arise.

The occurrence of hard cores in otherwise ripe fruit has been associated with CA storage (Testoni, 1992; Arpaia *et al.*, 1994a) or treatment with MCP (Gwanpua *et al.*, 2019). However, they can also occur in fruit not stored in CA or treated with MCP (Lallu *et al.*, 1989; MacRae *et al.*, 1989). In 'Hayward' fruit, hard cores were observed in early-harvested fruit, suggesting an association with maturation (Lallu *et al.*, 1989; MacRae *et al.*, 1989). More recently, with a wider range of genotypes being studied globally, it appears that hard cores arise from a temporal dissociation of the core and outer pericarp change to rapid (phase 2) softening, starting on the vine (J. Burdon, 2023, unpublished results). The risk of a hard core arises when the phase 2 softening occurs markedly (weeks) later in the core than in the outer pericarp. In addition, a more linear, softening pattern in the core than the outer pericarp has been reported for 'Hayward' fruit during storage (Burdon *et al.*, 2017). Thus, the temporal difference between start of core and outer pericarp softening for the particular genotype is the baseline from which the occurrence of hard cores may be exacerbated by early harvest or postharvest treatment. The softening of both the pericarp and core can be accelerated by ethylene. However, ethylene treatment shortly after harvest limits storage whilst treatment later results in an over-soft pericarp when the core is acceptably soft.

White core inclusions have been reported in 'Hayward' fruit exposed to high CO_2 and ethylene (Arpaia *et al.*, 1986). This disorder is most obvious where the core has darkened, possibly due to water-soaked tissues. The uncommon nature of the disorder is likely related to the specific conditions of the research: fruit with a high incidence of chilling damage, suggestive of a lack of maturity, exposed to a high CO_2 and ethylene. As research and commercial operations have progressed, this combination of circumstances is no longer seen.

A disorder described as pericarp granularity and translucency has been ascribed to CA storage (Gwanpua *et al.*, 2019), although this seems unlikely and chilling damage a more obvious explanation. In establishing CA conditions, it is common for CA operators to remove heat from the fruit before applying the CA, and over-cooling is a risk dependent on the fruit maturity.

The re-greening of the yellow-fleshed 'Hort16A' kiwifruit in response to ethylene has been proposed as a kiwifruit disorder (Pranamornkith *et al.*, 2012). This proposition seems biologically unlikely, and given the small difference in colour observed, may simply reflect the use of independent fruit samples when making destructive flesh colour measurements throughout storage.

20.10.3 Pathological

Storage rots (see Chapter 17) differ depending on the fruit-growing region and the fungal populations in the orchard. They may be divided into opportunistic rots, which need fruit damage, and those that can colonize sound fruit. Alternatively, they may be classified by whether they develop early in storage, or late, when the fruit are close to ripe (ripe rots). Ethylene produced by rotting fruit (Niklis *et al.*, 1997) will accelerate ripening of sound fruit.

20.11 Future Trends

The increasing number of commercially available cultivars is likely to have significant implications for both the scientific and commercial postharvest sectors (Burdon, 2018). For the major marketing companies, there is the basic requirement to offer proprietary green-, yellow- and red-fleshed cultivars and any other cultivars with a marketing advantage.

For scientists and technologists, as it becomes more apparent how different 'Hayward' is from other cultivars, with its late maturation and favourable softening pattern, the challenge is to understand new cultivars and develop predictable robust commercial postharvest systems. Understanding the fruit allows development of quality assurance systems essential for orderly and profitable marketing.

Inspection processes during grading can already be automated with on-line vision systems, and technology is likely to progress further as availability and cost of labour becomes a greater constraint. In addition, the development of new biological understanding and technological capabilities will enhance this process further, with the development of systems that progress beyond what can be achieved by simple compositional measures of the fruit or is immediately obvious to the human eye.

The long storage life of 'Hayward' allows for sea freight to all corners of the world for months after harvest. Sea freight may be more problematic for some of the less robust new cultivars, making the application of technology such as CA more common during shipping.

With sustainability as a common feature in all aspects of fruit production and marketing, there will be impacts throughout the postharvest sector from the types of packaging used to the speed at which fruit reach the market as a trade-off against fuel usage.

References

Ahmadi, E. (2012) Bruise susceptibilities of kiwifruit as affected by impact and fruit properties. *Research in Agricultural Engineering* 58, 107–113. DOI: 10.17221/57/2011-RAE

Alavi, M., Fullerton, C., Pidakala, P. and Burdon, J. (2021) Prediction of chilling injury risk in 'Zesy002' kiwifruit from softening early in storage. *New Zealand Journal of Crop and Horticultural Science* 50(2–3), 223–241. DOI: 10.1080/01140671.2021.1934041

Antunes, M.D.C. (2007) The role of ethylene in kiwifruit ripening and senescence. *Stewart Postharvest Review* 3(2), 9. DOI: 10.2212/spr.2007.2.9

Antunes, M.D.C. and Sfakiotakis, E.M. (1997) The effect of controlled atmosphere and ultra low oxygen on storage ability and quality of 'Hayward' kiwifruit. *Acta Horticulturae* 444, 613–618. DOI: 10.17660/ActaHortic.1997.444.83

Antunes, M.D.C. and Sfakiotakis, E.M. (2002) Ethylene biosynthesis and ripening behaviour of 'Hayward' kiwifruit subjected to some controlled atmospheres. *Postharvest Biology and Technology* 26(2), 167–179. DOI: 10.1016/S0925-5214(02)00040-6

Arpaia, M.L., Mitchell, F.G., Mayer, G. and Kader, A.A. (1984) Effects of delays in establishing controlled atmospheres on kiwifruit softening during and following storage. *Journal of the American Society for Horticultural Science* 109, 768–770.

Arpaia, M.L., Mitchell, F.G., Kader, A.A. and Mayer, G. (1985) Effects of 2% O_2 and varying concentrations of CO_2 with or without C_2H_4 on the storage performance of kiwifruit. *Journal of the American Society for Horticultural Science* 110, 200–203.

Arpaia, M.L., Mitchell, F.G., Kader, A.A. and Mayer, G. (1986) Ethylene and temperature effects on softening and white core inclusions of kiwifruit stored in air or controlled atmospheres. *Journal of the American Society for Horticultural Science* 111, 149–153.

Arpaia, M.L., Mitchell, F.G. and Kader, A.A. (1994a) Postharvest physiology and causes of deterioration. In: Hasey, J.K., Johnson, R.S., Grant, J.G. and Reil, W.O. (eds) *Kiwifruit Growing and Handling*. University of California Division of Agriculture and Natural Resources Publication 3344, Oakland, California, pp. 88–93.

Arpaia, M.L., Mitchell, F.G. and Mayer, G. (1994b) Cooling, storage, transportation and distribution. In: Hasey, J.K., Johnson, R.S., Grant, J.G. and Reil, W.O. (eds) *Kiwifruit Growing and Handling*. University of California Division of Agriculture and Natural Resources Publication 3344, Oakland, California, pp. 108–115.

Asiche, W.O., Mitalo, O.W., Kasahara, Y., Tosa, Y., Mworia, E.G. *et al.* (2018) Comparative transcriptome analysis reveals distinct ethylene-independent regulation of ripening in response to low temperature in kiwifruit. *BMC Plant Biology* 18, 47. DOI: 10.1186/s12870-018-1264-y

Atkinson, R.G., Gunaseelan, K., Wang, M.Y., Luo, L., Wang, T. *et al.* (2011) Dissecting the role of climacteric ethylene in kiwifruit (*Actinidia chinensis*) ripening using an ACC-oxidase knockdown line. *Journal of Experimental Botany* 62, 3821–3835. DOI: 10.1093/jxb/err063

Bauchot, A.D., Hallett, I.C., Redgwell, R.J. and Lallu, N. (1999) Cell wall properties of kiwifruit affected by low temperature breakdown. *Postharvest Biology and Technology* 16, 245–255. DOI: 10.1016/S0925-5214(99)00016-2

Biswas, P. and Brummell, D.A. (2019) Chilling injury. In: de Freitas, S.T. and Pareek, S. (eds) *Postharvest Physiological Disorders in Fruits and Vegetables*. CRC Press, Boca Raton, Florida, pp. 67–87. DOI: 10.1201/b22001

Botondi, R., Russo, V. and Mencarelli, F. (2011) Study of anaerobic metabolism of kiwifruits in ULO, LO, or regular CA at room temperature. *Acta Horticulturae* 913, 567–574. DOI: 10.17660/ActaHortic.2011.913.77

Boyd, L.M., Ferguson, I.B., Thorp, T.G., de Silva, N.H., Barnett, A.M. *et al.* (2006) Determining the relationship between fruit nutrient status and the development of physiological pitting in kiwifruit. *Acta Horticulturae* 721, 279–284. DOI: 10.17660/ActaHortic.2006.721.38

Brook, P. (1992) Botrytis stem-end rot and other storage diseases of kiwifruit – a review. *Acta Horticulturae* 297, 545–550. DOI: 10.17660/ActaHortic.1992.297.71

Burdon, J. (2015) Soluble solids revisited: a maturity or harvest index for kiwifruit. *Acta Horticulturae* 1096, 257–266. DOI: 10.17660/ActaHortic.2015.1096.28

Burdon, J. (2018) New cultivars: physiological challenges to commercial success. *Acta Horticulturae* 1218, 45–60. DOI: 10.17660/ActaHortic.2018.1218.6

Burdon, J. (2019a) Kiwifruit biology: the commercial implications of fruit maturation. *Horticultural Reviews* 46, 385–421. DOI: 10.1002/9781119521082.ch8

Burdon, J. (2019b) Kiwifruit quality: management in the supply chain. *Scripta Horticulturae* 20, 153–158.

Burdon, J. (2020) 18.7 Subtropical fruits: kiwifruit. In: Gil, M.I. and Beaudry, R. (eds) *Controlled and Modified Atmospheres for Fresh and Fresh-cut Produce*. Elsevier, London, UK, pp. 447–454. DOI: 10.1016/C2015-0-02025-1

Burdon, J. (2022) Understanding kiwifruit physiology: why we can predict aspects of postharvest performance. *Acta Horticulturae* 1332, 253–260. DOI: 10.17660/ActaHortic.2022.1332.34

Burdon, J. and Lallu, N. (2011) Kiwifruit (*Actinidia* spp.). In: Yahia, E.M. (ed.) *Postharvest Biology and Technology of Tropical and Subtropical Fruit. Vol. 3: Cocona to Mango*. Woodhead Publishing Limited, Cambridge, UK, pp. 326–360. DOI: 10.1533/9780857092885.326

Burdon, J. and Lallu, N. (2022) Minimising chilling injury in Italian-grown 'Hort16A' kiwifruit. *Acta Horticulturae* 1332, 343–349. DOI: 10.17660/ActaHortic.2022.1332.45

Burdon, J., McLeod, D., Lallu, N., Gamble, J., Petley, M. *et al.* (2004) Consumer evaluation of 'Hayward' kiwifruit of different at-harvest dry matter contents. *Postharvest Biology and Technology* 34, 245–255. DOI: 10.1016/j.postharvbio.2004.04.009

Burdon, J., Lallu, N., Francis, K. and Boldingh, H. (2007) The susceptibility of kiwifruit to low temperature breakdown is associated with pre-harvest temperatures and at-harvest soluble solids content. *Postharvest Biology and Technology* 43, 283–290. DOI: 10.1016/j.postharvbio.2006.09.011

Burdon, J., Pidakala, P., Martin, P., McAtee, P.A., Boldingh, H.L. *et al.* (2014a) Postharvest performance of the yellow-fleshed 'Hort16A' kiwifruit in relation to fruit maturation. *Postharvest Biology and Technology* 92, 98–106. DOI: 10.1016/j.postharvbio.2014.01.004

Burdon, J., Wohlers, M., Pidakala, P., Laurie, T., Punter, M. *et al.* (2014b) The potential for commonly measured at-harvest fruit characteristics to predict chilling susceptibility of 'Hort16A' kiwifruit. *Postharvest Biology and Technology* 94, 41–48. DOI: 10.1016/j.postharvbio.2014.03.005

Burdon, J., Punter, M., Billing, D., Pidakala, P. and Kerr, K. (2014c) Shrivel development in kiwifruit. *Postharvest Biology and Technology* 87, 1–5. DOI: 10.1016/j.postharvbio.2013.07.031

Burdon, J., Punter, M., Billing, D., Pidakala, P. and Kerr, K. (2015) Shrivel in kiwifruit. *Acta Horticulturae* 1096, 309–315. DOI: 10.17660/ActaHortic.2015.1096.34

Burdon, J., Pidakala, P., Martin, P., Billing, D. and Boldingh, H. (2016) Fruit maturation and the soluble solids harvest index for 'Hayward' kiwifruit. *Scientia Horticulturae* 213, 193–198. DOI: 10.1016/j.scienta.2016.10.027

Burdon, J., Pidakala, P., Martin, P. and Billing, D. (2017) Softening of 'Hayward' kiwifruit on the vine and in storage: the effects of temperature. *Scientia Horticulturae* 220, 176–182. DOI: 10.1016/j.scienta.2017.04.004

Burdon, J., Martin, P., Ireland, H., Schaffer, R., McAtee, P. *et al.* (2021) Transcriptomic analysis reveals differences in fruit maturation between two kiwifruit cultivars. *Scientia Horticulturae* 286, 110207. DOI: 10.1016/j.scienta.2021.110207

Burdon, J., Fullerton, C. and Billing, D. (2022) Degreening, softening and chilling sensitivity of early harvest 'Zesy002' kiwifruit under a high temperature controlled atmosphere. *Horticulturae* 8, 125. DOI: 10.3390/horticulturae8020125

Clark, C.J., McGlone, V.A., De Silva, H.N., Manning, M.A., Burdon, J. *et al.* (2004) Prediction of storage disorders of kiwifruit (*Actinidia chinensis*) based on visible-NIR spectral characteristics at harvest. *Postharvest Biology and Technology* 32, 147–158. DOI: 10.1016/j.postharvbio.2003.11.004

Crisosto, C.H. (1997) Final preconditioning guidelines for kiwifruit shippers. *Central Valley Postharvest Newsletter* 6, 1–4.

Crisosto, C.H. and Crisosto, G.M. (2001) Understanding consumer acceptance of early harvested 'Hayward' kiwifruit. *Postharvest Biology and Technology* 22, 205–213. DOI: 10.1016/S0925-5214(01)00097-7

Crisosto, G.U., Mitchell, F.G., Arpaia, M.L. and Mayer, G. (1984) The effect of growing location and harvest maturity on the storage performance and quality of 'Hayward' kiwifruit. *Journal of the American Society for Horticultural Science* 109(4), 584–587.

De Silva, H.N., Hall, A.J., Burdon, J., Lallu, N., Connolly, P. *et al.* (2007) Modelling the effect of holding temperature on flesh de-greening of 'Hort16A' (ZESPRI™ GOLD) kiwifruit. *Acta Horticulturae* 753, 769–775. DOI: 10.17660/ActaHortic.2007.753.101

Du, D., Wang, B., Wang, J., Yao, F. and Hong, X. (2019) Prediction of bruise susceptibility of harvested kiwifruit (*Actinidia chinensis*) using finite element method. *Postharvest Biology and Technology* 152, 36–44. DOI: 10.1016/j.postharvbio.2019.02.013

Feng, J., Maguire, K.M. and MacKay, B.R. (2003a) Factors affecting ethylene production of 'Hayward' kiwifruit. *Acta Horticulturae* 610, 203–209. DOI: 10.17660/ActaHortic.2003.610.27
Feng, J., MacKay, B.R. and Maguire, K.M. (2003b) Variation in firmness of packed 'Hayward' kiwifruit. *Acta Horticulturae* 610, 211–218. DOI: 10.17660/ActaHortic.2003.610.28
Gambi, F., Pilkington, S.M., McAtee, P.A., Donati, I., Schaffer, R.J. *et al.* (2018) Fruit of three kiwifruit (*Actinidia chinensis*) cultivars differ in their degreening response to temperature after harvest. *Postharvest Biology and Technology* 141, 16–23. DOI: 10.1016/j.postharvbio.2018.03.009
Gao, M., Guo, W., Huang, X., Du, R. and Zhu, X. (2021) Effect of pressing and impacting bruises on optical properties of kiwifruit flesh. *Postharvest Biology and Technology* 172, 111385. DOI: 10.1016/j.postharvbio.2020.111385
Gerasopoulos, D., Chlioumis, G. and Sfakiotakis, E. (2006) Non-freezing points below zero induce low-temperature breakdown of kiwifruit at harvest. *Journal of the Science of Food and Agriculture* 86, 886–890. DOI: 10.1002/jsfa.2429
Ghasemnezhad, M., Fullerton, C., Billing, D., Ampomah-Dwamena, C. and Burdon, J. (2022) Fruit degreening in different cultivars of kiwifruit. *Acta Horticulturae* 1332, 285–293. DOI: 10.17660/ActaHortic.2022.1332.38
Glowacz, M. (2020) Advances in the use of ozone in the disinfection of horticultural produce. In: Watkins, C.B. (ed.) *Advances in Postharvest Management of Horticultural Produce*. Burleigh-Dodds Science Publishing Ltd, Cambridge, UK, pp. 297–324. DOI: 10.19103/AS.2019.0055.12
Gong, H.J., Fullerton, C., Billing, D. and Burdon, J. (2020) Retardation of 'Hayward' kiwifruit tissue zone softening during storage by 1-methylcyclopropene. *Scientia Horticulturae* 259, 108791. DOI: 10.1016/j.scienta.2019.108791
Gorini, F. (1991) Skin pitting of kiwifruit during storage. *Acta Horticulturae* 297, 595–598. DOI: 10.17660/ActaHortic.1992.297.78
Gwanpua, S.G., Saeed, M., Jabbar, A. and Heyes, J. (2019) Kiwifruit. In: de Freitas, S.T. and Pareek, S. (eds) *Postharvest Physiological Disorders in Fruits and Vegetables*. CRC Press, Boca Raton, Florida, pp. 421–434. DOI: 10.1201/b22001
Hallett, I.C. and Sutherland, P.W. (2005) Structure and development of kiwifruit skins. *International Journal of Plant Sciences* 166, 693–704. DOI: 10.1086/431232
Harman, J. (1981) Kiwifruit maturity. *Orchardist of New Zealand* 54, 126–127, 130.
Harman, J.E. and McDonald, B. (1983) Controlled atmosphere storage of kiwifruit: effects on storage life and fruit quality. *Acta Horticulturae* 138, 195–201. DOI: 10.17660/ActaHortic.1983.138.22
Hopkirk, G. and Clark, C. (1992) Postharvest fruit losses in the New Zealand kiwifruit industry. *Acta Horticulturae* 297, 611–616. DOI: 10.17660/ActaHortic.1992.297.81
Hyodo, H. and Fukasawa, R. (1985). Ethylene production in kiwifruit. *Journal of the Japanese Society for Horticultural Science* 54, 209–215.
Jabbar, A. and East, A. (2016) Quantifying the ethylene induced softening and low temperature breakdown of 'Hayward' kiwifruit in storage. *Postharvest Biology and Technology* 113, 87–94. DOI: 10.1016/j.postharvbio.2015.11.002
Jiao, J., Guo, L., Liu, H., Wang, X., He, Y. *et al.* (2021) Effect of different packaging film thicknesses on chilling injury in postharvest 'Cuixiang' kiwifruit. *New Zealand Journal of Crop and Horticultural Science* 49, 168–181. DOI: 10.1080/01140671.2020.1824998
Jordan, R.B., Walton, E.F., Klages, K.U. and Seeleye, R.J. (2000) Postharvest fruit density as an indicator of dry matter and ripened soluble solids of kiwifruit. *Postharvest Biology and Technology* 20, 163–173.
Kader, A.A. (2002) *Postharvest Technology of Horticultural Crops*, 3rd edn. University of California Agriculture and Natural Resources Publication 3311, Oakland, California.
Kanellis, A.K., Tonutti, P. and Perata, P. (2009) Biochemical and molecular aspects of modified and controlled atmospheres. In: Yahia, E.M. (ed.) *Modified and Controlled Atmospheres for the Storage, Transportation and Packaging of Horticultural Commodities*. CRC Press, Boca Raton, Florida, pp. 553–567.
Kempler, C., Kabaluk, J.T. and Toivonen, P.M.A. (1992) Effect of environment and harvest date on maturation and ripening of kiwifruit in British Columbia. *Canadian Journal of Plant Science* 72, 863–869. DOI: 10.4141/cjps92-108
Kim, H.O., Hewitt, E.W. and Lallu, N. (1999) The role of ethylene in kiwifruit softening. *Acta Horticulturae* 498, 255–262. DOI: 10.17660/ActaHortic.1999.498.29
Lallu, N. (1997) Low temperature breakdown in kiwifruit. *Acta Horticulturae* 444, 579–586. DOI: 10.17660/ActaHortic.1997.444.89

Lallu, N. and Webb, D.J. (1997) Physiological and economic analysis of precooling kiwifruit. *Acta Horticulturae* 444, 691–697. DOI: 10.17660/ActaHortic.1997.444.106

Lallu, N., Searle, A.N. and MacRae, E.A. (1989) An investigation of ripening and handling strategies for early season kiwifruit (*Actinidia deliciosa* cv. 'Hayward'). *Journal of the Science of Food and Agriculture* 47, 387–400. DOI: 10.1002/jsfa.2740470402

Lallu, N., Rose, K., Wiklund, C. and Burdon, J. (1999) Vibration induced physical damage in packed 'Hayward' kiwifruit. *Acta Horticulturae* 498, 307–312. DOI: 10.17660/ActaHortic.1999.498.36

Lallu, N., Burdon, J., Yearsley, C.W. and Billing, D. (2003) Commercial practices used for controlled atmosphere storage of 'Hayward' kiwifruit. *Acta Horticulturae* 610, 245–251. DOI: 10.17660/ActaHortic.2003.610.32

Lallu, N., Burdon, J., Billing, D., Burmeister, D., Yearsley, C. *et al.* (2005) Effect of carbon dioxide removal systems on volatile profiles and quality of 'Hayward' kiwifruit stored in controlled atmosphere rooms. *HortTechnology* 15, 253–260. DOI: 10.21273/HORTTECH.15.2.0253

Li, H., Zhu, Y., Luo, F., He, H., Yuan, H. *et al.* (2015) Use of controlled atmospheres to maintain postharvest quality and improve storage stability of a novel red-fleshed kiwifruit (*Actinidia chinensis* Planch. var. *rufopulpa* [C.F. Liang et R.H. Huang] C.F. Liang et A.R. Ferguson). *Journal of Food Processing and Preservation* 39, 907–914. DOI: 10.1111/jfpp.12303

Lu, Q., Tang, M.-H., Cai, J.-R., Zhao, J.-W. and Vittayapadung, S. (2011) Vis/NIR hyperspectral imaging for detection of hidden bruises on kiwifruit. *Czech Journal of Food Science* 29(6), 595–602. DOI: 10.17221/69/2010-CJFS

Ma, Q., Suo, J., Huber, D., Dong, Z., Han, Y. *et al.* (2014) Effect of hot water treatments on chilling injury and expression of a new C-repeat binding factor (*CBF*) in 'Hongyang' kiwifruit during low temperature storage. *Postharvest Biology and Technology* 97, 102–110. DOI: 10.1016/j.postharvbio.2014.05.018

MacRae, E. and Redgwell, R. (1992) Softening in kiwifruit. *Postharvest News and Information* 3, 49N–52N.

MacRae, E.A., Bowen, J.H. and Stec, M.G.H. (1989) Maturation of kiwifruit (*Actinidia deliciosa* cv. Hayward) from two orchards: differences in composition of the tissue zones. *Journal of the Science of Food and Agriculture* 47, 401–416.

Manning, M. and Lallu, N. (1997) Fungal diseases of kiwifruit stored in controlled atmosphere conditions in New Zealand. *Acta Horticulturae* 444, 725–731. DOI: 10.17660/ActaHortic.1997.444.111

Manning, M., Burdon, J., De Silva, N., Meier, X., Pidakala, P. *et al.* (2016) Maturity and postharvest temperature management affect rot expression in 'Hort16A' kiwifruit. *Postharvest Biology and Technology* 113, 40–47. DOI: 10.1016/j.postharvbio.2015.10.012

Manning, M.A., Pak, H.A. and Beresford, R.M. (2010) Non-fungicidal control of Botrytis storage rot in New Zealand kiwifruit through pre- and post-harvest crop management. In: Prusky, D. and Gullino, M.L. (eds) *Plant Pathology in the 21st Century*, Vol. 2. *Post-harvest Pathology*. Springer Science + Business Media B.V., Berlin, Germany, pp. 183–196. DOI: 10.1007/978-1-4020-8930-5_13

McAtee, P.A., Richardson, A.C., Nieuwenhuizen, N.J., Gunaseelan, K., Hoong, L. *et al.* (2015) The hybrid ethylene and non-ethylene ripening response in kiwifruit (*Actinidia chinensis*) is associated with differential regulation of MADS-box transcription factors. *BMC Plant Biology* 15, 304. DOI: 10.1186/s12870-015-0697-9

McDonald, B. (1990) Precooling, storage, and transport of kiwifruit. In: Warrington, I.J., and Weston, G.C. (eds) *Kiwifruit: Science and Management*. Ray Richards Publisher in association with the New Zealand Society for Horticultural Science, Auckland, New Zealand, pp. 429–459.

McDonald, B. and Harman, J.E. (1982) Controlled atmosphere storage of kiwifruit. 1. Effect on fruit firmness and storage life. *Scientia Horticulturae* 17, 113–123. DOI: 10.1016/0304-4238(82)90003-6

McGlone, V.A. and Kawano, S. (1998) Firmness, dry matter and soluble solids assessment of postharvest kiwifruit by NIR spectroscopy. *Postharvest Biology and Technology* 13, 131–141. DOI: 10.1016/S0925-5214(98)00007-6

McMath, K.L., Paterson, V.J., Young, H., Ball, R.D. and MacRae, E.A. (1992) Factors affecting the sensory perception of sweetness and acidity in kiwifruit. *Acta Horticulturae* 297, 489–500. DOI: 10.17660/ActaHortic.1992.297.64

Mencarelli, F., Massantini, R. and Botondi, R. (1996) Influence of impact surface and temperature on the ripening response of kiwifruit. *Postharvest Biology and Technology* 8, 165–177.

Minas, I.S., Tanou, G., Krokida, A., Karagiannis, E., Belghazi, M. *et al.* (2018) Ozone-induced inhibition of kiwifruit ripening is amplified by 1-methylcyclopropene and reversed by exogenous ethylene. *BMC Plant Biology* 18, 358. DOI: 10.1186/s12870-018-1584-y

Minchin, P.E.H., de Silva, N., Snelgar, W.P., Richardson, A.C. and Thorp, T.G. (2003) Modelling of colour development in the fruit of *Actinidia chinensis* 'Hort16A'. *New Zealand Journal of Crop and Horticultural Science* 31, 41–53. DOI: 10.1080/01140671.2003.9514234

Montanaro, G., Dichio, B., Xiloyannis, C. and Lang, A. (2012) Fruit transpiration in kiwifruit: environmental drivers and predictive model. *AoB PLANTS* 2012, pls036. DOI: 10.1093/aobpla/pls036

Murakami, S., Ikoma, Y. and Yano, M. (2014) Low temperature increases ethylene sensitivity in *Actinidia chinensis* 'Rainbow Red' kiwifruit. *Journal of the Japanese Society for Horticultural Science* 83, 322–326. DOI: 10.2503/jjshs1.CH-104

Mworia, E.G., Yoshikawa, T., Salikon, N., Oda, C., Asiche, W.O. *et al.* (2012) Low-temperature-modulated fruit ripening is independent of ethylene in 'Sanuki Gold' kiwifruit. *Journal of Experimental Botany* 63, 963–971. DOI: 10.1093/jxb/err324

Niklis, N., Sfakiotakis, E. and Thanassoulooulos, C.C. (1997) Ethylene production by *Botrytis cinerea*, kiwifruit and botrytis rotted kiwifruit under several storage temperatures. *Acta Horticulturae* 444, 733–737. DOI: 10.17660/ActaHortic.1997.444.112

OECD (Organisation for Economic Co-operation and Development) (1992) International Standardisation of Fruit and Vegetables. Kiwifruit. OECD Publications Service, Paris, France.

Patterson, K., Burdon, J. and Lallu, N. (2003) 'Hort16A' kiwifruit: progress and issues with commercialisation. *Acta Horticulturae* 610, 267–273. DOI: 10.17660/ActaHortic.2003.610.35

Pennycook, S.R. and Manning, M.A. (1992) Picking wound curing to reduce botrytis storage rot of kiwifruit. *New Zealand Journal of Crop and Horticultural Science* 20, 357–360. DOI: 10.1080/01140671.2003.9514234

Pranamornkith, T., East, A. and Hayes, J. (2012) Influence of exogenous ethylene during refrigerated storage on storability and quality of *Actinidia chinensis* (cv. Hort16A). *Postharvest Biology and Technology* 64, 1–8. DOI: 10.1016/j.postharvbio.2011.09.011

Regiroli, G. and Vriends, P. (2007) SmartFresh^{SM} (1-methylcyclopropene) benefits for kiwifruit. *Acta Horticulturae* 753, 745–753. DOI: 10.17660/ActaHortic.2007.753.98

Richardson, A.C., Boldingh, H.L., McAtee, P.A., Gunaseelan, K., Luo, Z.W. *et al.* (2011) Fruit development of the diploid kiwifruit, *Actinidia chinensis* 'Hort16A'. *BMC Plant Biology* 11, 182. DOI: 10.1186/1471-2229-11-182

Ritenour, M.A., Crisosto, C.H., Garner, D.T., Cheng, G.W. and Zoffoli, J.P. (1999) Temperature, length of cold storage and maturity influence the ripening rate of ethylene-preconditioned kiwifruit. *Postharvest Biology and Technology* 15, 107–115. DOI: 10.1016/S0925-5214(98)00074-X

Saltveit, M.E. (2020) Biological basis for CA and MA. In: Gil, M.I. and Beaudry, R. (eds) *Controlled and Modified Atmospheres for Fresh and Fresh-cut Produce*. Elsevier, London, pp. 3–22. DOI: 10.1016/B978-0-12-804599-2.00002-8

Seager, N.G., Warrington, I.J. and Hewett, E.W. (1996) Maturation of kiwifruit grown at different temperatures in controlled environments. *Journal of Horticultural Science* 71, 639–652. DOI: 10.1080/14620316.1996.11515444

Sevillano, L., Sanchez-Ballesta, M.T., Romojaro, F. and Flores, F.B. (2009) Physiological, hormonal and molecular mechanisms regulating chilling injury in horticultural species: postharvest technologies applied to reduce its impact. *Journal of the Science of Food and Agriculture* 89, 555–573. DOI: 10.1002/jsfa.3468

Smith, G.S., Klages, K.U., Green, T.G.A. and Walton, E.F. (1995) Changes in abscisic acid concentration, surface conductance, and water content of developing kiwifruit. *Scientia Horticulturae* 61, 13–27. DOI: 10.1016/0304-4238(94)00733-V

Sommer, N.F., Suadi, J.E. and Fortlage, R.J. (1994) Postharvest storage diseases. In: Hasey, J.K., Johnson, R.S., Grant, J.G. and Reil, W.O. (eds) *Kiwifruit Growing and Handling*. University of California Division of Agriculture and Natural Resources Publication 3344, Oakland, California, pp. 116–122.

Spadaro, D., Galliano, A., Pellegrino, C., Gilardi, G., Garibaldi, A. *et al.* (2010) Dry matter, mineral composition, and commercial storage practices influence the development of skin pitting caused by *Cadophora luteo-olivacea* on kiwifruit 'Hayward'. *Journal of Plant Pathology* 92(2), 349–356. DOI: 10.4454/jpp.v92i2.176

Suo, J., Li, H., Ban, Q., Han, Y., Meng, K. *et al.* (2018) Characteristics of chilling injury-induced lignification in kiwifruit with different sensitivities to low temperatures. *Postharvest Biology and Technology* 135, 8–18. DOI: 10.1016/j.postharvbio.2017.08.020

Tabatabaekoloor, R., Hashemi, S.J. and Taghizade, G. (2013) Vibration damage to kiwifruits during road transportation. *International Journal of Agriculture and Food Science Technology* 4(5), 467–474.

Testoni, A. (1992) Influence of storage conditions on softening of the columella. *Acta Horticulturae* 297, 587–593. DOI: 10.17660/ActaHortic.1992.297.77
Testoni, A., Grassi, B., Quaroni, S., Saracchi, M. and Sardi, M. (1997) Pitting on kiwifruit in storage caused by *Phialophora* sp. *Acta Horticulturae* 444, 751–756. DOI: 10.17660/ActaHortic.1997.444.115
Thomai, T.K. and Sfakiotakis, E.M. (1997) Effect of low-oxygen atmosphere on storage behaviour of kiwifruit. *Advances in Horticultural Science* 11(3), 137–141.
Thompson, A.K., Prange, R.K., Bancroft, R.D. and Puttongsiri, T. (2018) *Controlled Atmosphere Storage of Fruit and Vegetables*. 3rd edn. CAB International, Wallingford, UK.
Thorp, T.G., Ferguson, I.B., Boyd, L.M. and Barnett, A.M. (2003) Fruiting position, mineral concentration and incidence of physiological pitting in 'Hayward' kiwifruit. *Journal of Horticultural Science and Biotechnology* 78, 505–511. DOI: 10.1080/14620316.2003.11511655
Tyson, J.L., Manning, M.A., Everett, K.R. and Fullerton, R.A. (2019) *Neofabraea actinidiae* in New Zealand kiwifruit orchards: current status and knowledge gaps. *New Zealand Plant Protection* 72, 75–82. DOI: 10.30843/nzpp.2019.72.268
Walsh, K.B., McGlone, V.A. and Han, D.H. (2020) The use of near infra-red spectroscopy in postharvest decision support: a review. *Postharvest Biology and Technology* 163, 111139. DOI: 10.1016/j.postharvbio.2020.111139
Wang, F., Yang, Q., Zhao, Q. and Zhang, X. (2020) Roles of antioxidant capacity and energy metabolism in the maturity-dependent chilling tolerance of postharvest kiwifruit. *Postharvest Biology and Technology* 168, 111281. DOI: 10.1016/j.postharvbio.2020.111281
Wang, R., Martin, P., McAtee, P.A., Schaffer, R.J. and Burdon, J. (2022) New insights into the storage performance of 'Hayward' kiwifruit from a comparison of three cultivars. *New Zealand Journal of Crop and Horticultural Science* 50(4). DOI: 10.1080/01140671.2022.2096083
Wang, Z., Künnemeyer, R., McGlone, A., Sun, J. and Burdon, J. (2021) Comparison of a dual-laser and a Vis-NIR spectroscopy system for detection of chilling injury in kiwifruit. *Postharvest Biology and Technology* 175, 111418. DOI: 10.1016/j.postharvbio.2020.111418
Wang, Z., Künnemeyer, R., McGlone, A., Sun, J., Burdon, J. *et al.* (2022a) Non-destructive detection of chilling injury in kiwifruit using a dual-laser scanning system with a principal component analysis – back propagation neural network. *Journal of Near Infrared Spectroscopy* 30(2), 67–73. DOI: 10.1177/09670335211061842
Wang, Z., Künnemeyer, R., McGlone, A., Burdon, J., Sun, J. *et al.* (2022b) Spectral discrimination of kiwifruit with chilling injury from fruit with other types of tissue damage in high-speed grading applications. *Acta Horticulturae* 1332, 277–284. DOI: 10.17660/ActaHortic.2022.1332.37
Wei, X., Xie, D., Mao, L., Xu, C., Luo, Z. *et al.* (2019) Excess water loss induced by simulated transport vibration in postharvest kiwifruit. *Scientia Horticulturae* 250, 113–120. DOI: 10.1016/j.scienta.2019.02.009
Xia, Y.X., Chen, T., Qin, G.Z., Li, B.Q. and Tian, S.P. (2016) Synergistic action of antioxidative systems contributes to the alleviation of senescence of kiwifruit. *Postharvest Biology and Technology* 111, 15–24. DOI: 10.1016/j.postharvbio.2015.07.026
Yanez Lopez, L., Galicia, R.M., Alonso, R., Cabrera, D., Rocha, J. *et al.* (1999) Effect of CO_2 enriched atmospheres on ethanol accumulation in different ripeness stages of kiwifruit. *Acta Horticulturae* 498, 293–298. DOI: 10.17660/ActaHortic.1999.498.34
Yang, Q., Rao, J., Yi, S., Meng, K., Wu, J. *et al.* (2012) Antioxidant enzyme activity and chilling injury during low-temperature storage of kiwifruit cv. Hongyang exposed to gradual postharvest cooling. *Horticulture, Environment and Biotechnology* 53(6), 505–512. DOI: 10.1007/s13580-012-0101-8
Yang, Q.-Z., Rao, J.-P and Wang, Y.-P. (2013a) Effect of different cooling modes on chilling injury, quality and active oxygen metabolism in harvested 'Xuxiang' kiwifruit. *Acta Horticulturae Sinica* 40(4), 651–662.
Yang, Q.-Z., Zhang, Z., Rao, J., Wang, Y., Sun, Z. *et al.* (2013b) Low-temperature conditioning induces chilling tolerance in 'Hayward' kiwifruit by enhancing antioxidant enzyme activity and regulating endogenous hormones levels. *Journal of the Science of Food and Agriculture* 93, 3691–3699. DOI: 10.1002/jsfa.6195
Yildirim, I. and Bayir, A. (2013) Effects of different storage techniques on the antioxidant capacity, total phenolics and flavonoids of 'Hayward' kiwifruit. *Acta Horticulturae* 1012, 807–813. DOI: 10.17660/ActaHortic.2013.1012.109
Yin, X., Allan, A.C., Xu, Q., Burdon, J., Dejnoprat, S. *et al.* (2012) Differential expression of kiwifruit ERF genes in response to postharvest abiotic stress. *Postharvest Biology and Technology* 66, 1–7. DOI: 10.1016/j.postharvbio.2011.11.009
Zhao, M., Bronlund, J.E. and East, A.R. (2015) Effect of cooling rate on kiwifruit firmness and rot incidence in subsequent storage. *Acta Horticulturae* 1079, 313–318. DOI: 10.17660/ActaHortic.2015.1079.38

21 Consumer Research into Fresh Kiwifruit: Fruit Quality, Consumer Methods and Future Trends

Denise C. Hunter[1]* and Anne White[2]
[1]*FORWARD Insight and Strategy, Auckland, New Zealand;* [2]*The New Zealand Institute for Plant and Food Research Ltd, Auckland, New Zealand*

21.1 Introduction

Despite vast genetic variation in the genus *Actinidia*, only a small number of species have significant commercial value. While the genetic variation represents opportunities for innovation, the ultimate success of commercial kiwifruit cultivars relies on consumer acceptance and repeat purchasing. Considering the investment required to bring a new kiwifruit cultivar to market, understanding the interactions between consumer variability in perceptions and preferences and fruit variability driven by on-orchard, postharvest and in-market management may improve the chances of success. The earlier this can be determined in the innovation pipeline, the better.

A consumer-based definition of food quality relates to the *perception* of all characteristics of a food, including sensory attributes, safety, convenience, cost and value, context and expectations (Cardello, 1995). When extended to kiwifruit quality specifically, this may also be true, although the value of select objective/instrumental measurements of fruit quality for predicting consumer responses should not be underestimated.

This chapter examines factors that contribute to fruit quality and have been demonstrated to influence consumer responses to kiwifruit. Methods used to evaluate sensory and consumer responses to kiwifruit are reviewed and, once in market, extrinsic factors that influence purchase decisions of kiwifruit are considered.

21.2 Environmental and Management Factors that Influence Consumer Responses to Kiwifruit

Consumers generally respond favourably to increasing ripe fruit soluble solids content (SSC) and/or dry matter (DM) (e.g. Harker *et al.*, 2009; Crisosto *et al.*, 2011; Jaeger *et al.*, 2011). SSC is primarily a measurement of soluble sugars, as well as soluble pectins, and organic, amino and ascorbic acids. DM is largely a measure of total carbohydrates in the fruit; sugars are accumulated and converted to starch during development on the vine (Okuse and Ryugo, 1981; MacRae *et al.*, 1989), and starch is then converted to soluble sugars as kiwifruit ripen (Matsumoto *et al.*, 1983). Ripe SSC (rSSC) is highly positively correlated with fruit DM (Jordan *et al.*, 1997)

*Corresponding author: denise@forward-insight.com

© CAB International 2024. *Kiwifruit: Botany, Production and Uses* (eds A.C. Richardson, J.N. Burdon and A.R. Ferguson)
DOI: 10.1079/9781800620933.0021

and the terms are often used interchangeably, particularly when referencing consumer responses. DM and rSSC are the main drivers of consumer responses to kiwifruit while other factors, such as fruit size or acidity, moderate the responses to DM. One possible exception is kiwifruit firmness, considering the concurrence of softening, conversion of starch to sugar and production of flavour volatiles during ripening, which lead to a smoother and juicier texture, sweeter taste and more complex flavour. While preferred firmness at eating ripeness varies among consumers (Stec *et al.*, 1989), it is crucial that firmness is controlled in consumer research to ensure preferences are not simply being driven by differences in fruit ripeness. Thus, while the influence of factors on consumer responses will be considered here, DM and firmness will be given special consideration owing to their substantial impact on preference.

21.2.1 Preharvest factors

Environmental and on-orchard factors that influence fruit quality (DM, rSSC, and/or size) include temperature (Richardson *et al.*, 2004; Snelgar *et al.*, 2007), manipulation of carbohydrate supply through canopy management techniques (Richardson and McAneney, 1990; Minchin *et al.*, 2010; Snelgar *et al.*, 2012; Le Lievre *et al.*, 2021), rootstock genotypes (Nitta and Ogasawara, 1999; Friend *et al.*, 2014; Richardson *et al.*, 2021) and time of harvest (Burdon *et al.*, 2004) (see Chapters 12 and 16 for more detail regarding DM manipulation on-orchard and starch metabolism in fruit, respectively).

Management of on-orchard factors is important because these factors have been demonstrated to influence consumer perceptions and preferences. For example, higher carbohydrate supply to fruit through canopy management resulted in fruit perceived to be sweeter than fruit from canes with low carbohydrate supply, reflecting the titratable acidity (TA), DM and SSC of the fruit (Le Lievre *et al.*, 2021). Higher DM is expected to accumulate in later-harvested fruit, and MacRae *et al.* (1989) demonstrated that later harvests of *A. chinensis* var. *deliciosa* 'Hayward' fruit were more acceptable than fruit from earlier harvests due to increased flavour intensity, fewer off-flavours and improved texture. However, early-harvested fruit with sufficient rSSC and relatively low ripe TA may still be acceptable to consumers, while kiwifruit with a combination of relatively low rSSC and high ripe TA may be rejected (Crisosto and Crisosto, 2001).

Care should be taken when interpreting consumer responses to different environmental and on-orchard factors if the DM of fruit evaluated by consumers is not controlled. For example, consumers preferred conventionally grown 'Hayward' fruit to organically grown fruit primarily due to increased flavour perception, reflecting the higher rSSC of conventionally grown fruit (Nunes-Damaceno *et al.*, 2013). However, with equivalent DM, conventionally and organically grown 'Hayward' kiwifruit were liked similarly by Japanese consumers (Harker *et al.*, 2009).

Factors that influence fruit size may influence both the texture and taste of kiwifruit. Small fruit from a range of genotypes were characterized by a firmer and more fibrous core while larger fruit types were perceived as juicier (Nardozza *et al.*, 2011). Smaller *A. chinensis* var. *chinensis* 'Zesy002' (marketed as Zespri SunGold™ Kiwifruit) fruit tend to have more acid, less sugar and a lower DM content than larger fruit of this cultivar (Seal *et al.*, 2018) and are thus less appealing to consumers (Harker *et al.*, 2009).

21.2.2 Postharvest storage

Cool storage is widely used to reduce softening of kiwifruit after harvest by delaying ripening. The temperature and length of storage can affect consumers' sensory perceptions of kiwifruit. Lower storage temperatures have been shown to result in fruit that is perceived as sweeter and less acidic than fruit stored at higher temperatures (Marsh *et al.*, 2004; Fisk *et al.*, 2006), reflecting the higher SSC and lower TA. However, storage at low temperature results in fruit with firmer cores than storage at 20°C (Jeffery *et al.*, 2013), and increased time in cool storage resulted in a significant reduction in aroma-related esters in *A. chinensis* var. *chinensis* 'Hort16A' and reduced retronasal odour and taste intensity scores (Günther *et al.*, 2015).

Use of controlled atmosphere (CA) and 1-methylcyclopropene (1-MCP) to increase storage life has little or no negative effect on fruit sensory characteristics (Harman and McDonald, 1989; Burdon *et al.*, 2007; Salazar *et al.*, 2019). When 1-MCP did negatively affect consumer responses, this was likely due to 1-MCP-treated fruit being firmer than control fruit (Deng *et al.*, 2015).

21.2.3 Ripening

As kiwifruit softens into the eating firmness range, SSC increases, and flavours develop. Fruit firmness is positively correlated with perceived acidity (Jordan and Seelye, 2009; Hunter *et al.*, 2018) and interacts with rSSC to predict edibility and consumer satisfaction (Harker *et al.*, 2019). As fruit softens from near ripe to overripe, total volatile production increases and some of the changes in volatile content could explain changes in fruit flavour detected by trained panellists (Wang *et al.*, 2011).

The different tissue zones in kiwifruit (core, inner pericarp and outer pericarp) have been shown to differ in softening patterns (Redgwell *et al.*, 1990, Hallett *et al.*, 1992), which can affect overall fruit firmness (Li *et al.*, 2016). These differences between tissue zones can manifest as the presence of hard cores in otherwise ripe fruit. Exposure to ethylene results in more rapid and synchronous ripening (Schröder and Atkinson, 2006) and reduces firmness differences between the core and outer pericarp (Zoffoli *et al.*, 2016). Ethylene treatment after cool storage has been shown to increase aroma-related esters in 'Hort16A' and a sensory panel found that 'tropical' and 'fruit candy' aromas were stronger and 'green' odour notes less intense in ethylene-ripened fruit (Günther *et al.*, 2015).

21.3 Current and Future Consumer Research Methods for Evaluating Kiwifruit

Developing new products from the *Actinidia* genus is resource-intensive. Consumer research is critical to maximize the chance of product success. Overall liking or preference has historically been the primary measure in consumer research based on the premise that higher liking will predict what consumers purchase and consume outside of the experimental setting, although this relationship is rarely validated (Tuorila *et al.*, 2008; Giacalone *et al.*, 2022). This section considers methods beyond liking, to extract maximal value from consumer research using *Actinidia* species.

21.3.1 Sensory profiling

Historically, consumer evaluation focused on preference/liking and product profiling was undertaken by trained or expert panellists. Recently, the relevance of this approach has been questioned, leading to the development and validation of methods generating sensory profiles by consumers, which allow for collection of information regarding preference and consumer-relevant sensory attributes in a single step.

'Check all that apply' (CATA) is one method that has been used by consumers to profile the sensory attributes of kiwifruit (e.g. Le Lievre *et al.*, 2021) and in programmes using innovative methods to explore consumers' perceptions of specific kiwifruit flavours (e.g. Zeng *et al.*, 2020). Researchers provide consumers with a list of predefined attributes expected to describe the products, and consumers select 'all that apply' after seeing/tasting the product. The method is versatile and efficient, but its success relies on having suitable numbers of consumers, and appropriate terms on the list. A minimum of 60–80 consumers is required to achieve data stability; more if the differences between samples are small (Ares *et al.*, 2014).

Selection of terms suitable for CATA use has received considerable attention. Terms integral to describing kiwifruit (e.g. 'sweet', 'juicy') should be included to prevent their perceived sensation being expressed using other similar, yet distinctly different, terms (e.g. 'fruity' or 'tropical') (Jaeger *et al.*, 2019a), known as 'dumping'. Acidity may also be considered integral to the eating experience of kiwifruit and has been demonstrated to discriminate between kiwifruit samples in most studies, although the specific term used to describe acidity may influence its discriminative capacity (e.g. sour/acidic, acid or acidic, sour, tangy, or sour fruity) (Jaeger *et al.*, 2019a). Which term to use may depend on where the research is taking place, as one expression may be used more in some cultures or translate more accurately. Other terms to consider may reflect general or specific flavours (e.g. 'fruity' vs. 'pineapple'), textures (e.g. 'crunchy', 'mushy'), and mouthfeel (e.g. 'prickly'/'tingly').

How CATA methodology is implemented in consumer research on kiwifruit depends on the objectives of the research and the sample set. For example, when evaluating a single kiwifruit cultivar where differences among the sample set may be minor and somewhat predictable, such as on-orchard manipulation of carbohydrate supply, a short list of terms may be sufficient (e.g. Le Lievre, 2017). Where the objective is to evaluate different types of kiwifruit (e.g. green-, red- or gold-fleshed kiwifruit) or the differences between samples cannot be predicted, the list of terms may need to be more encompassing. Jaeger *et al.* (2019a) have presented a range of terms applicable to kiwifruit and described their relative efficiency at discriminating between samples. Importantly, recent research suggests that citation frequencies (i.e. the number of participants selecting a given term) may be used as a proxy for perceived intensity (Vidal *et al.*, 2021).

Another sensory profiling approach that quantifies consumers' perceived attribute intensity is the 'just about right' (JAR) scale. Consumers rate the intensity of an attribute in relation to their internal perception of ideal (or near-ideal); is it too little, just about right, or too much? This question format is typically used for packaged goods, where products may be more easily reformulated, and this method has also proved helpful for differentiating kiwifruit samples. For example, in a sample set where on-orchard manipulation of carbohydrate supply did not significantly influence hedonic responses, larger proportions of consumers described the acidity as 'too much' and sweetness as 'not enough' in fruit with lower carbohydrate supply compared with control fruit or fruit with higher carbohydrate supply (Le Lievre, 2017). Furthermore, when used in conjunction with CATA questions, JAR scales can enhance product discrimination (Jaeger *et al.*, 2015).

21.3.2 Multi-factor interactions

While overall liking provides useful information regarding comparative preference in a product set, a multivariable approach to consumer evaluation of products provides a useful supplement to differentiate products, particularly those liked similarly (Jaeger *et al.*, 2017a, b). Additional measures may include emotion responses, familiarity or perceived uniqueness, and appropriateness and utility across consumption occasions.

Emotional associations

Measurement of consumers' emotion responses to products has been demonstrated to differentiate between similarly liked products or product attributes (e.g. Ristic *et al.*, 2019). Focus on research into global feelings of emotion has given rise to the circumplex-inspired emotion questionnaire, which includes 12 pairs of emotions that span the two dimensions of global emotions: valence (displeasure to pleasure) and arousal (inactivation to activation) (Jaeger *et al.*, 2021) (Fig. 21.1).

Using this circumplex approach in the fruit category, significant emotion linkages have been made with several sensory attributes using a diverse fruit sample set that included green-fleshed kiwifruit with relatively low and high DM (Jaeger *et al.*, 2019b). The strongest linkage across all fruit was between juicy and emotions associated with positive valence and high arousal (active/alert, energetic/excited, enthusiastic/inspired and happy/satisfied). Other attributes typically present in kiwifruit also tended to be positively linked with at least some of these terms, including sweet, sour and kiwifruit flavour. Kiwifruit flavour was linked with positive valence and low arousal (passive, quiet), potentially reflecting the milder flavour of lower DM kiwifruit. Negative valence emotions were also positively associated with sensory attributes that may be found in kiwifruit, such as overripe flavour, bland/lacking flavour and dry/not juicy, which were associated with low arousal, and sour, which was associated with higher arousal (jittery/nervous). These associations were defined using a broad fruit set. These findings suggest that fruit can stimulate different emotions, and kiwifruit with different sensory profiles may also be associated with different emotions.

Context – situation and occasion

In consumer research, it is generally accepted that situational appropriateness measures

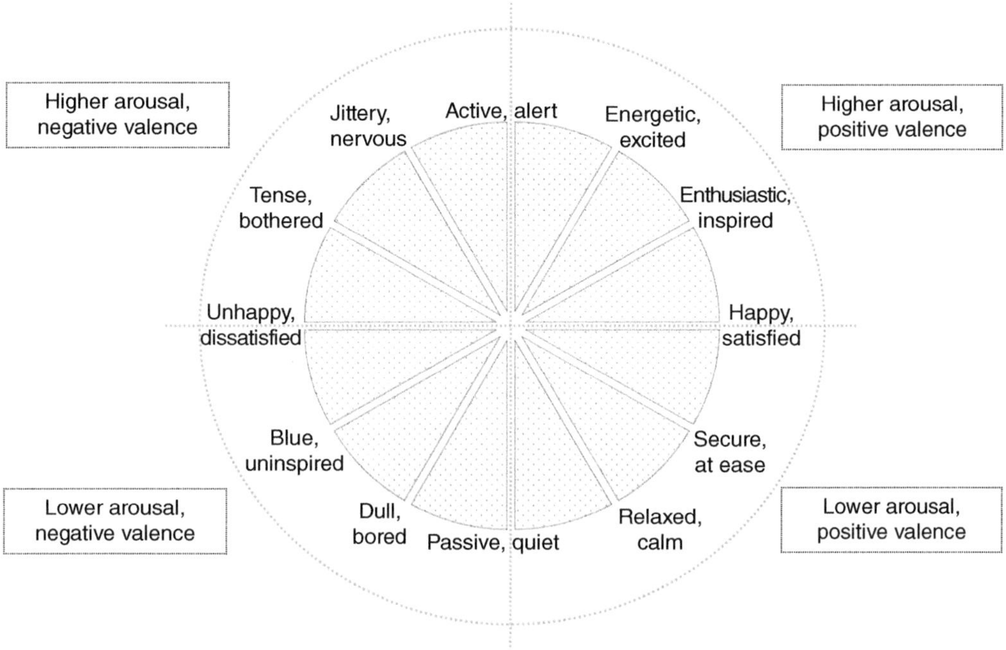

Fig. 21.1. Valence × arousal emotion circumplex framework. (Adapted from Jaeger *et al.*, 2021)

(i.e. the perceived fit between a product and its intended usage situation) are more predictive of consumer behaviour than simple measures of acceptance (Marshall and Bell, 2003; Giacalone and Jaeger, 2016).

A study of food diaries of New Zealand women revealed that kiwifruit were usually consumed earlier in the day and especially with breakfast (Bava *et al.*, 2012). However, more recent research indicates the situations in which kiwifruit are consumed may be more nuanced and influenced by factors such as familiarity and sensory profile. Indeed, consumer evaluation of conceptual kiwifruit products that ranged in flesh colour, flesh type and skin type/colour showed that more familiar kiwifruit types (green- and gold-fleshed kiwifruit) tended to be rated more often as appropriate for most stated situations than red- and pink-fleshed concepts (Giacalone and Jaeger, 2016). The more familiar green- and gold-fleshed types were considered appropriate by the largest proportion of consumers for use 'in a lunchbox' and 'with breakfast' (Giacalone and Jaeger, 2016). Therefore, lack of familiarity may be a barrier for new kiwifruit cultivars to overcome.

It is important to note that there may be interactions between these factors. For example, evoking (imagining) a kiwifruit consumption context in a research setting may influence the emotion responses. Evocation of the consumption situation (breakfast, lunch or dinner) during evaluation of gold-fleshed kiwifruit resulted in highly significant ($p < 0.001$) differences in the proportions of consumers indicating the emotion responses of active, energetic, enthusiastic, free and friendly (Piqueras-Fiszman and Jaeger, 2014). Evocation of situations also allows evaluation of the impact of extrinsic factors during kiwifruit consumption, such as the format of fruit (whole or cut), availability of utensils, location and social situation (Piqueras-Fiszman and Jaeger, 2015).

21.4 Extrinsic Factors that Influence Consumer Responses to Kiwifruit

Several extrinsic factors influence consumer purchasing and consumption of kiwifruit. Naturally, the marketing mix has a large impact on purchase likelihood and frequency, including the product, its branding and packaging, the places where kiwifruit is sold, where and how kiwifruit is displayed, the price and how it is promoted. Most of these are beyond the scope of this chapter but will be addressed as they relate to extrinsic factors specific to kiwifruit.

21.4.1 Convenience

Convenience is one of the major drivers or barriers of fruit consumption (Briz *et al.*, 2008), and describes both the convenience of the eating experience (i.e. easy to prepare, with little mess and waste) suitable for the entire family, and the consistency and availability of fruit for purchase (Harker *et al.*, 2007).

Some rejectors of kiwifruit do so because it is unsuitable to eat 'on the run' or to be included in a lunchbox and can be a 'hassle to peel and cut' (Harker *et al.*, 2007). Innovations in kiwifruit breeding, such as cultivars with an edible or peelable skin, may increase its appeal to those who would otherwise reject kiwifruit based on convenience. Indeed, the small size and edible skin of *A. arguta* 'Hortgem Tahi' were described as convenient, even by rejectors of kiwifruit (Harker *et al.*, 2007).

Consistency and availability can be barriers to consumption of kiwifruit because the skin provides few clues to ripeness (Harker *et al.*, 2007), unlike a banana, for example. Increased consumption of kiwifruit may, therefore, be achieved by effective ripening and quality management. The growth in demand for kiwifruit in the USA has been at least partly attributed to kiwifruit being 'more ready to eat when you buy it and provides a more consistent eating experience' (Rabobank, n.d.).

Growing demand has also been attributed to the supply of kiwifruit to the USA market 52 weeks of the year, which helps establish a regular consumption pattern (Rabobank, n.d.). Purchase convenience was found to be the main reason the majority (73%) of Chinese consumers purchased kiwifruit online, driven by access to a greater variety of products, convenience of home delivery and access to special offers/promotions (Tait *et al.*, 2018).

21.4.2 Health benefits

The health benefits of kiwifruit are many and varied (Skinner *et al.*, 2013), although not all are well known to consumers. Kiwifruit does appear to have a health halo in the eye of consumers, with more than 80% of respondents from Shanghai indicating in an online survey that they agreed with the statement, 'I eat kiwifruit mainly for the health benefits' (Tait *et al.*, 2018). Respondents in a Japanese online study showed health outcomes were important for almost 90% of regular kiwifruit consumers (Tait *et al.*, 2020).

21.4.3 Country of origin

Country of origin plays an important role in consumer choice of kiwifruit, at least in China (Tait *et al.*, 2018) and Japan (Tait *et al.*, 2020). Chinese consumers demonstrated a willingness to pay a premium for fruit from New Zealand, Chile and China, with the relative value reflected by this order (Tait *et al.*, 2018). Interestingly, participants stated that Zespri kiwifruit from New Zealand were important to them, and that they would purchase a similar amount of Zespri kiwifruit even if they originated from other countries, such as China, France and Italy. It is possible that consistent quality contributes to the high purchase intent for Zespri kiwifruit. However, although quality was an important attribute for participants, food safety was more important.

Japanese consumers also demonstrated a willingness to pay a premium for New Zealand kiwifruit, although a larger premium was demonstrated for kiwifruit grown in Japan (Tait *et al.*, 2020). However, New Zealand-grown kiwifruit was purchased most frequently, and 59% of respondents ranked New Zealand kiwifruit first for quality, compared with only 20% for Japanese-grown kiwifruit. While these two findings appear contradictory, most respondents (78%) agreed that supporting local producers was one of their most important considerations when buying food. Taste was also rated as important by Japanese respondents, closely followed by freshness, value for price and food safety certification. Again, this suggests that fruit quality and safety are more important than country of origin, although, increasingly, sustainability may be expected to influence consumer purchasing.

21.5 Current and Future Trends

The kiwifruit industry, as with most primary production, is facing an increasing challenge from consumers in terms of sustainability and environmental impact. A willingness by Chinese consumers to pay approximately 50% more for organic production has been demonstrated, with social responsibility, water use and pollution minimization valued slightly lower (Tait *et al.*, 2018). Japanese consumers (~50%) also considered reduced environmental impact and socially responsible production as being important when buying kiwifruit, with organic production and carbon neutral production also considered important (Tait *et al.*, 2020). While on-orchard management is the obvious target for improvement, the rich genetic resources of *Actinidia* species represent an opportunity for enhancing sustainability (Huang *et al.*, 2004). In this quest, consumer research will be critical in identifying selections that both promote environmental health and deliver an excellent eating experience.

The rich genetic resources of the *Actinidia* species also provide considerable opportunity for new product innovation, such as new (to consumers) flesh colours, flavour profiles or edible skin. Should gene editing become more widely accepted, these opportunities may be even greater as negative sensory attributes, such as skin bitterness, may be more easily avoided. As new flavour profiles are developed, it is possible that the strength of relationship between DM or rSSC and preference may weaken, and new tools for predicting consumer acceptance will be needed. This will extend to whether consumers accept a new type of kiwifruit produced using gene technologies and monitoring the perceptions of sustainability of kiwifruit production and supply chains and their impact on purchasing behaviours. The integration of modern consumer/sensory science and new product innovation will continue to provide the kiwifruit industry with a strong future.

References

Ares, G., Tárrega, A., Izquierdo, L. and Jaeger, S.R. (2014) Investigation of the number of consumers necessary to obtain stable sample and descriptor configurations from check-all-that-apply (CATA) questions. *Food Quality and Preference* 31, 135–141. DOI: 10.1016/j.foodqual.2013.08.012

Bava, C.M., Worch, T., Dawson, J., Marshall, D.W. and Jaeger, S.R. (2012) Characteristics of eating occasions that contain fruit. *British Food Journal* 114, 1361–1379. DOI: 10.1108/00070701211262966

Briz, T., Sijtsema, S.J., Jasiulewicz, A., Kyriakidi, A., Guàrdia, M.D. *et al.* (2008) Barriers to fruit consumption: driving forces behind consumer behaviour. *Scripta Horticulturae* 8, 7–18.

Burdon, J., McLeod, D., Lallu, N., Gamble, J., Petley, M. *et al.* (2004) Consumer evaluation of 'Hayward' kiwifruit of different at-harvest dry matter contents. *Postharvest Biology and Technology* 34, 245–225. DOI: 10.1016/j.postharvbio.2004.04.009

Burdon, J., Wang, M., Friel, E. and Lallu, N. (2007) Aroma volatiles of SmartFresh^{SM} treated 'Hayward' and 'Hort16A' kiwifruit. *Acta Horticulturae* 753, 763–768. DOI: 10.17660/ActaHortic.2007.753.100

Cardello, A.V. (1995) Food quality: relativity, context and consumer expectations. *Food Quality and Preference* 6, 163–170. DOI: 10.1016/0950-3293(94)00039-X

Crisosto, C.H. and Crisosto, G.M. (2001) Understanding consumer acceptance of early harvested 'Hayward' kiwifruit. *Postharvest Biology and Technology* 22, 205–213. DOI: 10.1016/S0925-5214(01)00097-7

Crisosto, C.H., Zegbe, J., Hasey, J. and Crisosto, G.M. (2011) Is dry matter a reliable quality index for 'Hayward' kiwifruit? *Acta Horticulturae* 913, 531–534. DOI: 10.17660/ActaHortic.2011.913.71

Deng, L., Jiang, C.-Z., Mu, W. and Wang, Q. (2015) Influence of 1-MCP treatments on eating quality and consumer preference of 'Qinmei' kiwifruit during shelf life. *Journal of Food Science and Technology* 52, 335–342. DOI: 10.1007/s13197-013-0986-y

Fisk, C.L., McDaniel, M.R., Strik, B.C. and Zhao, Y. (2006) Physicochemical, sensory and nutritive qualities of hardy kiwifruit (*Actinidia arguta* 'Ananasnaya') as affected by harvest maturity and storage. *Journal of Food Science* 71, S204–S210. DOI: 10.1111/j.1365-2621.2006.tb15642.x

Friend, A.P., Palmer, J.W., Seymour, S.M. and Diack, R.N. (2014) Potential of clonal rootstocks for devigoration and enhanced fruit characteristics in kiwifruit orchards. *Acta Horticulturae* 1058, 429–434. DOI: 10.17660/ActaHortic.2014.1058.52

Giacalone, D. and Jaeger, S.R. (2016) Better the devil you know? How product familiarity affects usage versatility of foods and beverages. *Journal of Economic Psychology* 55, 120–138. DOI: 10.1016/j.joep.2016.02.005

Giacalone, D., Llobel, F. and Jaeger, S.R. (2022) 'Beyond liking' measures in food-related consumer research supplement hedonic responses and improve ability to predict consumption. *Food Quality and Preference* 97, 104459. DOI: 10.1016/j.foodqual.2021.104459

Günther, C.S., Marsh, K.B., Winz, R.A., Harker, F.R., Wohlers, M.W. *et al.* (2015) The impact of cold storage and ethylene on volatile ester production and aroma perception in 'Hort16A' kiwifruit. *Food Chemistry* 169, 5–12. DOI: 10.1016/j.foodchem.2014.07.070

Hallett, I.C., MacRae, E.A. and Wegrzyn, T.F. (1992) Changes in kiwifruit cell wall ultrastructure and cell packing during postharvest ripening. *International Journal of Plant Sciences* 153, 49–60. DOI: 10.1086/297006

Harker, F.R., Jaeger, S.R., Lau, K. and Rossiter, K. (2007) Consumer perceptions and preferences for kiwifruit: a review. *Acta Horticulturae* 753, 81–88. DOI: 10.17660/ActaHortic.2007.753.7

Harker, F.R., Carr, B.T., Lenjo, M., MacRae, E.A., Wismer, W.V. *et al.* (2009) Consumer liking for kiwifruit flavour: a meta-analysis of five studies on fruit quality. *Food Quality and Preference* 20, 30–41. DOI: 10.1016/j.foodqual.2008.07.001

Harker, F.R., Hunter, D.C., White, A., Richards, K., Hall, M. *et al.* (2019) Measuring changes in consumer satisfaction associated with kiwifruit ripening: a new approach to understand human-product interactions with fruit. *Postharvest Biology and Technology* 153, 118–124. DOI: 10.1016/j.postharvbio.2019.03.010

Harman, J.E. and McDonald, B. (1989) Controlled atmosphere storage of kiwifruit: effect on fruit quality and composition. *Scientia Horticulturae* 37, 303–315. DOI: 10.1016/0304-4238(89)90141-6

Huang, H., Wang, Y., Zhang, Z., Jiang, Z. and Wang, S. (2004) *Actinidia* germplasm resources and kiwifruit industry in China. *HortScience* 39, 1165–1172. DOI: 10.21273/HORTSCI.39.6.1165

Hunter, D.C., Beresford, M.K., White, A., Jaeger, S.R., Wohlers, M. *et al.* (2018) Measuring perception of acidity and sweetness in kiwifruit. *Acta Horticulturae* 1218, 191–197. DOI: 10.17660/ActaHortic.2018.1218.26

Jaeger, S.R., Harker, F.R., Triggs, C.M., Gunson, A., Campbell, R.L. *et al.* (2011) Determining consumer purchase intentions: the importance of dry matter, size, and price of kiwifruit. *Journal of Food Science* 76(3), S177–S184. DOI: 10.1111/j.1750-3841.2011.02084.x

Jaeger, S.R., Hunter, D.C., Kam, K., Beresford, M.K., Jin, D. *et al.* (2015) The concurrent use of JAR and CATA questions in hedonic scaling is unlikely to cause hedonic bias, but may increase product discrimination. *Food Quality and Preference* 44, 70–74. DOI: 10.1016/j.foodqual.2015.04.001

Jaeger, S.R., Cardello, A.V., Chheang, S.L., Beresford, M.K., Hedderley, D.I. *et al.* (2017a) Holistic and consumer-centric assessment of beer: a multi-measurement approach. *Food Research International* 99, 287–297. DOI: 10.1016/j.foodres.2017.05.004

Jaeger, S.R., Cardello, A.V., Jin, D., Hunter, D.C., Roigard, C.M. *et al.* (2017b) Product uniqueness: further exploration and application of a consumer-based methodology. *Food Quality and Preference* 60, 59–71. DOI: 10.1016/j.foodqual.2017.03.013

Jaeger, S.R., Hunter, D.C., Vidal, L., Chheang, S.L., Ares, G. *et al.* (2019a) Sensory product characterization by consumers using check-all-that-apply questions: investigations linked to term development using kiwifruit as a case study. *Journal of Sensory Studies* 34, e12490. DOI: 10.1111/joss.12490

Jaeger, S.R., Lee P.-Y., Xia, Y., Chheang, S.L., Roigard, C.M. *et al.* (2019b) Using the emotion circumplex to uncover sensory drivers of emotional associations to products: six case studies. *Food Quality and Preference* 77, 89–101. DOI: 10.1016/j.foodqual.2019.04.009

Jaeger, S.R., Roigard, C.M. and Chheang, S.L. (2021) The valence × arousal circumplex-inspired emotion questionnaire (CEQ): effect of response format and question layout. *Food Quality and Preference* 90, 104172. DOI: 10.1016/j.foodqual.2020.104172

Jeffery, P.B., East, A.R., Jabbar, A., Heyes, J.A. and Bollen, A.F. (2013) Characterisation, measurement and factors influencing columella softening in kiwifruit. *Acta Horticulturae* 1012, 1487–1491. DOI: 10.17660/ActaHortic.2013.1012.201

Jordan, R.B. and Seelye, R.J. (2009) Relationship between taste perception, density and soluble solids concentration in kiwifruit (*Actinidia deliciosa*). *New Zealand Journal of Crop and Horticultural Science* 37, 303–317. DOI: 10.1080/01140671.2009.9687585

Jordan, R.B., Osborne, S.D., Künnemeyer, R. and Seelye, R. (1997) Harvest time prediction of eating time properties of kiwifruit using NIR transmission. In: *Sensors for Nondestructive Testing: Measuring the Quality of Fresh Fruits and Vegetables*. Proceedings from the Sensors for Nondestructive Testing International Conference and Tour, Orlando, Florida, February 18–21. Northeast Regional, Agriculture, and Engineering Service, Ithaca, New York, pp. 101–110.

Le Lievre, D.E. (2017) The development and taste of fruit of gold kiwifruit (*Actinidia chinensis* Planch. var. *chinensis* 'Gold3'). MSc thesis, University of Waikato, Hamilton, New Zealand.

Le Lievre, D.E., Anderson, R., Boldingh, H., Cooney, J., Seelye, R. *et al.* (2021) Modifying carbohydrate supply to fruit during development changes the composition and flavour of *Actinidia chinensis* var. *chinensis* 'Zesy002' kiwifruit. *Plants* 10, 1328. DOI: 10.3390/plants10071328

Li, H., Pidakala, P., Billing, D. and Burdon, J. (2016). Kiwifruit firmness: measurement by penetrometer and non-destructive devices. *Postharvest Biology and Technology* 120, 127–137. DOI: 10.1016/j.postharvbio.2016.06.007

MacRae, E.A., Bowen, J.H. and Stec, M.G.H. (1989) Maturation of kiwifruit (*Actinidia deliciosa* cv. Hayward) from two orchards: differences in the composition of the tissue zones. *Journal of the Science of Food and Agriculture* 47, 401–416. DOI: 10.1002/jsfa.2740470403

Marsh, K., Attanayake, S., Walker, S., Gunson, A., Boldingh, H. *et al.* (2004) Acidity and taste in kiwifruit. *Postharvest Biology and Technology* 32, 159–168. DOI: 10.1016/j.postharvbio.2003.11.001

Marshall, D. and Bell, R. (2003) Meal construction: exploring the relationship between eating occasion and location. *Food Quality and Preference* 14, 53–64. DOI: 10.1016/S0950-3293(02)00015-0

Matsumoto, S., Obara, T. and Luh, B.S. (1983) Changes in chemical constituents of kiwifruit during post-harvest ripening. *Journal of Food Science* 48, 607–611. DOI: 10.1111/j.1365-2621.1983.tb10800.x

Minchin, P.E.H., Snelgar, W.P., Blattmann, P. and Hall, A.J. (2010) Competition between fruit and vegetative growth in Hayward kiwifruit. *New Zealand Journal of Crop and Horticultural Science* 38, 101–112. DOI: 10.1080/01140671003781728

Nardozza, S., Gamble, J., Axten, L.G., Wohlers, M.W., Clearwater, M.J. *et al.* (2011) Dry matter content and fruit size affect flavour and texture of novel *Actinidia deliciosa* genotypes. *Journal of the Science of Food and Agriculture* 91, 742–748. DOI: 10.1002/jsfa.4245

Nitta, H. and Ogasawara, S. (1999) Characteristics of 'Hayward' kiwifruit vines grown on their own roots or grafted onto *Actinidia polygama* or *Actinidia rufa*. *Acta Horticulturae* 498, 319–324. DOI: 10.17660/ActaHortic.1999.498.38

Nunes-Damaceno, M., Muñoz-Ferreiro, N., Romero-Roderíguez, M.A. and Vázquez-Odériz, M.L. (2013) A comparison of kiwi fruit from conventional, integrated and organic production systems. *LWT – Food Science and Technology* 54, 291–297. DOI: 10.1016/j.lwt.2013.05.002

Okuse, I. and Ryugo, K. (1981) Compositional changes in the developing 'Hayward' kiwi fruit in California. *Journal of the American Society for Horticultural Science* 106, 73–76. DOI: 10.21273/JASHS.106.1.73

Piqueras-Fiszman, B. and Jaeger, S.R. (2014) Emotion responses under evoked consumption contexts: a focus on the consumers' frequency of product consumption and the stability of responses. *Food Quality and Preference* 35, 24–31. DOI: 10.1016/j.foodqual.2014.01.007

Piqueras-Fiszman, B. and Jaeger, S.R. (2015) The effect of product-context appropriateness on emotion associations in evoked eating occasions. *Food Quality and Preference* 40, 49–60. DOI: 10.1016/j.foodqual.2014.08.008

Rabobank (n.d.) Significant growth opportunities for NZ kiwifruit in US market – California-based analyst. Available at: https://www.rabobank.co.nz/media-releases/2019/191023-significant-growth-opportunities-for-nz-kiwifruit-in-us-market/ (accessed 25 April 2022).

Redgwell, R.J., Melton, L.D. and Brasch, D.J. (1990) Cell wall changes in kiwifruit following post harvest ethylene treatment. *Phytochemistry* 29, 399–407. DOI: 10.1016/0031-9422(90)85087-V

Richardson, A.C. and McAneney, K.J. (1990) Influence of fruit number on fruit weight and yield of kiwifruit. *Scientia Horticulturae* 42, 233–241. DOI: 10.1016/0304-4238(90)90085-S

Richardson, A.C., Marsh, K.B., Boldingh, H.L., Pickering, A.H., Bulley, S.M. *et al.* (2004) High growing temperatures reduce fruit carbohydrate and vitamin C in kiwifruit. *Plant, Cell & Environment* 27, 423–435. DOI: 10.1111/j.1365-3040.2003.01161.x

Richardson, A.C., Eyre, V., Kashuba, P., Ellingham, D., Jenkins, H. *et al.* (2021) Early shoot development affects carbohydrate supply and fruit quality of red-fleshed *Actinidia chinensis* var. *chinensis* 'Zes008'. *Agronomy* 11, 66. DOI: 10.3390/agronomy11010066

Ristic, R., Danner, L., Johnson, T.E., Meiselman, H.L., Hoek, A.C. *et al.* (2019) Wine-related aromas for different seasons and occasions: hedonic and emotional responses of wine consumers from Australia, UK and USA. *Food Quality and Preference* 71, 250–260. DOI: 10.1016/j.foodqual.2018.07.011

Salazar, J., Jorquera, C., Campos-Vargas, R., Jorgensen, C., Zapata, P. *et al.* (2019) Effect of the application timing of 1-MCP on postharvest traits and sensory quality of a yellow-fleshed kiwifruit. *Scientia Horticulturae* 244, 82–87. DOI: 10.1016/j.scienta.2018.09.028

Schröder, R. and Atkinson, R.G. (2006) Kiwifruit cell walls: towards an understanding of softening? *New Zealand Journal of Forestry Science* 36, 112–129.

Seal, A.G., Clark, C.J., Sharrock, K.R., de Silva, H.N., Jaksons, P. *et al.* (2018) Choice of pollen donor affects weight but not composition of *Actinidia chinensis* var. *chinensis* 'Zesy002' (Gold3) kiwifruit. *New Zealand Journal of Crop and Horticultural Science* 46, 133–143. DOI: 10.1080/01140671.2017.1365732

Skinner, S.J.M., Hunter, D.C., Cho, S. and Skinner, M. (2013) The potential health benefits of the subtropical fruits kiwifruit, feijoa and tamarillo. In: Skinner, M. and Hunter, D. (eds) *Bioactives in Fruit: Health Benefits and Functional Foods*. Wiley-Blackwell, Oxford, UK, pp. 169–195.

Snelgar, W.P., Hall, A.J., Richardson, A.C. and Currie, M.B. (2007) Influence of temperature on between-season variation in dry matter content of 'Hayward' kiwifruit. *Acta Horticulturae* 753, 383–387. DOI: 10.17660/ActaHortic.2007.753.48

Snelgar, W.P., Minchin, P.E.H., Blattman, P. and Hall, A.J. (2012) Sink priority on 'Hayward' kiwifruit vines. *New Zealand Journal of Crop and Horticultural Science* 40, 253–263. DOI: 10.1080/01140671.2012.681386

Stec, M.G.H., Hodgson, J.A., MacRae, E.A. and Triggs, C.M. (1989) Role of fruit firmness in the sensory evaluation of kiwifruit (*Actinidia deliciosa* cv Hayward). *Journal of the Science of Food and Agriculture* 47, 417–433. DOI: 10.1002/jsfa.2740470404

Tait, P., Rutherford, P., Driver, T., Li, X., Dalziel, P. *et al.* (2018) Consumer insights and willingness to pay for attributes: kiwifruit in Shanghai, China. AERU Research Report No. 346. Lincoln University: Agribusiness and Economics Research Unit. Available at: https://researcharchive.lincoln.ac.nz/handle/10182/10039 (accessed 23 February 2023).

Tait, P., Saunders, C., Dalziel, P., Rutherford, P., Driver, T. *et al.* (2020) Japanese kiwifruit consumer consumption behaviours and product preferences: a Latent Class Analysis. AERU Research Report No. 361. Lincoln University: Agribusiness and Economics Research Unit. Available at: https://researcharchive.lincoln.ac.nz/handle/10182/13868 (accessed 23 February 2023).

Tuorila, H., Huotilainen, A., Lähteenmäki, L., Ollila, S., Tuomi-Nurmi, S. *et al.* (2008) Comparison of affective rating scales and their relationship to variables reflecting food consumption. *Food Quality and Preference* 19, 51–61. DOI: 10.1016/j.foodqual.2007.06.007

Vidal, L., Ares, G. and Jaeger, S.R. (2021) Differences in citation proportions in CATA questions can be interpreted as differences perceived intensity of sensory attributes. *Journal of Sensory Studies* 36, e12695. DOI: 10.1111/joss.12695

Wang, M.Y., MacRae, E., Wohlers, M. and Marsh, K. (2011) Changes in volatile production and sensory quality of kiwifruit during fruit maturation in *Actinidia deliciosa* 'Hayward' and *A. chinensis* 'Hort16A'. *Postharvest Biology and Technology* 59, 16–24. DOI: 10.1016/j.postharvbio.2010.08.010

Zeng, Y., Wang, M.Y., Hunter, D.C., Matich, A.J., McAtee, P.A. *et al.* (2020) Sensory-directed genetic and biochemical characterization of volatile terpene production in kiwifruit. *Plant Physiology* 183, 51–66. DOI: 10.1104/pp.20.00186

Zoffoli, J.P., Flores, K. and D'Hainaut, D. (2016) Effect of ethylene on ripening of kiwifruit stored under controlled or modified atmosphere packaging and treated with 1-methylcyclopropene. *Journal of Berry Research* 6, 381–393. DOI: 10.3233/JBR-160117

22 Kiwifruit: Future Potential, Risks and New Directions

Annette C. Richardson[1]* and Jeremy N. Burdon[2]
[1]*The New Zealand Institute for Plant and Food Research Ltd, Kerikeri, New Zealand;* [2]*The New Zealand Institute for Plant and Food Research Ltd, Auckland, New Zealand*

22.1 Introduction

Global kiwifruit production has increased rapidly over the past four decades, from 81,500 tonnes in 1983 (Warrington, 1990) to 4.47 million tonnes (Mt) in 2021 (Statista, 2021). However, kiwifruit represents only 0.22% of the global fresh fruit bowl, which is dominated by bananas (124.98 Mt), apples (93.14 Mt), oranges (75.7 Mt) and grapes (73.52 Mt). Kiwifruit remains a specialized, high-value fresh fruit, while other mainstream fruit crops are commodities that may also be sold as processed foods and beverages. Kiwifruit has a short history of commercial production, over decades, compared with other fruit crops that have been commercially produced for centuries. The extensive genetic diversity within the genus *Actinidia* may provide enormous potential for market expansion based on commercialization of new cultivars.

Commercial kiwifruit production started in New Zealand from seed collected from China, and from there spread to Italy, France, Chile and the USA, based largely on *Actinidia chinensis* var. *deliciosa* 'Hayward' fruit. China is now the leading kiwifruit producer, although little fruit is exported, followed by the major exporting countries New Zealand, Italy, Greece, Iran and Chile. Production has increased rapidly in Iran and industries are becoming established in countries such as Turkey (see Chapter 2). Kiwifruit growing systems have rapidly become more sophisticated, increasing maximum production from 25 to 90 tonnes ha^{-1}. Development of new growing systems has tended to occur in premium growing locations, with suitable soil and climates, as well as available labour and resources. Recently, pest and disease problems such as *Pseudomonas syringae* pv. *actinidiae* (Psa) and the kiwifruit vine decline syndrome (KVDS) complex in Italy have devastated some crops, impacting worldwide kiwifruit production.

In future there will be many opportunities, as well as risks, in increasing global kiwifruit production. Here we cover these opportunities and risks under the general areas of consumer demand, breeding targets, environmental challenges, integrated growing technologies, fruit quality and maturation, labour supply and automation, and sustainability.

22.2 Consumer Demand

Kiwifruit is marketed predominantly as a fresh fruit; few kiwifruit-based processed products are

*Corresponding author: annette.richardson@plantandfood.co.nz

© CAB International 2024. *Kiwifruit: Botany, Production and Uses* (eds A.C. Richardson, J.N. Burdon and A.R. Ferguson)
DOI: 10.1079/9781800620933.0022

available. Consumer preferences for fresh fruit are based largely on the internal/external appearance of fruit and its flavour. Beyond these basics, the fruit must be of consistent quality, be available at a ready-to-eat, or close to, stage of ripeness and be at an acceptable price. Year-round supply from regional production helps with brand development within the marketplace. With the wider range of green-, yellow- and red-fleshed kiwifruit cultivars now available, along with kiwiberry, consumers have a much wider choice than when only the green-fleshed 'Hayward' was seasonally available. With their small size and smooth edible skins, kiwiberries are much more suitable as a convenient snack food than the large-fruited kiwifruit cultivars. As new kiwifruit cultivars develop, new sensory tools will be required to closely map consumer preferences and responses.

However, as with other fruit such as bananas and avocados, it is technically challenging to deliver ready-to-eat kiwifruit of any type, particularly early in the season, shortly after harvest. With the expansion of online direct marketing systems, the ability to reliably supply fruit of the correct ripeness becomes more important. Ripening and storage protocols tailored for each specific fruit cultivar are required to deliver fruit that will be perfectly ripe on arrival to the consumer.

Current trends suggest that other features of kiwifruit will more strongly influence future consumer choices. Kiwifruit is already recognized as having very nutrient-dense fruit with high vitamin C and fibre content as well as antioxidant and anti-inflammatory properties that may protect against heart disease, cancer, diabetes and central nervous system diseases (Hunter *et al.*, 2016). These health properties boosted sales of kiwifruit during the COVID-19 pandemic and are highly sought after by the ageing global population. Future cultivars and growing systems should continue to focus on maintaining or enhancing these health properties.

Consumers are increasingly environmentally conscious and in future will require that the fruit they buy is sustainably, possibly organically, produced and delivered with a low carbon footprint. Integrated systems that reduce chemical inputs, minimize water use, maintain water and soil quality, maintain biodiversity and reduce carbon emissions will be required. While gene-editing technologies may increasingly be used to meet industry and market targets, it is still uncertain how some consumers will respond to such fruit.

22.3 Breeding Targets

The global kiwifruit market started with 'Hayward' fruit and a few other early selections. In contrast, Chinese cultivars have mainly been sourced from the wild, although this is now changing with focused breeding programmes. 'Hayward' vines and fruit have been the subject of most scientific studies over the past four decades. This is rapidly changing as new cultivars are released; however, generally, data on only a few genotypes from a small number of *Actinidia* taxa have been published. A selection is first made from a single growing environment, and suitability in different environments should be checked before widespread planting. Therefore, it is impossible to generalize on the properties of and management required for new cultivars, and focused studies are required to develop their full potential.

There are now active kiwifruit breeding programmes in several countries, including China, New Zealand, Italy, Japan and Korea. These programmes have developed a diverse range of new *Actinidia chinensis* var. *deliciosa* (green), *Actinidia chinensis* var. *chinensis* (green, red and yellow) and *Actinidia arguta* (kiwiberry) cultivars. Many programmes are focusing on increased yield and fruit quality as well as resistance to biotic and abiotic stresses. The wide diversity of *Actinidia* germplasm offers significant opportunities to develop new fruit types, particularly through interspecific cultivars. However, the wide range of ploidy in *Actinidia* spp. remains a significant challenge for rapid progress. Maintenance of germplasm for future use is important, and this has been highlighted by losses of material during the global spread of Psa.

New kiwifruit cultivars are generally protected by plant variety rights (PVR) and are licensed to growers through propagation and marketing agreements. This limits both the breadth and availability of research on any cultivar with results from many studies remaining unpublished owing to commercial imperatives. As part of the integrated package required for future kiwifruit success, breeding programmes have expanded to include rootstock and male cultivars. To date, these areas have relied on material selected from female breeding programmes or germplasm collections. Rootstock breeding is targeting resistance to abiotic and biotic stresses, particularly soil waterlogging, soil nutrient limitations and disease-resistant material. The significant contributions of male cultivars to fruit properties and seasonal flowering overlap with females have become targets for the selection of new cultivars. The development of hermaphrodite kiwifruit that are self-setting and develop quality fruit when selfed will have a significant impact on productivity through co-location of both sexes within a single flower, thereby removing the necessity for separate male plants occupying orchard space. Some growers have already removed male vines and are relying on artificial pollination.

Rapid expansion of genetic knowledge and new genetic technologies have underpinned the recent development of breeding programmes but have yet to deliver new cultivars. Whole-genome maps and genome-wide association studies have helped to identify key genes controlling important breeding targets. High-throughput genotyping and phenotyping methods have been integrated with bioinformatic tools to aid discovery of new genes and pathways behind key target traits. Molecular markers for key genetic kiwifruit traits like sex, colour, composition, yield, and pest and disease resistance have been mapped. New technologies offer opportunities to stack desirable traits in new cultivars and speed up the traditionally prolonged breeding process. Genome editing (CRISPR/Cas9) has provided a non-transgenic means of modifying kiwifruit plants, e.g. to develop a compact, rapid flowering, hermaphrodite plant to speed up kiwifruit breeding. Rapid development and introduction of new cultivars need further attention.

22.4 Environmental Challenges

The natural environments where *Actinidia* are grown are highly variable, although most commercial cultivars are derived from material sourced from cool, temperate rainforest conditions. Many commercial orchards are situated in prime horticultural locations, but some current orchards have unfavourable conditions with temperatures of up to 40°C, or high pH and waterlogged soils, that can adversely affect vine and fruit growth. Unfavourable future changes to soils and climate may have a greater impact on where commercial kiwifruit production occurs.

In future, seasonal changes in temperature and water availability will affect vine growth, productivity, fruit quality and pest and disease susceptibility. In addition, adverse weather events with high winds, flooding and hail may severely damage crops or kill vines. To future-proof the kiwifruit industry, new cultivars, growing technologies and systems that mitigate adverse climatic changes and limit damage from extreme events will be required. Introduction and adaptation of techniques to protect vines against frost, hail and wind may be required for a greater number of plantings. Kiwifruit cultivars require significant water inputs and future limitations to water availability through reduced supply and greater community demand may limit kiwifruit production in some current growing regions. Climate change will offer new opportunities for kiwifruit, e.g. to develop in new growing regions or to develop more intensive and technologically advanced covers (hail netting, plastic or agrivoltaic coverings) and even indoor growing systems. This is likely to introduce further challenges including how to control vine vigour and development, manage altered pest and disease dynamics and achieve pollination indoors. The ability to grow kiwifruit in low-chill environments may be limited by the unavailability of some budbreak enhancers, e.g. hydrogen cyanamide which is already banned in Europe. Conversely, this provides an opportunity for breeding programmes to produce new low-chill cultivars.

Global demand for land to house and feed a growing population is likely to reduce the availability of premium land for kiwifruit orchards, particularly in areas close to markets or infrastructure. Therefore, new plantings are likely to be made in marginal sites, where the depth and physical, chemical and biological properties of soils, site topography and climate are less than optimal. Production at these sites may initially require considerable efforts to improve and maintain soil properties using a range of physical, chemical and biological ameliorations as well as careful water management to avoid further erosion, degradation or pollution. If growing systems move to controlled environments, it is possible that vines will be grown using soilless media. Major changes to plant traits will be required to maintain kiwifruit production using more challenging growing sites.

22.5 Integrated Growing Technologies

Many studies have been carried out, initially on 'Hayward' but more recently on newer cultivars, to understand and optimize the production and quality of kiwifruit. It is now clear that knowledge and production systems will need to be refined for each cultivar, and likely each cultivar/rootstock combination. A key consideration will be to understand how to use the significant potential of the canopy to capture carbon and effectively partition it and other key resources (water and nutrients) to ensure high production of quality fruit from vines each season. Rather than starting afresh with each new cultivar, a generic framework of underlying principles is needed as a starting point. A small amount of initial work has been done to develop vine and fruit models. However, this work has not progressed from the basic descriptions from early studies of 'Hayward'. Much more work is required before there are robust models of the whole vine that can then be used to test new growing strategies *in silico*. Further development will then be required to include new scion/rootstock cultivars, a range of environmental factors, new management technologies, pest and disease control methods, and other emerging constraints to develop a useful management model.

22.6 Fruit Quality and Maturation

Significant recent progress has been made in the understanding of fruit development, maturation and ripening. These have been key breeding targets and where the most rapid progress has been made in recent genomic and metabolic kiwifruit studies. This work has included the discovery of the metabolic controls of colour, starch, seed development, fruit maturation and ripening. These studies provide the potential for future cultivar development via gene editing. They also provide useful tools to understand how future environmental and management changes may affect fruit quality and maturation. 'Hayward' fruit has a storage life that is not matched by many newer cultivars. The rapidly softening kiwiberry has many desirable market attributes but its relatively short storage life (compared with other kiwifruit, although it stores well compared to many berry fruits) is a challenge. To ensure that fruit reaches consumers in optimum condition, new protocols will be required to obtain the best postharvest storage and shelf life for each new cultivar. New metrics beyond simple appearance and composition (soluble solids content, colour, firmness and dry matter) will be required to ensure consumers receive fruit in the best possible condition. Management of postharvest fruit quality will be closely linked to the sustainability of the kiwifruit industry through the impact of packaging and freight on the environment.

22.7 Labour Supply and Automation/mechanization

Increasing demand for food, including kiwifruit, can be met by expanding production. However, if current intensive production systems are used for kiwifruit, a significant increase in labour will be required. The kiwifruit industry already has a staffing crisis due to ageing of the current workforce, industry expansion and, in some countries, reliance on international workers. It has been difficult to attract people into horticultural careers due to the seasonal nature of the work and an outdated underestimation of the technical and management skills required. To attract new workers, the kiwifruit industry needs to promote itself as a highly innovative space in which there is continuous adoption of new research and technology. It is possible that further development and simplification of growing and postharvest systems will occur, so that a greater number of processes can be mechanized, and labour requirements decreased. Progress is already being made with robotic harvesting (Saunders,

2020) and packing systems (Hall, 2020) as well as optical and near infra-red sensors for fruit grading. With further technological progress, shortages of labour may no longer be so important.

22.8 Sustainability

Sustainable practices are becoming the target for the whole kiwifruit supply chain from the orchard to the consumer. In future, the sustainability of each step of the chain will come under greater scrutiny and modifications will be required to maintain, or even enhance, local environments as well as satisfying increasing consumer demand for environmentally friendly fruit. Water resources are under increased pressure and new technologies will be required to efficiently manage this, reduce the volume used, prevent run-off and leaching of nutrients and other agrichemicals into waterways, and share the resource equitably with local communities. More effective nutrient management will be closely aligned with efficient water management strategies using new sensors to monitor both soil and plant status. Further research is required to customize nutrient inputs for both vine and fruit development to optimize fruit yield and quality. New methods of promoting plant processes like budbreak, flowering and fruit growth will also be required. Many of the challenges of growing kiwifruit sustainably are likely to be met by developing new rootstock/scion options, new vine management strategies and more effective real-time monitoring tools. Our increasing knowledge of the interactions between genetics, the environment and vine management, combined with new technologies, is likely to facilitate advances in sustainable kiwifruit production.

The postharvest sector is already making changes to become more sustainable. One of the most common is the use of more environmentally friendly (biodegradable, recyclable) packaging materials that meet international food safety standards (Scion, 2023a). Yet the new packaging must still meet the requirements of protecting and unitizing the fruit, and this remains challenging.

Transportation impacts are a major consideration for the sustainability of all globally traded products. The transportation of exported kiwifruit, with its generally long storage life, is not as time-critical as it is for more perishable crops. Hence, sea freight is not problematic at present, with sustainability initiatives focused on the use of low sulphur fuels and a reduction in some vessel speeds to improve fuel use efficiency. However, new cultivars may not be as robust as 'Hayward' and refrigerated transport may need to be supplemented with technologies such as controlled atmospheres. Efforts to improve sustainability may also depend on marketing within the natural window of a cultivar and not simply expecting the same storage life as 'Hayward'. Some of the recently released red-fleshed cultivars clearly have reduced storage life compared with 'Hayward'. In the longer term, breeding programmes will address the issue of storage life, making new cultivars that are at least as robust as 'Hayward', if not better.

Large-scale production of premium fruits like kiwifruit can produce large waste streams. To improve the sustainability and efficiency of the crop, waste streams need to be reduced or eliminated throughout the supply chain. Alternative uses for waste are always well received (Renewable Carbon News, 2023); an example of this is the Biospife – a compostable spoon-knife, used to eat kiwifruit, made from kiwifruit waste (Scion, 2023b). The consumer wants to have safe and socially responsible food.

Increased global risk from the rapid spread of pests and diseases has been highlighted by the impact of Psa and, more recently, KVDS on kiwifruit production. As the global spread, number and severity of pests and diseases increases there will be a greater need for environmentally sustainable solutions to these problems. Maintaining a high level of pest and disease control with acceptable options is also essential for continued market access and consumer acceptance. There has been an industry trend towards reduced applications of chemical pesticides as well as more targeted and environmentally friendly control. Globally, the kiwifruit industry is a small market for agrichemical manufacturers and would benefit from the wider availability and use of generic chemicals for crop protection. Other options, like pest exclusion nets, may be beneficial if they can be combined with protected/indoor growing systems. Molecular tools have been widely used to understand plant defence systems and resistance genes for specific plant health problems and will be part of future control strategies. Alternative technologies that use the phytobiome, specific microorganisms and biomolecules have the potential to provide a large range of new control options, although biopesticides are not yet very effective. Further understanding of kiwifruit defence systems as well as the biology of pests and diseases will underpin the development of new diagnostic and control tools.

22.9 Conclusions

In future, new cultivars and technologies will be integrated into knowledge systems to support the growth of the global kiwifruit industry. Focus needs to remain on consumers to ensure their changing requirements are anticipated and met. Maintaining and enhancing the environment and meeting many regulatory challenges will also need regular consideration. Modelling of the basic biology of growing systems and the supply chain will integrate new findings with the wealth of existing knowledge. This will provide a solid basis to rapidly innovate with new scion and rootstock cultivars, alternative growing systems, and modified supply chains and marketing systems while moderating new biotic and abiotic risks. Increased real-time collection of data using remote sensing to acquire images and data will provide regular updates to validate models, provide options and assess outcomes. Increasing mechanization will support greater production with reduced labour requirements. Continued research will be required to fully understand the biology of kiwifruit throughout the continuum from production to consumption and to introduce new innovations for the kiwifruit industry to thrive.

References

Hall, C. (2020) Robots run EastPack kiwifruit cool storage site in a world first at Te Puke. Available at: https://www.nzherald.co.nz/bay-of-plenty-times/news/robots-run-eastpack-kiwifruit-cool-storage-site-in-a-world-first-at-te-puke/C6EWWSEK4AE7JF7DFCVSIYV5WY/ (accessed 27 April 2023).

Hunter, D.C., Skinner, M.A. and Ferguson, A.R. (2016) Kiwifruit and health. In: Watson, R.R. and Preedy, V.R. (eds) *Fruits, Vegetables, and Herbs*. Academic Press, Amsterdam, pp. 239–269. DOI: 10.1016/B978-0-12-802972-5.00012-3

Renewable Carbon News (2023) Scion institute created compostable bioplastics from kiwifruit waste. Available at: https://renewable-carbon.eu/news/scion-institute-created-compostable-bioplastics-from-kiwifruit-waste/ (accessed 10 April 2023).

Saunders, S. (2020) The robots that can pick kiwi-fruit. Available at: https://www.bbc.com/future/bespoke/follow-the-food/the-robots-that-can-pick-kiwifruit.html (accessed 27 April 2023).

Scion (2023a) Ticking all the boxes. Available at: https://www.scionresearch.com/about-us/about-scion/corporate-publications/scion-connections/current-issue/scion-connections-issue-42,-february-2023/ticking-all-the-boxes (accessed 24 April 2023).

Scion (2023b) The ZESPRI biospife. Available at: https://www.sciencelearn.org.nz/resources/1472-the-zespri-biospife#Identifying%20The%20Opportunity (accessed 10 April 2023).

Statista (2021) Global fruit production in 2021, by selected variety (in million metric tons). Available at: https://www.statista.com/statistics/264001/worldwide-production-of-fruit-by-variety/ (accessed 10 April 2023).

Warrington, I.J. (1990) Areas and trends of kiwifruit production in New Zealand and around the world. In: Warrington, I.J. and Weston, G. (eds) *Kiwifruit: Science and Management*. Ray Richards Publisher in association with the New Zealand Society for Horticultural Science, Auckland, New Zealand, pp. 511–525.

Index

Note: All index entries refer to Kiwifruit e.g. 'global demand' *(of kiwifruit)*, unless otherwise described. Page numbers in *italic* type refer to *figures;* page numbers in **bold** type refer to **tables.**

'74–49' cultivar (kiwiberry) **100**

'Abbott' cultivar 120, 151
ABC model, floral differentiation 239, *239*
abscisic acid (ABA) 234, 264, 279
'Ac 459 011' cultivar 58, 61
'AC 1536' cultivar 64, 130
acclimation conditions, frost survival 265
acetaldehyde 360
acibenzolar-S-methyl 327
Act d 1 and Act d 2 (allergenic proteins) 142
Act d 5, and Act d 11 (allergenic proteins) 142
Actinidia (genus) 1, 29–51
    cultivar list xv–xviii
    evolution 30, 39
    mihoutao shu (Chinese) 1
    nomenclature changes and reasons for 31–32
    species number 29
    taxonomic history 29–30
    vine architecture 185–186
    wild
        in China *see* China
        fruit diversity 42
        germplasm evaluation 9–10, 13
        insect pollination 211, 212
        interspecific crosses 30
        pests from 332
    as woody perennials 228, 237, 257
Actinidia (genus), species of xiii, 30–32
    chromosome numbers *33*, 33–34, *34*
    cultivation of species not *A. chinensis* 13
    difficulties for taxonomists 30–31
    diversity of, and importance 44–45
    geographic distribution 31–32, 32–33, 44
    geographic variation 30
    growth habit 34–36, *35*, 76
    habitat 30, 32, 33
    morphological characters 30, 32
    no boundaries between 30
    nomenclature changes 31–32
    ploidy *see* ploidy, *Actinidia*
    taxonomy ('splitters' and 'lumpers') 31
    variability 30, 31
    see also specific topics and species
Actinidia arguta 1, 13, 29, 54
    budbreak 229
    buds 37, 96
    canes, shoot growth and 243
    carbon transport (planteose) 260
    climatic requirements 44, 94
    cold tolerance 87, 266
    cultivars 98, *99*, **100**, 130
    cultivation, limitations 110
    distribution and variability 31
    flower odour 212
    forms/polymorphic 30
    frost tolerance, rootstocks 87
    fruit *see* kiwiberry
    in vitro propagation 122
    male cultivars for 80
    myo-inositol in fruit 280
    nomenclature changes 31
    origin/history 94

Actinidia arguta (*Continued*)
photosynthesis rates 257
ploidy, variation 34, *34*, 94, 96
Psa susceptibility and 106
pollen sources affecting fruit (xenia) 209, 210
pollen viability in storage 217
pollination 74, 214
Psa resistance 106, 143, 319
rootstocks, and graft compatibility 87, 88
seed number 211
self-setting 40
shoots and leaves 94, *95*, 96
similarities to other *Actinidia* species 94
varieties (two) 94
vegetative propagation 121
water deficit stress tolerance 263
waterlogging response/tolerance 264
see also kiwiberry
Actinidia arguta complex, nomenclature changes 31
Actinidia arguta 'Hortgem Tahi' 80, *99*, **100**, 380
Actinidia arguta 'Issai' *99*, **100**, 122
Actinidia arguta var. *arguta* 31, *34*, 44, 94
Actinidia arguta var. *giraldii* 31, 34, 94
Actinidia arguta var. *hypoleuca* 31, *34*
Actinidia arguta var. *purpurea* 31
Actinidiaceae 29
Actinidia CEN (*AcCEN*) genes 238
Actinidia chinensis 127
carpel number 39
in China 1, 31
discovery 1, 31
fruit skin colour 53
natural variation in wild populations 1
nomenclatural history 1–2, 31–32
ploidy variation 34
taxonomy 31–32
'under-pollinated' 210
as zhonghua mihoutao 1
Actinidia chinensis complex, nomenclature changes 31–32
Actinidia chinensis Planch 1, 32
Actinidia chinensis var. *chinensis* 1
A. chinensis var. *deliciosa* distinguished from 32
'Ac 459 011' cultivar 58, 61
Actinidia rufa cross with 44
allergens 142
association maps (genetic) 133
budbreak 228–229
buds 37
chilling sensitivity, storage 359
climatic requirements 44
CPPU effect 200
cultivars, new, vigour of 201
cultivar selection in China 10
cultivation in China 3, 9, 10
cultivation in other countries 3–4, 11–12
domestication place/timing 2
drought tolerance 88–89, 263
early cultivation attempts 4
first orchards 9
flowering time 37
flowers 38, *38*, *73*
development 239
odour 212
pistillate 32, *38*, *73*
staminate 5, 32, *38*, *73*
frost, response to 265
fruit
aroma 43–44
composition 274–275
diversity 42
flavour 43
protection 53
shape 32
size 41
fruit flesh colour 2, 32, 43, *53*, 53–54
green-fleshed ('Cuiyu') 59–60, 245
yellow-fleshed *see* yellow-fleshed kiwifruit
fruiting cultivars 52–71
ideal, characteristics 52, 53, 54, 55, 56
promotion 57
genetic diversity 45
genome sequencing 134
graft, 'Bounty71' rootstock *86*
growth habit 34–36, *35*
hairs on skin 2, *2*, 31, 32, 41
'Hort16A' cultivar *see* 'Hort16A' cultivar
in vitro propagation 122
iron deficiency, sensitivity 86, 89
iron-induced leaf chlorosis 86
in Italy 8
leaves 32, 37
locules from mature fruit *75*
photosynthesis rates 257
ploidy 34, 56
pollen viability in storage 218
pollination, insect *215*
Psa susceptibility 53, 318, 320
biovar 3 319
red pigmentation in 65
root-knot nematode resistance 90
roots 36
rootstocks, interspecific 87
seed set as measure of pollination 210, *210*
type specimen 2, *3*
vitamin C content 55
water deficit tolerance 88–89, 263
in waterlogged soils 88–89, 264
wild, cultivars raised from 10
winter chilling and endodormancy 233
xenia, pollen characteristics affecting fruit 209
'Zesy002' cultivar *see* 'Zesy002' cultivar

Actinidia chinensis var. *deliciosa*
Actinidia chinensis var. *chinensis* distinguished from 32
allergens 55, 142
'Allison' cultivar 233
'Bruno' seedling *vs* 'Bounty71' rootstock *87*
buds 37, *209*
chromosome squash/number *33*, 34
climatic requirements 44
clonal rootstocks for 85, 86, 87
graft compatibility 88
CPPU effect 200
cultivars
fruiting 52–71
hermaphrodite 40, 52, 68, 136–137
male cultivars 77
selection in China 10
selection in New Zealand 6
cultivation 3–8, 127
in China 9, 10
as a crop, development 13
cultivar selection 6, 10
early attempts 4, 5, 5–6
exports and consequences 6–7
female plants 5, 6
first plants outside China 3–4, 6
main cultivars grown 58–61
management practices, development 6, *7*, 184
in New Zealand 3, 5–6, 7
total area (New Zealand) 7
in USA, Europe and Japan 7–8
distribution and habitat 31–32
domestication place/timing 2, 129–130
drought tolerance 88–89, 263
endodormancy, winter chilling and 233
floral morphogenesis, stages **230**
flower bud, popcorn stage *209*
flower development 239
flowering shoots 37
flowering time 37
flowers *30*, 32, 37, 38, 39, 40, *73*
number in inflorescence 37
odour 212
frost, response to 265
fruit
aroma 43–44
composition 274–275, **279**
diversity 42
flavour 43
flesh colour 2, 32, 43, *53*
shape 32
size 41
skin, protective 53
structure 41, *41*, 274, **275**
weight at harvest 40
fruiting cultivars 52–71
gene linkage map 134
genetic diversity 45
hairs on skin 2, *2*, 31, 32, 41, 53
'Hayward' cultivar *see* 'Hayward' cultivar
inconstant female 40
insect pollination *213*
managed 213, 214
in vitro propagation 122
micro-grafting 121
leaves 32, 37
male cultivars 77
photosynthesis rates 257
ploidy 34, 56
pollen 208
grains, number and shape *74*, 208
kiwiberry pollination by 104
viability in storage 217, 218
propagation
grafting, graft formation 121
in vitro 121, 122
vegetative 120–121
Psa susceptibility 317, 318
Pukekohe dwarf mutant 239, *240*
red pigmentation in 65
root-knot nematode resistance 90
roots 36
salinity tolerance 266
seed germination, promoting 120
seedling rootstocks 85
seed number 211
seed structure 41
stems 36
taxonomy and nomenclature 31–32
vitamin C in fruit 42
water deficit stress tolerance 88–89, 263
in waterlogged soils 88–89, 151, 264
wild, in China 3, *4*
winter chilling 233
xenia, pollen characteristics affecting fruit 209
Actinidia chinensis var. *hispida* 31
Actinidia chinensis var. *setosa* 31, 34, 37
Actinidia chrysantha 119, 211
Actinidia deliciosa 31, 32
see also Actinidia chinensis var. *deliciosa*
Actinidia eriantha 13
budbreak 228–229
carpel number 39
cultivars, commercial 130
easy-to-peel 142
floral evocation, genes controlling 238
flowers
pistillate 39
size 38
staminate 39
fruit colour and chlorophyll level 138
fruiting cultivars 52–71

Actinidia eriantha (*Continued*)
genome sequencing 134
hairs on fruits 41, 54
interspecific cross with 130–131
male cultivars for 80
photosynthesis rates 257
pollen viability in storage 217, 218
rootstocks 87
seed number 211
shoot apex removal, flowering 231, *232*
waterlogged soils or drought tolerance 88–89, 264
wild, fruit diversity 42
Actinidia FT (*AcFT*) genes 238
Actinidia globose, fruit flavour 43
Actinidia grandiflora, fruit flavour 43
Actinidia guilinensis 90, 143
Actinidia hemsleyana 86, 87, 88, 89, 266
Actinidia henryi 42
Actinidia hypoleuca 31
Actinidia indochinensis 37
Actinidia kolomikta 13
cultivars, commercial 130
flowering time 37
fruit flavour 43
pollen viability in storage 218
rootstocks 87, *88*
staminate flowers 39
Actinidia latifolia
flowering time 37
flowers *39*
inflorescence *237*
vitamin C in fruit 42
Actinidia longicarpa 89, 266
Actinidia macrosperma
'Bounty71' rootstock *86*, 86–87, *87*
CPPU, response to 200
drought tolerance 89
fruit colour *42*
fruit flavour 43
fruits 41
in vitro propagation 122
Psa resistance 143
root pressure 262
waterlogged soil tolerance 89, 121, 152
Actinidia melanandra 31, 54, 200
fruit flesh colour 43
in vitro propagation 122
Actinidia polygama 13
budbreak 229
buds 37
climatic requirements 44
fruit colour 139
fruit flavour 43
insect pollination 211, 212
in vitro propagation 122
pollen viability in storage 218
Psa resistance 143
rootstocks 87, 88
waterlogging response 264
wind *vs* insect pollination 211
Actinidia purpurea 31
Actinidia rufa 44
rootstocks 87, 89
vegetative propagation 121
waterlogging or drought tolerance 88–89, 152
Actinidia sabiifolia 38, 39
Actinidia seed-borne latent virus (ASbLV) 304
Actinidia setosa 31, 32
Actinidia valvata
drought tolerance 89
fruit flavour 43
KR5 genotype 89
salinity tolerance 152, 266
waterlogged soil tolerance 89, 152, 264
actinidins 142
Adoxophyes orana (summerfruit tortrix) 334, *335*, *336*
ADP-glucose pyrophosphorylase 280
Adris tyrannus Guenée 334
AdZAT5 transcription factor 143
AGAMOUS (*AG*) gene 239, *239*, 240
ageing population, kiwifruit consumption 26
agrochemicals
fertilizers *see* fertilizers
generic 345
insecticides 333, *336*, 337, 389
pest management 344, 345, 389
alcohol acyltransferase 287
aldehydes 286–287
allergens 142
Actinidia chinensis var. *chinensis* 142
Actinidia chinensis var. *deliciosa* 55, 142
breeding to overcome impact 142–143
proteins 142
yellow-fleshed kiwifruit 55, 142
'Zesy002' cultivar 55
allergies, to kiwifruit 55, 142
Allison, Alexander 5
'Allison' cultivar 233
allopatric speciation 30
alloploids 34
alpha-amylase 2 (*AMY2* gene) 280
Alpha male cultivar
axillary bud development **230**
budbreak 228
floral morphogenesis, stages **230**
flower bud development **230**
pre-formed nodes and shoot development 229
winter chilling and endodormancy 233
Alternaria alternata **108**, 308
Alternaria species 308, 366
alternate year cropping (AYC) 195, *195*
1-aminocyclopropane-1-carboxylic acid oxidase (ACO) 285, 361

1-aminocyclopropane-1-carboxylic acid synthase (ACS) 285
Amphipsalta zelandica (chorus cicada) *107*, *338*, *339*, 340, 341
Amphitetranychus viennensis (hawthorn spider mite) 334
AMY2 gene 280
amylases 286
  genes 280
amylopectin 279
amylose 279
anaerobic metabolism 360–361
'Ananasnaya' cultivar (kiwiberry) *99*, **100**
Anastatus bifasciatus 341
Anastrepha fraterculus (South American fruit fly) 344
androdioecious flowers 38
anoxia 88, 89, 151, 179, 263
  see also waterlogged soils
anthers
  blossom blight 302
  colours 38, 96
  development, timing 238–239
  drying, pollen harvesting 217
  on female flowers, sterile pollen 96
  pollen release, bee pollination 214
  Psa infection *324*
  removal, pollen harvesting 77, 217
anthocyanins 43, 139, 281–282
  biosynthetic pathway *141*, 282
  localization in fruit 282
  metabolism 282
  synthesis, regulatory genes 139
anthracnose **108**, 308
anti-hail nets *see* hail netting
antioxidants 263, 361, 364
APETALA1 (AP1) gene 235, 238
APETALA2 (*AP2*) gene 239, *239*, 240, *240*
APETALA3 (*AP3*, *AP3-1*, *AP3-2*) genes 239
apical buds 229
Apis cerana 212
Apis dorsata 212
Apis mellifera (western honey bee) 212–214, *213*
  stocking rates of hives 213
Apis spp. *see* honey bees (*Apis* spp.)
APL4 (AGPase large subunit 4) gene 280
apoplast/apoplastic space 262, 325
Apple stem grooving virus (ASGV) 302–303
Arabidopsis 136
Armillaria species (white root rot) **108**, 295–296
Armillariella mellea **108**, 295–296
Armobreak® 247, **247**
Armoured scale **107**
aroma, of fruit *see* fruit (kiwifruit), aroma
arthropods 332
  see also pests of kiwifruit
artificial pollination *see* pollination, artificial
ascorbic acid *see* vitamin C
association maps 133
'AU Authur' 233
'AU Fitzgerald' 233
'Autari' (Ac 12.255) male cultivar **79**, 80
automation 388–389
  grading of fruit 358, 368
  packing of fruit 359, 389
autoploids 34
autotrophic leaves 259
autotrophic shoots 259
autumn
  Psa infection 318, 320, *321*, 326
  temperature for cultivated vines 156–157
auxins 278–279
axillary buds 228–233
  development 228–229, *229*, **230**, 230–231
    comparison with other fruit trees 228
    future perspectives 247
    'Hayward' *vs* Alpha (male) **230**, 230–231
    inflorescence meristem development 230, *231*, 237
    schematic diagram *232*
    timing 228, *229*, 230–231, 237
  dormancy in winter 228, *229*, 231, 232
  endodormancy 232–235
  floral productivity 231
  inflorescence meristem appearing 237
  in vitro cultures 122
  second-order 231
  spring growth 36, 228, *229*, *232*
  see also bud(s); shoot(s)

Bacillus thuringiensis 339
bacteria, increasing diversity, organic orchards 153
bacterial blossom blight 123, 301–302, *302*
bacterial canker, cause *see Pseudomonas syringae* pv. *actinidiae* (Psa)
bacterial disease 295
  blossom rot (bacterial blossom blight) 123, 301–302, *302*
  Psa virulence activated 327
bagging, fruit *66*, 199–200, *200*, 356
bags, modified atmosphere 358–359
BAM9 gene 280
BBCH scale *277*, 278
bees, pollination by 215
  kiwiberry 104
  wild *Actinidia* species 212
  see also bumble bees; honey bees (*Apis* spp.)
beetles
  as pests 334
  pollination by 216
'Belén' male cultivar **78**, 80
Belgium, as importer and re-exporter 20
Bencao Yanyi 3
Benefit® treatment 201

6-benzylaminopurine (BAP) 122, 123
berries, *Actinidia* fruit defined as 41, 273
beta-amylase 9 (*BAM9* gene) 280
BFT genes 234, *236*, 238
bicoloured kiwifruit *142*
'Bingo' cultivar (kiwiberry) *99*, **100**
bins, for harvested fruit 356
Biodiversity Treaty of Rio 127
biofilm, Psa 320
biological control 390
  pests 339–340, 345
  Psa 327
biomass (plant), production 166
  development (annual), growing season 166–167, *167*
  nutrient requirements 166
  whole-canopy photosynthetic rate and 259
biopesticides 339–340, 345
biosecurity 20
Biospife 389
black-lyre leafroller (*Cnephasia jactatana*) 337, *338*
black mould 301
black rot (*Alternaria alternata*) **108**, 308
black spot 301
blossom-end rots (BER) 305
blossom rot (bacterial blossom blight) 123, 211, 301–302, *302*, 305
blue net 198
BluPrins® 247, **247**
body rots (BR) 305
Boeremia exigua (*Phoma exigua*) 308
Boerica®, 'Hayward' cultivar 59, 130
Bombus terrestris (buff-tailed bumble bee) *215*
boron 153, 165
  toxicity 152
Botryosphaeria species 297–298, *305*, 306
  B. dothidea (side rot, dimple rot) 306
Botryotinia fuckeliana 304, *304*
Botrytis cinerea (botrytis rot, stem-end rot, grey mould) **108**, 187–188, 304, *304*, 305
  fruit curing 356–357
  reduced by ozone in storage 361
  wound rot 308, *309*
'Bounty71' rootstock *86*, 86–87, *87*, 91, 143
  KVDS-inducing soil, survival in 90
'Boverlig 02', 'Hayward' bud mutation 59, 130
'box' males 189
brand identity 22
branding of kiwifruit 56–57, 68
brands of kiwifruit 24–25, 56, 57
  Zespri *see under* Zespri Group Ltd, New Zealand
breeding (kiwiberry), targets 111
breeding/breeding programmes 119, 127, 386
  A. chinensis complex 130
  Actinidia species (not *A. chinensis*) 130
  constraints for breeders 128–129
    dioecy 40, 128
    large area for cross-progeny 129
    long juvenility 128–129
    ploidy 128, 386
  future perspectives 144
  genetically modified kiwifruit 135
  genetic diversity importance 44–45
  genome-based *see* genome-based breeding
  goals 44, 127–128, 386–387
  'Hort16A' cultivar development 12
  integrated package for success 386, 387
  interspecific hybrids *see* interspecific crosses
  male cultivars 80–81, 387
  new technologies 135–136, 144
  in New Zealand 130, 131, 386
  in other countries 130, 131, 386
  plant variety rights 57, 132, 386–387
  population genetics underlying *see* population genetics
  selection procedures 131–132
    controlled crosses 131, 132
    demonstration orchards 132
    female cultivars 131–132
    field observations 131
    fruiting selections 131–132
    male genotypes (pollenizers) 132
    multivariate function assessment 131–132
    pollenizers 132
    rootstocks 132, 387
    seedling growth 131
    storage trials 132
  somatic hybridization 131
  for specific objectives 136–144, 386–387
    carbohydrate accumulation 139
    fruit flesh colour 137–139, *139*
    fruit storage 143, 389
    hermaphrodite cultivars, genes 136–137, *137*, 387
    overcoming allergen impact 142–143
    pest/disease resistance 143, 144
    rootstocks 143–144, 387
    skin peelability 142, 380
    sweetness 139
  for sustainable industry 144
  targets 386–387
  traditional 129–130
    domestication *see* domestication of *Actinidia*
    pre-breeding 129–130
  see also fruiting cultivars (kiwifruit)
Brevipalpus chilensis (false spider mite) 344
BROTHER OF FT AND TFL1 (BFT) gene 234, *236*, 238
brown-headed leafroller (*Ctenopseustis herana*) 340
brown-headed leafroller (*Ctenopseustis obliquana*) 337, *338*, *339*
brown marmorated stink bug (*Halyomorpha halys*) 20, 333, *335*, *336*, 341, *342*, 344
brown spot (*Alternaria alternata*) **108**, 308

'Bruce' male cultivar *75*, *77*, 240
'Bruno' cultivar 60, 85
    daylength as endodormancy driver 232
    fruit weight at harvest 40
    grafting method and success 121
    light-saturated photosynthesis rate 257
    seed for rootstocks 85, *87*
    seed germination, promoting 120
    'Tomuri' cross, germination experiment 120
    vegetative propagation 120
bud(s) (kiwiberry) 96
bud(s) (kiwifruit) 36–37, *231*
    apical 229
    axillary *see* axillary buds
    capacity maintenance, winter chilling 233
    cold tolerance 232
    development 228, *229*
        management 242–247
        over two seasons 228, *229*, 255
    dormant, winter 228, *229*
    ecodormancy 232, 234
    endodormancy *232*, 232–235, 245
        abscisic acid levels 234
        cues inducing 232
        depth and duration 233
        management 245
        mechanisms 233–235, *235*, *236*
        starting 233
        transcriptome genes 234, 235
        transcriptome profiles 234, 235, *235*
        transition to ecodormancy 232, 234
        winter chilling for 232–233, *234*
    formation 228, *229*
    frost effect 265
    low temperature effect, damage 265
    meristems, winter chilling resetting 232, 233
    numbers, setting in winter pruning 196
    opening 228, *229*
    overwintering, bud scales on 32
    Psa infection *324*
    respiration, endodormancy onset 233
    respiration rates, flower number prediction 245
    second-order basal, spiral phyllotaxy *231*
    swelling before budbreak 228, *229*
    winter, optimum number 188
    winter chilling and 44, 232–233
budbreak (in spring) 37, 44, 56, 154–156, 228–229, *229*
    buds swelling before 228, *229*
    carbohydrate reserves for 244
    days after (DAB), floral differentiation 238–239, 255
    duration 229
    enhancers 154, 200, 245, 246–247, **247**, 387
    environmental conditions 229
    in first season, reasons 231
    floral differentiation stages 238–239
    frost damage 154–155, 265
    by genotype 228–229
    hydrogen cyanamide effect 200, 246, 387
    ideal male cultivars 74–75
    improvement, techniques 246–247
    kiwiberry 101
    predicting 245–246
    reduction, after very low temperature 265
    in second season 228, *229*, 231
    sepal and petal development 238
    temperatures for 154, 155–156, 229, 245
    timing 56, 228, 229, **230**, *256*
        predicting 245–246
    upward *vs* downward-facing buds 229
    vine biomass development after 166–167
    winter chilling requirement 229, 233
    winter management and 245
    xylem sap composition 262
bud scales 32, 230
budswell 228, 229, *229*
    inflorescence development 238
    shoot development 229, *229*
    see also budbreak (in spring)
bumble bees
    colonies, stocking rate of hives 214
    open-field 214
    pollen release induced by 208
    pollination by (kiwiberry) 104
    pollination by (kiwifruit) 214, *215*
        attracting, factors affecting 214
        managed pollination 212, 214
        timing and climate effect 214
        visit number, honey bee comparison 214
        wild *Actinidia* species 212
butterflies, pollination by 216

C3 plants 257
CA *see* controlled atmosphere (CA)
Cadophora luteo-olivacea (pitting, side rot) 307, 366
Cadophora species 297, 298
calcium (Ca) 87, 165
    Ca:Mg ratio 165
    deficiency 152
    depletion, Psa symptom increase 326
    excess, nutrient uptake and 165
    fruit concentration 171, *171*, 262
        high levels, fruit quality 170
        long *vs* short shoots 262
        low levels, disorders and 170
        seasonal accumulation pattern 170–171, *171*, 262–263
        vapour pressure deficit and 171, *172*
    K:Ca ratio, fruit quality 172
    leaf:fruit ratio 262
    pistillate pollen 207
    plant organ content in mature orchard 166, **166**

calcium (Ca) (*Continued*)
- soil chemistry changed by 165
- transport in xylem 262–263
- transport to fruit 171, 262
- transport to leaves 170, 262
- upper *vs* deep soil layers 170
- uptake by fruit 170–171
- uptake by roots 262

calcium carbonate 105
callus, graft formation 121
canes 185
- budbreak timing and 229
- carbohydrate reserves 244, 245
- fruiting 186, 187, 188, 190
- girdling 197
- kiwiberry 103
- large, winter pruning and 188
- productivity (one-year-old) 243
- pruning *see* pruning
- replacement 7, 185, 186, *186*, 187, 188, 192, 193, 195
- size (one-year-old) 243
- spacing, canopy density 188
- stringing systems and 193, *194*, 195, *195*
- summer pruning 188
- training *see* pergola support systems; T-bar training system
- vigour and 187, 188
- vine architecture 185–186
- winter pruning 186, 188
- *see also* shoot(s)

cankers 320, *321*, *322*
- exuding 325, 326
- Psa *see Pseudomonas syringae* pv. *actinidiae* (Psa)

canopy 255, 256–260
- area
  - soil-water loss by evaporation 176
  - water-use efficiency and 176
- for carbon capture 257, 259, 388
- in China, traditional *vs* commercial orchards 193
- dense, disease increase 187–188
- density, cane spacing and bud density 188
- depth, photosynthetic rates 259
- development 255, 256, 266
  - rate, timing and seasons 256
  - root death during 256, 259
  - temperature dependence 256
  - young plants 261
  - *see also* shoot(s)
- diseases 187–188, 300–302
- divided 188, 193
  - secondary canopies 193, *194*
  - strip-male system 188, *189*
- establishment in spring 256
- fruiting
  - alternate year cropping system 195, *195*
  - density 188
- fruiting, formation 186, *186*, *187*
  - management in New Zealand *187*, 188
  - semi-permanent 187, *187*
- growth in spring 188, 256
- leaf function 255, 256–257
- low solar radiation, as assimilate sink not source 176
- low-vigour 255
- photosynthesis *see* photosynthesis
- removal, root turnover reduced 261
- secondary 193, *194*
- source/sink relationships 176, 259–260
- wagon-wheel arrangement 192, *192*
- whole-canopy carbon acquisition 257, 259, 388

canopy management 184–196
- carbohydrate supply to fruit 376
- innovative growing systems 193–195, *194*, *195*, 201–202
  - alternate-year cropping 195, *195*
  - secondary canopies 193, *194*
  - stringing systems 193, *194*
- in New Zealand *187*, 188
- pruning *see* pruning
- training vines *see* training of vines
- vegetative regrowth control 195–196
- vine architecture and 185–186
- vine physiology 186–187

carbohydrate
- accumulation 256
  - breeding for 139
  - in fruit *see below*
  - root pruning timing 198
  - trunk girdling effect 197, 256, 260
- cell wall 143, 280, 286
- competition 260
  - floral abortion 240
  - fruit/shoot *vs* root growth 261
  - fruit *vs* woody storage organs 245
  - growing *vs* storage 260, 261
  - shoot development affected 244
  - trunk girdling effect 197, 256, 260
- in fruit 256, 274, 275, 279–280
  - composition by genotype **276**, **279**
  - metabolism during fruit growth 279–280
  - starch *see* starch
  - weight **276**
- metabolism, during fruit growth 279–280
  - soluble carbohydrates 280
  - starch 279–280
  - structural carbohydrates 280
- partitioning, in water-deficit stress 263
- reserves 244, 245, 260
  - light/shading effect on 244, 245
  - partitioning 245
- root content, recovery from waterlogging 264
- soluble 280
- source/sink relationships 260

storage 260
competition, growth and 260, 261
by roots 244, 260, 264
by shoots 243, 244
structural 280
supply to fruit, canopy management 376
see also starch
carbon
acquisition 259
increased shoot growth 256
rates, models 259
whole-canopy 257, 259, 388
assimilation per unit water transpired 176
capture 156, 257, 259, 388
fixation 256–257, 259, 263
requirements (total vine) 259
sink 259, 260
fruit growth, allocation 186, 259
leaf growth 156
removal, effect on photosynthesis 260
shoot growth 156, 259
sink/source relationships 259–260
soil organic (SOC) 152, 153, 177
starvation
effect on anthocyanin synthesis 282
effect on cytokinin synthesis 278
storage and mobilization 186, 259–260, 279
sucrose, transport in phloem 259–260
supply for vines/fruit, indicators 156
use/mobilization patterns, spring 156, 259
carbon dioxide (CO_2)
assimilation rate *258*
controlled atmosphere for storage 360
reduced chlorophyll loss with 361
high, for storage of fruit 360
intercellular, in leaves *258*, 264, 265
photosynthetic response of leaves to 257
release rates 257
β-carotene 53, *140*
carotenoids 138–139, 282
biosynthetic pathway *140*
carpels 39, 41, *41*
carpenter bees 208
category reputation 24
caterpillars 337
cellulose 143, 280
CEN genes 129, 136, 234, 238, *241*
CEN4 gene 238, *241*
Central China Agricultural College (Yichang, Hubei), China 10
CENTRORADIALIS (*CEN*) genes 129, 136, 234, 238, *241*
Ceratitis capitata (Mediterranean fruit fly) 343, 345
Ceratocystis fimbriata (ceratocystis wilt) *298*, 298–299
Ceroplastes rusci (fig wax scale) *343*, 344
'Chang Jiang No. 1' cultivar (kiwiberry) **100**
'Chaohong' male cultivar 37, 231
'check all that apply' (CATA) 377–378
'Chengxiang', *A. arguta* from the wild 40
'Cheongsan' cultivar (kiwiberry) **100**
Cherry leaf roll virus (CLRV) 303, *304*
'Chiak' cultivar (kiwiberry) **100**
'Chico' cultivar 58
see also 'Hayward' cultivar
'Chico-Male' (California Male) cultivar 77
'Chieftain' male cultivar 77, **78**, 80
Chile 57
'Hayward' cultivar 8
kiwifruit exports 20
pests in 344
chilling, winter *see* winter chilling
chilling damage 354, 358, 359, 363–364, *365*
low-temperature breakdown 363–365
chilling requirements 44
see also winter chilling
chilling sensitivity, postharvest 357–358, 359
'Hayward' tolerance 359
China
A. arguta varieties 31
A. chinensis origin 1, *2*
A. chinensis var. *chinensis* 1–2, 31
cultivation 9, 13
selections/cultivars 10
A. chinensis var. *deliciosa* 2, 31–32
cultivation 9
selections/cultivars 10
A. chinensis varieties 1–2, *2*, 31
accessions, genetic studies 127, 130
artificial pollination 216, 219
budbreak time prediction 245
consumer preferences, reasons 380
cultivars developed in 13, 57, 62–63, 130–131, 386
from interspecific hybrids 13, 62–63, 130–131
from the wild, selection 9–10, 62, 63, 386
cultivars grown commercially 3, 10, 11, 18, 57, 59–61, 192
changes in 11, 18, **19**, 59, 68
green-fleshed cultivars 18, 59–61
most widely growth (green-fleshed) 59, 60, 61
red-fleshed cultivars 10, 11, 13, 18, 65–67
regions/provinces 59–60, 62–63
yellow-fleshed cultivars 10, 13, 61, 62–63
demand for kiwifruit 24, 26
New Zealand fruit consumption 380, 381
early cultivation attempts 3, 130
expansion of kiwifruit industry 10–11, *11*, 13, 68
first kiwifruit orchards 9
fruit bagging 199–200

China (*Continued*)
fruit harvest time 44, 55
fruit quality *vs* yield 193
geographic distribution of *Actinidia* 31, 32–33
germplasm resources/repositories 9, 10, 144
habitat of *Actinidia* 32, 33
as largest kiwifruit producer 16–17, *17*, 385
local consumption *vs* exports 11, 20, 385
male cultivars used 77, 80
netting and plastic coverings 198, *199*
orchards 9, *11*
total area 10–11, 18
traditional *vs* commercial 192–193
pests of kiwifruit *see* pests
production volume 16, 17, *17*, **18**
Psa biovar 3 origin 319
rain-out shelters 198, *199*
training systems 192–193
wild mihoutao 1, *2*, 3, *3*, *4*, 31
wild populations of *Actinidia* 1–2, 3, 13, 31–32, 130, 184
germplasm evaluation 9–10, 13
'Chinese gooseberry' 4, 6, 7
change of name to kiwifruit 7, *8*
Chinese mihoutao 1, 3, *4*, 31
chloride ions 266
chlorine
deficiency 152
kiwiberry vines 106
chlorophyll 53, 54, 137, 138
accumulation 138
degradation in yellow-flesh fruit 43, 53, 54, 138–139, 281, 355
not ripening-related *139*, 281
rates, cultivars 281
temperature sensitivity/dependent 281, 357
degradation pathway 138, *138*, 282
degradation with ripening 138, *139*, 274, 282
controlled atmosphere storage 361
cytokinin role 282
slow, green-flesh colour 138, *139*, 281
fruit bagging time effect on 199
metabolism 282
retention, green-flesh coloured fruit 43, 138, *139*, 281
chloroplasts 41, 43, 274, 281
carbon fixation 257, 263
degeneration, degreening of fruit 274, 282
inner *vs* outer pericarp 41
starch storage 279
chlorosis, leaf
in high-pH soils 266
iron-induced 86, 89
chorus cicada (*Amphipsalta zelandica*) **107**, *338*, *339*, 340, 341
chromoplasts 282
chromosome numbers *33*, 33–34, *34*, 40, 134
'Chuhong' cultivar 65
Cicadas **107**, *338*, *339*, 340, 341
cincturing (trunk girdling) 196–198, *197*, 260
cineole 287
cisgenesis 144
citric acid 280
citrus flatid planthopper (*Metcalfa pruinosa*) 325, *342*, 342–343, *343*, 344
Cladosporium species 308
Clematoclethra 29
climacteric fruit 273, 275, 284, 353
atypical, kiwifruit 275, 284, 353
definition 353
climacteric–non-climacteric continuum 284
climate, for commercial kiwifruit production 37, 153–157, 158
hail 157
requirements 44
temperature *see* temperature
wind 153, *154*
climate change x, 158, 164, 387
crosses adapted to warm conditions 44
'Hayward', impact on 55–56, 81, 154
ideal cultivars and 55–56
kiwiberry distribution increase 94
lack of winter chilling, effect 154, 387
new cultivars for 387
opportunities for kiwifruit 387
risk to production 20
vine productivity/profitability 247
water availability 387
climbing plants 34–36, *35*
clonal propagation *see* micro-propagation; propagation of kiwifruit plants
clonal rootstocks 85–87, 119
Clone 8, Clone K and Clone Maeba, 'Hayward' cultivar 59
Cnephasia jactatana (black-lyre leafroller) 337, *338*
cold storage *see* storage of kiwifruit
cold tolerance 245, 265, 266
collar rot (*Fusarium solani* species) 296–297
collar rot (*Phytophthora* species) 296
Colletotrichum species **108**, *305*, 308
colour
fruit flesh *see* fruit flesh colour
fruit skin 53
petals 38
columella 39, 41
combining ability, general and specific 133
commercial fruit production
global *see* global production/supply
kiwiberry *see* kiwiberry
site selection for *see* site selection for orchards
soil and climate for *see* climate; soil
tonnage increase 385
see also management systems/practices; orchard (kiwifruit)

commercially grown cultivars 3, 8, 9, 57, 248, 332, 368, 386
in China *see* China
green-fleshed kiwifruit *see Actinidia chinensis* var. *deliciosa*; 'Hayward' cultivar
innovation/new cultivars 381
investment required for 375, 377
in Italy 19, **20**, 64
new, lack of familiarity as barrier 379
in New Zealand 12, 18–19, **19**, 58, 65, 68, 332
red-fleshed kiwifruit 12–13
see also 'Zes008' cultivar
selection criteria, assessment 131–132
yellow-fleshed kiwifruit 11–13
see also 'Hort16A' cultivar; 'Zesy002' cultivar
compatibility
graft *see* graft compatibility
male and female cultivars 76, 189
pollen and pistil 76
competition
for carbohydrate *see* carbohydrate
for honey bee pollinators 213–214
for land 158
between producers (countries) 57
for resources, floral abortion and 240, 241
compitum (pollen tube distributor cap) 39
compost 168, 170
compound dichasia 239
compression damage to fruit 363, *363*
consortia, plant propagation rights 57
consumer demand 385–386
global *see* global demand
consumer preferences/decision-making 24, 25–26, *26*, 68, 354, 375, 378–379, 390
ageing population 26
consistent quality 380
convenience 26, 54, 380
cool storage of fruit 376
country of origin 380–381
delivering preferences 22
determining *see* consumer research
eating experience 377
eating occasions 26, 378–379
eating quality and harvest indices 354, 355
environmental factors 386
familiarity, as factor 379
fruit attributes 52, 54, 377
acidity 54, 139, 276, 376, 378
colour *53*, 53–54, 386
emotional associations 378
firmness 54, 376, 377
flavour/taste 54, 139, 355, 376, 378, 380
juiciness 54, 376, 378
ready to eat 380, 386
ripeness 54, 376, 377, 380
sizes 376
sweetness 54, 62, 139, 276, 376, 378
texture 54, 376
'Hayward' cultivar, conventional *vs* organic 376
health 25–26, 42, 55, 380, 386
negative emotions, causes 378, *379*
new cultivar fruit 381
product innovation 26, 381
rejection, reasons 380
retail channels 26
sensory profiling 377–378
supply for 52 weeks/year 380, 386
sustainability 26
time for eating 379
consumer research 375–384, 377–379
evaluation of kiwifruit 377–379, 381
'check all that apply' (CATA) 377–378
circumplex approach 378, *379*
context (situation/occasion) 378–379
eating, time of 379
emotional associations 378, *379*
'just about right' (JAR) scale 378
multi-factor interactions 378–379
sensory profiling 377–378
consumer responses to kiwifruit 375–384
current/future trends 381
environmental/management factors, influence 375–377, 381
on-orchard management 376, 381
postharvest storage 376–377
preharvest factors 376
extrinsic factors affecting 379–381
ripe SSC (rSSC) and dry matter (DM) 375–376
controlled atmosphere (CA) 353, 356, 360–361
chilling sensitivity, loss 358
degreening of yellow-fleshed fruit 357, 358
pre-grading treatment 357, 358
storage of fruit 360–361, 377
consumer response 377
fruit ripening slowed 360
fruit softening, effect on 360
hard cores 367
low-temperature breakdown reduced by 364
pericarp granularity and translucency 367
convenience factor 26, 54
Convention on Biological Diversity (1992) 127
'Convi 97,001', 'Hayward' bud mutation 59, 130
cooling of fruit, storage 359
see also chilling sensitivity, postharvest; storage of kiwifruit
copper, for kiwiberry vines 106
copper-based compounds, Psa control 326–327
cordons (leaders) 103, 186, *186*
core of fruit *see* fruit (kiwifruit), core
coronatine **318**, 319, 326
Cortinicara gibbosa (minute scavenger beetle) 335
cossid moths 334
Cossus cossus L. 334

cost(s)
- agrochemical development 345
- artificial pollination 219
- labour and on-orchard 21
- T-bar training system 187

COVID-19 pandemic 26
CPPU, effect on fruit development 200–201
CRISPR/Cas9-mediated genome editing 129, 136, 234, *236*, 387
- hermaphrodite cultivar 240, *241*

crop coefficient value (K_c) 174, 175, *176*, 178
crop demand 165–166
- *see also* global demand

crop evotranspiration 174–175, *175*
crop irrigation *see* irrigation
crop load 38
- fruit size affected by 53, 196
- setting 196

crop management 196–198
- girdling practices 196–198, *197*
- root pruning 198
- setting crop load 196

crop yield 55, 56
- alternate year cropping 195, *195*
- fruit number *vs* weight 196
- girdling practices and 196–198, *197*
- 'Hayward' cultivar 184, 193, 196
- improved by reduced tillage 170
- improved by soil organic matter 170
- irrigation and water availability benefits 174
- secondary canopies 193

crosses
- controlled 131, 132
- interploid, single species 128
- interspecific *see* interspecific crosses
- land required 129, *129*
- selection procedures 131–132

crown pathogens 295–297
'crush tipping' (tip-squeezing) 195–196
cryopreservation 121–122
cryotherapy 123
cryptic dioecy 96
Cryptosporiopsis actinidiae 306–307, 366
Ctenopseustis herana (brown-headed leafroller) 340
Ctenopseustis obliquana (brown-headed leafroller) 337, *338*, *339*
'Cuixiang' cultivar 18, 59, 130, 364
'Cuiyu' cultivar 59–60, 245
cultivars xv–xviii
- commercially grown *see* commercially grown cultivars
- fruiting (female) *see* fruiting cultivars (kiwifruit)
- male (pollenizer) *see* male cultivars
- *see also* breeding/breeding programmes; *specific cultivars*

curing of kiwifruit 356–357
cutin 273
cuttings of kiwifruit cultivars 120–121
cylindrocarpon rot 307–308
Cylindrocarpon species 297, 299, 307–308
cymes 75, 96
cytokinin
- 6-(3-hydroxybenzyl) adenine (mT) 122–123
- chlorophyll breakdown and 282
- control of levels in kiwifruit 278
- floral sex expression 240
- forchlorfenuron (CPPU) 200–201
- fruit growth 273, 278
- increase with fruit ripening 278
- seed development 283
- synthesis, carbon starvation effect 278

'D1 Vitroplant'® 86, 143
Dactylonectria species 297
damage to fruit *see* fruit, damage
darkness, seed germination and 120
daylength, endodormancy driver 232
decaploids 34, 94, 128
deciduous nature, of *Actinidia* 37
defence systems, kiwifruit 390
defoliation 231
degreening 138, 274, 281, 282, 355, 357
- chloroplast degeneration 274, 282
- controlled atmosphere storage 357, 358, 361
- quantification 355
- rates 281
- red-fleshed fruit 139, *139*, 282
- slow, green-fleshed fruit 139, *139*, 281, 355
- temperature sensitivity, postharvest 357, *357*, 359
- yellow-fleshed fruit 138, 281, 355, 357
- *see also* chlorophyll, degradation

'Deliwoong' male cultivar **79**, 80
demand for kiwifruit *see* global demand
demonstration orchards 132
description of kiwifruit 1
Diaporthe species *305*, 305–306, *309*
diaspidids 337, 340
dichasia (dichasium) 75, 239
- compound 239

dieback of vines 297
dietary fibre 110, 386
diffusion, nutrient absorption 165
dimple rot (*Botryosphaeria dothidea*) 306
dioecy/dioecious *Actinidia* flowers *38*, 38–39, *39*, 40, 72, *73*, 240
- all commercial cultivars 52, 72
- breeding programme problems 40
- consequences 40, 132
- crosses between genotypes of different sex 128, 132
- dependency on pollination 210–211
- female-to-male ratio 128

fruiting or inconstant males 40, 136
genes controlling 136, 137, *137*
kiwiberry cultivars 96
sex expression, flower development 240
transition to hermaphroditism 81
diploid *Actinidia* 32, 33, 34, 56, 73, 128
chilling requirements 44
flowering time 37, 73–74
flowers *73*
genetic maps 133
genome size 33, 134
pollen yield and fertility 75
disease(s) *see* fruit diseases; postharvest disease; vine disease
disease resistance
breeding for 143
cultivar selection 52, 53
ideal male cultivars 76, 81
importance of genetic diversity 44
Psa *see Pseudomonas syringae* pv. *actinidiae* (Psa)
of rootstocks 90, 91
DNA sequencing 133, 134
domestication of *Actinidia* 2–3, 85, 129–130
Actinidia chinensis var. *chinensis* 2
Actinidia chinensis var. *deliciosa* 2, 129–130
in New Zealand 2–3, 5–6, 85, 129–130
'Donghong' cultivar 11, 18, 65, 130
dormancy, winter 245, 255
axillary buds 228, *229*, 231, 232
chemicals breaking 56, 245, 246
budbreak enhancers 154, 200, 245, 246–247, **247**, 387
ecodormancy 232, 234
endodormancy *see* bud(s), endodormancy
DORMANCY-ASSOCIATED MADS-box genes (*DAM*) 234
Drosophila suzukii (spotted-wing drosophila) **107**, 343–344
drought
air embolism in xylem 263
tolerance 150, 263
screening of rootstocks for 88–89
water deficit stress 263
xylem issues, hydraulic conductance 261
see also water, deficit
dry matter (DM) 186, 375, 376
composition 275
consumer response 375–376
definition 375, 376
higher, later-harvest fruit 376
kiwiberry fruit at-harvest 109, **276**
percentage in *A. chinensis* fruit 274–275
in plant organs of mature orchard 166, **166**
pollen source/characteristics affecting (xenia) 209
ripe fruit eating quality prediction 356
ripe soluble solids content (rSSC) and 276, 375–376
Dynamic Chill Model 233, 245
'Earp' (MC79) male cultivar 77, **78**
'easy-to-peel' fruit 142
eating occasions, for kiwifruit 26
ecodormancy 232, 234
e-commerce 26
economic losses/impact
of pests 332
of Psa 317
economic recession 24
Edwardsiana salicicola 344
effector proteins 319
elephantiasis, kiwifruit 297
embryo growth 283
embryo rescue 131
'Emihoutao No. 1' cultivar 60
emotional responses to fruit 378
Empoasca vitis (leafhopper) 344
Endoclita excrescens 334
endodormancy *232*, 232–235
see also bud(s), endodormancy
environment, natural, for *Actinidia* 387
environmental challenges 387–388
environmental factors/issues
affecting foliar fertilization 168
agronomic practices protecting soils 170
bud/shoot and flower development 247
consumer preferences 386
effect on photosynthesis 257, *258*
integrated systems for production 386
pest management challenges 344
sustainability of kiwifruit 26
environmental pollution 169
environmental stress responses 263–266
frost 265–266
high-pH soils 266
high temperatures 264–265
low temperatures 265–266
salinity 266
water deficit stress *see* water, deficit
waterlogging *see* waterlogged soils
epicarp *see* fruit (kiwifruit), skin (epicarp)
Epicoccum species 308
Epiphyas postvittana (light brown apple moth) 340
Erger® 247, **247**
Erthesina fullo (yellow-spotted stink bug) 333, 334, *335*, *336*
esca-like disease 298
esters, production 287
ethanol 360–361
Ethrel® 283
ethylene 273, 284, *285*, 353, 361
after cool storage, fruit aroma increase 377
after early harvesting 355
biosynthesis, enzymes 285
contamination, risks 284, 353
development of fruit response to 284, 362

ethylene (*Continued*)
endogenous production by fruit 273, 284, 285, *285*, *286*, 353, 354
reduced by low O_2 in storage 361
exogenous, controlled ripening 284, 353, 355, 361–362, 377
coordination of all fruit 286, 362
fruit sensitivity to 284, 285, 353–354, 361
low storage temperature increasing 285, 360
reducing, methods 361
fruit softening 284, 285, 286, *286*, 353, 368
fruit unresponsive to, MCP bound to receptors 361
Psa virulence and 326
receptors 285, 361
regreening of yellow-fleshed fruit 368
removal, CA/MA storage 359, 360, 361
removal/avoidance in coolstore 361
respiratory activity increase 284, 353
rotting fruit producing 368
volatile production in fruit 287, 377
ethylene response factor transcription factors 285, 361
ethylene response genes 285
Europe
history of kiwifruit cultivation 7–8
'Hort16A' cultivar in 12
importer and re-exporter 20
see also France; Greece; Italy
European and Mediterranean Plant Protection Organization (EPPO) 317, 320
evapotranspiration 174, 177
reference value (ET_0) 174, 176, **179**
annual (Italy) 174, *175*
see also transpiration
evocation, floral *see* floral evocation
evolution
Actinidia species 30, 39
genome duplication 134
kiwifruit as a crop 13
exopolysaccharide (bacterial), Psa *322*
explants, for *in vitro* propagation 122
exporters of kiwifruit 16, 19–20
China (domestic consumption greater) 11, 20, 385
export value **21**, 21–22
global volume 20, **20**, 21, **21**
Greece and Chile 20
Italy 20
New Zealand 6–7, 12, 19–20, 22, 57, 354
elements driving increase 22–24
yellow-fleshed fruit 12
see also global producers

F_1 populations 42, 128
false spider mite (*Brevipalpus chilensis*) 344
fatty acid desaturase gene (*AdFAD1*) 287
Fe chelate treatment 89
female cultivars *see* fruiting cultivars (kiwifruit)
female flowers *see* flower(s), pistillate (female)
female-to-male ratio 128
female vines 72
male vines planted among 72, 76
rows, strip-male system 188, *189*
'Fengyue' cultivar 62, 245
'Ferguson' (MC110) male cultivar 77, **78**, 80
fertigation 168–170
nitrate concentration, seasonal 169, *169*
plan **169**
fertilizers
application, timing 167
fertigation *see* fertigation
foliar fertilization 105, 168
impact on environment 164
kiwiberry production 105
NPK 105
organic 170
Psa virulence affected 326
recommendations, soil types 168–169
slow-release solid 169
solid 169
distribution in soil 167
vs fertigation, nitrate levels 169, *169*
field bins 356
field capacity (FC) 177, 179
fig wax scale (*Ceroplastes rusci*) *343*, 344
filaments, pistillate flowers 39
flavonoids *141*, 282
biosynthetic pathway *141*
decline with storage 361
flavonols *141*
flavour of fruit 4, 43–44, 286
consumer preferences 54, 139, 355, 376, 378, 380
flesh colour association 54, 58
flies, pollination by 216
floral abortion 86, 240–241, *242*
floral constancy 76, 213
floral differentiation 238–240, *239*
genetic regulation *239*, 239–240
hydrogen cyanamide improving 246, *246*
model 239, *239*
timing and stages 238–239
floral evocation 228, *232*, *235*, 237–238
genetic regulation 238
prevention, shoot girdling 237
timing 237–238
floral primordia 239
floral productivity, increasing 246–247
floral sex expression 240
'florigen' 238
flow cytometry 33, *34*, 128
flower(s) (kiwiberry) 96, *98*
flower(s) (kiwifruit) *30*, 32, *38*, 38–40, *39*, 207–210

abortion (floral) 86, 240–241, *242*
Actinidia chinensis var. *chinensis* 32, 38, *38*, *208*
Actinidia chinensis vs. deliciosa 32, 38
bisexual 136
bud development **230**
climate affecting 81, 207, 208
collection for pollen harvesting 217
constancy 76, 213
development 228, *229*, 237–242, 255
    anthers 238–239
    floral differentiation 238–240, *239*, 246
    floral evocation (commitment to) 228, *232*, *235*, 237–238
    future perspectives 247
    genetic regulation 238
    initiation, mechanism 237
    over two seasons 228, *229*, *232*, 237
    primordia shape/development 239
    sepals and petals 238
    sex expression 240
    stigma, and ovary 238–239
    see also floral differentiation; floral evocation
duration 81, 207
each setting a fruit 56
excess, removal 196
failure to, temperature effect 243
female *see* flower(s) (kiwifruit), pistillate (female)
king (terminal) *vs* lateral 56, 75
lateral 37, 56, 75, 86, 196
    development 239
    removal/abortion 37, 56, 75, 86, 196, 246
male *see* flower(s) (kiwifruit), staminate (male)
number 207, 228
    climate change effect 81
    crop yield and 56
    factors affecting 207–208
    improvement, techniques 246–247
    male cultivars and 74–75, 207
    predicting 245–246
    shoot orientation effect 243
    timing of development in previous season 243
opening, timing 208
phenology 207–208
pistillate (female) 38, *38*, *39*, *73*, 128, 207, *208*, *324*
    bud development **230**
    in controlled crosses 131
    development 39, 240
    honey bees visiting/preference 76, 213
    pollen (sterile) 39, 72, *74*, 128, 131, 207
    pollen grain, appearance *74*
    pollination *see* pollination
    Psa infection/multiplication *324*, 325
    structure 39, 72, 128
pollen from *see* pollen
popcorn stage, pollen 208, *209*, 217
primordia 239
production 207–208
Psa epiphytic growth 320
Psa infection cycle *324*, 325, 326
scent 39–40, 212
self-grooming 56
sex expression 240
single, preference for 37, 56
sizes 32, 38
staminate (male) 5, 38, *38*, 72, *73*, 81, 128, *208*, *324*
    abundance, honey bee preference and 213
    bud development **230**
    in controlled crosses 131
    development 240
    honey bees visiting 213
    number 81, 207
    pollen 207, 208
    sizes 32
    structure 38, 39, 72
structure *38*, 38–40, *39*, 72, 128
terminal, development 238–239
on terminated spurs *vs* canes 243
thinning 196
flowering
    asynchronous (male/female) 216
    commitment to (evocation) *see* floral evocation
    improvement, techniques 246–247
    initiation 237
    regulation, genetic 238
    repressor gene, editing 278
    time/duration 81, 207
        artificial pollination and 77
        heritability 80–81
        juvenile phase duration and 128–129
        ploidy and 37, 73–74
        several in growing season 231
        short, commercial cultivars 207–208
FLOWERING LOCUS C (FLC) gene 136, 235
flowering shoots 37–38, 243
flowers per winter bud (FWB) 246, **247**
flyspeck **108**
foliar fertilization 105, 168
food allergies 55
food diaries 379
food labelling 142
food webs, organic orchards 153
forced-air cooling 359
forchlorfenuron (CPPU) 200–201
France
    interspecific crosses 131
    pruning 189–192
    vine diseases 298
Fraser, Isabel 5, *5*, 129–130
free-spindle training system 190

freezing damage 364, *365*
freezing wounds, Psa invasion 322
frequency domain reflectometry (FDR) *178*
fresh fruit market (all fruit)
  consumer demand 385–386
  size and revenue 21, **21**, 385
  total production *vs* kiwifruit production 16, *17*, **20**
friction damage to fruit 363, *363*
Friendly Boy (FrBy) gene 137, *137*, 240, *241*
frost
  avoidance, tall T-bar system (Pergoletta Alta) 191
  damage 154, 157, *158*, 265
  preharvest 157
  protection 155
    kiwiberry *101*
  response to 265–266
  sensitivity
    kiwiberry vines 101, *101*
    kiwifruit vines 154, *156*, 265
  tolerance 265
frost-tolerant rootstocks 87
fructokinase 280
fructose **276**, 280
fruit (kiwiberry) *see* kiwiberry, fruit
fruit (kiwifruit) 40–41, 266, 273–294, 353–354
  acidity 110, 211, 276, 280–281
    consumer preferences 54, 139, 276, 376, 378
    firmness correlation 377
    improving fruit 211
    pollen source/characteristics affecting (xenia) 209
    titratable (TA) 42, 376
  appearance 40, 42–43, 53–54
    see also specific cultivars
  aroma 43–44, 286
    reduced by cool storage 376
  bagging *66*, 199–200, *200*, 356
  biology 353–354
  biomass accumulation ('Hayward') 167, *167*
  bruising 362, 363, *363*
  calcium transport to, xylem 171
  calcium uptake/supply 170–171, *171*
  carbohydrate accumulation *see* carbohydrate; starch
  carbohydrate metabolism 279–280
    soluble carbohydrates (glucose, sugars) 280
    starch 279–280
    structural carbohydrates 280
  cell wall components 143, 280
    carbohydrates 143, 280, 286
  characteristics affected by interploid pollination 74
  Classes (UNECE) 53
  colour *see* fruit flesh colour
  composition 42, 43–44, 273, 274–276, **276**, **279**
    calcium levels and quality 170
    carbohydrate 275, **276**, 279–280, 286
    dry matter, % by fruit structure 274–275
    by genotype **276**, **279**
    glucose 280
    by growth stage *277*, 278
    minerals 170–173, **276**
    organic acids 280–281
    organic acids by growth stage *277*, 278
    starch 260, 274, 275, 279–280
    sugar:acid balance 43, 276
    sugars **276**, **279**, 280
    vitamins 42, **276**
    weight of each component **276**
    see also carbohydrate; starch; vitamin C
  compression 363, *363*
  core (columella)
    hard 54, 367–368
    structure 39, 41, 58, 274, **274**, *274*
    white inclusions 368
  damage
    bruising 362, 363, *363*
    chilling causing/symptoms 363–364, *365*
    compression 363, *363*
    cultivars to reduce 53
    freezing 364
    fruit-to-fruit 37, 53, 56
    impact, signs of 362, *363*
    by packaging 359, 362
    pedicel stub causing 356
    by pests *336*
    physical, causes 362
    rot development 308
    rubbing 363
    sensitivity to 362
    by vibration, friction 363
    weather and fungal disease 300
  defects 362
    see also under quality of kiwifruit
  development 273
    calcium supply 170–171
    CPPU effects 200–201
    temperatures for 156
    tomato similarities 275
    water supply 171
  discoloured
    chilling damage 364, *365*
    pitting 366
    storage stain 366, *367*
  disorders, mineral imbalance 170
  diversity 41–44, *42*, *43*
  drop 38, 56
    premature, pests causing 335
    Sclerotinia sclerotiorum 300
  dropped shoulders 211
  dry weight (DW) 276, 277

epicarp *see* fruit (kiwifruit), skin
fasciated 53
firmness 284, *285*, 358
    acidity correlation 377
    consumer response 54, 376, 377
    pollen sources affecting (xenia) 209, 283
    quality assessment 362
    storage temperature 376
'flat' 53
flavour *see* flavour of fruit
flesh colour *see* fruit flesh colour (kiwifruit)
foliar fertilization 'target' 168
fresh weight (FW) 276, 277
from 'fruiting' males 136, *136*
future research targets 287
grading *356*, 358, 368
growth 255–256, 276–279
    auxins and 278–279
    carbohydrate metabolism 279–280
    cell division stage 277, *277*, 278
    cell enlargement stages 277, *277*, 278
    cessation at maturation 282, 284
    by cultivar *277*, 278
    cytokinin levels 273, 278
    dry weight (DW) 276, 277
    fresh weight (FW), sigmoidal pattern increase 277
    organic acids during 280–281
    phytohormones and 278–279
    rapid, in stage I *277*, 278
    slow, in stage II 277, *277*, 278
    stages 255–256, *277*, 277–278
    transition to maturation 283
    water supply, sensitivity 171
handling 52, 53, 356
harvest *see* harvest/harvesting
indentations, damage 363, *363*
    at stylar end 53
low-temperature breakdown (LTB) 363–364
maturation 282–284
    CPPU effect 201
    growth cessation 282, 284
    harvest after 353
    phase 1 softening 285
    ripening difference 282–283
    seed coat colour change 283
    seed development 283, **284**
    soluble solids accumulation 283, 354–355, *355*
    temperature affecting 156
    time 54
    transition from fruit growth 283
    transition to ripening 285
    triggering 283
mature (unripe) 282–283
morphology, CPPU effect 201
nutritional value 25
osmotic pressure 275, 279–280
packing *see* packaging/packing of fruit
parthenocarpic 40
picking *see* harvest/harvesting
pitting 364, 366, *366*
    fungal 366
pollen source/characteristics affecting (xenia) 209–210
postharvest *see* postharvest
quality *see* quality of kiwifruit
removal, storage/sink relationship study 260
ripe 43, 54, 273
    composition 273, **276**
    esters 287
    flesh colour *see* fruit flesh colour (kiwifruit)
    skin colour 43, 53, *281*, 281–282
    soluble solids content 59, 61, 108, 109, 281, 283
    texture 41–42, 54, 376
    volatiles 286, 287
ripeness
    assessment difficulty 380
    consumer responses 376, 377, 380
ripening 284–287, 353, 361–362, 377
    chilling damage risk reduced 359
    climacteric 273, 275, 284, 353
    climacteric–non-climacteric continuum 284
    control 284–285, *285*
    criterion for 284
    ethylene application *see* ethylene
    ethylene production 273, 284, *285*, *286*, 353–354, 361
    genetic control 285, 287, 354
    maturation *vs* 282–283
    mechanism 275
    slowed by controlled atmosphere 360
    slowed by ozone 361
    starch conversion to sugars 275–276, *355*
    starting, time 44
    transition, Phase 1 to Phase 2 softening 285, *286*, 354
    transition from maturation 285
    triggering 283, 284
    volatile production 286–287
set 255
    honey bee pollination and 213
setting rate 210
shape 58
    ideal cultivars 53
    variety *chinensis vs. deliciosa* 32
    see also specific cultivars
shrinkage 284
shrivel 366–367
sizes 40–41, 72
    Benefit® treatment 201
    carbohydrate availability increasing 256

fruit (kiwifruit) (*Continued*)
consumer preferences 376
estimation method 276–277
factors affecting 211
flower quality and pollination 211
forchlorfenuron (CPPU) enhancing 200–201
ideal cultivars 53
increase over time 276
large 53
LDD (length, smallest/largest diameters) 276–277, 284
pollen grain number and 210
root pruning and 198
seed number and 211, 283
trunk girdling effect 197
sizing, in breeding 132
skin (epicarp) 41, 273, *274*, 362
colour 53
damage 56, 362
discolouration, low-temperature breakdown 364
genetic control 273
hairs *see* hairs, on kiwifruit skin
ideal cultivars 53
lenticel spots 366, *366*
lifting, discolouration and 362, *362*
markings 364, 366, *366*, *367*
peelability, breeding for 142, 380
permeance 367
pitting 364, 366, *366*
raised spots 366, *366*
softening 283, 285–286, 353, 377
breeding to reduce 143
control 143, 284, 353
controlled/modified atmosphere effect 359, 360
ethylene unresponsiveness 361
hard cores and 367–368
inhibition by MCP 361
key events in 143, *286*
measurement 285
pattern and phases 143, 285, *286*, 354, *354*, 377
Phase 1 285, *286*, 354, *354*, 360
Phase 2 285–286, *286*, 354, *354*, 360, 361, 368
Phases 3 and 4 286, *286*, 354, *354*, 361
rate of, temperature and 285, 359–360, *360*
specific tissue zones 377
starch metabolism 286
temperature-induced 285, 359–360, *360*
triggers 143, 285, 286, *286*, *354*
volatile production 286–287, 377
see also fruit, ripening
storage *see* storage of kiwifruit
structural carbohydrates 280
structure 273–275, *274*
contribution of each tissue to weight 274
core (columella) 39, 41, 58, 274, **274**, *274*
epicarp *see* fruit, skin (epicarp)
hairs *see* hairs, on kiwifruit skin
inner pericarp *see* pericarp, inner
morphology, CPPU effect 201
outer pericarp *see* pericarp, outer
periderm 273
red pigment site 54, *281*, 281–282
softening of pericarp/core 286
tissue zones 273–274, *274*
vascular bundles 274
sun damage 44
taste 43–44, 355
consumer preferences 54, 139, 355, 376, 378, 380
organic acid impact 280
temperature affecting 156
fruit softening 285, 359–360, *360*
high temperatures 265
low-temperature breakdown 363–364
thinning 196
timing of shoot development and 243–244
transpiration *see* transpiration
transport, sea freight 368, 389
unacceptable quality 358
unripe 43, 353–354
vitamin C *see* vitamin C (ascorbic acid)
volume, measurement 277
water and nutrient transport to 171
weight
contribution of each tissue type 274
diversity 42
dry weight (DW) 276–277
factors influencing 196
fresh weight (FW) 276–277
fruit number relationship 196
at harvest, seed number and 40, 72
increase from setting to harvest 277
ploidy correlation 56
pollination importance 72
wound rots 308, *309*
yields *see* crop yield
fruit diseases 304–308
global risks 20, 389
of kiwiberries 106, **108**
Psa *see Pseudomonas syringae* pv. *actinidiae* (Psa)
see also postharvest disease
fruit flesh colour (kiwifruit) 2, 42–43, *281*, 281–282
anthocyanin metabolism *see* anthocyanins
bicoloured *142*
breeding for 137–139, *139*
carotenoid metabolism *140*, 282
changes late in development 12, 43
chlorophyll degradation *see* chlorophyll
CPPU effect 201

degreening *see* degreening
diversity *42, 43*
flavour association 54
green *see* green-fleshed kiwifruit
harvest indices and 355
for ideal cultivar *53*, 53–54
measurement 355
in peridermal *vs* epidermal skin 281
pollen sources affecting (xenia) 209–210
red *see* red-fleshed kiwifruit
skin lifting and discolouration 362, *362*
variety *chinensis vs. deliciosa* 32
yellow *see* yellow-fleshed kiwifruit
fruit fly 343–344, 345
FRUITFUL (FUL) 235
fruiting cultivars (kiwiberry) 98, *99*
fruiting cultivars (kiwifruit) 44, 52–71, 184
breeding programmes aim 44
choosing 68
climatic requirements 44
current and future trends ix, 68
developed in China 57
fruit quality prediction 68
green-fleshed 58–61
see also green-fleshed kiwifruit
hermaphrodite 40, 52
see also hermaphrodite cultivars
key attributes 2–56
convenience 54
cultivar promotion 56–57
fruit flavour and texture 54
fruit flesh colour *53*, 53–54
fruit size and shape 53
fruit skin and handling 53
harvesting 54–55
health attributes 55
novelty 56
ploidy 56
Psa tolerance 52, 53
storage 55
vine attributes 55–56
list, taxon, gender and tradenames xv–xviii, 57
main commercial ones *53*, 57–58
ploidy 34
promotion of 56–57
red-fleshed 65–67
see also red-fleshed kiwifruit
risks (commercial) associated 52
yellow-fleshed 61–65
see also yellow-fleshed kiwifruit
fruiting males 136, *136*
fruitlets, excess, removal 196
fruit-piercing moths 334
fruit-to-fruit damage 37, 53, 56
FT gene and FT protein 238
FUL gene 239
FUL-like gene 238, 239, *239*
fungal diseases 295
canopy diseases 300–302
kiwiberry **108**, 110
leaf diseases 301
pitting due to 306–307, 366, *367*
root pathogens 295–297
specific diseases
Armillaria species (white root rot) 295–296
Botryosphaeria species 297–298
Cadophora species 297, 298
Fusarium avenaceum 298
Fusarium solani species complex 296–297
Neonectria microconidia 299
Phytophthora species 296
Pseudocercospora actinidiae 301
rust fungi 301
Sclerotinia sclerotiorum 76, 300, *300*
Verticillium species 297
trunk diseases 297
fungicides 296, 358
Fusarium avenaceum 298
Fusarium solani species complex 296–297
Fusarium species 308, *309*
future potential/directions 385–390
breeding and breeding programmes 144
consumer responses to kiwifruit 381
fruiting cultivars ix, 68
global industry (kiwifruit) 25–26, 385
irrigation 179–181
male cultivars 81
postharvest, trends 388
propagation of kiwifruit plants 123–124
rootstocks 90–91
yellow-fleshed kiwifruit 68

galactose 280, *286*
L-galactose pathway 281
'Ganmi No. 2 ' ('Kuimi') cultivar 62
'Ganmi No. 3' ('Jinfeng') cultivar 62
garlic oil 247
gender
genetic control 137
see also flower(s) (kiwifruit), pistillate; flower(s) (kiwifruit), staminate
gender inconstancy 40, 136
gene(s)
allelic variations 134
bud endodormancy control 234, *235*
expression change with hydrogen cyanamide 246
floral differentiation, regulation *239*, 239–240
floral evocation regulation 238
number 134
Psa biovar 3 resistance 319
Psa genome 318–319, 320
Psa susceptibility 327

gene annotation 134
gene editing 81, 129, 135–136, 381, 387, 388
  flowering repressor gene 278
  fruit development research 136, 287
gene linkage maps 134
genetically modified organisms 135
genetic control (of)
  bud endodormancy 234, *235*
  ethylene biosynthesis 285
  floral differentiation *239*, 239–240
  floral evocation regulation 238
  fruit ripening 285
  fruit skin 273
  fruit starch accumulation 280
  sex expression 137, *137*
  soluble solids content (SSC) 354–355
  waterlogging stress response 264
genetic diversity 127
  importance 40, 44–45
genetic improvement 127–149, 387
  see also breeding/breeding programmes
genetic maps 133–134
genetic recombination 40
genetic transformation 135
'Geneva' cultivar (kiwiberry) *99*, **100**
Geneva Double Curtain (GDC) system 191, *191*
genome ix, *134*
  Psa 318
genome-based breeding 133–135, 144, 387
  gene annotation 134
  genome editing *see* gene editing
  genome sequencing 134
  molecular markers, genetic maps 133–134
genomic selection (GS) 135, *135*, 144
geographic distribution, *Actinidia* 31–32, 32–33, 44
germplasm 2, 3, 247, 386
  fruit colour diversity 138–139
  screening, of rootstocks 132
  wild *Actinidia*, evaluation in China 9–10, 13
gibberellin 264, 279
girdling practices 196–198, *197*
  rationale/actions 197, 245, 256, 260
  shoot girdling 237
  timing 197–198
  trunk girdling 196–198, *197*, 260
global brands 24–25, 56, 57, 237
  see also Zespri Group Ltd, New Zealand
global demand (for kiwifruit) 7, 8, 16, 21–24, 385–386
  China 24, 25
  global risks to 24
  key elements driving increases 22–24
    future trends/opportunities 24, 25–26, *26*
  market size 21, **21**, 127
global demand (for land) 387
global industry (kiwifruit) 16–28, 385–390
  brands 24–25, 56, 57, 237
  cultivar changes 18–19
  current trends 386
  demand *see* global demand
  early, in New Zealand 6–7
  exporters *see* exporters of kiwifruit
  future trends/opportunities 25–26, 385, 390
  global supply *see* global production/supply
  growth increase/volume 16, *17*, 21, **21**, 385
  importer and re-exporter (Belgium) 20
  processed kiwifruit products 24, 25
  producers/suppliers *see* global producers
  risks to 20–21, 24, 52
  trade value *see* global trade, value
  see also China; Italy; New Zealand
global producers 16–18, *17*
  China 16–17
  Greece 17–18
  Italy 8, 17
  New Zealand 17, *17*, **18**, **19**
  production (Mt) 17, **18**
global production/supply 16–21, 385
  commercial cultivar changes 18–19
  compared to total fresh fruit 16, *17*
  growth by country (2016-2020) 17–18, **18**
  growth by year (2001-2020) 16, *17*, 385
  growth of exports (2000, 2020) 21, **21**, **22**
  growth (%) of top producers 17–18, **18**
  optimizing, future trends 388
  risks affecting 20–21, 24
  volumes/area in New Zealand 17, *17*, **18**, **19**
global sales 22, *23*
global trade, value **21**, 21–22
  countries contributing to 22
  global risks to 24
  key elements driving increases **22**, 22–24
  Zespri Group Ltd case study 22
glucose **276**
  rapid fruit growth (stage I) *277*, 278, 280
glutathione S-transferase 246
'Golden Dragon' cultivar 233
'Golden Sunshine' cultivar 233
gooseberries, kiwifruit flavour likened to 4
grading of fruit *356*, 358, 368
graft compatibility 87–88, 121, 132
  incompatibility 88, 121, 132, 143
grafted cultivars 36, 119
  see also rootstocks
graft incompatibility 88, 121, 132, 143
grafting 121, 124
  cleft 121
  graft formation, callus formation 121
  interspecific 87–88
  methods 121
  rootstocks *see* rootstocks
granularity, pericarp 368
grape clearwing moth (*Paranthrene regalis*) 334
grapevine, training system for kiwifruit 191–192, *192*
grapevine moth (*Lobesia botrana*) 343

Greece 385
kiwifruit exports 20
kiwifruit pests 344
kiwifruit production 17–18, **18**
greedy scale (*Hemiberlesia rapax*) 337, *338*, *339*
green-fleshed kiwifruit 42–43, *43*, *53*, 53–54, 138, *139*, 281, *281*
breeding 138
chlorophyll accumulation 138
chlorophyll retention 43, 138, *139*, 281
cultivars 58–61, 130
China 59–61, 130
global 58–59
Italy 61
Japan 61
New Zealand 61
cytokinin signal and 278
harvest indices 354, 355
processed products 25
see also Actinidia chinensis var. *deliciosa*; 'Hayward' cultivar
green-headed leafroller (*Planotortrix* spp.) 340
greenhouse cultivation 81
greenhouse thrips (*Heliothrips haemorrhoidalis*) *338*, 340, 344
green/red-fruit flesh colour 139, *139*, *142*
green vegetable bug (*Nezara viridula* (L.)) 341–342, *342*, *343*
grey mould *see Botrytis cinerea* (botrytis rot, stem-end rot, grey mould)
growing degree days (GDD) 154
growing strategies, new, testing 388
growth
active, minerals absorbed from soil 165, 166
annual (by season), biomass accumulation 166–167, *167*
axillary buds 36, 228, *229*, *232*
canopy, in spring 188, 256
cessation, endodormancy 232–235
embryo 283
fruit *see* fruit (kiwifruit)
interruption, juvenility phase reduction 129
leaves 156, 175, 231
nutrient requirements/removal and uptake 166
roots *see* root(s)
shoots *see* shoot(s)
growth chambers 131
growth cycle 228, *229*, 255–256
flower development over two seasons 228, *229*, *232*, 237
key stages ('Hayward') 255, *256*
growth habit 34–36, *35*, 242
male cultivar choice 76
vigorous climbing 34–36, *35*
Guangxi Forest Institute, China 10
Gui-1 rootstock 90
'Guichang' cultivar 10, 59, 60, 130
gynoecium *30*, 39
habitat, *Actinidia* 30, 32, 33
'Haegeum' cultivar 130, 150
'Haewon' cultivar 130
hail damage 157
hail netting 157, 189–190, *190*, 198
honey bee *vs* bumble bee pollination and 214
hairs, on kiwifruit skin 2, *2*, 31, 41, 273
A. chinensis var. *chinensis* 2, *2*, 31, 32, 41
A. chinensis var. *deliciosa* 2, *2*, 31, 32, 41, 53
A. eriantha 41, 54
convenience and 54
foliar fertilization and 168
Psa epiphytic growth 320, *323*
hairs, on leaves 32, 37, 168
'Halla Gold' cultivar 130
Halyomorpha halys (brown marmorated stink bug) 20, 333, *335*, *336*, 341, *342*, 344
handling, of kiwifruit 52, 53, 356
haploid kiwifruit, chromosome number 33, 134
hardwood cuttings 121
harvest/harvesting (kiwiberry) 106, 108–110
harvest/harvesting (kiwifruit) 54–55, 354–356
costs, tall T-bar system (Pergoletta Alta) 191
firm fruit, and allowed to ripen 54
fruit assessment before 358
indices 354–356
nutrients removed from orchard by 166
optimal soil water content after 177
optimum harvest index 354
pests on harvested fruit 341
picking fruit 356, 357
practice/methods 356
robotic 388–389
time 44, 54, 55, 353, 354–355
consumer responses 376
SSC and dry matter content 376
too early 355
see also postharvest
hawthorn spider mite (*Amphitetranychus viennensis*) 334
'Hayward Behi' 120
'Hayward' cultivar 1, 3, 11, *58*, 127
axillary bud development **230**
biomass accumulation in growing season 166–167, *167*
budbreak 228, 229
enhancer action 247
improvement techniques 246
time prediction 245
bud endodormancy 233–234
bud mutations 59, 130
cane size, shoot number 243
carbohydrate reserves 244
cell wall carbohydrates 280
chilling damage 364, *365*
chilling tolerance 359
chlorophyll loss, CA affecting 361
climate change impact 55–56, 81, 154
clonal rootstocks 85, 87

'Hayward' cultivar (*Continued*)
'Bruno' seedling rootstocks *vs* 85
drought tolerance 89
graft compatibility 88
waterlogged soil intolerance 89
clones, advantages 59
cold tolerance and bud endodormancy 232
consumer preferences 376
CPPU effect 201
crop yield 58, 184, 193, 196
cultivar most widely grown 3, 11, 58–59, 68, 127, 184
dominance being reduced 12, 13, 18, 19
cultivar not protected 56–57
cultivation in China 59
cultivation in Greece 17–18
cultivation in New Zealand *see* New Zealand
cultivation in USA/Europe/Japan/Chile 8
differences from other cultivars 58, 59, 368
drought tolerance 89
endodormancy 233
exports from New Zealand *7*, 23, 58
first cultivation, New Zealand 6, *6*, 58, 129–130
floral abortion 240
floral differentiation timing 238
floral evocation timing 237–238
floral morphogenesis, stages **230**
flower bud development **230**
flowering time 38, 58
fruit *58*
appearance 58, *58*, *59*
bagging 199–200
composition **276**, **279**
damage *363*
development stages *277*
flavour 286
flesh colour *53*, 58, *59*
growth stages *277*, 277–278
hard cores 367
harvesting 44, 54–55
maturation (late) 54–55, *277*, 355, 368
ripening 279, 284
shrivel 367
softening, temperature-induced *360*
structure **275**
syncarpous ovary *41*
transpiration and VPD 171, *172*
weight, tissue types 274
weight and seed number 72
weight at harvest 40
fruit storage 55, 283, 356, 368, 388
anaerobic metabolism 360
controlled atmosphere effect 360, 361
storage life 55, 388, 389
temperature management 359
growth stages 255, *256*
harvest index 354
'Hort16A' cultivar comparison 12
K_c (crop coefficient value) 174
kiwifruit synonymous with 1, 3, 11, 18, 58
leaf area index 186
leaf development, temperatures for 156
limitations/disadvantages 58–59
low-temperature breakdown 364, *365*
male cultivars for 77
management practice development 6, *7*, 184
management/pruning, response to 56
mutant selection 59, 130
nomenclature 32
ovary size/shape 239
ovule number 210
on pergola support system *186*
pests 333, 337, 340, 341
photosynthetic rates 257, 259, 263
pollen grain number for seed set and fruit size 210
pollination 209, 210, 211
temperature effect 209
pre-formed nodes and shoot development 229
productivity affected by development timing 243–244
productivity of orchards 184
propagation
clonal, Psa spread 123
by grafting 121
vegetative 120, 121
Psa infection 123, 317
Psa susceptibility 320
respiration rate of buds 245
root systems and depth 150, 173, *173*, 260
seed number 72, 119
seed viability, factors affecting 119
shading, effect on storage reserves 244, 245
shoot growth and vigour 243–244
shoot orientation effect 243
temperature effect 156, 243, 264
shoots, basal nodes 230
soil depth for 150
soil nutrient deficiencies 152
strip-male vine system *189*
training systems, Italy *191*, 191–192, *192*
trunk girdling *197*, 245
vine growth 58
vitamin C content 55, 58, 281
water-deficit stress 263
waterlogged soil intolerance 89, 151, 264
whole-canopy carbon acquisition 259
winter chilling 154, 233
'Hayward mark' 58
health 55, 380, 386
attributes for ideal cultivar 55
kiwiberry fruits, actions 110, 386
nutritional value of kiwiberries 110
nutritional value of kiwifruit 25–26, 42, 386

hedgerow system, kiwiberry fruit production 101, *103*
Heliothrips haemorrhoidalis (greenhouse thrips) *338*, 340, 344
Hemiberlesia lataniae (Latania scale) 340
Hemiberlesia rapax (greedy scale) 337, *338*, *339*
hemicelluloses 280
herbarium specimens 30
heritability of traits 132–133, **133**
hermaphrodite cultivars 40, 52, 136, *241*
  Actinidia chinensis var. *deliciosa* 40, 52, 68, 136–137
  advantages 81, 387
  breeding for, genes 136–137, *137*, *241*, 387
  commercial 40, 68
  development 240
  FrBy expression in female cultivar 240, *241*
  gene editing for 81
  planting design 387
  SyGI silencing or *FrBy* gene on X chromosome 137, *137*
heterozygosity 40
2-hexanal, and n-hexanal 286
hexaploid plants 32, 34, 56, 73, 128
  chilling requirements 44
  flowering time 37, 74
  flowers 39, *73*
  genetic maps 133–134
  male cultivars 77
  pollen yield and fertility 75
hexokinase 280
hexose sugars 233, 262
  see also glucose
history of kiwiberries 13
history of kiwifruit 1–15, 127, 385
  Actinidia species cultivation 13
    see also specific species
  Chinese gooseberry, name change to kiwifruit 7, *8*
  as a crop, evolution 13
  cultivation in China *see* China
  cultivation in New Zealand 2–3, 5–6, *7*
  cultivation in USA/Europe/Japan 7–8
  domestication *see* domestication of *Actinidia*
  first cultivation attempts 3–4, 5–6, 130
    'Hayward', in New Zealand 6, *6*, 58, 129–130
  first orchard 6, *6*
  first plants in New Zealand 4, 5–6
  first plants outside China 3–4
  fruit exports 6–7
  management practices 6
    training, vertical fences/pergolas 184–185, *185*, 187
  nomenclatural 1–2
  origin of name 1, 7
  as ornamental plant 4
  red-fleshed kiwifruit 10, 12–13
  type specimen (*A. chinensis* var. *chinensis*) 2–3
  wild mihoutao in China 1, *2*, 3, *3*, *4*, 31
  yellow-fleshed kiwifruit 10, 11–12
hives 213, 214
holistic approach 202
honey bees (*Apis* spp.) 212–214
  attracting, methods 214
  lipids from pollen 207
  pollination by 76, 212–214
    bumble bee comparison 214
    flower constancy 213
    kiwiberry 104
    managed pollination 212–214, *213*
    plant competitors for 213–214
    staminate *vs* pistillate flowers 212, 213
  Psa dispersal 325
  species 212
'Hongshi No. 2' cultivar 66
'Hongyang' (Rainbow Red) cultivar 10, 11, 13, 54, 65–66, 68
  advantages 10, 11, 65, 66
  in China 18, 57, 65–66, 130
  consumer preference 54
  development of cultivar 65–66
  fruit appearance *66*, *67*
  genome sequencing 134
  as hypoallergenic cultivar 142–143
  in Italy 12, 67
  low actinidin content 142–143
  low-temperature breakdown 364
  pests 333, 334
  protection of fruit by bagging *66*
  Psa susceptibility 53, 65
  rootstock transferring Psa resistance 90
  storage, CA effect 360, 361
  storage temperature, anthocyanins and 282
  temperature effect on pigment 282
hormones, plant 264
'Hort16A' cultivar 12, 13, *64*
  alternate year cropping 195, *195*
  budbreak enhancer action 247
  bud endodormancy, mechanism 234, 235
  carbohydrate supply limited 245
  CPPU effect 201
  degreening 355, 357, *357*
  description, fruit *64*, 64–65
  development in New Zealand 12, 13, 64, 77, 127, 130
  diseases 12, 13, 20, 53, 65
  floral abortion 240, *242*
  floral evocation, genes 238
  fruit
    carbohydrates **279**
    chilling damage 364, *365*
    ester production 287
    growth stages *277*, 278
    inner pericarp tissue 274

'Hort16A' cultivar (*Continued*)
locules *75*
regreening, ethylene causing 368
ripening 284
sharp 'beak' (stylar protrusion) 53
shrinkage 284
size after Benefit® 201
softening, temperature-induced *360*
storage, CA effect 360
structure **275**
fruit skin
discolouration, low-temperature breakdown 364
lifting 362, *362*
markings 366, *367*
harvest indices 355
inflorescence *237*
interspecific rootstocks for 87, *88*
in vitro propagation, and viruses 53
leaf area index 186
leaf chlorosis, high-pH soils 266
male cultivars for 77
management techniques 12
novelty, consumer appeal 56
orientation of shoots and vigour 243
pests 337, 340, 341
photosynthesis, environmental factors and 257, *258*
pruning *187*
Psa infection 12, 13, 20, 53, 65, 317
root pruning 198
temperature for degreening 357, *357*
'Hortgem Rua' cultivar (kiwiberry) *99*, **100**
'Hortgem' series, female cultivars 80
'Hortgem Tahi' cultivar (kiwiberry) *99*, **100**
host-pathogen recognition 143
'Huate' cultivar 60
'Huayou' cultivar 62
humidity
fruit transpiration and 171
kiwiberry fruit production 101
photosynthetic rate decline 257, *258*
pollen storage and 217
pollen viability reduced with 208
hydrogen cyanamide 196, 200, 241, 246–247
application method/timing 246
budbreak improvement 81, 200, 246, *246*
flower number improvement 81, 200, 246, 247
mechanism of action 246–247
hydroponic trial 260
hydrostatic pressure (root pressure) 261, 262
generation 261–263
6-(3-hydroxybenzyl) adenine (meta-Topolin, mT) 122–123
hypersensitive response, to Psa biovar 3 effectors 319
hypodermis 41
ice nucleation 322
ideal cultivars *see under* fruiting cultivars (kiwifruit)
Ilyonectria species 297
immunity, plant 143
impact sprinklers 105
importers of kiwifruit 20
indole-3-butyric acid (IBA) 120, 122
indole-acetic acid, waterlogging and 264
industry (kiwifruit) *see* global industry
inflorescences *39*, 56, 75, 96, 196, 237, *237*
abortion 240–241, *242*
development 230, 231, *232*, 237, 238, 239
timing 238
flower number 37, *237*
meristems, development *see* meristems
mixed, hermaphrodite cultivars 136
number per shoot 240
infructescence 56
innovative growth systems *see under* canopy management
insecticides 333, *336*, 337, 389
insect pollination 40, 72, 77, 211, 219
challenges 220
dry pollination 217
failure 216
flower odour/scent 212
kiwiberry 104
managed 212–214, 219
bumble bees 212, 214
honey bees 212–214, *213*
pollen as only reward 212
pollinators 211–216
bees (unmanaged) 215–216
beetles 216
bumble bees *see* bumble bees
butterflies and moths 216
flies 216
honey bees *see* honey bees (*Apis* spp.)
male cultivar attractiveness 76–77
other 214–216
Psa dispersal by 325
seed number and fruit size 211
timing (morning) 211
unmanaged 214
wild *Actinidia* species 211, 212, 216
integrated growing technologies 388
integrative conjugative element (ICE), Psa 319
internodes 229, 230
interploidy crosses 128
interspecific crosses 37, 68, 128, 130–131, 144
adapted to warm conditions, climate change and 44
breeding programmes 130–131, 144
in China 13, 63, 130–131
dioecy complicating 128
'Jinyan' development 13, 63, 130

large area required for progeny 129, *129*
in other countries 131
ploidy complicating 128
poor seed set after 119
in the wild 30
interspecific grafting 87–88
interspecific rootstock 86, 87
invertase 280
in vitro germination tests 75, 76
in vitro micro-grafting 121
in vitro propagation 85, 121–123, 124
bacteria/disease transfer 123, 124
explants for, culture initiation 122
genetic integrity concerns 123
macro- and micronutrients for 122, 124
media for explant growth 122
plant growth regulators 122–123
plant performance, factors affecting 121–122
root initiation protocol 122–123
shoots transfer, media for 122
Iran, kiwifruit production 18, **18**, 385
iron 165
deficiency 86, 89, 152
pistillate pollen 207
iron chelate treatment 89
iron-induced leaf chlorosis 86, 89
irrigation 157, 164, 173–181
benefits 174
daily 178
drip, for kiwiberry 105
drip systems (kiwifruit) 177
excess 157
waterlogging 179
fertigation *see* fertigation
frequency, seasonal 177–178
frost protection of young vines *156*
future directions 179–181
high frequency, small amounts 177–178
kiwiberry production 104–105, *105*
management 174, 176–178, 179
precision approaches *178*, 178–179
optimal 177–178
precision scheduling, remote/proximal sensors 178
Psa infection outcome 326
quality (high) 157
scheduling 177–178, 179
soil volume affected by 176
soil-water balance method 179, **179**
sprinkler 176
sustainable 178–179
sustained/regulated, water stress due to 177
time to start, identifying 177
timing, Psa spread reduction 326
volume 157, 176, 177, 179
calculation 174, 176
monthly averages (Italy) **179**
soil-moisture probes to assess 179
soil-water balance method 179, **179**
water relations and hydraulic features 173–174
irrigation method efficiency (Eff) 176
isopentenyladenine 278
'Issai' cultivar (kiwiberry) *99*, **100**
Italy
artificial pollination 216, 219
breeding programmes 130
challenges to production 17
commercial cultivar changes 19, **20**, 64
consortia and cultivar propagation rights 57
controlled crosses in 131
'D1 Vitroplant'® 86
evotranspiration (ET_0) 174, *175*
green-fleshed cultivars 61
hail netting use 189–190, *190*
harvesting time 55
interspecific crosses 131
Jingold™ brand 12, 24
kiwifruit industry 8, *9*, 19, 20, **20**
male cultivars used 80
production volumes 8, *17*, **18**, 385
pruning 189–192
rainfall, annual trends 174, *175*
rain-out shelters 198
red-fleshed cultivars 11, 19, **20**, 67
total area planted 8
trade value 22
water requirements and deficits 174, *175*
yellow-fleshed cultivars 12, 19, **20**, 57, 61, 64

James Veitch & Sons Ltd 4
Japan
breeding programmes 130
consumer preference, country of origin 380, 381
green-fleshed cultivars grown 61
history of kiwifruit cultivation 7–8
insect pollination of wild *Actinidia* 212
male cultivars 80
rootstocks for kiwifruit 89
yellow-fleshed cultivars grown 64
Japanese beetle **107**
jasmonic acid 279
'Jecy Gold' cultivar 130, 231
'Jenny' cultivar 52, 68
'Jinfeng' ('Ganmi No. 3') cultivar 62
Jingold™ 12, 24, 57
'Jinkui' cultivar 60, 74
'Jinmei' cultivar 62
'Jinshi No. 1' cultivar 62
'Jintao' cultivar 12, 57, 62–63
fruit appearance 62–63, *63*
in Italy 12, 57, 64
Psa infection 317

'Jinyan' cultivar 10, 13, 61, 63
crosses with 62
development 13, 63, 130–131
fruit appearance 63, *63*
in Italy 64
'Jinyuan' cultivar 61, 62, 63
juice, kiwiberry 110–111, *111*
'Jumbo' cultivar (kiwiberry) *99*, **100**
Just, Bruno 6
'just about right' (JAR) scale 378
juvenility phase, long 128–129

'Kaimai' interspecific rootstock 86, 87, 143
K_c (crop coefficient value) 174, 175, *176*, 178
'Ken's Red' cultivar (kiwiberry) *99*, **100**
kiwellin (Act d 5 allergenic protein) 142
kiwiberry 1, 94–118
attractive features 94
botany 94–98
breeding targets 111
buds 37, 96
commercial fruit production 99, 101–110, 111
growing conditions 103
harvest 106, 108–110
irrigation and water supply 104–105, *105*
maintenance 103
male:female plant ratio 103
nutrient management 105–106, **106**
orchard site selection 99, 101
pests and diseases 101, 106, **107**, **108**
planting density 101, 103
planting design 101, *102*, 103
pollination *see below*
soil pH and fertility 101
T-bar system, cordons and canes 103
temperature 99, 101
trellis structure 101, *102*, 103
consumer preference, reasons 380, 386, 388
cultivars 98, *99*, **100**
fruiting 98, *99*, **100**
male 98
cultivation, history 13
diseases 101, 106, **108**
distribution (geographic) 94
elevation range 94
female plants 96, 103–104
flowering period duration 96
flowers 96, *98*, 104
frost sensitivity and protection 101, *101*, 111
fruit
cold storage 109
development, timing 96
diversity 96, *97*
dried, uses 110–111
hairless, convenience 54
harmful compounds in 110
harvest indicator methods 108–109
harvesting 106, 108–110
harvest method, manual 109
health, effects 111
juices 110–111, *111*
maturation 96, 106, 108
maturity, indication 106, 108, 109
nutritional composition 110
postharvest decline 109
postharvest handling 106, 108–110, 111
processed products from 110–111, *111*
quality and bioactive composition 110
ripening 96, 98, 106
shelf life 98, 106, 109, 110
size/weight 96, 104, 211
skin (epicarp) 94, 96, 110, 362
softening 98, 108, 109
SSC and at-harvest dry matter 109
storage *see* storage of kiwiberries
taste and sensory attributes 109
temperature and perishability 109
uses 110–111, *111*
variations in cultivars 98, *99*
water loss postharvest 109
wind damage 96
fruit flesh colour 31, 43, 96, *97*, 108, 281
changes, harvest indicator 109
fruit set 104
future perspectives 111
growth period 99
leaves 94, *95*, 96
nutrient ranges 106
male plants 96, 103, 104
distribution among female plants 103
morphological variability 94, *95*, 96
nutrient requirements 105–106, **106**
pests 106, **107**, 343
pollen 104
pollination 74, 103–104
artificial 104
honey bees and bumble bees 104
wind 104
pruning 96
roots 105
seed colour, fruit maturity 108
shoots 94, *95*, 96, *98*, 101
vigour 96, 101, 103
vine maintenance 103
water deficiency and requirements 104
wild 94
see also Actinidia arguta
Kiwifruit Breeding Centre, New Zealand 24, 64, 130
kiwifruit neonectria canker 299, 307–308
kiwifruit vine decline syndrome (KVDS) 17, 86, 90, 295, 299–300, 385, 389
control 299–300
rootstocks resistant to 90, 91

symptoms and causes 151
waterlogging and 151, 299
Kiwigreen, programme 344
Kiwi Passion™ (Italy) 25
'Koryoku' cultivar 61, 130
'Kuimi' ('Ganmi No. 2') cultivar 62
'Kuliv' cultivar (kiwiberry) *99*, **100**

labelling, food 142
labour costs 21
labour supply 388–389, 390
Lactobacillus plantarum 327
Lasioglossum (sweat bees) 215
Latania scale (*Hemiberlesia lataniae*) 340
'LD133' cultivar (kiwiberry) **100**
leaching of minerals 169
leader pruning 188
leaf area index (LAI) 175–176, *176*
'Hayward' and 'Hort16A' 186
leaf drop 157
leaf:fruit ratio 256
leafhoppers 334, 344
leaf initials 231
leafrollers **107**, 337, 340
management 337, 339–340
leaf spot and blight **108**
leaf spotting, Psa 318, 320, 325, 326
LEAFY (LFY) gene 238
leaves (kiwiberry) *see* kiwiberry
leaves (kiwifruit) 30, 32, 37
A. chinensis, variety *chinensis vs. deliciosa* 32
area, in orchard 168, 175–176, *176*
after budbreak, seasonal 175–176
variations in early years 176
autotrophic 259
biomass accumulation ('Hayward') 167, *167*
calcium transport to 170
carbon transport from 260
characteristics affecting foliar fertilization 168
chlorosis, in high-pH soils 266
development/growth 156, 175, 231
exposed (to sun), transpiration rate 176
fertilization (foliar) 168
functions 256–257
fungal diseases 301
hairs on 32, 37, 168
high temperatures effect 264, 265
hydraulic conductance 261
male/female plants and variations 30
maturation, photosynthetic rate 257
maximum area, orchard 175
metabolism, floral evocation affected 237
nutrient absorption, foliar fertilizers 168
nutrient accumulation 166
pests of 334
photosynthesis *see* photosynthesis
planteose transport from 260
pre-formed and neo-formed 230, **231**
Psa epiphytic growth 320
Psa infection/invasion 322, *323*, 325–326
removal, root growth rate decrease 261
salinity affecting 266
shaded, transpiration rate 176
shape, size and texture 37
structure, tissues *323*
transitional 230
transpiration *see* transpiration
water-deficit stress 263
water-use efficiency (WUE) and 176
wind damage 153
Leiocarpae 30, 88, 131
Leioproctus spp. 215
lenticels 29, 366
lenticel spots 366, *366*
lianas 36, 242, 266
light
effect on vine reserves 244
increase, reflective ground covers *194*, 195, 198–199
photosynthesis rates 257, 265
seed germination 120
shoot growth and vigour 243
light brown apple moth (*Epiphyas postvittana*) 340
light-compensation point 257
lime, soil chemistry change 165
limonene 287
linkage disequilibrium 135
linkage maps 134
lipids, in pollen 207
litter moths 341
Lobesia botrana (grapevine moth) 343
locules 41
logistics, difficulties 21
'Longcheng No. 2' cultivar (kiwiberry) *99*, **100**
loop-mediated isothermal amplification (LAMP) assays 320
low-temperature breakdown (LTB) 363–364
'Lushanxiang' cultivar 63
Lycorma delicatula (spotted lanternfly) 334, *335*, *336*

'M36' male cultivar **79**
'M43' male cultivar **79**
'M91' male cultivar **78**
'MacLoughlin' (MC352) male cultivar 77, **78**
macronutrients
plant organ content 166, **166**
see also specific minerals
MADS-box genes 136, 234, *235*, 239
magnesium (Mg) 87, 165
Ca:Mg and Mg:K ratios 165
deficiency, 'Hayward' vines 152
fruit quality and 172
for kiwiberry vines 106
plant organ content, mature orchard 166, **166**

male cultivars (kiwiberry) 98
male cultivars (kiwifruit) 52, 72–84
breeding 80–81, 387
fruiting (inconstant) 136, *136*
future prospects for 81
genetic contribution to fruit characteristics 128
high-performing, importance 72, 73
history, New Zealand 77
key attributes 73–77, 80, 81–82
attractiveness to pollinators 76–77
climate change effect 81
compatibility 76, 189
flower number (high) 74–75, 81, 239
heritability of 81
lateral flowers 75
ploidy, and pollination 73–74
pollen yield and fertility 75, 77
resistance to pests/diseases 76, 81
suitability for artificial pollination 77
suitability for growing systems 76
synchrony with females 72, 73, 77
timing of flowering 189
vigour/growth habit 76
list xv–xviii, 77–80, **78–79**
pollenizer evaluation 132
production system changes 81
selection 72, 77, 80
from wild 80–81
senescent flower tissues retained 76
significance of males 72–73
male flowers *see* flower(s) (kiwifruit), staminate (male)
male vines 72, 73
management, trained on trellis system 76, 82
in matrix layouts ('box' males) 189
percentage in commercial orchard 40
planting among female vines 72
kiwiberry 103
pollen grains *74*, 75, 77
pruning 76
rows, strip-male system 188, *189*
spatial distribution and size 76
malic acid 280
management systems/practices 36, 152, 184–206, 385
budbreak enhancers 154, 200.246–247, 245, **247**, 387
canopy management *see* canopy management
changes, effect on male cultivars 81
crop management *see* crop management
development, in New Zealand 6, 7, 184
disease reduction 53, 389
future directions 201–202
'Hayward' cultivar 6, *7*, 184
'Hort16A' cultivar 12
innovative growing systems *see under* canopy management
in Italy 8, *9*
kiwiberry 103
male vines 76
modified environments 198–200
on-orchard factors, consumer response 376, 381
to optimize productivity 184, 388
organic orchards 152–153
pest management 345
plant growth bioregulators 200–201
pruning *see* pruning
Psa infection control 326–327
soil management systems 152–153
stringing systems 193, *194*
training *see* training of vines
vegetative regrowth control 195–196
vine protection, in climate change 387
water-use efficiency (WUE) and 176
manganese (Mn) 165
deficiency 106, 152
depletion, Psa symptom increase 326
for kiwiberry vines 106, **106**
pistillate pollen 207
manure, organic 153, 168
marker-assisted selection (MAS) 135, 138
market access
pests disrupting 345
restrictions 24
marketing 368
kiwiberries 111
new cultivars 56
online direct 386
market size 21, **21**, 127
see also global production/supply
MAS (marker-assisted selection) 135, 138
mating designs, population genetics 132
'Matua' male cultivar 77, **78**, 80
frost, survival 265
rooting depth 260
vegetative propagation 120–121
winter chilling and endodormancy 233
maturation of fruit *see* fruit (kiwifruit), maturation
1-MCP *see* 1-methylcyclopropene (1-MCP)
Mealybug **107**
mechanization 388–389, 390
Mecinus pascuorum (plantain weevil) *339*, 340
Mediterranean fruit fly (*Ceratitis capitata*) 343, 345
Megachile rotundata (alfalfa leafcutting bee) 212
megachilid bees 212
melatonin 263
Meloidogyne spp. 90
'Meris', 'Hayward' cultivar 59
meristems 185, 228, 230
apical, evolution, gene control *137*
axillary 185, 230
bud, sucrose levels 245
bud endodormancy mechanism and 232, 233–234
cork (phellogen) 273
development 230, *231*, 233, 237

floral 230, 233, 238, *239*
floral abortion 240
gene control *239*
inflorescence
development 230, *231*, 237
timing of development 237
undeveloped in winter 228
lateral and terminal 185, 238, 239
shoot apical (SAM) 237
winter chilling resetting 232, 233
meta-Topolin (mT) 122–123
Metcalfa pruinosa (citrus flatid planthopper) 325, *342*, 342–343, *343*, 344
'Meteor' (CK3) male cultivar 77
1-methylcyclopropene (1-MCP) 109, 285, 353, 361
hard cores and 367
treatment 361, 367
consumer response 377
microbial diversity, organic orchards 153
microclimate modification 158
micro-grafting 121
micronutrients
for *in vitro* propagation 122
for kiwiberry production 106
micro-propagation 85, 120
disease spread 123
micro-sprinklers 105, *105*
Mighties® brand 25
mihoutao (monkey peach) 1, *2*, 3
wild, in China 1, *2*, 3, *3*, *4*, 31
'Miliang No. 1' cultivar 10, 60, 245
'Miliang' rootstock 86
minerals
absorption from soil 165, 166
accumulation in plant organs 165–167, **166**
deficit, high-pH soils 266
foliar fertilization increasing 168
fruit, composition 170–173, **276**
imbalances, uptake affected by 165
leaching of 157, 169
mobility in soil 169, 170
pollen composition 207
requirements 166
solutions, foliar fertilization 168
transport in xylem 262
weight in fruit **276**
see also nutrients/nutrition (vine)
Minolta Chroma Meter 355
minute scavenger beetle (*Cortinicara gibbosa*) 335
miRNA 139, 239
'Mitsuko' cultivar (kiwiberry) *99*, **100**, 200
models, fruit and vine 388, 390
modified atmosphere (MA) 360–361
fruit softening, effect on 360
low-temperature breakdown reduced by 364
packaging 358–359
storage of fruit 360–361
modified environments 198–200
fruit bagging *66*, 199–200, *200*
overhead netting *see* netting
rain-out shelters 198, *199*
reflective ground covers *194*, 195, 198–199
molecular approaches *see* genome-based breeding
monkeys, kiwifruit eaten by 1, *2*, 3
'Monty' cultivar 40, 233
moria (Moria del kiwi) *see* kiwifruit vine decline syndrome (KVDS)
'Moshan No. 4' male cultivar 77, **78**
mosquitoes 216
moth flies 216
moths
as pests 334, 340, 341
pollination by 216
'M Series' male cultivars 77, **78**, **79**
MS medium 122, 123
mucilaginous cells 41
Mucor species 308, *309*
'Muller' (MC482) male cultivar 77, **78**
MYB10 282
MYB gene family 139
MYB transcription factors 139, 273
myo-inositol 280

NAC class transcription factors 285
Nagoya Protocol of 2010 127
name (of kiwifruit), origin 1
naphthalene acetic acid (NAA) 201
near infra-red (NIR) spectroscopy 358
nectar, lacking 38, 40, 76, 104, 207, 211, 212, 219
nectarthrodes 322
Nectria haematococca 296–297
nematodes, root-knot 90
Neodryinus typhlocybae 342–343
Neofabraea actinidiae *305*, 306–307, 366
Neonectria microconidia 299, 307–308
netting
blue *vs* red or grey 198
hail *see* hail netting
overhead 198
pest management 345
uses and benefit 198
New Zealand
A. chinensis var. *deliciosa* cultivation 2–3, 5–6, 7, 58, 130
artificial pollination 216, 219
Bay of Plenty *7*, 8
breeding programmes for new cultivars 56
budbreak time prediction 245
Chinese gooseberry, name change to kiwifruit 7, *8*
commercial cultivar changes 12, 18–19, **19**, 58, 65, 68, 332
crop yields 184, 193, 196

New Zealand (*Continued*)
- domestication of *Actinidia* 2–3, 5–6, 85, 129–130
- export of kiwifruit 6–7, 12, 19–20, 22, 57, 354
  - China/Japan consumers 380, 381
- export success 7, 12, 17, 57
- global trade value increase 22
  - key elements driving 22–24, **23**, *23*, 56
- green-fleshed cultivars (new) grown 61
- harvest indices 354–355
- 'Hayward', main cultivar 58, 127
  - crop yields 184, 193, 196
  - productivity of orchards 184
- history of *Actinidia* in 2–3, 5–6, 85, 127, 385
  - first plants 4, 5–6
- 'Hort16A' cultivar development 12, 13, 64, 77, 127, 130
- 'Hort16A' cultivar growth and disease 12, 64–65
- insect pollination
  - managed pollination by bees 212
  - wild *Actinidia* 212
- kiwifruit industry 3, 12, 13, 57, 58
- male cultivars 77
- management practices, development 6, *7*, 184
- orchards, total area 7, 11
- production volumes 17, *17*, **18**, **19**, 57
- productivity of orchards 184
- Psa biovar 3 319
- Psa impact 317
- red-fleshed cultivars 12, 19, 67
- root pruning 198
- secondary canopies 193–194
- training and pruning systems 6, *7*, 184, *186*, *187*, 187–189, *189*
  - early (history), vertical fences/pergolas 184–185, *185*, 187
- trunk girdling 196–197, *197*
- Whanganui, first orchard 6, *6*
- wind/storm damage to vines 153
- yellow-fleshed cultivars developed in 65
- yellow-fleshed cultivars grown 12, 18–19, **19**, 61, 64–65, 68
- Zespri Group Ltd *see* Zespri Group Ltd, New Zealand
- 'Zesy002' cultivar 12, 18–19, 23, 61, 65, 68

Nezara viridula (L.) (green vegetable bug) 341–342, *342*, *343*
'Nijitaro' male cultivar **78**, 80
nitrates
- roots *vs* shoots 262
- soil 169, *169*

nitrogen (N)
- application, kiwiberry 105, 106, **106**
- availability, root zone 169
- deficiency, 'Hayward' vines 152
- fruit quality and 170, 172–173
- fruit storage life and 172–173
- mobility in soil, and leaching 169
- plant organ content in mature orchard 166, **166**
- Psa virulence affected 326
- shoot growth enhanced 173
- slow-release solid fertilizers 169
- soil 152, 153, 166, 169
- uptake by roots 169, 262

'No. 1 Maoxiong' male cultivar 80
nodes and internodes 229–230, **231**
nomenclature, changes 1–2
- *Actinidia arguta* complex 31
- *Actinidia chinensis* complex 31–32

'Nongda Jinmi' cultivar 63, 266
'Nongda Mixiang' cultivar 60
'Nongda Yuxiang' cultivar 60
NON-RIPENING (*NOR*) genes 285
North Carolina Mating II (NCM2) design 132
novelty, for ideal cultivar 56
nucellus 283
nutrients/nutrition (vine) 164–173
- absorption, by diffusion 165
- availability in soil 164, 165
- balance (output and input) 165, 166, 167, 168
- Ca:Mg and Mg:K ratios 165
- deficiencies, in soil 152, 165, 168
- estimation/measurement 167
- fertigation supplying 168–170
- fertilizers supplying *see* fertilizers
- foliar fertilization increasing 168
- of fruit, quality affected by 170
- inputs (manure/compost) 168, 170
- for *in vitro* propagation 122
- management 165, 167–170
  - fertilizer application *see* fertilizers
  - future trends 389
  - kiwiberry production 105–106, **106**
  - organic matter and compost 170
  - tillage reduction to preserve root integrity 170
- mobility in soil 166, 169
- organic orchards 153
- 'plant annual nutrient demand' 166
- pre-planting, analysis 168
- removed from orchards in fruit 166
- requirements 165–167, **166**
- reserves, accumulation 167, *167*
- seasonal mineral accumulation 165–167, **166**
- supply in arid climates 169
- transport and translocation 164–165, *165*, 171
  - *see also* transpiration
- uptake by plants 164–165, 167, 168
- *see also* minerals; *individual nutrients/minerals*

nutritional value
- kiwiberries 110
- kiwifruit 25–26

Nysius huttoni White (wheat bug) 341

octaploids 34, 128
on-orchard costs 21
on-orchard factors, consumer response 376, 381

optimizing production 184, 388
Oraesia excavata Butler 334
orange-coloured fruit 138, 139
orchard (kiwifruit) *7*, 180
canopy management *see* canopy management
in China 9, 11, 192–193
climate affecting *see* climate
demonstration, selection of cultivars 132
factors affecting consumer response 376, 381
first, in New Zealand 6, *6*
irrigation *see* irrigation
in Italy (covered orchard) 8, *9*
land for, challenges 387
male vine percentage 40
management *see* management systems/practices
at marginal (less suitable) sites 387–388
in New Zealand, total area, productivity 7, 11, 184
organic 152–153
planting design/density *see* planting density; planting design
position and aspect and slope 157, *158*
shelter for (from high winds) 153, *154*, *155*
site selection 99, 150, 157, 158
space for growing seedlings by breeders 129, *129*
three years old, arthropod variability 332
total area
China 10–11, 18
Europe 8, 10, 11
Italy 8
New Zealand 7, 11
orchard machinery 157, 193
organic acids 275, 280–281
organic matter, soil 170, 177
organic orchards 152–153, 376
Oscar® (France) 25
Osmia bees (mason bees) 212
osmotic flow, xylem 261, 262
osmotic pressure, fruit 275, 279–280
outcrossing 40
ovary 39, 41
development, timing 238–239
Psa infection *324*
size and shape variations 239
syncarpous 39, *41*, 211
overhead netting 198
overhead sprinklers 105, *105*
overwintering
buds (kiwifruit) 32
Pseudomonas syringae pv. *actinidiae* (Psa) *321*, *322*, 325, 326
scale, *Pseudaulacaspis pentagona* (white peach scale) 333
ovules 39, 41
initiation 283, **284**
number 207
pollen-to-ovule ratio 211
oxalates, in kiwiberries 110
oxidative metabolism 364
oxidative stress 364
oxygen, controlled atmosphere for storage 360
ozone, storage of fruit 361

packaging/packing of fruit 356, 358–359
automated 359, 389
bruising of fruit 362
fruit damage and 359, 362, 363
kiwiberry 109–110
kiwifruit 358–359
sustainable materials 389
Paranthrene regalis (grape clearwing moth) 334
parasitoids 341
parthenocarpic fruit 40
passionvine hopper (*Scolypopa australis*) **107**, *338*, *339*, 340
patents, plant 57
PCR assay, Psa detection 320
PEBP genes 234, *235*
pectins 143, 280, *286*
peelability, breeding for 142, 380
Pelargonium zonate spot virus (PZSV) 303
Penicillium species 308, *309*, *361*
pentaploids 128
pentatomids 333
peptone 123
'percent solubilized' concept 356
pergola support systems 76, 184–185, *185*, *186*
cane replacement 189
China 193
cost, T-bar systems *vs* 187
kiwiberry 101, *102*, 103
New Zealand 187
orientation of shoots and vigour 243, *244*
stringing systems 193, *194*, 195
strip male system and 188, *189*
vine planting distances 187
water use efficiency (WUE) and 176
whole-canopy photosynthetic rate 259
Pergoletta Alta system 191
pericarp 41
granularity and translucency 368
inner 41, *41*, 42, *67*, *139*, 274, **274**, *274*, 286
structure 41, 274, *274*
water-soaked, low temperature breakdown 364
outer 41, *41*, 42, *139*, 171, 274, *274*, 286
colour 42–43, *67*, *139*
damage, water-soaking 362, 363, *363*
granular, low-temperature break-down 364
hard cores and 367–368
structure 41, 274, *274*
white line, impact damage 362, *363*
softening 286

periderm 41, 53, 273, 362
 origin, layers 273
 suberized, of fruit 273
 trunk girdling effect 197
Pestalotiopsis species **108**, 308, *309*
pesticides 389
pests of kiwiberry 106, **107**, 343
pests of kiwifruit 20, 332–352
 in Argentina 344
 in Chile 344
 in China 333–335, *335*, *336*, 344, **345**
  major pests 333–334, *335*, *336*
  minor pests 334–335, *336*
  from wild *Actinidia* 332
 economic impact 332
 global risks 389
 in Greece and Turkey 344
 increase with increasing production 332
 increase with orchard age 332
 in Italy 341–344, **345**
  major pests 341–342, *342*, *343*
  minor pests *342*, 342–344, *343*
 life cycles 333, 334, 337
 major *vs* minor 332
 management
  biological control 339–340, 345
  challenges and restrictions 344–345
  of citrus flatid planthopper 342–343
  of greedy scale 337
  of leafrollers 337, 339–340
  of white peach scale 333, 341
 in New Zealand 335, 337–341, **345**
  major pests 335, 337–340, *338*, *339*
  management 344
  minor pests 340–341
 range expansion 345
 resistance
  breeding for 143
  ideal male cultivars 76
  of rootstocks 90
 taxonomy of major pests **345**
 see also individual pests
petals, of flowers 38, 238
Peyracchia system 190
pH (soil)
 kiwiberry production 101
 kiwifruit production *see* soil (for kiwifruit production)
Phaeoacremonium species 297
phellem (cork) and phelloderm 273
phenolic compounds 362
 decline with storage 361
 released, low-temperature breakdown 364
phenology 255–256
 flowers 207–208, 255
 vines 228, 245, 255–256
phloem *165*, 259–260, *323*
 trunk girdling effect 197
Phoma species *305*, 308, *309*
Phomopsis species 305–306
phosphatidylethanolamine-binding protein (PEBP) genes 234, *235*
phosphonates 296
phosphorus (P) 165
 for kiwiberry vines 106, **106**
 plant organ content in mature orchard 166, **166**
photoinhibition 265
photosynthesis
 C3 pathway 256–257
 canopy depth and 259
 decrease, trunk girdling 197
 downregulation, by root pruning 260
 effects of treatments or cultivar differences 257, 259
 environmental factors affecting 257, *258*
 light effect 257, 264–265
 portable measurement systems 259
 rate
  high temperature effect 264, 265
  light-saturated (Pmax) 257, 265
  reduced in water-deficit stress 263
  respiration exceeding 259
  response to waterlogging 264
  shading in high temperatures 265
 regulation 260, 265
 removing sink carbon, effect 260
 temperature (high) affecting 264–265
 whole-canopy (Ac) 259
photosynthetically active radiation (PAR) 244, 257, *258*
photosystem II (PSII) stress 263
Phyllactinia species 301
phytohormones, fruit growth 278–279
Phytophthora species **108**, 296
Phytopythium helicoides 89
Phytopythium vexans 89
pitting, fruit *366*, *367*
 fungal 307, 366, *367*
 storage of fruit 364, 366, *366*
planar cordon systems 81
Planotortrix spp. (green-headed leafroller) 340
plantain weevil (*Mecinus pascuorum*) *339*, 340
'plant annual nutrient demand' 166
plant-based biostimulants 200
plant bioregulators 200–201
 budbreak 154, 200
  see also hydrogen cyanamide
 forchlorfenuron (CPPU) 200–201
 in vitro propagation 122–123
 shoot growth retardants 201
plant breeding rights 57, 68
plant collectors, early 1, 5, 5, 31
planteose 260, 273, 280

planthoppers 344
citrus flatid planthopper (*Metcalfa pruinosa*) 325, *342*, 342–343, *343*
planting density
kiwiberry 101, 103
kiwifruit 187, 188
planting design
'box' males (matrix layout) 189
hermaphrodite cultivars 387
for kiwiberry fruit production 101, *102*, 103
male and female vine rows alternated 72
strip male system 188, 189, *189*
planting distances, vines 187
plant variety rights (PVR) 57, 132, 386–387
plastic covers 198, *199*
plastic rain-out shelters 198
ploidy, *Actinidia* 33–34, *34*, 40, 56, 128
A. arguta (kiwiberry) 34, *34*, 94, 96
breeding constraints 128, 386
chilling requirements variation 44
effect on anthocyanin accumulation 282
estimation, flow cytometry 33, *34*, 128
flowering times and 37, 73–74
flower size and 38
fruit weight correlation 56
graft (rootstock) compatibility 88
for ideal cultivar 56, 73–74
interploidy crosses 128
male cultivar, attribute 73–74
manipulation 128
pollen yield association 75
see also specific ploidy races
'Pohwa' male cultivar 75, **78**, 80
pollen (kiwiberry plants) 96, 104
pollen (kiwifruit plants) *74*
application in artificial pollination 217
availability and honey bee abundance 213
bee-collected, harvesting 217
composition, minerals 207
daily release, measurement 208
development 39
diluents, for dry pollination (artificial) 218
donors, factors influencing choice 72–73
dose-response relationship to fruit setting 210–211
early-season, storage 217
excess, effect 211
extraction, controlled crosses 131
fertility, ideal male cultivar 75, 77, 80
harvesting for artificial pollination 217, *218*
infected pollen 219
international trade and Psa spread 319
vacuuming from dehiscent flowers 217, *218*
hollow, non-fertile *74*, 75
lipids 207
maturation failure 39
nutritional value to pollinators 207
origin, affecting seed development 283
popcorn stage flower bud 208, *209*, 217
production 208–209
Psa contamination/infection 319, *324*, 325, 326
release, timing 208, 212
release by explosive dehiscence 208
seed set variation and 72–73
staminate *vs* pistillate flowers 207
sterile, in female flowers 39, 72, *74*, 128, 131, 207
storage
for artificial pollination 77, 217–218
fertility 77
temperature 217
storage life 217–218
suspension medium for 218
viability 75, 77, 208–209
climate change effect 81
factors affecting 208–209
in storage 77, 217
xenia 209–210, 283
yield 75
ideal male cultivar 75, 77, 80
maximizing, artificial pollination 217
ploidy and 75
pollen dispensers 214
pollen grains
from female vines *74*
kiwiberry 96
from male vines *74*, 75, 77
number (kiwifruit) 208, 210
requirement, different *Actinidia* species 210, 211
pollenizers 132
pollenkitt, lacking from pollen 211
pollen-to-ovule ratio 211
pollen tube(s) 39, 76, 209
germination, temperature (kiwiberry) 104
syncarpy and 211
pollen tube distributor cap (compitum) 39
pollination (kiwiberry flowers) 74, 103–104
pollination (kiwifruit flowers) 39, 40, 72, 207–227
artificial 216–219
application time and costs 219
bee pollination comparison 216
broadcast systems 219, **220**
dry pollination 217, 218, **220**
electrostatic charge 217, 218, 219
by hand *see below*
interploidy cross 128
kiwiberry 104
male cultivar attribute 77
methods 219, **220**
pollen application 218–219, **220**
pollen application, pass number 217
pollen application systems 219, **220**

pollination (kiwifruit flowers) (*Continued*)
pollen diluents for 218
pollen from hexaploid males 74
pollen harvesting 217, *218*
pollen suspension medium 218
pollen yield 217
requirements 217
risks and benefits 219, 220
robotic 219
seed set variation 76, 211
semi-targeted 219, **220**
targeted 219, **220**
targeted autonomous 219, **220**
targeted broadcast 219, **220**
technologies 218–219, **220**
tractor-mounted/-pulled 219, **220**
wet pollination 217, 218–219, **220**
attractiveness of flowers 76–77
challenges to 220
'deceptive' 207
dependency on 210–211
dry 217
effectiveness, importance 72
effective period for 209
failure, ploidy effect 128
by fans, or bees in orchards 72
fruit size and quality 211
by hand 72, 76, 80, 219, **220**
controlled crosses 131
pollen excess, reduced seed number 211
seed set/number 76, 128, 211
see also pollination, artificial
high winds affecting 153
by honey bees *see* honey bees (*Apis* spp.)
by insects *see* insect pollination
interploid 74
mechanical 40, 72, 77
not required for fruit set 40
requirement, different *Actinidia* species 210, 211
seed set as measure of 210, *210*
successful, factors in 207, 210
temperatures affecting 156
timing 209
'under-pollinated', effect 210
wet 217
by wind *vs* insect 40, 72, 211
pollution, leaching of minerals from soil 169
polyploidy 33, 34, 40
genetic maps 133–134
polysaccharides, fruit softening 286
polyuronide 280
population genetics 132–133
combining ability and progeny testing 133
heritability of traits 132–133, **133**
mating designs 132
positive chill accumulation 233
postharvest (kiwiberry)
handling 106, 108–110
treatments 109–110
postharvest (kiwifruit) ix, 353–374
chilling sensitivity 357–358, 359
current/future trends 388
fruit softening pattern 143, 285, *286*, 354, *354*
handling of fruit 52, 53, 356
packing of fruit *see* packaging/packing of fruit
pre-grading treatments 356–358
curing 356–357
fungicide application 358
temperature conditioning 357–358
quality of kiwifruit 170–173
storage *see* storage of kiwifruit
sustainability 389
systems and process 356, *356*
postharvest disease 87, 304–308
Botrytis cinerea (botrytis rot) **108**, *187–188*, 304, *304*
ripe rots *305*, 305–308
minor pathogens 308
wound rots 308, *309*
potassium (K)
deficiency, 'Hayward' vines 152
fruit quality relationship 172
impact on Fe and Mn uptake 165
K:Ca ratio 172
for kiwiberry vines 106, **106**
once-a-season application 165
plant organ content in mature orchard 166, **166**
powdery mildew **108**, 301
pre-grading treatment *see under* postharvest (kiwifruit)
preharvest factors, consumer response 376
proanthocyanidins *141*
processed products
kiwiberry 110–111, *111*
kiwifruit 24, 25, 385
product innovation 27
production, global *see* global production
Proeulia auraria 344
Proeulia chrysopteris 344
programmed cell death 143
prolepsis 185
proline 246
propagation of kiwifruit plants 119–126
clonal 85, 120, 124
disease transfer 123
rootstocks 85–86, 119
see also micro-propagation; *vegetative (subentry below)*
disease-free plants 123, 124
future prospects 123–124
grafting *see* grafting
in vitro see in vitro propagation

micro- *see* micro-propagation
seed 119–120, 124
vegetative 120–121, 124
hardwood cuttings 121
semi-hardwood cuttings 120–121
softwood cuttings 120
protected cultivars 12
protein effector, Psa 319
protoplast fusion 131
pruning (kiwiberry vines) 103
pruning (kiwifruit vines) 184, 187–193, 201
cane replacement
benefits 184–185, 187
in China 193
in Italy and France 189
in New Zealand 187, *187*, 188
in China 193
cultivar response 56
delayed 188
detailed *vs* less precise 187
excessive in summer 188
flower production increase 247
in France 192
Geneva Double Curtain system 191, *191*
in Italy 189–192
leader 188, 191, *191*
long-cane
in Italy/France 190
in New Zealand 188, 190
male vines 76
in matrix layout ('box' males) 188, 189, *189*
in strip-male system 189, *189*
mixed-age
in Italy 190
in New Zealand 187, *187*, 188
new shoot growth after 186, 187, 188
in New Zealand *187*, 187–189, *189*
Peyracchia system 190
practices, genotypes/cultivars 188
reasons for 186
root 198
short-cane, in Italy/France 190
spring decisions for summer 188
spur 187
of strip male vines 189, *189*
summer 188, 189
time taken for 191
timing 187, 188
total cane replacement 187, *187*, 188
winter 188, 196
zero-leaf 196
Psa *see Pseudomonas syringae* pv. *actinidiae* (Psa)
Pseudaulacaspis pentagona (white peach scale) 333, *335*, *336*, 341, *342*
Pseudocercospora actinidiae 301
Pseudocercospora hangzhouensis 301
Pseudomonas fluorescens 123
Pseudomonas species 301–302, *302*
Pseudomonas syringae pv. *actinidiae* (Psa) ix, 12, 17, 20, 90, 201, 317–331, 385, 389
biovar 1 and biovar 2 317, 320
biovar 3 (aggressive) 317, **318**, 319
copper-resistant 327
detection method 320
effectors and pathogenicity 319
PacICE1, PacICE2, PacICE3 319
biovar 4 320
biovar 5 and biovar 6 318, 320
biovars 317
characteristics **318**
detection methods 320
genome sequences 318
colonization (leaf, flower, buds) 320, *321*, *323*
comparative genomics 318–320
control 184, 326–327
biological 327
copper resistance 327
cultivars susceptible to 12, 13, 20, 53, 68
'Hort16A' cultivar 12, 13, 20, 53, 65
red-fleshed cultivars 53, 65
cultivars tolerant of (resistant to) 44, 52, 53, 56, 65, 68, 319
breeding for 143
inheritance of resistance 81
male cultivar choice 76, 81
resistance genes 319
detection 320
disease cycle 325–326
distribution (countries) and biovars 317, 319
economic losses/impact 317
effectors (protein) 319
epidemiological cycle 320, *321*
epiphytic population 320, *321*, *323*, 326
frost damage increasing 265
future trends 327, 389
genes 318–319
resistance (biovar 3) 319
susceptibility 327
genome
accessory/flexible 319
core 318–319
infection process 320–325, 326
acropetal and basipetal migration 325
endophytic migration *324*, 325, 326
epiphytic growth 320, *321*, *323*, 326
flower cycle and *324*, 325, 326
invasion via stomata 322, *323*, 326
leaf cycle and 322, *323*, 325–326
migration via xylem *322*, *323*, 325, 326
motility-based chemotaxis 325
seasonal changes 320, *321*, 325–326
systemic invasion 325
infection threshold 322
isolation and isolation media 320

Pseudomonas syringae pv. *actinidiae* (*Continued*)
kiwiberry infection 106
long-distance spread 325
management systems to minimize 184, 327
plastic shelters 198, *199*
molecular mechanisms, Psa–fruit interactions 327
as most important pathogen 295
origin 318–320
overwintering *321*, *322*, 325, 326
pathovars 319
phylogenetic tree 318
Psa-free phytosanitary certificate 317
reactivation in spring 325, 326
resistance
cultivars *see above*
genes 319
rootstock 90
risk increase by frosts 155
rootstock infection/resistance 90
spread, conditions associated 325, 326
symptoms 218, 318, 325, 326
toxins 319
transfer, kiwiberry pollination 104
transfer by artificial pollination 219, 319
transfer by *in vitro* propagation 123
transfer to clonally propagated plants 123
vectoring/vectors 325, 342
virulence 326, 327
whole-genome sequences 318
wound infection 322, *323*, 325
Pseudomonas syringae pv. *actinidifolium* 320
Pseudomonas syringae pv. *syringae* 302, 322
Pseudomonas viridiflava (blossom rot) 123, 211, 301–302, *302*, 305
Pukekohe dwarf mutant 239, *240*
purple fruit variant 31, 43

'Qihong' cultivar 66–67, 266
'Qinmei' cultivar 10, 59, 60, 263, 266
fruit 60, *61*, 274
quality of kiwiberries 110
quality of kiwifruit 362–368
consistent, consumer preference 380
consumer-based definition 375
crop load and 196
current/future trends 388
eating quality 354, 355
environmental and on-orchard factors 376
evaluation 358, 362
flower quality affecting 211
girdling practices 196–198, *197*
improved by better pollination 211
mineral composition affecting 170–173
calcium levels 170–171
K:Ca ratio and 172
nitrogen levels 172–173
potassium levels 172
optimizing, future trends 388
pathological defects 368
physical defects 362–363, *363*
see also fruit, damage
physiological defects 363–368
chilling damage 363–364, *365*
other disorders 367–368
shrivel 366–368
skin markings 364, 366, *366*, *367*
postharvest 170–173
premium, Zespri system 22
soil-water balance and irrigation 174
unacceptable 358
quantitative trait loci (QTL) 273
quarantine, Psa 317
quinate dehydrogenase (QDH) 280–281
quinic acid 280

rain
flower opening and 208
Psa spread 326
Rainbow Red *see* 'Hongyang' cultivar
rainfall
annual trends (Italy) 174, *175*
monthly averages (Italy) **179**
rain-out shelters 198
raised spots 366, *366*
raphides 41
RAW *see* water, readily available
reactive oxygen species (ROS) 326
red-brown rot 296
red-fleshed kiwifruit 42–43, *43*, *53*, 54, 65–67, *281*, 281–282
anthocyanins causing colour 43, 139, 281, 282
block red 54, *281*, 281–282
breeding 139
consumer appeal 56, 65
cultivars grown 12–13, *53*, 54, 65–67, 130, *281*
in China 10, 11, 13, 18, 65–67, 130
in Italy 12, 19, **20**, 67
in New Zealand 12, 19, 67
in South Korea 12
current and future trends 68
degreening 139, *139*, 282
during development 282
genomic selection 135, *135*
green/red-fruit flesh colour 139, *139*, *142*
pattern of red pigment 54, 65, *281*, 281–282
pigment synthesis, control 139
Psa susceptibility 65
'starburst' *53*, 54, *67*, 281, 282
see also 'Hongyang' cultivar; 'Zes008' cultivar

Red Sun No. 1 (RS1) *see* 'Hongyang' cultivar
'Redvita' cultivar 12, 67
reflective ground covers *194*, 195, 198–199
'Renact' cultivar 52, 68
reproductive biology 128–129
research, limited, impact of 45
respiration
buds
endodormancy onset 233
rate, flower number prediction 245–246
controlled atmosphere effect 360
exceeding leaf canopy photosynthetic rate 259
increase, ethylene response 284
modified atmosphere effect 360
retail channels 26
revenue in fruit sector 21, **21**
rhizosphere 153
Ribgrass mosaic virus 303–304
ribulose-1,5-bisphosphate carboxylation 257
Ricania simulans *343*, 344
ripe fruit *see* fruit (kiwifruit), ripe
ripening
kiwiberry 96, 98, 106
kiwifruit *see* fruit (kiwifruit), ripening
RIPENING INHIBITOR-like genes (*RIN*) 285
ripe rots *305*, 305–308, 307
robotic systems x, 219, 388–389
root(s), *Actinidia* 36, 260–263
anatomy 36
anchorage role 260
anoxic 88, 89, 151, 179
calcium uptake 170, 262
carbohydrate reserves 244, 260
cortex 261
damage, by frost and low temperatures 265
death or loss 256, *256*, 259
depths 150, 260
development 173–174
diseases/pathogens 295–297
distribution 173, 260–261
'feeder', depth 260
fibrous roots 36, 173, 260
fine roots 36, 173, 260
growth 260–261
carbohydrate competition for 261
cessation after budbreak 256
cessation in waterlogging 264
changes, water uptake rate change 261
control (girdling/pruning) 150
leaf removal affecting 261
organic matter effect 170
phases 256, *256*, 261
starting in winter 255, *256*
temperature for 156
young plants 261
hydraulic conductance 261, 262
hypoxia 179
initiation, *in vitro* propagation 122
kiwifruit vine decline syndrome 151
lateral extension/spread 260
long, thickened 36, 173, 260, 261
moisture content of soil and 170
nitrogen uptake 262
nutrient absorption 165, 170, 262
older, structure 36
oxygen consumption 151
pattern 36
pressure 261–263
pruning 198, 260
removal, storage/sink relationship study 260
rootstock and scion interaction affecting 261
root system characterization 173–174
root system structure 36, 173, *173*, 260–261
root system weight 173, **173**
salinity affecting 266
in sandy loam soils 260
sap flow from 261, 262
secondary 173, 260
to shoot transport 261, 262
soil depth 150, 151
soil salinity tolerance 152
soil volume explored by 36, 173, **173**, 176–177
soil water reservoir for 179
solute movement in xylem 261, 262
in upper soil layers, preserving integrity of 170
water deficit 174
water flow in xylem 261, 262
in waterlogged soils 151, 174, 263
water requirement *see* soil (for kiwifruit production); water
water uptake rates 261
white rot (*Armillariella mellea*) 295–296
rooting, method to increase rate 120, 121
root-knot nematodes 90
root pressure 261–263
root rot (*Fusarium solani* species) 296–297
root rot (*Phytophthora* species) 296
rootstocks 85–93, 119, 132
Actinidia species potential 86, 87
'Bounty71' *86*, 86–87, 91
breeding for 143–144, 387
clonal 85–87, 119
'D1 Vitroplant' 86
frost-tolerant 87
future research 90–91
graft compatibility 87–88, 121, 132
graft incompatibility 88, 121, 132, 143
graft survival rate 88
interspecific 86, 87–88, *88*, 91
'Kaimai' 86
light-saturated photosynthesis rates (Pmax) 257
low-vigour 87, *88*, 90–91

'Miliang' 86
pest/disease resistance 90
kiwifruit vine decline syndrome 90, 91
Psa 90
root-knot nematodes 90
root growth and distribution 261
screening for difficult soils and drought 88–90, 91, 132
alkaline calcareous soils 89–90, 91
drought tolerance 88–89
iron deficiency tolerance 89
root rot tolerance 89
waterlogged soils 88–89
seedling 85, 119, 121
selection 132
vigour 87, *88*, 90–91
winter chilling requirement, effect on 233
'Z1 Vitroplant' 86
rot
blossom (bacterial blossom blight) 123, 301–302, *302*, 305
collar (*Fusarium solani* species) 296–297
collar (*Phytophthora* species) 296
cylindrocarpon 307–308
opportunistic 368
pitting, side (*Cadophora luteo-olivacea*) 307, 366
ripe *305*, 305–308, 368
root (*Fusarium solani* species) 296–297
root (*Phytophthora* species) 296
Sclerotinia **108**
side (*Botryosphaeria dothidea*) 306
side (*Cadophora luteo-olivacea*) 307
stem-end *see* stem-end rot
storage 368
white root (*Armillaria*) **108**, 295–296
wound 308, *309*
'Ruiyu' cultivar 60–61
'Russell' male cultivar 77, **78**, 81
russeting, kiwiberries 110
rust fungal infections 301

salicylic acid pathway 327
salinity, of soil 152
susceptibility to 266
tolerance 266
'Sanuki Gold' cultivar 64, 80, 89, 130
'Sanuki kafunriki' male cultivar **78**, 80
sap flow 261, 262
Saurauia 29
scarab beetles 334
scent, of flowers 39–40, 212
Schizothyrium pomi **108**
scion cultivars 85, 86, 87
graft compatibility with rootstock 87–88, 121, 132
rootstocks for *see* rootstocks
sclerotinia blossom blight 300, *300*
Sclerotinia nivalis **108**
Sclerotinia rot **108**
Sclerotinia sclerotiorum 76, 300, *300*
Scolypopa australis (passionvine hopper) **107**, *338*, *339*, 340
screens, Psa spread reduction 326
sea freight 368, 389
secondary canopies 193, *194*
seed(s)
BBCH scale of fruit growth *277*, 278
coat colour change 283
development 283, **284**, 287
stages 283, **284**
dispersal 41
dormancy 119–120, 124
protocols to overcome 120
fruit maturation and 283, **284**
germination 119–120, 124, 131
in breeding programmes 131
conditions for, experiment 120
poor, endangered status of species 119
promoting, methods 120
rate 119
lacking, parthenocarpic fruit 40
low-temperature stratification 120
number
commercial pollination success and 210
different *Actinidia* species 210, 211
fruit quality improvement and 211
fruit size and 211, 283
fruit weight and 40, 72
propagation by 119–120, 124
seed characteristics 119
seed set 132
compatibility and 76
hand pollination reducing 128, 211
male cultivar choice and 72, 76
as measure of pollination 72–73, 76, 210, *210*, 211
poor, after interspecific hybridization 119
variations 72–73, 76
size and distribution 72
sowing 131
storage 119
structure 41
viable, ploidy affecting in interploidy cross 128
seedling(s) 131
area for and spacings 129, *129*
from controlled crosses, growth 131
growing conditions 120, 131
growth and transplanting 131
long juvenility, reducing 128–129
ploidy after interploidy cross 128
seedling rootstocks 85, 119, 121
selection in breeding *see* breeding/breeding programmes

self-setting species 40
semi-hardwood cuttings 120–121
sensory profiling 377–378
SEPALLATA (SEP) genes 239, *239*
sepals, of flowers 38
sex determination 81
sex-determining region (SDR) 240
sex expression, genetic control 137, *137*
shade nets 198
shading
  effect on storage reserves 244, 245
  in high temperatures 265
shelf life
  kiwiberry 98, 106, 109, 110
  kiwifruit 55, 170, 388
shelter(s) 153
  artificial 153, *154*
  natural (living, trees) 153, *155*
  rain-out 198
shelterbelts 153
shelter nets 198
shikimate dehydrogenase (SDH) 280–281
shikimic acid pathway 280–281
shock cooling, kiwiberry 109
shoot(s) (kiwiberry) 94, *95*, 96, 101
  terminated and non-terminated 96, *98*
shoot(s) (kiwifruit)
  apex, removal, budbreak 231, *232*
  apical ends, photosynthesis rates 257
  apical meristem, flower development 237
  autotrophic 259
  biomass accumulation ('Hayward') 166–167, *167*
  blind, pruning 188
  branching, acrotonic 185
  carbohydrate storage 243, 244
  control methods 195–196
  development 228, *229*, 229–230, 237, 242, 255
    carbohydrate competition 244
    future perspectives 247
    late *vs* early, effect on fruit 243–244
    management 242–247
    orientation, effect on vigour 243
    schematic diagram *232*
    temperature for 156
    water stress effect 177
  embryonic 230, *231*
  floral, kiwiberry 96
  flowering 37–38, 243
  girdling, floral evocation prevention 237
  growth 243–244, 255, 256, 266
    carbon acquisition and 256
    cessation, factors affecting 230
    control 195–196
    dynamics 243
    enhanced by nitrogen 173
    fruit removal effect 260
    number, cane thickness 243
    rates of 242–243, 256
    seasons/timing *229*, 242–243
    temperature affecting 243, 256, 264
    vigour and 243–244, 256, 266
  growth retardants 201
  limiting, tip-squeezing 195–196
  long upright 230, *231*
    see also canes
  medium 230, **231**
  orientation, effect on vigour 243
  photosynthesis rates in leaves along 257
  productivity, previous development timing 243–244
  regrowth after pruning 186, 187, 188, 196, 201
    prevention/retardants 188, 201
  reserve storage 243, 244–245
    light effect on 244, 245
  root-to-shoot transport 261
  secondary 185–186
  second-order basal 231
  shading, effect on storage reserves 244, 245
  short 229, 230, **231**, 257
  spurs, flower production 243
  terminated and non-terminated 188, 229–230
  types 230, **231**, 257
shoot-tip necrosis 185
SHORT VEGETATIVE PHASE (SVP) repressor 234
shrivel 366–367
Shy Girl (SyGI) gene 137, *137*, 240
side rot (*Botryosphaeria dothidea*) 306
side rot (*Cadophora luteo-olivacea*) 307
single nucleotide polymorphism (SNP) 133, 318
site selection for orchards 99, 150, 157, 158
  position and aspect 157
  slope 157, *158*
  water supply 157
skin *see* fruit (kiwifruit), skin
sloping sites 157, *158*
'sluicing' 296
snacking, healthy 26, 386
sodium ions (Na^+), accumulation in roots, high salinity 266
sodium toxicity 152, 266
softening of kiwifruit *see* fruit (kiwifruit), softening
softwood cuttings 120
soil (for kiwiberry production) 101, 104
soil (for kiwifruit production) 150–153, 158
  acidic 152
  agronomic practices protecting 170
  alkaline 266
  alkaline calcareous, tolerance 89–90, 91
  analyses before pre-planting fertilization 167
  available water (AW) 177, 179
  biological properties 152–153, 158
  carbon 152, 153, 177
  chemical properties 152, 158, 165

soil (for kiwifruit production) (*Continued*)
  depth 150, 151, 177
  drainage, poor 151
  evapotranspiration 174
  fertilizer (solid) distribution 167
  heavy subsoil/hard pain 151–152
  high-porosity, root system 173, *173*
  impaction, effect on root development 174
  improvement 151–152
  irrigation wetting, depth 177
  management systems 152–153, 170, 174
    water retention improvement 177
  microporosity improvement 152
  mineral imbalance 165
  mineral leaching 169
  mineral mobility 169
  minerals in upper *vs* deep layers 170
  moisture measurement probes *178*, 178–179
  nitrates, optimal 169, *169*
  nutrient absorption by roots 165, 166
  nutrient availability 164, 165
  nutrient toxicities/deficiencies 152, 165
  organic carbon (SOC) and nitrogen 152, 153
  organic matter 170, 177
  organic orchards 152–153
  pH 152
    high (alkaline soil) 89–90, 91, 266
  physical properties 151–152, 158
  potential rooting depth 150
  readily available water (RAW) *see under* water
  root development and 174
  rootstocks tolerant of 88–90, 91
  salinity 152
  sandy loam, rooting depth 260
  volume affected by irrigation 176
  volume explored by root system 173, **173**, 176–177
  water, irrigation volume 176, 177
  water availability reduction 263
  water balance 174
    method, sustainable irrigation 179, **179**
  water content 177, 179
    monitoring probes 178, *178*
    optimal after harvest 177
    seasonal trend 179, *180*
    upper *vs* deep layers 170
  water-holding 150, 151–152, 177
  waterlogging *see* waterlogged soils
  water loss 176, 177
  as water reservoir 177, 179
  water retention capacity/curves 177
solar radiation, transpiration rate and 176
solitary bees 215
soluble solids (SS), increase/accumulation *355*
soluble solids content (SSC) 42, 54, 96
  accumulation and starch breakdown 283
  accumulation during maturation 283
  consumer acceptability and 355, 375–376
  fruit assessment, grading 358
  genetic control 354–355
  harvest index 354
  kiwiberry 96, 108, 109
  pollen source/characteristics affecting (xenia) 209
  prediction from dry matter in fruit 276
  ripe (rSSC) 355–356, 375–376, 377
  sigmoidal pattern of accumulation 283, 354–355, *355*
somaclonal variation 123
somatic doubling 34
somatic embryogenesis 131
somatic hybridization 130–131
sooty mould/sooty splotch **108**, *339*, 341, 344
sooty spot 301
'Soreli' cultivar 64, 130
South Africa, flower production improvement 247
South American fruit fly (*Anastrepha fraterculus*) 344
South Korea
  artificial pollination 216, 219
  breeding programmes 130
  male cultivars 80
  red-fleshed cultivars grown 67
  subsoiling using pressured air 151–152
  yellow-fleshed cultivars grown 64
'Sparkler' (CK2) male cultivar 77
spotted lanternfly (*Lycorma delicatula*) 334, *335*, *336*
spotted-wing drosophila (*Drosophila suzukii*) **107**, 343–344
spring
  axillary bud growth 36, 228, *229*, *232*
  budbreak *see* budbreak (in spring)
  canopy growth 188, 256
  carbon use/mobilization patterns 156, 259
  Psa (bacterial canker) infection 318, 320, *321*, 325–326
  temperatures 154–156
  xylem vessels refilling 261–262
sprinkler systems 105, 177
spur systems 189
SSC *see* soluble solids content (SSC)
stamens
  kiwiberry flowers 96
  kiwifruit flowers *38*, 39, 72, *73*, 128, *208*
Standard medium 122
starch 279–280
  accumulation in fruit 142, 274, 275, 279–280, 355, *355*
    anthocyanin accumulation and 282
    fruit growth stages *277*, 278, *355*
    by genotype **279**
    prediction of rSSC 355–356
    structures/sites (outer pericarp) 274, *274*, 279
    timing **279**, 280, *355*

accumulation in roots 260
accumulation in vines 260
breakdown 280, *355*
fruit ripening 284, 355
fruit softening 286, *286*
SSC accumulation 283, 355, *355*
conversion to sugars in ripening fruit 275–276, 375
formation from sugars 375
metabolism 286
kiwifruit as model 275–276
structure 279
type-A and type-B polymorphs 279
see also carbohydrate
Stathomopoda horticola Dugdale 341
Stathomopoda skelloni 341
stem(s), structure 36
stem-end rot (SER)
Botrytis cinerea see Botrytis cinerea (botrytis rot, stem-end rot, grey mould)
Neonectria microconidia 307–308
stigma
development, timing 238–239
Psa infection/multiplication *324*, 325
viability 209
stinkbug species
in China 333
in Italy 341
stomata 168, 174, *323*
behaviour, comparable with other C3 species 257
conductance 88, 89, 257
reduced in water deficit 174, 263
reduced in waterlogging 264
root pruning effect 260
waterlogged soil assessment 88, 89
conductance values 257
density 174
foliar nutrient uptake 168
gas exchange, photosynthesis control 265
Psa invasion via 322, *323*, 326
storage life 55
'Hayward' cultivar 55, 388, 389
kiwiberries 98, 109
pollen sources affecting (xenia) 209
storage of kiwiberries
diseases 110
methods 109–110
storage life 98, 109
treatments prior to cold storage 109
storage of kiwifruit 55, 353, 359–361
assessment 55
conditions for 55
consumer responses 376–377
controlled ripening with ethylene 361–362
early experiments 6
fruit aroma reduced 376
'Hayward' *see* 'Hayward' cultivar
quality (fruit)
high calcium levels 170
nitrogen levels affecting 172–173
responses to atmosphere 360–361
carbon dioxide levels 360
controlled/modified atmosphere 360–361, 364, 368
ethylene 361
ozone 361
see also controlled atmosphere (CA)
rots 368
skin discolouration 366
skin markings 364, 366, *366*, *367*
storage life *see* storage life
temperatures 109, 283, 353, 354, 359–360
consumer responses 376
curing 357
ethylene sensitivity after 285
forced-air cooling 359
fruit firmness 376
fruit responses to 359–360, *360*
low-temperature breakdown 363–364
management 359
pitting associated 366
rate of cooling, effect 359
softening induced by 359–360
temperature conditioning 357–358
trials, for new cultivars 132
storage stain 366, *367*
storm damage 153
'stratification thermal time' model 120
stress responses *see* environmental stress responses
stringing systems 193, *194*
strip male system 188, *189*
pruning of vines 189
secondary canopies 193
stubbing 196
styles (of flowers) 39
suberin 273
succinic acid 266
sucrose **276**
bud endodormancy onset 233
in fruit 280
transport in phloem 259–260
SuF gene 137, *137*
sugars
consumer response 375
in fruit **276, 279**
translocated 280
see also specific sugars
summer
pruning *see* pruning (kiwifruit vines)
Psa (bacterial canker) infection 318, 320, *321*, 326
temperatures 156
'Summerfaenza' cultivar 80
summerfruit tortrix (*Adoxophyes orana*) 334, *335*, *336*

'Summerkiwi 3373' cultivar 61, 80, 130
sun damage 44
'superfood' 110
support structures *see* pergola support systems; T-bar training system
SUPPRESSOR OF OVEREXPRESSION OF CONSTANS1 (SOC1) 235
surfactants, foliar fertilization and 168
sustainability 26, 144, 247, 389–390
  challenges, resolving 389
  fertilizer management 164
  irrigation strategy 178–179
  organic fertilizers 170
  pest management challenges 344
  postharvest 368, 389
  as target 389
  water and nutrients for vines 389
SVP genes 234
sweat bees (*Lasioglossum*) 215
Sweeki® (Italy) 25
sweetness of fruit 54, 62, 139, 276, 376, 378
  breeding for 139
syncarpous ovaries 39, *41*, 211
syrphid flies 216

talc, pollen diluent 218
taste, of fruit
  kiwiberry 109
  kiwifruit *see* fruit (kiwifruit), taste
taxonomy, *Actinidia* species 31
T-bar training system 187
  advantage, hail netting compatibility 189, *190*
  cane replacement 189
  costs 187, 191
  in France 192
  Geneva Double Curtain system 191, *191*
  increase of winter buds, limitation 193
  in Italy 189, 190, 191
  kiwiberry fruit production 101, *102*, 103
  in New Zealand 187
  orientation of shoots and vigour 243
  tall (Pergoletta Alta) 191
  whole-canopy photosynthetic rate 259
technology, new, impact 81, 82
temperature
  for budbreak 154, 155–156, 229, 245
  canopy development 256
  chilling, fruit damage *see* chilling damage
  climate change effect *see* climate change
  conditioning, postharvest 357–358
  for cultivated vines 44, 153–157, 158
    autumn 156–157
    spring 154–156
    summer 156
    winter 81, 154
  degreening yellow-fleshed fruit 357, *357*, 359
  effect on anthocyanin accumulation 282
  as endodormancy driver 232, 233
  frost sensitivity 154
  fruit bagging effect 199
  fruit softening and 285, 359–360, *360*
  high, vine response 264–265
  kiwiberry fruit production 99, 101
  kiwiberry perishability 109
  for leaf development/growth 156
  low
    chilling damage *see* chilling damage
    response to 265–266
    storage 353, 359
    see also under storage of kiwifruit
  low-temperature breakdown (LTB) 363–364
  low-temperature stratification, seed dormancy 120
  orchard shelters affecting 153
  pollen storage 217
  pollen tube germination (kiwiberry) 104
  pollen viability reduced with 208
  seed germination requirements 120
  shoot growth and vigour 156, 243, 256, 264
  sloping orchards 157
  SSC accumulation 283
  storage of fruit, responses 283, 353, 359–360, *360*
    see also storage of kiwifruit
  in winter, climate change impact 81
  winter chilling 44, 55, 153, 154, 233
tendone support structure 191–192, *192*
terpenes 287
tetraploid plants 32, 33, 34, 56, 73, 128
  chilling requirements 44
  flowering time 37, 74
  flowers *73*
  genetic maps 133–134
  male cultivars 77, 80
  pollen yield and fertility 75
texture of fruit 41–42
  consumer preferences 54, 376
thinning of fruit 86
thrips **107**, 340
tillage 170
tip-squeezing 195–196
tissue culture 85, 121
  see also in vitro propagation
'Tomua' cultivar 61
'Tomuri' male cultivar 74, 77, **78**
  'Bruno' cross, germination experiment 120
tortricids 334, *336*, 343
total area of orchards *see* orchard (kiwifruit)
trace elements, foliar fertilizers replacing 168
trade of kiwifruit 19–20
  exporters *see* exporters of kiwifruit
  global value *see* global trade, value

importers 20
re-exporters 20
risks 20–21
traditional medicine, kiwiberries 111
training of vines 56, 184, 187–193
by country
China 193
Italy and France 189–192
New Zealand *186*, *187*, 187–189, *189*
cultivar response 56
Geneva Double Curtain system 191, *191*
history (vertical fences) 184–185, *185*, 187
on pergolas *see* pergola support systems
stringing systems 193, *194*
T-bar *see* T-bar training system
tendone support structure 191–192, *192*
wagon-wheel arrangement 192, *192*
training population, genomic selection 135, *135*
transcription factors 143, 273, 285
see also specific transcription factors
transcriptome, bud endodormancy mechanism 234, 235, *235*
transpiration *165*, 170, **171**
calcium transport to leaves 170
cessation in winter 261
fruit 170–171
daily fluctuations, humidity 171, *172*
high vapour pressure deficit and 171, *172*
humidity and 171, *172*
seasonal pattern 170–171, *171*
leaves *vs* fruit 170, **171**
mineral uptake from soil and 164, 165, *165*, 170
rate 257, 261
leaf area index and 175–176
shaded or exposed (sun) leaves 176
total area, water requirements and 175
water-use efficiency (WUE) and 176
xylem flow driven by 165, *165*, 261
transportation 389
sea freight 368, 389
trees, as shelter for orchards 153, *155*
trellis structure 157
for kiwiberry fruit production 101, *102*, 103
see also T-bar training system
trials, semi-commercial 68
trichomes *see* hairs, on kiwifruit skin
trunk girdling 196–198, *197*, 260
'Tsechelidis' 40, 130
Turkey, kiwifruit pests 344
Turnip vein-clearing virus 304
twig blight 300, *300*
twiners 34
tyloses 325
Type III secretion system 319

Uncinula actinidiae **108**
Uromyces species 301
uronic acid 280
USA
history of kiwifruit cultivation 7–8
retail e-commerce 26
Utah (Richardson) chill units 233, **234**, 245

vapour pressure deficit (VPD)
fruit transpiration and 171, *172*
high, fruit shrivel 367
photosynthetic rate and 257, *258*
water requirements and 176
vascular bundles, in fruit 274
vectors, for Psa 325, 342
vegetative growth 177
high planting density and 187
reduced, for productive vines 186–187
water stress effect 177
vegetative propagation *see* propagation of kiwifruit plants
vegetative regrowth, controlled 195–196
tip-squeezing 195–196
ventilation
ethylene reduced in coolstore by 361
Psa spread, effect on 326
Verona vine decline *see* kiwifruit vine decline syndrome (KVDS)
Verticillium dahliae 297
Verticillium species 101, **108**, 297
Verticillium wilt 101, **108**
vibration damage to fruit 363, *363*
vigour
canes 187, 188
control, spring sap flow and 87
interspecific scion/rootstock combinations 87–88, *88*
kiwiberry 96, 101, 103
low
canopy 255
male cultivar choice 76
rootstocks 87, *88*, 90–91
new *Actinidia chinensis* var. *chinensis* cultivars 201
orientation of shoots and 243
shoot growth and 243–244, 256, 266
vine(s), *Actinidia* 34–36, *35*
age, pollen viability 209
architecture 185–186
area occupied by 129
attributes for ideal cultivar 55–56
canopy *see* canopy
commercial production *see* commercial fruit production; kiwiberry; orchard (kiwifruit)
from cuttings *vs* grafting 86

vine(s), *Actinidia* (*Continued*)
female *see* fruiting cultivars (kiwifruit)
growth habit *see* growth habit
kiwiberry 96
lianas 36, 242, 266
male *see* male cultivars (kiwifruit)
management and pruning *see* management systems/practices
physiology 186–187
site selection for *see* site selection for orchards
structure 36–44
vigour *see* vigour
as woody perennials 228, 237
see also specific structures
vine disease 295–304
bacterial 123, 295, 301–302, *302*
canopy diseases 300–302
development with commercial orchards 295
fungal *see* fungal diseases
global risks 20
kiwiberry 106, **108**
KVDS *see* kiwifruit vine decline syndrome (KVDS)
root and crown pathogens 295–297
transmitted by artificial pollination 219
trunk diseases 297–299
viral diseases 302–304, *303*
waterlogged soil impact 151, 299
vine management systems *see* canopy management; management systems/practices
vine wilt, *Ceratocystis fimbriata* (ceratocystis wilt) *298*, 298–299
viral diseases 295, 302–304, *303*
kiwiberry 106
reduction, *in vitro* propagated plants 123
vision systems, automated grading of fruit 358
vitamin(s) **276**
in kiwiberries 110
vitamin C (ascorbic acid) 25, 42
content (fruit) 25, 42, 55, 281, 386
during fruit growth 280, 281
'Hayward' cultivar 55, 58, 281
pollen source/characteristics affecting (xenia) 209
yellow-fleshed fruit 62
kiwiberries 110
levels by genotype 281
synthesis 281
volatile organic compounds (VOC) 39–40, 43–44
changes induced by Psa infection 325
fruit softening 286–287, 377
pollination by honey bees 76–77
V-trellis system 190

waste streams 389
water
availability
benefits to vines 174
climate change effect 387
available (AW), soil 177, 179
budget 176–177
chemical composition 157
consumption, leaf area index and 175–176, *176*
deficit 88–89, 174, 263
kiwiberry growth poor 104
kiwifruit sensitivity 157
Psa infection outcome 326
stress due to 263
see also drought
evaporation 168
fruit skin permeance 367
irrigation *see* irrigation
loss 176
kiwiberry fruit postharvest 109
reduction by packaging liners 358
shrivel of fruit 366–367
from soil 176, 177
mobilization of reserves 262
quality, for pollen suspension medium 218
readily available water (RAW) 165, 177, 179
irrigation volume needed 177
time to start irrigation 177
relations 173–174, 180–181
requirements 174–176, 261
high vapour pressure deficit 176
kiwiberry production 104
kiwifruit vines 157
leaf area index and 175–176, *176*
resources, management 389
root hydraulic conductance 261
saving, sustainable irrigation 179, **179**
soil content *see* soil (for kiwifruit production)
soil retention of 177
storage
roots 262
soil depth and 150
stress 263–264
effects 174, 177, 263
short-term, recovery from 263
symptoms 263
tolerance 263
see also water, deficit; waterlogged soils
supply for orchards 157
fertigation with 168–170
fruit sensitive to 171
kiwiberry production 104–105
see also irrigation
transpiration, 'Hayward' 170, **171**
see also transpiration
transport 174
to fruit, xylem 171
to leaves, xylem 174, 261
for wet pollination 218–219

waterlogged soils 151–152, 174, 263–264
disease impact relationship 151, 296
duration 264
excess irrigation causing 179
genes involved in response to 89
kiwifruit vine decline syndrome and 90
response, duration-dependent 264
root sensitivity 263
symptoms (for vines) 151
tolerance 88–89, 121, 264
genetics and genotypes 264
screening for new rootstock 88–89
vine sensitivity to 151, *151*, 263
vine survival, factors affecting 264
water potential 265
water-potential gradient 262
water use efficiency (WUE) 176, 265
'Weiki' cultivar (kiwiberry) *99*, **100**
western honey bee (*Apis mellifera*) 212–214, *213*
wheat bug (*Nysius huttoni* White) 341
white core inclusions 368
white peach scale (*Pseudaulacaspis pentagona*) 333, *335*342, *336*, 341
white root rot (*Armillaria* species) **108**, 295–296
whole-orchard system 202
wild *Actinidia* vines *see Actinidia* (genus), wild
'Wilkins Super', 'Hayward' cultivar 59, 130
Wilson, E.H. 1, 4, *5*
wilting point 177, 263
wind
damage 153
orchard position/aspect 157
Psa spread 326
wind pollination 40, 72, 211
kiwiberry 104
winter 44, 154
Actinidia cultivar response 44, 154
bud endodormancy 232–235
cold, cultivars growing in areas of 37, 44
dormancy *see* dormancy, winter
hail netting folded away 190
Psa (bacterial canker) infection 318, 320, *321*, *322*, 326
temperatures 154
winter chilling 44, 55, 153, 154, 232–233, 247
budbreak and 44, 229
time prediction 245
bud endodormancy 232–233, *234*
bud meristem resetting 232, 233
climate change effect 154, 387
management 245
temperature models 233, **234**
woody perennials, *Actinidia as* 228, 237, 257
workers, shortage 388–389, 390
wound(s), Psa infection/invasion 322, *323*, 325
wound rots 308, *309*
Wright, Hayward 6, *6*

'Xangji' 40
xanthophylls 138, 282
X chromosome 240
xenia 209–210, 283
'Xuxiang' cultivar 18, 59, 61, 130, 266, 364
vibration treatment, damage 363
xylem 36, 165, *165*, 171, 261, *323*
air embolisms 261, 263
calcium transport 262–263
osmotic flow 261, 262
pressures 262, 326
Psa migration *322*, *323*, 325, 326
refilling in early spring 262
sap, composition changes 262
solute concentrations 262
solute movement 261, 262
substances/minerals transported 171, 262
trunk girdling effect 197
vessel diameter 174, 261
water transport to leaves/fruit 171, 261
xyloglucanase 286
xyloglucans 280, *286*
XY sex-determining system 240

'Yangshi Jinhong No. 50' cultivar 67
yangtao 1
Y chromosome 137, *137*, 240
yellow-fleshed kiwifruit 12, 42–43, *43*, 61–65, 281
allergens 55
breeding for 138–139
carotenoids and xanthophylls 138–139
chilling sensitivity 357–358
chlorophyll degradation *see under* chlorophyll
countries growing 61–62
cultivars commercially grown 11–12, 61–65, 130
in China 10, 13, 18, 61, 62–63
in Italy 12, 19, **20**, 57, 61, 64
in Japan 64
in New Zealand 12, 18–19, **19**, 61, 64–65, 68
in other countries (not China) 11–12, 18–19, 61, 64–65
in South Korea 64
current and future trends 68
cytokinin signal and 278
degreening 138, 281, 355, 357
first marketed 12
fruit colour development 12, 43, 281
harvest indices 355
'Hort 16A' *see* 'Hort16A' cultivar
'Jintao' *see* 'Jintao' cultivar
male cultivars for 77
management techniques 12
processed products 25
production volume and areas (by year) 18–19, **19**, **20**

yellow-fleshed kiwifruit (*Continued*)
Psa infection 317
re-greening, ethylene response 368
total area planted in China 10
'Zesy002' *see* 'Zesy002' cultivar
see also Actinidia chinensis var. *chinensis*
yellow-red fruit flesh colour *142*
yellow-spotted stink bug (*Erthesina fullo*) 333, 334, *335*, *336*
yields *see* crop yield
Y-trellis 243

'Z1 Vitroplant' (Vip Zedone®) 86
zeatin 278
zeatin riboside 264
zero-leaf pruning 196
'Zes006' cultivar 278, 279, **279**
'Zes007' cultivar 61, 243
'Zes008' cultivar 12, 19, 23, 130
fruit flesh colour *53*, *67*, *67*
light-saturated photosynthesis rate 257
male cultivar for pollinating 77
in New Zealand 67, 130
shoot growth and vigour 243
'Zesh004' cultivar 19, 23, 58, 61
Zespri GOLD™ Kiwifruit 12
Zespri Group Limited, New Zealand 8, 17, 22, 57, 64, 130
brand (Zespri) identity 12, 22, 23, 24, 25, 57
consumer preferences 22, 23, 380
case study 22, **22**, *23*
current/predicted revenue 23, *24*
export growth 22, **22**, *23*
global sales 22, *23*
increase demand for health 26
opportunities to increase demand 23
premium quality fruit 22
product mix (two cultivars) 22–23
volume and value of exports 23, **23**
Zespri Organic Green™ Kiwifruit 22
Zespri RubyRed™ Kiwifruit 19, 23
Zespri SunGold™ Kiwifruit 12, 17, 23
Zespri Sweet Green™ Kiwifruit 22
'Zesy002' cultivar 12, 13, 17, 64
allergens 55
artificial pollination 77
budbreak enhancer action 246, *246*, 247
budbreak time, prediction 245
bud endodormancy, mechanism 234, 235
damage to fruit *363*
chilling damage 364, *365*
development, ploidy cross 12, 128, 130
fruit
composition **269**, **276**
flesh colour *53*, 65, 361
growth stages *277*, 278
skin discolouration, low-temperature breakdown 364
softening 285
storage, CA effect 360, 361
structure 41, **275**
vitamin C levels 281
grafted, 'Bounty71' rootstock *86*
male cultivar for 77
orientation of shoots and vigour 243
ovary size/shape 239
pests 337, 341
pollen grain number, seed set and 210
production in Italy 19, 64
production in New Zealand 12, 18–19, 23, 61, 65, 68
pruning 188
Psa susceptibility 320
root system 173, *173*
seed number, fruit weight and 72, 210
shoot growth/orientation, vigour 243, *244*
'Zesy003' cultivar 65, 367
Zhengzhou Fruit Institute, China 10
'Zhongmi No. 2' cultivar 61
zinc 165
deficiency 152
for kiwiberry vines 106
Zorion guttigerum 216
'Zuva' male cultivar **78**, 80

This book is published by **CABI**, an international not-for-profit organisation that improves people's lives worldwide by providing information and applying scientific expertise to solve problems in agriculture and the environment.

CABI is also a global publisher producing key scientific publications, including world renowned databases, as well as compendia, books, ebooks and full text electronic resources. We publish content in a wide range of subject areas including: agriculture and crop science / animal and veterinary sciences / ecology and conservation / environmental science / horticulture and plant sciences / human health, food science and nutrition / international development / leisure and tourism.

The profits from CABI's publishing activities enable us to work with farming communities around the world, supporting them as they battle with poor soil, invasive species and pests and diseases, to improve their livelihoods and help provide food for an ever growing population.

CABI is an international intergovernmental organisation, and we gratefully acknowledge the core financial support from our member countries (and lead agencies) including:

UKaid from the British people

Ministry of Agriculture People's Republic of China

Australian Government Australian Centre for International Agricultural Research

Agriculture and Agri-Food Canada

Ministry of Foreign Affairs of the Netherlands

Schweizerische Eidgenossenschaft Confédération suisse Confederazione Svizzera Confederaziun svizra
Swiss Agency for Development and Cooperation SDC

Discover more

To read more about CABI's work, please visit: **www.cabi.org**

Browse our books at: **www.cabi.org/bookshop**,
or explore our online products at: **www.cabi.org/publishing-products**

Interested in writing for CABI? Find our author guidelines here:
www.cabi.org/publishing-products/information-for-authors/

Save **10%** at www.cabi.org

Also available from CABI

Crop Production Science in Horticulture

This book series examines economically important horticultural crops selected from the major production systems in temperate, subtropical and tropical climatic areas.

Written by internationally renowned experts with extensive practical experience, each volume covers all aspects of production, from physiology, breeding, propagation and planting, through husbandry, crop protection, harvesting, handling and storage.

This series provides a set of succinct and readable resources for horticulture students and staff, growers, extension workers and industry personnel.

Read more and order with **10% discount** at

www.cabi.org/bookshop

*10% discount is available on individual purchases when ordering direct from www.cabi.org

Printed and bound by CPI Group (UK) Ltd, Croydon, CR0 4YY

11/01/2026

14804836-0004